Lecture Notes in Computer Science 4688

Commenced Publication in 1973
Founding and Former Series Editors:
Gerhard Goos, Juris Hartmanis, and Jan van Leeuwen

Kang Li Minrui Fei
George William Irwin Shiwei Ma (Eds.)

Bio-Inspired Computational Intelligence and Applications

International Conference on
Life System Modeling and Simulation, LSMS 2007
Shanghai, China, September 14-17, 2007
Proceedings

Springer

Volume Editors

Kang Li
George William Irwin
Queen's University Belfast
School of Electronics, Electrical Engineering and Computer Science
Ashby Building, Stranmillis Road, BT9 5AH Belfast, UK
E-mail: {K.Li, g.irwin}@ee.qub.ac.uk

Minrui Fei
Shiwei Ma
Shanghai University
School of Mechatronics and Automation
Shanghai, China
E-mail: {mrfei, swma}@staff.shu.edu.cn

Library of Congress Control Number: 2007933811

CR Subject Classification (1998): F.2.2, F.2, E.1, G.1, I.2, J.3

LNCS Sublibrary: SL 1 – Theoretical Computer Science and General Issues

ISSN 0302-9743
ISBN-10 3-540-74768-0 Springer Berlin Heidelberg New York
ISBN-13 978-3-540-74768-0 Springer Berlin Heidelberg New York

Springer is a part of Springer Science+Business Media

springer.com

Printed in Germany

Typesetting: Camera-ready by author, data conversion by Scientific Publishing Services, Chennai, India
Printed on acid-free paper SPIN: 12118496 06/3180 5 4 3 2 1 0

Preface

The International Conference on Life System Modeling and Simulation (LSMS) was formed to bring together international researchers and practitioners in the field of life system modeling and simulation as well as life system-inspired theory and methodology.

The concept of a life system is quite broad. It covers both micro and macro components ranging from cells, tissues and organs across to organisms and ecologic niches. These interact and evolve to produce an overall complex system whose behavior is difficult to comprehend and predict. The arrival of the 21st century has been marked by a resurgence of research interest both in arriving at a systems-level understanding of biology and in applying such knowledge in complex real-world applications. Consequently, computational methods and intelligence in systems, biology, as well as bio-inspired computational intelligence, have emerged as key drivers for new computational methods. For this reason papers dealing with theory, techniques and real-world applications relating to these two themes were especially solicited.

Building on the success of a previous workshop in 2004, the 2007 International Conference on Life System Modeling and Simulation (LSMS 2007) was held in Shanghai, China, September 14–17, 2007. The conference was jointly organized by The Shanghai University, Queen's University Belfast together with The Life System Modeling and Simulation Special Interest Committee of the Chinese Association for System Simulation. The conference program offered the delegates keynote addresses, panel discussions, special sessions and poster presentations, in addition to a series of social functions to enable networking and future research collaboration.

LSMS 2007 received a total of 1,383 full paper submissions from 21 countries. All these papers went through a rigorous peer-review procedure, including both pre-review and formal referring. Based on the referee reports, the Program Committee finally selected 333 good-quality papers for presentation at the conference, from which 147 were subsequently selected and recommended for publication by Springer in one volume of *Lecture Notes in Computer Science* (LNCS) and one volume of *Lecture Notes in Bioinformatics* (LNBI). This particular volume of *Lecture Notes in Computer Science* (LNCS) includes 84 papers covering 6 relevant topics.

The organizers of LSMS 2007 would like to acknowledge the enormous contributions made by the following: the Advisory Committee and Steering Committee for their guidance and advice, the Program Committee and the numerous referees worldwide for their efforts in reviewing and soliciting the papers, and the Publication Committee for their editorial work. We would also like to thank Alfred Hofmann, from Springer, for his support and guidance. Particular thanks are of course due to all the authors, as without their high-quality submissions and presentations, the LSMS 2007 conference would not have been possible.

Finally, we would like to express our gratitude to our sponsor – The Chinese Association for System Simulation, – and a number of technical co-sponsors: the IEEE United Kingdom and Republic of Ireland Section, the IEEE CASS Life Science Systems and Applications Technical Committee, the IEEE CIS Singapore Chapter, the

IEEE Shanghai Section for their technical co-sponsorship and the Systems and Synthetic Biology (Springer) for their financial sponsorship. The support of the Intelligent Systems and Control group at Queen's University Belfast, Fudan University, the Shanghai Institute for Biological Sciences, the Chinese Academy of Sciences, the Shanghai Association for System Simulation, the Shanghai Association for Automation, Shanghai Association for Instrument and Control, the Shanghai Rising-star Association, the Shanghai International Culture Association, Shanghai Medical Instruments Trade Association is also acknowledged.

June 2007

Bohu Li
Guosen He
Mitsuo Umezu
Min Wang
Minrui Fei
George W. Irwin
Kang Li
Luonan Chen
Shiwei Ma

LSMS 2007 Organization

Advisory Committee

Kazuyuki Aihara, Japan
Panos J. Antsaklis, USA
John L. Casti, Austria
Zongji Chen, China
Aike Guo, China
Roland Hetzer, Germany
Alfred Hofmann, Germany
Huosheng Hu, UK
Okyay Kaynak, Turkey
Frank L. Lewis, USA
Iven Mareels, Australia
Kwang-Hyun Park, Korea
Xiaoyuan Peng, China
Shuzhi Sam Ge, Singapore
Eduardo Sontag, USA
Steve Thompson, UK
Yishan Wang, China
Paul Werbos, USA
Stephen Wong, USA
Zhenghai Xu, China
Hao Ying, USA
Minlian Zhang, China
Xiangsun Zhang, China
Guoping Zhao, China
Yufan Zheng, Australia
Mengchu Zhou, USA

Steering Committee

Joseph Sylvester Chang, Singapore
Kwang-Hyun Cho, Korea
Seung Kee Han, Korea
Tom Heskes, Netherlands
Yan Hong, HK China
Jenn-Kang Hwang, Taiwan China
Fengju Kang, China
Young J Kim, Korea
Gang Li, UK
Yixue Li, China
Zaozhen Liu, China
Zengrong Liu, China
Sean McLoone, Ireland
David McMillen, Canada
Yi Pan, USA
Dhar Pawan, Singapore
Chen Kay Tan, Singapore
Kok Kiong Tan, Singapore
Stephen Thompson, UK
Svetha Venkatesh, Australia
YuguoWeng, Germany
Tianyuan Xiao, China
Jianxin Xu, Singapore
Wu Zhang, China
Tianshou Zhou, China
Quanmin Zhu, UK

Honorary Chairs

Bohu Li, China
Guosen He, China
Mitsuo Umezu, Japan

General Chairs

Min Wang, China
Minrui Fei, China
George W. Irwin, UK

International Program Committee

IPC Chairs

Kang Li, UK
Luonan Chen, Japan

IPC Local Chairs

Luis Antonio Aguirre, Brazil
Yongsheng Ding, China
Orazio Giustolisi, Italy
Xingsheng Gu, China
Pheng-Ann Heng, HK, China
Nicolas Langlois, France
WanQuan Liu, Australia
Zhijian Song, China
Shu Wang, Singapore
T.C. Yang, UK
Jun Zhang, USA

IPC Members

Akira Amano, Japan
Vitoantonio Bevilacqua, Italy
Weidong Cai, Australia
Ming Chen, China
Zengqiang Chen, China
Wushan Cheng, China
Xiaochun Cheng, UK
Minsen Chiu, Singapore
Sally Clift, UK
Patrick Connally, UK
Rogers Eric, UK
Haiping Fang, China
Huijun Gao, China
Xiaozhi Gao, Finland
Zhinian Gao, China
Ning Gu, China
Weihua Gui, China
Lingzhong Guo, UK
Liqun Han, China
Jiehuan He, China
Liangjian Hu, China
Guangbin Huang, Singapore
Sunan Huang, Singapore
Peter Hung, Ireland
Ping Jiang, UK
Prashant Joshi, Austria
Tetsuya J Kobayashi, Japan
Abderrafiaa Koukam, France
Xuecheng Lai, Singapore
Ziqiang Lang, UK
Keun-Woo Lee, Korea
Raymond Lee, UK
Donghai Li, China
Jun Li, Singapore
Shaoyuan Li, China
Wanqing Li, Australia
Xiaoou Li, Mexico
Yunfeng Li, China
Paolo Lino, Italy
Guoqiang Liu, China
Han Liu, China
Julian Liu, UK
Junfeng Liu, USA
Mandan Liu, China
Wei Lou, China
Zuhong Lu, China
Guido Maione, Italy
Fenglou Mao, USA
Kezhi Mao, Singapore
Marco Mastrovito, Italy
Marion McAfee, UK
Carlo Meloni, Italy
Zbigniew Mrozek, Poland
Antonio Neme, Mexico
Manamanni Noureddine, France
Philip Ogunbona, Australia
Jianxun Peng, UK
Girijesh Prasad, UK
Yixian Qin, USA
Wei Ren, China

Qiguo Rong, China
Ziqiang Sun, China
Nigel G Ternan, UK
Bing Wang, UK
Ruiqi Wang, Japan
Xiuying Wang, Australia
Guihua Wen, China
Lingyun Wu, China
Qingguo Xie, China
Jun Yang, Singapore
Ansheng Yu, China
Jingqi Yuan, China
Jun Zhang, USA
Cishen Zhang, Singapore
Yisheng Zhu, China
Da Ruan, Belgium
Sanjay Swarup, Singapore
Shanbao Tong, China
Jihong Wang, UK
Ruisheng Wang, Japan
Yong Wang, Japan
Peter A. Wieringa, Netherlands
Xiaofeng Wu, China
Meihua Xu, China
Tao Yang, USA
Weichuan Yu, HK China
Dong Yue, China
Yi Zhang, China
Xingming Zhao, Japan
Chenxi Shao, China
Shin-ya Takane, Japan
Gabriel Vasilescu, France
Ning Wang, China
Xingcheng Wang, China
Zhuping Wang, Singapore
Guangqiang Wu, China
Hong Xia, UK
Zhenyuan Xu, China
Maurice Yolles, UK
Wen Yu, Mexico
Zhoumo Zeng, China
Zuren Zhang, China
Huiyu Zhou, UK

Secretary General

Shiwei Ma, China
Ping Zhang, China

Co-Secretary-General

Li Jia, China
Qun Niu, China
Banghua Yang, China
Lixiong Li, China
Yang Song, China
Xin Li, China
Ling Wang, China

Publication Chairs

Xin Li, China
Sanjay Swarup, Singapore

Special Session Chair

Hai Lin, Singapore

Organizing Committee

OC Chairs

Jian Wang, China
Yunjie Wu, China
Zengrong Liu, China
Yuemei Tan, China

OC Co-chairs

Tingzhang Liu, China
Shiwei Ma, China
Weiyi Wang, China
Xiaojin Zhu, China

OC Members

Jian Fan, China
Zhihua Li, China
Zhongjie Wang, China
Weiyan Hou, China
Hai Lin, Singapore
Lisheng Wei, China
Aimei Huang, China
Xin Sun, China
Xiaolei Xia, UK

Reviewers

Jean-Francois Arnold
Xiaojuan Ban
Leonora Bianchi
Mauro Birattari
Ruifeng Bo
Jiajun Bu
Dongsheng Che
Fei Chen
Feng Chen
Guochu chen
Hang Chen
Mingdeng Chen
Lijuan Chen
Zengqiang Chen
Cheng Cheng
Guojian Cheng
Jin Cheng
Maurizio Cirrincione
Patrick Connally
Marco Cortellino
Jean-Charles Creput
Shigang Cui
Dan Diaper
Chaoyang Dong
Guangbo Dong
Shuhai Fan
Lingshen Fang
Dongqing Feng
Hailin Feng
Zhanshen Feng
Cheng Heng Fua
Jie Gao
Padhraig Gormley
Jinhong Gu
Lan Guo
Qinglin Guo
Yecai Guo
Yu Guo
Dong-Han Ham
Zhang Hong
Aimin Hou
Yuexian Hou
Jiangting Hu
Qingxi Hu
Wenbin Hu
Xianfeng Huang
Christian Huyck
George W. Irwin
Yubin Ji
Li Jian
Shaohua Jiang
Guangxu Jin
Hailong Jin
Xinsheng Ke
Mohammad Khalil
Yohei Koyama
Salah Laghrouche
Usik Lee
Chi-Sing Leung
Gun Li

Honglei Li
Kan Li
Kang Li
Ning Li
Xie Li
Yanbo Li
Yanyan Li
Zhonghua Li
Xiao Liang
Xiaomei Lin
Binghan Liu
Chunan Liu
Hongwei Liu
Junfang Liu
Lifang Liu
Renren Liu
Wanquan Liu
Weidong Liu
Xiaobing Liu
Xiaojie Liu
Xuxun Liu
Yumin Liu
Zhen Liu
Zhiping Liu
Xuyang Lou
Tao Lu
Dajie Luo
Fei Luo
Suhuai Luo
Baoshan Ma
Meng Ma
Xiaoqi Ma
Quentin Mair
Xiong Men
Zhongchun Mi
Claude Moog
Jin Nan
Jose Negrete
Xiangfei Nie
Xuemei Ning
Dongxiao Niu
Jingchang Pan
Paolo Pannarale
Konstantinos Pataridis
Jianxun Peng
Son Lam Phung
Xiaogang Qi
Chaoyong Qin
peng qin
Zhaohui Qin
Lipeng Qiu
Yuqing Qiu
Yi Qu
Qingan Ren
Didier Ridienger
Giuseppe Romanazzi
R Sanchez
Jesus Savage
Ssang-Hee Seo
Tao Shang
Zichang Shangguan
Chenxi Shao
JeongYon Shim
Chiyu Shu
Yunxing Shu
Vincent Sircoulomb
Anping Song
Chunxia Song
Guanhua Song
Yuantao Song
Yan Su
Yuheng Su
Suixiulin
Shibao Sun
Wei Sun
Da Tang
Pey Yuen Tao
Shen Tao
Keng Peng Tee
Jingwen Tian
Han Thanh Trung
Callaghan Vic
Ping Wan
Hongjie Wang
Kundong Wang
Lei Wang
Lin Wang
Qing Wang
Qingjiang Wang
Ruisheng Wang
Shuda Wang
Tong Wang
Xiaolei Wang
Xuesong Wang
Ying Wang
Zhelong Wang
Zhongjie Wang
Hualiang Wei
Liang Wei
Guihua Wen
Qianyong Weng
Xiangtao Wo
Minghui Wu
Shihong Wu
Ting Wu
Xiaoqin Wu
Xintao Wu
Yunna Wu
Zikai Wu
Chengyi Xia
Linying Xiang
Xiaolei Xia
Yougang Xiao
Jiang Xie
Jun Xie
Xiaohui Xie
Lining Xing
Guangning Xu
Jing Xu
Xiangmin Xu
Xuesong Xu
Yufa Xu
Zhiwen Xu
Qinghai Yang
Jin Yang
Xin Yang
Yinhua Yang
Zhengquan Yang
Xiaoling Ye
Changming Yin
Fengqin Yu
Xiaoyi Yu
Xuelian Yu
Guili Yuan
Lulan Yuan
Zhuzhi Yuan
Peng Zan
Yanjun Zeng
Chengy Zhang
Kai Zhang
Kui Zhang

Hongjuan Zhang
Hua Zhang
Jianxiong Zhang
Limin Zhang
Lin Zhang
Ran Zhang
Xiaoguang Zhang
Xing Zhang
Haibin Zhao
Shuguang Zhao
Yi Zhao
Yifan Zhao
Yong Zhao
Xiao Zheng
Yu Zheng
Hongfang Zhou
Huiyu Zhou
Qihai Zhou
Yuren Zhou
Qingsheng Zhu
Xinglong Zhu
Zhengye Zhu
Xiaojie Zong

Table of Contents

The Third Section: Ant Colonies and Particle Swarm Optimization and Application

The Fourth Section: Fuzzy, Neural, Fuzzy-Neuro Hybrids and Application

The Fifth Section: Intelligent Modeling, Monitoring, and Control of Complex Nonlinear Systems

The Sixth Section: Biomedical Signal Processing, Imaging and Visualization

An Algorithm Based on Nonlinear PCA and Regulation for Blind Source Separation of Convolutive Mixtures

Liyan Ma and Hongwei Li

School of Mathematics and Physics, China University of Geosciences, Wuhan 430074, China
{Liyan Ma,liyan_ma}@126.com
{Hongwei Li,hwli}@cug.edu.cn

Abstract. This paper proposes a method of blind separation which extracts independent signals from their convolutive mixtures. The function is acquired by modifying a network's parameters so that a cost function takes the minimum at anytime. Firstly we propose a regulation of a nonlinear principle component analysis (PCA) cost function for blind source separation of convolutive mixtures. Then by minimizing the cost function a new recursive least-squares (RLS) algorithm is developed in time domain, and we proposed two update equations for recursively computing the regularized factor. This algorithm has two stages: one is pre-whitening, the other is RLS iteration. Simulations show that our algorithm can successfully separate convolutive mixtures and has fast convergence rate.

1 Introduction

Blind source separation is one of the rising techniques in signal processing at present, which has potential in many applications. Blind source separation can be divided into instantaneous mixtures and convolutive mixtures by the mixing type. Convolutive mixtures are mutual convolutive mixing by multi-signal in complicated circumstance. In practical, the sensors obtained usually are convolutive mixing signals due to the complexity of mixing system and the delay of signal transmission; hence numerous pursuers are attracted by blind source separation of convolutive mixtures [1, 3-10].

There are two main kinds of methods for blind source separation of convolutive mixtures: time-domain method [1, 5-6] and frequency-domain method [3-4, 7, 9]. The frequency-domain method firstly transforms the signals to frequency-domain by applying Fourier transforms. Since the time-domain convolutive mixtures are instantaneous mixtures at every frequency point in frequency-domain, we can apply the algorithm of instantaneous mixtures of plural signals to separate signals at every frequency point. However, there will be some problems, which are inherent in blind source separation: the permutation problem and the scale problem [2]. To reconstruct source signals from frequency-domain data, we must solve these problems. However, it is too difficult. Such problems are not occurring in time-domain method. Certainly the time-domain method also has disadvantages: it takes time to computing when the separated filter is too long. The algorithm proposed in this paper is a time-domain

K. Li et al. (Eds.): LSMS 2007, LNCS 4688, pp. 1–9, 2007.

method. Our method estimates the unknown transfer function by modifying the parameters of an adaptive network. We extended the algorithm presented in [5] by regulation. Then the learning rule of the network's parameters is derived from the minimization of cost function. Simulations show that this algorithm can successfully separate convolutive mixtures and has fast convergence.

This paper is organized as follows. Section 2 describes the convolutive network's structures and algorithm's model. Section 3 proposes a novel cost function based on nonlinear PCA and regulation for separating convolutive mixtures and two update equations for recursively computing the regularized factor. A novel RLS method is presented in section 4. Section 5 shows the simulations. There is a conclusion in section 6.

2 The Convolutive Network's Structures and Algorithm's Model

The general model of convolutive mixtures can express as:

$$\boldsymbol{x}(t) = \boldsymbol{A} * \boldsymbol{s}(t) = \sum_{p=-\infty}^{+\infty} \boldsymbol{A}_p \boldsymbol{s}(t-p), \quad t = 1, 2, \cdots \tag{1}$$

where $\boldsymbol{s}(t) = [s_1(t), \cdots, s_n(t)]^{\mathrm{T}}$ is a vector of n mutual statistical independent source signals, $\boldsymbol{x}(t) = [x_1(t), \cdots, x_m(t)]^{\mathrm{T}}$ is m convolutive mixtures, in this paper, we let $m = n$, $\boldsymbol{A}$ is the mixing filter, " $*$ "denotes convolutive operator. To separate the source signals from the observed signals, we use an adaptive feed forward network. The input signals of the network are the observed signals $\boldsymbol{x}(t)$. The network's outputs are given by the following equation:

$$\boldsymbol{y}(t) = \sum_{p=-\infty}^{+\infty} \boldsymbol{W}_p \boldsymbol{x}(t-p) \tag{2}$$

Here $\boldsymbol{y}(t) = [y_1(t), \cdots, y_n(t)]^{\mathrm{T}}$ is a column vector of dimension n , $\{\boldsymbol{W}_p, \ -\infty \le p \le \infty\}$ is a series of $n \times n$ dimensional coefficient matrices. In z transform domain, the input and output system are

$$\boldsymbol{X}(z) = \boldsymbol{A}(z)\boldsymbol{S}(z) \tag{3}$$

and

$$\boldsymbol{Y}(z) = \boldsymbol{W}(z)\boldsymbol{X}(z) = \boldsymbol{W}(z)\boldsymbol{A}(z)\boldsymbol{S}(z) = \boldsymbol{C}(z)\boldsymbol{S}(z) \tag{4}$$

where

$$\boldsymbol{W}(z) = \sum_{p=-\infty}^{\infty} \boldsymbol{W}_p z^{-p}, \quad \boldsymbol{A}(z) = \sum_{p=-\infty}^{\infty} \boldsymbol{A}_p z^{-p}, \quad \boldsymbol{C}(z) = \boldsymbol{W}(z)\boldsymbol{A}(z) \tag{5}$$

If $\boldsymbol{y}(t)$ is the estimation of $\boldsymbol{s}(t)$, then

$$\boldsymbol{C}(z) = \boldsymbol{P}\boldsymbol{D}(z) \tag{6}$$

where $\boldsymbol{P}$ is any permutation matrix, $\boldsymbol{D}(z)$ is a non-singular diagonal matrix, its diagonal element is $c_i z^{-\Delta i}$, $i=1,\cdots,n$, c_i is non-zero coefficient, Δi is a real number.

3 A Novel Cost Function Based on Nonlinear PCA and Regulation

We consider finite filter in this paper, the corresponding mixing and de-mixing system are

$$\boldsymbol{x}(t)=\sum_{p=0}^{L_1-1}\boldsymbol{A}_p\boldsymbol{s}(t-p) \tag{7}$$

$$\boldsymbol{y}(t)=\sum_{p=0}^{L-1}\boldsymbol{W}_p\boldsymbol{x}(t-p) \tag{8}$$

Let $\boldsymbol{B}(z)$ be a $n\times n$ whitening filter, it makes

$$\boldsymbol{V}(z)=\boldsymbol{B}(z)\boldsymbol{X}(z) \tag{9}$$

such that

$$\mathrm{E}\left[\boldsymbol{v}(t)\boldsymbol{v}^{\mathrm{T}}(t-j)\right]=\boldsymbol{I}\delta(j) \tag{10}$$

The cost function for convolutive mixtures in [5] is

$$\min\ J(\boldsymbol{W}(t))=\sum_{i=1}^{t}\beta^{t-i}\left\|\boldsymbol{v}(i)-\sum_{p=0}^{L-1}\boldsymbol{W}_p^{\mathrm{T}}(t)\mathrm{f}\left(\boldsymbol{y}(i+p)\right)\right\|^2 \tag{11}$$

where β is the forgetting factor, $\boldsymbol{v}$ is the whitening signals of mixing signals, $\boldsymbol{y}(t)=\sum_{l=0}^{L-1}\boldsymbol{W}_l(t)\boldsymbol{v}(t-l)$, $\sum_{l=0}^{L-1}\boldsymbol{W}_l\boldsymbol{W}_{l+i}^T\approx\boldsymbol{I}\delta(i)$, f is a nonlinear function. When the source signal is sub-Gaussian, $\mathrm{f}(y)=\tanh(y)$; when the source signal is supper-Gaussian, $\mathrm{f}(y)=y-\tanh(y)$.

Since the de-mixing filter is inverse of mixing filter, $\boldsymbol{W}(\mathrm{z})$ should be infinite. However, in practical we can but compute finite, therefore there will exist certain error. We introduce regulation owning to the regulation allow of some error [6]. Then we get a novel cost function as follows

$$\min\ J(\boldsymbol{W}(t))=\sum_{i=1}^{t}\beta^{t-i}\left\|\boldsymbol{v}(i)-\sum_{p=0}^{L-1}\boldsymbol{W}_p^{\mathrm{T}}(t)\mathrm{f}\left(\boldsymbol{y}(i+p)\right)\right\|^2+\sum_{l=0}^{L-1}\eta_l\left\|\boldsymbol{W}_l(t)\right\|^2 \tag{12}$$

where $\left\|\boldsymbol{W}_l(t)\right\|$ is defined as $\left\|\boldsymbol{W}_l(t)\right\|^2=Tr(\boldsymbol{W}_l(t)\boldsymbol{W}_l^{\mathrm{T}}(t))$, here Tr denotes the trace of matrix, η_l is regularized factor.

Enlightened by [13], we give two update equations for recursively computing the regularized factor η_l in the following.

Utilizing the 'projection approximation' proposed in [11, 13], we get

$$\boldsymbol{y}(t)=\sum_{l=0}^{L-1}\boldsymbol{W}_l(t)\boldsymbol{v}(t-l)\approx\sum_{l=0}^{L-1}\boldsymbol{W}_l(t\text{ -}1)\boldsymbol{v}(t-l) \tag{13}$$

Then the derivative of $J(\boldsymbol{W}(t))$ in expression (12) at $\boldsymbol{W}_p(t)$ is

$$\frac{\partial J(\boldsymbol{W}(t))}{\partial \boldsymbol{W}_p(t)}=-2\sum_{i=1}^{t}\beta^{t-i}\left(z(i+p)\boldsymbol{v}^{\mathrm{T}}(i)-z(i+p)\sum_{p_1=0}^{L-1}z^{\mathrm{T}}(i+p_1)\boldsymbol{W}_{p_1}(t)\right)+\eta_p\boldsymbol{W}_p(t) \tag{14}$$

where $z(i)=\mathrm{f}\left(\boldsymbol{y}(i)\right)$. Let $\frac{\partial J(\boldsymbol{W}(t))}{\partial \boldsymbol{W}_p(t)}=0$, then we have

$$\eta_p\boldsymbol{W}_p(t)=2\sum_{i=1}^{t}\beta^{t-i}\left(z(i+p)\boldsymbol{v}^{\mathrm{T}}(i)-z(i+p)\sum_{p_1=0}^{L-1}z^{\mathrm{T}}(i+p_1)\boldsymbol{W}_{p_1}(t)\right) \tag{15}$$

Computing Frobenius norm at both sides, we obtain

$$\eta_p\left\|\boldsymbol{W}_p(t)\right\|_F=\left\|2\sum_{i=1}^{t}\beta^{t-i}\left(z(i+p)\boldsymbol{v}^{\mathrm{T}}(i)-z(i+p)\sum_{p_1=0}^{L-1}z^{\mathrm{T}}(i+p_1)\boldsymbol{W}_{p_1}(t)\right)\right\|_F \tag{16}$$

In order to update η_p more adaptively, we compute it as follows

$$\eta_p\left\|\boldsymbol{W}_p(t)\right\|_F=\left\|z(t+p)\boldsymbol{v}^{\mathrm{T}}(t)-z(t+p)\sum_{p_1=0}^{L-1}z^{\mathrm{T}}(t+p_1)\boldsymbol{W}_{p_1}(t)\right\|_F \tag{17}$$

Then we consider two cases: ① all of η_l are equal, namely $\eta_0=\cdots=\eta_{L-1}=\eta$; ②every η_l has each iterative formula.

① To equation (17), we plus the equations for $p=0,\cdots,L-1$, so we obtain

$$\eta\sum_{l=0}^{L-1}\left\|\boldsymbol{W}_l(t)\right\|_F=\sum_{l=0}^{L-1}\left\|z(t+l)\boldsymbol{v}^{\mathrm{T}}(t)-z(t+l)\sum_{p_1=0}^{L-1}z^{\mathrm{T}}(t+p_1)\boldsymbol{W}_{p_1}(t)\right\|_F \tag{18}$$

then

$$\eta=\frac{\sum_{l=0}^{L-1}\left\|z(t+l)\boldsymbol{v}^{\mathrm{T}}(t)-z(t+l)\sum_{p_1=0}^{L-1}z^{\mathrm{T}}(t+p_1)\boldsymbol{W}_{p_1}(t)\right\|_F}{\sum_{l=0}^{L-1}\left\|\boldsymbol{W}_l(t)\right\|_F} \tag{19}$$

② Dividing both sides of (17) by $\left\|\boldsymbol{W}_p(t)\right\|_F$, we get

$$\eta_p=\frac{\left\|z(t+p)\boldsymbol{v}^{\mathrm{T}}(t)-z(t+p)\sum_{p_1=0}^{L}z^{\mathrm{T}}(t+p_1)\boldsymbol{W}_{p_1}(t)\right\|_F}{\left\|\boldsymbol{W}_p(t)\right\|_F} \tag{20}$$

Looking that denominator $\|W_p(t)\|_F$ may be zero or close to zero, then it will result in ill condition, so that we modify the above expressions to get

$$\eta_p = \frac{\left\| z(t+p)v^{\mathrm{T}}(t) - z(t+p)\sum_{p_1=0}^{L-1} z^{\mathrm{T}}(t+p_1)W_{p_1}(t) \right\|_F}{\|W_p(t)\|_F + \gamma} \tag{21}$$

where γ is a small positive real number.

4 A Novel RLS Algorithm

4.1 Whitening Algorithm

Some of the algorithms for blind source separation need pre-whitening, which can speed up the computation. The whitening algorithm for instantaneous mixtures can get whitening signals through singular value decomposition. However, it is not the same with convolutive mixtures on account of its mixing type. For the moment, most of the whitening for convolutive mixtures is achieved by translating the convolutive mixtures to instantaneous mixtures; certainly the separated algorithm utilizes the algorithms for instantaneous mixtures in such case. Ohata et.el.[8] recently obtained the whitening signals by defined the cost function of whitening from the construct of convolutive mixing type.

In this paper, we use the whitening algorithm proposed in [8], let the whitening filter $\boldsymbol{B}(z) = \sum_{\tau=-K}^{K} \boldsymbol{B}_\tau z^{-\tau}$, the update expressions is that

$$\boldsymbol{v}(t-k) = \sum_{\tau=-K}^{K} \boldsymbol{B}_\tau \boldsymbol{x}(t-k-\tau) \tag{22}$$

$$\Delta \boldsymbol{B}_\tau = \alpha \left\{ \boldsymbol{B}_\tau - \sum_{\tau'=-K}^{K} \boldsymbol{v}(t-3K)\boldsymbol{v}(t-3K-\tau+\tau')\boldsymbol{B}_{\tau'} \right\} \tag{23}$$

$$\boldsymbol{B}_\tau = \frac{1}{2}\left(\boldsymbol{B}_\tau + \boldsymbol{B}_{-\tau}^{\mathrm{T}}\right) \tag{24}$$

where $\alpha (0 < \alpha < 1)$ is iterative step size.

4.2 RLS Algorithm

Let $\frac{\partial J(\boldsymbol{W}(t))}{\partial \boldsymbol{W}_p(t)} = 0$ (Eq.14), we can obtain the optimized matrix at time t as

$$\boldsymbol{W}_p(t) = \left(\sum_{i=1}^{t} \beta^{t-i} z(i+p) z^{\mathrm{T}}(i+p) + \eta_p \boldsymbol{I} \right)^{-1} \left(\sum_{i=1}^{t} \beta^{t-i} \left(z(i+p)\boldsymbol{v}^{\mathrm{T}}(i) - z(i+p)\boldsymbol{zp}(i) \right) \right) = \boldsymbol{R}^{-1}(t)\boldsymbol{C}(t) \tag{25}$$

where $\boldsymbol{I}$ is identity matrix

$$\boldsymbol{zp}(i)=\sum_{\substack{p_1=0\\p_1\neq p}}^{L-1} z(i+p_1)^{\mathrm{T}}\boldsymbol{W}_{p_1}(t) \tag{26}$$

$$\boldsymbol{R}(t)=\sum_{i=1}^{t}\beta^{t-i}z(i+p)z^{\mathrm{T}}(i+p)+\eta_p\boldsymbol{I} \tag{27}$$

$$\boldsymbol{C}(t)=\sum_{i=1}^{t}\beta^{t-i}\left(z(i+p)\boldsymbol{v}^{\mathrm{T}}(i)-z(i+p)\boldsymbol{zp}(i)\right) \tag{28}$$

Let $\boldsymbol{P}(t)=\boldsymbol{R}^{-1}(t)$, we utilize the 'projection approximation' once more, get

$$\boldsymbol{zp}(t)=\sum_{\substack{p_1=0\\p_1\neq p}}^{L-1} z(t+p_1)^{\mathrm{T}}\boldsymbol{W}_{p_1}(t)\approx\sum_{\substack{p_1=0\\p_1\neq p}}^{L-1} z(t+p_1)^{\mathrm{T}}\boldsymbol{W}_{p_1}(t-1) \tag{29}$$

Since

$$\boldsymbol{R}(t)=\beta\boldsymbol{R}(t-1)+(z(t+p)z^{\mathrm{T}}(t+p)+\eta_p\boldsymbol{I})\boldsymbol{I} \tag{30}$$

$$\boldsymbol{C}(t)=\beta\boldsymbol{C}(t-1)+z(t+p)\boldsymbol{v}(t)-z(t+p)\boldsymbol{zp}(t) \tag{31}$$

then $\boldsymbol{W}_p(t)$ can be recursively computed by applying the sum of matrices inversion lemma [12], yielding a new adaptive RLS algorithm for blind source separation of convolutive mixtures. We summarized it as follows:

$$\boldsymbol{y}(t+p_1)=\sum_{l=0}^{L-1}\boldsymbol{W}_l(t-1)\boldsymbol{v}(t+p_1-l) \tag{32}$$

$$z(t+p_1)=\mathrm{f}\left(\boldsymbol{y}(t+p_1)\right) \tag{33}$$

$$\boldsymbol{zp}(t)=\sum_{\substack{p_1=0\\p_1\neq p}}^{L-1} z(t+p_1)^{\mathrm{T}}\boldsymbol{W}_{p_1}(t-1) \tag{34}$$

$$\boldsymbol{Q}(t)=\frac{\boldsymbol{P}(t-1)(z(t+p)z^{\mathrm{T}}(t+p)+(\eta_p-\eta_p\beta)\boldsymbol{I})}{\beta+\boldsymbol{P}(t-1)(z(t+p)z^{\mathrm{T}}(t+p)+(\eta_p-\eta_p\beta)\boldsymbol{I})} \tag{35}$$

$$\boldsymbol{P}(t)=\frac{1}{\beta}\left[\boldsymbol{P}(t-1)-\boldsymbol{Q}(t)\boldsymbol{P}(t-1)\right] \tag{36}$$

$$\boldsymbol{W}_p(t)=\boldsymbol{W}_p(t-1)+\left[\boldsymbol{P}(t)\left(z(t+p)\boldsymbol{v}^{\mathrm{T}}(t)-z(t+p)\boldsymbol{zp}(t)\right)-\boldsymbol{Q}(t)\boldsymbol{W}_p(t-1)\right] \tag{37}$$

5 Simulations

Experiment 1

In order to verify the effectiveness of the algorithm proposed in this paper, we consider the separation of the following sub-Gaussian source signals $\sin(t/811)\cos(t/51)$, $\sin(t\pi/800)$, $\sin(t\pi/90)$. Mixing filter $\boldsymbol{A}$ is randomly generated in each run, its order $L_1 = 4$, the order of the whitening filter $K = 7$, the step size of the whitening algorithm $\alpha = 5\times10^{-7}$, the order of the separated filter $L = 8$, the forgetting factor $\beta = 0.999$, $\gamma = 0.01$, the nonlinear function $f(y) = \tanh(y)$, $\boldsymbol{P}(0) = \boldsymbol{I}$, $\boldsymbol{W}_p(0) = \delta(p)\boldsymbol{I}$, here $\delta(p)$ is a Kronecker function. The length of source signals data is 50000. Fig. 1 shows (a) the source signals, (b) mixing signals and (c) separated signals (η_p updates as (19)). We can see that the source signals are successfully separated.

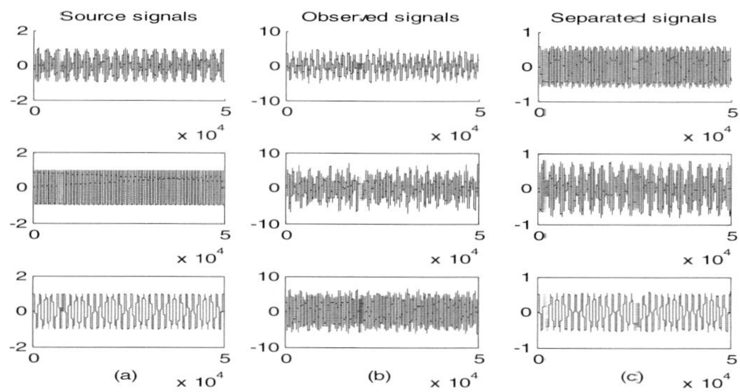

Fig. 1. (a)Source signals, (b)mixing signals and (c)separated signals (η_p takes Eq.19)

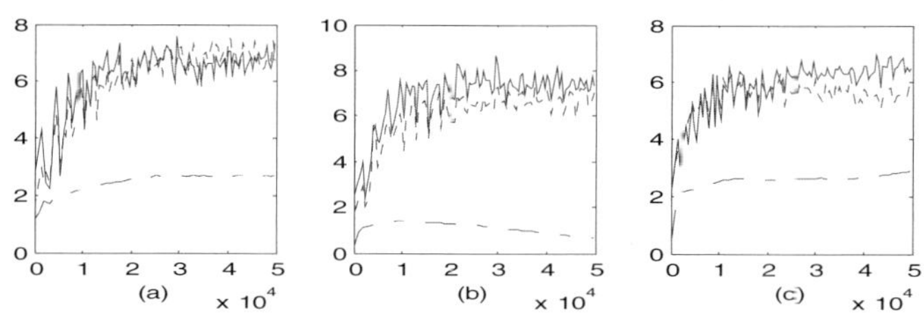

Fig. 2. The comparisons of the separated signals' performance of the algorithm we proposed (η_p updates as (19)), the algorithm in [5](*dot-line*) and the algorithm in [1] (*dot-and dash line*), (a), (b), (c) respectively show SR_1, SR_2, SR_3.

Experiment 2

For comparison, we run the presented algorithms, the algorithm of Amari[1] and the algorithm in [5] over 100 independent times utilizing the source signals in experiment 1. The performance of the separated algorithm is defined as[10]

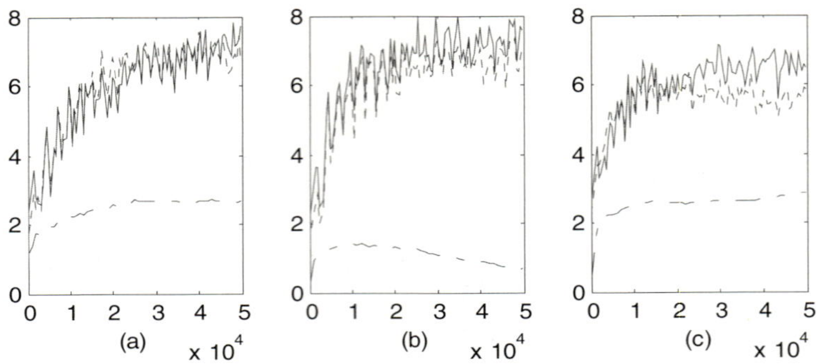

Fig. 3. The comparisons of the separated signals' performance of the algorithm we proposed (η_p updates as (21)), the algorithm in [5](*dot line*) and the algorithm in [1] (*dot-and dash line*), (a), (b), (c) respectively show SR_1 , SR_2 , SR_3 .

$$SR_i = 10\log_{10} \frac{c_{i,i}}{\sum_{\forall j \neq i} c_{i,j}} \tag{38}$$

where $c_{i,j} = \mathrm{cor}(s_i(t), \tilde{s}_j(t))$, $\mathrm{cor}(\cdot)$ denotes correlation coefficient, $\tilde{s}_j(t)$ is the estimation of the jth source signal. SR_i is larger when the ith signal is better separated. In Fig.2-3, (a), (b), (c) respectively show SR_1 , SR_2 , SR_3 . We can see that, the performances of the algorithms we proposed (real line) and algorithm in [5] (dot line) are far better than the algorithm in [1] (dot-and-dash line), and our algorithm has more fast convergence and better performance.

6 Conclusion

We proposed an algorithm for blind source separation of convolutive mixtures in this paper. By regularizing the cost function in [5], we got a novel algorithm based on nonlinear PCA and regulation. In simulations, we compared our algorithm with Amari's algorithm [1] and the algorithm in [5], simulations shows that our algorithm has more fast convergence and better performance. However, the performance curve has a little vibration, and how to improve the robust of algorithm is our aim in the future.

Acknowledgments. The authors would like to thank the anonymous reviewers for their helpful comments and suggestions. The work is supported by the National Science Foundation of China under Grant No.60672049.

References

1. Amari, S., Douglas, S.C., Cichocki, A., Yang, H.H.: Multi-channel blind deconvolution and equalization using the natural gradient. In: Proc. IEEE Workshop on Signal Processing Advances in Wireless Communication, Paris, France, April 1997, pp. 101–104. IEEE Computer Society Press, Los Alamitos (1997)

2. Comon, P.: Independent component analysis, A new concept? Signal Processing 36(3), 287–314 (1994)
3. Dapena, A., Castedo, L.: A novel frequency domain approach for mixtures of temporally-white signals. Digital Signal Processing 13(2), 301–316 (2003)
4. Dapena, A., Servitere, C., Castedo, L.: Inversion of the sliding Fourier transform using only two frequency bins and its application to source separation. Signal Processing 83(2), 453–457 (2003)
5. Ma, L.Y., Li, H.W.: An algorithm based on nonlinear PCA for blind separation of convolutive mixtures. Acta Electronica Sinica (Chinese) (submitted)
6. Matsuoka, K., Nakashima, S.: A robust algorithm for independent component analysis. In: SICE Annual Conference, Fukui, Japan, August 2003, pp. 2131–2136 (2003)
7. Murata, N., Ikeda, S., Ziehe, A.: An approach to blind source separation based on temporal structure of speech signals. Neurocomputing 41(1-4), 1–24 (2001)
8. Ohata, M., Matsuoka, K., Mukai, T.: An adaptive blind separation method using para-Hermitian whitening filter for convolutively mixed signals. Signal Processing 87(1), 33–50 (2007)
9. Smaragdis, P.: Blind separation of convolved mixtures in the frequency domain[J]. Neurocomputing 22(1-3), 21–34 (1998)
10. Smaragdis, P.: Convolutive speech bases and their application to supervised speech separation. IEEE Trans. on Audio, Speech and Language Processing 15(1), 1–12 (2007)
11. Yang, B.: Projection approximation subspace tracking. IEEE Trans. Signal Processing 43(1), 95–107 (1995)
12. Zhang, X.D.: Matrix analysis and application (Chinese). Tsinghua University Press, China, Beijing (2004)
13 Zhu, X.L., Zhang, X.D., Su, Y.T.: A fast NPCA algorithm for online blind source separation. Neurocomputing 69(7-9), 964–968 (2006)

An Edge-Finding Algorithm on Blind Source Separation for Digital Wireless Applications

Jie Zhou [1], Keyou Yao[1], Ying Zhao[1], Yiyue Gao[1], and Hisukazu Kicuchi[2]

[1] College of Electronic and Information Engineering of Nanjing University of Information Science and Technology, P.R. China

[2] Department of Electrical and Electronic Engineering Faculty of Engineering Niigata University, Japan

yao1776@hotmail.com

Abstract. In this paper we discuss the problem of blind estimating multiple digital co-channel communication signals using an antenna array. A new approach is proposed to the problem, which is to find all the independent edges of X base on analysis of the projections of **X** onto coordinates. Through the simulation results of separating two blind audio signals, we can find out that blind signals could be separated with the algorithm quickly and rightly.

1 Introduction

In order to cope with the situation of rapid growth of wireless cellular radio communication, we have to increase the frequency re-use factor of cells. A largely used way to achieve this increase is by employing smart antenna array at the cell base sites [1] [2]. On one hand, with the antenna array, we can process the receive data so as to reject co-channel interference from neighboring cells which is mainly generated by the large delay multi-path, and thus can increase the frequency re-use factor. On the other hand, the array can even distinguish signals from different location in a single cell. This important ability allows us to exploit spatial separation to modulate and demodulate user signals. This source separation problem is known as the blind source separation, and is the essential to the Space-Division Multiple Access (SDMA).

The authors in previous works [3][4]have probed into and exploited the following two properties in blind SDMA problem. Firstly, the signals from different users share a known finite symmetric digital signaling alphabet. Secondly, the signals have linearly independent, but unknown, spatial signatures as measured at the antenna array. They actually not only have given a model which can be interpreted as an geometric-algebraic model: **X= AS**, where **X**, **A**, **S** are matrices stated in Section 2 of our paper, but also made a study of the geometric aspect and propose an "hyperplane algorithm" to solve the blind SDMA problem[5]. To solve this complex problem, instead of finding the hyperplanes of the underlying parallelotope **X**, we propose a new approach to the problem, which is to find all the independent edges of **X** based on analysis of the projections of **X** onto coordinates. From the simulation, we can find out that blind signals could be separated with the algorithm quickly and rightly.

K. Li et al. (Eds.): LSMS 2007, LNCS 4688, pp. 10 – 17, 2007.

This paper is organized as follows. Section 2 restates the mathematical model of the problem and the outlines of our idea for deal with the problem in the no-noise channel. Section 3 provides the geometric basis the algorithm relies on. The algorithm is summarized in Section 4. In this section, we also present simulation results of separating two blind audio signals. Finally, Section 5 concludes our paper.

2 Mathematic Model and New Approach

2.1 Mathematic Model

Let $S(t) = [s_1(t),\cdots,s_d(t)]^T$ be the input vector of the antenna array at time *t* from *d* users, where each $s_i(t)$, the *i*th signal from the *i*th user, takes value from the *L*-level alphabet $\mathbb{A}=\{\pm1, \pm3,\cdots,\pm(2L-1)\}$, where *L* is an positive integer. The output vector of the antenna at time *t* is $X(t) = [x_1(t),\cdots,x_m(t)]^T$, where *m* is the number of sensors of the antenna array and $d \leqslant m$. And let $V(t) = [v_1(t), \cdots, v_m(t)]^T$ be the additive noise. Then

$$X(t) = [a_1,\cdots,a_d]S(t) + V(t) = AS(t) + V(t) \tag{1}$$

where a_i is the spatial signature of the *i*th signal marked by antenna array. Let $t = 1, \cdots, N$, and concatenating *N* snap shots yields:

$$\begin{aligned} X &= [x(1),\cdots,x(N)] \\ &= A[s(1),\cdots,s(N)] + [v(1),\cdots,v(N)] \\ &= AS + V \end{aligned} \tag{2}$$

Where **X** and **V** are $m \times N$, **A** and **S** are $m \times d$ and $d \times N$ matrices, respectively.

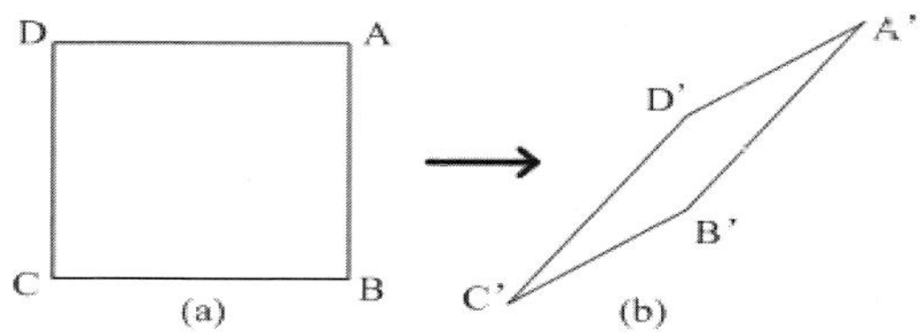

Fig. 1. (a): Outputs of antenna array **X** (digital source signals); (b): Inputs of antenna array **S** (received signals converted by matrix A)

Ignore the additive noise, (2) becomes

$$X = AS \tag{3}$$

In this paper we assume:

(1)**S** is drawn from a finite alphabet $\mathbb{A}$

(2) $N >> d$ and all the possible columns will occur in **S**.
(3)**X** is known.

The blind digital co-channel communication problem is: give only **X** in (1) or (2) and the knowledge of that **S** is drawn from a finite alphabet A = {±1, ±3, ···, ± (2*L*-1)}, where *L* is a positive integer, estimating the matrix A and the transmitted signal **S**.

2.2 Sketches to Problem

In this paper, we first deal with the problem where the entries in matrix **S** are composed of 1´s and -1´s. If $L = 1$, the inputs of antenna array **S** is a square as displayed in Fig.1. Because of the influence of the communication matrix **A**, the outputs of antenna array **X** is a parallelogram as displayed in Fig.1. So the problem is that how to separate the input signals **S** without knowing anything except **X**.

In our approach, both **X** and **S** are viewed as $d \times 2$ matrix, whose columns are just all the vertices of a parallel-tope and a cube respectively in Euclidean space $\mathbf{R}^d$, and **A** is a non-degenerate linear transformation on $\mathbf{R}^d$ such that **X=AS**.

Let $e_1 = (1,0,\cdots,0)^T$, $e_2 = (0,1,\cdots,0)^T$, ··· , $e_d =(0,0,\cdots,1)$ be the natural basis of $\mathbf{R}^d$. If two columns of **X** (two points in $\mathbf{R}^d$), say P and Q, from an edge PQ of the parallelotope **X**, then PQ is an image of an edge of the corresponding cube **S**. Thus, PQ = $\pm 2Ae_i$ for some *i* because each edge of the cube **S** is parallel to some coordinate axis and its length is equal to 2. Hence the *i*-th column of **A** is $\mathbf{A}e_i = \pm (Q-P)/2$.

Thus, if we know all the independent edges of the parallel-tope **X** (where "independent" can be explained as "no pair of edges are parallel") then we will know all the columns of **A** up to signs and order. Therefore the aim in our method is to find all independent edges of the given parallel-tope **X**. Basically, our method of finding the edges is a projection method, i.e. a method for selecting edges by investigating the projections on the coordinates.

Our algorithm is easy to be understood and lower complexity. All algorithms exploit only addition, subtraction and comparison of two real numbers, copying and deleting columns. There is no need to modify matrix **X** by linear or orthogonal transformations.

3 Basic Concepts and Facts About X=AS

To solve **X=AS**, let $P = (p_1, p_2, \cdots, p_d)$ and $Q = (q_1, q_2, \cdots, q_d)$ are two points in space $\mathbf{R}^d$, the line PQ through them is the set of the solutions satisfying equations:

$$\frac{x_1 - p_1}{q_1 - p_1} = \frac{x_2 - p_2}{q_2 - p_2} = \cdots = \frac{x_d - p_d}{q_d - p_d} \tag{4}$$

Where x_i is between p_i and q_i for all i = 1, 2, ···, d. P is a hyperplane. A line and a hyperplane are said to be orthogonal if this line is orthogonal to any line in the hyperplane. In order to simplify the presentation, we use the notations as follows:

$E_i*(P)$: the i-th row of a given parallelotope P which is actually the set of projections of the vertices of P onto axis e_i.

$V_i(y, P)$ or $V_i(y)$: the set of vertices (i.e. columns) of a given parallelotope P whose i-th component (i.e. projection on e_i) is equal to y.

$V_i(Y, P)$ or $V_i(Y)$: the set of vertices (i.e. columns) of a given parallelotope P whose i-th component is equal to some $y \in Y$.

P−Q: the set of columns obtained from P after deleting all columns of Q.

$E_i*(P-Q)$: the ith row of P−Q.

$|Q|$, $|V_i(y, P)|$, $|V_i(Y, P)|$, $|P-Q|$: the numbers of vertices in Q, $V_i(y, P)$, $V_i(Y, P)$

Thus the following facts are easy to be understood.

Fact (I): The edges of cube S are parallel to $e_1, e_2, \cdots, e_c$ respectively.

Fact (II): Suppose $P'Q'$ is an edge of cube **S**, where P' and Q' are two vertices (i.e. columns) of S, then $\mathbf{A}(P'Q')$ is an edge of parallelotope **X**, and each edge of **X** is an image of some edge of **S**. Now, from the fact above, $P'Q' = Q' - P' = \pm 2e_i$, for some $i \in \{1, 2, \cdots, d\}$, hence $\mathbf{A}(Q' - P') = \pm 2\mathbf{A}e_i$. Thus if we find two columns P and Q of **X** such that PQ is an edge of parallelotope **X**, then $\mathbf{A}e_i = \pm(Q - P)/2$, hence we know the i-th column of **A** up to a sign. In this way we will know **A** up to signs and order of columns if we find all the d independent edges of **X**, where “independent” can be explained as “no pair of edges is parallel”.

Fact (III): The ith-row E_i* (P) of P is the projection of the vertices of parallelotope P on e_i.

Fact (IV): The projection of vertices of **X** on e_i consists of at least two points for any e_i.

Fact (V): If P and Q are two different vertices of P, then their projections are different on at least one e_i, for some $i \in \{1, 2, \cdots, d\}$.

Let E_i* (P) be ith row of a given m-dimensional parallelotope P, $c = \min(E_l*(P))$ the minimum element in E_i* (P). Then $|V_i(c)| = 2^k$ for some $k \in \{1, 2, \cdots, m\}$ and $V_i(c)$ (i.e. $V_i(c, P)$) form a k-dimensional facet of P (it is itself a k-dimensional parallelotope denoted by $V_i(c)$also), which is orthogonal to e_i.

Let $E_i*(P)$ be a row of a given m-dimensional parallelotope P, $c = \min(E_i*(P))$, $y = \min(E_i*(P - V_i(c)))$. If $|V_i(c)| = 1$, then for any vertex $Q \in V_i(y)$, PQ is an edge of P, where P is the only vertex in $V_i(c)$.

4 Algorithms and Simulation Results

4.1 A New Algorithms

Let P be an m-dimensional parallelotope in $\mathbf{R}^d$, where $m \le d$. In this section we will firstly give two algorithms for finding an edge and an $(m-1)$-facet of given P. The P is represented as a set or a matrix consisting of column vectors in Euclidean space $\mathbf{R}^d$. Finally we give a main algorithm of this paper for solve the problem.

From the point of mathematic view, all the data of **X** is float (real) numbers with some error. Therefore all the notations involving in comparison of two real numbers need to modify. We do not intend to concern the detail involved in these error problem for it will make the paper have complicated outlook.

Let $\mathtt{P}$ be a given *m*-dimensional parallelotope in $\mathbf{R}^d$. Then we can find an edge of $\mathtt{P}$. We can get the pseudo code in Algorithm 4.1 below.

Algorithm 4.1 Edge (P)
Input: An *m*-dimensional parallelotope $\mathtt{P}$ in $\mathbf{R}^d$.
Output: Two columns P, Q of $\mathtt{P}$ such that PQ is an edge of $\mathtt{P}$.
1. $\mathtt{P}_0 \leftarrow \mathtt{P}$, $k \leftarrow 1$;
2. $c_1 \leftarrow \min(E_1{}^*(\mathtt{P}_0))$;
3. $\mathtt{U} \leftarrow V_1(c_1, \mathtt{P}_0)$;
4. while $|\mathtt{U}| > 1$
5. $\quad \mathtt{P}_k \leftarrow \mathtt{U}$;
6. $\quad k \leftarrow k+1$;
7. $\quad c_k \leftarrow \min(E_k{}^*(\mathtt{P}_{k-1}))$;
8. $\quad \mathtt{U} \leftarrow V_k(c_k, \mathtt{P}_{k-1})$;
9. end while
10. $y \leftarrow$ the second minimum real number in $(E_k{}^*(\mathtt{P}_{k-1})$;
11. $P \leftarrow$ the only column in $\mathtt{U}$; $Q \leftarrow$ any column in $V_k(y, \mathtt{P}_{k-1})$;
12. Return P, Q.

Let P, Q, P′ be three columns of parallelotope $\mathtt{P}$ in $\mathbf{R}^d$ and Q′ be a column vector in $\mathbf{R}^d$ such that PQ is an edge of P. Then P′Q′ is an edge parallel to PQ (hence Q' is in $\mathtt{P}$) if and only if $E_j(Q)-E_j(P) = E_j(Q')- E_j(P')$ or $E_j(Q)-E_j(P) = -(E_j(Q')- E_j(P'))$ for all $j = 1,2,\cdots,d$.

If we know two columns P,Q of $\mathtt{P}$ form an edge PQ of $\mathtt{P}$. Then we can find a (*m-1*)-facet $\mathtt{F}$ of $\mathtt{P}$ such that P is in it but Q is not. Thus we can design the following Algorithm to find required (*m-1*)-dimensional facet of $\mathtt{P}$.

Algorithm 4.2 Facet ($\mathtt{P}$, P, Q)
Input: An m-dimensional parallelotope $\mathtt{P}$ in R^d and two vertices P and Q (columns) of $\mathtt{P}$ such that PQ is an edge of $\mathtt{P}$.
Output: $\mathtt{F}$ – an (*m-1*)-dimensional facet of $\mathtt{P}$ such that containing P but not Q.
1. $\xi \leftarrow \mathtt{P}$, $\mathtt{F} \leftarrow \varphi$;
2. Columns P and Q is given; we can find an axes e_i such that $E_i^*(P) \neq E_i^*(Q)$;
3. $\mathbf{q} = (E_1(Q)-E_1(P), \cdots, E_d(Q)-E_d(P))$;
4. while $\xi \neq \varphi$
5. $\quad a \leftarrow \min(E_i^*(\xi))$;
6. $\quad V \leftarrow V_i(a, \xi)$, $\xi \leftarrow \xi - V$;
7. $\quad \mathtt{U} \leftarrow \{P' + \mathbf{q} \mid P' \in V\}$, $\xi \leftarrow \xi - \mathtt{U}$;
8. $\quad \mathtt{F} \leftarrow \mathtt{F} \cup V$;
9. end while
10. return $\mathtt{F}$

So let P be a given m-dimensional parallelotope in R^c, then we can find all independent m edges of P. Now return the problem. If PQ is an edge of **X**, then **A**ei = $\pm$ (Q-P)/2 for some $i \in \{1, 2, \cdots, d\}$. So, after preparations above, we can design the main algorithm below, to find **A** up to signs and the order of columns of **A**.

Main Algorithm Main(X)
Input: A d-dimensional parallelotope **X** in $\mathbf{R}^d$.
Output: A $d \times d$ matrix **A**.

```
P ← X;
while k ≤ d
      P, Q ← Edge(P);
      Ae_k ← (P -Q)/2 (Ae_k is the k-th column of A);
      if k < d then
            P ← Facet (P, P, Q);
            k = k +1;
      end if
end while
return A = (A_1e_1, ···, A_de_d).
```

Note that, in above algorithm, d-k+1 is the dimension of P. Thus we have decided the matrix **A** up to the order of columns and signs for each column. **S** can be obtained as $\mathbf{A}^{-1}\mathbf{X}$.

4.2 Simulation Results

In this section, we describe the simulation results of a MIMO (Multiple Input and Multiple Output) antenna system. In the experiment, we had d = 2 digital source transmitting over the communication channels. We also use M = 2 antennas. We assume the two voices: s1, s2 as Fig.2 displays. First, convert them to digital signals. Secondly, we assume the communication matrix **A** as our talk as. Because **X** = **AS,** we can work out the outputs of arrays. The outputs of antenna array **X** are completely different from the inputs **S**. We find all the independent edges of **X** first. Then on the base of $\mathbf{A}e_i = \pm$ (Q-P)/2, we work out **A** and **S**. At last, we convert the digital signals to analog signals.

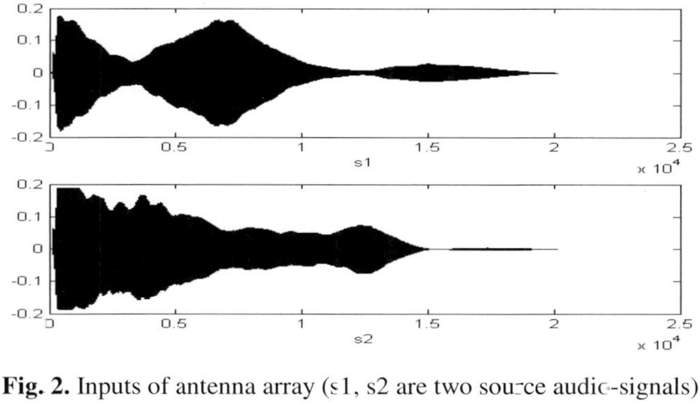

Fig. 2. Inputs of antenna array (s1, s2 are two source audio-signals)

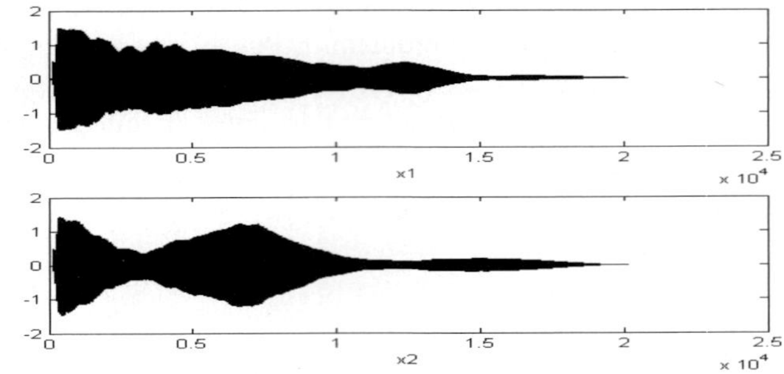

Fig. 3. Outputs of antenna array(x1, x2 are received audio-signals transmitted by matrix A)

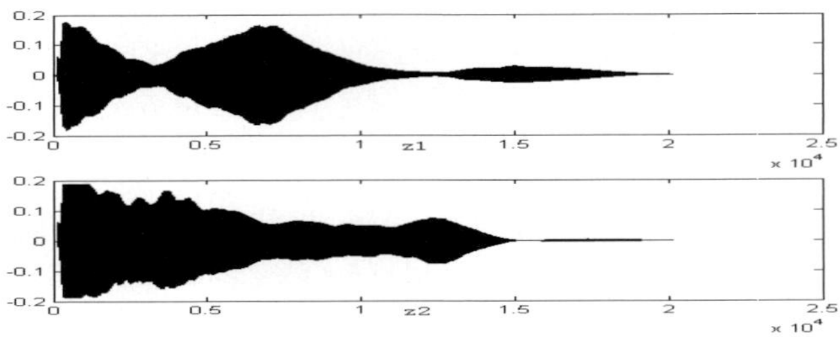

Fig. 4. Separated signals (z1,z2 are separated audio-signals using the algorithm)

Fig.3 displays the outputs of antenna array. They are different from the inputs. Fig.4 displays the separated signals. The separated signals are the same as the inputs of antenna array. This shows that our algorithm is valid. In this simulation, we may require more data samples to make the test valid.

In order to have a qualitative analysis of our algorithm, we use the relativity ζ_{ij} of source and separated signals to evaluate the effect of our algorithm.

$$\zeta = \zeta(z_i, S_j) = \frac{|\sum z_i(t) s_j(t)|}{\sqrt{\sum z_i^2(t) \sum s_j^2(t)}} \tag{5}$$

We work out the ζ_{ij} as follows:

$$\zeta = \begin{bmatrix} 0.9532 & 0.0467 \\ 0.0416 & 0.9504 \end{bmatrix} \tag{6}$$

The elements of main diagonal are approximate to 1 and the elements of assistant diagonal are approximate to 0. That is the separated signals are similar to the source signals.

4.3 Extension to Large-Scale Cases

Our method is also can be extended to Large-Scale case, where **S** is drawn from alphabet A = {±1, ±3, ···, ±(2*L*-1)}, L ≥ 1. Here we only give its sketch as follows.

The columns of **X** and **S** can also be viewed as points in Euclidean space $\mathbf{R}^d$, the columns of **S** define all the vertices of (2*L*-1) adjacent congruent small hyper cubes of edge length 2 in $\mathbf{R}^d$ and these hyper cubes make a big hypercube of edge length 2(2*L*-1) in $\mathbf{R}^d$, and, correspondingly, the columns of **X** define all the vertices of $(2L - 1)^n$ adjacent small parallel topes in R and these parallel topes form a big parallelotope.

If we know an edge PQ of a small parallelotope of length 2. it is easy to find an edge of the big parallelotope PQ^b, where $Q^b = P + (2L - 1)(P - Q)$. Thus program in general case is nearly the same as the case of L = 1; however we omit the mathematic proofs because it is very cumbersome.

Since we perform the operations on the field of real number, all the data are float number. We do not discuss complexity problem, however the efficiency is greater than the case of L = 1. Since there are $(2L)^d$ rows in **X** from beginning. Each time after use Algorithm 4.2, the number of columns is reduced from $(2L)^k$ to $(2L)^{k-1}$. Finally we have not discussed the case when the noise **V** presented in equation (1). It is our future works.

5 Conclusions

In this paper, we have shown that we can blindly separate the digital signals transmitting over multiple channels with an edge-finding algorithm. Instead of finding the hyper planes of the underlying parallelotope **X**, we proposed a new approach to the problem, which is to find all the independent edges of **X** base on analysis of the projection of **X** onto coordinates. We then developed a fast and simple algorithm. We have also presented the simulation results to illustrate the effectivity of the new approach.

References

1. Hanson, L., Xu, G.: A Hyper-plane Based Algorithm for the Digital Co-Channel Communication Problem. IEEE Transactions on Information Theory 43(5) (1997)
2. Hansen, L.K., Xu, G.: A Geometric Properties of the Blind Digital Co-Channel Communication Problem. In: Proc. of ICASSP'96, Atlanta, GA, pp. 1085–1088 (1996)
3. Liu, H.: Smart Antenna Application to Wireless System. Ph D. thesis, U. of Texas at Austin, Austin, TX (1995)
4. Liu, H., Xu, G.: Deterministic Approach to Blind System Estimation. IEEE Signal Processing Letter 1(12), 204–205 (1994)
5. Talwar, S., Vilberg, M., Pautraj, A.: Blind Estimation of Multiple Co-Channel Digital Signals Using an Antenna Array. IEEE Signal Processing Letter 1(12), 29–31 (1994)
6. Torlak, M., Hansen, L., Xu, G.: A Geometric Approach to Blind Source Separation for Digital Wireless Application. IEEE Signal Processing, 153–167 (1999)
7. Nakayama, K., Hirano, A., Sakai, T.: An adaptive nonlinear function controlled by kurtosis for blind source separation, Honolulu, Hawai (May 2002)
8. Mansour, A., Puntonet, C.G.: Blind multi-users separation of instantaneous mixture algorithm base on geometrical concepts. Signal Processing 82(8), 1155–1175 (2002)

Research of Immune Neural Network Model Based on Extenics

Xiaoyuan Zhu, Yongquan Yu, and Hong Wang

Faculty of computer, Guang Dong University of Technology,
Guangzhou Guangdong P.R. China 510090
kobe_zhuxiaoyuan@163.com

Abstract. In order to conquer the disadvantage of the traditional Immune Neural Network (INN), the paper presents INN model which is based on extenics. With matter-element analysis, the model can solve the problem that antibody identifies and memorizes antigen in immune system. The model also can make correct judgments about activation or control of nerve cell. Consequently, the structure design of INN can be optimized. And then, the new model is applied in experiment which is used for solving the problem of nonlinearity function. Based on experiment results, the model is compared with the traditional neural network. Simulation results indicate that the new model has better convergence and stability.

1 Introduction

Immune system of organism can resist large numbers of viruses with finite resource. Immune algorithm is a new algorithm which is designed based on revelation of immune system. Recently, immune algorithm is applied in neural network. Because immune algorithm can identify one and others, and has mechanism that antibody resists antigen. A new research field, which is immune neural network (INN) [7], is inaugurated.

Along with development of Genetic Algorithm (GA), people can design and optimize INN with GA [10]. GA has capability of the whole convergence, so it can make the process of optimization close to the whole optimization. But GA has shortage. For example, the results easily get into local optimization. The process of optimization is difficult to control.

Extenics was established in 1983 by Professor Wen Cai. Extenics is a original subject which has high value. It is consisted by three theoretical structures: the theory of matter-element [1], the theory of extension set, the logic of extension [3]. Extenics investigates extensibility of matter and exploits new rule. It also can solve the problem of contradiction [4].

Based on the merit of extenics, the paper integrates INN, GA and the theory of extenics. This paper presents INN model which is based on extenics. The model can conquer the disadvantage of the traditional INN.

K. Li et al. (Eds.): LSMS 2007, LNCS 4688, pp. 18–27, 2007.

2 Theory of Extenics

2.1 Characteristic of Extension Set

$R = (N, c, v)$ is regarded as essence-element which describes matter. It is called matter-element. N expresses matter[4]. c expresses the name of characteristic. v expresses the variables of N with regard to c. They are called three elements of matter-element.

When U is theory field, if u is any element in U, means $u \equiv U$. There is a real number $K(u) \in (-\infty, +\infty)$, which is correlative with u. Therefore, $\tilde{A} = \{(u, y) \mid u \in U, y = K(u) \in (-\infty, +\infty)\}$ is the extension set of the theory field U. And $y = K(u)$ is dependent function, $K(u)$ is dependent degree of u about $\tilde{A}$.

The definitions are listed as follows.

(1) $A = \{(u, y) \mid u \in U, y = K(u) \geq 0\}$ is plus field of $\tilde{A}$.

(2) $\bar{A} = \{(u, y) \mid u \in U, y = K(u) \leq 0\}$ is minus field of $\tilde{A}$.

(3) $J_o = \{(u, y) \mid u \in U, y = K(u) = 0\}$ is zero field of $\tilde{A}$.

Clearly, $u \in Jo$ implies $u \in A$ and $u \in \bar{A}$.

2.2 Matter-Element Model of the Identified Object Formulas

Suppose that $\mathrm{P_o}$ is a subset of P, means $\mathrm{P_o} \subset \mathrm{P}$. We try to judge whether any p belongs to $\mathrm{P_o}$. The degree that p belongs to $\mathrm{P_o}$ is computed. The method is as follows.

$$R_o = \begin{bmatrix} \mathrm{P_o}, & \mathrm{c}, & \mathrm{V}_{o1} \\ & \mathrm{c}_2, & \mathrm{V}_{o2} \\ & \cdots & \cdots \\ & \mathrm{c}_n, & \mathrm{V}_{on} \end{bmatrix} = \begin{bmatrix} \mathrm{P_o}, & \mathrm{c}_1, & < a_{o1}, b_{o1} > \\ & \mathrm{c}_2, & < a_{o2}, b_{o2} > \\ & \cdots & \cdots \\ & \mathrm{c}_n, & < a_{on}, b_{on} > \end{bmatrix}$$

V_{oi} is the variable value which is ensured by characteristic variables $c_i (i = 1,2,\ldots,n)$ of $\mathrm{P_o}$. So V_{oi} belongs to $< a_{oi}, b_{oi} >$ which is a classical field[5].

$$R_p = \begin{bmatrix} \mathrm{P}, & \mathrm{c}_1, & \mathrm{V}_{p1} \\ & \mathrm{c}_2, & \mathrm{V}_{p2} \\ & \cdots & \cdots \\ & \mathrm{c_n}, & \mathrm{V}_{pn} \end{bmatrix} = \begin{bmatrix} \mathrm{P}, & \mathrm{c}_1, & < a_{p1}, b_{p1} > \\ & \mathrm{c}_2, & < a_{p2}, b_{p2} > \\ & \cdots & \cdots \\ & \mathrm{c}_n, & < a_{pn}, b_{pn} > \end{bmatrix}$$

P expresses all gather. $V_{\rho i}$ is the variable value which is ensured by characteristic variables of P . So $V_{\rho i}$ belongs to $< a_{\rho i}, b_{\rho i} >$ which is a modulation field [5].

2.3 Compute the Dependent Degree of Object p

The Identified Object can be expressed as follows.

$$R = \begin{bmatrix} p, c_1, v_1 \\ c_2, v_2 \\ \dots \quad \dots \\ c_n, v_n \end{bmatrix}$$

Dependent function can be expressed as follows.

$$k(v_i) = \begin{cases} -\dfrac{\rho(v_i, V_{oi})}{|V_{oi}|}, & v_i \in V_{oi} \\ \dfrac{\rho(v_i, V_{oi})}{\rho(v_i, V_{pi}) - \rho(v_i, V_{oi})}, & v_i \notin V_{oi} \end{cases}, \quad (i = 1,2,\dots,n) \tag{1}$$

In equation (1),

$$\rho(v_i, V_{oi}) = | v_i - \frac{a_{oi} + b_{oi}}{2} | - \frac{b_{oi} - a_{oi}}{2}, \quad (i = 1,2,\dots,n) \tag{2}$$

$$\rho(v_i, V_{pi}) = | v_i - \frac{a_{pi} + b_{pi}}{2} | - \frac{b_{pi} - a_{pi}}{2}, \quad (i = 1,2,\dots,n) \tag{3}$$

2.4 Decide Weight Coefficient

Firstly we can decide every weight coefficients $\alpha_1, \alpha_2, \dots \alpha_n$. The degree that p belongs to P_o is computed as follows.

$$K(p) = \sum_{i=1}^{n} \alpha_i k_i(v_i) \tag{4}$$

(1) When $K(p) \geq 0$, $p \in P_o$.
(2) When $-1 \leq K(p) \leq 0$, $p \in P_o$, $p \notin P_o$.
(3) When $K(p) \leq -1$, $p \notin P_o$.

3 Immune Neural Network Model

3.1 Biology Immune System

Immune apparatus, immune organization and lymphocyte cell are made of biology immune system. Lymphocyte cell mainly includes T-cell and B-cell[9]. When antigen invades body, B-cell can destroy antigen by excreting antibody. After antigen be identified, T-cell activates B-cell to produce antibody. When concentration of antibody reaches the degree of controlling antibody, T-cell controls the propagation of B-cell.

3.2 Matter-Element Model of Immune System

We replace two lymphocytes by two matter-element models:

$$R_T = (\text{T-lymphocyte}, \text{antibody}, \text{T-cell}) = \begin{bmatrix} \text{T-lymphocyte}, & \text{antibody1}, & \text{damage T-cell} \\ & \text{antibody2}, & \text{control T-cell} \\ & \text{antibody3}, & \text{assistance T-cell} \end{bmatrix} = \begin{bmatrix} R_{T1} \\ R_{T2} \\ R_{T3} \end{bmatrix}$$

$$R_B = (B-\text{lymphocyte}, \text{antibody}, \text{B-cell}) = \begin{bmatrix} \text{B-lymphocyte}, & \text{antibody1}, & \text{B-cell} \\ & \text{antibody2}, & \text{cytoplast cell} \end{bmatrix} = \begin{bmatrix} R_{B1} \\ R_{B2} \end{bmatrix}$$

Here, we replace memory by a matter-element model $R = (\text{antigen}, \text{structure}, \text{V})$. P is set up as all antigens which have invaded.

So $$R_\rho = \begin{bmatrix} \text{antigen}, & \text{structure1}, & V_{\rho 1} \\ & \text{structure2}, & V_{\rho 2} \\ & \dots & \dots \\ & \text{structuren}, & V_{\rho n} \end{bmatrix} = \begin{bmatrix} \text{antigen}, & \text{structure1}, & < a_{\rho 1}, b_{\rho 1} > \\ & \text{structure2}, & < a_{\rho 2}, b_{\rho 2} > \\ & \dots & \dots \\ & \text{structuren}, & < a_{\rho n}, b_{\rho n} > \end{bmatrix}.$$

P_o is set up as a subset of P.

So $$R_o = \begin{bmatrix} \text{antigen}, & \text{structure1}, & V_{o1} \\ & \text{structure2}, & V_{o2} \\ & \dots & \dots \\ & \text{structuren}, & V_{on} \end{bmatrix} = \begin{bmatrix} \text{antigen}, & \text{structure1}, & < a_{o1}, b_{o1} > \\ & \text{structure2}, & < a_{o2}, b_{o2} > \\ & \dots & \dots \\ & \text{structuren}, & < a_{on}, b_{on} > \end{bmatrix}.$$

For an invasive antigen p, there is $$Rp = \begin{bmatrix} \text{antigen}, & \text{structure1}, & V_1 \\ & \text{structure2}, & V_2 \\ & \dots & .. \\ & \text{structuren}, & V_n \end{bmatrix}.$$

Finally, $K(p)$ is acquired from equation (1), (2), (3), (4). It expresses degree that antigen p belongs to P_o.

3.3 Structure of Immune Neural Network

INN has 3 layers. The first layer is the input layer. It is made of the management cell which takes over the input signal. The second layer is the management layer. Based on the input signal, it decides whether the management cell is in activation, control or uncertainty, and how to manage the input signal. The third layer is the output layer. Based on the results of the second layer, it decides the output type of the management cell. The structure is shown in Fig.1.

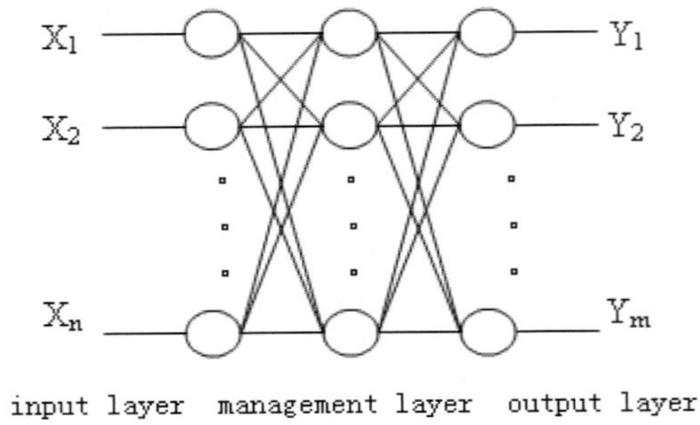

Fig. 1. Structure of INN

Suppose that the management cells have n input nerve cell and m output nerve cell. $X_1, X_2, \ldots, X_n$ are the input of the nerve cell. θ_i is the threshold value of the nerve cell i $(i = 1,2,\ldots,m)$. $W_{i1}, W_{i2}, \ldots, W_{in}$ are respectively the weight coefficients of $X_1, X_2, \ldots, X_n$ about the nerve cell i . $Y_1, Y_2, \ldots, Y_m$ are the output of the nerve cell. The transfer function of the nerve cell can be expressed as follows.

$$Y_i = f[\sum_{j=1}^{n} W_{ij} X_j - \theta_i], (i = 1,2,\ldots,m) \tag{5}$$

3.4 Mechanism of Activation and Control

We can structure extension dependent function $K(u) = \sum_{j=1}^{n} W_{ij} X_j - \theta_i$, and compute dependent degree $K(u)$.

When $\sum_{j=1}^{n} W_{ij} X_j$ exceed the threshold value θ_i , means $K(u) \geq 0$, and the nerve cell immediately comes into the state of activation. Now the input signal can be taken over, means that it can be memorized.

When $\sum_{j=1}^{n} W_{ij} X_j$ is under the threshold value θ_i, means $K(u) \le 0$, the nerve cell is in the state of control. Now the input signal can not be taken over, means that it can not be memorized.

When $\sum_{j=1}^{n} W_{ij} X_j$ is equal to the threshold value θ_i, means $K(u) = 0$, the nerve cell is in the state of uncertainty. If the affinity between the antibody and the antigen is strong, then the nerve cell is activated. In this way, the antigen can be taken over, means that it can be memorized. If the affinity is weak, then the nerve cell is controlled. In this way, the antigen can be thrown away, means that it can not be memorized.

4 Design and Implement of Model

Model algorithm is as follows:

Step1 The objective function, which optimizes network, is input as antigen.
Step2 In order to structure the whole neural network, antibody colony is initialized.
Step3 In order to structure the fitness function, the affinity between the antibody and the antigen is computed.
Step4 The concentration of every antibody is computed, so antibody can be operated by activation or control.
Step5 Cross and mutation are operated.
Step6 If finish condition is satisfied, the algorithm is over. Otherwise return step2.

4.1 Recognition of Antigen

The objective function, which optimizes network, makes the total error of sample least. The objective function is used for the antigen of INN. The definition is as follows:

$$F = \sum_{i=1}^{n} (Y_i - \widehat{Y}_i)^2 \tag{6}$$

In equation (6), n expresses the total of sample.

Y_i expresses the output value of actual network.

$\widehat{Y}_i$ expresses the anticipant output value .

4.2 Initialization of Antibody Colony

Every antibody corresponds with a structure of neural network. The number of latent node and weight value are coded by mixed real number. Every antibody is shown in Fig.2.

number of latent node N	weight value correlated with latent node 1th	weight value correlated with latent node 2th	...	weight value correlated with latent node Nth

Fig. 2. Antibody code

4.3 Structure of Fitness Function

Antibody is randomly selected from colony as the latent cell. When the whole neural network is structured, we can test the objective function of network. We have repeatedly structured the network, until every antibody is tested for certain times. We can compute the affinity between the antibody $Ab(k)$ and the antigen Ag as follows.

$$AY_{Ab(k),\,Ag} = \frac{1}{1+\min\{F_1, F_2, \ldots, F_n\}} \quad (k = 1,2,\ldots,m) \tag{7}$$

In equation (7), n expresses the times that antibody participates in the structure of network.

$Fi\ (i = 1,2,\ldots,n)$ expresses the objective function of the ith network.

$\min\{F_1, F_2, \ldots, F_n\}$ expresses the least value which is in the objective function of n networks.

The affinity between the antibody and the antigen reflects the degree that every antibody matchs the objective function. So the affinity between the antibody $Ab(k)$ and the antigen Ag is defined as the fitness of the antibody $Ab(k)$, means $f(k) = AY_{Ab(k),\,Ag}\ \ (k = 1,2,\ldots,m)$.

4.4 Colony Selection Based on Concentration

The affinity between the antibody and the antibody reflects the similar degree of antibodies. When antibodies are similar, the affinity is strong. Or else, the affinity is weak. The affinity between the antibody $Ab(k)$ and the antibody $Ab(d)$ is computed as follows.

$$AY_{Ab(k),\,Ab(d)} = \frac{1}{1+H_{k,d}} \tag{8}$$

$H_{k,d}$ is the distance between two antibodies. It is expressed by Euclid distance.

The concentration $C(k)$ of the antibody $Ab(k)$ is computed as follows.

$$C(k) = \frac{\sum_{d=1}^{L} AY_{Ab(k),\,Ab(d)}}{L} \tag{9}$$

In equation (9), L expresses the number of antibody.

Immune system controls the antibody which has strong concentration and weak affinity, and activates the antibody which has weak concentration and strong affinity. Based on such characteristic, we can acquire the equation of the probability selection as follows, which is based on the concentration of the antibody $Ab(k)$.

$$P(k) = a * f(k) * (1 - C(k)) \tag{10}$$

In equation(10), a is a adjustable parameter, $a \in [0,1]$.

4.5 Cross and Mutation

In the paper, we adopt cross of two points and mutation of gauss Mutation of gauss is the mutation that every antibody is decoded as the structure of the corresponding network. All weight values of the network are modified as follows.

$$W_{kj} = W_{kj} + \partial * \sqrt{f(k)} * \mu(0,1) \quad (k = 1,2,...,m) \ (j = 1,2,...,n) \tag{11}$$

In equation (11), ∂ is a adjustable parameter, $\partial \in (-1,1)$. $\mu(0,1)$ is a gauss operator. After mutation, all latent node and weight values are used for forming a new antibody.

5 Experiment Results and Analysis

In order to test the performance of the INN which is based on extenics, the paper simulates testing function. The simulation is operated by MATLAB.

Algorithm (1) expresses the algorithm of the INN which is improved based on extenics. Algorithm (2) expresses the algorithm of the INN which has not been improved. Output (1) expresses the output results of the sample. Output (2) expresses the output results which are improved based on extenics. Output (3) expresses the output results which have not been improved.

The example of nonlinearity function is as follows.

$$f(x_1, x_2) = x_1 * |\sin(x_2)| + (x_1 + x_2)^2, x_1, x_2 \in [1, \pi] \tag{12}$$

Algorithm (1) and algorithm (2) are separately used for closing the above nonlinearity function. The most fitness of two algorithms is shown in Table 1.

Table 1. Data of the most fitness

Evolution Time	Algorithm(1)	Algorithm(2)
0	0.15	0.12
50	0.80	0.47
100	0.856	0.73
150	0.86	0.785
200	0.865	0.79

The change is shown in Fig.3. By all appearances, the algorithm of the INN, which is improved, has a obvious improvement on the speediness and the stability of the convergence.

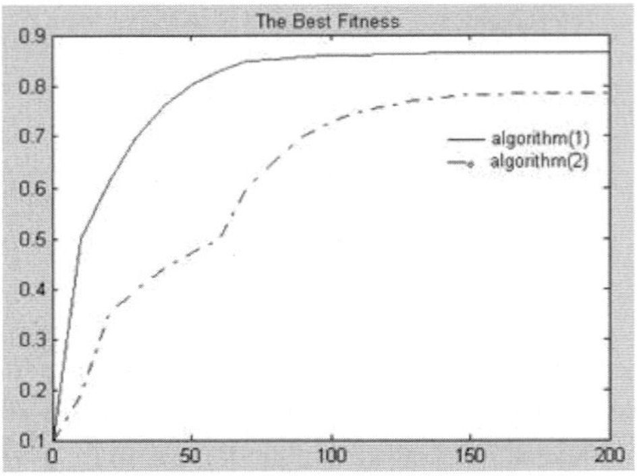

Fig. 3. Change of the most fitness

After learning of 200 generations, we can acquire the comparative perfect results. The output results are shown in Fig.4.

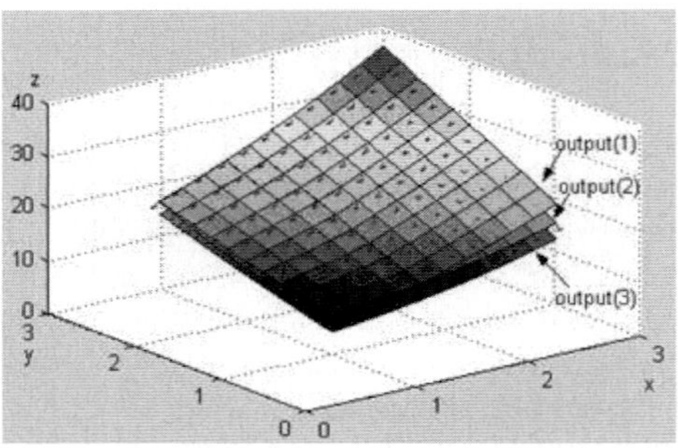

Fig. 4. Simulation results

6 Conclusions

The paper presents a new INN model which is based on extenics. With matter-element analysis, the model can solve the problem that antibody identifies and memorizes antigen in immune system. The model also can make correct judgments about activation or control of nerve cell. Consequently, the structure design of INN can be optimized. Experiment results show that the model not only saves the whole convergence of GA, but also well control the process of network optimization. The structure of the new model effectively integrates the theory of extenics, INN and GA. It also provides a new method for INN which extenics is applied in. So the new model has great reference value.

Acknowledgements. The authors are grateful to National Natural Science Foundation of China (60272089) and Guangdong Provincial Natural Science Foundation of China (980406) (04009464) for decisive support.

References

1. Cai, W.: The extension set and incompatible problem. Science Exploration (in Chinese) 3(1), 83 (1983)
2. Yu, Y.Q.: The extended detecting technology. Engineering Science (In Chinese) 3(4) (April 2001)
3. Cai, W.: The Extension theory and its application. Chinese Science Bulletin 44(17), 1538–1548 (1999)
4. Cai, W.: Matter-element Models and Their Application (in Chinese). Science and Technology Documentation Publishers, Beijing (1994)
5. Cai, W., Yang, C.Y., Lin, W.C.: Methods of Extension Engineering (in Chinese). Science Press, Beijing (1997)
6. Yang, C.Y.: Event element and its application. The theory of System Project and Practice (1998)
7. Watkins, A., Timmis, J.: Artificial immune recognition system (AIRS): revisions and refinements [A]. In: Timmins (ed.) Artificial Inunune System, Berlin (2003)
8. De Castro, L.N., Zuben, F.J.: An evolutionary immune network for data clustering [J]. In: IEEE Brazilian Symposium on Artificial Neural Networks, vol. 1, pp. 84–89. IEEE Computer Society Press, Los Alamitos (2000)
9. Zak, M.: Physical model of immune inspired computing[J]. Infollmtion Sciences 129(1-4), 61–79 (2000)
10. Nikolaev, N.Y., Iba, H.: Learning Polynomial Feed-forward Neural Networks by Genetic Programming and Back-propagation [J]. IEEE Trans.on Neural Networksl 4(2), 337–350 (2003)
11. Watkins, A., Boggess, L.: A new classifier based on resource limited artificial immune systems[A]. In: Proc. of Congress on Evolutionary Computation, Part of the World Congress on Computational Intelligence[C], Springer, Heidelberg (2002)
12. De Castro, L.N., Von Zuben, F.J.: Immune and neural network models: Theoretical and empirical comparisons [J]. International Journal of Computational Intelligence an Application (IJCIA) 1(3), 239–257 (2001)
13. Hofmeyr, S.A., Forrest, S.: Architecture for an Artificial Immune System. Evolutionary Computation 8(4), 443–473 (2000)

A Novel ANN Model Based on Quantum Computational MAS Theory*

Xiangping Meng[1], Jianzhong Wang[2], Yuzhen Pi[1], and Quande Yuan[1]

[1] Department of Electrical Engineering, Changchun Institute of Technology
Changchun, Jilin Province, China, 130012
xp_meng@ccit.edu.cn
[2] Department of Information Engineering, Northeast Dianli University
Jilin, Jilin Province, China, 132012
wangjianzhong681@163.com

Abstract. Artificial Neural Networks (ANNs) are powerful computational modeling tools, however there are still some limitations in ANNs. In this paper, we construct a new artificial neural network, which based on MAS theory and quantum computing algorithm. All nodes in this new ANN are presented as quantum computational (QC) agents, and these QC agents have learning ability via implementing reinforcement learning algorithm. This new ANN has powerful parallel-work ability and its training time is shorter than classic algorithm. Experiment results show this method is effective.

1 Introduction

Artificial Neural Networks (ANNs) are powerful computational modeling tools, which is found extensive acceptance in many disciplines for modeling complex real-world problems. ANN may be defined as structures comprised of densely interconnected adaptive simple processing-elements (called artificial neurons or nodes), which are capable of performing massively computations for data processing and knowledge representation[1,2]. Although ANNs are drastic abstractions of the biological counterparts, the idea of ANNs is not to replicate the operation of the biological systems but to make use of what is known about the functionality of the biological networks for solving complex problems. The ANNs have gained great success, however there are still some limitations in ANNs. Such as:

- Most of the ANNs are not really distribute, so it's nodes or neurons cannot parallel-work.
- Training time is long.
- The nodes number is limited by the capability of computer.

To solve these problems we try to reconstruct the ANN using Quantum Computational Multi-Agent System (QCMAS)[3], which consists of quantum computing agents.

* Supported by Key Project of the Ministry of Education of China for Science and Technology Researchment(ID:206035).

K. Li et al. (Eds.): LSMS 2007, LNCS 4688, pp. 28–35, 2007.

Multi-agent technology is a hotspot in the recent study on artificial intelligence. The concept of Agent is a natural analogy of real world. Quantum computing is capable of processing huge numbers of quantum states simultaneously, in parallel. Quantum searches can be proven faster compared with classical searches. In certain cases, QCMAS can be computational more powerful than any MAS by means of properly designed and integrated QC agents.

In this paper, we give a new method to construct artificial neural network, which based on QCMAS theory and reinforcement learning algorithm. All nodes in this new ANN are presented as QC agents, and these QC agents have learning ability via implementing reinforcement learning algorithm. We use QCMAS reinforcement learning algorithm as this new neural network's learning rules.

2 Related Work

2.1 Quantum Computing

A. Quantum Bits

The unit of quantum information is the quantum bit (qubit). The qubit is a vector in a two-dimensional complex vector space with inner product, represented with quantum state. It is an arbitrary superposition state of two-state quantum system[4,5]:

$$|\psi\rangle = \alpha|0\rangle + \beta|1\rangle,\ |\alpha|^2 + |\beta|^2 = 1 \tag{1}$$

Where α and β are complex coefficients. $|0\rangle$ and $|1\rangle$ correspond to classical bit 0 and 1. $|\alpha|^2$ and $|\beta|^2$ represent the occurrence probabilities of $|0\rangle$ and $|1\rangle$ respectively when the project $|\psi\rangle$ is measured, the outcome of the measurement is not deterministic. The value of classical bit is either Boolean value 0 or value 1, but qubit can simultaneously store 0 and 1, which is the main difference between classical and quantum computation.

B. State Space

The hadamard transform (or hadamard gate) is one of the most useful quantum gates and can be represented as [6]:

$$H \equiv \frac{1}{\sqrt{2}}\begin{bmatrix} 1 & 1 \\ 1 & -1 \end{bmatrix} \tag{2}$$

Through the Hadamard gate, a qubit in the state $|0\rangle$ is transformed into a superposition state of two states, i.e. $H|0\rangle = \frac{1}{\sqrt{2}}|0\rangle + \frac{1}{\sqrt{2}}|1\rangle$; Similarly, a qubit in the state $|1\rangle$ is transformed into the superposition state $H|1\rangle=\frac{1}{\sqrt{2}}|0\rangle-\frac{1}{\sqrt{2}}|1\rangle$, i.e. the magnitude of the amplitude in each state is $1/\sqrt{2}$ but the phase of the amplitude in the state $|1\rangle$ is inverted. In classical probabilistic algorithms the phase has no analog since the amplitudes are in general complex numbers in quantum mechanics. Now consider a quantum system described by n qubits, it has 2^n possible states. To prepare an equally weighted superposition state, initially let each qubit lie in the state $|0\rangle$, then we can perform the transformation H on each qubit independently in sequence and thus change the state of the system. The state transition matrix representing this operation will be of $2^n \times 2^n$ dimensions and can be implemented by n shunt-wound Hadamard gates. This process can be represented into:

$$H^{\otimes n}\left|\overbrace{00\cdots0}^{n}\right\rangle=\frac{1}{\sqrt{2^n}}\sum_{a=00\cdots0}^{\overbrace{11\cdots1}^{n}}|a\rangle \tag{3}$$

C. Grover's Searching Algorithm

Grover's searching algorithm is well known as searching an unstructured database of $N = 2^n$ records for exactly one record which has been specifically marked.

Classically, this process would involve querying the database, at best once and at worst N times, so on average $N/2$ times. In other words the classical search scales $O(N)$ and is dependent upon structuring within the database [7]. Grover's algorithm offers an improvement to $O(\sqrt{N})$ and works just as well in a structured database as a disordered one.

As we know, the probability to get a certain vector in a superposition in a measurement is the square of the norm of its amplitude. Luk Grover's idea was to represent all numbers in $\{0, 1\}$ via a superposition ϕ of n qubits, and shift their amplitudes so that the probability to find an element out of L in a measurement is near 1. To achieve this, a superposition of n qubits in which all elements having the same amplitude is created. Then the following operations are applied alternately:

- Amplitudes of all elements $\in L$ are reflected at zero.
- Amplitudes of all elements are reflected at the average of all amplitudes.

Those two operators together are called the Grover-operator. The other fundamental operation required is the conditional phase shift operation which is an important element to carry out Grover iteration. According to quantum information theory, this transformation may be efficiently implemented on a quantum computer. The conditional phase shift operation does not change the probability of each state since the square of the absolute value of the amplitude in each state stays the same.

2.2 Quantum Computational Multi-Agent System

A. QC Agents
A Quantum Computational agent extends an intelligent agent by its ability to perform both classical agent, and quantum computing to accomplish its goals individually, or in joint interaction with other agents.

QCAs are supposed to exploit the power of quantum computing to reduce the computational complexity of certain problems where appropriate.

B. QC Multi-Agent System (QCMAS)
A quantum computational multi-agent system is a multi-agent system that consists of QC agents that can interact to jointly accomplish their goals. QCMAS can be computational more powerful than any MAS by means of properly designed and integrated QC agents.

2.3 Multi-Agent Q-Learning

Learning behaviors in a multi-agent environment is crucial for developing and adapting multi-agent systems. Reinforcement learning (RL) has been successful in finding optimal control policies for agents operating in a stationary environment, specifically a Markov decision process (MDP)[8,9].

Q-learning is a standard reinforcement learning technique[10,11]. Multi-agent environments are inherently non-stationary since the other agents are free to change their behavior as they also learn and adapt. In multi-agent Q-learning, the Q-function of agent i is defined over states and joint-action vectors $\vec{a}=(a^1,a^2,\cdots,a^n)$, rather than state-action pairs. The agents start with arbitrary Q-values, and the updating of Q value proceed as following:

$$Q^i_{t+1}(s,\vec{a})=(1-\alpha)Q^i_t(s,\vec{a})+\alpha\left[r^i_t+\beta\cdot V^i(s_{t+1})\right] \tag{4}$$

Where $V^i(s_{t+1})$ is state value functions, and

$$V^i(s_{t+1})=\max_{a^i\in A^i} f^i(Q^i_t(s_{t+1},\vec{a})) \tag{5}$$

In this generic formulation, the keys elements are the learning policy, i.e. the selection method of the action $\vec{a}$, and the determining of the value function $V^i(S_{t+1})$, $0\le\alpha<1$. The different action selection / value function computation methods generate different multi-agent learning algorithms.

We just let

$$\pi^i(s_{t+1},a^i)=\arg\max_{\pi^i} Q(S,\vec{a}) \tag{6}$$

3 A New Artificial Neural Network

ANNs can be successfully employed in solving complex problems in various fields of mathematics, engineering, medicine, economics, meteorology, neurology and many others[12].

The fundamental processing element of a neural network is a neuron (node). Artificial Neural networks are the clustering of the artificial neurons. This clustering occurs by creating layers, which are connected to one another. Basically, all artificial neural networks have a similar structure or topology as shown in Fig. 1: input layer, output layer, and possibly one or more intermediate layers of neurons, called hidden layer. In the structure some of the neurons interface to the real world to receive their inputs. Other neurons provide the real world with the network's outputs. All the rest of the neurons are hidden from view in the hidden layers receive the signals from all of the neurons in the input layer. After a neuron performs its function, it passes its output to all of the neurons in the output layer, providing a feed forward path to the output.

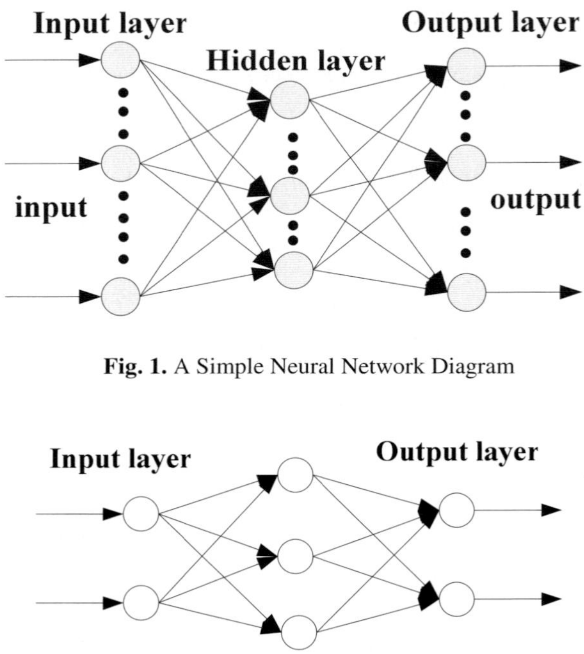

Fig. 1. A Simple Neural Network Diagram

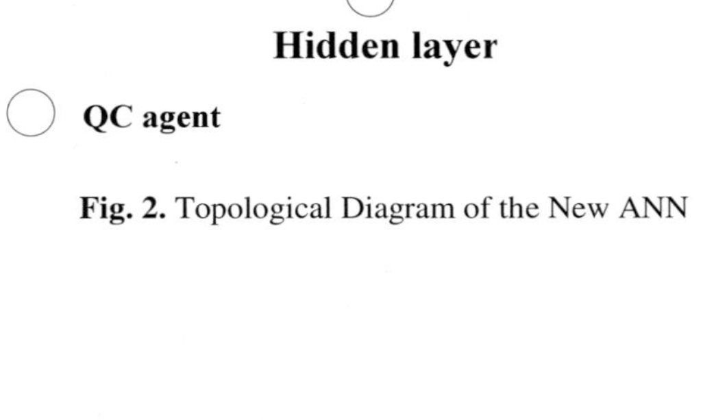

Fig. 2. Topological Diagram of the New ANN

Each neuron in the ANNs can process information in a dynamic, interactive, and self-organizing way. So the neuron have the intelligent characteristic just like agent and ANNs just as multi-agent system that can be trained to solve problems that are difficult for conventional computers or human beings.

Now we can construct the ANN using QCMAS theory: each neuron or node is a QC agent. We constructed a simple three layers ANN, whose topological diagram is shown in Fig. 2. There are three types of QC agents: input nodes QC agents, output nodes QC agents, and hidden layer QC agents. These QC agents can be not only on same computer but also different ones. The other QC agent can find the node QC agents that it need to link. The QCMAS can be distributed, that's to say QC agents can be running on different computers, so the number of QC agents is not limited. And every agent has learning ability via implementing reinforcement learning.

Succeed to the characteristic of the quantum computing and MAS, the new ANN is really distributed and it's nodes or neurons can organize dynamically which can process information in parallel. The new ANN has more powerful computing ability than the classic ANNs.

4 A Novel ANN Training Method

As discussed above, the new ANN's training can be regard as the QCMAS's reinforcement learning.

4.1 Define States and Actions

According to the above method, we can define the states and actions in quantum computational multi-agent reinforcement learning system, whose states may lie in a superposition state:

$$\left|s^{(m)}\right\rangle = \sum_{s=00\cdots0}^{\overbrace{11\cdots1}^{m}} C_s \left|s\right\rangle \tag{7}$$

The mapping from states to actions is $f(s)=\pi : S \to A$, and can be defined as following:

$$f(s) = \left|a_s^{(n)}\right\rangle = \sum_{s=00\cdots0}^{\overbrace{11\cdots1}^{n}} C_a \left|a\right\rangle \tag{8}$$

Where C_s and C_a is probability amplitudes of state $\left|s\right\rangle$ and action $\left|a\right\rangle$, respectively.

4.2 The Process of QCMAS Q-Learning

The process of QCMAS Q-learning is shown in Table 1.

Table 1. The Process of QCMAS Q-Learning

1. Initialize:

(1) Select initial learning rate α and discount factor β and let t=0;

(2) Initialize the state and action to the equal superposition state, respectively;

$$\left|s^{(m)}\right\rangle = \sum_{s=00\cdots0}^{\overbrace{11\cdots1}^{m}} C_s \left|s\right\rangle,\ f(s) = \left|a_s^{(n)}\right\rangle = \sum_{s=00\cdots0}^{\overbrace{11\cdots1}^{n}} C_a \left|a\right\rangle;$$

(3) For all states $\left|s^{(m)}\right\rangle$ and actions $\left|a_s^{(n)}\right\rangle$,

Let $Q_0^i(s^{(0)}, a_s^{(1)}, a_s^{(2)}, \cdots, a_s^{(n)}) = 0$;

$$\pi_0^i(s^{(0)}, a_s^{(i)}) = \frac{1}{n}, \overline{\pi}_0^i(s^{(0)}, a_s^{(i)}) = \frac{1}{n}, C^i(s) = 0;$$

2. Repeat the following process (for each episode)

(1) For all states, observe $\left|a_s^{(n)}\right\rangle$ and get $\left|a\right\rangle$, using formula (8);

(2) Execute action $\left|a\right\rangle$, observe reward $(r_t^{(1)}, \cdots, r_t^{(n)})$ and new states $\left|s_t^{(m)}\right\rangle$;

(3) Update Q_t^i using formula (6) using Grover- operator.

(4) Update probability amplitudes, i.e., explore next action.

Until for all states $|\Delta V(S)| < \varepsilon$

5 Experiment and Results

We construct a simple three layer networks to training it to learn XOR.

The experiment results are shown in Fig. 3.

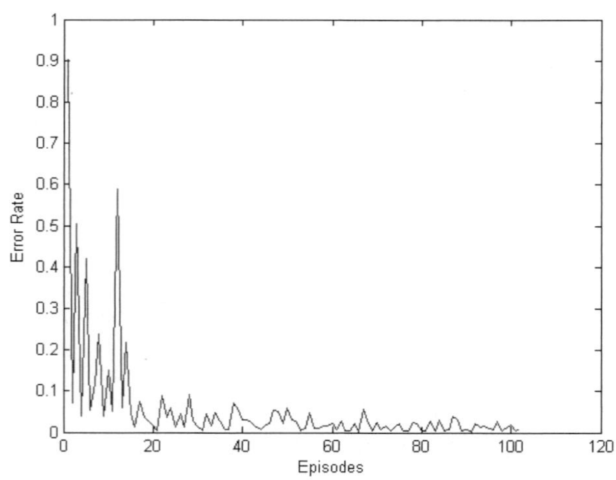

Fig. 3. The error rates vs number of training episodes

6 Conclusions

In this paper, we introduced quantum theory, multi-agent reinforcement learning and ANN respectively, and proposed a new method to construct artificial neural networks with the QC agents based on the theories of quantum computation and agent. We regard the new ANN as a QCMAS and use the QCMAS Q-learning to train the ANN. We adopt quantum-searching algorithm in agents' action selection policy of multi-agent reinforcement learning, which can speed up the learning of the new ANN. The results show this method is effective.

The combination of quantum computation methods and multi-agent reinforcement learning used in ANN is an attempt. With the development of quantum computation, MAS and ANN, we will continue attending this aspect, and more research will be done in the future work.

References

1. Hecht-Nielsen: Neuro computing. Addison-Wesley, Reading, MA (1990)
2. Schalkoff, R.J.: Artificial Neural Networks. McGraw-Hill, New York (1997)
3. Klusch, M.: Toward Quantum Computational Agents. In: Nickles, M., Rovatsos, M., Weiss, G. (eds.) AUTONOMY 2003 LNCS (LNAI), vol. 2969, Springer. Heidelberg (2004)
4. Aharonov, D.: Quantum Computation. LANL Archive quant-ph/981203 (1998)
5. Benenti, G., Casati, G., Strini, G.: Principles of quantum computation and information, vol. 1, pp. 144–150. World Scientific, Singapore (2005)
6. Nielsen, M.A., Chuang, I.L.: Quantum Computation and Quantum Information. Cambridge University Press, Cambridge (2000)
7. Grover, L.: A Fast Quantum Mechanical Algorithm for Database Search. In: Proc. 28th Annual ACM Symposium on Theory of Computation, pp. 212–219. ACM Press, New York (1996)
8. Shoham, Y., Powers, R., Grenager, T.: Multi-agent reinforcement learning: a critical survey. Technical report, Computer Science Department, Stanford University, Stanford (2003)
9. Sutton, R.S., Barto, A.G.: Reinforcement Learning. MIT Press, Cambridge, MA (1998)
10. Watkins, C.J.C.H., Dayan, P.: Q-learning. Machine Learning 8(3/4), 279–292 (1992)
11. Hu, J., Wellman, M.: Multiagent Q-learning. Journal of Machine Learning (2002)
12. Basheer, I.A., Hajmeer, M.: Artificial neural networks: fundamentals, computing, design, and application. Journal of Microbiological Methods 43, 3–31 (2000)

Robust Speech Endpoint Detection Based on Improved Adaptive Band-Partitioning Spectral Entropy

Xin Li[1,2,3], Huaping Liu[3], Yu Zheng[4], and Bolin Xu[1]

[1] Department of Electronic and Information Engineering, Nanjing University, 210093 Nanjing, China

[2] State Key Lab of Pattern Recognition, Institute of Automation, Chinese Academy of Sciences, 100080 Beijing, China

[3] School of Electromechanical Engineering and Automation, Shanghai University, 200072 Shanghai, China

[4] School of Computer Engineering and Science, Shanghai University, Shanghai 200072, China

Abstract. The performance of speech recognition system is often degraded in adverse environments. Accurate Speech endpoint detection is very important for robust speech recognition. In this paper, an improved adaptive band-partitioning spectral entropy algorithm was proposed for speech endpoint detection, which utilized the weighted power spectral subtraction to boost up the signal-to-noise ratio (SNR) as well as keep the robustness. The idea of adaptive band-partitioning spectral entropy is to divide a frame into some sub-bands which the number of it could be selected adaptively, and calculate spectral entropy of them. Although it has good robustness, the accuracy degrades rapidly when the SNR are low. Therefore, the weighted power spectral subtraction is presented for reducing the spectral effects of acoustically added noise in speech. The speech recognition experiment results indicate that the recognition accuracy have improved well in adverse environments.

1 Introduction

The speech endpoint detection is aim to distinguish the speech and non-speech phase, it plays the key role in the speech signal processing. Inaccurate endpoint detection causes the recognition ratio lower and increases the computation amounts. The research shows that, the probability of false recognition of isolating word is over half cause by unreliable endpoint detection [1]. The method of the ideal detection should have the traits of reliability, stability, accuracy, adaptability and simplicity, which process at any time and needn't prior knowledge. So, it makes sense to explore the proper endpoint detection method for robust speech recognition systems. During the last decades, a number of endpoint detection methods have been developed. We can categorize approximately those methods into two classes [2]. One is based on thresholds. Generally, this kind of method first extracts the acoustic features for each frame of signals and then compares these values of features with preset thresholds to classify each frame. The other is pattern-matching method that needs estimate the model

K. Li et al. (Eds.): LSMS 2007, LNCS 4688, pp. 36 – 45, 2007.

parameters of speech and noise signal. The method based on pattern-matching [3]-[5] has the traits of high accuracy, but the disadvantages are model dependency, high complexity and enormous computation. It is difficult to apply for the real-world speech signal processing system.

Compared with pattern-matching method, threshold-based method is simpler and faster since it does not need to keep much training data and train models. Traditional short-time energy and zero-crossing rate method is part of this sort[6]-[8], but it is sensitive to varies types of noise and cannot fully specify the characteristics of a speech signal, the detection effect will become worse in the adverse environment. Several other parameters including linear prediction coefficients (LPCs), Cepstral coefficients, pitch, time-frequency (TF) and adaptive time-frequency (ATF) [9]-[13] faced same situation.

J. L. Shen et al. [14] used the entropy for endpoint detection under adverse conditions. Entropy is a metric of uncertainty for random variables, thus it is definite that the entropy of speeches is different from that of noise signals because of the inherent characteristics of speech spectrums. Bing-Fei Wu, Kun-ching Wang advanced the adaptive band-partitioning spectral entropy (ABSE) concept[15], which introduces the refined adaptive band selection (RABS) method[16] to adaptively select useful bands. Experiment results show the ABSE parameter strengthens the boundary between speech and noise more clearly at poor SNRs. Although this parameter has stability for almost kinds of noises, the effect of endpoint detection is still not satisfied as the decline of SNR. The accuracy reaches 94.8% based on ABSE endpoint detection algorithm when babble noises SNR is 40dB, and it glides down to only 81.8% when the SNR is decline to 0dB.

In this paper, a speech endpoint detection algorithm based on improved adaptive band-partitioning spectral entropy is proposed, which is directed against the disadvantages of speech endpoint detection based on adaptive band-partitioning spectral entropy. Weighted power spectral subtraction is applied to adaptive band-partitioning spectral entropy process for enhancement, while the result of endpoint detection is used to update every frame's background noise estimation for spectral subtraction.

This paper is organized as follows. Section 2 will introduce the theory of spectral entropy and the algorithm of adaptive band-partitioning spectral entropy. In Section 3, weighted power spectral subtraction is derived in detail, and then the procedure for implementing the proposed improved adaptive band-partitioning spectral entropy algorithm is outlined in Section 4. Section 5 discusses the performance of the proposed algorithm under various noise conditions and compares its performance with that of other algorithms. Finally, Section 6 summarizes the results from experiments and discusses future works.

2 Adaptive Band-Partitioning Spectral Entropy

2.1 Spectral Entropy

Entropy is cited by informative theory by Shannon, the basic theory of spectral entropy is the spectrum is first obtained for each frame by fast Fourier transform (FFT). Then the probability density function (pdf) for the spectrum can be calculated by normalizing every frequency component over all frequency components of one frame:

$$P_i = \frac{Y_m(f_i)}{\sum_{k=0}^{N-1} Y_m(f_k)} \qquad i = 1,2,\cdots N \tag{1}$$

where N is the total number of frequency components in FFT, $Y(f_i)$ is the spectral energy of the frequency component f_i, P_i is the corresponding probability density. Since most of the energy of speech is in the region between 250Hz and 3750Hz, the empirical constraint to enhance the discriminability of this pdf between speech and non-speech signals was defined as follow:

$$Y(f_i) = 0, \quad \text{if} \quad f_i < 250HZ \quad \text{or} \quad f_i > 4500HZ \tag{2}$$

After applying the above constraint, the negative spectral entropy for each frame can be calculated:

$$H_m = -\sum_{k=1}^{N} P_k \log P_k \tag{3}$$

The validity of the bad-partitioning spectral entropy as a feature used on endpoint detection is indicated in the following Figure 1 and Figure 2. The entropy of speech signals is different from that of most noise signals because of the intrinsic characteristics of speech spectrums and the different probability density distributions. The spectral entropy is not influenced by the total energy in theory if the spectral distribution keeps unchanged. In practice, the distribution is changed by the actual pronunciation so that the entropy becomes different. However, the change of entropy is small compared with that of the energy. Fig. 1 shows that although the energy of the two waveforms different, the change of the spectral entropy is unconspicuous.

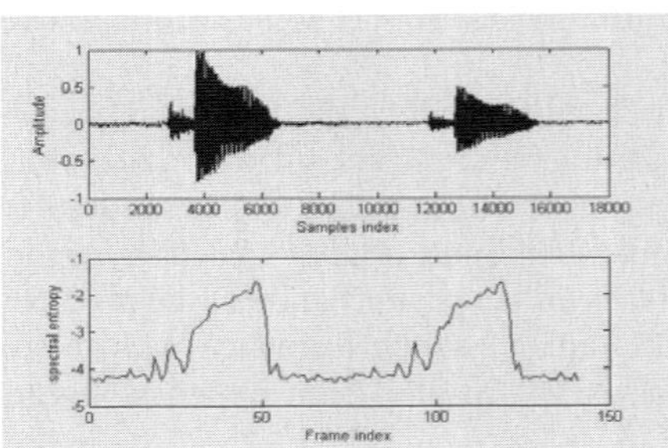

Fig. 1. Waveform and entropy

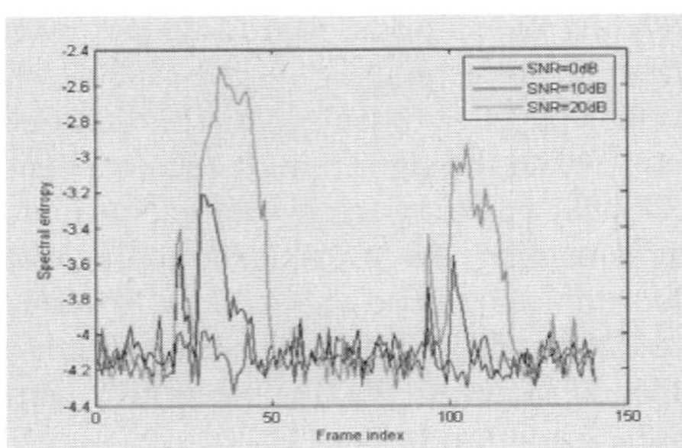

Fig. 2. Entropy under different SNR (0-20dB) conditions with white noise

The band-partitioning spectral entropy is robust to noise to some extent. For example in Fig. 2, each line represents the entropy under different SNR conditions with white noise. With the drop of the SNR, the shape of negative entropy is almost keeping.

Nevertheless, the negative entropy decreases so that endpcint detection becomes more difficult if the SNR decreases. It shows that the spectral entropy parameter needs to be improved.

2.2 Adaptive Band-Partitioning Spectral Entropy

By Analyzing the trait of the band-partitioning spectral entropy form the above, even under the very lower SNR condition, speech still has keeping high entropy, but noises haven't. Shen J. L., et al. [14] raised an improved band-partitioning spectral entropy for endpoint detection which

$$X(k,l) = \sum_{n=1}^{N} H(n) \bullet S(n,l) \bullet \exp(-j2\pi kn / N) \qquad 1 \le k \le N \tag{4}$$

$X(k,l)$ is the fast Fourier transform (FFT) of signal $S(n,l)$, By normalization, the corresponding spectral entropy for a given frame is defined as follows:

$$H(l) = \sum_{i=1}^{N/2} P(i,l) \bullet \log[1 / P(i,l)] \tag{5}$$

Where $H(l)$ is the spectral entropy of the i-th frame. The foregoing calculation of the spectral entropy parameter implies that the spectral entropy depends only on the variation of the spectral energy but not on the amount of it. Consequently, the spectral entropy parameter is robust against changing level of noise.

However, the magnitude associated with each point in the spectrum is easily contaminated by noise and then the performance of endpoint detection would be degraded seriously under low SNRs. This study addresses the multi-band analysis of noisy speech signals to overcome the sensitivity of the spectral magnitude in noisy environments. The band energy of each band for a given frame is described as follows:

$$E_b(m,l) = \sum_{k=1+(m-1)*4}^{k=1+(m-1)*4+3} X_{energy}(k,l) \qquad 1 \le m \le N_b \tag{6}$$

Where N_b is the total band size of each frame(N_b =32), and $E_b(m,l)$ represents the band energy of the m-th band. By using the set of weighting factors $w(m,l)$, the proposed spectral entropy parameter is finally represented as follows:

$$H_b(l) = \sum_{m=1}^{N_b} w(m,l) \bullet P_b(m,l) \bullet \log[1 / P_b(m,l)] \tag{7}$$

Bing-Fei Wu, et al.[15] applied the adaptive band-partitioning to adapt the essence of speech signal by making the every frame Nb different. The experiments of adaptive band-partitioning spectral entropy under kinds of noise envircnment and different SNRs show that, this algorithm has good stability and high accuracy for speech endpoint detection under adverse environments.

3 Weighted Power Spectral Subtraction

3.1 Power Spectral Subtraction

Power spectral subtraction [17] is widely used for speech enhancement. It is easy and the effect of noise-removing is good. It bases on the following basic hypothesis: the speech signals and the noise are not interrelated with each others, and they are additional in frequency domain. Since to subtract the estimate noise form the signal can decline the noise. The deduction is as follows:

Assume that a windowed noise signal *n(i)* has been added to a windowed speech signal *s(k)*, with their sum denoted by *x(i)*. Then

$$x(i) = s(i) + n(i) \tag{8}$$

Taking the Fourier transform， $x(i) \leftrightarrow X(\omega)$, $s(i) \leftrightarrow S(\omega)$, $n(i) \leftrightarrow N(\omega)$, We have：

$$X(\omega) = S(\omega) + N(\omega) \tag{9}$$

$$|X(\omega)|^2 = |S(\omega)|^2 + |N(\omega)|^2 + S(\omega)\overline{N(\omega)} + \overline{S(\omega)}N(\omega) \tag{10}$$

Where， $|X(\omega)|^2$, $|S(\omega)|^2$ and $|N(\omega)|^2$ denote power spectral entropy. $\overline{S(\omega)}, \overline{N(\omega)}$ is the conjugate of $s(\omega)$ and $N(\omega)$ respectively. On the assumption that *s(i)* is irrelevant with the noise *n(i)*,

$$|X(\omega)|^2 = |S(\omega)|^2 + |N(\omega)|^2 \tag{11}$$

$$|S(\omega)|^2 = |X(\omega)|^2 - |N(\omega)|^2 \tag{12}$$

It shows that the power spectral entropy of speech signal can derived from the subtraction of the power spectral entropy of speech signal and noise. Since only the speech with noise we can get under most situations, so $|N(\omega)|^2$ in the formula (3-5) can not to be acquired directly. Usually, the average power spectral entropy before the speech frame $E[\,|N(\omega)|^2]$ is used to estimate $|N(\omega)|^2$:

$$|S(\omega)| = \left(|X(\omega)|^2 - E\left[|N(\omega)|^2\right]\right)^{\frac{1}{2}} \tag{13}$$

then by taking the inverse Fourier transform of $|S(\omega)|$□the enhancement of speech signal *s(i)* can be obtained.

3.2 Weighted Power Spectral Subtraction

The main defect of the traditional power spectral subtraction is by using the noise estimation of several beginning frames instead of the whole phase noise. It is not practical in real world. Since the frequency and amplitude of noise spectral entropy are changing.

Liu XiaoMing, et al. [18] raised the weighted power spectral entropy subtraction to alleviate this problem. The estimation of noise $E[\,|N(\omega)|^2]$ is updating for each frame by detecting the endpoint of every frame.

In [18], the short-time energy-zero is the parameter to be used for endpoint detection. The experiment results show that the weighted power spectral entropy subtraction has some extent effects in restraining the noise, but under low SNR circumstance, the update of the voiceless phase is ineffective because of the weak robustness of the short-time energy-zero parameter.

Therefore, the adaptive band-partitioning spectral entropy is used to replace the short-time energy-zero in this paper, and it obviously better theoretically[2].

4 Improved Adaptive Band-Partitioning Spectral Entropy

The adaptive band-partitioning spectral entropy (ABSE) is robust than other parameters such as short-time energy, zero-crossing rate or ATF, it is reasonable to choose it for endpoint detection. Since the effects of ABSE are influenced by the SNRs, we apply the weighted power spectral subtraction for enhancement to solve this problem in this paper. On the other hand, the adaptive band-partitioning spectral entropy is used to replace the short-time energy-zero parameter in the traditional weighted power spectral subtraction. It makes our approach combines all the merits of these two methods. The steps of the algorithm are as follows:

Step1: Using the beginning voiceless phases estimate speech background noise.

$$\mu = \frac{1}{5}\sum_{l=1}^{5} H_b(l) \tag{14}$$

$$\sigma_n = \frac{1}{5-1}\sum_{l=1}^{5} (H_b(l) - \mu_n)^2 \tag{15}$$

$$T_s = \mu + \alpha * \sigma \tag{16}$$

Where α is an empirical constant.

Step 2: Speech enhancement.

The voiceless phase usually has several frames, so it needs to calculate the FFT of each frame, then to compute the average value of those frames. After the average power spectral $E[\,|N(\omega)|^2]$ of voiceless phase has obtained, the spectral entropy of speech signal can be computed by using formula (13). Then the enhanced speech signal *s(i)* of current frame has got by doing IFFT.

Step 3: Speech endpoint detection by using adaptive band-partitioning spectral entropy.

Step 4: Update the voiceless phase by weighting.

If the step 3 shows the current frame still in the voice phase, fetch the data of next frame, turn to step 2. If it shows the current frame in the phase of voiceless, fetch the data of current frame, average it with the data of last frame, then update the estimation of noise of the voiceless phase before turn to Step 2. When all the data have been fetched, the stage of speech endpoint detection is finished.

5 Experiment Results

5.1 Parameter Selection

The speech we used in the experiment is a recorded English number ("Two'), the additional white noise at 5 dB comes from NOISE-92 database. Fig. 3 and Fig. 4 show the enhancement result between the traditional spectral subtraction and weight spectral subtraction. The voiceless sound or fricative of the speech was cut by the traditional spectral subtraction speech enhancement, while it could be remained in the weight spectral subtraction.

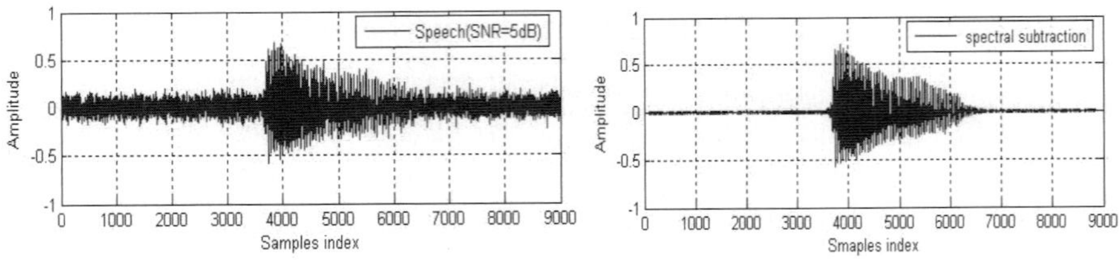

Fig. 3. Tradition power spectral speech enhancement

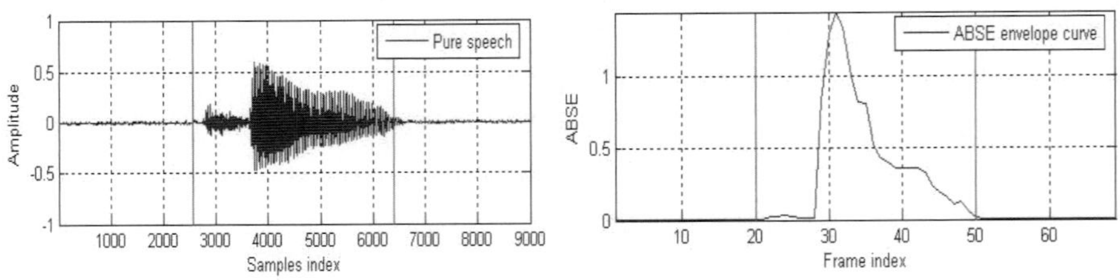

Fig. 4. Weighted power spectral speech enhancement

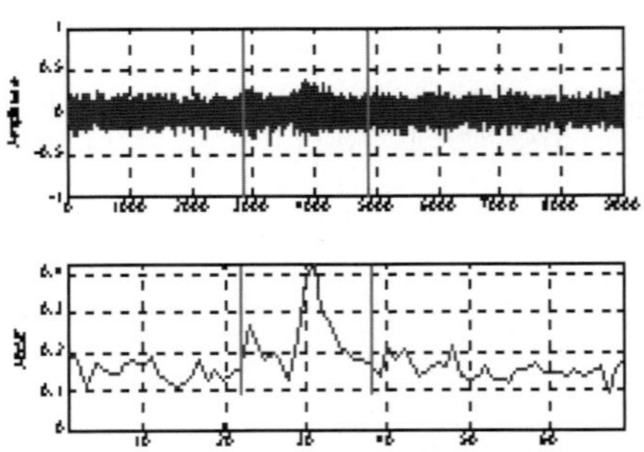

Fig. 5. Endpoint detection based on ABSE for clean speech

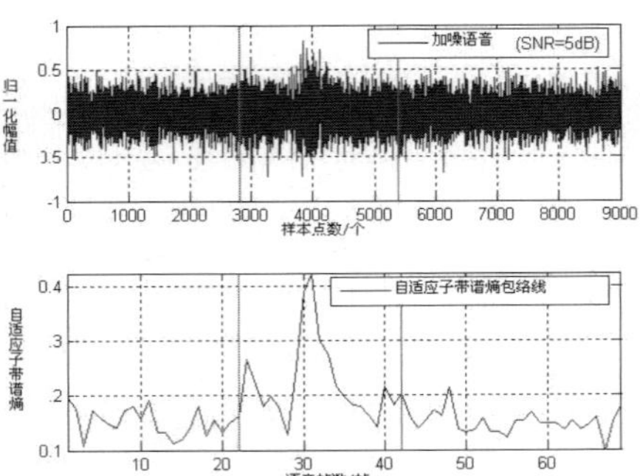

Fig. 6. Endpoint detection based on ABSE for noised speech (SNR=5dB)

Fig 5 and Fig.6 are endpoint based on adaptive band-partitioning spectral entropy without noise or with noise at SNR=5dB. It shows the adaptive band-partitioning spectral entropy under the noise would be great corrupted than using the clean speech. Thus the parameter needs to be improved by enhancement.

Fig 7 is the graph of speech endpoint detection based on improved adaptive band-partitioning spectral entropy SNR=5dB.The two vertical red line separately represent the starting point and ending point for detecting to speech, from the simulation comparative graph, it shows that the feasibility of algorithmic in this article.

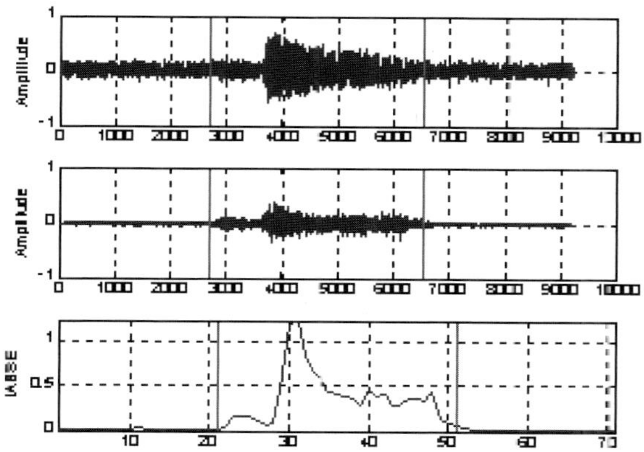

Fig. 7. Endpoint detection based on IABSE for noised speech (SNR=5dB)

5.2 Speech Recognition Experiment

The speech database used in the experiments here contains the isolated words in Mandarin Chinese produced by specific speaker. Every word has 10 samples, where three among them were taken the testing words while the others were used to train a speech model. 12 order mel-frequency cepstrum coefficients and the corresponding 12 delta cepstrum coefficients were used as the feature parameters. First order, left-to-right continuous density hidden Markov models (CHMM) were trained for each word with four states for each model. The recognition accuracy was evaluated by the average of the three testing samples. A variety of types of noise were collected from the NOISEX-92 noise-in-speech database including white noise, babble and volvo noise (car noise). Furthermore, signal to noise ratios are set to 40dB, 15dB, 10dB, 0dB respectively.

Table 1 shows the comparative experiment results. By using traditional endpoint detection parameter such as short-time zero-cross and energy (ZCR\AMP), it works under high SNRs. It is completely not able to detect endpoint (N/A) when SNRs degrade. The ABSE algorithm could detect endpoint under lower SNRs, but the performance declined rapidly. For example, the recognition rate of ABSE with white noise is as high as 89.93% at 40dB, when it reduces to 50.36% at 0dB. However, 96.97% and 69.24% recognition rates can be achieved respectively by using the proposed algorithm (IABSE).

Table 1. Speech recognition accuracy

		ZCR/AMP(%)	ABSE(%)	IABSE(%)
clean		95.6	96.97	100
White	SNR=40dB	74.24	83.33	92.4
	SNR=15dB	N/A	89.35	92.64
	SNR=10dB	N/A	71.21	83.36
	SNR=0dB	N/A	50.36	69.24
Babble	SNR=40dB	N/A	84.45	94.45
	SNR=15dB	N/A	81.82	84.85
	SNR=10dB	N/A	71.21	74.24
	SNR=0dB	N/A	N/A	60.73
Volvo	SNR=40dB	93.94	95.6	97.85
	SNR=15dB	92. 24	93. 94	94. 45
	SNR=10dB	89. 93	90. 90	91. 67
	SNR=0dB	N/A	52. 16	65. 34

6 Conclusion

The improved Adaptive Band-Partitioning Spectral Entropy combined the weighted power spectral subtraction and the adaptive band-partitioning spectral entropy. It detects the endpoint while doing enhancement. The proposed algorithm remains the voiceless sound and fricative after enhancement, thus the accuracy of speech endpoint detection increased. The robustness of speech recognition was improved by the procedure of enhancement.

The experiment also indicates that using spectral entropy for endpoint detection with babble noise is lower than that with other noises. The recognition rates at same SNR are differ nearly 10 percent. The reason is the babble spectral entropy is similar with speech spectral entropy [19]. For solving this problem, the discriminative features [20] could be further improvement.

Acknowledgments. This work was supported by Shanghai Leading Academic Disciplines (under Grant No.T0103).

References

1. Bush, K., Ganapathiraju, A., Kornman, P.: A Comparison of Energy-based Endpoint Detectors for Speech Signal Processing [C]. In: MS State DSP Conference, pp. 85–98 (1995)
2. Jia, C., Xu, B.: An Improved Entropy based Endpoint Detection Algorithm[C]. In: ISCSLP, p. 96 (2002)
3. En-qing, D., He-ming, Z., Ya-tong, Z., Xiao-di, Z.: Applying Support Vector Machines to Voice Activity Detection. Journal of China Institute of Communications 24(3), 3 (2003)
4. Sohn, J., Kim, N.S., Sung, W.: A Statistical Model-Based Voice Activity Detection. IEEE Signal Processing Letters 6(1), 1–3 (1999)

5. Qi, Y., Hunt, B.R.: Voiced-unvoiced-silence classification of speech using hybrid features and a network classifier. IEEE Tran. Speech Audio Processing 1, 250–255 (1993)
6. Lamel, L., Labiner, L., Rosenberg, A., Wilpon, J.: An improved endpoint detect for isolated word recognition. IEEE ASSP Mag. 29(4), 777–785 (1981)
7. Savoji, M.H.: A robust algorithm for accurate endpointing of speech. Speech Commun. 8, 45–60 (1989)
8. Ney, H.: An optimization algorithm for determining the endpoints of isolated utterances. In: Proc. ICASSP, pp. 720–723 (1981)
9. Rabiner, L.R., Sambur, M.R.: Voiced-unvoiced-silence detection using the Itakura LPC distance measure. In: Proc. ICASSP, pp. 323–326 (May 1977)
10. Haign, J.A., Mason, J.S.: Robust voice activity detection using cepstral features. In: Proc. IEEE TEN-CON, pp. 321–324. IEEE Computer Society Press, Los Alamitos (1993)
11. Chengalvarayan, R.: Robust energy normalization using speech/nonspeech discriminator for German connected digit recognition. In: Proc. Eurospeech, pp. 61–64 (September 1999)
12. Junqua, J.C., Mak, B., Revaes, B.: A robust algorithm for word boundary detection in the presence of noise. IEEE Trans. Speech Audio Process. 2(4), 406–412 (1994)
13. Wu, G.D., Lin, C.T.: Word boundary detection with mel-scale frequency bank in noise environment. IEEE Trans. Speech Audio Process. 8(3), 541–554 (2000)
14. Shen, J.L., Hung, J.W., Lee, L.S.: Robust entropy-based endpoint detection for speech recognition in noisy environments. In: ICSLP (1998)
15. Wu, B.-F., Wang, K.-C.: Robust Endpoint Detection Algorithm Based on the Adaptive Band-Partitioning Spectral Entropy in Adverse Environments. IEEE Transactions On Speech and Audio Processing 13(5(9)), 762–775 (2005)
16. Wu, G.-D., Lin, C.-T.: Word Boundary Detection with Mel-Scale Frequency Bank in Noisy Environment. IEEE Transactions On Speech and Audio Processing 8(5(09)), 541–554 (2000)
17. Boll, S.F.: Suppression of acoustic noise in speech using spectral subtraction. IEEE Transaction on Acoustics, Speech and Signal Proc. 27, 113–120 (1979)
18. XiaoMing, L., Sheng, Q., ZongHang, L.: Simulation of Speech Endpoint Detection. Journal of system simulation 17(8(8)), 1974–1976 (2005)
19. Hua, L.-s., Yang, C.-h.: A Novel Approach To Robust Speech Endpoint Detection In Car Environments. In: ICASSP, pp. 1751–1754 (2000)
20. Yamamoto, K., Jabloun, F., Reinhard, K., Kawamura, A.: Robust endpoint detection for speech recognition based on discriminative feature extraction. In: ICASSP, pp. 805–808 (2006)

A Novel Neural Network Based Reinforcement Learning

Jian Fan[1,2], Yang Song[1], MinRui Fei[1], and Qijie Zhao[3]

[1] Shanghai Key Laboratory of Power Station Automation Technology, Collgeg of Mechatronics Engineering and Automation, Shanghai University, Shanghai 200072, China
[2] Operations Research Center, Nanjing Army Command College, Nanjing, Jiangsu Province 210045, China
[3] Collgeg of Mechatronics Engineering and Automation, Shanghai University, Shanghai 200072, China
jfan@mail.shu.edu.cn,
simonsongcn@163.com,{mrfei,zqj}@staff.shu.edu.cn

Abstract. Many function-approaching methods such as neural network, fuzzy method are used in reinforcement learning methods for solving its huge problem space dimensions. This paper presents a novel ART2 neural network based reinforcement learning method (ART2-RL) to solve the space problem. Because of its adaptive resonance characteristic, ART2 neural network is used to process the space measurement of reinforcement learning and improve the learning speed. This paper also gives the reinforcement learning algorithm based on ART2. A simulation of path planning of mobile robot has been developed to prove the validity of ART2-RL. As the complexity of the simulation increased, the result shows that the number of collision between robot and obstacles is effectively decreased; the novel neural network model provides significant improvement in the space measurement of reinforcement learning.

1 Introduction

Reinforcement learning RL[1] is an important machine learning method, also called strengthen learning or encourage learning. Its essence is an asynchronous dynamic planning method based on Markov Decision Process and always used to acquire self-adaptability response behavior in dynamic environment. Because of its ability of implementing the decision optimization through alternating with environment, RL has extensive application value in solving complex optimization control problem.

Adaptive resonance theory based ART2 neural network[2]performs steady classification through competitive learning and self-stabilization mechanism; it is widely used in pattern recognition and pattern classification. For instance, ART2 neural network is applied in [3] for face recognition; Li Ming, Yan Chaohua[4] proposed an ART2 neural network with more rigorous vigilance test criterion for higher exact recognition proportion.

Under continuous state space, RL is often disabled by huge problem space dimensions, so it's very important to measure its state space through simple methods. This paper presents a novel RL solution ART2-RL based on ART2 neural network.

K. Li et al. (Eds.): LSMS 2007, LNCS 4688, pp. 46–54, 2007.

Because ART2 has the good characteristic of classifying and planning, so we can use ART2 to solve the adaptability and agility problem for implementing the space measurement of RL, improve the learning speed and obtain satisfactory learning effect.

2 ART2-RL Reinforcement Learning System

Shown as Fig.1, ART2-RL reinforcement learning system is composed of ART2 neural network module, RL module and behavior selected module. The space measurement problem of RL is solved through these three modules.

The function of RL module is to find a logical behavior through disposing measured space, which can make learning effect favorable. Subsequently, environment produces an evaluation signal based on the behavior, and then RL modifies the goal value based on the evaluation signal. The behavior selected module selects behavior based on the goal value of RL through the optimization policy.

Usually, the input signal of learning system is presented as continuous transformation which constitutes continuous transformable input space. In continuous input space, the learning search area is complicated, so it is needed to transform continuous space to discrete space for reducing the computing complexity. The measurement methods include BOX, neural network and fuzzy method and so on. This paper uses adaptive resonance characteristic of ART2 neural network to implement the measurement of RL input space.

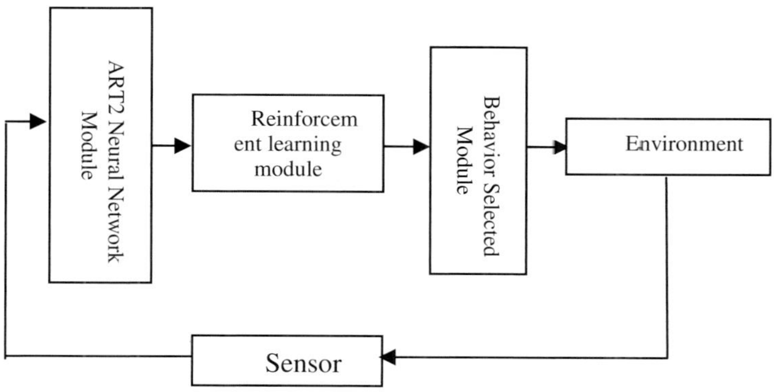

Fig. 1. ART2-RL reinforcement learning system architecture

Adaptive Resonance Theory is proposed by Grossberg in 1976 [5-6] and ART neural network is proposed by him later. Developed as yet, ART neural network can be classified as ART1, ART2 and ART3. ART1 is used to handle binary value, ART2 for arbitrary simulation and binary value, ART3 for classification search. So it is a feasible solution to use adaptive resonance theory to solve the area planning problem of space measurement of RL.

3 The Implementation of ART2-RL

3.1 ART2 Neural Network Architecture

ART2 networks consists of three fields: a pre-processing field F_0, a layer of processing units called feature representation field F_1,and a layer of output units called category representation field F_2, F_1 and F_2 are fully connected in both directions via weighted connections called pathways. The set of pathways with corresponding weights is called an adaptive filter. The weights in the adaptive filter encode the long-term memory (LTM) traces. Patterns of activation of F_1 and F_2 nodes are called short-term memory (STM) traces. The connections leading from F_1 to F_2, and from F_2 to F_1 are called bottom-up and top-down adaptive filters, respectively. There are also corresponding bottom-up weights and top-down weights.

Fig.2 illustrates the ART2 architecture used in this paper. F_0 field has the same structure as that of F_1、F_2 field and both of them have M nodes. The *ith* node in F_0 field only connects to the corresponding *ith* F_1 node in one direction, namely, from F_0 to F_1. The F_0 field pre-processes input patterns. Each input pattern can be represented as a vector I which consists of analog or binary components

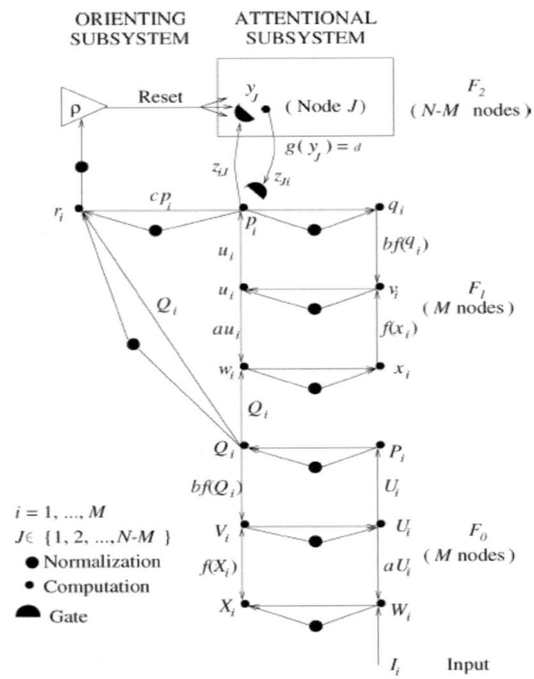

Fig. 2. ART2 architecture (adapted from [7])

$I_i(i=1,...,M)$. The nodes of F_2 field are numbered as $J=(M+1,...,N)$, that is, the F_2 field consists of N-M nodes which are fully connected among themselves in both directions [7].

The fields F_1 and F_2, as well as the bottom-up and top-down adaptive filters constitute the attentional subsystem of ART2 networks. An auxiliary orienting subsystem becomes active when a bottom-up input to F_1 fails to match the learned top-down expectation (vector Z_J is composed of $N-M$ components, Z_{Ji} ($i=1,...,N-M$) readout by the active category representation at F_2).In this case, the orienting subsystem is activated and causes rapid reset of the active category representation in F_2. This reset event automatically induces the attentional subsystem to proceed with a parallel search. Alternative categories are tested until either an adequate match is found or a new category is established.

3.2 The Learning Flow

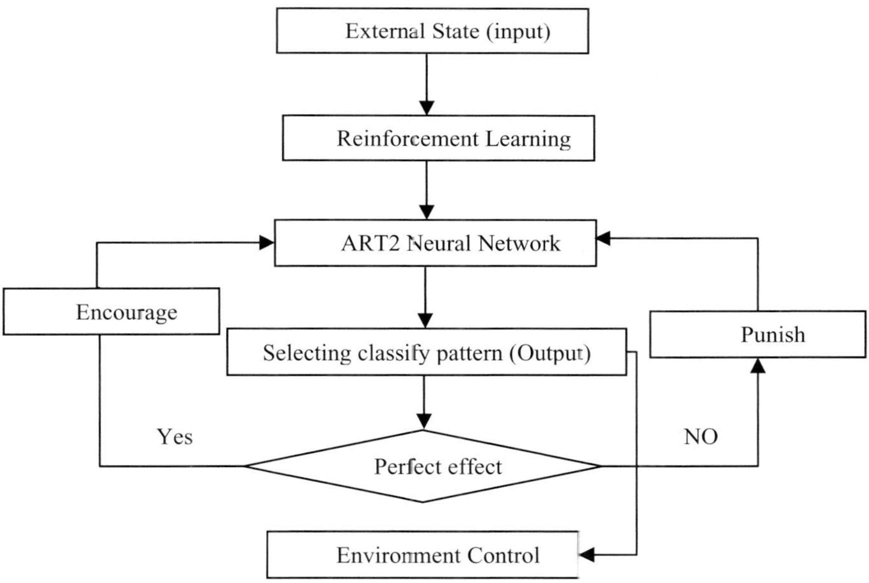

Fig. 3. The learning flow of ART2-RL

The flow of ART2-RL is shown as Fig.3.

1. When reinforcement learning get the external states, reinforcement learning pass the external states to ART2, then ART2 selects a corresponding classified pattern through inner competition and applies the pattern to environment control.
2. Then the evaluation module of RL evaluates the running effect of the classified pattern.

3. If the running result is ideal, then reinforcement learning encourages the classified pattern, viz. increases the corresponding weights of ART2, else punishes the pattern, and viz. decreases the corresponding weights of ART2.
4. When the neural network completed certain learning times, it will turn around to select other classified patterns. So ART2 neural network can be possessed comparative distinguish probability through interactive learning with environment and RL space measurement is solved by this way.

3.3 The Learning Algorithm

RL can be implemented by different neural networks. Its input signal is measured state vector $s_t = \{s_1(t), s_2(t), \ldots, s_n(t)\}$; every output of the neural network is corresponded with the goal value of a behavior. The learning algorithm is the key of implementing RL with neural network.

The goal function definition of RL is shown as equation (1):

$$G(s_{t+1}) = G(s_t) + \alpha(r_{t+1} + \gamma G(s_{t+1}) - G(s_t)) \tag{1}$$

Where α is the learning rate, θ_t is the evaluation value at time t, γ is the influence factor, t is current time.

Equation (1) is come into existence only under the precondition of optimization policy. In the learning phase, the error signal is

$$\Delta \mathrm{G} = \theta_{\mathrm{t}} + \gamma G(s_{t+1}) - G(s_t) \tag{2}$$

Where $G(s_{t+1})$ is the goal value of next state. The error $\Delta \mathrm{G}$ can be decreased through modifying the neural network weights. When RL is implemented by ART2 neural network, the modification of ART2 weights is

$$\Delta \mathrm{W}_{\mathrm{i}} = \alpha(\theta_{\mathrm{t}} + \gamma G(s_{t+1}) - G(s_t))\mathrm{e}_{\mathrm{t}} \tag{3}$$

Qualification mark e_t is:

$$\mathrm{e}_t = \sum_{K=1}^{t} \lambda^{t-k} \nabla w G(s_k) \tag{4}$$

Where λ is the discount factor.

$$\nabla w G(s_t) = \frac{\partial G(s_i)}{\partial W_t} \tag{5}$$

The modification of ART2 weights is:

$$z_{Ji} = \mathrm{a}du_i + (1 + ad(d-1))z_{Ji} + \Delta W_i \tag{6}$$

$$z_{iJ} = \mathrm{a}du_i + (1 + ad(d-1))z_{iJ} + \omega \Delta W_i \tag{7}$$

Where ω is the influence factor between weight z_{Ji} and z_{iJ}. a and d are ART2 parameters, u_i is the compute equation of ART2 unit U_i.

In order to test the validity of ART2-RL, ART2-RL is used in the simulation experiment of path planning of mobile robot. Robot state space and CABs (collision avoidance behavior) are stored by ART2-RL. When robot R observers that it will collide with obstacle O, then it selects one CAB based on present state in ART2-RL to avoid the collision and evaluate the collision avoidance result. If every CAB is effective, then R can arrive at the destination without collision.

Teambots [8] is open source software which is used as the foundation of the experiment simulation. All kinds of data are got by average computing in 100 epochs and 300 cycles. Q-Learning is choosed as the RL method. Simulation environment is shown as Fig.4, the object grading as 0 is robot R, which movement velocity is 0.5. The objects grading as 1 and 2 are moving obstacles O, the moving directions of two Os are $\pi/2$ and $3\pi/2$ denoted as polar coordinates, their velocities are 0.3. The object similar as long-pole is the goal area and other objects are static obstacles. The evaluation value of ART2-RL learning algorithm is shown as equation (8).

$$\theta = \begin{cases} 0.2 & \text{if without collision} \\ -0.1 & \text{if collision} \end{cases} \tag{8}$$

In simulation experiment, the maximum deflection angle of R is ±20^0, and the maximum acceleration is ±0.02 unit. In the behavior space table, R can only change its angle in 5^0, and acceleration interval is 0.005 units. Each obstacle's state can be described as four parameters (velocity, direction, x coordinate, y coordinate). F_1 layer of ART2 neural network has 12 nodes corresponding to 12 state variables of R and two Os, F_2 layer has 65 nodes corresponding to R's 65 CABs.

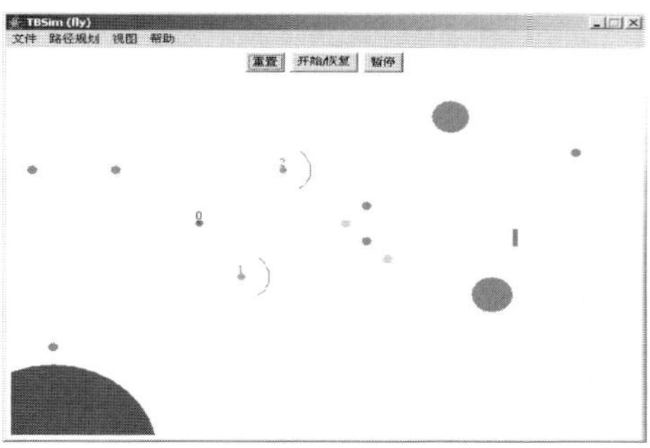

Fig. 4. The setting of simulation environment

For effect contrast, BP neural network which is often used to measure RL state space is also used in the simulation. The parameters of ART2-RL and BP-RL are shown as:

$$a=b=10,\ c=0.1,\ d=0.9,\ \rho=0.9,\ e=0.05,\ \omega=0.1,\ \alpha=0.9,\ \gamma=0.7$$

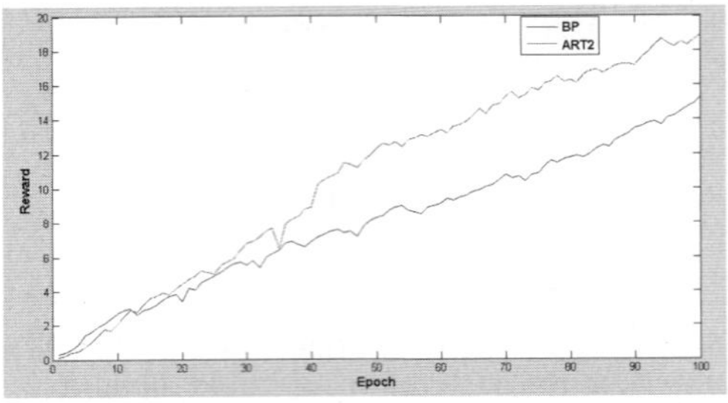

Fig. 5. Total reward in each epoch

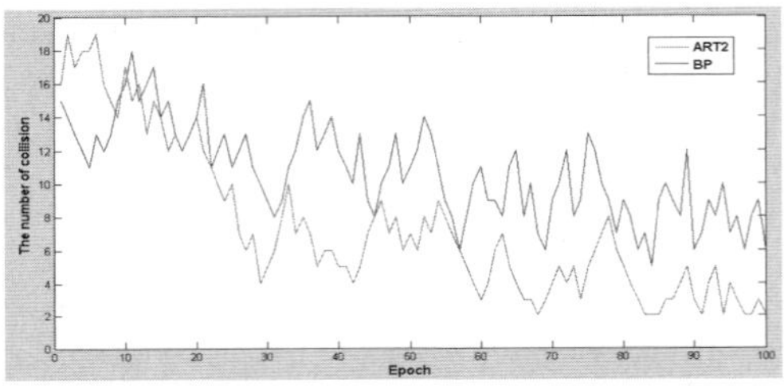

Fig. 6. The number of Collision the robot with obstacles

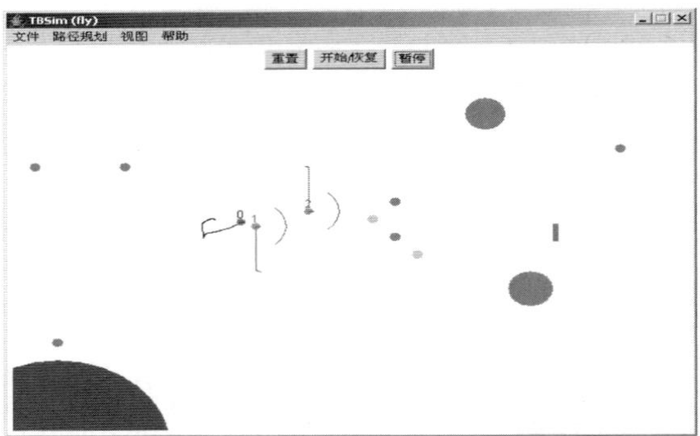

Fig.7. The sketch map of robot path planning before avoiding collision

Through analysing Fig.5, it can be got that ART2 has smaller reward than BP in learning initial stage, but with the deep learning, the reward became more and more large. Also in Fig.6, we can find the same phenomena and educe that the number of collision between R and O is effective decreased. The whole process of path planning is shown as Fig.7, Fig.8 and Fig.9.

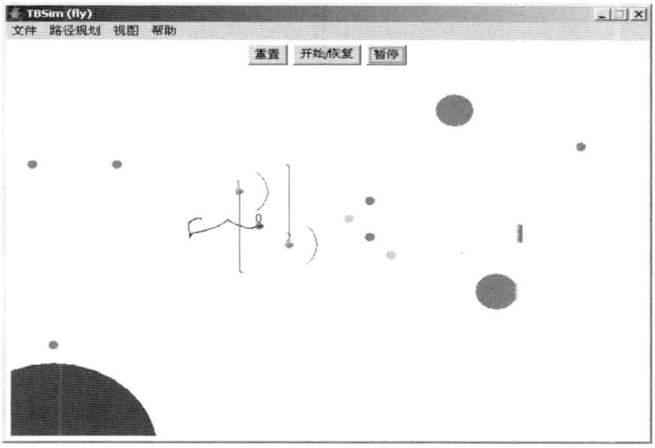

Fig. 8. The sketch map of robot path planning after avoiding collision

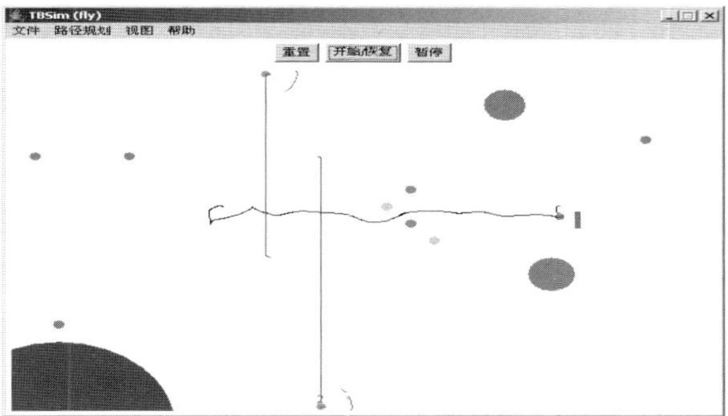

Fig. 9. Path planning of the mobile robot

4 Conclusions

A novel ART2 neural network based reinforcement learning ART2-RL is been successfully proposed in this paper. Because of the good ability of classifying and planning by ART2 adaptive resonance characteristic, so it's feasible to solve the adaptability and agility problem of the space measurement of reinforcement learning, improve learning speed and obtain satisfactory learning effect by using ART2. And a simulation of path planning of mobile robot is been developed to prove the validity of ART2-RL. Also in the simulation experiment, ART2 is compared with BP neural network. The simulation result indicates that ART2-RL has larger reward and the number of collision between robot and obstacle is effectively decreased. ART2-RL makes favorable path planning effect.

Acknowledgement. This work was supported by Shanghai Key Laboratory of Power Station Automation Technology and Shanghai Leading Academic Disciplines under grant T0103.

References

1. Richard Sutton, S., Andrew Barto, G.: Reinforcement Learning: An Introduction. MIT Press, Cambridge (March 1998)
2. Bruce, K.B., Cardelli, L., Pierce, B.C.: Comparing Object Encodings. In: Ito, T., Abadi, M. (eds.) TACS 1997. LNCS, vol. 1281, pp. 415–438. Springer, Heidelberg (1997), Carpenter, G.A., Grossberg. S.: ART2: Stable self-organization of category recognition codes for analog input patterns. Applied Optics 26, 4919–4930 (1987)
3. XiaoHua, L., et al.: Face recognition using adaptive resonance theory. In: Proceedings of the 2003 International Conference on Machine Learning and Cybernetics, vol. 5, pp. 3167–3171 (2003)
4. Ming, L., Chao-hua, Y., Gao-hang, L.: ART2 Neural Networks with More Vigorous Vigilance Test Criterion. Journal of Image and Graphics 6, 81–85 (2001)
5. Grossberg, S.: Adaptive pattern classification and universal recoding, I: Parallel development and coding of neural feature detectors. Biol. Cybern. 23, 121–134 (1976)
6. Grossberg, S.: Adaptive pattern classification and universal recoding, II: Feedback, expectation, olfaction, and illusions. Biol. Cybern. 23, 187–202 (1976)
7. Carpenter, G.A., Grossberg, S., Rosen, D.B.: Fuzzy ART: Fast stable learning and categorization of analog patterns by an adaptive resonance system. Neural Netw. 4, 759–771 (1991)
8. Tucker Balch: http://www.cs.cmu.edu/ trb/TeamBots
9. NanFeng, X., Nahavandi, S.: A Reinforcement Learning Approach for Robot Control in an Unknown Environment. In: 2002 IEEE International Conference on Industrial Technology, vol. 2, pp. 1096–1099. IEEE Computer Society Press, Los Alamitos (2002)

A Design of Self Internal Entropy Balancing System with Incarnation Process

JeongYon Shim

Division of General Studies, Computer Science, Kangnam University
San 6-2, Kugal-Dong, Kihung-Gu,YongIn Si, KyeongKi Do, Korea
Tel.: +82 31 2803 736
mariashim@kangnam.ac.kr

Abstract. In this paper, adopting the concept of autonomic nervous system, we design Self Internal Entropy Balancing System with incarnation focused on the self maintaining function, we define self type and self internal entropy as a property of system. This system checks SEG(Surviving Energy Gauge) periodically and make a balance by adjusting the treatment. In the case of the situation to survive, it keeps the system by processing the incarnation process. It is applied to knowledge network system of virtual memory and tested with sample data.

1 Introduction

A human being encircled with the environment which has dynamic, complex and dangerous characteristics maintains his life systems by self balancing system. A self balancing system makes a living thing keep a life adjusting the balancing factors,protecting a body from stimulus of the externals and making the balance of the system.

Especially, autonomic nervous system takes a part of controlling the essential functions for keeping a life automatically. It also takes part in self-repairing functions. A human being cannot be conscious of this autonomic controlling system. Human controlling system consists of conscious controlling part and autonomic nervous controlling part.

As the information technology develops rapidly and the complexity of information environment is getting high, the requirements of high qualified intelligent system are becoming more and more. The design of the intelligent system adopting a life concept can be a good approach for solving this problem. For these recent decades, concepts of a life have been successfully applied and built.

In this paper, adopting the concept of autonomic nervous system, we design Self Internal Entropy Balancing System with incarnation focused on the self maintaining function, we define self type and self internal entropy as a property of system. This system checks the balancing factor periodically and make a balance by adjusting the treatment. In the case of the situation to survive, it

K. Li et al. (Eds.): LSMS 2007, LNCS 4688, pp. 55–62, 2007.

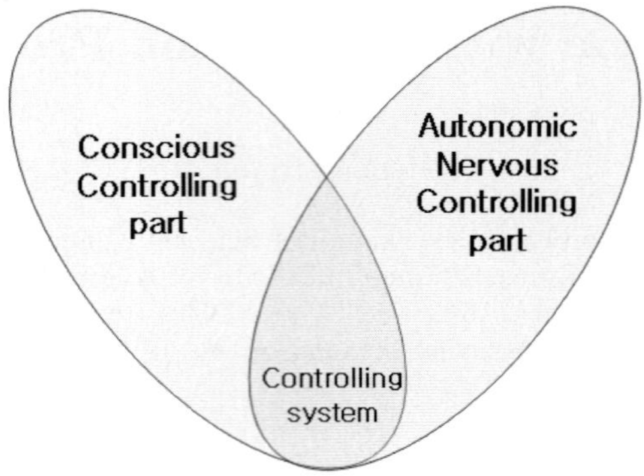

Fig. 1. The controlling system in the body

keeps the system by processing the incarnation process. It is applied to knowledge network system of virtual memory and tested with sample data.

2 A Structure of Intelligent Knowledge Management with Self Internal Entropy Balancing System

A proposed Intelligent Knowledge Management system has conscious part and Self Internal Entropy Balancing. The conscious part consists of Knowledge acquisition, Inference and Retrieval module and their functional processes are described in detail by author in the previous paper[1]. In this paper we focus on Self Internal Entropy Balancing System. As shown in figure 2, Self Internal Entropy Balancing System controls whole system and keeps the balance by checking the SEG(Surviving Entropy Gauge) and incarnation process.

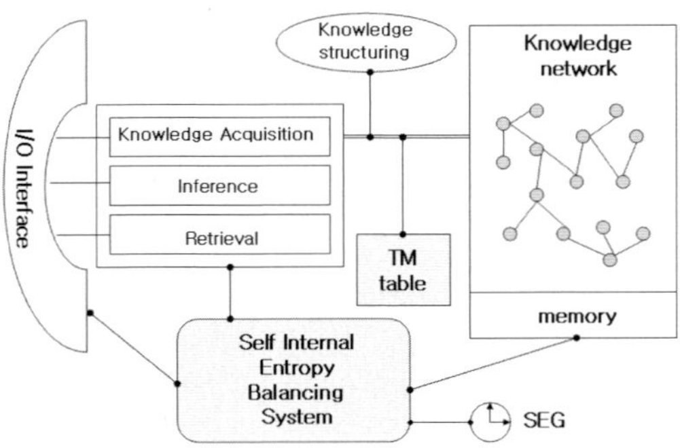

Fig. 2. A structure of Intelligent Knowledge management System with Self Internal Entropy Balancing

3 Entropy Balancing Strategy and Incarnation Process of Self Internal Entropy Balancing System

3.1 A Design of a Knowledge Cell

The acquired knowledge by knowledge acquisition process is stored in the form of knowledge network of memory by knowledge structuring mechanism. The structured knowledge network is used for the processing of Inference and Retrieval process. Knowledge network consists of a collection of cells and relations. It is represented as a knowledge associative list in memory for an efficient access. Figure 3 represents the structure of a knowledge cell which is composed of ID,Self Type,cell Internal Entropy and contents.

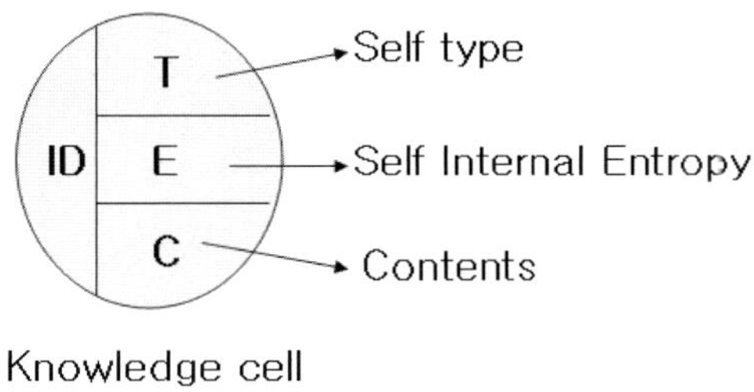

Fig. 3. A structure of Knowledge Cell

ID represents a symbol of identifying a knowledge cell And self type is its own property of a knowledge cell[2].

A knowledge cell is an atomic element of knowledge network in the memory.

Knowledge associative list
Because knowledge network is a logical concept, it is necessary to transform this to the simple form for more efficient computational processing. As shown in the figure, the form of knowledge network is transformed to Knowledge associative list. One record of Knowledge associative list consists of Flag, knowledge node i, IE(Internal Entropy), Rel(Relation) and connected knowledge node i+1. IE represents Internal Entropy of itself and Flag which represents knowledge state. Flag has one of four states as described before. The value of Relation is calculated by eq(1).

$$R_{ij} = P(K_j|K_i) \tag{1}$$

3.2 Knowledge Network Management Strategy

Knowledge network management strategy system takes charge of maintaining efficient memory by discriminating dead cells and removing dead cells. Main function of Knowledge network management strategy is pruning the dead cells or negative cells and maintain the new state of memory constantly. In this process, the system checks if there is a negative cell and calculates Internal entropy of whole memory. If it finds a dead cell, it sends a signal and starts the pruning process. In the case of a negative cell, it determine if a negative cell should be destroyed. During destroying a negative cell, memory configuration is made. After removing, new associative relation should be made by calculating the new strength. In this case of following example in Figure 4, node k_j is removed because it is a dead cell. After removed, k_i is directly connected to k_k with new strength. The new strength is calculated by eq(2).

$$R_{ik} = R_{ij} * R_{jk} \tag{2}$$

For the more general case as shown in Figure 5, if there are n number of dead cells between k_i and k_j,the new strength R_{ij} is calculated by eq(3).

$$R_{ij} = \prod R_{i,i+1} * R_{j-1,j} \tag{3}$$

3.3 Self Internal Entropy and Surviving Energy Gauge

Self Internal entropy,E,is defined as a internal strength of cell calculated by entropy function(eq.4). The calculated value is normalized between -1.0 and 1.0. A plus value means a positive value, a minus value means negative entropy and zero depicts the state of no-entropy.

$$E_i = \frac{1 - exp(-\sigma x)}{1 + exp(-\sigma x)} \tag{4}$$

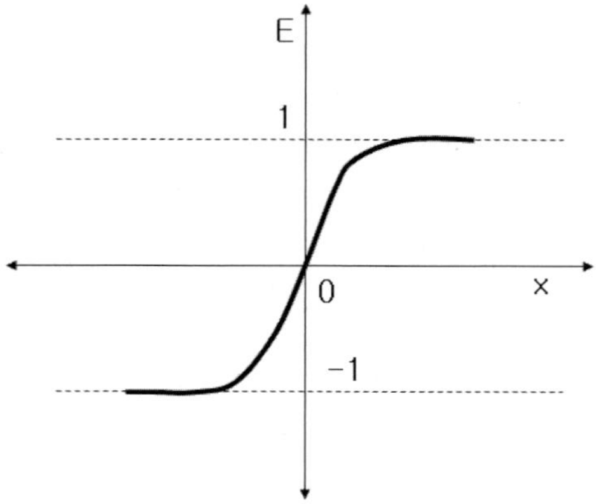

Fig. 4. An Entropy Function

Table 1. Knowledge associative list of initial knowledge network

Type	Knowledge i	E	Rel.	Knowledge $i+1$
M	K_1	0.30	1.0	K_2
M	K_1	0.30	0.7	K_5
M	K_1	0.30	0.8	K_8
M	K_1	0.30	0.9	K_9
F	K_2	-0.35	0.9	K_3
S	K_3	-0.71	0.4	K_4
M	K_4	0.21	0.0	NULL
S	K_5	0.61	0.3	K_3
S	K_5	0.61	0.6	K_6
F	K_6	-0.91	0.7	K_7
F	K_7	-0.29	0.0	NULL
E	K_8	0.30	0.2	K_7
E	K_8	0.30	0.9	K_{10}
M	K_9	0.40	0.5	K_{10}
M	K_9	0.40	0.7	K_{12}
F	K_{10}	-0.40	0.9	K_{11}
F	K_{11}	-0.41	0.3	K_{14}
M	K_{12}	0.52	0.7	K_{13}
S	K_{13}	0.80	0.8	K_{15}
S	K_{14}	0.41	0.9	K_{15}
M	K_{15}	0.30	0.0	NULL

We define SEG(Surviving Entropy Gauge) as a value of whole energy representing the state of surviving as the following equation. the value of SEG is used for Entropy Balancing.

$$SEG(t) = \frac{\sum_{i=1}^{n} E_{ti}}{n} \tag{5}$$

3.4 Entropy Balancing and Incarnation

For keeping Entropy Balancing, the system calculates the value of SEG periodically and the essential functions as energy supplementation or Incarnation process. The process of energy supplementation is receiving the new knowledge cells for supplementing the surviving energy of whole system. Incarnation process starts when the state of starvation continues for the limited steps and the system cannot be survived any more. The state of starvation can be checked by the value of SEG that is zero. Incarnation process is as following algorithm 1:

Algorithm 1. Incarnation process

begin
STEP1 calculate the value of SEG.
STEP2 IF(abs(SEG) $\leq \theta$)
  Then start the Incarnation process
  Else IF((SEG $\leq$ 0) or (SEG $\leq \gamma$))
  Then Supplementation process; Else Goto STEP3
Incarnation process:
.Select the type which has maximum number in Input pool.

.Change Self type to the selected type;
Supplementation process:
.Execute knowledge acquisition and knowledge structuring process
STEP3 Stop

Knowledge cell	Self type	Initial	Incarnation
K1	M	0.300000	0.000000
K2	F	-0.350000	0.000000
K3	S	-0.710000	-0.710000
K4	M	0.210000	0.000000
K5	S	0.610000	0.610000
K6	F	-0.910000	0.000000
K7	F	-0.290000	0.000000
K8	E	0.700000	0.700000
K9	M	0.400000	0.000000
K10	F	-0.400000	0.000000
K11	F	-0.410000	0.000000
K12	M	-0.520000	0.000000
K13	S	0.800000	0.800000
K14	S	0.410000	0.410000
K15	M	0.300000	0.000000
SEG		0.009333	0.120667

Fig. 5. The changed value of entropy by Incarnation

4 Experiments

The knowledge cell information of virtual memory was applied for representing the incarnation process. The following table shows a simple form of knowledge associative list applied to the experiments. Figure 5 depicts the variation

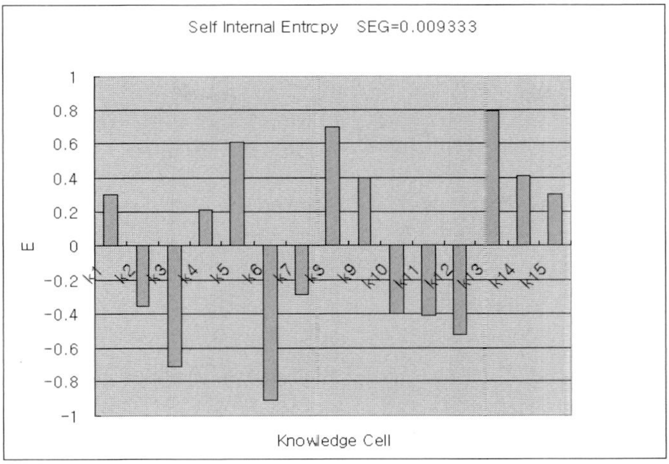

Fig. 6. Initial state

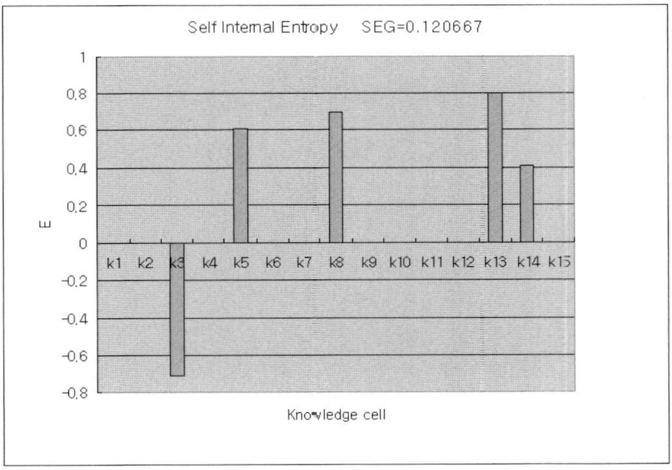

Fig. 7. Changed state

of the Internal Entropy after incarnation process. This table includes the information of knowledge ID, Type,Internal Entropy. The graphs shown in Figure 6 and figure 7 also represents the changed states during the incarnation process. In initial graph,it shows the starvation state of which SEG indicates 0.009333(threshold=0.050000). After Incarnation process,the value of SEG was changed to 0.120667 and the difficult situation of starvation was successfully removed.

5 Conclusions

Self Internal Entropy Balancing System with Incarnation process was designed for making the self maintenance system. Self Internal Entropy function and SEG was defined and Incarnation process was proposed. We applied this proposed system to virtual memory consisting of knowledge network and tested the results. As a result of testing , we could find that the memory was successfully updated by memory cleaning mechanism. The proposed system was also can be usefully applied to many areas requiring the efficient memory management.

References

1. Shim, J.-Y.: Knowledge Retrieval Using Bayesian Associative Relation in the Three Dimensional ModularSystem. In: Yu, J.X., Lin, X., Lu, H., Zhang, Y. (eds.) APWeb 2004. LNCS, vol. 3007, pp. 630–635. Springer, Heidelberg (2004)
2. Shim, J.-Y.: Automatic Knowledge Configuration by Reticular Activating System. LNCS (LNAI), pp. 1170–1178. Springer, Heidelberg (2005)
3. Anderson, J.R.: Learning and Memory. Prentice-Hall, Englewood Cliffs

Cooperative Versus Competitive Coevolution for Pareto Multiobjective Optimization

Tse Guan Tan, Hui Keng Lau , and Jason Teo

Center for Artificial Intelligence,
School of Engineering and Information Technology,
Universiti Malaysia Sabah, Malaysia
tseguantan@gmail.com , {hklau,jtwteo}@ums.edu.my

Abstract. In this paper, we propose the integration between Strength Pareto Evolutionary Algorithm 2 (SPEA2) with two types of coevolution concept, Competitive Coevolution (CE) and Cooperative Coevolution (CC), to solve 3 dimensional multiobjective optimization problems. The resulting algorithms are referred to as Strength Pareto Evolutionary Algorithm 2 with Competitive Coevolution (SPEA2-CE) and Strength Pareto Evolutionary Algorithm 2 with Cooperative Coevolution (SPEA2-CC). The main objective of this paper is to compare competitive against cooperative coevolution to ascertain which coevolutionary approach is preferable for multiobjective optimization. The competitive coevolution will be implemented with *K*-Random Opponents strategy. The performances of SPEA2-CE and SPEA2-CC for solving tri-objective problems using the DTLZ suite of test problems are presented. The results show that the cooperative approach far outperforms the competitive approach when used to augment SPEA2 for tri-objective optimization in terms of all the metrics (generational distance, spacing and coverage).

1 Introduction

Coevolutionary learning [14] occurs at two different levels, individual level and population level. At the individual level, the coevolution concept may be applied to the evolutionary process within a single population. The fitness of each member depends on other members within the same population. At the population level, the coevolution concept may be applied to evolutionary process with more than one population. The fitness of each member depends on other from different population.

Generally, the coevolution model can be divided into competitive coevolution and cooperative coevolution. The competitive coevolution involves individuals that compete against each other for dominance in the population. Competitive coevolution have been implemented with some famous competitive fitness strategies such as Bipartite Tournament [4], Single Elimination Strategy [1], *K*-Random Opponents [7] and Hall of Fame [10] to solve the problems of sorting network [4] and Tic-Tac-Toe [1]. On the other hand, cooperative coevolution [9] involves a number of individuals working together to solve the problems.

K. Li et al. (Eds.): LSMS 2007, LNCS 4688, pp. 63 – 72, 2007.

Based on the literature, some references were found in which coevolution model is applied to solve multiobjective optimization problems. Parmee and Watson [8] introduced a competitive coevolution in evolutionary multiobjective optimization for bi-objective and tri-objective problems to design the airframes. Lohn et al. [6] presented a competitive coevolution genetic algorithm to solve bi-objective problems. In his approach, the tournament is held between two populations, namely the trial population and target population. Keerativuttitumrong et al. [5] proposed Multiobjective Cooperative Coevolutionary Genetic Algorithm (MOCCGA), which integrates the cooperative coevolutionary concept with Multiobjective Genetic Algorithm (MOGA). Each species represents a single decision variable. In order to evaluate individuals in any species, collaborators will be selected from other species to form a complete solution. Then this complete solution is mapped into objective vector by the objective function. The evolution of these populations is controlled through MOGA. Coello Coello and Sierra [2] introduced a multiobjective evolutionary algorithm that integrated the cooperative coevolutionary concept, named CO-MOEA. This algorithm separates the search space into several regions and later focuses on the promising regions. The promising regions are measured by an analysis of the current Pareto front. The evolutionary process of this algorithm was divided into four distinct stages. The number of generations will determine the change of stages.

In our proposed algorithms, Competitive Coevolution and Cooperative Coevolution approaches were integrated with Strength Pareto Evolutionary Algorithm 2 (SPEA2) in solving tri-objectives problems. The two modified SPEA2s were referred to as SPEA2-CE and SPEA2-CC. The CE was implemented with *K*-Random Opponents strategy. The results of the SPEA2-CE were benchmarked against the SPEA2-CC for performance using the DTLZ [3] suite of test problems.

2 Strength Pareto Evolutionary Algorithm 2 with Competitive Coevolution (SPEA2-CE)

Generally, the framework for the SPEA2-CE is similar to the framework for SPEA2 [16] with the exceptions of two additional methods, Opponents_selection and Reward_assignment. In the evolution process, after finishing the calculation of the raw fitness value for each individual in the population, the Opponents_selection method will randomly select individuals as the opponents based on the K-Random Opponents strategy [7] from the same population without repeating the identical opponents and prohibits self-play. The K is tested with the values of 10, 20, 30, 40, 50, 60, 70, 80 and 90. After that, each individual will compete against the K number of opponents. During the tournament, the reward value will be calculated for each competition based on the reward function. The reward value will be summed up as the fitness score for the individual up to the K number of competitions, using Reward_assignment. The reward function is defined as follow. I represent the participating individual, while O represents the opponent. R is the raw fitness value,max (R) is the maximum raw fitness value and the min (R) is the minimum raw fitness value. The range for values in this function is within [-1, 1]. If Reward(I, O) = 0, it corresponds to the competition as draw.

$$\mathrm{Re}\,ward(I,O) = \frac{R(O) - R(I)}{\max(R) - \min(R)} \tag{1}$$

3 Strength Pareto Evolutionary Algorithm 2 with Cooperative Coevolution (SPEA2-CC)

The CC [9] architecture allowed for cooperating subpopulations to evolve in parallel and when combined together, they produced a complete solution. Each subpopulation was genetically isolated from the others and evolved on its own. For evaluation, each subpopulation was considered in turn. In our approach, method for selecting collaborators is by selecting the current best individuals from alternative subpopulations to generate a complete chromosome for fitness evaluation [13]. The general framework for SPEA2-CC is shown below.

```
gen = 0
for each species s do begin
          Pops(gen) = randomly initialized population
          Fitness_assignment Pops(gen)
          Environmental_selection Pops(gen)
          end
while termination = false do begin
          gen = gen + 1
          for each species s do begin
          Mating_selection Pops(gen)
          Variation Pops(gen)
          Fitness_assignment Pops(gen)
          Environmental_selection Pops(gen)
          end
end
```

4 Optimization Results and Discussion

In this section, the results from applying the SPEA2-CE and the SPEA2-CC to solve DTLZ1 to DTLZ7 are presented. In order to have a fair comparison, all runs considered are implemented with the same real-valued representation, simulated binary crossover (SBX), polynomial mutation and tournament selection. The number of evaluations in each run is fixed at 60,000. Table 1 lists all the parameter setting for each evolutionary multiobjective optimization algorithm. In the experiment, each algorithm was run 30 times for each test function.

Table 1. Parameter Settings for SPEA2-CE and SPEA2-CC

Parameter	SPEA2-CE	SPEA2-CC
Population size	100	100
Number of decision variables per generation	12	12
Number of objectives	3	3
Number of generations	600	50
Mutation probability	0.08	0.08
Crossover probability	1	1
Polynomial mutation operator	20	20
SBX crossover operator	15	15
Number of repeated runs	30	30
Population size per species	100	100
Total population size per generation	100	1200
Number of species per generation	1	12

The results for test problems from DTLZ1 to DTLZ7 with respect to generational distance and spacing are summarized in Table 2 to Table 8. The graphical presentations in box plots of these two metrics are shown in Fig. 1. The leftmost box plot relates to SPEA2-CC while from the second left box plot to the rightmost box plot relate to SPEA2-CE with K-Random Opponents. The dark dash is the median, the top of the box is the upper quartile, and the bottom of the box is the lower quartile. The results for the test problems in terms of the coverage metric are shown in Fig. 2. Each rectangle contains seven box plots representing the distribution of the C value; the leftmost box plot relates to DTLZ1 while the rightmost box plot relates to DTLZ7. The scale of 0 means no coverage at the bottom while 1 means total coverage at to top per rectangle.

Generational Distance [12]: Overall, SPEA2-CC performed much better than SPEA2-CE in most of the DTLZ test problems, particularly in DTLZ3, DTLZ5 and DTLZ6. However, only in DTLZ4, SPEA2-CE had a better performance. The results showed that the solutions obtained with SPEA2-CC are very close to the true Pareto front.

Spacing [11]: SPEA2-CC performed better than SPEA2-CE in all of the DTLZ test problems, especially in DTLZ1, DTLZ2, DTLZ3, DTLZ5 and DTLZ6. Based on the results, the SPEA2-CC had a very good distribution for nondominated solutions in the objective space.

Coverage [15]: Again, in all the test problems, SPEA2-CC clearly outperforms SPEA2-CE. Basically, none of the nondominated solutions in SPEA2-CC were covered by the set SPEA2-CE as depicted in Fig. 2.

Table 2. Comparison of Results between SPEA2-CE against SPEA2-CC for DTLZ1

		DTLZ1 (60, 000 evaluations)					
Algorithm	Opponents Size	Metric	Mean	St Dev	Best	Median	Worst
SPEA2-CC		GD	0.01610	0.01325	0.00043	0.01293	0.04365
		SP	0.0850	0.0777	0.0085	0.0637	0.4284
SPEA2-CE	10	GD	2.306	0.782	1.238	2.141	4.395
		SP	9.519	4.736	3.782	8.461	22.183
	20	GD	1.519	0.818	0.353	1.560	3.324
		SP	6.225	3.949	0.386	5.855	15.636
	30	GD	1.948	1.529	0.379	1.531	7.911
		SP	7.748	5.287	0.462	6.582	21.345
	40	GD	1.519	2.002	0.151	1.203	10.979
		SP	5.577	4.375	0.101	5.225	18.328
	50	GD	1.309	1.124	0.347	0.940	6.129
		SP	5.791	4.235	0.713	4.204	16.576
	60	GD	1.114	0.868	0.216	0.894	4.244
		SP	4.890	4.263	0.232	3.452	16.471
	70	GD	1.394	1.642	0.276	0.801	7.051
		SP	5.255	5.456	0.260	2.987	19.580
	80	GD	1.046	1.114	0.238	0.889	6.312
		SP	3.684	3.363	0.163	2.754	9.492
	90	GD	0.7256	0.4287	0.2059	0.5931	1.8454
		SP	3.486	4.046	0.149	1.679	15.172

Table 3. Comparison of Results between SPEA2-CE against SPEA2-CC for DTLZ2

		DTLZ2 (60, 000 evaluations)					
Algorithm	Opponents Size	Metric	Mean	St Dev	Best	Median	Worst
SPEA2-CC		GD	0.00173	0.00230	0.00099	0.00114	0.01104
		SP	0.02756	0.01385	0.02098	0.02491	0.09943
SPEA2-CE	10	GD	0.00225	0.00026	0.00174	0.00222	0.00301
		SP	0.05649	0.00658	0.04731	0.05462	0.07387
	20	GD	0.00205	0.00019	0.00167	0.00207	0.00244
		SP	0.05788	0.00925	0.03872	0.05716	0.08877
	30	GD	0.00201	0.00021	0.00166	0.00200	0.00245
		SP	0.05398	0.00751	0.03702	0.05505	0.06664
	40	GD	0.00204	0.00027	0.00162	0.00201	0.00264
		SP	0.05872	0.01050	0.03080	0.05914	0.08281
	50	GD	0.00197	0.00022	0.00160	0.00192	0.00264
		SP	0.05544	0.00762	0.04201	0.05482	0.07509
	60	GD	0.00194	0.00022	0.00154	0.00192	0.00248
		SP	0.05623	0.00796	0.04238	0.05619	0.06979
	70	GD	0.00195	0.00022	0.00148	0.00194	0.00237
		SP	0.05493	0.00700	0.03489	0.05591	0.06830
	80	GD	0.00196	0.00023	0.00153	0.00194	0.00255
		SP	0.05556	0.00981	0.04188	0.05252	0.08213
	90	GD	0.00189	0.00026	0.00147	0.00189	0.00280
		SP	0.05319	0.00789	0.04028	0.05226	0.07952

Table 4. Comparison of Results between SPEA2-CE against SPEA2-CC for DTLZ3

DTLZ3 (60, 000 evaluations)							
Algorithm	Opponents Size	Metric	Mean	St Dev	Best	Median	Worst
SPEA2-CC		GD	0.06090	0.06000	0.00110	0.03860	0.20580
		SP	0.22390	0.32560	0.02340	0.12190	1.58040
SPEA2-CE	10	GD	6.941	4.228	2.165	5.886	23.583
		SP	21.28	7.80	8.55	19.70	42.83
	20	GD	7.545	4.646	1.009	5.680	22.620
		SP	20.24	10.85	5.57	21.97	50.79
	30	GD	5.625	4.256	0.887	4.183	17.510
		SP	16.46	10.33	1.30	15.02	43.09
	40	GD	6.056	4.071	0.762	4.542	15.502
		SP	16.83	10.42	2.25	14.17	38.35
	50	GD	7.12	5.74	0.84	6.39	26.21
		SP	20.68	9.64	6.47	20.47	41.06
	60	GD	6.262	4.028	1.022	5.099	16.856
		SP	18.44	9.18	2.15	19.93	35.13
	70	GD	6.673	5.469	1.514	5.352	25.970
		SP	19.74	10.91	1.79	17.88	48.71
	80	GD	7.371	4.611	0.826	6.817	17.713
		SP	19.12	10.13	1.04	17.09	38.67
	90	GD	7.569	5.444	0.514	5.155	20.625
		SP	18.63	9.13	0.28	20.01	37.72

Table 5. Comparison of Results between SPEA2-CE against SPEA2-CC for DTLZ4

DTLZ4 (60, 000 evaluations)							
Algorithm	Opponents Size	Metric	Mean	St Dev	Best	Median	Worst
SPEA2-CC		GD	0.00326	0.00326	0.00104	0.00125	0.01180
		SP	0.03108	0.01857	0.00331	0.02687	0.09188
SPEA2-CE	10	GD	0.00195	0.00051	0.00114	0.00210	0.00347
		SP	0.03989	0.02282	0.00727	0.05396	0.06883
	20	GD	0.00178	0.00044	0.00110	0.00186	0.00262
		SP	0.03681	0.02675	0.00000	0.03344	0.07703
	30	GD	0.00194	0.00042	0.00110	0.00199	0.00270
		SP	0.04822	0.02233	0.00000	0.05566	0.07372
	40	GD	0.00176	0.00040	0.00112	0.00184	0.00279
		SP	0.03804	0.02663	0.00766	0.03551	0.08973
	50	GD	0.00192	0.00039	0.00113	0.00197	0.00251
		SP	0.05030	0.02436	0.00000	0.05895	0.10106
	60	GD	0.00180	0.00039	0.00112	0.00189	0.00240
		SP	0.04280	0.02589	0.00000	0.05324	0.07547
	70	GD	0.00187	0.00033	0.00117	0.00195	0.00236
		SP	0.05111	0.02404	0.00724	0.05968	0.08936
	80	GD	0.00190	0.00034	0.00114	0.00198	0.00245
		SP	0.04443	0.02480	0.00000	0.05757	0.07355
	90	GD	0.00182	0.00035	0.00113	0.00192	0.00225
		SP	0.04389	0.02490	0.00000	0.05697	0.07686

Table 6. Comparison of Results between SPEA2-CE against SPEA2-CC for DTLZ5

		DTLZ5 (60, 000 evaluations)					
Algorithm	Opponents Size	Metric	Mean	St Dev	Best	Median	Worst
SPEA2-CC		GD	0.00002	0.00000	0.00002	0.00002	0.00003
		SP	0.00468	0.00052	0.00389	0.00467	0.00593
SPEA2-CE	10	GD	0.00050	0.00010	0.00032	0.00049	0.00073
		SP	0.01599	0.00508	0.00806	0.01536	0.02822
	20	GD	0.00041	0.00008	0.00029	0.00039	0.00060
		SP	0.01606	0.00869	0.00774	0.01296	0.04513
	30	GD	0.00049	0.00029	0.00025	0.00044	0.00160
		SP	0.01809	0.01084	0.00649	0.01371	0.04787
	40	GD	0.00036	0.00007	0.00023	0.00036	0.00051
		SP	0.01545	0.00741	0.00716	0.01308	0.04129
	50	GD	0.00041	0.00014	0.00027	0.00039	0.00101
		SP	0.01616	0.00495	0.00868	0.01578	0.03059
	60	GD	0.00035	0.00007	0.00023	0.00035	0.00051
		SP	0.01559	0.00671	0.00677	0.01446	0.04018
	70	GD	0.00036	0.00006	0.00026	0.00036	0.00049
		SP	0.01627	0.01122	0.00654	0.01246	0.06378
	80	GD	0.00039	0.00015	0.00025	0.00034	0.00087
		SP	0.01404	0.00586	0.00733	0.01343	0.03917
	90	GD	0.00037	0.00009	0.00026	0.00035	0.00065
		SP	0.01682	0.00770	0.00726	0.01468	0.03770

Table 7. Comparison of Results between SPEA2-CE against SPEA2-CC for DTLZ6

		DTLZ6 (60, 000 evaluations)					
Algorithm	Opponents Size	Metric	Mean	St Dev	Best	Median	Worst
SPEA2-CC		GD	0.00041	0.00172	0.00002	0.00002	0.00918
		SP	0.00851	0.01716	0.00390	0.00493	0.09752
SPEA2-CE	10	GD	0.01503	0.00104	0.01346	0.01503	0.01822
		SP	0.02961	0.00831	0.02064	0.02766	0.05343
	20	GD	0.01448	0.00104	0.01268	0.01431	0.01728
		SP	0.02660	0.00737	0.01967	0.02406	0.05685
	30	GD	0.01415	0.00112	0.01213	0.01418	0.01693
		SP	0.02970	0.01657	0.01769	0.02556	0.10925
	40	GD	0.01452	0.00123	0.01236	0.01461	0.01683
		SP	0.02528	0.00480	0.01822	0.02445	0.03811
	50	GD	0.01421	0.00097	0.01184	0.01422	0.01645
		SP	0.02467	0.00473	0.01688	0.02398	0.03888
	60	GD	0.01380	0.00112	0.01203	0.01380	0.01621
		SP	0.02597	0.01034	0.01581	0.02483	0.07576
	70	GD	0.01387	0.00107	0.01180	0.01361	0.01602
		SP	0.02535	0.00609	0.01458	0.02440	0.04640
	80	GD	0.01416	0.00134	0.01164	0.01404	0.01796
		SP	0.02408	0.00473	0.01689	0.02359	0.03802
	90	GD	0.01356	0.00073	0.01250	0.01343	0.01520
		SP	0.02418	0.00593	0.01746	0.02194	0.04175

Table 8. Comparison of Results between SPEA2-CE against SPEA2-CC for DTLZ7

DTLZ7 (60, 000 evaluations)							
Algorithm	Opponents Size	Metric	Mean	St Dev	Best	Median	Worst
SPEA2-CC		GD	0.00504	0.01908	0.00097	0.00122	0.10596
		SP	0.06600	0.18000	0.02680	0.03260	1.01860
SPEA2-CE	10	GD	0.00161	0.00045	0.00072	0.00156	0.00315
		SP	0.06087	0.01847	0.00289	0.06406	0.08786
	20	GD	0.00165	0.00085	0.00104	0.00141	0.00546
		SP	0.06861	0.01311	0.03834	0.07072	0.08925
	30	GD	0.00142	0.00049	0.00056	0.00127	0.00290
		SP	0.06802	0.01350	0.01887	0.06810	0.09251
	40	GD	0.00200	0.00175	0.00100	0.00135	0.00848
		SP	0.06690	0.01159	0.04631	0.06722	0.09113
	50	GD	0.00169	0.00131	0.00098	0.00129	0.00746
		SP	0.06715	0.01279	0.04147	0.06685	0.10080
	60	GD	0.00289	0.00298	0.00104	0.00150	0.01167
		SP	0.06320	0.01392	0.03063	0.06332	0.09278
	70	GD	0.00213	0.00189	0.00099	0.00139	0.00884
		SP	0.06569	0.01489	0.02159	0.06733	0.09698
	80	GD	0.00195	0.00160	0.00098	0.00138	0.00655
		SP	0.07024	0.01112	0.05233	0.06910	0.09488
	90	GD	0.00273	0.00252	0.00093	0.00171	0.01052
		SP	0.06450	0.01721	0.00000	0.06664	0.09645

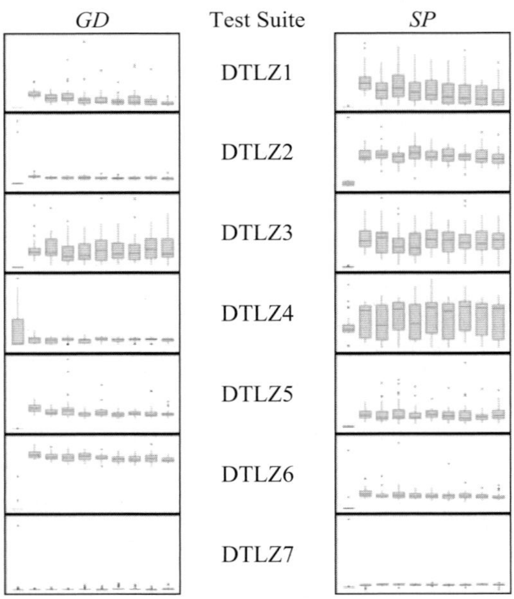

Fig. 1. Box plots for Generational Distance (GD) and Spacing (SP)

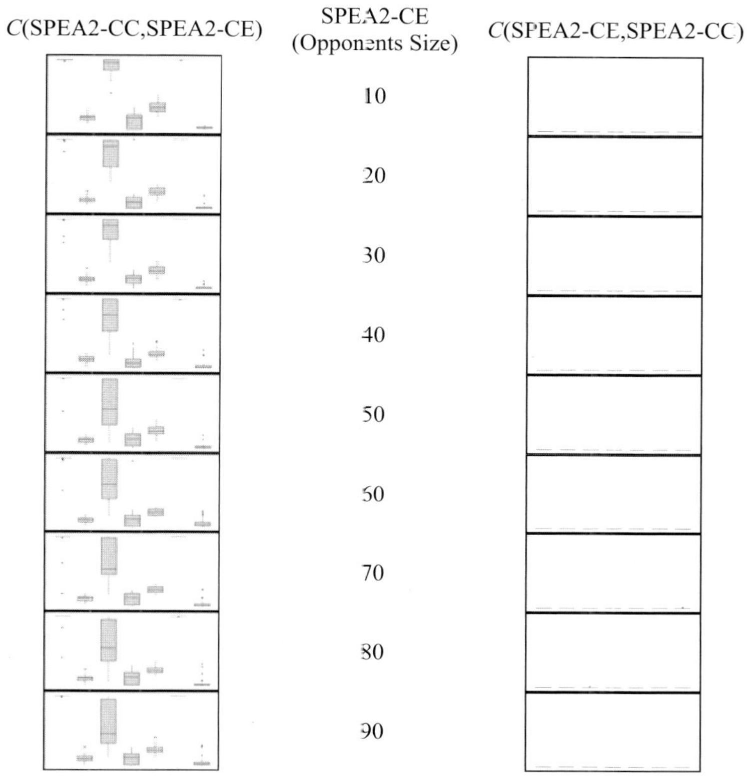

Fig. 2. Box plots of function C (X, Y which is covered by X) indices for each pair of SPE2-CE and SPEA2-CC

5 Conclusions and Future Work

Two new modified algorithms, SPEA2-CE and SPEA2-CC are introduced in this paper. The SPEA2-CE is benchmarked against the SPEA2-CC using seven well-known test problems, DTLZ1 to DTLZ7, using three objectives. It has been shown that SPEA2-CC has a much better performance in terms of the average distance of the contaminated solutions to the true Pareto front. In addition, the results show that the SPEA2-CC has an excellent performance in terms of the even distribution of solutions across the solution space and also the solutions found by SPEA2-CC practically totally covered the solutions found by the SPEA2-CE.

As future work, it would be interesting to investigate whether the SPEA2-CE using different competitive fitness strategies would be able to improve the performance. It would also be highly beneficial to conduct further tests of scalability to higher dimensions for SPEA2-CE and SPEA2-CC.

References

1. Angeline, P.J., Pollack, J.B.: Competitive Environments Evolve Better Solutions for Complex Tasks. In: Forrest, S. (ed.) Proc. 5th International Conference on Genetic Algorithm, pp. 264–270. Morgan Kaufmann, San Francisco (1993)
2. Coello Coello, C.A., Reyes Sierra, M.: A Coevolutionary Multi-Objective Evolutionary Algorithm. Evolutionary Computation 1, 482–489 (2003)
3. Deb, K., Thiele, L., Laumanns, M., Zitzler, E.: Scalable Test Problems for Evolutionary Multi-Objective Optimization. KanGAL Report 2001001, Kanpur Genetic Algorithms Laboratory (KanGAL), Department of Mechanical Engineering, Indian Institute of Technology Kanpur, India (2001)
4. Hillis, W.D.: Co-evolving Parasites Improve Simulated Evolution as an Optimization Procedure, pp. 228–234. MIT Press, Cambridge (1991)
5. Keerativuttitumrong, N., Chaiyaratana, N., Varavithya, V.: Multi-objective Co-operative Co-evolutionary Genetic Algorithm. In: Guervós, J.J.M., Adamidis, P.A., Beyer, H.-G., Fernández-Villacañas, J.-L., Schwefel, H.-P. (eds.) PPSN VII. LNCS, vol. 2439, pp. 288–297. Springer, Heidelberg (2002)
6. Lohn, J., Kraus, W., Haith, G.: Comparing a Coevolutionary Genetic Algorithm for Multi-objective Optimization. In: Fogel, D., et al. (eds.) Proc. 2002 Congress on Evolutionary Computation (CEC'02), pp. 1157–1162. IEEE Computer Society Press, Los Alamitos (2002)
7. Panait, L., Luke, S.: A Comparative Study of Two Competitive Fitness Functions. In: Langdon, W.B., et al. (eds.) Proc. Genetic and Evolutionary Computation Conference (GECCO 2002), pp. 503–511. Morgan Kaufmann, San Francisco (2002)
8. Parmee, I.C., Watson, A.H.: Preliminary Airframe Design Using Co-evolutionary Multiobjective Genetic Algorithms. In: Banzhaf, W., Daida, J., Eiben, A.E., Garzon, M.H., Honavar, V., Jakiela, M., Smith, R.E. (eds.) Proc. Genetic and Evolutionary Computation Conference (GECCO 99), vol. 2, pp. 1657–1665. Morgan Kaufmann, San Francisco(1999)
9. Potter, M.A., DeJong, K.A.: A Cooperative Coevolutionary Approach to Function Optimization. In: Davidor, Y., Männer, R., Schwefel, H.-P. (eds.) PPSN III. LNCS, vol. 866, pp. 249–257. Springer, Heidelberg (1994)
10. Rosin, C.D., Belew, R.K.: Methods for Competitive Co-evolution: Finding Opponents Worth Beating. In: Eshelman, L. (ed.) Proc. 6th International Conference on Genetic Algorithms, pp. 373–380. Morgan Kaufmann, San Francisco (1995)
11. Schott, J.R.: Fault Tolerant Design Using Single and Multicriteria Genetic Algorithm Optimization. Master's thesis, Department of Aeronautics and Astronautics, Massachusetts Institute of Technology, Cambridge, Massachusetts (1995)
12. Van Veldhuizen, D.A., Lamont, G.B.: On Measuring Multiobjective Evolutionary Algorithm Performance. Evolutionary Computation 1, 204–211 (2000)
13. Wiegand, R.P., Liles, W.C., DeJong, K.A.: An Empirical Analysis of Collaboration Methods in Cooperative Coevolutionary Algorhtms. In: Spector, L., et al. (eds.) Proc. Genetic and Evolutionary Computation Conference, pp. 1235–1242. Morgan Kaufmann, San Francisco (2001)
14. Yao, X.: Evolutionary Computation. In: Sarker, R., Mohammadian, M., Yao, X. (eds.) Evolutionary Optimization. International Series in Operations Research and Management Science, pp. 27–46. Kluwer, United States (2002)
15. Zitzler, E., Deb, K., Thiele, L.: Comparison of Multiobjective Evolutionary Algorithms: Empirical Results. Evolutionary Computation 8(2), 173–195 (2000)
16. Zitzler, E., Laumanns, M., Thiele, L.: SPEA2: Improving the Strength Pareto Evolutionary Algorithm. Technical Report 103, Computer Engineering and Network Laboratory (TIK), Swiss Federal Institute of Technology (ETH) Zurich, Switzerland (2001)

On the Running Time Analysis of the (1+1) Evolutionary Algorithm for the Subset Sum Problem

Yuren Zhou [1], Zhi Guo [1], and Jun He [2]

[1] School of Computer Science and Engineering, South China University of Technology, Guangzhou 510640, China
yrzhou@scut.edu.cn
[2] School of Computer Science, University of Birmingham, Birmingham, B15 2TT, UK

Abstract. Theoretic researches of evolutionary algorithms have received much attention in the past few years. This paper presents the running time analysis of evolutionary algorithm for the subset sum problems. The analysis is carried out on (1+1) EA for different subset sum problems. It uses the binary representation to encode the solutions, the method "superiority of feasible point" that separate objectives and constraints to handle the constraints, and the absorbing Markov chain model to analyze the expected runtime. It is shown that the mean first hitting time of (1+1) EA for solving subset sum problems may be polynomial, exponential, or infinite.

1 Introduction

The subset sum problem (SSP) is one of the most frequently encountered combinatorial optimization problems. It is a problem that offers many practical applications in computer science, operations research, cryptography and management science. Subset sum problem belongs to a special class of knapsack problems, and it is the NP-complete problem whose computational complexity increases exponentially as n increases. Hence in its general form, it is a difficult problem to solve. The most accepted algorithms in the literatures are the pseudo polynomial algorithm and the fully polynomial time approximation scheme [1-2]. Different heuristic algorithms, including evolutionary algorithms, have been proposed for this problem [3- 6].

Evolutionary algorithms are population based iterative stochastic techniques for solving optimization problems by modeling the process of natural selection and genetic evolution [3]. Over the last 20 years they have been used to solve a wide range of science research and engineering optimization problems. However, theoretical researches of evolutionary algorithms (EAs) receive much attention recently [7]. A major part of the theoretical result on discrete search spaces is the runtime analysis of the algorithms, which estimates the average optimization time of evolutionary algorithms on various problems [8-11]. In this paper, we present the running time analysis of evolutionary algorithm for the subset sum problem, and the rest of this paper is organized as follows: Section 2 introduces the concept of time complexity of evolutionary algorithms; section 3 analyzes the mean first hitting time of (1+1) EA for different subset sum problems and section 4 concludes this paper.

K. Li et al. (Eds.): LSMS 2007, LNCS 4688, pp. 73 – 82, 2007.

2 Evolutionary Algorithms and Their Time Complex Analysis

We concentrate on the minimization of so-called Pseudo-Boolean function:

$$f: \{0,1\}^n \rightarrow R \tag{1}$$

Denote the EA using mutation and selection approach with group size of 1 as (1+1) EA. Its general description is:

Algorithm 1 (1+1) EA

1. Initialization: Choose randomly an initial bit string $\mathbf{x} \in \{0,1\}^n$.
2. Mutation: $\mathbf{y}$: =mutate($\mathbf{x}$).
3. Selection: If f($\mathbf{y}$) < f($\mathbf{x}$), $\mathbf{x}$: = $\mathbf{y}$.
4. Stopping Criterion

If the stopping criterion is not met, continue at line 2.

(1+1) EA generally uses two kinds of mutation which are called local mutation and global mutation respectively:

1. LM randomly chooses a bit x_i $(1 \leq i \leq n)$ from the individual $\mathbf{x}$ $=(x_1 \dots x_n) \in \{0,1\}^n$ and flips it.
2. GM flips each bit of individual $\mathbf{x} =(x_1 \dots x_n) \in \{0,1\}^n$ independently with probability $\frac{1}{n}$.

(1+1) EA is a simple but effective random hill-climbing EA. It is a basic model to analyze EA's time complexity. The methodologies and results of analyzing (1+1) EA's time complexity are of theoretically importance as well as instructiveness to more general and complicated theoretical analysis.

Generally, the population sequence of the evolutionary algorithm without adaptive mutation can be modeled by a homogenous Markov chain. The population stochastic process introduced by (1+1) EA is an absorbing Markov chain and its absorbing set is the optimal solution. Basic knowledge of absorbing Markov chain is introduced as follows. Its details can be found in literature regarding "Random Process" (e.g. [12]).

Let (X_t ; t = 0, 1, …) denote a discrete homogenous absorbing Markov chain in a finite state space S. T is the transient state set, H = S – T is the absorbing set. The canonical transition matrix is:

$$\mathbf{P} = \begin{pmatrix} \mathbf{I} & \mathbf{O} \\ \mathbf{R} & \mathbf{T} \end{pmatrix}$$

where **I** is an identity matrix, the region **O** consists entirely of 0's, the matrix **R** concerns the transition from transient to absorbing states, and the matrix **T** concerns the transition from transient to transient states. By the nature of the power of **P** we know that the region **I** remains **I**. This corresponds to the fact that once an absorbing state is entered, it can't be left.

Definition 1. Let (X_t ; t = 0, 1, …) be an absorbing Markov chain. The first hitting time from status i (i∈ S) to the absorbing status set H is: τ_i = min { t: t≥0, $X_t \in H$ | X_0 = i }.If the right side is an empty set, let $\tau_i = \infty$.

When i∈ H, it is obvious that τ_i =0. The first hitting time τ_i is a random variable, what we are interested is its expectation value E[τ_i].

The following theorem can be used to estimate the mean hitting time of evolutionary algorithms when they are modeled by homogenous absorbing Markov chain.

Theorem 1. [12]: Let (X_t ; t = 0, 1, …) be an absorbing Markov Chains, τ_i is the first hitting time of transient state i. Let m_i = E[τ_i], $\mathbf{m} = [m_i]_{i \in T}$, then:

$$\mathbf{m} = (\mathbf{I} - \mathbf{T})^{-1}\mathbf{1}$$

where **1** represents vector $(1,\ldots,1)^t$.

3 The Mean First Hitting Time of (1+1)EA for Subset Sum Problems

The subset sum problem (SSP) is generally described as follows:

$$\text{Minimize } f(\mathbf{x}) = -\sum_{i=1}^{n} a_i x_i \,.$$

$$\text{subject to } g(\mathbf{x}) = \sum_{i=1}^{n} a_i x_i \le b;\ x_i = 0 \text{ or } 1,\ i = 1,2,\ldots,n,$$

where all the coefficient a_i and the right-hand-value b are positive integers.

The subset sum problem is a special case of non-linear programming(NLP): **x** is a binary string. Several methods for handling NLP by EAs are available in the literature. In the survey article [13], Coello divides these EAs into five categories: Penalty functions, Special representations and operators, Repair algorithms, Separation of objectives and constraints, and Hybrid methods. The most common approach in the EA community to handle constraints is to use penalties. A drawback of penalty function method is the large number of parameters. In this paper we focus the method "superiority of feasible point" that separate objectives and constrains.

Deb [14] suggested a non-parameter penalty function evolutionary algorithm for constrained optimization problem in which an individual is evaluated using:

$$\text{fitness}(\mathbf{x}) = \begin{cases} f(\mathbf{x}), & \text{if } \mathbf{x} \text{ is feasible,} \\ f_{worst} + \sum_{j=1}^{m} g_j(\mathbf{x}), & \text{if } \mathbf{x} \text{ is infeasible.} \end{cases} \quad (2)$$

where f_{worst} is the objective function value of the worst feasible solution in the population, and $g_j(\mathbf{x})$ refers to inequality constraints.

When two individuals based on fitness function (2) are compared using binary tournament selection, where two solutions are compared at a time, the following criteria are applied:

1. A feasible solution is superior to an infeasible solution;
2. Between two feasible solutions, the one having a better objective function value is preferred;
3. Between two infeasible solutions with unequal constraint violations, the one having smaller constraint violation is preferred.

In the following, we use the binary representation to encode the solutions and the fitness function (2) to handle the constraints.

The time complexity analysis of EAs for solving general knapsack problem is very complicated and difficult. This paper only focuses on the following three examples of subset sum problem.

SSP1. Minimize $f_1(\mathbf{x}) = -((n-1)x_1 + \ldots + x_n)$,

subject to $g_1(\mathbf{x}) = (n-1)x_1 + \ldots + x_n - (n-1) \leq 0$,

where $x_i = 0$ or 1, i = 1, 2, …, n

The optimal solution of SSP1 is (1, 0, …, 0) or (0, 1, …, 1). SSP1 is the concerned with the NP-complete set partition problem [1]. This problem can be formulated as follows: Given a set of n integers W={ w_1, w_2, …, w_n }, we want to partition W into two subsets W_1 and W_2 so that the sum of the integers in W_1 is equal to the sum of the integers in W_2. This is one of the easiest-to-state NP-complete problems. There already are some average-case analyses of classical greedy heuristics [15], and worst-case and average-case of (1+1)EA for this problem[6]. Following Garey and Johnson[1] the set partition problem can be transformed to SSP as follows:

$$\text{Minimize } -\sum_{i=1}^{n} w_i x_i ,$$

$$\text{subject to } \sum_{i=1}^{n} w_i x_i \leq \sum_{i=1}^{n} w_i / 2 ,$$

where $x_i = 0$ or 1, i = 1, 2, …, n.

SSP2. Minimize $f_2(\mathbf{x}) = -(x_1 + \ldots + x_n)$,

subject to $g_2(\mathbf{x}) = x_1 + \ldots + x_n - m \leq 0$,

where $x_i = 0$ or 1, i=1, 2, … , n, m $\leq$ n.

The optimal solution of SSP2 is $\mathbf{x} = (x_1 \ldots x_n)$, where $\mathbf{x}$ satisfy the condition $x_1 + \ldots + x_n = m$.

SSP3 Minimize $f_3(\mathbf{x}) = -((\sum_{i=2}^{n} a_i + 1) x_1 + a_2 x_2 + \ldots + a_n x_n)$,

subject to $g_3(\mathbf{x}) = (\sum_{i=2}^{n} a_i + 1) x_1 + a_2 x_2 + \ldots + a_n x_n - (\sum_{i=2}^{n} a_i + 1) \le 0$,

where a_i (i=2, … , n) is positive integer, x_i =0 or 1, i=1, … , n.

SSP3 is similar to the deceptive problem, and the optimal solution is (1, 0, …, 0).

The followings derive the mean runtimes of the (1+1) EA with local mutation operator and global mutation operator for SSP1, SSP2 and SSP3 respectively.

For $\mathbf{x} = (x_1, \ldots, x_n) \in \{0,1\}^n$, denote $|\mathbf{x}| := x_1 + \ldots + x_n$, and $S_k := \{\mathbf{x} \in \{0,1\}^n \mid |\mathbf{x}| = k\}$, k=0, 1, …, n. Let $S_k^0 := \{\mathbf{x} \mid \mathbf{x} = (0, x_2, \ldots, x_n), |\mathbf{x}|=k\}$(k=0, …, n-1) and $S_k^1 := \{\mathbf{x} \mid \mathbf{x} = (1, x_2, \ldots, x_n), |\mathbf{x}|=k+1\}$(k=0, …, n-1) be the partition of the search space $S=\{0,1\}^n$.

Proposition 1. The expected runtime of local (1+1)EA on SSP1 is O(n*log*n), i.e. the mean first hitting time of transient state i

$$m_i = \begin{cases} 1+(n-1)(1+\ldots+\frac{1}{n-2}), & i \in S_C^0 \\ n(1+\ldots+\frac{1}{n-(k+1)}), & i \in S_k^0 (1 \le k \le n-2), \\ 0, & i \in S_{n-1}^0, \\ 0, & i \in S_0^1 \\ \frac{n}{k+1}(k+\sum_{h=1}^{k}\sum_{j=1}^{n-(h+1)}\frac{1}{j}), & i \in S_k^1 (1 \le k \le n-2), \\ (n-1)+\sum_{h=1}^{n-2}\sum_{j=1}^{n-(h+1)}\frac{1}{j}, & i \in S_{n-1}^1. \end{cases}$$

Proof. Let X_t (t=0, 1, …) be the population sequence introduced by the local (1+1) EA on SSP 3. The transition probabilities of the population stochastic process X_t can be described as follows:

When k=0, for i$\in S_k^0$,

$$P(X_{t+1} \in S_k^1 \mid X_t = i) = \frac{1}{n}.$$

$$P(X_{t+1} \in S_{k+1}^0 \mid X_t = i) = 1 - \frac{1}{n}.$$

When $1 \leq k \leq n-2$, for any $i \in S_k^0$,

$$P(X_{t+1} \in S_{k+1}^0 \mid X_t = i) = \frac{n-(k+1)}{n},$$

$$P(X_{t+1} \in S_k^0 \mid X_t = i) = \frac{k+1}{n}.$$

When $k=n-1$, for $i \in S_k^0$,

$$P(X_{t+1} \in S_k^0 \mid X_t = i) = 1.$$

Similarly, when $k=0$, for $i \in S_k^1$,

$$P(X_{t+1} \in S_k^1 \mid X_t = i) = 1.$$

When $1 \leq k \leq n-1$, for any $i \in S_k^1$,

$$P(X_{t+1} \in S_k^0 \mid X_t = i) = \frac{1}{n},$$

$$P(X_{t+1} \in S_{k-1}^1 \mid X_t = i) = \frac{k}{n},$$

$$P(X_{t+1} \in S_k^1 \mid X_t = i) = 1 - \frac{k+1}{n}.$$

Introduce an auxiliary homogeneous Markov chain Z_t (t=0, 1, …) with the state space $\{ z_0^0, z_1^0, \ldots, z_{n-1}^0, z_0^1, z_1^1, \ldots, z_{n-1}^1 \}$, the transition probabilities are defined by

$$P(Z_{t+1} = z_h^v \mid Z_t = z_k^u) = P(X_{t+1} \in S_h^v \mid X_t \in S_k^u),$$

where u, v=0 or 1, and h, k = 0, …, n-1.

Then Z_t is an absorbing Markov chain with the absorbing state z_0^1 and z_{n-1}^0, and for any $i \in S_k^u$ (u=0 or 1, k = 0, …, n-1), the mean first hitting time m_i equals $m_{z_k^u}$.

According to Theorem 1, the mean first hitting time of stochastic process Z_t is given by

$$\begin{cases} -\frac{n-1}{n} m_{z_{k+1}^0} + m_{z_k^0} = 1, & k=0, \\ -(1-\frac{k+1}{n}) m_{z_{k+1}^0} + (1-\frac{k+1}{n}) m_{z_k^0} = 1, & k=1,\ldots,n-2, \\ m_{z_k^0} = 0, & k=n-1, \\ m_{z_k^1} = 0, & k=0, \\ -\frac{k}{n} m_{z_{k-1}^1} - \frac{1}{n} m_{z_k^0} + \frac{k+1}{n} m_{z_k^1} = 1, & k=1,\ldots,n-1. \end{cases}$$

The above linear equations can be solved as

$$\begin{cases} m_{z_0^0} = 1 + (n-1)(1 + \ldots + \frac{1}{n-2}), & \\ m_{z_k^0} = n(1 + \ldots + \frac{1}{n-(k+1)}), & 1 \le k \le n-2, \\ m_{z_{n-1}^0} = 0, & \\ m_{z_0^1} = 0, & \\ m_{z_k^1} = \frac{n}{k+1}(k + \sum_{h=1}^{k} \sum_{j=1}^{n-(h+1)} \frac{1}{j}), & 1 \le k \le n-2, \\ m_{z_{n-1}^1} = (n-1) + \sum_{h=1}^{n-2} \sum_{j=1}^{n-(h+1)} \frac{1}{j}. & \end{cases}$$

This completes the proof. □

Proposition 2. For SSP 1, the expected runtime of global (1+1)EA is O(n*log*n).

Proof. For any i∈ S_k^0 (k=0, 1, …, n-2), i is a feasible solution, then the possibility of the global (1+1)EA leading i to some j∈ $S_{k+1}^0 \cup \ldots \cup S_{n-1}^0$ is greater than (n-k-1)$\frac{1}{n}(1-\frac{1}{n})^{n-1} \ge \frac{n-k-1}{e \times n}$.

This leads an upper bound on the expected runtime of the algorithm, i.e.

$m_i \le$ e×n (1+…+$\frac{1}{n-k-1}$)= O(n*log*n) (i∈ S_k^0, k=0, 1, …. n-2).

For any i∈ S_k^1 (k=1, …, n-1), i is an infeasible solution, then the possibility of the global (1+1)EA leads i to some j∈ $S_{n-1}^0 \cup \ldots \cup S_0^0 \cup S_{k-1}^1 \cup \ldots \cup S_0^1$ is greater than (k+1)$\frac{1}{n}(1-\frac{1}{n})^{n-1} \ge \frac{k+1}{e \times n}$.

Therefore we have $m_i \le$ e×n (1+…+$\frac{1}{k+1}$)+ O(n*log*n) = O(n*log*n) (i∈ S_k^1, k=1, …, n-1). This proves the claim. □

Proposition 3. For any i∈ S_k = {x∈ $\{0,1\}^n$ | |**x**|= k }(k=0, 1, …, n), the expected runtime of local (1+1)EA on SSP 2 is

$$m_i = \begin{cases} n\sum_{h=k}^{m-1} \frac{1}{n-h} = O(n\log n), & \text{if } k < m, \\ 0, & \text{if } k = m, \\ n\sum_{h=m+1}^{k} \frac{1}{h} = O(n\log n), & \text{if } k > m. \end{cases}$$

Proof. The proof can be done in the same way as the proof of Proposition 1. □

Proposition 4. For SSP 2, the expected runtime of global (1+1) EA is O(n*log*n).

Proof. The proof can be done in the same way as the proof of Proposition 2. □

Proposition. 5. For any i∈ S=$\{0,1\}^n$, the expected runtime of local (1+1) EA on SSP 3 is:

$$m_i = \begin{cases} 0, & \text{if } i \in S_0^1, \\ +\infty, & \text{otherwise.} \end{cases}$$

Proof. Let X_t (t=0, 1, …) be the population sequence introduced by the local (1+1) EA on SSP 3. The transition probabilities of the population stochastic process X_t can be described as follows:

When k=0, for any i∈ S_k^0,

$$P(X_{t+1} \in S_1^0 \mid X_t = i) = \frac{n-1}{n},$$

$$P(X_{t+1} \in S_0^1 \mid X_t = i) = \frac{1}{n}.$$

When 1≤k<n-1, for any i∈ S_k^0

$$P(X_{t+1} \in S_{k+1}^0 \mid X_t = i) = \frac{n-1-k}{n},$$

$$P(X_{t+1} \in S_k^0 \mid X_t = i) = \frac{1+k}{n}.$$

This implies that the local (1+1) EA starting from i∈ S_k^0 (1≤k≤n-1) will never reach the optimal solution (1, 0, …, 0). Thus, we have m_i =+∞ (i∈ S_k^0, k = 0, 1, …, n-1).

Similarly, when 1≤k≤n-1, for any i∈ S_k^1,

$$P(X_{t+1} \in S_{k-1}^1 \mid X_t = i) = \frac{k}{n},$$

$$P(X_{t+1} \in S_k^0 \mid X_t = i) = \frac{1}{n},$$

$$P(X_{t+1} \in S_k^1 \mid X_t = i) = 1 - \frac{k+1}{n}.$$

Therefore we have m_i =+∞ (i∈ S_k^1, k = 1, …, n-1). This completes the proof. □

Proposition 6. For any $0<\varepsilon<\frac{1}{2}$, with a probability of $\frac{1}{2}(1-e^{-\Omega(n)})$, the expected runtime of global (1+1)EA on SSP3 is $\Omega(n^{(\frac{1}{2}-\varepsilon)n})$.

Proof. By Chernoff bounds (see R.Motwani and P.Raghavan [16]), for any $0<\varepsilon<\frac{1}{2}$, the probability that the initial string has at least $(\frac{1}{2}-\varepsilon)n$ of ones equals $(1-e^{-\Omega(n)})$, i.e. exponentially close to one. So with a probability of $\frac{1}{2}(1-e^{-\Omega(n)})$, the initial string has the property $\mathbf{x}^0$ = (0,*...*) and $|\mathbf{x}^0|\geq(\frac{1}{2}-\varepsilon)n$. The probability that such string reach optimal solution (1,0...0) by mutation is at most

$$(\frac{1}{n})^{1+(\frac{1}{2}-\varepsilon)n}(1-\frac{1}{n})^{n-1-(\frac{1}{2}-\varepsilon)n}\leq(\frac{1}{n})^{(\frac{1}{2}-\varepsilon)n}.$$

This proves the claim. □

From propositions 1 - 6, the time complexity of (1+1) EA for solving subset sum problems may be polynomial, exponential, and infinite.

4 Conclusions

As an important branch of the modern heuristic algorithms, evolutionary algorithms are successfully applied to a number of real-world optimization problems but the theory of these is not well developed. The theoretical study of EAs is still in its infantile stage, the analysis of the time complexity for evolutionary algorithms is far behind the analysis of classical deterministic algorithms. While current theoretical research of EAs generally focuses on some so-called artificial function, more work need to be done on study of time complexity of EAs for real-world and complicated problems in the future.

Acknowledgments. This work was supported in part by the National Natural Science Foundation of China under Grant No.60673062, in part by the Natural Science Foundation of Guangdong Province of China under Grant No.06025686, and in part by the Science and Technology Plan Project of Guangdong Province of China under Grant Nos. 2005B10101048, 2006B11201003.

References

1. Garey, M.R., Johnson, D.S.: Computers and intractability-A guide to the theory of NP-completeness. Freeman, New York (1979)
2. Hochbaum, D.S.: Approximation algorithms for NP-Hard problems. Wadworth publish company (1997)
3. Michalewicz, Z.: Genetic algorithms + data structures = Evolution programs. Springer, Heidelberg (1992)
4. Wang, R.: A genetic algorithm for subset sum problem. Neurocomputing, 463–468 (2004)
5. Lin, F.T., Kao, C.Y., Hsu, C.C.: Applying the genetic approach to simulated annealing in solving some NP-hard problems. IEEE transactions on system, man, and cybernetics 23, 1752–1767 (1993)
6. Witt, C.: Worst-case and average-case approximations by simple randomized search heuristics. In: Diekert, V., Durand, B. (eds.) STACS 2005. LNCS, vol. 3404, pp. 44–56. Springer, Heidelberg (2005)
7. Beyer, H.-G., Schwefel, H.-P., Wegener, I.: How to analyze evolutionary algorithms. Theoretical Computer Science, 101–130 (2002)
8. Garnier, J., Kallel, L., Schoenauer, M.: Rigorous hitting times for binary mutations. Evolutionary Computation, 167–203 (1999)
9. Droste, S., Jansen, T., Wegener, I.: On the analysis of the (1+1)-evolutionary algorithm. Theoretical Computer Science, 51–81 (2002)
10. Wegener, I., Witt, C.: On the analysis of a simple evolutionary algorithm on quadratic pseudo-boolean function. Journal of discrete algorithms 61–78 (2005)
11. He, J., Yao, X.: Towards an analytic framework for analyzing the computation time of evolutionary algorithms. Artificial Intelligence, 59–97 (2003)
12. Iosifescu, M.: Finite Markov processes and their applications. John Wiley & Sons, Chichester (1980)
13. Coello, C.A.C.: Theoretical and numerical constraint-handling techniques used with evolutionary algorithms: a survey of the state of the art. Computer methods in applied mechanics and engineering 19, 1245–1287 (2002)
14. Deb, K.: An efficient constraint handling method for genetic algorithms. Computer methods in applied mechanics and engineering 186(2-4), 311–338 (2000)
15. Coffman Jr., W.W.: Recent asymptotic results in the probabilistic analysis of schedule makespans. Scheduling Theory and its applications, pp. 15–31. Wiley, Chichester (1995)
16. Motwani, R., Raghavan, P.: Randomized algorithms. Cambridge University Press, Cambridge (1995)

Parameter Identification of Bilinear System Based on Genetic Algorithm

Zhelong Wang and Hong Gu

Dalian University of Technology
Dalian, Liaoning, 116024, China
zhelong.wang@dunelm.org.uk

Abstract. The paper presents a method for the identification of bilinear system parameters by using an improved Genetic Algorithm. Good results could still be obtained when the system output was influenced by Gaussian noise in the simulation. By comparing with RLS and COR through a simulation experiment to a SISO bilinear system, it is found that the method can get better result than the other two methods. Through a simulation experiment to a MIMO bilinear system, the method can get reasonably good results too. These simulations show that the method is simpler and can get better results than RLS and COR. Through a simulation study to an MIMO bilinear system, good results can still be got. In the last section, the paper describes that a hybrid GA, the combination of Genetic Algorithm and nonlinear Least Square, was developed to identify bilinear system structure and parameters simultaneously.

1 Introduction

Many objects in engineering, economics, ecology and biology etc. can be described by using a bilinear system.[1] The bilinear system is the simplest and most similar nonlinear system to a linear system in the form. It has many advantages, such as its special variable structure property which makes it advantageous on system optimization and modeling. The research of system identification and parameter estimation of the bilinear system have been carried out and received extensive attentions. Several identification methods have been reported, such as Recursive Least Squares (RLS), Recursive method of instrumental variable (RIV) and Correlative Function method (COR), etc. RLS is simple to use and costs little computation but it gives large estimation error when the system is influenced by colorful noises. Although COR can acquire more accurate parameter estimation but the estimation error is still not satisfing. RIV can get accurate parameter but it costs large amount of complex computation. Most of methods used in the system identification are based on the assumption that the searching space of identified system model is continuous and differentiable. If the searching space is undifferentiable or parameter span is nonlinear, traditional recurrent methods can not gain the global optimization.[2] However, Genetic Algorithm (hereafter referred to as GA) does not need to assume that the searching space is differentiable or continuous. GA employs random, yet directly, search for locating the

K. Li et al. (Eds.): LSMS 2007, LNCS 4688, pp. 83–91, 2007.

globally optimal solution and can search discrete, noisy and multimodal space.[3] In this paper, an improved GA was applied to identify SISO bilinear system, which can be described as the following equation (1).

$$y(k)+\sum_{i=1}^{n} a_i y(k-i)=\sum_{i=1}^{n} b_i u(k-d-i)+ \sum_{i=1}^{n}\sum_{j=0}^{n} c_{ij} y(k-j)u(k-d-j)+v(k) \tag{1}$$

In the equation, $\{y(k)\}$ is an input series and $\{u(k)\}$ is an output series. The integer d is time lag. $\{v(k)\}$ is a stochastic noise series. The system order is n. $\sum_{i=1}^{n}\sum_{j=0}^{n} c_{ij} y(k-j)u(k-d-j)$ is a bilinear term. Parameters which are needed to be identified are a_i, b_i and c_{ij}.

2 Research Methodology

2.1 Genetic Algorithm (GA)

GA is a robust search and optimization technique, which simulates heritage mechanism in the nature and natural evolutionary. When it is applied to a practical problem, parameters of the problem need be encoded into strings. Several strings form an individual and many individuals form a population. Reproduction, Crossover and Mutation techniques are applied to evolve and locate global optimum of the parameters in the population. In addition, a fitness function needs to be formed and control parameters of GA such as crossover probability P_c; mutation probability P_m and population N will be selected.

Given parameters to be identified $a_1, a_2, \cdots, a_n$ are encoded into strings as following,

a_1	a_2	$\cdots$	a_n

The length of each parameter is set according to actual requirement of research objects. For example, the length of each parameter is L, the whole length of the string is K=L×N.

Fitness function gives a fitness value for each individual in the population and its selection is related with the researched problem. The variance of fitness function usually is required to correspond to the variance of objective function and then theindividual with the best fitness will be the best solution. The working parameters of GA are usually set as: crossover probability $P_c \in [0.25,1.0]$, mutation probability $P_m \in [0.005,0.05]$, the size of population $N \in [30,100]$.

2.2 Adaptive Genetic Algorithm (AGA)

An Adaptive Genetic Algorithm was applied to do the parameter identification of a bilinear system. AGA takes advantage of adaptive crossover probability and mutation probability. [1] The structure of AGA is shown as follows.

1) Adaptive crossover probability p_c and adaptive mutation probability p_m

$$P_c = \begin{cases} k_1 (f_{\max} - f')/(f_{\max} - \bar{f}) & ,f' \ge \bar{f} \\ k_2 & ,f' < \bar{f} \end{cases}$$

$$P_m = \begin{cases} k_3 (f_{\max} - f)/(f_{\max} - \bar{f}) & ,f \ge \bar{f} \\ k_4 & ,f < \bar{f} \end{cases}$$

constants $k_1, k_2, k_3, k_4 \le 1.0$; $f_{\max}$ is a maximum fitness value in each population; $\bar{f}$ is an average fitness value in each population. f' is the bigger fitness value within the selected two strings to do a crossover operation. f is the maximum fitness value in the selected strings to do a mutation operation.
2) Crossover in two positions
3) Crossover positions are selected in unequal probabilities. [4]
4) Elitist Strategy
5) The convergence criterion is that fitness value of each individual is equal to average fitness value.

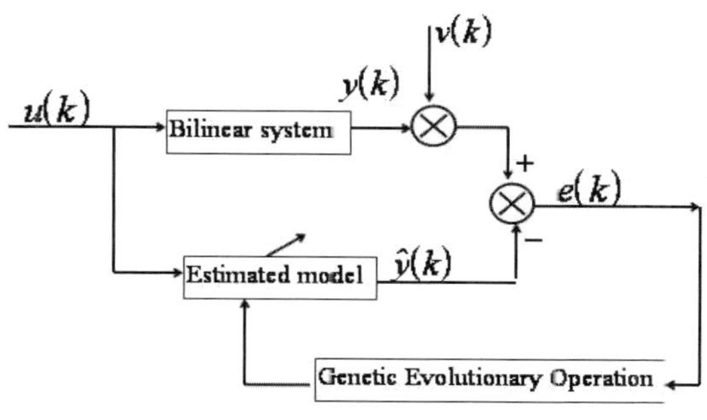

Fig. 1. A diagram of bilinear system identification by using GA

2.3 Parameter Identification of a Bilinear System by Using AGA

Fig. 1 shows that GA is applied to identify the parameters of a bilinear system.
The procedure of the identification algorithm is as follows:

1) Set the encoding method and decide the binary encoding length of each parameter length and the length of the formed individual.
2) Set the working parameter of GA. Randomly produce N binary strings and initialize population.
3) Decode binary strings to parameters of the bilinear system.
4) Calculate the fitness of each group of parameters and evaluate each individual.

A parameter identification problem actually is a minimization problem and the objective function J needs to be minimized. Hence, the objective function needs to be converted to a fitness function when GA is applied. Define:

$$f(k)=\begin{cases} C_{\max}-J(k) & J(k)<C_{\max} \\ 0 & Other \end{cases},$$

Where $J(k)=\sum_{j=0}^{s} e^2(k-j)$, the sum of squared errors of system outputs.

5) Perform reproduction, crossover and mutation to generate the next population.
6) Repeat the 3), 4), 5) steps until the convergence is reached.
7) Parameter value represented by the individual which has the best fitness in the population is the result of identified system parameter.

3 Simulations

A bilinear system is given as equation (2) and simulations were carried out respectively when the bilinear system is added without noise or with Gaussian noise. [4]

$$\begin{aligned} y(k+1)&=0.898y(k-1)-0.106y(k)u(k) \\ &+0.284u(k) \end{aligned} \tag{2}$$

Case 1: The System Was Not Added With Noise.

The input signal is M series and its amplitude is 1. Its period is 127. Working parameters of GA are population $N=80$; adaptive crossover possibility $k_1=k_2=0.95$ and mutation possibility, $k_3=k_4=0.05$; Crossover at two different positions; the length of chromosome is 20; S =40; Parameters of identified system are varied in the area [-0.2,1.0]; The system was sampled every three genetic operations. The convergence was reached after 232 genetic operations. The simulation result is shown in Table. I and the sum of squared parameter errors is 0.000005.

Table 1. Identification of the Bilinear System by Using GA (without noise)

Algorithm	a	b	c
AGA	0.897268	-0.107984	0.284729
Real value	0.898	-0.106	0.284

Case 2: The System Was Added With Zero-Mean Gaussian Noise.

The inputs of RLS, RIV and COR are evenly distributed random value in $|u(t)| \le 0.5$. Under the same signal-noise-ratio, the inputs of AGA are evenly distributed random value in $|u(t)| \le 4.0$. The input signal is M series and its amplitude is 1. Its period is 127. Working parameters of AGA are population $N = 100$; adaptive crossover possibility $k_1 = k_2 = 0.76$ and mutation possibility, $k_3 = k_4 = 0.05$; Crossover at two different positions; the length of chromosome is 15; S =30; Parameters of identified system are varied in the area[-0.2, 1.0]; The system was sampled every three genetic operations. The convergence was reached after 1851 genetic operations when the signal-noise-ratio is 50%. The simulation result is shown as follows.

a =0.897 b =-0.096 c =0.284

The convergence was reached after 3310 genetic operations when the signal-noise-ratio is 100%. The simulation result is shown as follows.

a =0.903 b =-0.103 c =0.219

Table 2. Comparison of Different Identification Methods

signal-noise-ratio	Method	a	b	c	The sum of squared parameter errors
50%	RLS	0.712	-0.089	0.301	0.035174
	RIV	0.894	-0.106	0.298	0.000212
	COR	0.899	-0.265	0.241	0.027131
	AGA	0.897	-0.096	0.264	0.000501
100%	RLS	0.438	-0.043	0.325	0.235482
	RIV	0.873	-0.141	0.290	0.001886
	COR	0.891	-0.352	0.210	0.066041
	AGA	0.903	-0.103	0.219	0.004259
Real value		0.898	-0.106	0.284	

It is shown in the Table II that RLS has the biggest error and COR has the second largest error. RIV has the smallest error. Although AGA's simulation result is not as accurate as RIV, its result is still reasonably well and much better than RLS and COR. Furthermore, it is much easier to use than RIV and it will be a good choice if the requirement to the identification accuracy is not very high.

Case 3: a MIMO Bilinear System

A MIMO bilinear system was given as follows:

$$\begin{bmatrix} x_1(k+1) \\ x_2(k+1) \end{bmatrix} = \begin{bmatrix} -0.2 & 0 \\ 0 & 0.35 \end{bmatrix} \begin{bmatrix} x_1(k) \\ x_2(k) \end{bmatrix} + \begin{bmatrix} 0.61 & 0.1 \\ 0.5 & 0 \end{bmatrix} \begin{bmatrix} x_1(k) \\ x_2(k) \end{bmatrix} u_1(k) + \begin{bmatrix} -0.31 & 0 \\ 0.53 & -0.4 \end{bmatrix} \begin{bmatrix} x_1(k) \\ x_2(k) \end{bmatrix} u_2(k) + \begin{bmatrix} 0.02 & 0.5 \\ 0 & -0.2 \end{bmatrix} \begin{bmatrix} u_1(k) \\ u_2(k) \end{bmatrix}$$

The input signal is M series and its amplitude is 1. Its period is 2^9-1. Working parameters of GA are population $N=70$; adaptive crossover possibility $k_1=k_2=0.95$ and mutation possibility, $k_3=k_4=0.05$; Crossover at multiple different positions and inverse operation is also applied; the length of chromosome is 14; S =35; The system was sampled every three genetic operations. The convergence was reached after 2782 genetic operations. The simulation result is shown in Table III and the sum of squared parameter errors is 0.003101.

Table 3. Identification of Mimo Bilinear system by using AGA

	Results of AGA	Real value
1	-0.2041	-0.2
2	0.2307	0.25
3	0.5925	0.61
4	0.0832	0.1
5	-0.3075	-0.31
6	-0.0096	0
7	0.0469	0.02
8	0.5155	0.5
9	0.0876	0.1
10	-0.1650	-0.18
11	0.3263	0.31
12	0.3584	0.35
13	0.4887	0.5
14	0.0104	0
15	0.5202	0.53
16	-0.3978	-0.4

Case studies above show that reasonably good results could be gained when GA was applied to identify bilinear system parameters.

4 Combined Genetic Algorithm

In this section, a combined algorithm of GA and nonlinear Least Square was applied to identify the order and parameters of a bilinear system simultaneously. The nonlinearity of bilinear system makes the searching space of GA show strong nonlinearity and hence increases the calculation amount and the convergence time. So, the combined algorithm takes advantage of GA's global search capability and the reliable convergence of nonlinear Least Square. The performance of combined algorithm is much better than GA and nonlinear Least Square.

4.1 The Procedure of Combined Algorithm

1) Set the encoding method and decide the binary encoding length of each parameter length and the length of the formed individual.
2) Set the working parameter of GA. Randomly produce N binary strings and initialize population.
3) Decode binary strings to parameters of the bilinear system.
4) Apply nonlinear Least Square method to estimate parameters of each model in the population and simplex search method are applied as an optimization method.
5) Calculate system errors and the order and parameter fitness of each model. Evaluate each model by using fitness value.
6) Perform reproduction, crossover and mutation to generate the next population.
7) Repeat the 3), 4) ,5) and 6) steps until the convergence is reached.
8) Parameter value represented by the individual who has the best fitness in the population is the result of identified system parameters and order.

4.2 Simulation

A bilinear system is given as follows

$$y(k) = 0.106y(k-2)u(k) + 0.898y^2(k-1)$$
$$+0.284u(k) + v(k)$$

In the simulation,

1) Population N =100;

2) Adaptive crossover possibility $k_1 = k_2 = 0.95$ and adaptive mutation possibility $k_3 = k_4 = 0.05$; Crossover needs to be done at two different positions since parameters to be identified are n_0, n_1, n_2.

3) The length of chromosome is 14 so the length of each individual L=14×3=42; S =40; Parameters of identified system are varied in the area[0,3];

4) Elitist strategy $R = E$; The best current individual will be brought in the next generation.

5) Initialize the vertices of the simplex, expansion coefficient $\mu = 1.4$, contraction coefficient $\lambda = 0.75$, allowable error $\varepsilon = 0.01$.

The input signal u is M series and its amplitude is 1. Its period is 127. Output noise is zero mean white noise with a variance of 0.01. The system was sampled every three genetic operations. The convergence was reached after 886 genetic operations. The simulation result is shown in Table IV and V. The sum of squared parameter errors is 0.005080.

Table 4. Order Parameters

	n_0	n_1	n_2
Combined algorithm	1	2	0
Real value	1	2	0

Table 5. Coefficient Parameters

	a_0	a_1	a_2
Combined algorithm	0.166072	0.321654	0.905334
Real value	0.106	0.284	0.898

The above simulation shows the feasibility that the combined algorithm of GA and nonlinear Least Square is applied to identify the order and parameters of bilinear system.

5 Conclusions

The paper describes a method for parameter identification of bilinear system by using an improved Genetic Algorithm. Simulations to a bilinear system have been carried out to investigate the effectiveness of proposed method. It is shown that good results could still be obtained when the bilinear system was influenced by Gaussian noise in the simulation. By comparing with RLS and COR through a simulation experiment to a SISO bilinear system, it is found that the method can get more accurate result than the two methods. Through a further simulation experiment to a MIMO bilinear system, the proposed method can get reasonably accurate results too. These simulations show that the proposed method is simpler and can get more accurate results than RLS and COR. Through a simulation study to an MIMO bilinear system, good results can still be acquired. Finally, the paper describes that a hybrid GA—the combination of Genetic Algorithm and nonlinear Least Square was developed to identify bilinear system structure and parameters simultaneously. The simulation result shows that the hybrid GA is feasible and effective to identify bilinear system structure and parameters.

Acknowledgments. The authors would like to thank to National Science Foundation of China for their funding support to this research work.

References

1. Hua, X.: Modeling and Control of Bilinear System. Northeast Chemical Engineering Institute Press (1990)
2. Goldkberg, D.: Genetic Algorithms in Search, Optimization and Machine Learning. Addison-Wesley, Reading, MA (1989)
3. Kristinsson, K.: System Identification and Control Using Genetic Algorithms. IEEE Trans. Syst., Man, Cyber. 12(5), 1033–1046 (1992)
4. Zhang, G., Liu, G.: Selecting crossover site with unequal probability in genetic algorithms. Information and Control 26, 53–60 (1997)
5. Srinivas, M., Patnaik, L.: Adaptive probabilities of crossover and mutation in genetic algorithm. IEEE Trans. Syst., Man and Cyber. SMC-24(4), 656–666 (1994)
6. Fang, C., Xiao, D.: Process Identification. TsinghuaPress, Beijing (1988)
7. Wang, Z.: Nonlinear System Identification Based on Genetic Algorithm. Dalian University of technology, M.Sc. thesis (1996)
8. Wang, Z., Gu, H.: Grey Identification of Cobb-Douglas Production Function Model Parameters. System Engineering and Electronic Technology 7, 19–21 (1999)
9. Castillo, O., Melin, P., Montiel, O., Sepúlveda, R.: Application of A Breeder Genetic Algorithm For System Identification In An Adaptive Finite Impulse Response Filter. Evolvable Hardware, 146–156 (2001)
10. Konak, A., Smith, A.: A hybrid genetic algorithm approach for backbone design of communication networks. In: Proceedings of the 1999 Congress on Evolutionary Computation, Washington DC, pp. 112–117. IEEE, Los Alamitos (1999)
11. Castillo, O., Melin, P., Montiel, O., Sepúlveda, R.: Evolutionary Non-Linear System Identification. IC-AI, 98–104 (2004)
12. Akramizadeh, A., Akbar Farjami, A., Khaloozadeh, H.: Nonlinear Hammerstein Model Identification Using Genetic Algorithm. In: ICAIS'02, pp. 755–764 (2002)
13. Castillo, O., Melin, P., Montiel, O., Sepúlveda, R.: The evolutionary learning rule for system identification. Appl. Soft Comput. 3, 343–352 (2003).

A Random Velocity Boundary Condition for Robust Particle Swarm Optimization

Jian Li, Bo Ren*, and Cheng Wang

HuBei Key Laboratory of Digital Valley Science and Technology,
Huazhong University of Science and Technology
430074 Wuhan, China
fibin@263.net

Abstract. The particle swarm optimization (PSO) is a stochastic evolutionary computation technique based on the behavior of swarms that can be used to optimize objects with complex search spaces. However, it has been observed that its performance varies duo to the dimensionality of the object and the location of the global optimum in the search space. This paper introduces a "random" velocity boundary condition to address the problem, where the velocity boundary alters randomly to prevent the velocity of a particle from stopping on a same boundary during the evolution. Simulation results on two benchmark functions with 30 and 300 dimensionalities and three types of locations of the global optimum solutions in the search spaces have shown that with the proposed "random" velocity boundary condition, a highly competitive optimization performance can be obtained for PSO regardless of the dimensionality and the location of the global optimum solution.

1 Introduction

The particle swarm optimization (PSO) is a stochastic search algorithm first proposed by Kennedy and Eberhart [1]. Its basic idea was rooted in the study and action simulation of the simplified social model of bird population. The performance of the PSO varies depending on both the dimensionality of the object [2] and the distance between the optimum solution and the boundary of the search space, which could prevent the PSO from providing effective performance for high dimensional objects in which the location of the global optimum with respect to the search space boundary is unknown [3].

In most optimizations, the position of each particle is restricted in a range. Without any boundaries or limits on the velocity, a particle could essentially fly out of the physically meaningful solution space. One approach to address this problem is to simply enforce a maximum allowed velocity called $v_{\max}$. And $v_{\max}$ was set to a certain percentage of the dynamic range of each dimension in [4], [5]. The velocity boundary conditions in [4] and [5] are fixed conditions, where $v_{\max}$ is a constant during the evolution. Experience has shown that $v_{\max}$ do not always confine the particles within

* Corresponding author.

K. Li et al. (Eds.): LSMS 2007, LNCS 4688, pp. 92–99, 2007.

the solution space. To address this problem Robinson and Rahmat-Samii imposed three different boundary conditions [6] namely absorbing walls, reflecting walls and invisible walls. Huang and Sanagavarapu Mohan proposed a hybrid damping boundary condition that combines the characteristics offered by the absorbing walls and reflecting walls boundaries [3]. To solve constrained optimization problems, Takahama and Saka [7] proposed to divide a swarm into some groups with various v_{max} and control v_{max} adaptively by comparing the movement of particles in each group. There are alterable velocity boundary conditions in [3], [6], [7], except the invisible walls, in which there is no velocity boundary. In the absorbing, reflecting and damping boundary conditions there is no explicit velocity boundary, but the position will be changed if a particle tries to fly out of the feasible space, which can be regarded as the velocity is clamped in a range depending on the position boundary and the former position of the particle.

In this paper, to combine the characteristics of the fixed and the alterable velocity boundary conditions, we propose a random velocity boundary condition, where the velocity boundary consists of a fixed component and a random component, and the random component alters during the evolution. The effectiveness of the proposed velocity boundary condition is shown by comparing with various boundary conditions on some well-known problems, and the results have shown its robust and consistent performance for high dimensional optimization objects regardless of where the global optimum is located in the search space. In Section 2, the PSO technique with circle topology is introduced. Section 3 introduces some previous work on boundary conditions, after which the random velocity boundary condition is described. Section 4 presents the simulation results to validate the effectiveness offered by the proposed boundary condition. The conclusions of this paper are given in Section 5.

2 Particle Swarm Optimization with Circle Topology

The particle swarm optimization consists of a swarm of particles flying through the search space. Each particle i in the swarm contains parameters for position (X_i^t) and velocity (V_i^t) where $X_i^t \in R^n$, $V_i^t \in R^n$ ($t \in N$, $i \in N$), and n is the dimensionality of the search space, t is the number of current iteration. The position of each particle represents a potential solution to the optimization problem. At each iteration step, each particle adjusts its velocity based on its momentum and the influence of its best position (P_i^t) tracked and the best position found by the neighbors (P_G^t) so far, after which it computes a new position to examine. The PSO formulas are given as follows.

$$V_i^t = \omega \cdot V_i^{t-1} + c_1 \cdot rand_1() \cdot (P_i^{t-1} - X_i^{t-1}) + c_2 \cdot rand_2() \cdot (P_G^{t-1} - X_i^{t-1}) \tag{1}$$

$$X_i^t = X_i^{t-1} + V_i^t \tag{2}$$

Where c_1 and c_2 are constants knows as the cognitive and social acceleration coefficients and set to be 2.0 in this paper, respectively, ω is the inertia weight and set to be 0.9 in this paper, $rand_1()$ and $rand_2()$ are random numbers uniformly distributed between 0 and 1, and the position is clamped as $X_i^t \in [x_{\min}, x_{\max}]$.

The first part of (1) represents the previous velocity, which provides the necessary momentum for particles to fly across the search space. The second part is known as the "cognitive" component, which represents the personal thinking of each particle. This component encourages the particles to fly toward their own best positions found so far ($pbest$). The third part is known as the "social" component, which represents the collaborative effect of the particles in finding the global optimum. This component always pulls the particles toward the best positions found by their neighbors so far [8].

There are four neighborhood topology types were proposed in [9] viz., circles ($lbest$), wheels, starts ($gbest$) and random edges, and our approach adopts the circle topology. In the circle topology, each particle communicates with its k immediate neighbors. In this paper $k = 2$, a particle communicates with its immediately adjacent neighbors. The neighborhood is determined through an index label assigned to all particles. In this version of the PSO, particles can traverse larger area of the search space [10], [11].

3 Random Velocity Boundary Condition

In most optimizations, the position of each particle is restricted in a range $X_i^t \in [x_{\min}, x_{\max}]$. To prevent the particle from flying out of the physically meaningful solution space, its velocity is clamped in a range as $V_i^t \in [-v_{\max}, v_{\max}]$. Eberhart and Kennedy [5] noted that the PSO performed better if $v_{\max}$ in each dimension is set equal to the dynamic range of that dimension. Robinson and Rahmat-Samii imposed three different boundary conditions [6] namely absorbing walls, reflecting walls and invisible walls. Huang and Sanagavarapu Mohan proposed a hybrid damping boundary condition that combines the characteristics offered by the absorbing walls and reflecting walls boundaries [3]. The feature of each boundary condition is illustrated in Fig.1 (a)-(d).

It was observed that with the three former the searching performance of the PSO will vary significantly depending on the problem dimensionality and the location of the global optimum with respect to the search space boundary. And the damping boundary condition is reported to provide robust performance regardless the dimensionality and the location.

This paper proposes a random velocity boundary condition based on the fixed velocity boundary condition [5], and the results of the PSO with the condition will be compared to those with the fixed velocity boundary condition, the absorbing walls boundary condition and the damping boundary condition in Section 4.

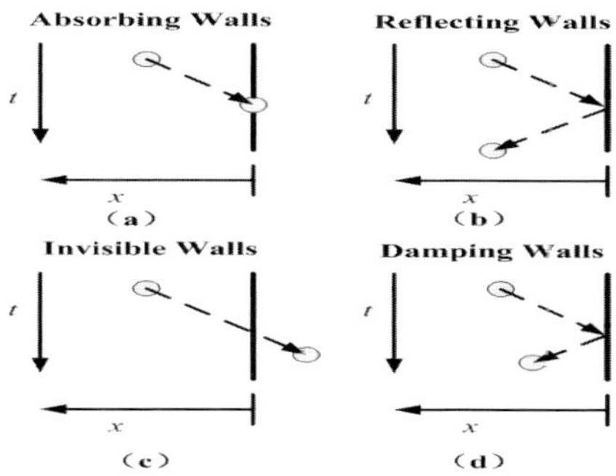

Fig. 1. (a)For the absorbing boundary the velocity of the particle is zeroed and the particle is stopped at the boundary. (b)For the reflecting boundary the sign of the velocity is reversed and the particle is reflected back to the search space after the impact. (c)For the invisible boundary the particle is allowed to escape the boundary of the search space and is ignored by the fitness evaluator. (d)For the damping boundary part of the velocity is absorbed by the boundary during the impact and the particle is then reflected back with a lesser velocity of a reversal of sign.

With the fixed velocity boundary condition, the velocity is restricted in a fixed range during the evolution. If any dimension of the velocity calculated by (1) tries to exceed any boundary of the dimension, it will be set to value of the boundary of the dimension. This boundary condition will reduce the diversity of the swarm potentially, because some dimensions of the velocity will be set to the value of the upper or the lower boundary at many iteration steps during the evolution. In order to address the problem, this paper introduces a random velocity boundary condition, where the upper and the lower velocity boundaries keep on altering during the whole evolution. The boundaries are calculated by the equations as follows.

$$v_{max}^{t} = (a \times R_1^t + b) \times (x_{max} - x_{min}) \tag{3}$$

$$v_{min}^{t} = (a \times R_2^t + b) \times (x_{min} - x_{max}) \tag{4}$$

Where R_1^t and R_2^t are random numbers uniformly distributed between 0 and 1, a and b are constants to adjust the random and the fixed components of the boundaries, and are set to be 2 and -0.5 in this paper, respectively, which were empirically derived after a set of experiments. And then the boundaries will alter at each iteration step and the particles can fly faster at some steps to enhance global searching ability and fly slower to enhance local searching ability at the others. With the random velocity boundary condition, a particle still has the chances to escape the boundary of the search space, if it occurs the particle is stopped at the boundary of the search space in this paper

4 Simulation Results

To compare the PSO performance with different boundary conditions, two benchmark functions, viz., Rastigrin and Rosenbrock functions were employed [2]. The two functions are defined as follows.

Rastigrin function

$$f_1(x) = \sum_{n=1}^{N} [x_n^2 - 10\cos(2\pi x_n) + 10] \tag{5}$$

Rosenbrock function

$$f_2(x) = \sum_{n=1}^{N-1} [100(x_{n+1} - x_n^2)^2 + (x_n - 1)^2] \tag{6}$$

Where N is the dimension of each function. The global minimum of both functions is equal to zero, where $f_1(0) = 0$ and $f_2(1) = 0$.

In this paper, in order to compare the results of the boundary conditions with different dimension and different location of the global optimum solution, each function is optimized of 30 and 300 dimensions with three types of search spaces, whose size in terms of the Euclidean distance between the minimum and maximum boundaries are the same, but the search range in each dimension is different. For the Rastigrin function, the three search ranges are [0, 10], [-5, 5] and [-2, 8], and for the Rosenbrock function, the three search ranges are [1, 11], [-4, 6] and [-2, 8], and then the global optimum of each function is located at the lower boundary, at the middle and at a generic position, respectively.

In this paper, the population of the swarms is set to be 30, and a set of 50 different initial populations for each type of the search space of 30 and 300 dimensions were created. For each of the boundary condition mentioned earlier, 50 simulations were performed using the same initial population set created earlier. Each simulation consists of 2000 iterations, and the results are averaged at the end of 50 simulation runs. The average fitness value versus the number of iterations of Rastigrin and Rosenbrock functions are plotted in Fig.2 and Fig.3, respectively.

For the functions, referring to Fig.2 and Fig.3, the PSO performance obtained by each boundary condition is affected by both the increase in the dimension of the functions and the variation in the location of the global optimum with respect to the search space boundary. And the solutions obtained by the proposed boundary conditions are much better than those by the other boundary conditions.

It can be seen that when the global optimum is at the boundary of the search space, the PSO with the proposed boundary converged to the global optimum very soon, while the average fitness values of the PSO with the fixed and the absorbing boundary conditions decreased sharply in the former phases and represented slight improvement in the latter phases. And the damping boundary condition represented the second-best performance except for the 300 dimension case of the Rosenbrock function.

For the cases that the global optimum is at the middle and at a generic position in the search space, the proposed boundary condition has represented highly competitive performance, with which the PSO has found much better solutions than with the others in all the cases. And it also can be seen that when the solution is at a generic position in the search space, the results of the fixed, the absorbing and the damping boundary conditions are similar, which improved slightly in the latter phases of the evolutions.

With the increase of the dimensionalities, the result obtained by each boundary condition deteriorated, while the curves obtained by the proposed random velocity boundary condition of 300 dimensionality cases represented the same characteristic as those of 30 dimensionality cases, which shows the consistence of the proposed boundary.

The results have shown that with the proposed random velocity boundary condition the PSO offers a robust and consistent performance for optimizing a high dimensional bench mark function with any location of the global optimum with respect to the search space.

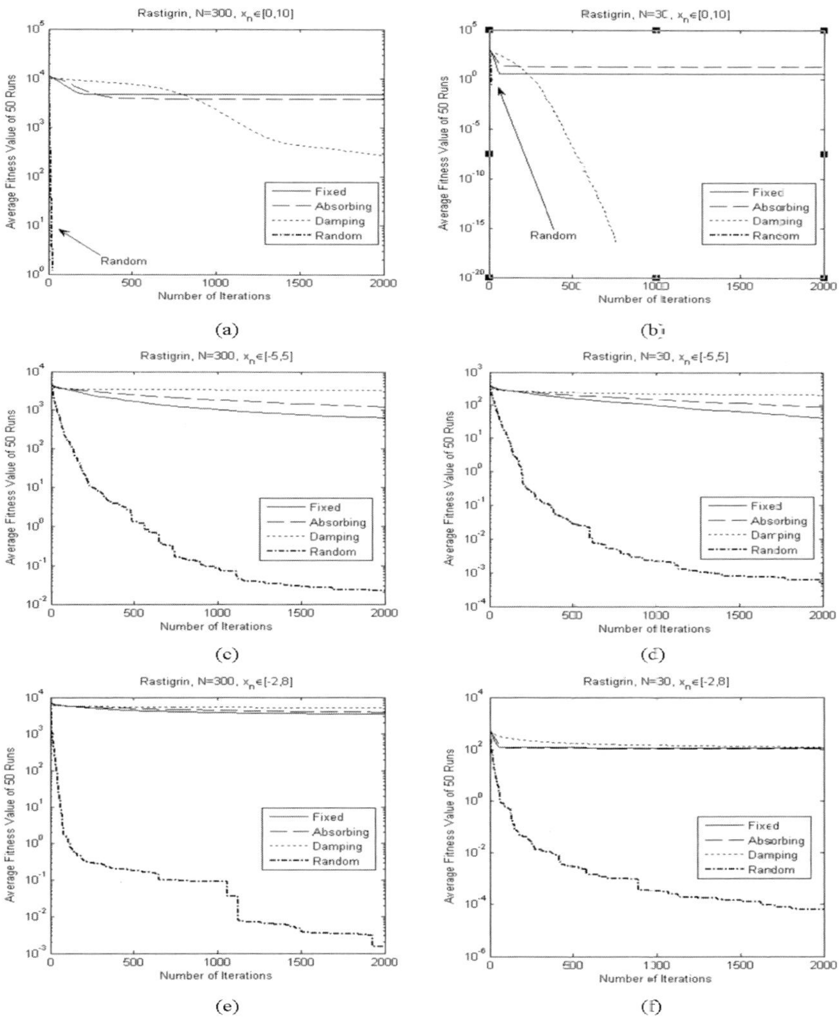

Fig. 2. The average fitness value obtained versus the number of iterations for the 50 independent PSO runs of Rastigrin function of 300 and 30 dimensionalities

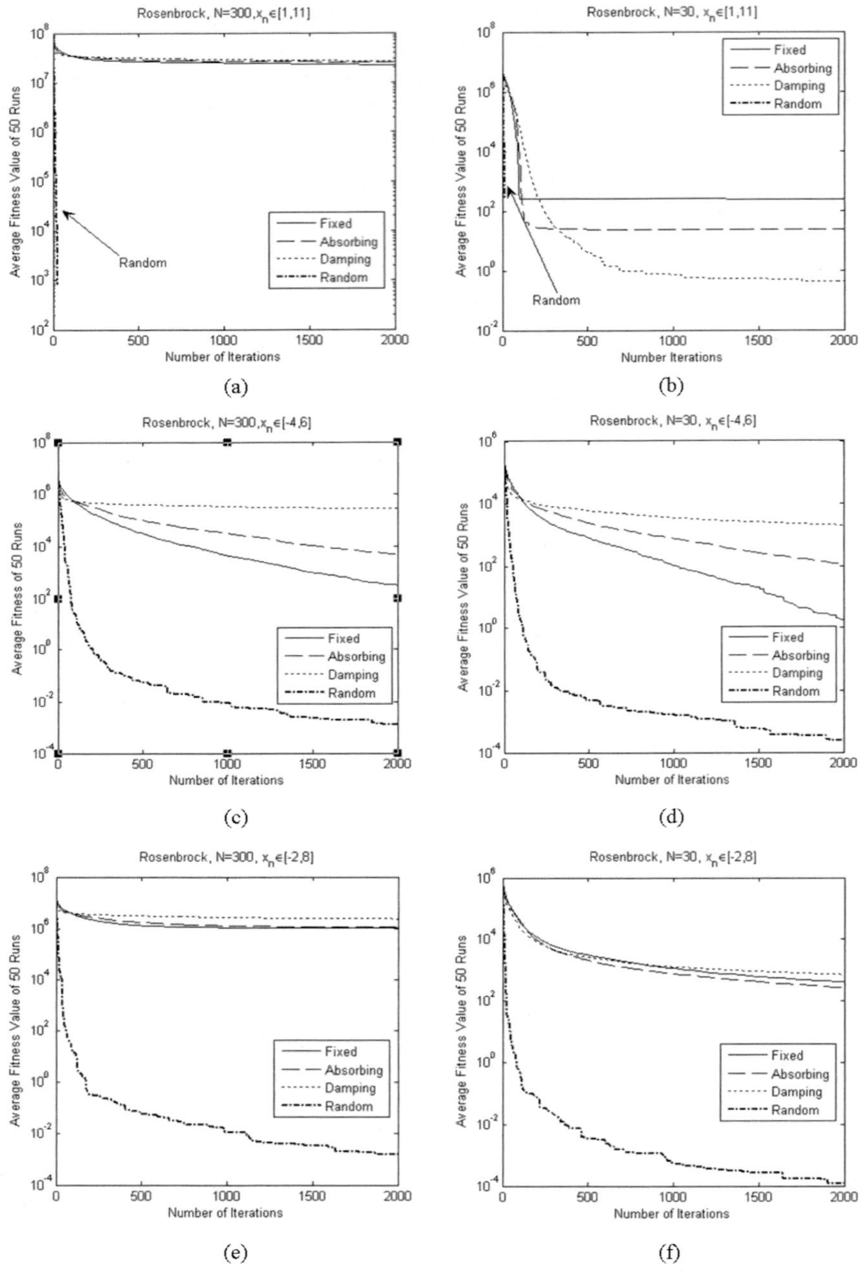

Fig. 3. The average fitness value obtained versus the number of iterations for the 50 independent PSO runs of Rosenbrock function of 300 and 30 dimensionalities

5 Conclusion

In this paper, we have introduced the random velocity boundary condition for PSO with circle topology structure, where the velocity boundary condition alters randomly during the evolution. This boundary condition is simple and easy to implement, besides it can maintain the diversity of the particles by providing different velocity boundaries for particles in different iterations. As compared to the existing boundary conditions, the simulation results have shown that the proposed boundary condition can provide much better solutions with different dimensionalities and locations of the global optimum solutions in the search spaces. Thus, the proposed random velocity boundary condition can be implemented to PSO to optimize various dimensionality functions without known the location of the global optimum in the search spaces.

References

1. Kennedy, J., Eberhart, R.: Particle Swarm optimization. In: Proceedings of IEEE International Conference on Neural Network, pp. 1944–1948. IEEE Computer Society Press, Los Alamitos (1995)
2. van den Bergh, F., Engelbrecht, A.P.: A Cooperative approach to particle swarm optimization. IEEE Transactions on Evolutionary Computation 8, 225–239 (2004)
3. Huang, T., Mohan, A.S.: A Hybrid Boundary Condition for Robust Particle Swarm Optimization. IEEE Antennas and Wireless Propagation Letters 4, 112–117 (2005)
4. Carlisle, A., Doizier, G.: An off-the-shelf PSO. In: Proceedings of the Workshop on Particle swarm optimization, Indiannapolis, IN, pp. 1–6 (2001)
5. Eberhart, R., Shi, Y.: Particle Swarm Optimization: Developments, Applications and Resources. In: Proceedings of the 2001 Congress on Evolutionary Computation, vol. 1, pp. 81–86 (2001)
6. Robinson, J., Rahmat-Samii, Y.: Particle swarm optimization in electromagnetics. IEEE Transactions on Antennas and Propagation 52, 397–407 (2004)
7. Takahama, T., Sakai, S.: Solving constrained optimization problems by the epsilon constrained particle swarm optimizer with adaptive velocity limit control. In: The 2nd IEEE International Conference of Cybernetics and Intelligent System, pp. 683–689. IEEE Computer Society Press, Los Alamitos (2006)
8. Ratnaweera, A., Halgamuge, S.: Self-Organizing Hierarchical Particle Swarm Optimizer with Time-Varying Acceleration Coefficients. IEEE Transaction on Evolutionary Computation 8, 240–255 (2004)
9. Kennedy, J.: Small Worlds and Mega-minds: Effects of Neighborhood Topology on Particle Swarm Performance. IEEE Congress on Evolutionary Computation, 1931–1938 (1999)
10. Shi, Y., Eberhart, R.C.: Parameter Selection in Particle Swarm Optimization. In: Porto, V.W., Waagen, D. (eds.) Evolutionary Programming VII. LNCS, vol. 1447, pp. 591–600. Springer, Heidelberg (1998)
11. Clerc, M., Kennedy, J.: The Particle Swarm-explosion, Stability, and Convergence in a Mutildimensional Complex Space. IEEE Transaction Computation 6, 58–73 (2002)

Kernel-Based Online NEAT for Keepaway Soccer

Yun Zhao, Hua Cai, Qingwei Chen, and Weili Hu

Department of Automation, Nanjing University of Science and Technology, Nanjing, Jiangsu, China, 210094
Zhaoyun83@163.com
hwatsai@hotmail.com
cqw1002@sina.com
hwl1002@mail.njust.edu

Abstract. This paper presents a kernel-based online neuroevolutionary of augmenting topology (KO-NEAT) algorithm, which borrowing the selection mechanisms used in temporal difference (TD) algorithms and combining the kernel function approximator for individual fitness initiation. KO-NEAT can improve evolution's online performance of NEAT and learns more quickly. Empirical results in keepaway soccer problem demonstrate that KO-NEAT can substantially improve the original algorithm.

1 Introduction

Keepaway soccer[1] is a subtask of Robocup soccer, in which one team, the keepers, tries to learn the best behavior policy to maintain possession of the ball as long as possible. Each robot has only a partial world view at any given moment and also has noisy sensors and actuators. In addition, the environment is uncertainty with large state space and action space. Since then it has been put forth as an important benchmark problem for multiagent reinforcement learning (RL), and bring huge challenge to research.

Neuroevolution of augmenting topologies (NEAT) [2] is an efficient RL policy search algorithm, which evolves network topologies and weights simultaneously using Genetic Algorithms (GA). It has had substantial success in RL domains like Pole-Balancing Problem [2], Game Playing [3] and Robot Control [4]. Taylor [5] learned robot keepway with NEAT and compared with currently best methods-linear Sarsa with radial basis function approximation [1]. Results demonstrated that NEAT could learn better policies in this task, though it required more evaluations to do so. The first reason for learning more slowly of NEAT is that re-evaluations are needed to cope with noisy fitness evaluations. Another reason is that evolution process does not exploit the information gained earlier in the generation sufficiently, which also restricts NEAT to on-line scenarios [6].

For pre-mentioned shortcomings of NEAT for keepaway soccer task, this paper presents a kernel-based online NEAT (KO-NEAT), which borrowing the selection mechanism of temporal difference (TD) method[7], another efficient RL method, and combining kernel function approximation for individual fitness initiation. Empirical results in 3 vs. 2 keepaway soccer problem illustrate that proposed approach can substantially accelerate and improve the on-line performance of the standard NEAT.

K. Li et al. (Eds.): LSMS 2007, LNCS 4688, pp. 100–107, 2007.

2 Keepaway Soccer

Keepaway soccer is a sub-problem of Robocup simulated soccer [10], in which one team, the keepers, tries to maintain possession of the ball within a limited region, while the opposing team, the takers, attempts to gain possession. Whenever the takers take possession or the ball leaves the region, an episode ends and the players are reset for next episode [1]. Figure 1 shows a starting configuration for a 3 vs.2 keepaway soccer episode in a 20m*20m region, with the keepers being given possession. Figure 2 shows a middle episode, K1 is the keeper with the ball, K2 and K3 are respectively the next closet and third closest keeper to the ball. T1 and T2 are respectively the closet and next closest taker to the ball.

For comparison, this paper adopts the same setting as document [1], in which, robots choose from higher level macro-actions based closely on skills used in the CMUnited-99 team [8]. Figure 3 shows keepers' and takers' policy space.

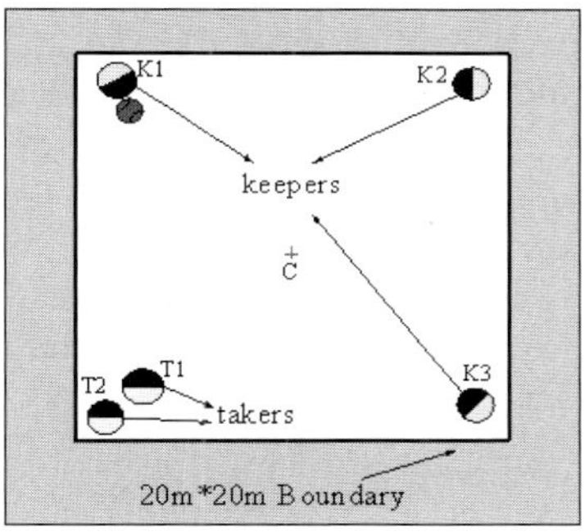

Fig. 1. Starting configuration

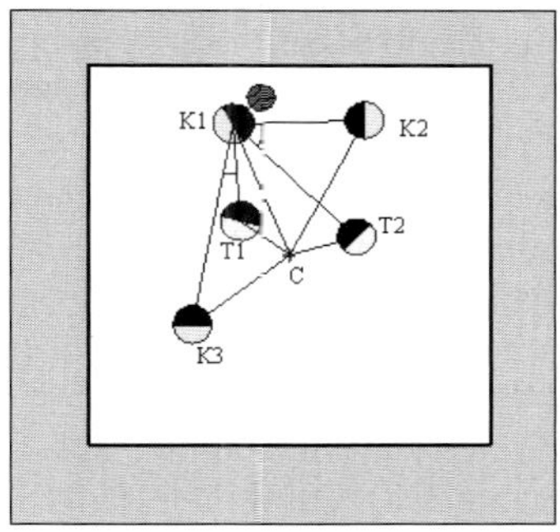

Fig. 2. Middle episode

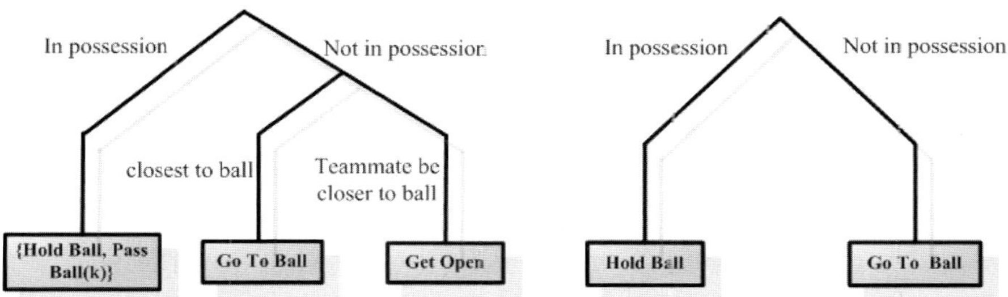

a. Keepers' policy space **b.** Takers policy space

Fig. 3. Keepers' and takers' policy space

Where: Hold ball: Remain stationary;
Pass ball (k): Kick the ball directly towards keeper k;
Get open: Move to a safe position and open for a pass from the ball's current position;
Go to ball: move directly towards the ball.

Keepers and takers choose macro-actions following respective policy space. The keepers should learn to choose from {Hold ball; Pass ball (2); Pass ball (3)} when in possession of the ball, based on 13 state variables, while the takers behave following a certain policy.

3 Kernel-Based Online NEAT for Keepaway Soccer

NEAT is an effective reinforcement learning approach. It searches for the best policy from the policy space of RL problem to maximize accumulated reward agent accrued along the learning process.

Because the noisy environment of keepaway soccer task induces noisy fitness evaluation. Standard NEAT re-evaluates and averages the fitness so that NEAT needs longer time to learn. Re-evaluating every individual for exactly the same number of episodes is also unadvisable. In addition, to excel in on-line scenarios, a learning algorithm must effectively balance two competing objectives: searching for better policy (exploration) and accruing maximal reward (exploitation)[9]. But standard NEAT is purely exploratory within each generation, and limited to offline learning problems [6].

In the light of defects of standard NEAT for keepaway soccer, this paper presents kernel-based online NEAT (KO-NEAT) for keepaway soccer. Proposed method boosts evolution's on-line performance by borrowing an action selection mechanism from TD-Interval Estimation[6], exploiting the fitness information gained earlier in generations to systematically give more evaluations to the most promising individuals and avoid re-evaluating the weak ones. Doing so allows NEAT to increase the reward accrued during learning. KO-NEAT also introduces kernel function approximator to compute initial fitness of neonatal individuals without episodic re-evaluations, thus accelerates learning speed.

3.1 Design of KO-NEAT for Keepaway Soccer

Primary contents of improved algorithm include: genetic encoding, population's initialization and speciation, individual fitness initialization and updating, crossover and mutation. The remainder of this section provides a detailed description of KO-MEAT for keepaway soccer.

Genetic encoding. Each neural network is encoded as an individual in the population, representing a candidate policy in the form of macro-action selector. The inputs to the network describe keeper's (in possession) current 13 state variables. There is one output for each of 3 macro-actions; the keeper takes whichever action has the highest activation.

For corresponding genes to be easily lined up when two genomes crossover during mating, network is encoded by linear genetic encoding scheme. Each genome includes a list of node genes and a list of connection genes. Each node gene specifies one node information. Each connection gene consists of the in-node, out-node, connection weight, an enable bit (whether or not the connection is expressed) and innovation number (historical marking about the connection's appearance for the first time).

Relation between any individuals in a topologically diverse population is established based on innovation number of each gene. The connection genes in both genomes with the same innovation numbers must represent the same structures (although possibly with different weights), called *matching* genes. Genes that do not match are either *disjoint* or *excess*, depending on whether they occur within or outside the range of the other genome's innovation number.

Population initialization. Networks in initial population have uniform simple structures without hidden nodes and only 13 inputs connected to 3 outputs.

Population speciation. In most case, individuals with larger structures evolve more slowly. Population speciation protects the individuals with large structures in the generations so as to prevent being eliminated before maturation. Speciation method divides the population into several species by the compatibility distance between two network encodings such that similar topologies are in the same species. Then individuals compete primarily within their own species rather than the population at large. The compatibility distance d is defined as follows:

$$d = \frac{a \times E}{N} + \frac{b \times D}{N} + c \times \overline{W} \tag{1}$$

Thereinto: three factors $E, D, \overline{W}$ are respectively the number of *excess* genes, the number of *disjoint* genes and the average weight differences of *matching* genes. The three coefficients, *a*, *b* and *c*, allows us to adjust the importance of the three factors. Factor *N*, the number of genes in the larger genomes, normalizes for genome size.

Individual is compared to each species orderly. Once the distance to a randomly chosen individual is less than compatibility threshold δ_t; it is placed into that species. Species number is restricted not more than 5, by adjusting δ_t dynamically.

Kernel-based individual fitness initialization. KO-NEAT adopts $(\mu + \lambda)$ evolutionary strategy idea to generate next generation's parent population in every species. In each generation, the population members for the next generation are chosen from the combined μ parents and λ children population according to their fitness, with the size of λ.

The selection of members for next generation depends on individuals' fitness. KO-NEAT introduces kernel function approximatior to compute children's fitness as their initial fitness. In this way, needed number of re-evaluations for accurate fitness is reduced and learning steps out accordingly.

Supposedly, parents population in the species is $\{p_1, p_2, \ldots p_\mu\}$, children population is $\{c_1, c_2, \ldots c_\lambda\}$, $d(p_i, c_j)$ specifies the distance between parent p_i and child c_j. Then initial fitness of child c_j is computed as follows:

$$f(c_j) = \frac{1}{z(c_j)} \sum_{i=1}^{\mu} \phi\left(\frac{d(p_i, c_j)}{\beta}\right) \cdot f(p_i) \quad (j = 1, 2, ...\lambda) \tag{2}$$

Where $z(c_j) = \sum_{i=1}^{n} \phi\left(\frac{d(p_i, c_j)}{\beta}\right)$ normalizes the weights. We use Gaussian kernel function to compute these weights, so $\phi(x) = e^{-x^2}$. The bandwidth parameter β thus corresponds to the standard deviation of each Gaussian.

Introducing interval estimation and fitness updating. Interval estimation is an efficient action selection mechanism in TD, which naturally balance exploration and exploitation. KO-NEAT borrows the mechanism to choose policies for evaluation in species. This approach is described as follows:

$$\text{Return } p = \arg\max_{p \in P_e} [f(p) + \frac{\sigma(p)}{\sqrt{e(p)}} \cdot u_{\frac{\alpha}{2}}]$$

Where, $f(p), \sigma(p), e(p)$ are the fitness, standard deviation and re-evaluation number respectively, for policy p. $u_{\frac{\alpha}{2}}$ is two-side 100α fractile under the standard deviation.

This strategy encourages exploitation by using fitness information gained from early generations. Instead of giving the same number of episodic re-evaluations to every policy, KO-NEAT favors policies with high fitness and also focuses exploration on the most promising but with inaccurate fitness policies. Thereby, introducing interval estimation technique improves evolution's on-line performance in keepaway.

A candidate policy p is selected to evaluate by allowing the correspond network to control all the keepers' macro-action selection when they are in possession during the episode. The policy's fitness is simply the episode duration time, also called accumulated reward accrued during the episode. After the policy has been re-evaluated $e(p)$ times, its fitness is updated as the average of the $e(p)$ episodic fitness. The max $e(p)$ is limited to not more than parameter *max-indi-e*.

Crossover. When crossing over, the node genes are simply lined up, connection genes are randomly chosen from either parent at matching genes, whereas all excess or disjoint genes are always inherited from the fitter parent. This way allows crossover using linear genomes without the need for expensive topological analysis.

Mutation. Figure 4 exhibits two types of structural mutation: node mutation and connection mutation, both of which enlarge the genomes gradually. In evolution, only the offspring with effective expanding can survive so that time-consuming to optimize offspring with bad expanding is avoided.

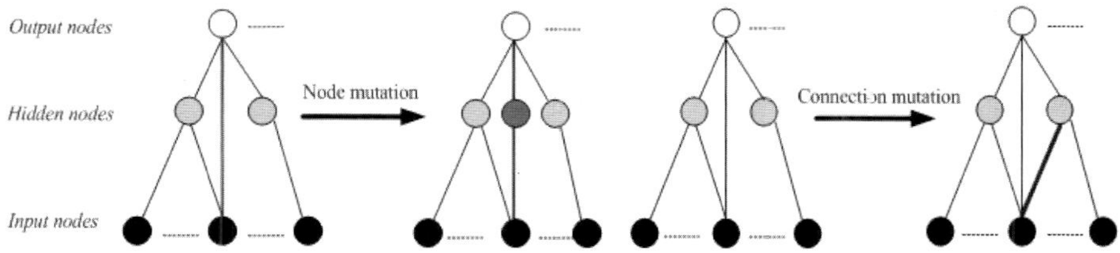

Fig. 4. Two types of structural mutation

3.2 Learning Flow of KO-NEAT

Figure 5 shows the learning flow of proposed KO-NEAT for keepaway soccer.

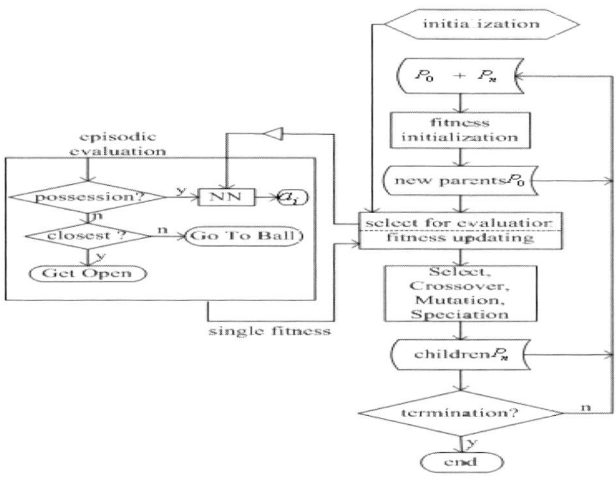

Fig. 5. KO-NEAT for keepaway soccer

4 Data Simulation

Proposed KO-NEAT and standard NEAT were tested on the 3 vs. 2 keepaway soccer platform and record the max fitness gained in each generation. The population size was 100. Coefficients for adjusting compatibility were $a = 1.0, b = 1.0, c = 2.0$. The compatibility threshold δ_t was adjusted dynamically to maintain a target of 5 species. Max re-evaluation number for each individual, *max-indi-e*, was 100. Since the keepaway task is stochastic and the evaluations are noisy, it is difficult to establish a priori the optimal number of episodes to evaluate each generation. To set the parameter, the number of keepaway episodes per generation was tried as one of {3000, 4500, 6000, 8000} and found that 4500 episodes per generation for KO-NEAT and 6000 episodes per generation for standard NEAT yielded the best performance respectively. Other parameters were the same as document [2]. Relevant two learning curves were compared in figure 6.

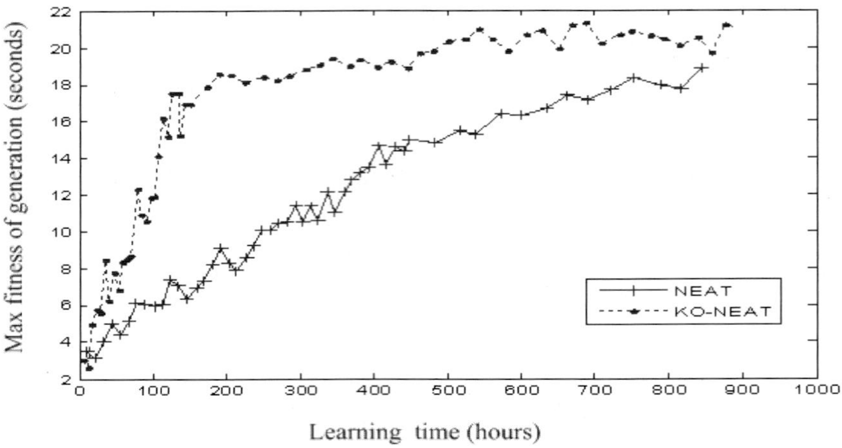

Fig. 6. Learning curves of KO-NEAT and NEAT

From the figure 6, we can see, firstly, proposed KO-NEAT plateaued much sooner than standard NEAT. At simulator time about 480 hours, the learning curve of KO-NEAT tended to plateau; mean episodic duration of best individual achieved 20 seconds, while learning curve of standard NEAT had not leveled off and the mean fitness of best individual to date was only 14 seconds. Secondly, KO-NEAT accrued more rewards than standard NEAT during the same number of episodic evaluations. When proposed KO-NEAT implemented about 189000 episodic evaluations, viz. 42 generations, the curve plateaued and accumulated rewards were 480 simulator hours. With roughly the same number of episodic estimations, viz. 31 generations, standard NEAT gained accumulated rewards of about 360 simulator hours. Finally, KO-NEAT can learn better policy than standard NEAT for keepaway soccer task. Until simulator time reached 900 hours, the max fitness to date of KO-NEAT and NEAT were 21 seconds and 19 seconds respectively and curve's holistic level of KO-NEAT was upper than that of standard NEAT all along. Overall, proposed KO-NEAT substantially improves standard NEAT's on-line performance, boosts learning efficiency and accelerates its learning speed.

5 Conclusion

Aiming at the shortcomings of standard NEAT for keepaway soccer task, like learning slowly and being limited to off-line scenarios, this paper presents an improved algorithm, named KO-NEAT. Proposed method borrows an action-selection mechanism of another efficient RL algorithm (TD) and combines kernel-based individual fitness initialization. Empirical results in 3 vs.2 keepaway soccer demonstrate that KO-NEAT can substantially improve the evolution's on-line performance and accelerate the learning speed. Further work will focus on transferring the KO-NEAT into larger task space.

References

1. Stone, P., Kuhlmann, G., Taylor, M.E., Liu, Y.: Keepaway Soccer: From Machine Learning Testbed to Benchmark. In: Bredenfeld, A., Jacoff, A., Noda, I., Takahashi, Y. (eds.) RoboCup 2005. LNCS (LNAI), vol. 4020, pp. 93–105. Springer, Heidelberg (2006)
2. Stanley, K.O., Miikkulainen, R.: Evolving Neural Networks through Augmenting Topologies. Evolutionary Computation 10(2), 99–127 (2002)
3. Stanley, K.O., Miikkulainen, R.: Evolving a Roving Eye for Go. In: Proceeding of the genetic and evolutionary computation conference, pp. 1226–1238 (2004)
4. Stanley, K.O.: Competitive Coevolution through Complexification. Journal of artificial intelligence research 21, 63–100 (2004)
5. Kaelbling, L.P., Littman, M.L., Moore, A.W.: Reinforcement learning: a survey. Journal of artificial intelligence research, 237–285 (1996)
6. Tayor, M.E., Whiteson, S., Stone, P.: Comparing Evolutionary and Temporal Difference Methods in a Reinforcement Learning Domain. In: Proceeding of the genetic and evolutionary computation conference (GECOO-2006), Seattle, Washington, USA, pp. 1321–1328 (2006)
7. Whiteson, S., Stone, P.: On-line Evolutionary Computation for Reinforcement Learning in Stochastic Domains. In: Proceeding of the genetic and evolutionary computation conference, pp. 1577–1584 (2006)
8. Sutton, R.S.: Learning to Predict by the Methods of Temporal Differences. Machine Learning, pp. 9–44. Kluwer Academic Publishers, Boston (1988)
9. Stone, P.: Layered Learning in Multiagent Systems: A Winning Approach to Robotic Soccer. MIT Press, Cambridge (2000)
10. Tong, L., Lu, J.: Overview of Robotsoccer Learning Methods. Computer Simulation 21(6), 1–5 (2004)
11. Zhang, R., Gu, G., Liu, Z., Wang, X.: Reinforcement Learning Theory, Algorithms and Its Application. Control Theory and Application 17(5), 637–642 (2000)

Immune Clonal Strategy Based on the Adaptive Mean Mutation

Ruochen Liu and Licheng Jiao

Institute of Intelligent Information Processing, Xidian University, Xi'an, 710071, China
ruocheenliu@yahoo.com.cn

Abstract. Based on the clonal selection theory, the main mechanisms of clone are analyzed in this paper. A novel strategy algorithm based on Artificial Immune System-Immune Clonal Strategy Algorithm base on the Adaptive Mean Mutation (ICSAMM), is presented, in which the Gauss mutation in the Classical Evolutionary Strategies algorithm (CES) is replaced by the Adaptive Mean one, and the size of algorithm step can be adjusted adaptively. Compared with CES, ICSAMM is shown to be an evolutionary strategy capable of avoiding prematurity, increasing the converging speed and keeping the variety of solution in the simulation. Using the theories of Markov Chain, its convergence is proved.

1 Introduction

Evolutionary computation (including genetic algorithms, evolutionary programming and evolutionary strategies) is an artificial intelligent technicality which uses ideas and gets inspirations from natural evolutionary process, due to its intrinsic parallelism and some intelligent properties such as self-organizing, adaptation and self-learning, evolutionary computation has enormous potential supplying novel methods to solve complex problems [1]. According to Darwinism and Mendelism, biologic evolution can be realized through propagation, mutation, competition and selection, The simulation of evolutionary computation on the biologic reproduce mechanism focus on the sexual one [2,3], the reasons lie in that the sexual reproduction realize the information communication between generations with emphasis on the producing new information, however, When exposed to antigens, immune cells which can recognize and eliminate the antigens can be selected in the body and mount an effective response against them, this process depend on the clone-- a set of cells that are the progeny of the single cell to stimulate antibody to proliferate and divide so as to increase the numbers of antibody[4,5], the idea has been extensively applied in some fields like computer programme[6,7], system control[8], interactive parallel simulation[9], etc. Based on clonal selection theory, a novel clonal operator- immunity clonal operator is proposed in this paper.

Evolutionary strategies were firstly developed by Rechenberg and Schwefel in the 1960s for optimization in hydrodynamical problems [10] and have been applied to many engineering fields. The simple classical evolutionary strategies are evolved

K. Li et al. (Eds.): LSMS 2007, LNCS 4688, pp. 108–116, 2007.

mainly through the mutation and selection with the crossover unused and the mutate mechanism is Gauss mutation; we refer to this method as Classical Evolutionary Strategies (CES). In 1996, inspired by the fast stimulate annealing, C. Kappler proposed a new mutate strategy, Cauchy mutation and applied this method to $(1+1)-\mathrm{ES}$, the results shows it is fast and robust to optimize one-dimensional functions. Recently, this idea is developed by X. Yao and used for $(\mu,\lambda)-\mathrm{ES}$ [11]; series of good results are obtained. Each mutation strategy has their strength and weakness; Gaussian mutation is better at search in a small local neighborhood while Cauchy mutation is very good at search in a large neighborhood, In order to combine their merits, Chellaiapilla proposed combined mutation called Adaptive Mean Mutation Operator (AMMO)[12], in this paper, the mutate manipulation in the clonal operator adopts Adaptive Mean Mutation, we call the algorithm as Immunity Clonal Strategy Algorithm based on the Adaptive Mean Mutation (ICSAMM).

This paper is organized as follows. Section 2 defines the Clonal operator, and a novel algorithm of Artificial Immune System, Immune Clonal Strategy Algorithm based on the Adaptive Mean Mutation (ICSAMM) is put forward. In Section 3, we use the theory of Markov chain to prove the convergence of ICSAMM. Section 4 demonstrates a special implementation of ICSAMM used to compare with the Classical Evolutionary Strategies (CES), Immune Clonal strategy Algorithm based on Gaussian mutation and Cauchy mutation, they are denoted as ICSAGM and ICSACM respectively, ICSAMM, ICSAGM and ICSACM are called by a joint name as ICSA. The numerical experiment shows the new algorithm can enhance the diversity of the population and avoid the prematurity to some extend. Finally Section 5 gives some conclusions.

2 Clonal Operator and ICSAMM

The clonal selection theory (F. M. Burnet, 1978) is used by the immune system to describe the basic features of an immune response to an antigenic stimulus; it establishes the idea that only those cells that recognize the antigens proliferate, thus being selected. Upon exposed to antigens, an immune cell with an appropriate receptors will divide. Each daughter cell will have a receptor with the same receptor with the same specificity. In other words, an antigen specific cell will undergo clonal expansion. As a result, there will be many more cells capable of reacting with antigen, sufficient to mount a strong response, Some of this clones will differentiate into effector cells, other will remain and will be ready to give rise to effector cells upon re-exposure of the host to the same antigen. We will simulate the clone selection theory in following several points in this paper

(1) the new cells (clone) are copies of their parents, subject to a mutation mechanism with high rates;
(2) elimination of newly differentiated lymphocytes carrying self-reactive receptors;
(3) proliferation and differentiation on contact of mature cells with antigens;
(4) the persistence of forbidden clones, resistant to early elimination by self-antigens, as the basis of the autoimmune disease.

In order to use for reference this mechanism in artificial intelligent research fields, we propose a new operator—immunity clonal operator in this paper.

2.1 Clonal Operator

The clonal operator is an antibody random map induced by the affinity including four steps: clone, clonal recombination, clonal mutation and clonal selection. The state transfer of antibody population is denoted as follows:

$$C\text{: } \mathrm{A(k)} \xrightarrow{\text{clone}} \mathrm{B(k)} \xrightarrow{\text{clonalrecombination}} \mathrm{C(k)} \xrightarrow{\text{clonalmutation}} \mathrm{D(k)} \xrightarrow{\text{clonalselection}} \mathrm{A(k+1)}$$

here antibody, antigen, the affinity between antibody and antigen are similar to the definitions of the objective function and restrictive condition, the possible solution, match between solution and the fitting function in AIS respectively. According to the affinity function f (*), a point $a_i = \{x_1, x_2, \cdots x_m\}, a_i(k) \in A(k)$ in the solution space will be divided into q_i different points $a_i'(k) \in A'(k)$, by using clonal operator, a new antibody population is attained.

If the antibody population is an n-dimension vector $A = \{a_1, a_2 \cdots a_n\}$, Clonal Operator can be described as follows:

Clone Θ : The clonal operator Θ is defined as

$$\Theta(\mathrm{A}) = [\Theta(\mathrm{a}_1) \quad \Theta(\mathrm{a}_2) \quad \cdots \quad \Theta(\mathrm{a}_n)]^{\mathrm{T}} \cdot \tag{1}$$

where $\Theta(a_i) = I_i \times a_i \quad i = 1,2 \cdots n$, and I_i is q_i dimension row vectors.

$$\mathrm{q_i} = \mathrm{g}(\mathrm{N_c}, \mathrm{aff}(\mathrm{a_i})) \cdot \tag{2}$$

Generally, q_i is given by:

$$\mathrm{q_i} = \mathrm{N_c} * \frac{\mathrm{aff}(\mathrm{a_i})}{\sum_{j=1}^{n} \mathrm{aff}(\mathrm{a_j})}, \quad \mathrm{i} = 1, 2 \cdots \mathrm{n}. \tag{3}$$

$N_c > n$ is a given integer relating to the clonal scale, it is easy to see that the clonal scale is regularized automatically. After cloning, the antibody population becoming the following equation:

$$\mathrm{B} = \{\mathrm{A}, \mathrm{A}_1', \mathrm{A}_2', \cdots, \mathrm{A}_n'\} \tag{4}$$

Where:

$$\mathrm{A}_i' = \{\mathrm{a}_{i1}, \mathrm{a}_{i2}, \cdots, \mathrm{a}_{iq_i-1}\}, \ \mathrm{a}_{ij} = \mathrm{a}_i, \quad \mathrm{j} = 1, 2, \cdots, \mathrm{q}_i - 1. \tag{5}$$

Clonal Recombination T_r^C : unlike the general recombination operator in GA , the clonal recombination is unused to $A \in A'$, namely:

$$\mathrm{T}_c^C(\mathrm{a}_i, \mathrm{a}_t) = \mathrm{a}_i', \quad \mathrm{a}_i \in \mathrm{A}_j', \mathrm{j} = 1, 2, \cdots \mathrm{n}, \quad \mathrm{a}_t \in \mathrm{A}. \tag{6}$$

Clonal Mutation T_m^C **:** similar to the clonal mutation, in order to save the information of the original population, the clonal mutation is unused to $A \in A'$ and we apply Adaptive Mean mutation here, namely, component x_i of individual a becomes the equations below after clonal mutation:

$$\begin{aligned} \eta_{1i}'(j) &= \eta_{1i}(j)\exp(\tau' N(0,1) + \tau N_i(0,1)) \\ \eta_{2i}'(j) &= \eta_{2i}(j)\exp(\tau' N(0,1) + \tau N_i(0,1)) \end{aligned} \tag{7}$$

$$x_i'(j) = x_i(j) + \eta_{1i}'(j)N(0,1) + \eta_{2i}'(j)C(0,1) \tag{8}$$

Where $x_i(j), x_i'(j), \eta_{1i}(j), \eta_{1i}'(j), \eta_{2i}(j), \eta_{2i}'(j)$ denote the jth component of the vector $x_i, x_i', \eta_{1i}, \eta_{1i}', \eta_{2i}, \eta_{2i}'$ respectively, $N_i(0,1)$ denotes a normally distributed one-dimensional random number with mean zero and standard deviation one, $c(0,1)$ is thy Cauchy variable with mean 0 and $t = 1$, τ and τ' are set commonly to $(\sqrt{2\sqrt{n}})^{-1}$ and $(\sqrt{2n})^{-1}$ [11].

Clonal Selection T_s^C **:** $\forall i = 1,2,\cdots n$, if $b = \max\{f(a_{ij}) \mid j = 2,3,\cdots q_i - 1\}$, let:

$$f(a_i) < f(b), \quad a_i \in A. \tag{9}$$

Then b replaces the antibody a_i in the original population. So the antibody population is updated, and the information communication between generations can be realized.

It is easy to find that the essential of the clonal operator is producing a variation population around the parents according to their affinity. Then the searching area is enlarged, ulteriorly, the clonal operator maps a problem in a low dimension space (n dimensions) to a high one (N_c dimensions), and then projects the results to the original space after solving. Thereby the problem can be solved better.

2.2 Immunity Clonal Strategy Algorithm Based on the Adaptive Mean Mutation Algorithm (ICSAMM)

Before introduction of the algorithm, we firstly describe the one-dimensional Cauchy density centered at the origin and Gaussian density function with mean zero and standard deviation one

(1) Cauthy density function

$$f_{cauchy(o,t)}(x) = \frac{1}{\pi}\frac{t}{t^2 + x^2}, \qquad x \in R \tag{10}$$

Where t is the scale parameter, the corresponding distributions is:

$$F_t(x) = \frac{1}{2} + \frac{1}{\pi}\operatorname{arctg}(x/t) \tag{11}$$

(2) Gaussian density function

$$f_{Gaussian(0.1)}(x) = \frac{1}{\sqrt{2\pi}} e^{-\frac{x^2}{2}} \quad (12)$$

The shape of $f_{Cauchy(0,t)}(x)$ is similar to the Gaussian density function $f_{Gaussian(0,1)}(x)$, the difference is that the former inclines to x axis infinitely but the latter intersects it, which can be find in figure1, it is easy to find that the larger random values are more likely to occur in Cauchy distributions than in Gaussian distributions which implies Cauthy mutation more likely escape from the local optimal. In order to combine their merits, an improved evolutionary strategies algorithm–immunity clonal strategy algorithm based on the adaptive mean mutation is presented, Concretely, the clonal operator mentioned above replaces the generic operator in the classical evolutionary strategies algorithm, and ICSAMM weights each offspring before summing the two distribution thereby allowing adaptation of not only the parameters of the distribution, but also its shape.

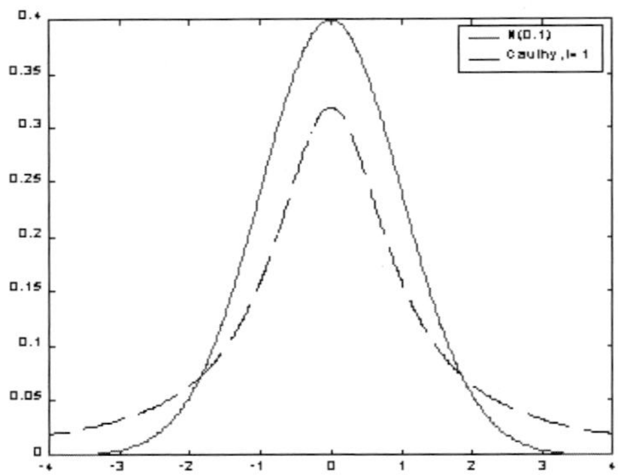

Fig.1. The compares ion between Cauchy density and Gaussian density

$f: R^m \to R$ is the optimized object function, without loss of generality, we consider the minimum of affinity function the affinity function$\Phi: I \to R$ where $I = R^m$ is the individual space, n is population size, m is the number of the optimized variable, namely, $a = \{x_1, x_2, \cdots x_m\}$. The main step of the algorithm is described as follows:

$k = 0$

initial the antibody population:

$A(0) = \{a_1(0), a_2(0), \cdots a_\mu(0)\} \in I^\mu$

Calculate the affinity:

$A(0): \{\Phi(A(0))\} = \{\Phi(a_1(0)), \Phi(a_2(0)), \cdots \Phi(a_\mu(0))\}$

while there is no satisfied candidate solution in $A(k)$, do

Clone: $B(k) = \Theta(A(k)) = [\Theta(a_1(k)), \Theta(a_2(k)) \cdots, \Theta(a_\mu(k))]^T$

```
Clonal recombination: C(k) = T_r^C(B(k))
Clonal mutation: D(k) = T_m^C(C(k))
Calculate the affinity: D(k) : {Φ(D(k))}
Clonal selection: A(k+1) = T_s^C(D(k))
Calculate the affinity:
A(k+1) : {Φ(A(k+1))} = {Φ(a_1(k+1)), Φ(a_2(k+1)), ···Φ(a_μ(k+1))}
k = k+1
end
```

The halt conditions are defined as the restricted iterative number or the time when the solutions are not improved at successively iterations or the two methods blending.

3 Convergence of the Algorithm

Theorem 3.1. The population series of ICSAMM $\{A(n),\quad n\geq 0\}$ is finite homogeneous Markov chain.

Proof. Similar to the evolutionary strategies, the state transfer of ICSAMM are processed on the finite space, therefore, population is finite, since

$$A(k+1)=T(A(k))=T_s\circ T_m\circ T_r\circ\Theta(A(k)) \tag{13}$$

T_s, T_m, T_r, Θ have no relation with n, so A (k+1) just relates with A(k), namely, $\{A(n),\quad n\geq 0\}$ is finite homogeneous Markov chain.

The size of the initial population of the algorithms is μ and the size of the middle population is N, all the possible solutions in the initial population can be seen as a point in the state space $S^1 := X^{\mu}$ and all the possible solutions in the middle population can be seen as a point in the state space $S^2 := X^N$, when it is not necessary to distinguish S^1 and S^2, we denote the state space as S, $s_i \in S$ denote the ith state in S, let f is the affinity function defined on the hunting space X. Let:

$$s^* = \{x \in X \mid f(x) = \max_{x_i\in X} f(X)\} \tag{14}$$

Then we can define the convergence of the algorithms as follows.

Definition 1. Suppose for arbitrary initial distribution, the following equation satisfies:

$$\lim_{k\to\infty}\sum_{s_i\cap s^*\neq\varnothing} p\{A_k^i\}=1 \tag{15}$$

Then we call the algorithm is convergent.

The definition of the convergence means that if the algorithm implements for enough iteration, then the probability with which the population contains the optimal individual will verge on 1, thus the definition shown as the above is usually called the convergence with probability 1.

Theorem 3.2. ICSAMM is convergent with the probability 1.

4 Simulations and Results

Similar to classical evolutionary strategies, the immune clonal strategies algorithm based adaptive mean mutation can be used to solve complex machine learning tasks, like numerical optimization. The following test functions [2, 10, 11] are used to evaluation the performance of ICSAMM. The propriety of test functions are shown in Table1.

$$f_1(x) = x^2 + y^2 - 0.3\cos 3\pi x + 0.3\cos 4\pi y + 0.3 \tag{16}$$

$$f_2(x) = 4x^2 - 2.1x^4 + \frac{1}{3}x^6 + xy - 4y^2 + 4y^4 \tag{17}$$

$$f_3(x) = \sum_{i=1}^{30}[x_i^2 - 10\cos(2\pi x_i) + 10] \tag{18}$$

$$f_4 = \sum_{i=1}^{30}\left(\lfloor x_i + 0.5 \rfloor\right)^2 \tag{19}$$

$$f_5 = -20\exp\left(-0.2\sqrt{\frac{1}{N}\sum_{i=1}^{N} x_i}\right) - \exp\left(\frac{1}{N}\sum_{i=1}^{N}\cos(2\pi x_i)\right) + 20 + e,\ (N = 30) \tag{20}$$

The results of 20 times experiments with different initial antibody population are presented in Table 1. The population size used in CES is100 and n=50, N_c=100 in ICSA, p_m=0.1, P_c=0.7, the maximum evolutionary generation is 500 for f_1 and f_2; for f_3–f_5, the maximum evolutionary generation is 2000. The number in the bracket is the optimal accuracy. Additionally, because the CES is immersed in the local optimal values many times, some statistical parameters cannot be calculated (denotated as "/" in Table 1).

ICSAGM also run into the local optimal value (two times from ten), but the results is very close to the global optimum. ICSA is superior to CES in ten tests distinctly, Obviously, the convergent speed of ICSAMM is faster than that of ICSACM and

Table 1. Comparing performance on functions for all algorithms

(Mean number of iterations for convergence)

algorithm / function	CES	ICSAGM	ICSACM	ICSAMM
$f_1(10^{-3})$	90	24	17	13
$f_2(10^{-3})$	98	66	40	35
$f_3(10^{-3})$	/	1171	935	979
$f_4(10^{-3})$	/	679	602	557
$f_5(10^{-3})$	/	592	582	544

ICSAGM when optimizing functionsf_1, f_2, f_4, f_5; when optimizing f_3, ICSACM is better slightly than ICSAMM, but ICSAMM is more stable, Additionally, it is demonstrated that bigger clonal size can improve the diversity ulteriorly, as a result, the prematurity can be avoided effectively, but mean time performing each generation will be increased.

After a further analysis of the simulation results, it is ease to find that ICSA can hunt more local optimas than CES. The results of ICSA are more like to find a set of the local optimal value including the global optimal one score the characteristic of the objective function more accurately, this can be seen from Fig.2. ICSAMM can hunt more local optimas than ICSAGM and ICSACM, and it can search the global optima every simulation.

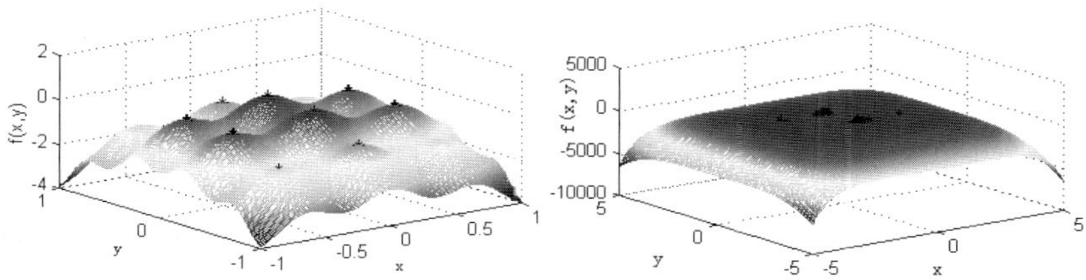

(a) $f_1(x,y)$ (b) $f_2(x,y)$

Fig. 2. ICSA simulation results of $f_1(x,y)$ and $f_2(x,y)$

5 Conclusions

The mechanize of the antibody clonal selection is discussed systematically in this paper, clonal operator is proposed, By applying the clonal operator and the adaptive Mean mutation to the evolutionary strategies, Immune clonal Strategy based on the Adaptive Mean Mutation is put forward. We find the essential of the clonal operator is producing a variation population around the parents according to their affinity, then the searching area is enlarged, compared with CES, ICSAMM can enhance the diversity of the population and avoid the prematurity to some extent, and have the high convergence speed.

References

1. Bentley, P.J., Wakefield, J.P.: Overview of Generic Evolutionary Design Systems. In: Proceedings of the 2nd On-Line World Conference on Evolutionary Computation (WEC2) (1996)
2. Guoliang, C., Xifa, W., Zhenquan, Z.: Genetic Algorithms and Application. People Telecommunication Press, Beijing (1996)
3. Guangyan, Z.: Principles of Immunology. Shang Hai Technology Literature Publishing Company (2000)
4. Fogel, D.B., Atmar, J.W.: Comparing Genetic Operators with Gaussian Mutations in Simulated Evolutionary Processes Using Linear Systems. Biological Cybernetics 63, 111–114 (1993)

5. Schwefel, H.P.: Evolutionary Optimum Seeking. John Wiley&Son, New York (1995)
6. Szu, H.H., Hartley, R.L.: Nonconvex Optimization by Fast Simulated Annealing. Proceeding of IEEE 75, 1538–1540 (1987)
7. Kappler, C.: Are Evolutionary Algorithms Improved by Larger Mutations. In: Ebeling, W., Rechenberg, I., Voigt, H.-M., Schwefel, H.-P. (eds.) PPSN IV. LNCS, vol. 1141, pp. 346–355. Springer, Heidelberg (1996)
8. Yao, X.: A New Simulated Annealing Algorithms. Int. J. of Computer Math. 56, 162–168 (1995)
9. Bäck, T., Schwefel, H.P.: An overview of Evolutionary Algorithms for Parameter Optimization. Evol. Comput. 1, 11–23 (1993)
10. Yao, X., Liu, Y., Lin, G.: Evolutionary Programming Made Faster. IEEE Transactions on Evolutionary Computation 3(2), 82–102 (1999)
11. Chellapilla, K.: Combining Mutation Operators in Evolutionary Programming. IEEE Transactions Computation 2(3), 91–96 (1998)
12. Baldonado, M., Chang, C., Gravano, C.K., Paepcke, L.: The Stanford Digital Library Metadata Architecture. Int. J. Digit. Libr. 1, 108–121 (1997)
13. Bruce, K.B., Cardelli, L., Pierce, B.C.: Comparing Object Encodings. In: Ito, T., Abadi, M. (eds.) TACS 1997. LNCS, vol. 1281, pp. 415–438. Springer, Heidelberg (1997)
14. van Leeuwen, J. (ed.): Computer Science Today. LNCS, vol. 1000. Springer, Heidelberg (1995)
15. Michalewicz, Z.: Genetic Algorithms + Data Structures = Evolution Programs, 3rd edn. Springer, New York (1996)

An Agent Reinforcement Learning Model Based on Neural Networks

Liang Gui Tang[1,2], Bo An[3], and Dai Jie Cheng[1]

[1] College of Computer Science, Chongqing University, Chongqing, P.R. China
[2] College of Computer Science, Chongqing Technology and Business University, Chongqing, P.R.China
[3] Dept. of Computer Science, University of Massachusetts, Amherst, USA
tglj@126.com, ban@cs.umass.edu, djcheng@cqu.edu.cn

Abstract. This paper thoroughly analyzes the transfer and construction of the state-action space of the agent decision-making process, discusses the optimal strategy of agent's action selection based on Markov decision-making process, designs a neural networks model for the agent reinforcement learning, and designs the agent reinforcement learning based on neural networks. By the simulation experiment of agent's bid price in Multi-Agent Electronic Commerce System, validated the Agent Reinforcement Learning Algorithm Based on Neural Networks has very good performance and the action impending ability.

1 Introduction

Reinforcement Learning (RL) is a kind of machine learning methods, which finds out the optimal decision-making strategy by trial-and-error and gaining the feedback through interaction with environments. The main characteristic of RL is making use of uncertain reward values of environments to find the optimal action strategy. In recent years, the research of RL has gained great achievements in theory and applications [1-3].

At present, some reinforcement learning algorithms have been put forward, such as TD(λ) [4], Q-Learning [5], Sarsa-Learning [6], and so on. Most of them are based on Markov Decision-making Process (MDP) and value function iterative computing based on dynamic programming. Generally by use of the value function mechanism, such as polynomial basis function, decision tree or multilayer feedforward neural networks, these algorithms could resolve the Curse of Dimensionality problem of large-scale continuous state space's computation, and implement its generalization to a certain extent.

The research of RL based on neural networks mostly focuses on the following two aspects [3, 10, 11]. Firstly, the reinforcement learning system is regarded as a frame of control systems, which uses neural networks to adjust the input parameters of reinforcement learning systems, so as to reinforce and optimize the learning action and ability. Secondly, it employs Q-value iterated algorithm's ideas of RL to adjust and optimize the joint weights of neural networks, especially to optimize the output layer weights by learning, so as to improve the approaching ability of neural networks.

In Multi-Agent Systems (MAS), an agent may know little about others because information is distributed. Even when an agent has some prior information about others,

K. Li et al. (Eds.): LSMS 2007, LNCS 4688, pp. 117–127, 2007.

the behavior of others may change over time. Therefore, besides continuously enhancing its own ability and maximizing its income, a single agent still needs to cooperate with other agents as a consociation colony during the process of resolving problems. Namely, to improve efficiency and action approaching ability, an agent should conjecture other agents' action strategies and the union state trends, through the current state, the tasks to be solved and the rewards of environment in competing and cooperating. It is therefore naturally to use reinforcement learning. According to the cybernetic model of an agent, we can use neural networks to approach an agent's reinforcement learning action.

2 The Agent Behavior Choice Strategy Based on MDP

Definition 1: Single Agent's Markov Decision-making Process (SAMDP). Assume that the system has continuous state space and discrete action space, then the SAMDP is defined as a quadruple <S,A,R,T>, where S is continuous state space, A is finite discrete action space, $r \in R : S \times A \to R$ is reward function, where R is the set of reward functions, and $p \in P : S \times A \to P$ is transfer function, where P is the set of probability distributions over state space S. In general, under the strategy π of action selection, the probability distributions P may probably change at any moment, i.e., the action strategy is not balanced.

Definition 2: The values function of a state. Assume that π denotes the action selection strategy of an agent, s_t denotes the system state, $r_t \in R$ denotes the reward under state s_t, and $\beta \in [0,1]$ denotes the discount factor. The agent value function of state $v(s_t, \pi)$ is defined as:

$$v(s_t, \pi) = \sum_{t=0}^{\infty} \beta^t E(r_t / \pi, s_t) \tag{1}$$

If there exists an act $a_t \in A$ under a specified action selection strategy π, such that the system state is shifted from s_t to s_{t+1}, then the equation (1) may be rewritten as:

$$v(s_t, \pi) = r(\pi(s_t)) + \beta \sum_{s_{t+1} \in S} p(s_t, a_t, s_{t+1}) v^{\pi}(s_{t+1}) \tag{2}$$

According to the theory of dynamic programming, there at least exists an optimization action selection strategy π^* which makes the following equation holds:

$$v(s_t, \pi^*) = \max_a \left\{ r(\pi(s_t)) + \beta \sum_{s_{t+1}} p(s_t, a_t, s_{t+1}) v(s_t, \pi^*) \right\}, \forall s_t, s_{t+1} \in S \tag{3}$$

where $v(s_t, \pi^*)$ is called the optimal value function of the state.

Definition 3: The values function of action. Suppose $\pi*$ denotes the action selection strategy of an agent, at time t, $a_t \in A \subseteq \pi*$, the system state is s_t. When the action a_t is executed, the system state is transferred to s_{t+1}, and gains the reward r_t. $\beta \in (0,1)$ is discount factor. The value function of action $Q(s_{t+1}, a_t)$ can be defined as:

$$Q(s_{t+1}, a_t) = r_t + \beta \sum_{s_{t+1} \in S} P(s_t, a_t, s_{t+1}) v(s_{t+1}, \pi*) \tag{4}$$

Assume that the system executes action a_t under state s_t, and then it takes action $a_{t'} (t' \neq t)$ according to the optimal strategy $\pi*$ in the state space S, so the total accumulative discounted reinforcing value can be defined as:

$$Q(s_{t+1}, a_t) = r_t + \beta \max_{a \in A \subseteq \pi*} \{Q(s_{t+1}, a)\} \tag{5}$$

From the previous definitions, we have known that the equation (5) holds only if the optimal strategy has been acquired, and in the course of learning, the equation (5) doesn't hold. So according to the Temporal Difference Method (TDM), an agent could update its Q-values as follows:

$$Q(s_t, a_t) = \begin{cases} \eta_t \left[r_t + \beta \max_{a \in A} \{Q(s_{t-1}, a)\} - Q(s_{t-1}, a_{t-1}) \right] \\ \qquad + Q(s_{t-1}, a_{t-1}), \ s=s_t \in S, a=a_t \in A \\ Q(s_{t-1}, a_{t-1}), \quad \textit{the others} \end{cases} \tag{6}$$

where s_t is the current state, a_t is the selected action, η_t is the learning factor.

According to the analysis above, the selected action element from the set of action strategies should have the biggest probability and maximal action value of the function $Q^*(s,a)$, so action selection probability isn't continuous to the changes of value function estimation. Generally the adopted action selection strategy is Boltzman distributions, which are approximately greedy and continuously differentiable.

Assume that, the system has an action set $A = \{a_i, i = 1,2,...,n\}$, and $Q(s_t, a_i)$ is the estimation of action value function under state s_t, then the action selection probability could be defined as:

$$p(s_t, a_i) = \frac{e^{Q(s_t,a_i)/T}}{\sum_{a \in A} e^{Q(s_t,a)/T}} = \frac{1}{\sum_{a \in A} e^{\frac{Q(s_t,a) - Q(s_t,a_i)}{T}}} \tag{7}$$

where T>0 is temperature parameter or energy parameter, while $T \to 0$, the action selection strategy defined in equation (7) approximates to a greedy strategy.

3 Agent Reinforcement Learning Based on Neural Networks

Reinforcement learning is to learn what should be done -how to map situations to actions - so as to maximize a numerical reward signal. It is defined in the framework of optimal control model of Markov decision processes.

3.1 Agent RL Neural Network Model and Algorithm

Assume that the number of the elements in agent action set based on MDP is N , and we adopt n-layer feed forward neural networks to design the agent reinforcement learning model, which is composed of N Learning Units (LU_i, i=1, 2, …, N), where each learning unit is used to approach every action value function $Q(s_i,a_i)$ (i=1,2,…,N) respectively. The inputs of each learning unit contain the state s_i , the action a_i , the reward of environment r_i and the Q_{i-1} $(i=1,...,n)$. The output is the value of action value function $Q(s_i,a_i)$.

For that, the neural network model of single agent reinforcement learning could be designed as shown in Fig.1, where LU_i is a learning unit.

The inputs of each learning unit LU_i(i=1,…,n) are instantaneous state s_i which is composed by the state vector $X=\{x_1,x_2,...,x_m\}$. Assume that the weights of hidden layers of learning unit are $q_{ij}(i=1,2,...,m;\ j=1,2,...,l)$, where m is the dimension of inputs. l is the number of hidden layer units, then the input vector of LU_i is $Xrq=\{x_1,x_2,...,x_m;r_i,a_i,Q_{i-1}\}$. If the output vector is y_i , and the output layer weight vector is w_i , then the structure of Learning Unit i shown in Fig.2.

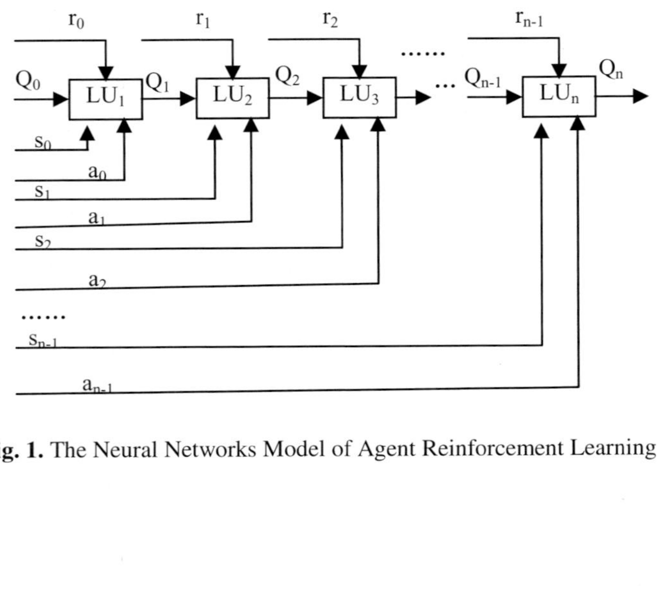

Fig. 1. The Neural Networks Model of Agent Reinforcement Learning

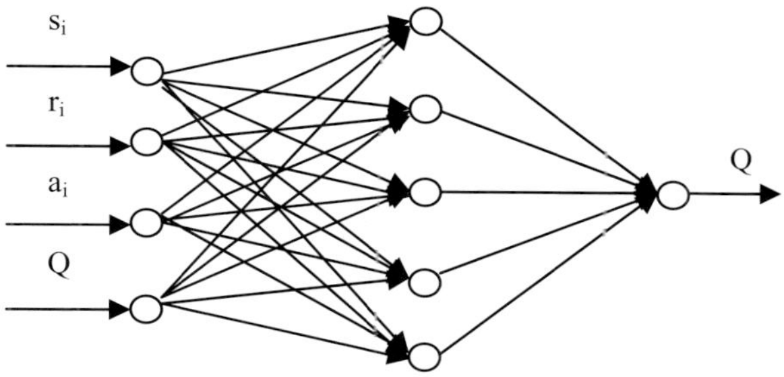

Fig. 2. The structure of Learning Unit$_i$ (LU_i)

The estimation of the single agent action value function may be represented in the following form:

$$Q(s_i, a_i) = (w_i)^T y_i = \sum_{j=1}^{n} w_i^j y_i^j \tag{8}$$

According to the definition of Q-function, the optimal action Q-value could be gained only under the optimal strategy. In the process of learning, Q-values iteration exists errors. Considering the method of the Temporal Difference, the TD error may be defined as:

$$\delta_t = r_t + \beta \max_{a\in A}\{Q(s_{t-1}, a)\} - Q(s_{t-1}, a_{t-1}) \tag{9}$$

To ensure that agent selection is the optimal action, the TD error should be minimized. Therefore we can achieve the target by minimizing the square sum of Bellman residues.

Assume the probability of action selection is $p(s_i, a_i)$. E[·] is the mathematical expectation that is defined over probability distributions P, then the square sum of Bellman residues is defined as follows:

$$\tilde{E} = \frac{1}{2n}\sum_{s_i}\sum_{a_i} E[r(s_i, a_i) + \beta \sum_{j=i+1} p(s_j, a_j) Q(s_j, a_j) - Q(s_i, a_i)]^2 \tag{10}$$

Using the algorithm of random gradient descent to minimize the $\tilde{E}$, then we have:

$$\frac{\partial \tilde{E}_t}{\partial w_i} = p(s_{t+1}, a_{t+1})\delta_t[p(s_{t+1}, a_{t+1})\frac{\partial \delta_t}{\partial w_i} + \delta_t \frac{\partial p(s_{t+1}, a_{t+1})}{\partial w_i}] \tag{11}$$

Where,

$$\frac{\partial \delta_t}{\partial w_i} = \beta \frac{\partial Q(s_{t+1}, a_{t+1})}{\partial w_i} - \frac{\partial Q(s_t, a_t)}{\partial w_i} \tag{12}$$

$$\frac{\partial p(s_{t+1},a_{t+1})}{\partial w_i}=\frac{-1}{Tp^2(s_{t+1},a_{t+1})}\times \sum_{a\in A}[e^{(Q(s_{t+1},a)-Q(s_{t+1},a_{t+1}))/T}\left(\frac{\partial Q(s_{t+1},a)}{\partial w_i}-\frac{\partial Q(s_{t+1},a_{t+1})}{\partial w_i}\right)] \tag{13}$$

$$\frac{\partial Q(s_{t+1},a)}{\partial w_i}=\begin{cases} y_i, & \text{if } a=a_i \\ 0, & \text{if } a\neq a_i \end{cases} \tag{14}$$

With the algorithm of the random gradient descent, we gain the iterative equation of the neural network joint weights:

$$\Delta w_i=-\eta_t\frac{\partial \tilde{E}_t}{\partial w_i} \tag{15}$$

So the reinforcement learning algorithm based on neural networks of single agent can be described as follows:

Algorithm 1. The agent reinforcement learning algorithm based on neural networks

```
For all LUi, i<=N
  Initialize the joint weight value w
  Initialize learning parameters s0, r0, Q0;
  t=0;
  While t<MaxNum
      Suppose current state is st
      Choose action ati based on p(st, a_t^i )
      Observe the state st+1 and reward r(st, at)
      Chose action a_{t+1}^i based on p;
      Update joint weight value While j<=M
             Computing ∂Ẽt/∂wi ;
           Update every output layer joint weight value
            by
               Δwi = −ηt ∂Ẽt/∂wi ;
      j=j+1
   End While
   Computing hidden layer joint weight value by BP;
   t=t+1;
  End While
  i=i+1
End For
```

Where, N is the number of learning units, $MaxNum$ is the algorithm's terminal condition based on observations of states, M is the number of the joints of output layer, $\eta_t \in (0,1)$ is the learning-rate parameter.

Using the algorithm 1, an agent can update its own Q-values, and as well as amend its belief about other agents. According to the property of neural networks that it can approach any value functions, so the algorithm ANNRL can reduce the dimension of the union state and action space, and improve the efficiency of multi-agent cooperation.

3.2 Convergence of the Algorithm 1

These algorithms can converge to an optimal and equilibrium value Q*(s, a) under certain assumptions in learning.
Convergence Theorem [8, 6]: Assume that all the state-action pairs are visited infinitely, and the learning-rate parameter $\eta_n(s_i, a_i)$ has its boundary conditions, such that $\eta_n(s_i, a_i)$ satisfies the condition

$$\sum_{n=0}^{\infty}\eta_n(s_i,a_i)=\infty \text{ and } \sum_{n=0}^{\infty}\eta_n^2(s_i,a_i)<\infty \text{ for all } (s_i,a_i),$$

Then, the sequence of Q-values $\{Q(s_i,a_i)\}$ generated by the agent reinforcement learning algorithm converges with probability 1 to the optimal value Q*(s, a) for all state-action pairs (s_i, a_i) as n approaches infinity, where n is the number of iterations.

To guarantee the convergence of the algorithm, the time-varying learning-rate parameter can be defined as follows:

$$\eta_n(s_i,a_i)=\frac{\alpha}{\lambda+visits_n(s,a)} \tag{16}$$

where $visits_n(s,a)$ denotes the total times of an agent's observation, α and λ are positive.

The value of learning-rate parameter $\eta_n(s_i,a_i)\in[0,1]$ in equation (18) will be reduced with the increase of $visits_n$. So it could diminish the extent of Q-values' updating gradually, and make sequence $\{Q(s_i,a_i)\}$ converge to a right Q function.

Proof. According to the assumptions, every state-action pair will be executed one time at least. Suppose $\hat{Q}_n$ is the estimation of Q_n, then we should prove that the max error is diminished according to discount factor β. Let ΔE is the maximum error of estimation, namely:

$$\Delta E \equiv \max_{s,\,a}\left|\hat{Q}_n(s,a)-Q(s,a)\right| \tag{17}$$

Now, in the $n+1$th iteration, the error of estimation $\hat{Q}_{n+1}(s,a)$ could be denoted as follows:

$$\left|\hat{Q}_{n+1}(s,a)-Q(s,a)\right|=\left|\left(r+\beta\max_{a'}\hat{Q}_n(s',a')\right)-\left(r+\beta\max_{a'}Q(s',a')\right)\right|$$

$$=\beta\left|\max_{a'}\hat{Q}_n(s',a')-\max_{a'}Q(s',a')\right|\leq\beta\max_{a'}\left|\hat{Q}_n(s',a')-Q(s',a')\right|$$

$$\leq\beta\max_{s'',a'}\left|\hat{Q}_n(s'',a')-Q(s'',a')\right|\leq\beta\Delta E$$

$$\because\ \Delta E\equiv\max_{s,a}\left|\hat{Q}_n(s,a)-Q(s,a)\right|,\ \left|\hat{Q}_n(s,a)-Q(s,a)\right|\leq\Delta E$$

So the error of updated $\hat{Q}_{n+1}(s,a)$ for all s and a is the β times of the maximum error ΔE. Because the values of $\hat{Q}_0(s,a)$ and $Q(s,a)$ are bounded for any s and a, the initial maximum error ΔE_0 is bounded. So the maximum error is $\beta\Delta E_0$ in the first zone in which all s and a have been visited. The maximum error for k zones is $\beta^k\Delta E_0$.

Because every state-action pair is visited infinitely, and the number of these zones is infinite, while $n\to\infty$, $\Delta E_n\to 0$. ❑

Note that, in the proof, we have used the following Lemma:

Lemma. For any functions f_1 and f_2, the following holds

$$\left|\max_a f_1(a)-\max_a f_2(a)\right|\leq\max_a\left|f_1(a)-f_2(a)\right|$$

Note that, in the proof, we introduce a new variable s'', and maximize Q-value based on s''. When an additive variable is allowed to change, the maximum value could be bigger or be equivalent at least.

4 Experiment Simulation and Analysis

In this section, we apply the ANNRL algorithm to a Multi-Agent Electronic Commerce Trade-off System (MAECTS). The Electronic Commerce Trade-off Model is a trade mode under some restrictive strategies based on Web. It satisfies every dealer's demands on their benefits to a certain extent. At the same time, it normalizes participators' bargaining behaviours according to some rules, so as to ensure that the trade and the deals are effective and fair.

In practical system, the trade environment in MAECTS generally is a virtual market, which is constituted of multi intelligent agents. The state space of the virtual market sometimes is continuous with high dimensions. The neural network model of multi-agent reinforcement learning can map the high dimension space of state and action into a computable lower dimension space, and optimize actions of agents, so as to improve the ability and efficiency of approaching, and to improve the accuracy of estimating future action strategies and market trends in complex environments.

We simulate the agent bid price actions, and compare the results with that of the Regress Neural Networks (RNN) and the general Reinforcement Learning (RL) algorithms.

In a bargaining process about some commodities, we assume that the bid price action of the agent$_i$ (i=1, 2,..., N) is a no-balance time sequence $\{Q_{i,j}; i=1,2,...,N; j=1,2,...,M\}$, where N is the number of agents, M is the times of bid price.

In practical applications, we often use statistic methods to approach a model, and evaluate its performance. In this paper, we use the Mean of the Root Mean Square Relative Error (MRMSRE) to appraise the effect of ANNRL. According to the definition of the root mean square error, MRMSRE can be defined as follows:

$$MRMSRE = \frac{1}{N}\sum_{i=1}^{N}\sqrt{\frac{1}{M}\sum_{j=1}^{M}\left(\frac{x_i - y_{j,l}}{x_i}\right)^2} \quad (18)$$

where, $y_{j,l}$ is the l steps' forecast value of x_i, N is the number of agents participating in biding price, M is the time of a bid.

In these experiments, we gained 157 quoted prices of 10 agents about some goods, and according to these data, we computed the conjectures of the future bid price action strategy of each agent through algorithms ANNRL, RNN and RL.

The experiment results are shown in Table 1 and Fig.3. The data in Table 1 is the averages of the estimation value's MRMSRE of 10 agents' previous six steps forecast by ANNRL, RNN and RL algorithms respectively.

Table 1. Forecast Step and The MRMSRE

Forecast Steps	ANNL	RNN	RL
1	0.0015	0.0025	0.0033
2	0.0026	0.0387	0.0417
3	0.0562	0.0869	0.1169
4	0.1188	0.1321	0.2352
5	0.1297	0.2572	0.3061
6	0.1125	0.3829	0.4727

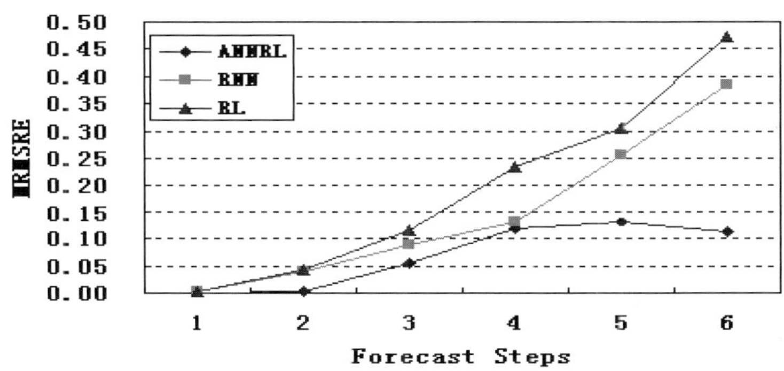

Fig. 3. The forecast results OF ANNRL, RNN, RL

These results show that the algorithm ANNRL is effective. The degree of accuracy reaches 89.89% only in six steps in the simulation, and it can converge to a limited value. This is what we expected. Because each agent updates its Q-values, at the same time it takes advantage of its belief about the other agents and the market.

5 Conclusions

We design a neural network model of single agent reinforcement learning, and discuss the modification mechanism of multi-agents' beliefs for the agents' cooperation with each other, and then realize the algorithm of the Agent Reinforcement Learning Model Based on Neural Networks (ANNRL). We have proved that it can converge to optimal Q-values under certain assumptions. Because neural networks can approach any value function, and the higher dimension space of information can be transformed into lower dimension space by the mapping of the value function. So the ANNRL algorithm could reduce the VC-dimensions of the state-action space computation to a certain extent. And by using random gradient descent algorithm to minimize the square sum of Bellman residues, we can gain a higher convergence speed and efficiency of the algorithm. We implement the ANNRL algorithm in Multi-Agent Electronic Commerce Trade-off System (MAECTS). The results of the simulation experiment of agent's bid prices in MAECTS have shown that the ANNRL algorithm is effective.

Acknowledgments. This research is supported by the Key Science Technology Project of Chongqing under Grant No. CSTC-2005AC2090, The Science Technology Project of Chongqing Municipal Education Commission under Grant No. KJ060704, and the Science Foundation of Chongqing under Grant No. CSTC-2004BB2167 and No. CSTC-2006BB2249.

References

1. Rubo, Z., Guochang, G., et al.: Reinforcement Learning Theory,Algorithm and Its Application[J]. Control Theory and Applications 5(17), 637–641 (2000)
2. Burnsa, T.R., Gomolinskab, A.: Socio-cognitive mechanisms of belief change Applications of generalized game theory to belief revision, social fabrication, and self-fulfilling prophesy[J]. Journal of Cognitive Systems Research 2, 39–54 (2001)
3. Kaelbling, L.P., Littman, M.L., Moore, A.W.: Reinforcement learning: A survey[J]. Journal of Artificial Intelligence Research 4, 237–285 (1996)
4. Tsitsiklis, J.N., Roy, B.V.: An analysis of temporal difference learning with function approximation[J]. IEEE Transactions on Automatic Control 42, 674–690 (1997)
5. Watkins, C.J., Dayan, P.: Q-learning[J]. Machine Learning 8, 279–292 (1992)
6. Singh, S.P., et al.: Convergence results for single-step on strategy reinforcement learning algorithms[J]. Machine Learning 38, 287–308 (2000)
7. Boutilier, C.: Planning, learning and coordination in multi-agent decision processes[J]. In: Proceedings of the Sixth Conference on Theoretical Aspects of Rationality and Knowledge[C], pp. 195–210. Morgan Kaufmann, San Francisco (1996)

8. Haykin, S.: Neural Networks: A Comprehensive Foundation[M]. Prentice Hall Tsinghua University Press (2001)
9. Tang, L., Li, S., Cheng, D.: The Research on Multi-Agent Based Matchmaking Tradeoff Model and Algorithm. Computer Engineering and Applications(in Chinese) [J] 23(46), 145–147, 228 (2003)
10. Tang, L., Cheng, D.: Structure of Multi-agent Making Match Tradeoff System. Computer Engineering(in Chinese) [J] 15(29), 166–168 (2003)
11. Yang, G., Shi-fu, C., et al.: Rearch on reinforcement learning technology: A Review. Acta Automatica Sinica[J] 30(1), 86–100 (2004)
12. Wallace, R.J., Freuder, E.C.: Constraint-based reasoning and privacy/efficiency tradeoffs in multi-agent problem solving. Artificial Intelligence[J], 209–227, 161(2005)
13. Yu, Z., Guo-chang, G., et al.: Surey of distributed reinforcement learning algorithms in multi-agent systems. Control Theory & Applications[J] 20(3), 317–322 (2003)
14. Sutton, R.S., Barto, A.G.: Reinforcement Learning: An Introduction [M]. MIT Press, Cambridge, MA (1998)

A Novel Agent Coalition Generation Method Based on Fuzzy Set Theory

Chunhua Yu[1], Xueqiao Du[1], and Na Xia[2]

[1] School of Sciences, Hefei University of Technology
[2] School of Computer and Information Science, Hefei University of Technology
Hefei City, Anhui Province, 230009, PR China
yxylucky@mail.china.com

Abstract. In Multi-Agent System, the generation of agent coalition is a key topic. It mainly researches how to generate the task-oriented optimal agent coalition. This paper proposes a novel coalition generation method based on fuzzy set theory. Defines a membership function, by which the membership degree of an agent to a coalition is calculated as a value between 0 and 1, and then the task-oriented coalition can be generated by λ matrix theory. An example is presented to illustrate the validity of the method.

1 Introduction

In Multi-Agent System, it is an important cooperation method for Agents to form a coalition [1]. In recent years, Agent coalition generation problem has been receiving more and more research attention [2, 3].

Agent coalition generation problem can be described as follows:

Agent set $\{A_1, A_2, \cdots, A_n\}$, task set $\{T_1, T_2, \cdots, T_m\}$. Define a coalition C as a non-empty subset of Agent set. The problem is how to form m coalitions $C_1, C_2, \cdots, C_m$ for m tasks, so that these tasks can be accomplished in optimal manner.

2 Related Work

Sandholm [4], Sen [5], and Hu [6] researched this problem using "coalition structure". Define a coalition structure CS $=\{C_1, C_2, \cdots, C_m\}$ as a complete partition of Agent set. Each coalition C_i corresponds to a task T_i. Then, the Agent coalition generation is converted to the searching for the optimal CS^* in the coalition structure map. To search for the CS^*, some evolutionary computing algorithms were adopted, such as Genetic Algorithm [7], Ant Colony Optimization [8], and Particle Swarms Optimization [9] etc. These coalition structure based methods restrict one Agent to join in multi-coalition, which is to say by these methods, one Agent can only join in one coalition.

K. Li et al. (Eds.): LSMS 2007, LNCS 4688, pp. 128–134, 2007.

Accordingly, some scholars have put forward the research content of one Agent joining in multi-coalition. In the literature [10], Zhang proposed an algorithm to generate Agent coalition in serial manner. The residual ability of Agent in the generated coalition will be put into the process of next coalition generation. Consequently, one Agent may join in several coalitions when all coalitions are generated. This method is too ideal, because the accurate measurement and division of Agent's ability are not practical. It should be in a fuzzy manner for Agent to join in multi-coalition. An Agent can join in any coalition, but its membership degree to different coalitions may be not equal.

This paper introduces the Fuzzy Set Theory to solve the Agent coalition generation problem, defines a membership function of an agent to a coalition, and then generates the task-oriented coalition by λ matrix theory.

3 Fuzzy Set Theory

A fuzzy set A of a field U is defined as follows:

$\forall\ u \in U$, a value $\mu_A(u) \in [0, 1]$ represents the membership degree of u to A.

$$\mu_A : U \to [0, 1],$$
$$u \mapsto \mu_A(u),$$

$\mu_A(u)$ is named membership function of A.

A fuzzy relation R between set X and Y is defined as a fuzzy set R on field $X \times Y = \{(x, y) | x \in X, y \in Y\}$.

$$\mu_R : X \times Y \to [0, 1],$$
$$(x, y) \mapsto \mu_R(x, y).$$

R can be expressed by a fuzzy relation matrix $(r_{ij})_{n\times m}$. The r_{ij} represents the membership degree of x_i to y_j, ($i = 1,2,\cdots,n$; $j = 1,2,\cdots,m$).

Given a threshold $\lambda \in [0, 1]$,

$$r_{ij}^{\lambda} = \begin{cases} 1, & if\ \ r_{ij} \geq \lambda, \\ 0, & if\ \ r_{ij} < \lambda. \end{cases}$$

The Boolean matrix $(r_{ij}^{\lambda})_{n\times m}$ is called λ matrix of R, signing R_λ.

4 Agent Coalition Generation Based on Fuzzy Set Theory

4.1 Establish Several Matrixes

(1) Agent matrix

Agent set $\{A_1, A_2, \cdots, A_n\}$. Every Agent has r kinds of abilities, composing an ability vector $A_i = \langle a_i^1, a_i^2, \cdots, a_i^r \rangle$, $(i = 1,2,\cdots,n)$. Where, $a_i^k \in [0,\ 1]$ is the fuzzy evaluation to the kth ability of Agent i. Then, n Agents having r kinds of abilities can be expressed by a matrix:

$$A_{n\times r}=\begin{bmatrix} a_1^1 & a_1^2 & \cdots & a_1^r \\ a_2^1 & a_2^2 & \cdots & a_2^r \\ \vdots & & & \\ a_n^1 & a_n^2 & \cdots & a_n^r \end{bmatrix} \quad (1)$$

Definition 1. Matrix (1) is Agent matrix. It shows the similarities and differences of all Agents on ability. Where, r is a variable number. The bigger it is, the more in detail the Agent is analyzed, and the evidence to generate coalitions is more sufficient.

(2) Task matrix

Task set $\{T_1, T_2, \cdots, T_m\}$. Every task needs r kinds of abilities, composing an ability demand vector $T_j=\langle b_j^1, b_j^2, \cdots, b_j^r\rangle$, $(j=1,2,\cdots,m)$. Then, m tasks needing r kinds of abilities can be expressed by a matrix:

$$T_{m\times r}=\begin{bmatrix} b_1^1 & b_1^2 & \cdots & b_1^r \\ b_2^1 & b_2^2 & \cdots & b_2^r \\ \vdots & & & \\ b_m^1 & b_m^2 & \cdots & b_m^r \end{bmatrix} \quad (2)$$

Definition 2. Matrix (2) is task matrix. It combines all the tasks together, and shows t heir special and common demand.

(3) Agent-task fuzzy relation matrix

n Agents performing m tasks can be considered as a fuzzy relation R between Agent set and task set. Every task is a fuzzy set of the Agent field. Thereby, the fuzzy relation matrix can be established as follows:

$$R=\begin{bmatrix} \mu_{T1}(A_1) & \mu_{T2}(A_1) & \cdots & \mu_{Tm}(A_1) \\ \mu_{T1}(A_2) & \mu_{T2}(A_2) & \cdots & \mu_{Tm}(A_2) \\ \vdots & & & \\ \mu_{T1}(A_n) & \mu_{T2}(A_n) & \cdots & \mu_{Tm}(A_n) \end{bmatrix} \quad (3)$$

This matrix sets up the relationship between Agents and tasks, so as to judge the fitness degree for an Agent to perform certain task, in other words, to judge the membership degree of an Agent to a coalition. According (3), Agent coalitions for tasks can be generated.

4.2 Design the Membership Function

The value in matrix (3) is calculated by the membership function $\mu_{Tj}(A_i)$ $(i=1,2,\cdots,n;\ j=1,2,\cdots,m)$.

Definition 3.

$$A_i \otimes T_j = \langle a_i^1 b_j^1, a_i^2 b_j^2, \cdots, a_i^r b_j^r \rangle = \langle a_{i,j}^1, a_{i,j}^2, \cdots, a_{i,j}^r \rangle$$

It represents the weighed ability vector of A_i to T_j.

Definition 4. Membership function

$$\mu_{Tj}(A_i) = \frac{1}{r}\sum_{k=1}^{r} e^{-\left(a_{i,j}^k - b_j^k\right)^2 / s_k^2}$$

It represents the membership degree of A_i to T_j. Where, s_k^2 is a proper positive value. $\mu_{Ti}(A_i) \in [0, 1]$.This definition integrates the attribute of Agent A_i and the demand of task T_i.

Matrix (3) sets up the relationship between n Agents and m tasks, and provides conversion evidence from fuzzy measurement to accurate generation of coalitions.

4.3 Generate Agent Coalitions

Selecting a proper threshold $\lambda \in [0, 1]$, the matrix R can be converted to another matrix R':

$$R' = \begin{bmatrix} \mu_{T1}'(A_1) & \mu_{T2}'(A_1) & \cdots & \mu_{Tm}'(A_1) \\ \mu_{T1}'(A_2) & \mu_{T2}'(A_2) & \cdots & \mu_{Tm}'(A_2) \\ \vdots & & & \\ \mu_{T1}'(A_n) & \mu_{T2}'(A_n) & \cdots & \mu_{Tm}'(A_n) \end{bmatrix} \quad (4)$$

Where, $\mu_{Tj}'(A_i) = \begin{cases} 1, & if\ \mu_{Tj}(A_i) \geq \lambda, \\ 0, & if\ \mu_{Tj}(A_i) < \lambda. \end{cases}$

$\mu_{Tj}'(A_i) = 1$, it means that A_i is assigned to perform T_j, in other words, A_i is included in coalition C_j. The matrix R' represents the result of coalition generation.

The value of λ influences the result directly. The bigger it is, the fewer Agents will be included in coalitions. On the contrary, more Agents will join in coalitions. The selection of λ is flexible. Considering a coalition must include an Agent at least,

$\lambda \le \min_j \{\max_i \{\mu_{Tj}(A_i)\}\}$. The fuzzy clustering method is applied to covert the fuzzy relation matrix to common matrix.

5 Example

Agent set $\{A_1, A_2, \cdots, A_6\}$□Task set $\{T_1, T_2, T_3\}$, their ability demand vectors are $\langle 9,7,0,3,1 \rangle$, $\langle 0,2,8,4,6 \rangle$, and $\langle 2,3,1,5,8 \rangle$. Generate the 3 task-oriented coalitions.

Using the method of this paper, the course of coalition generation is as follows:

1) Analyze the Agents' attribute, evaluate their five kinds of ability, and establish the Agent matrix as follows:

$$D_{6\times5} = \begin{bmatrix} 0.1 & 0.7 & 0.4 & 0.0 & 0.9 \\ 0.4 & 0.8 & 0.8 & 0.2 & 0.4 \\ 0.5 & 0.5 & 0.1 & 0.7 & 0.1 \\ 0.1 & 0.5 & 0.2 & 0.7 & 0.6 \\ 0.1 & 0.4 & 0.2 & 0.3 & 0.2 \\ 0.2 & 0.1 & 0.6 & 0.8 & 0.5 \end{bmatrix}$$

2) The ability demand vectors of the 3 tasks are known, and the task matrix is as follows:

$$T_{3\times5} = \begin{bmatrix} 9 & 7 & 0 & 3 & 1 \\ 0 & 2 & 8 & 4 & 6 \\ 2 & 3 & 1 & 5 & 8 \end{bmatrix}$$

3) Calculate the weighed ability vector of every Agent with different tasks by definition 3, calculate the membership degree of every Agent to different coalitions by definition 4, and establish the Agent-task fuzzy relation matrix as follows:

$$R = \begin{bmatrix} 0.400 & 0.479 & 0.342 \\ 0.368 & 0.386 & 0.379 \\ 0.378 & 0.321 & 0.205 \\ 0.459 & 0.322 & 0.155 \\ 0.308 & 0.247 & 0.121 \\ 0.495 & 0.313 & 0.260 \end{bmatrix}$$

4) Generate the Agent coalitions. According the matrix R above, $\min_j \{\max_i \{\mu_{Tj}(A_i)\}\} = 0.379$, the threshold $\lambda \le 0.379$, set $\lambda = 0.379$ as well, then the matrix R' is as follows:

$$R' = \begin{bmatrix} 1 & 1 & 0 \\ 0 & 1 & 1 \\ 0 & 0 & 0 \\ 1 & 0 & 0 \\ 0 & 0 & 0 \\ 1 & 0 & 0 \end{bmatrix}$$

Thus, the generated coalitions are as follows:

T_1-oriented coalition: $C_1 = \{A_1, A_4, A_6\}$

T_2-oriented coalition: $C_2 = \{A_1, A_2\}$

T_3-oriented coalition: $C_3 = \{A_2\}$

As we see, the method in this paper can generate the task-oriented coalitions with fuzzy clustering strategy, so that the tasks can be accomplished in the optimal manner. At the same time, this method realizes an Agent joining in multi-coalition, for example above, A_1 joining in C_1 and C_2 simultaneously, but the membership degrees of A_1 to these two coalitions are different.

The performances comparison of the methods in [5], [10] and this paper is shown in Table 1.

Table 1. Performances comparison

	One Agent in multi-coalition	Agent's ability allocation	Fitness of agent to task	Effect of task accomplished
Method in [5]	illegal	/	unknown	in suboptimal manner
Method in [10]	legal	accurate	unknown	in suboptimal manner
Method in this paper	legal	fuzzy	fit	in optimal manner

6 Conclusions

This paper analyzes the fuzzy characteristic in Agent coalition generation, and proposes a novel method based on fuzzy set theory. Defines a membership function, by which the membership degree of an agent to a coalition is calculated as a value between 0 and 1, and then generate task-oriented coalition by λ matrix theory. Finally, this method with fuzzy clustering strategy is illustrated to be valid and superior to related methods.

Acknowledgments. This work is supported by NSFC, National Natural Science Foundation of China, No. 60474035, and Natural Science Foundation of Anhui Province, No. 070412035.

References

1. Shehory, O., Kraus, S.: Task allocation via coalition formation among autonomous agents [A]. In: Proc. of IJCAI-95 [C], pp. 655–661. Morgan Kaufmann Publishers, Los Angeles, CA, USA (1995)
2. Jiang, J.G., Xia, N., Qi, M.B., et al.: Ant colony algorithm based multi-task coalition serial generation algorithm [J]. Acta Electronica Sinica 33(12), 2178–2182 (2005)
3. Jiang, J.G., Xia, N., Yu, C.H.: The Coalition Formation Strategy Based on Capability Vector Contribution-Rate and Auction [J]. Acta Electronica Sinica 32(12A), 215–217 (2004)
4. Sandholm, T., Larson, K., Andersson, M., et al.: Anytime coalition structure generation with worst case guarantees [A]. In: Proc. of the National Conference on Artificial Intelligence [C], Madison, WI, pp. 46–53 (July 1998)
5. Sen, S., Dutta, P.: Searching for optimal coalition structures [A]. In: Proc. of the 4th ICMAS [C], pp. 287–292. IEEE Press, Boston, MA, USA (2000)
6. Hu, S.L., Shi, C.Y.: An Anytime Coalition Structure Generation Algorithm [J]. Journal of Software 12(5), 729–734 (2001)
7. Yang, J.A., Luo, Z.H.: Coalition formation mechanism in multi-agent systems based on genetic algorithms [J]. Applied Soft Computing 7(2), 561–568 (2007)
8. Xia, N., Jiang, J.G., Wei, X., et al.: Searching for agent coalition for single task using improved ant colony algorithm [J]. Computer Research and Development 42(5), 734–739 (2005)
9. Zhang, G.F., Jiang, J.G., Xia, N.: Solutions of Complicated Coalition Generation Based on Discrete Particle Swarm Optimization [J]. Acta Electronica Sinica 35(2), 323–327 (2007)
10. Zhang, G.F., Jiang, J.G., Xia, N., et al.: Particle Swarms Cooperative Optimization for Coalition Generation Problem [A]. In: Wang, T.-D., Li, X., Chen, S.-H., Wang, X., Abbass, H., Iba, H., Chen, G., Yao, X. (eds.) SEAL 2006. LNCS, vol. 4247, pp. 166–173. Springer, Heidelberg (2006)

Application of the Agamogenetic Algorithm to Solve the Traveling Salesman Problem

Yinghui Zhang, Zhiwei Wang, Qinghua Zeng, Haolei Yang, and Zhihua Wang

Chengdu Neusoft Institute of Information,
611844 Chengdu, China
{zhangyh,wangzhw,zeng-qh,yang-hl,wangzhih}@neusoft.com

Abstract. Based on the biologic agamogenesis mechanism, the agamogenetic algorithm (AGA), which is similar to classical genetic algorithms except the agamogenesis and highly converging speed, is presented to solve the optimization problems. In order to solve the traveling salesman problem (TSP) by AGA, one agamogenetic operator R-Edge and one mutation operator NI-Dot are given by introducing the conception of the relative distance between cities. The validity of the AGA to solve the traveling salesman problem is shown by simulative experiments.

1 Introduction

The genetic algorithms (GA) have been researched widely and deeply in recent years, and have been applied to solve optimization problems. Many valuable results in both theory and practical application have been obtained. However, the converging speed and the global searching capability of GA are always inconsistency in practical application when solve complex or large-scalar optimization problems, for example, the traveling salesman problem (TSP) with the number of cities over 100. Some algorithms have quicker converging speed, but they could easily result in the "premature convergence" phenomenon because of local optimization. Other algorithms have stronger global searching capability, yet converging speed could be very slow as a result of larger solution space ([2], [3]).

In the classical GA to solve larger TSP, two individuals were selected to cross at a time. Since the chromosomes are too long, vast gene orders were rearranged by the crossover operator. In the early stages of genetic process, the converging speed of GA is rapid due to low fitness of the chromosomes. However in the late stages of genetic process, the converging speed begins to slow down and it is difficult to seek the optimal individual because large numbers of the chromosomes in the population have high fitness.

Agamogenesis is a genetic mode of the low-grade biology, with short genetic cycle in each generation. Its genetic process is simpler than that of gamogenesis. During the genetic process, the vice of genes should be continually self-modified to fit the changes of environment and to raise the fitness.

Based on agamogenetic mechanism, the agamogenetic algorithm (AGA) is developed to solve TSP in this paper. AGA is similar to the classics GA except that:

K. Li et al. (Eds.): LSMS 2007, LNCS 4688, pp. 135 – 143, 2007.

1) The select-operator selects one individual at a time instead of two individuals; 2) The agamogenetic operator replaces the crossover operator.

Hereinafter, the basic structure of AGA is given. In order to solve TSP, in section 3, the concept of relative distance between cities is introduced to picture ruleless distribution of cities; an agamogenetic operator R-Edge and a mutation operator HI-Dot are presented based on the concept. Several simulative experimenter results are given in section 4. The simulative experimenter shows that AGA can raise the converging speed and can void the premature convergence.

2 Basic Structure of Agamogenetic Algorithm

The basic structure of the agamogenetic algorithm is displayed as follows:

```
Input:     N = Population size;
           MaxGen = Maximum number of generation;
           pc = crossover probability
           pm = mutation probability
Output:    The best chromosome;    // Near-optimal
Begin
           g = 0;  // g is current number of generation
           Initiating population P(g);
           Evaluating P(g);
           While ( g<MaxGen )
             Begin
               Agamogenetic operation for P(g);
               Mutation operation for P(g);
               Evaluating P(g);
               g = g+1;
             End
           Output the result;
End.
```

AGA is very similar to the classics GA. The major differences are:

1) To select one individual to generate one new individual of next generation instead of two individuals at a time;

2) To use the agamogenetic operator to self-improve the fitness of the chromosome instead of the crossover operator to cross the genome.

3 Agamogenetic Algorithm to Solve TSP

The traveling salesman problem (TSP) is a classical difficult problem with NP-complete complexity. For conveniently describe AGA to solve TSP, some symbols are introduced as follows.

$X = \{1, 2, ..., n\}$ denotes a set of codes of n cities in an area;

$e(v_i, v_j)$ denotes the edge or direct line between city v_i and city v_j, $v_i, v_j \in X$;

$d(v_i, v_j)$ denotes the length of $e(v_i, v_j)$ or the distance (Euclidean) between city v_i and city v_j.

So, TSP can be described as seeking a permutation $P(v_1v_2...v_n)$, $v_i \in X$, $i = (1, 2, ..., n)$, to minimize

$$L(v_1 v_2 \cdots v_n) = d(v_1, v_n) + \sum_{i=1}^{n-1} d(v_i, v_{i+1}) \tag{1}$$

In order to simplify the depiction later, for the permutation $P(v_1v_2...v_n)$, e_i (i=1, 2, ..., n-1) denotes the edge $e(v_i, v_{i+1})$, e_n denotes the edge $e(v_n, v_1)$.

3.1 Relative Distance in TSP

Generally, the distributing of cities is different in thousands ways in TSP. In order to depict the relative distributing of distance between cities, the conception of relative distance between cities is introduced as follows

Define: Set $v_i \in X$, $E(v_i)$ as the expected distance of all cities v_j ($v_j \in X, j \neq i$) to v_i,

$$E(v_i) = \frac{1}{n-1} \sum_{j=1, j \neq i}^{n} d(v_i, v_j) \tag{2}$$

then,

$$r(v_i, v_j) = \frac{d(v_i, v_j)}{E(v_i)} \tag{3}$$

is called the relative distance of v_j to v_i. Specially, r_i (i = 1, 2, ..., n-1) denotes $r(v_i, v_j)$, and r_n denotes $r(v_n, v_1)$.

The properties as follows are distinct.

1. $r(v_i, v_i) = 0$;
2. $r(v_i, v_j) \neq r(v_j, v_i)$;
3. For any two edges $e(v_i, v_j)$ and $e(v_s, v_t)$, the inequation $r(v_i, v_j)>r(v_s, v_t)$ can not always be deduced by $d(v_i, v_j)>d(v_s, v_t)$, in such a manner, $r(v_i, v_j)<r(v_s, v_t)$ can not always be deduced by $d(v_i, v_j)<d(v_s, v_t)$ as well;
4. $\sum_{j=1}^{n} r(v_i, v_j) = n-1$.

Example: In the TSP shown in figure 1

$$\text{Matrix of distance} = \begin{bmatrix} 0 & 3 & 5.5 & 3.6 \\ 3 & 0 & 2.5 & 2 \\ 5.5 & 2.5 & 0 & 3.2 \\ 3.6 & 2 & 3.2 & 0 \end{bmatrix}, \tag{4}$$

$$\text{Matrix of relative distance} = \begin{bmatrix} 0 & 0.74 & 1.36 & 0.89 \\ 1.2 & 0 & 1 & 0.8 \\ 1.47 & 0.67 & 0 & 0.86 \\ 1.23 & 0.68 & 1.09 & 0 \end{bmatrix}, \tag{5}$$

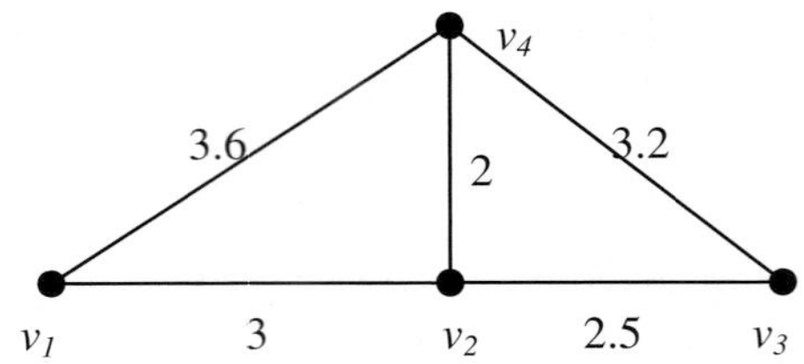

Fig. 1. Example of TSP with 4 cities

3.2 Agamogenetic Operator R-Edge

The basic function of R-Edge (Rupturing Edge) operator is rupturing the edge which has a large relative distance, taking the edge to be replaced by a sub-path whose two end-cities are near the two cities of the edge. The operator is stated in detail as follows.

Set the permutation $P(v_1 v_2 ... v_n)$ be a chromosome selected to be operated.

Step1: Select an relatively longer edge e_i according to probability p_i.

$$p_i = \frac{r_i}{\sum_{k=1}^{n} r_k} \tag{6}$$

Step2: Select a city v_s relatively near v_i and a city v_t relatively near v_{i+1} according to probability p_{is} and $p_{i+1,t}$, respectively.

$$p_{is} = \begin{cases} 0, & r(v_i, v_s) \geq 1; \\ \dfrac{r(v_i, v_s)^{-1}}{\sum\limits_{k=1, k \neq i, r(v_i, v_k) < 1}^{n} r(v_i, v_k)^{-1}}, & r(v_i, v_s) < 1; \end{cases}$$
$$p_{i+1,t} = \begin{cases} 0, & r(v_{i+1}, v_t) \geq 1; \\ \dfrac{r(v_{i+1}, v_t)^{-1}}{\sum\limits_{k=1, k \neq i+1, r(v_{i+1}, v_k) < 1}^{n} r(v_{i+1}, v_k)^{-1}}, & r(v_{i+1}, v_t) < 1; \end{cases} \tag{7}$$

Step3: Insert the path $(v_s...v_t)$ into between v_i and v_{i+1}. The operate process is as follows (generally to suppos $i < \min(s, t)$):

1. If $s<t$, insert the path $(v_s \cdots v_t)$ into the edge between v_i and v_{i+1} directly, i.e. translate (8) into (9).

$$(v_1 \cdots \underline{\underline{v_i v_{i+1}}} \cdots v_{s-1} \underline{v_s v_{s+1} \cdots v_{t-1} v_t} v_{t+1} \cdots v_n) ; \tag{8}$$

$$(v_1 \cdots \underline{\underline{v_i}\, v_s v_{s+1} \cdots v_{t-1} v_t \underline{v_{i+1}}} \cdots v_{s-1} v_{t+1} \cdots v_n) . \tag{9}$$

2. If $t<s$: reverse the path $(v_s \cdots v_t)$ into $(v_t \cdots v_s)$, then insert $(v_t \cdots v_s)$ into the edge between v_i and v_{i+1}, i.e. translate (10) into (11).

$$(v_1 \cdots \underline{\underline{v_i v_{i+1}}} \cdots v_{t-1} \underline{v_t v_{t+1} \cdots v_{s-1} v_s} v_{s+1} \cdots v_n) ; \tag{10}$$

$$(v_1 \cdots \underline{\underline{v_i}\, v_s v_{s-1} \cdots v_{t+1} v_t \underline{v_{i+1}}} \cdots v_{t-1} v_{s+1} \cdots v_n) . \tag{11}$$

3.3 Mutation Operator: NI-Dot

The basic function of NI-Dot (Nearby-inserting Dot) operator is to randomly choose one city, insert it into a nearby edge. The operator is stated in detail as follows.

Set the permutation $P(v_1 v_2 ... v_n)$ be a chromosome selected to be mutated.

Step1: Randomly select a city v_i a;

Step2: Select a city v_t relatively near v_i according to the probability p_{it},

$$p_{it} = \begin{cases} 0, & r(v_i, v_t) \geq 1; \\ \dfrac{r(v_i, v_t)^{-1}}{\sum\limits_{k=1, k \neq i, r(v_i, v_k) < 1}^{n} r(v_i, v_k)^{-1}}, & r(v_i, v_t) < 1; \end{cases} \tag{12}$$

Step3: Insert v_i into the edge e_t. i.e. translate (14) into (15).

$$(v_1 \cdots v_{i-1} \underline{\underline{v_i}} v_{i+1} \cdots \underline{v_t v_{t+1}} \cdots v_n) ; \tag{13}$$

$$(v_1 \cdots v_{i-1} v_{i+1} \cdots \underline{v_t \underline{\underline{v_i}} v_{t+1}} \cdots v_n) . \tag{14}$$

4 Simulative Experimenters

In order to validate the validity of AGA, the basic structure of AGA in section 2 is implemented by using the genetic operators in section 3. Table 1 and Figure 2 respectively show the optimal results and the converging process of solving several TSPs from ref. [4]. Figure 3~5 respectively show their optimal path.

Table 1. Optimal results of several TSPs by AGA. Column 3 lists the optimal values known in ref. [4]. Column 4 and 5 respectively list optimal values and the generations where optimal values were found by AGA. Column 6 and 7 respectively list optimal values and the generations by the classical GA.

Name of TSP	Num. of Cities	Opt. val. given	AGA		Classical GA	
			Optimal	Gens	Optimal	Gens
KroA100	100	21,282	21,685	9,450	48,412	19,854
KroA150	150	26,524	27,018	17,715	103,180	19,765
KroA200	200	29,368	30,441	18,385	135,508	18,753

The strategies and methods to solve these TSPs in Table 1 by using AGA and the classical GA as shown as follows:

1. Elitist selection genetic strategy: For each generation, if the maximum fitness in current generation less than the maximum fitness in earlier generations, the chromosome which has having minimum fitness in current generation should be replaced by the chromosome which has having maximum fitness in earlier generations.
2. The individual fitness function equal to the reciprocal of path length, i.e.

$$f(v_1 v_2 \cdots v_n) = \frac{1}{L(v_1 v_2 \cdots v_n)}. \tag{15}$$

3. The variable r_k in the probability formula 6, 7, 12 is replaced by r_k^2.
4. The crossover probability $p_c = 0.85$, the mutation probability $p_m = 0.2$. The size of a population = 4× the number of cities, the maximum generation is 20,000.
5. In AGA, an individual is selected at a time, and then it is operated by the R-Edge operator to produce a new individual of next generation. While in the classical GA, two individuals are selected at a time, and then they are operated by the order crossover operator ([1], [3]) to produce two new individuals.
6. In both of AGA and the classical GA, the mutation operator HI-Dot is employed to mutate the chromosomes. Moreover, the shorter one of the two paths that are gained by inserting v_i into e_t and inserting v_i into $e_{t\text{-}1}$ is selected for next generation population.

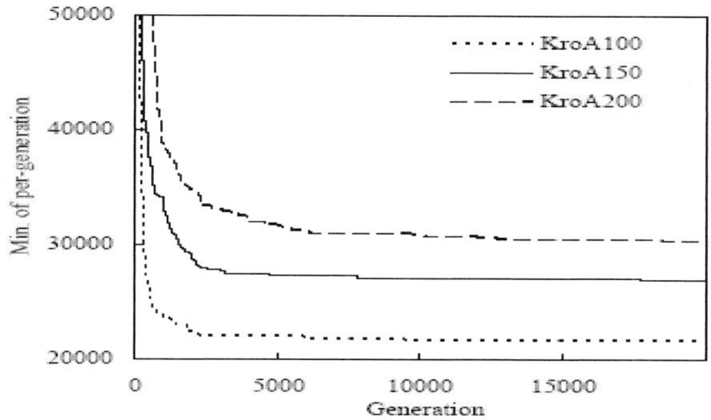

Fig. 2. Converging process of AGA for several TSPs. Minimum of Initial generation respectively is: KroA100= 145,439; KroA1500= 225,615; KroA200= 302,888. They all can rapidly converge down to below of 50,000.

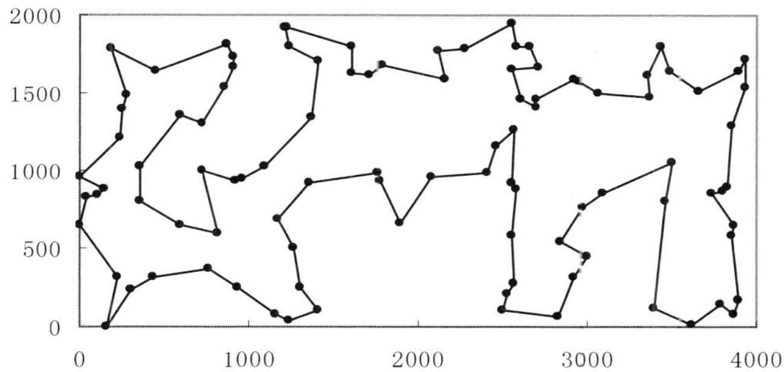

Fig. 3. Approximate optimal path of KroA100 100 cities TSP. The length of this path is 21,685.

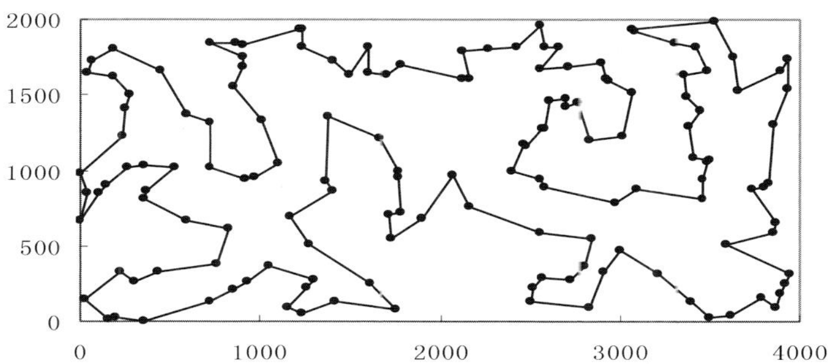

Fig. 4. Approximate optimal path of KroA150 150 cities TSP. The length of this path is 27,018.

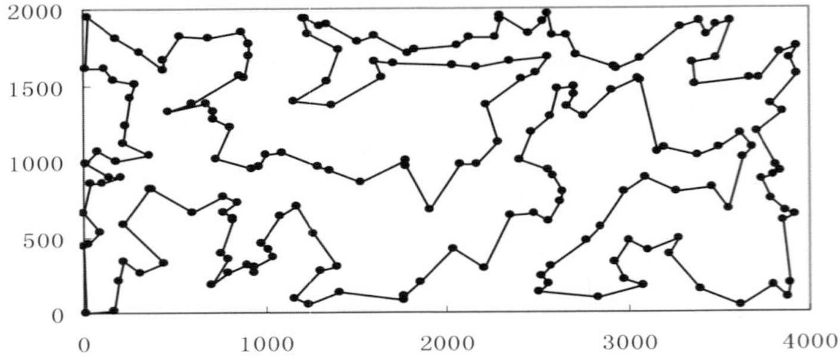

Fig. 5. Approximate optimal path of KroA200 200 cities TSP. The length of this path is 30,441.

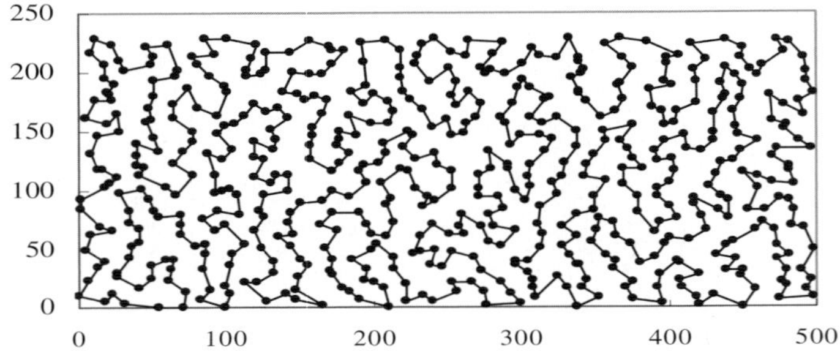

Fig. 6. Approximate optimal path of rat575 575 cities TSP. The length of this path is 6,870.

5 Conclusions

Simulative Experimenters show that the AGA can be used to solve large-scale TSPs. Although the number of generations of convergence is huge, the total time consumed by using the AGA is less than classical gamogenic algorithms because of the genetic operations are simple.

Figure 6 shows the approximate optimal path of rat575 575 cities TSP by AGA. The optimal value known by literatures is 6773 [4].

This paper aims at discussing the applications of AGA. There are many theoretical problems requiring further study.

The AGA can also be used to solve other optimization problems. The key is to adopt appropriate coding methods, and to construct corresponding genetic operators.

References

1. XiaopinWan, L.C.: Genetic Algorithms: Theory, Application and Software Implement. Xi an Jiaotong University Press, Xi An (2002)
2. Zhang, Y., Wang, X., Liu, J., Li, H.: An Effective Adaptive Model of Probabilities in Genetic Algorithms. Journal of Tsinghua University (Science and Technology) S2, 110–113 (1998)

3. Zhang, Y., Wang, Z., Zeng, Q.: The Genetic Algorithm Based on Mixed Genetic Operators. Computer Science 4, 222–224 (2007)
4. TSPLIB[EB/OL] (2003), www.iwr.uni-heidelberg.de/groups/comopt/software/TSPL IB95

An Artificial Life Computational Model for the Dynamics of Agro-ecosystem

Li Sa[1,3], Fanlun Xiong[2], and Yongsheng Ding[1]

[1] College of Information Sciences and Technology, Donghua University, 201620 Shanghai, China
sali@ustc.edu, ysding@dhu.edu.cn
[2] Institute of Intelligent Machines, CAS, 230031 Hefei, China
flxiong@tom.com
[3] College of Computer Engineering, Jimei University, 361021 Xiamen, China

Abstract. Artificial life computational model can be used to simulate the dynamic behavior of certain agro-ecosystem. In this paper, we present an agent-based artificial agro-ecosystem model to simulate the interaction between pest movements and trap crop physical design. The result of simulation shows that artificial agro-ecosystem model has a broad application prospect in the field of agro-ecosystem management.

1 Introduction

Artificial ecosystems have been defined in the literature [1] as a system in which a self contained environment is populated by independent entities (agents) representing some level of organization. They are a class of the artificial life model, which is an abstraction of natural living systems, preserving some characteristics of those systems that are the subject of study. The agents in an ecosystem interact locally with each other and with their environments. It has been argued that traditional simulation models are inadequate to implement artificial ecosystems [2]. They have always some sort of embedded constraints due to the model itself, forcing the system to behave according to the strict rules the model is based upon. An alternative is the use of agents. The rationale behind the modeling through agents [1] is that each of them implements the level of abstraction necessary to represent an entity while leaving aside aspects considered less important. Agent-based models consist of a collection of individuals, each of which possesses the information needed to behave autonomously, according to a set of rules, and to interact locally with neighboring agents. Interactions are simple and local, yet can lead to complex patterns at the cropland or global scale [3-5].

Modeling provides an alternative method for studying the interplay between the physical design of trap crop systems and pest population processes. Modeling can be used to reveal patterns produced by the different population dynamics, to provide insight about the importance of the different processes, and to generate testable predictions about pest distribution in different situations. A field-scale testing of trap crop models had been used to limit field experiments to the trap crop designs [6].

K. Li et al. (Eds.): LSMS 2007, LNCS 4688, pp. 144 – 151, 2007.

In this paper, we use a modeling approach to get more insight in the mechanisms underlying the efficacy of diversification strategies and to give an indication of the reliability of agro-ecosystem diversification as a pest management strategy. Agent Based Modeling as the technology to implement the model and Swarm as the tool [7].

2 Model Description

An artificial agro-ecosystem simulation models consist of three components: a cropland, pests, and a set of behavioral rules. The cropland is typically a two-dimensional grid of cells containing a heterogeneous distribution of crop plant or non-crop plant of interest to the pests. The pests are rational actors that move about on the cropland, interacting with the cropland and other pests. Together, the pests comprise the population to be studied.

2.1 The Cropland

The cropland is a two-dimensional grid with dimensions x × y cropland is distinguished by its (x, y) position with (x, y) as illustrated in Fig. 1. The Cropland class encapsulates all states and attributes for an individual cropland grid. Included in this class are member functions for modifying the cell states; also included are functions for retrieving the values of the states and attributes.

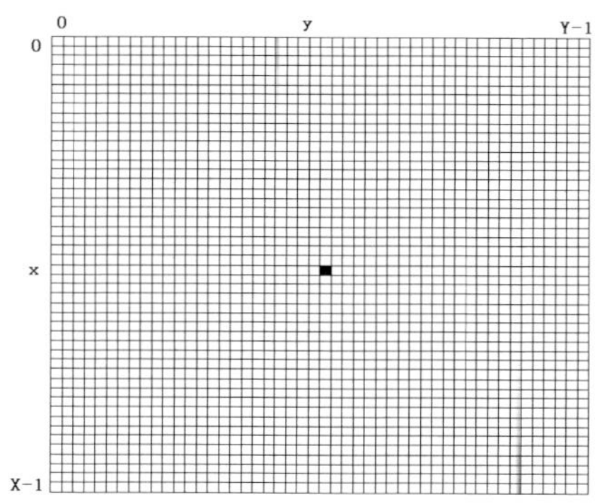

Fig. 1. Two-dimensional grid with dimensions x×y

2.2 The Pests

The pests are the inhabitants of the cropland. They harvest their "environment" for eats to grow, and die off when their lives go to the end. All pest interactions in the model are purely local. The pests have no global view of the model and cannot interact with any pest or cell on the cropland outside their local neighborhood. They have a number of attributes (with their variable names in brackets):

The efficient accumulated temperature level (levelOfEAT), the current energy level of the pest;

The pest's life (pestLife), measured in number of running times;

Current position on the cropland (x and y), this information is not available to the pest itself, and is used only by the modeler;

The birthrate of pest (pestBirthRate), set at the start of simulation;

The density of pest (pestDensity), set at the start of simulation;

The pest's colour (colour), to be used as a diagnostic tool. This will give the user a view of the state of the population as the model runs.

The field of view (FOV) attribute quantifies how far a pest can move from the current (x, y) location as illustrated in Fig. 2.

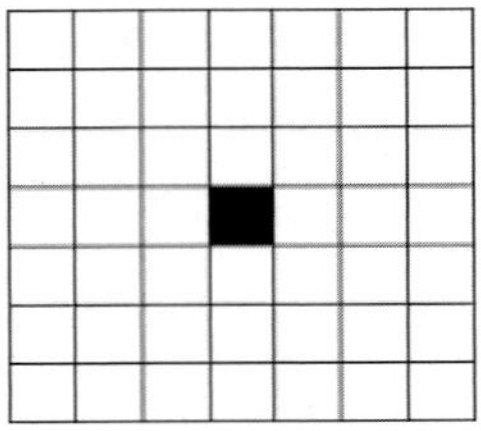

Fig. 2. The FOV for a pest if m = 3

2.3 Movement Rule

At each time step each pest is allowed to move. Population number in trap crop area depends on the movement speed, and the movement speed depends on the pest motility (m). Initial distribution of pests were placed at random at the beginning of simulation. Pest will make a short movement, no further than the neighboring plant, or initiate an emigration from the currently occupied cell. An emigration is initiated when a randomly generated number between 0 and 1 is below the emigration probability of the current position, which is initiated according to the type and state of the plant. The agent movement rule as follows.

- Look at all grids within the FOV.
- Select the closed grid with resource and move to that grid.
- If there is preferable resource in the closed grid, move to it and stay there for a long time.

3 Model Implementation

The entire model is implemented using Swarm, and programming in Java. The class capability available in Java naturally lends itself to implementing the artificial agro-ecosystem model. Consequently, we utilize separate class representations for cropland cells, for the entire cropland, for agents, and for maintaining a list of future simulated events.

3.1 The Pests

The PestAgent class encapsulates all states and attributes for a pest agent. Included in this class are member functions that implement the pest movement, birth, and death events. In addition to the states and attributes. The basic class structure for a pest agent follows.

```
class PestAgent
{public:
            void move();
            void death();
            void birth();
private:
            int FieldOfView;
            int AgentReproEnergy;
            int AgentMetab;
            int AgentAge;
            int AgentMaxLife;
            int AgentReproAge;}
```

3.2 Implementing the List of Events

In the implementation of a next-event simulation model, a suitable data structure is chosen to store the list of future events in simulated time. The simulation engine drives the time evolution of the model by appropriately accessing events in this list. The basic class structure of the event list follows.

```
public void PestStep()
{int pestEnergy = pestInitEnergy;
++pestAge;
          if (pestEnergy <= 0 || pestAge >= pestMaxLife)
          {
pestSpace.putObject$atX$Y(null, x, y);
           modelSwarm.pestDeath(this);
          }
           if (foodSpace.getValueAtX$Y(x, y) ! = 0)
           {
            pestEnergy = pestEnergy + foodCalorie(x, y);
             if (pestEnergy >= pestReproEnergy &&
pestAge >=
pestReproAge && Globals.env.uniformDblRand.
getDoubleWithMin$withMax(0.0, 1.0) <= pestDensity)
           {
            modelSwarm.addpests();
            pestEnergy /= 2;
           }
            }
            pestEnergy = pestEnergy - pestMetab;}
```

The ordering of cropland related events is shown below in Table. 1.

Table 1. The ordering and Description of the Model Schedules

Function call	Description	Order
ObserverSwarm._update_	Draws cropland and pests	1
ObserverSwarm._updateGraphs	Updates all graphs	2
ProbeDisplayManage.update	Updates all Object Probes	3
Cropspace.updateLattice	Performs Cellular Automata	4
ModelSwarm.death	Removes dead pest	5
ModelSwarm.born	Activates newborn pests	6
AgentStep calls	Runs all Pest routine	7

The model is initialized by placing on cropland a random distribution of agents, each with characteristic attributes and initial states. Consistent with the behavioral rules, as time evolves agents move about the cropland interacting with one another and with the cropland. Statistics are gathered during the model's lifetime to provide data for analyzing the resulting macro-scale behavior of the agents.

4 A Case Study

Simulating experiment had been done on this model to understand the mechanisms underlying pest population response to diversified agro-ecosystems, with a strong emphasis on the behavioral ecology of pests. A case study here is made to model the interaction between pest movements and trap crop physical design in a situation where the pest moves by a random walk with spatially variable mobility.

(a) (b) (c)

Fig. 3. Different designing of trap crop systems in field. All the fields are similar in the total area of the field, the total crop area (white), and the total trap crop area (green). The crop patch dimensions are (a) 50 x 50m, (b) 100 x 25m, (c) 200 x 12.5m.

The configuration of the environment may have an impact on the speed of the redistribution process, and thus it may also influence the efficiency of trap crops. Here, the aim was to define how crop patch size and shape influence pest redistribution from crop to trap crop. The pest moves by a random walk with spatially variable mobility. Situations were considered where fields are composed of rectangular crop patches surrounded by trap crops (Fig. 3), and where the initial density of pest individuals is the same in all parts of the field.

Parameters in the simulation model as follow: Environment size (Size), Pest type (Species), Mix plant probability (MixPro), Pest density size (PestDen), Max simulate time(MaxTime) (Table. 2). At each time step each pest is allowed to move. The amount of pest population in trap crop area depends on the motility of pests (MotilityOfPest m^2/day), and the motility of pests depends on the FOV of each type of pest as shown in Table. 3.

Table 2. Parameters in the simulation model

Size	Species	MixPro	PestDen	MaxTime
250×130	5	5:8	0.02	30

Table 3. The motility of pest and coloration

Species type	Type1	Type 2	Type 3	Type 4	Type 5
MotilityOfPest	1	2	3	4	5
Coloration	Red	Blue	Yellow	Orange	Purple

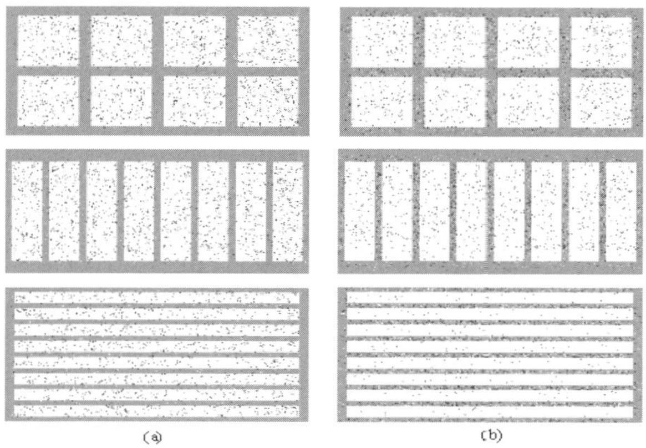

Fig. 4. System state graphs showing pests aggregation in the trap crop (a) time (days) = 1, (b) time (days) = 30

Model snapshot is shown in Fig. 4. Initial distribution of pests was placed at random at the beginning of simulation (Fig. 4a). The most of pests aggregated in the preferred trap crop after 30 days (Fig. 4b).

5 Discussion

Our simulation results show that species-specific mobility is an important factor determining the control efficacy of diversified agro-ecosystems Fast moving species aggregate more rapidly in the trap crop than slow moving species (Fig. 5). These results demonstrate that the configuration of the environment can have a dramatic impact on the pests' redistribution process, and thus on the efficiency of trap crops. These results suggest that maximizing the perimeter to area ratio of crop patches may be useful if a rapid reduction in pest population is important. Patch perimeter to area ratio can be increased by decreasing patch size or by increasing patch perimeter (i.e. by using elongated instead of square patches) (Fig. 3c).

The maximum rates of decay in pest population in the crop for the studied motilities are presented in Fig. 6. For example, consider that the aspired reduction in pest population in the crop is 50% in 15 days. If the field is divided into 200×12.5m crop patches, the required reduction in pest density can be achieved using a trap crop only if the motility of pest in the crop is at least 4m^2/day (Fig. 6).

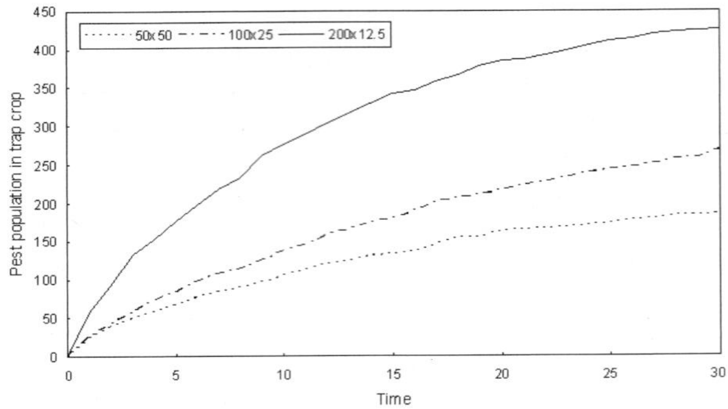

Fig. 5. The changing of the pest population emigrating to the trap crop in three different crop patch dimensions. Movement speed was varied by setting the motility (m) per time step.

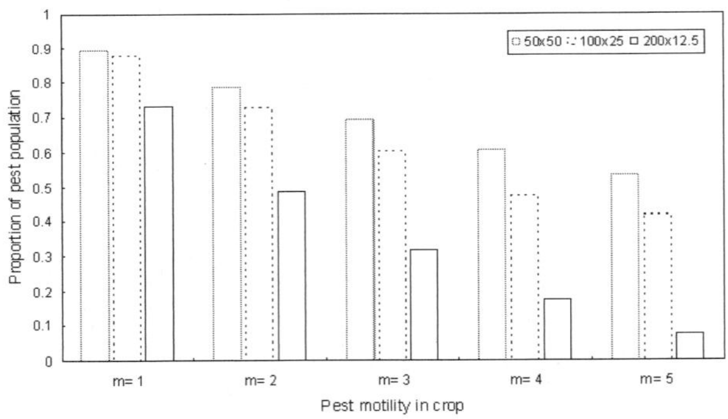

Fig. 6. The proportion of the pest population in the crop in three different crop patches at time (day) 15. Pest motility (m) is 1m^2/day, 2m^2/day, 3m^2/day, 4m^2/day and 5m^2/day respectively.

For a pest whose motility in the crop is smaller than this, no trap crop can reduce the pest population efficiently enough. If the field is divided into 200×12.5m crop patches, a similar reduction in pest population can be achieved only if the motility of pest in the crop is greater than 2m^2/day. This suggests that, when parameterized with pest motility observed on a crop species, the analytical model could be used to predict the maximum speed of pest redistribution from the crop to the trap crop. Trap crops with such wide variation in pest motilities may have the potential to reduce the pest density in the crop very quickly in others. Thus finding a trap crop that the pest distinctly prefers over the crop appears to be crucial for developing efficient trap crop systems. Designing field layouts to increase the perimeter to area ratio of crop patches may be beneficial.

6 Conclusions

There are very few studies of population dynamics that systematically vary both behavioral ecological parameters of pest and the composition and spatial arrangement of crop as we have done here. Swarm platform offers distinct advantages for collaborative modeling. The model discussed above has shown that it can be used to simulate the dynamic behavior of certain agro-ecosystem. Object-oriented, agent-based modeling is a powerful and flexible way to bridge the gap between individual behavioral ecology and population dynamics. Thus, these models are perfectly suited to use information from small-scale experiments to identify relevant mechanisms of population dynamics at a field-scale and allow extrapolation to novel conditions, for example to predict the response of pest to agro-ecosystem diversification.

References

1. Olson, R.L., Sequeira, R.A.: Emergent Computation and the Modeling and Management of Ecological Systems. Computers and Electronics in Agriculture, 183–209 (1995)
2. Ladanyi, M., Horvath, L., Gael, L.M.: Hufnagel. An agro-ecological simulation model system. Applied ecology and environmental research 1(1-2), 47–74 (2003)
3. Potting, R.P.J., Perry, J.N., Powell, W.: Pest behavioural ecology and other factors affecting the control efficacy of agro-ecosystem diversification strategies. Ecological Modeling 182, 199–216 (2005)
4. Berec, L.: Techniques of spatially explicit individual-based models: construction, simulation and mean-field analysis. Ecol. Model 150, 55–81 (2002)
5. Riley, J.R., Drake, V.A.: Technologies for pest movement and migration research. Comput. Electron. Agric. 35, 57–62 (2002)
6. Hannunen, S.: Modeling the interplay between pest movement and the physical design of trap crop systems. Agricultural and Forest Entomology 7, 11–20 (2005)
7. SDG.: Swarm Development Group. Swarm [EB/OL] (September 5, 2002), available at: www.swarm.org

A New Image Auto-Segmentation Algorithm Based on PCNN

Zhihong Zhang [1], Guangsheng Ma [1], and Zhijiang Zhao [2]

[1] Dept. of Computer Science and Engineering, Harbin Engineering University, 150001 Harbin, China
zzh_hrb@163.com

[2] Dept. of Information and Communications Engineering, Harbin Engineering University, 150001 Harbin, China
zzjrjtech@163.com

Abstract. Pulse Coupled Neural Networks (PCNN) is applied to image segmentation efficiently. Although the segmentation result with classical PCNN depends on the suitable concerned parameters, many experiments have shown that the segmentation result changed periodically with the calculation cyclic iteration times, N, after other parameters had been set. Therefore, how to decide the best iteration times N, is the key of applying PCNN to automated image segmentation. This paper brought forward a new edge-statistic algorithm based on calculation of connected regions, in order to automatically get the optimized value of N. An Edge-pixel Criterion was raised, and with it the algorithm calculated the valid edge pixels during the iteration process, and it meant that the maximum edge pixels were accordant with the best iteration times N, thereby the best segmentation result was achieved. The experiments show that the improved PCNN algorithm can promote the segmentation ability and has much better sensitivity than those methods based on image entropy or edge operator, and also has much stronger robustness of image noisy.

1 Introduction

The classification attribute of Artificial Neural Networks is the basis to implement image edge detection and region segmentation. And PCNN, achieved by the modification of Eckhorn's neuron model, is a kind of artificial neuron model, which is gained by observation on the cat visual cortex neurons and simulation of the neurons' activity. Since this comes directly from the study about visual characteristics of mammals, in terms of the physiological function, it is much more efficient to be applied to process, especially to segment and smooth all kinds of images [1-8]. PCNN is a single-layer model neural networks and its structure is superiority in parallel processing information, so it is perfectly suitable for real-time image process. The document [1] proposed the conditions for the perfect segmentation of partial overlapped image of PCNN. The document [2] accomplished the analysis of histograms in image segmentation and smoothing with PCNN, and discussed the influence of the inhibition item, and pointed out that finding out a suitable adjustment principle of segmentation

K. Li et al. (Eds.): LSMS 2007, LNCS 4688, pp. 152 – 161, 2007.

was vital to PCNN segmentation processing. The document [3] recalled the present situation of PCNN model development and its application. The document [4] analyzed the characteristics between PCNN and wavelet transform.

In short, compared with the previous classical Artificial Neural Networks model, PCNN can be applied to image segmentation without the pre-practice process. However, it needs to properly set the various threshold parameters, and the attenuation time constancy, and the weighted factor, and the connection coefficient, etc. of its mathematical model, and then can achieve the optimum segmentation. The present theories have a hard time to explain the relationship between the PCNN mathematical model parameters and the effect of image segmentation. Its theory discussions and application study is just in progress. In general, to achieve better segmentation effect need to do many experiments to choose those parameters. In condition of the certain parameters, the iteration time directly affects the segmentation effect, while the judgment needs the observation and analysis of human eyes, thus it causes the manual intervention.

To automatically perform the process, this essay proposes a kind of evaluation algorithm based on the statistic of image connected region and solves the problems of setting the iteration times when automatically segmenting the images of common scene, blood cells, Lena. It provides an evaluation principle, in order to make the segmentation free from the manual intervention and to automatically get the results accordant with the human subjective feelings. Here we use the connected region calculation as the rule to decide the image edges. The experiments prove that the algorithm has much better sensitivity, lower edge mismatch ratio and stronger anti-noisy capability.

2 The Principle of PCNN Image Segmentation

One neuron mathematical model of the PCNN achieved by the alteration of Eckhorn neuron model is shown in Figure1[7]. In it, F is the feeding input $F_{ij}[n]$ of the neuron (i, j) for the n time, $S_{ij}[n]$ is the input stimulus signal (here is the gray scale of the image pixel (i, j) in the pixel matrix), β is the connected coefficient, $L_{ij}[n]$ is the link item. $\theta_{ij}[n]$ and $Y_{ij}[n]$ are respectively the dynamic threshold and the pulse output of PCNN. And $U_{ij}[n]$ is the internal active item. Their relationships are defined as follows:

$$F_{ij}(n) = e^{-\alpha_F} F_{ij}(n-1) + S_{ij} + V_F \sum_{k,l} M_{ijkl} Y_{kl}(n-1) \tag{1}$$

$$L_{ij}(n) = e^{-\alpha_L} L_{ij}(n-1) + V_L \sum_{k,l} W_{ijkl} Y_{kl}(n-1) \tag{2}$$

$$U_{ij}(n) = F_{ij}(n)(1 + \beta L_{ij}(n)) \tag{3}$$

$$\theta_{ij}(n) = e^{-\alpha_\theta} \theta_{ij}(n-1) + V_\theta Y_{ij}(n-1) \tag{4}$$

$$Y_{ij}(n) = \begin{cases} 1, U_{ij}(n) > \theta_{ij}(n) \\ 0, U_{ij}(n) \le \theta_{ij}(n) \end{cases} \tag{5}$$

Here, Mijkl of the internal connected matrix M and Wijkl of W (generally M = W) are separately the weighted coefficient of $Y_{kl}[n]$ in $F_{ij}[n]$ and $L_{ij}[n]$. α_F, α_L and α_θ are respectively the attenuation time constancy of $F_{ij}[n]$, $L_{ij}[n]$ and $\theta_{ij}[n]$, and V_F, V_L and V_θ are respectively the electric potential. Shown in Figure1, PCNN neuron first receives the feedback input $\theta_{ij}[n]$ and link input $L_{ij}[n]$, and then forms the internal active item $U_{ij}[n]$ in the neuron activity system. When $U_{ij}[n]$ is bigger than the dynamic threshold $\theta_{ij}[n]$, PCNN outputs the pulse sequence $Y_{ij}[n]$ in time order.

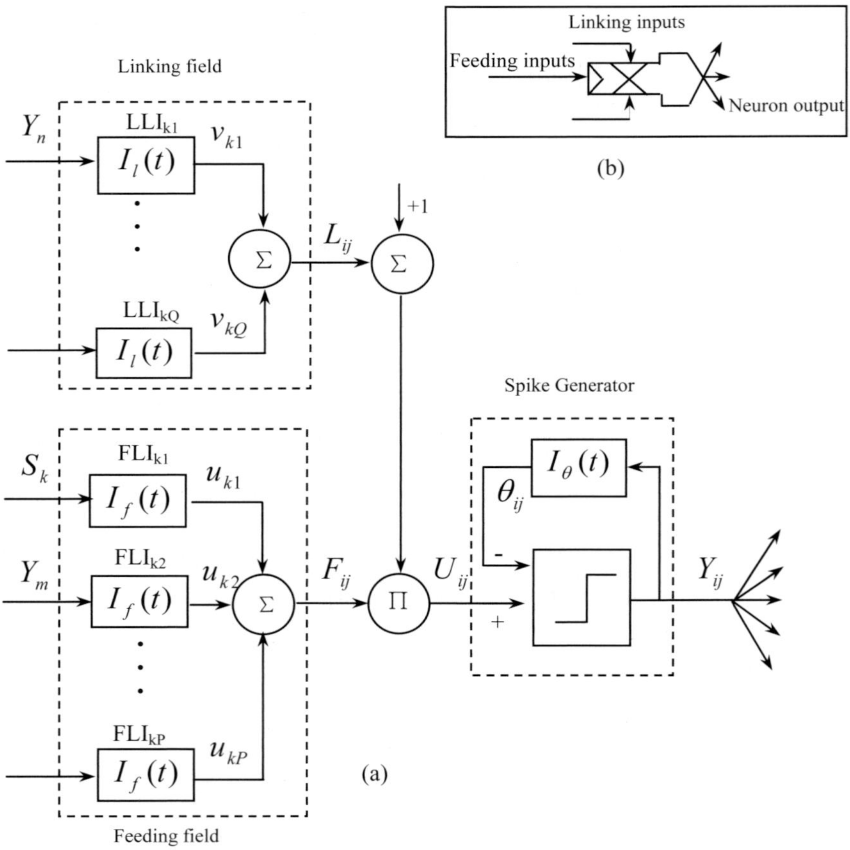

Fig. 1. Neuron model of PCNN

(a) the neuron's mathematics model (b) the action representation of the neuron

Each pixel in an image is considered as a neuron. The gray-scale of the pixel P_{ij} is put into the network as the external stimulus S_{ij} of network neuron N_{ij}. Suppose the original status values of all neuron are zero. During the first iteration, the internal activation U_{ij} of a neuron equals the external stimulus S_{ij}. The threshold value θ_{ij} of every neuron begins to attenuate from the original value $\theta_{ij}(0)$. When the threshold value θ_{ij} of some neuron is smaller than or equals the corresponding internal activity U_{ij}, the neuron fires, that is, it outputs a pulse, Y_{ij}=1. At the same time, the threshold value θ_{ij} of this neuron increases sharp. When θ_{ij} is bigger than U_{ij} again, the pulse output stops. Then θ_{ij} decays along with the time exponent. When the threshold value is smaller than or equals to the corresponding internal activation U_{ij}, the neuron fires again. So this course repeats periodically and the neuron produces a pulse signal sequence. During the production of pulse, the fired neurons will cause their surrounding neurons to fire by the interaction of the neurons, and they will cause the around neurons to fire in the same way, thus it can produce a pulse in the activated domain to spread outside. In the image, the domain with similar gray scale are usually corresponding with the neuron groups with similar threshold values. In the neuron group, any neuron's fire will fire the whole group. So we can quickly carry out the image segmentation by use of such synchronous characteristic.

3 The Edge Statistic Algorithm Based on the Calculation of the Connected Region

The result of image segmentation with PCNN is a binary image, which includes not only the image information but also a great deal of noisy. In order to get the edge statistic result satisfied with subjective evaluation, it is important to discriminate between the real edge and image noisy as much as possible. Hence, based on the calculation of connected regions, we propose the following statistic criterion about edge-pixels.

Edge-pixel Criterion: it must satisfy the following conditions to determine an image pixel as an edge-pixel:

(1) There must have two or more pixels whose values are 1 in its neighborhood.
(2) The values of the connected regions both of its neighborhood and of the inverse matrix must be 1.

In order to validate this criterion, firstly we get a series of exact edges images by means of the objective adjustment. Considering the universality, we take the 3×3 neighborhood of an arbitrary image pixel p in [C] as a matrix Z. Since the edge images [C] are binary images, the matrix Z has 2^9=512 kinds of different value in all. Without regard to the directivity, we combine the matrixes with the same shapes but different directions and get 140 template matrixes with different shapes. For edge pixel in the edge image group [C], take the histogram statistics of its 3×3 neighborhood matrix in the corresponding sample image group [A], each kind of template's ratio is shown in Figure 2(The templates have been arranged in order of proportion.). The proportion of the number of template matrixes to the total amount of the edges is shown in Figure 3, which says that more than 95% of edges are made up of the minority of the template matrixes, defined as edge template in this paper. With analysis, there are about over 20 kinds of edge templates, taking up about 14% of the total, and the values of the connected region of them and their inverse matrixes are both 1. Furthermore, among the 140 kinds of templates, 20 templates are proved that their values of connected region of matrix Z and its inverse Z^{-1} are both 1, and these 20 kinds are completely the same as the edge templates got in the statistic experiment. By further analysis, because the real edge is continuous but the noisy is discrete, this paper adopts the above criterion to judge whether an image pixel is an edge pixel and decide the segmentation result including the most edge pixels as the optimum segmentation result. In order to compare the accuracy of the decided edge, we respectively apply the Canny edge operator and the method in the paper to mark the segmentation results of the Lena image with noisy. The results are shown in Figure 4 and Figure 5. The experiment shows that the above criterion can retain the image edge and constrain the noisy much better.

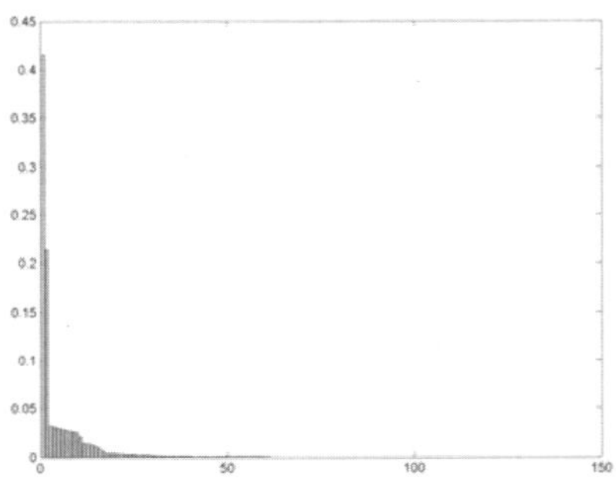

Fig. 2. The statiscal diagram of the edge templates

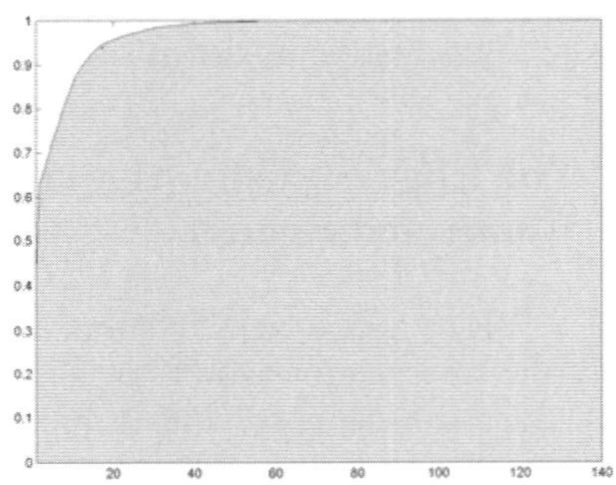

Fig. 3. The proportion of the edge templates to the total

Fig. 4. The marked edges result gained with Canny operator (edge pixels marked with l)

Fig. 5. The marked edges result gained with the algorithm proposed in this paper (edge pixels marked with +)

Table 1. The general parameters of PCNN

General Parameter	Value
α_F	693
α_L	0.693
α_T	0.069
V_F	0.01
V_L	1
V_T	0.2
β	0.2

Table 2. The statistic values of the edges gained during the segmentation process of image Lena

n	P(n)
2	405
3	843
4	1137
5	1321
6	1569
7	2037
8	2759
9	2872
10	2950
11	3009
12	3106
13	3039

Table 3. The statistic values of the edges gained during the segmentation process of blood cells micrograph with noisy

n	P(n)
2	3895
3	4227
4	2844
5	2612
6	2639
7	2750
8	2903
9	2985
10	3067
11	3166
12	3216
13	3142

4 Experiments

With lots of experiments, the PCNN parameters used in this paper are shown in Table 1. The input images are A (original image Lena), B (the blood cells micrograph added with 5% Gauss noise and 5% white noise), and C (the complex indoors-view image added with 5% Gauss noise and 5% white noise). We evaluate the segmentation result by the edge statistics algorithm based on the connected region. The results of P(n) of each iteration of the image A and B and C are in Table 2, 3 and 4. The comparing results of the method and the one based on entropy of the image and the one based on Canny edge operator are respectively shown in Figure 6, Figure 7 and Figure 8.

Table 4. The statistic values of the edges gained during the segmentation process of complex indoors-view image with noisy

n	P(n)
2	2381
3	2327
4	2555
5	2723
6	2859
7	2984
8	3071
9	3292
10	3429
11	3360
12	2936
13	2355

5 Discussion of the Experiment Results

1. For images without noise, from Figure 6, we can see that the evaluation curves respectively got with our method, Canny edge operator and image entropy, are similar and the evaluation results are the same and also meet very well with the objective evaluation. The reason is that the essence of the algorithm based on connected region is a kind of edge detection method, and the statistic number of edges is in direct proportion to the image details, which is just coherent with the definition of image entropy. Thus the three methods are nearly the same without the interference of noisy.

2. For images with noise, the method based on edge operator is greatly disturbed, and the curve decreased sharply, so it cannot be used as a reference any more (shown in Figure 7(c), Figure 8(c)). The methods based on the entropy of image and the algorithm of this paper are hardly disturbed and they can still show the correct evaluation curves and results. It demonstrates that these two methods have stronger robustness of noisy. Compared with the method based on the entropy of image, the evaluation curve of our method paper has bigger slope, steeper peak and better test sensitivity.

3. The locally optimum result of the evaluation curve of our method usually corresponds with a segmentation result, which can be accepted by human subjective (See Figure 7(c) and Figure 9). When n=2, P(n)=2381, it produces a middle solution. In Figure 9, we can see that the segmentation result is very fine. Comparatively, the

method based on the entropy of image can only get one solution during an iteration period. Since the evaluation about images has great subjectivity, whatever method to use, the result is a very close approximation to the subjective result. Our method can not only be used in image auto-segmentation, but also in the pre-search process, during which it can submit all the possible results at the same time and so extraordinarily increase the speed of image segmentation by means of man-computer interactive.

4. Because the segmentation process with PCNN runs in the method of cyclic iteration, the output object changes from zero to many and to zero again during the process. Suppose the iteration time in each period is a, the total iteration periods is n, so the total iteration times are a×n. Only take one period for example in the experiment, if you want to get a better segmentation result, you can properly increase the periods number, n. But if n is too large, it will cost a great deal of calculation time and the segmentation quality may become worse, so n can't be larger than five cycles in general.

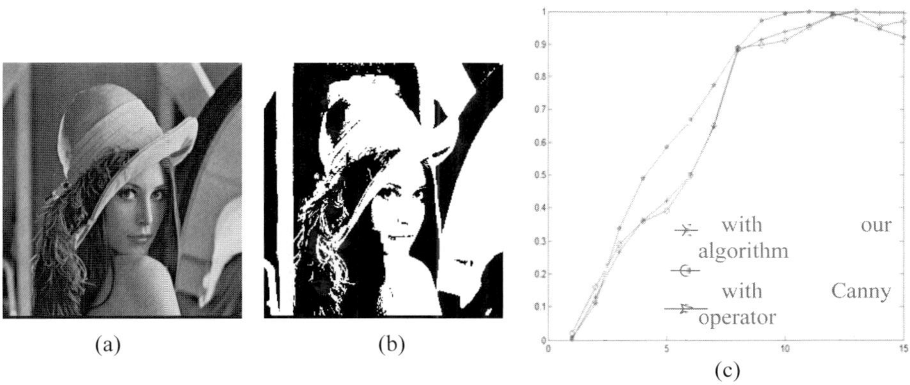

Fig. 6. (a) Image A : original Lena image (b) The optimum segmentation result (c) The iteration statistical diagram

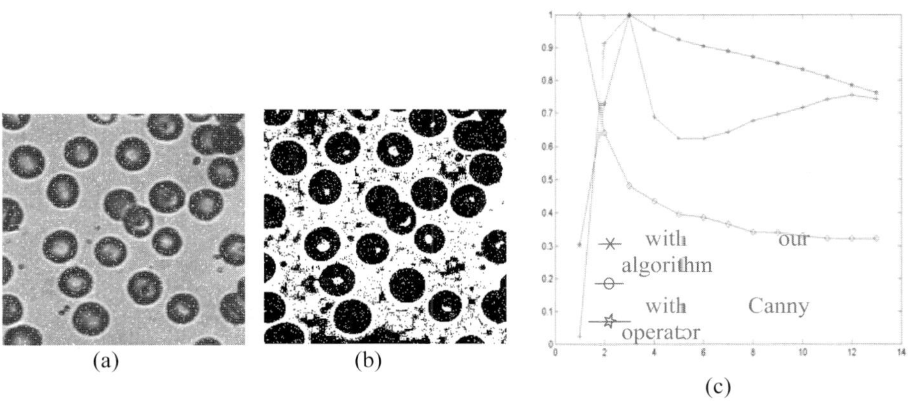

Fig. 7. (a) Image B: the blood cells micrograph with 5% Gauss noise and 5% white noise (b) The optimum segmentation result (c) The iteration statistical diagram

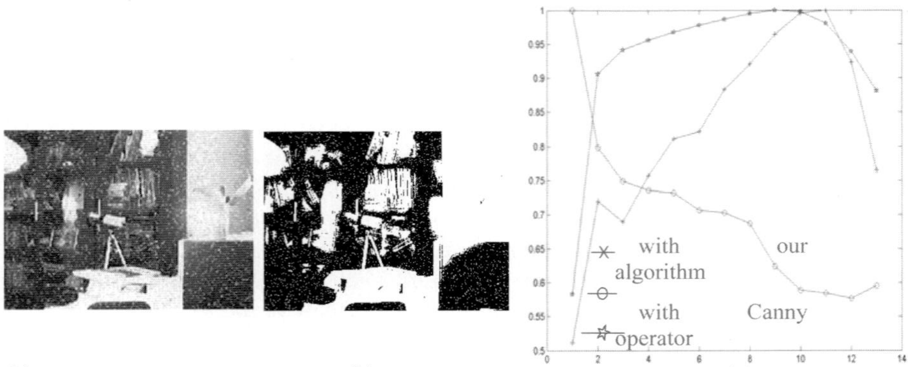

Fig. 8. Image C: the complex indoors-view image with 5% Gauss noise and 5% white noise(b) The optimum segmentation result (c) The iteration statistical diagram

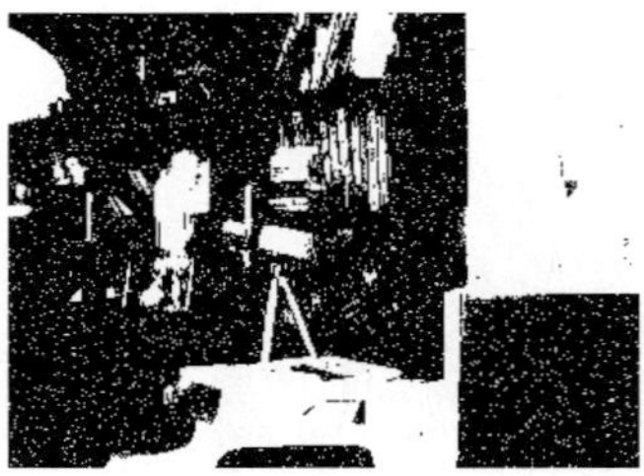

Fig. 9. The segmentation result of Image C at the first peak value when n=2

6 Conclusions

The research on PCNN is a leading field in the world at present and people often evaluate the segmentation result in the way of man-computer interaction with subjective judgment. But in practice, after we get the proper parameters by repeated experiments on the target image, to decide the optimum iteration times N automatically is the key point of the automatic process free from human intervener. The edge statistic algorithm based on the calculation of connected region is a try to solve the above problem. The advantages of the method are stronger robustness of noisy and higher test sensitivity. Its disadvantage is that it needs a bit lot of calculation. But we can satisfy the real-time need while processing 256×256 gray-scale images on the computer with 256MB memory and Pentium IV CPU. If one uses the quick algorithm of connected region calculation or some strategies with multiple-resolution from roughness to exquisiteness, the speed can be greatly increased. Besides, many experiments further prove that, the image segmentation based on PCNN have no obvious relationship with the brightness, contrast and noisy of image, but the image's internal properties can directly influence the choice of the parameters and the segmentation result. The area smoothing connection restrictions should be expressed during the decision of the optimum value of N. Some related theories and practice still need further study.

References

1. Kuntimad, G., Ranganath, H.S.: Perfect image segmentation using pulse coupled neural networks. IEEE Trans. on Neural Networks 10, 591–598 (1999)
2. Skourikhine, A.N.: Pulse couple neural network for image smoothing and segmentation[A]. In: International Symposium on Computational Intelligence[C], Kosice, Slovakia (2000)
3. Johnson, J.L., Padgett, M.L.: PCNN models and applications. IEEE Trans. on Neural Networks 10, 480–498 (1999)
4. Lindbiad, T.: Inherent features of wavelets and pulse coupled neural networks. IEEE Trans. on Neural Networks 10, 607–617 (1999)
5. Johnson, J.L.: Pulse-coupled Neural Networks, Adaptive Computing: Mathematics, Electronics and Optics. In: Proc. of a conference, Orlando, FL, April 4-5, 1994, vol. CR55 (1994)
6. Johnson, J.L.: Waves in Pulse-coupled Neural Networks. In: Proc. of the World Congress on Neural Networks, vol. 4, INNS Press (1993)
7. Ranganath, H.S., Kuntimad, G.: Pulse Coupled Neural Networks for Image Processing. IEEE, Los Alamitos (1995)
8. Lindblad, T., Kinser, J.M.: Image Processing Using Pulse-coupled Neural Networks[M]. Springer, Heidelberg (1998)

Dynamic Structure-Based Neural Network Determination Using Orthogonal Genetic Algorithm with Quantization

Liling Xing[1] and Wentao Zhao[2]

[1] College of Information System and Management, National University of Defense Technology, 410073, Changsha, China
xinglining@gmail.com
[2] Chengdu Institute of Computer Applications, Chinese Academy of Sciences, 610041 Chengdu, China
wtzhao.nn@gmail.com

Abstract. It proposed a novel dynamic structure-based neural network determination approach using orthogonal genetic algorithm with quantization in this paper. Both the parameter (the threshold of each neuron and the weight between two neurons) and the transfer function (the transfer function of each layer and the network training function) of the dynamic structure-based neural network were optimized in this proposed approach. In order to satisfy the dynamic transform of the neural network structure, the population adjustment operation was introduced into orthogonal genetic algorithm with quantization for dynamic modification of the population's dimensionality. A mathematical example was applied to evaluate this proposed approach. The experiment results suggested that this proposed approach is feasible, correct and valid.

1 Introduction

Neural networks (NNs) have been successfully applied to different kind of research fields and applications due to their model-free approximation capability to complex decision making process [1]. Compare to the real life nerve system, the structure of the conventional NNs are too simple and form in regular alignment patterns. Although conventional NNs are very powerful to solve many engineering problems, the structure of NNs has a potential constraint that differs from the real life nerve systems. In many applications, NNs are fully connected. However, the performance of a fully connected NN may not be better than that of a partly connected one with the same number of hidden nodes [2]. It is because some links in a NN could be redundant.

Recently, dynamic structure-based neural networks (DSNN) are being extensively applied in many fields of science and engineering [3-6]. Although DSNN is very efficacious tool for solving many complex practical problem, but its training is rather complicated. There are three difficulties in the DSNN's training.

❶ The existing training algorithm (such as, the back-propagation learning algorithm) is not available to DSNN because of its different structure.

K. Li et al. (Eds.): LSMS 2007, LNCS 4688, pp. 162 – 171, 2007.

❷ The training algorithm for DSNN should suffice to the dynamic modification of the solution's dimensionality because of dynamic transform of its structure.
❸ The decision variable of DSNN's training problem is more comprehensive than that of standard NN. For example, the network training function is a decision variable in DSNN's training, but it is predefined in the standard NN's training.

For this reason, it proposed a novel DSNN determination approach using orthogonal genetic algorithm with quantization (OGA/Q) [7] in this paper. This paper is organized as follows. Some related works are reviewed in section 2. The proposed approach is presented in section 3. A mathematical example is used to evaluate our method in section 4. Finally, the conclusion is summarized in section 5.

2 Related Work Review

There are two parts in this section: (1) dynamic structure-based neural network review and (2) neural network structure determination review.

2.1 Dynamic Structure-Based Neural Network

LING proposed a novel variable structure NN [5]. The proposed variable structure NN consists of a Neural Network with a Node-to-Node Relationship (N^4R) and a Network Switch Controller (NSC). In the N^4R, two different activation functions are used in neurons of the hidden layer, and a node-to-node relationship is established. Switches are introduced in some links of the N^4R to facilitate the tuning of the network structure. The on/off states of the link switches are controlled by the NSC, which is a 3-layer feed-forward NN.

Hsiao [8] developed an automatic NN generating algorithm and its learning algorithm. The initial generated NN has a randomly generated structure and weights. All generated neurons shall be deployed in a virtual three-dimensional (3-D) space. According to the distance between two neurons, the neural connections are generated by Bozeman distribution to decide which two neurons should be connected. Moreover, the neuron-regeneration mechanism according to the neuron-regeneration technique of the real life nerve system in the territory of medical science, are proposed to automatically reconstruct the NN for achieving higher performance.

2.2 Neural Network Structure Determination

Boozarjomehry [9] presents a new method for the auto-design of NNs, which is based on GA and Lindenmayer Systems [10-11]. The method is less computationally intensive than the existing iterative design procedures, and it can be applied to the automatic design of NNs for complex processes. Kottathra [12] proposes a novel prescriptive solution to decide the optimum number of neurons in the hidden-layer of multilayer feed-forward NNs. His approach uses the unconstrained mixed integer nonlinear multi-criteria optimization technique. Rovithakis [13] presents an algorithmic approach to determine the structure of high order NNs, to solve function approximation problems. The method is based on a genetic algorithm and is equipped with a stable update law to guarantee parametric learning.

Henrique [14] proposes a new procedure for model structure determination in feedforward NNs. Mao [15] presents a supervised network structure determination algorithm. Na [16] developed a fuzzy NN with an optimal structure constructor to achieve a reliable and efficient sensor monitoring system. Tsai [17] applied a hybrid Taguchi-genetic algorithm (HTGA) to solve the problem of tuning both network structure and parameters of a feedforward NN. Leung [2] presented the tuning of the structure and parameters of a NN using an improved genetic algorithm. Mao [18] proposed to select hidden layer neurons based on data structure preserving criterion.

3 The Proposed Determination Approach

3.1 Uniform Expression of the Dynamic Structure-Based Neural Network

In general, the uniform expression of DSNN can be drawn as follows.

$$DSNN = \{X, Y, I_1, O_1, I_2, O_2, \cdots I_m, O_m\} \tag{1}$$

Where, X denotes the inputs of DSNN, Y denotes the outputs of DSNN; m denotes the amount of sub-systems in the DSNN; I_i denotes the inputs of the ith sub-system, O_i denotes the outputs of the ith sub-system.

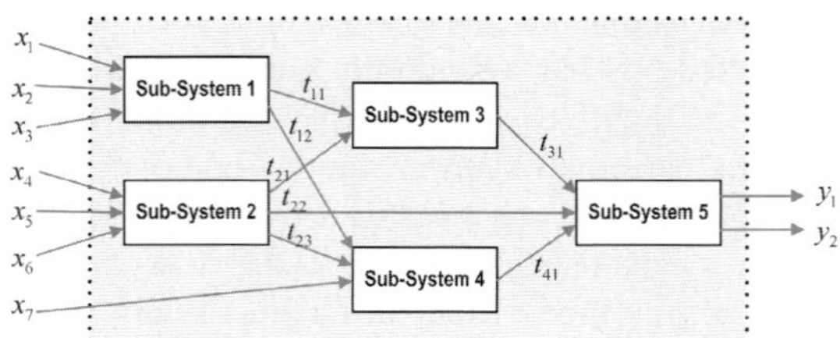

Fig. 1. Hierarchy structure and inputs-outputs relationship of a practical optimization system

Fig. 1 describes the hierarchy structure and inputs-outputs relationship of a practical optimization system. We can apply the DSNN drew as Fig. 2 to simulate the given system. The uniform expression showed as formula (2) is applied to express the given DSNN. With this simple uniform expression, both the hierarchy structure and inputs-outputs relationship of the given DSNN were displayed distinctly.

$$DSNN = \{X, Y, I_1, O_1, I_2, O_2, I_3, O_3, I_4, O_4, I_5, O_5\} \tag{2}$$

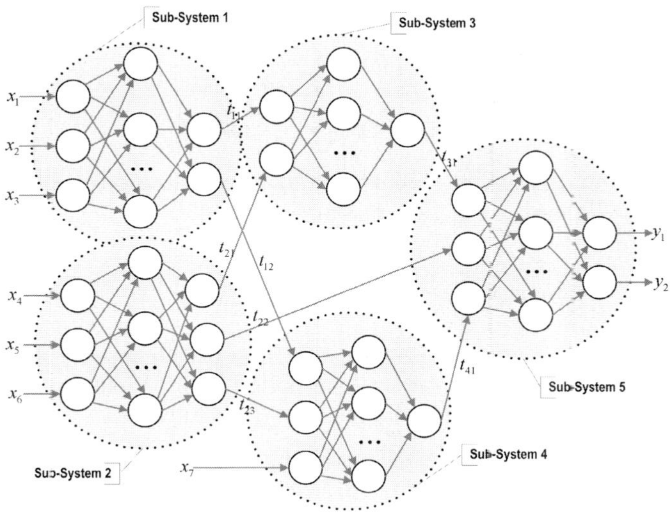

Fig. 2. DSNN model for simulating the given system drew as Fig. 1

Where

$$X=\{x_1,x_2,x_3,x_4,x_5,x_6,x_7\},\quad Y=\{y_1,y_2\};$$
$$I_1=\{x_1,x_2,x_3\},\quad O_1=\{t_{11},t_{12}\};$$
$$I_2=\{x_4,x_5,x_6\},\quad O_2=\{t_{21},t_{22},t_{23}\};$$
$$I_3=\{t_{11},t_{21}\},\quad O_3=\{t_{31}\};$$
$$I_4=\{t_{12},t_{23},x_7\},\quad O_4=\{t_{41}\};$$
$$I_5=\{t_{31},t_{22},t_{41}\},\quad O_5=\{y_1,y_2\}.$$

3.2 Error Computation of the DSNN

To a given DSNN, its error means the approximation error. Let $\{X_L,Y_L\}$ denotes the learning samples, $DSNN(P)$ denotes a known DSNN, P denotes the parameter set of the DSNN. $\hat{Y}$ denotes the outputs of $DSNN(P)$ when the inputs are X_L. In this paper, the error of DSNN is measured according to the mean of squared errors.

$$ERR\left(DSNN(P)\right)=MSE\left(Y_L,\hat{Y}\right)=\sqrt{\frac{\sum_{i=1}^{n}\sum_{j=1}^{k}\left\{\omega_j\left[Y_L(i,j)-\hat{Y}(i,j)\right]^2\right\}}{n-1}} \tag{3}$$

Where, n denotes the account of learning samples, k denotes the account of performance indexes of the optimization problem, ω_j denotes the weight of the jth performance index, $Y_L(i,j)$ denotes the output value of the jth performance index of the ith learning sample, $\hat{Y}(i,j)$ denotes the computation output value of the jth performance index of the ith learning sample using the given DSNN.

3.3 Determination Approach of the DSNN

Determination of DSNN can be summarized to an optimization problem as follows. Find an optimal parameter set P^*, satisfied that

$$ERR\left(DSNN\left(P^*\right)\right) = \min \quad ERR\left(DSNN\left(P\right)\right) \tag{4}$$

The DSNN with the parameter set P^* has the most excellent performance for simulating the given practical model. Before the optimization of parameter set P, let us analyze the detailed element of the parameter set P carefully.

$$P = \begin{bmatrix} WI_1 & TI_1 & TFI_1 & WH_1 & TH_1 & NTF_1 \\ WI_2 & TI_2 & TFI_2 & WH_2 & TH_2 & NTF_2 \\ \cdot & \cdot & \cdot & \cdot & \cdot & \cdot \\ WI_i & TI_i & TFI_i & WH_i & TH_i & NTF_i \\ \cdot & \cdot & \cdot & \cdot & \cdot & \cdot \\ WI_m & TI_m & TFI_m & WH_m & TH_m & NTF_m \end{bmatrix} \tag{5}$$

Where, m denotes the sub-system amount of the giving DSNN, WI_i denotes the input layer weights of the ith sub-system, TI_i denotes the input layer thresholds of the ith sub-system, TFI_i denotes the input layer transfer function of the ith sub-system, WH_i denotes the hidden layer weights of the ith sub-system, TH_i denotes the hidden layer thresholds of the ith sub-system, NTF_i denotes the network training function of the ith sub-system. Fig. 3 depicts the flow chart of this proposed determination approach of the DSNN.

Step 1. **Population Initialization.** Determines the neuron number of hidden layer of DSNN based on the amount of learning samples. The experiential formula $m = \log_2 num$ can be applied to determine the neuron number, where m denotes the neuron number of hidden layer, num denotes the amount of learning samples. Then generate a group of initial population using population initialization mechanism of OGA/Q. Obviously, the single individual of initial population is a set of parameter setting of DSNN.

Step 2. **Fitness Evaluation.** Calculate the fitness values of current population using the error computation formula of DSNN.

Step 3. **Population Evolvement.** Evolve the current population using crossover operator and mutation operator of OGA/Q.

Step 4. **Termination Condition Estimation.** The proposed algorithm is terminated when one of the following criteria is satisfied: ① maximum preset search times are exhausted; ② minimum preset optimization error is achieved (error between the current achieved optimal solution and the known optimal solution is smaller than the minimum preset optimization error).

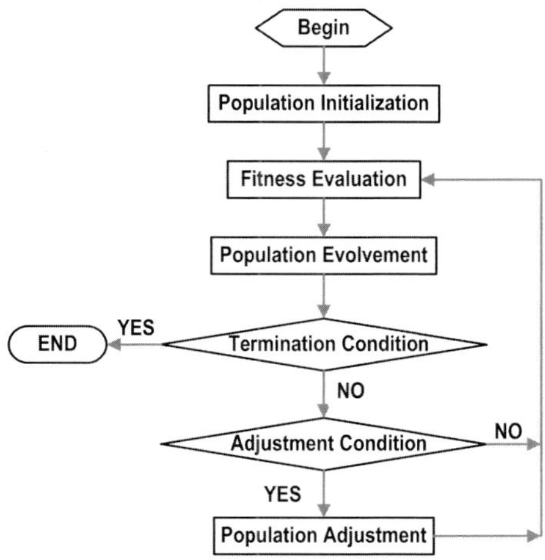

Fig. 3. Flow chart of the determination approach of the dynamic structure-based neural network

Step 5. **Adjustment Condition Estimation.** Our algorithm will implement the population adjustment when all of the solutions generated in five continuous iterations are not better than the globally best solution from the beginning trials.

Step 6. **Population Adjustment.** When the adjustment condition is satisfied, the neuron number of hidden layer of DSNN will be increases gradually.

3.4 Detailed Implementation of the Proposed Approach

(1) Population Initialization. The details for generating an initial population using the orthogonal design and quantization technology are introduced in reference [7]. After the operation, it will select G chromosomes having the greatest fitness value as the initial population.

(2) Fitness Evaluation. To each individual of current population, a group of parameter of DSNN, the error of DSNN will be applied to evaluate the fitness of this individual. The fitness evaluation function of this paper can be expressed as following.

$$FITNESS(pop) = ERR(DSNN(P)) = MSE(Y_L, \hat{Y}) \tag{6}$$

Where pop denotes the individual of current population, P denotes a group of parameter of DSNN corresponding to the individual pop, Y_L denotes the output value of the learning sample, $\hat{Y}$ denotes the computation output value of the learning sample using the given DSNN.

(3) Crossover Operation. On one hand, we want to obtain enough potential offspring to improve the current population; on the other hand, each pair of parents should not produce too many potential offspring in order to avoid large numbers of function evaluations during selection. It is critical to select a small but representative sample of points as the potential offspring. For this purpose, the orthogonal crossover with quantization [7] was employed as the crossover operator.

(4) Mutation Operation.

Step 1. Select one chromosome according to the mutation probability as parent individual.

Step 2. Choice one gene (independent variable) for mutation operation randomly.

Step 3. Apply the micro-perturbation method to mutate the selected parent chromosome. It is to say, modify the selected gene θ of this parent chromosome as $(1-2\sigma)\theta$, $(1-\sigma)\theta$, $(1+\sigma)\theta$ and $(1+2\sigma)\theta$, then we will achieve four offspring chromosomes after this mutation, where σ is a design parameter.

Step 4. Among parent chromosome and these offspring chromosomes, select one chromosome having the smallest cost as the results of this mutation operation.

Step 5. If the maximum preset mutation times are reached, then stop the mutation operation; else, go to ***Step 1***.

(5) Population Adjustment. The key idea of population adjustment is to increase the neuron number of hidden layer of DSNN gradually. The detailed adjustment scheme mainly expands the dimension of following variables: input layer weights, input layer thresholds, hidden layer weights and hidden layer thresholds. We will apply a simple example to explain this population adjustment. Just as Fig. 4(a), it describes a simple NN structure, its design variables and its corresponding individual (chromosome) structure. After population adjustment, this simple NN structure, its design variables and its corresponding individual (chromosome) structure will be modified as Fig. 4(b). Obviously, the dimension of every single individual among current population will be enlarged correspondingly after the population adjustment.

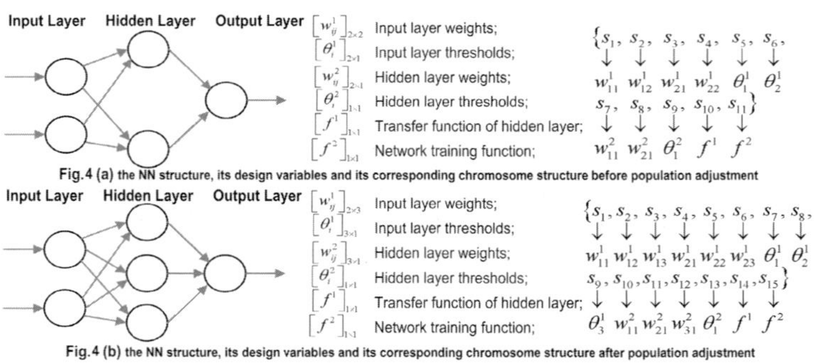

Fig. 4. A simple example of population adjustment

4 Experimental Results

In this section, the proposed approach is demonstrated by a practical optimization problem. In order to simplify subsequent computations, we only consider single objective simulation optimization problems. In fact, the appropriate approach can transform multi-objective simulation optimization problems to single objective simulation optimization problems. Given the case of intercommunion at future, a complex function is used to replace the input-output relationships of the known simulation system (sub-system). The experiments were performed on a Pentium IV 2.4 GHz personal computer with a single processor and 512M RAM. To avoid the randomness in the optimization process, each experiment was run 20 times. The parameter settings for the OGA/Q used in this paper are listed in Table 1.

Table 1. Parameter Settings of OGA/Q used in This Paper

Parameter	Signification	Setting
Q_1	Number of quantization levels	$Q_1 = \begin{cases} 5 & N \in (0,100] \\ 3 & N \in (100,\infty) \end{cases}$
B	Number of subspaces	1
G	Population size	20
Q_2	Number of quantization levels	3
F	Number of factors	$F = \begin{cases} 4 & N \in (0,100] \\ 8 & N \in (100,\infty) \end{cases}$
P_c	Crossover probability	0.60
P_m	Mutation probability	0.10
$MaxGen$	Maximum iterative	100
$MinErr$	Minimum optimization error	0.01

The input-output relationship of the giving simulation optimization problem is shown in Fig. 5. The complex function relationship of the giving simulation sub-systems ate listed in Table 2. The feasible space (searching space or definition space) of each complex numerical function was described in Table 2 too. Here, $Fix(x)$ rounds the elements of x to the nearest integers towards zero.

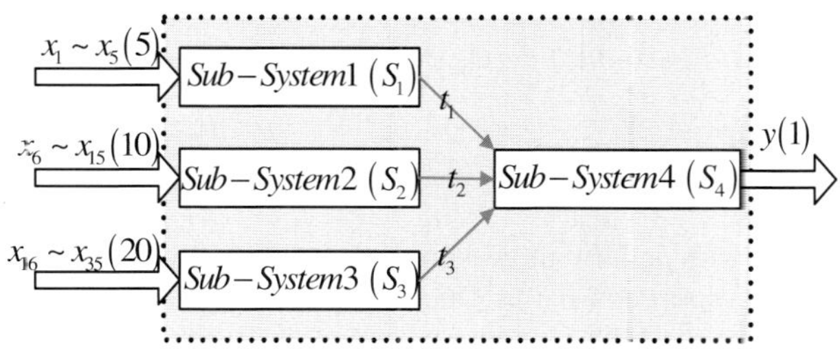

Fig. 5. Input-output relationships of the simulation optimization problem

This paper discusses a hybrid approach that combines EAs with NNs for solving simulation optimization problems. It applies the DSNN to replace the known simulation model for evaluating subsequent solutions. To demonstrate the efficiency of this hybrid approach, improved genetic-annealing algorithm, genetic algorithm-simplexes, new immune GAs, OGA/Q and this hybrid approach were applied to solve the simulation optimization problem. The experiment results produced using the five different approaches for solving the simulation optimization problem are listed as Table 3.

Table 2. The Complex Function Relationship of the Known Simulation Sub-systems

	Input-Output Relationship	Feasible Space	Simulation Time (S)
Sub-System 1	$S_1 = \sum_{i=1}^{5} Fix(x_{1i})$	$0 \le x_{1i} \le 20$	5
Sub-System 2	$S_2 = \sum_{i=1}^{10} \left(x_{2i}^2 - A\cos(2\times\pi\times x_{2i})\right)$	$A = 20$ $-1 \le x_{2i} \le 1$	10
Sub-System 3	$S_3 = \frac{1}{20}\sum_{i=1}^{20} \left(x_{3i}^4 - 16x_{3i}^2 + 5x_{3i}\right)$	$0 \le x_{3i} \le 5.5$	20
Sub-System 4	$S_4 = \max\left\{\sum_{i=1}^{3} S_i \sin(\sqrt{\lvert S_i \rvert})\right\}$	$0 \le S_i \le 100$	3

Table 3. The Experiment Results Produced using the Five Different Approaches

Name of Different Optimization Approach	Optimization Result	Simulation Time (h)
Improved genetic-annealing algorithm	127.75	794.39
Genetic algorithm-simplexes	179.63	151.62
New immune GAs	185.79	170.54
OGA/Q	186.56	255.07
This proposed hybrid approach	190.54	28.82

These results show that the hybrid approach is clearly superior to the other four approaches. The results show drastic reductions in computing time using the hybrid approach and that it outperforms the other, recently developed, algorithms for simulation optimization. The proposed hybrid approach is a valuable addition to the tools available for solving simulation optimization problems, being not only feasible, correct and valid, but also fast and efficient.

5 Conclusions

The contributions of this paper can be summarized as following. It proposed a novel DSNN determination approach using OGA/Q in this paper. Both the parameter and the transfer function of the DSNN were optimized in this proposed approach. In order to satisfy the dynamic transform of the neural network structure, the population adjustment operation was introduced into OGA/Q for dynamic modification of the population's dimensionality. The experiment results suggested that this proposed approach is feasible, correct and valid.

References

1. Avi-Itzak, H.I., Diep, T.A., Garland, H.: High Accuracy Optical Character Recognition using Neural Networks with Centroid Dithering. IEEE Transactions on Pattern Analysis and Machine Intelligence 17, 218–224 (1995)

2. Leung, F.H.F., Lam, H.K., Ling, S.H., Tam, P.K.S.: Tuning of the Structure and Parameters of Neural Network using an Improved Genetic Algorithm. IEEE Transaction on Neural Network 14, 79–88 (2003)
3. Loo, C.K., Mandava, R., Rao, M.V.C.: A Hybrid Intelligent Active Force Controller for Articulated Robot Arms Using Dynamic Structure Neural Network. Journal of Intelligent and Robotic Systems 40, 113–145 (2004)
4. Hung, C.P.: Integral Variable Structure Control of Nonlinear System Using a CMAC Neural Network Learning Approach. IEEE Transaction on System, Man and Cybernetics - Part B: Cybernetics 34, 702–709 (2004)
5. Ling, S.H., Lam, H.K., Leung, F.H.F., et al.: A Genetic Algorithm Based Variable Structure Neural Network. In: The Annual Conference of the IEEE on Industrial Electronics Society, IEEE, Roanoke, VA, United States (2003)
6. Ei-Sadek, M.Z., Ei-Saady, G., Abo-Ei-Saud, M.: A Variable Structure Adaptive Neural Network Static VAR Controller. Electric Power Systems Research 45, 109–117 (1998)
7. Leung, Y.W., Wang, Y.P.: An Orthogonal Genetic Algorithm with Quantization for Global Numerical Optimization. IEEE transaction on evolutionary computation 5, 41–53 (2001)
8. Siao, Y.T., Chuang, C.L., Jiang, J.A.: A Novel Dynamic Structural Neural Network with Neuro-regeneration and Neuro-degeneration. In: Proceedings of 9th International Workshop on Cellular Neural Networks and Their Applications, IEEE, Hsinchu, Taiwan (2005)
9. Boozarjomehry, R.B., Svrcek, W.Y.: Automatic Design of Neural Network Structures. Computers and Chemical Engineering 25, 1075–1088 (2001)
10. Lindenmayer, A.: Mathematical Models for Cellular Interactions in Development I. Filaments with One-sided Inputs. Journal of Theoretical Biology 18, 280–300 (1968)
11. Lindenmayer, A.: Mathematical Models for Cellular Interactions in Development II. Simple and Branching Filaments with Two-sided Inputs. Journal of Theoretical Biology 18, 300–316 (1968)
12. Kottathra, K., Attikiouzel, Y.: A Novel Multi-criteria Optimization Algorithm for the Structure Determination of Multilayer Feedforward Neural Networks. Journal of Network and Computer Applications 19, 135–147 (1996)
13. Rovithakis, G.A., Chalkiadakis, I., Zervakis, M.E.: High-Order Neural Network Structure Selection for Function Approximation Applications Using Genetic Algorithms. IEEE Transaction on System, Man and Cybernetics - Part B: Cybernetics 34, 150–158 (2004)
14. Henrique, H.M., Lima, E.L., Seborg, D.E.: Model Structure Determination in Neural Network Models. Chemical Engineering Science 55, 5457–5469 (2000)
15. Mao, K.Z., Tan, K.C., Ser, W.: Probabilistic Neural-Network Structure Determination for Pattern Classification. IEEE Transactions on Neural Network 11, 1009–1016 (2000)
16. Na, M.G., Sim, Y.R., Park, K.H., Lee, S.M.: Sensor Monitoring Using a Fuzzy Neural Network with an Automatic Structure Constructor. IEEE Transactions on Nuclear Science 50, 241–250 (2003)
17. Tsai, J.T., Chou, J.H., Liu, T.K.: Tuning the Structure and Parameters of a Neural Network by Using Hybrid Taguchi-Genetic Algorithm. IEEE Transactions on Neural Network 17, 69–70 (2006)
18. Mao, K.Z., Huang, C.B.: Neuron Selection for RBF Neural Network Classifier Based on Data Structure Preserving Criterion. IEEE Transaction on Neural Network 16, 1531–1540 (2005)

Simulation-Based Optimization Research on Outsourcing Procurement Cost

Jiangming Jia, Zhengxiao Wang, and Xiaohong Pan

Institute of Modern Manufacture Engineering, Zhejiang University, Hangzhou, 310027, P.R. China
{Jarky,Wangzhengxiao}@zju.edu.cn

Abstract. To solve the cost control problem of outsourcing procurement inherent in the complexness and difficulty for mathematical analysis, this paper established a simulation-based cost Optimization framework. Firstly, we adopted Poisson process to simulate multi-supplier outsourcing market environment, the procurement cost expectations objective function and constraints of the model was introduced. Secondly, taking decision strategy as feedback variables, procurement cost optimization model was proposed. Thirdly, several key technologies for algorithm implement were described. Finally, based on the history demand of an aircraft model factory, numerical simulation is carried out to show that this optimization model can effectively reduce the outsourcing procurement total cost and outsourcing parts inventory fluctuation.

1 Introduction

With the rapid development of globalization, more and more firms produce by means of globally orders. These orders are not only changeful batch and exact delivery time, but also the key components of these orders (outsourcing parts) are special and the price of the key components changes quickly. MTO (Make To Order) is a manufacturing mode to respond quickly to market demand, i.e., firms gain orders from customers and procure raw materials and parts, and then they will assemble the products to customers. In this manufacturing mode, manufacturing firms can arrange the firm-wide production activities reasonably, however, as a result of uncertainty in outsourcing parts supply, how to control this uncertainty effectively and complete orders perfectly is an urgent problem.

In essence, the problem is how to optimize the amount of suppliers and the batch decision, i.e., firms make sure how many suppliers and how much batch of outsourcing parts when receiving the orders to ensure a perfect implementation of the orders. Multi-supplier and less batch strategy easily guarantee stable supply with the complex management and high cost; concentrated suppliers and large batch strategy is not only the price promotion and the achievement of the seller's and the buyer's credit with easily subjected to several main suppliers. Cheng-liang et al. established multi-member supply chain decision-making model in uncertain demand and prices environment, and the model attempts to ensure to minimize cost under condition of

K. Li et al. (Eds.): LSMS 2007, LNCS 4688, pp. 172–181, 2007.

matching demand, however, they did not take the supply uncertainty into account [1]. Kevin et al. studied the short production cycle product orders decisions in uncertain lead time and uncertain demand environment, especially, lead time uncertainty was described [2]. For relative simple problem, analytical techniques [3] can be applied. As the problem gets more complicated, those analytical techniques become inapplicable. In recent years, the rise of simulation technology provides a new idea for resolving this problem [4]. The details can be found in several articles [5][6]. This paper attempts to use Simulation-Based Optimization to solve the decision-making optimization of outsourcing procurement. Simulation-Based Optimization method has a common defect, i.e., the search in whole problem space is not focused effectively. It will cost a lot of computing resources, and it cannot high-intensity simulate on the presumable space [7].

This paper studies to determine how to ensure the amount of suppliers and replenishment quantities dynamically, when making the minimum cost of the procurement logistics. The model adopts improved simulation technology, i.e., guiding the simulation search space by the results of optimization [8]. Finally, it produces decision variable as feedback on outsourcing replenishment, suitable to the constantly adjust of the delivery plan (the dynamic optimization of whole plan period).

2 Optimization Modeling

2.1 Problem Analysis

In MTO circumstances, the procurement decision-making problem can be described: firms purchase parts to suit the demand ω_t in the coming period of T, and then choose suppliers set G (including k suppliers) from the candidate Suppliers (S_1, S_2 , ..., S_K). At period $t=m(0\leqslant m\leqslant T)$, they were asked to replenish outsourcing parts inventory $\{P_i\}(i\in G)$.

After suppliers received the message of replenishment, due to its own inventory, production level, production planning and uncertainty of natural environmental

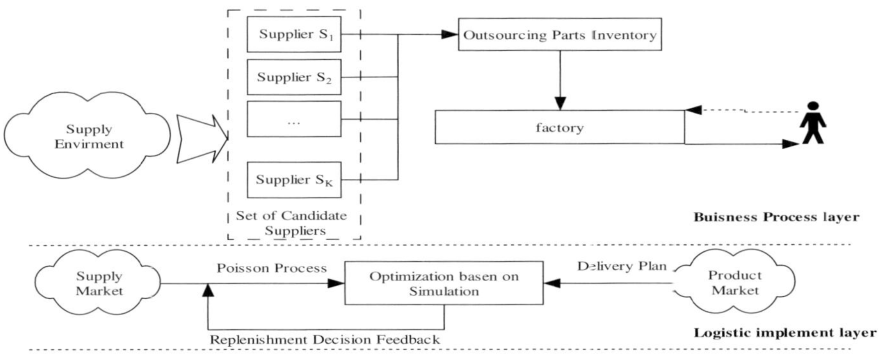

Fig. 1. Optimization of Outsourcing procurement cost

conditions, the supplier's delivery process is a random process at the firms view. When time went by, the inventory of available outsourcing parts under the suppliers' replenishment will increase to become the random cumulative process. The probability of Suppliers' replenishment batch reached is proportional to a very short time interval Δt, and the probability of more outsourcing batch is higher-order infinitesimal to Δt [7]. Poisson process is a random independent cumulative incremental process. So long as the random events in the intersected interval period are independent and it only occurs once within the considerable short interval period, the accumulative total number of the random events is a Poisson process. Therefore, accumulative outsourcing batch process $\{N(t), t \geqslant 0\}$ can be considered as a random Poisson process [9].

2.2 Objective Function

In this paper, the outsourcing parts demand is mapped to the product demand by BOM. By calculating the probability of out of outsourcing parts stock under orders delivery, we improve on Fu and Healy's experimental model [10] via expectation function. Then, we determine the amount of suppliers and replenishment batch to minimize the objective function. With changing the replenishment decision-making u(decisions of the amount of suppliers and outsourcing parts batch), we get the procurement cost objective function at period t=0-m($0 \leqslant m \leqslant T$):

$$
\begin{aligned}
&\min \sum\nolimits_0^m C_{total} \\
&\sum\nolimits_0^m C_{total} = \sum\nolimits_0^{m-1} C_{total} + E(C_{total})|_{t=m} \\
&E(C_{total})|_{t=m} = E(C_{\log})|_{t=m} + C_{cha}|_{t=m} + C_{los}|_{t=m} \\
&E(C_{\log})|_{t=m} = C_{tran}|_{t=m} + E(C_{inv})|_{t=m} = kC_{truf} + \sum_{i \in k} p_i C_{truu} + (w_{t+1} + E(R(t)))C_u|_{t=m} + C_{fix} \\
&\qquad = kC_{truf} + \sum_{i \in G} p_i C_{truu} + (w_t + E(N(t+1)) - E(N(t))|_{t=m} - \varpi_t|_{t=m} + E(R(t)))|_{t=m} C_u + C_{fix} \\
&\qquad = kC_{truf} + \sum_{i \in G} p_i C_{truu} + (w_t - \varpi_t + \frac{\lambda}{k} \sum_{i \in G} p_i + E(R(t)))|_{t=m} C_u + C_{fix} \\
&C_{cha}|_{t=m} = \begin{cases} C_{\sup} & k|_{t=m} \neq k|_{t=m-1} \\ C_{bat} & p_i|_{t=m} \neq p_i|_{t=m-1} \\ 0 & ortherwise \end{cases} \\
&Y_t = \begin{cases} 0 & if\,(\varpi_t - w_t) \geq 0 \\ 1 & if\,(\varpi_t - w_t) < 0 \end{cases} \\
&C_{los}|_{t=m} = Y_t C_{los} \\
&Q_t = \frac{\sum_{i=0}^{t} Y_t}{t} \\
&s.t. \\
&\mathrm{p_i} \leq \mathrm{P_i}, \mathrm{w_t} \leq \mathrm{W_{max}}, k \leq K, Q_t \leq 2\%, E(C_{total})|_{t=0} = 0 \\
&k, p_i, w_t, \varpi_t, \lambda \geq 0
\end{aligned}
\tag{1}
$$

Where, $E(C_{total})_m$ is the total procurement cost at period t=0-m, $E(C_{total})|_{t=m}$ is the increased cost at period t=m. The procurement logistics cost includes the total cost of inventory C_{inv} and the total cost of transportation C_{tran}. The total cost of inventory C_{inv}

includes the fixed costs of inventory C_{fix} and the variable cost of inventory C_u. The total cost of transportation C_{tran} includes the fixed cost of transportation C_{truf} and the variable cost of transportation C_{truu}. Y_t is the flag of out of outsourcing parts stock and Q_t is cumulative ratio of out of outsourcing pats stock at period 0—t.

Table 1 is notation and its significance in this paper.

Table 1. Notation table

Not.	Significance	Not.	Significance
K	Amount of candidate suppliers	$N(t)$	Process of bach reached at period 0-t
Pi	Max batch of suppliers S_i	Q_t	out ratio of stock at period 0-t
k	Amount of suppliers`	H_t	Status of system at period t
p_i	Batch of supplier S_i	x_0	Initial parameter of system
C_{truu}	Variable cost of transportation	μ_t	Decision of control at period t
C_{truf}	Fixed cost of transportation	u_t	Variable of control at period t
λ	Parameter of Poisson distribution	C_{inv}	Total cost of inventory
C_{tran}	Total cost of transportation	C_{total}	Total cost
C_{sup}	Cost of supplier changed	C_{bat}	Cost of batch changed
W_{max}	Max capacity of inventory	w_t	Amount of inventory at period t
C_{fix}	Fixed costs of inventory	C_u	Variable cost of inventory
T	Total plan period	T_{max}	Longest replenishment time
ω_t	Demand of outsourcing parts at period t	Y_t	Flag of out of stock at period t
G	Set of suppliers	Z_t	Est. of objective function at period t

3 Optimization Algorithm

Eqs.(1) is a nonlinear, discrete and mathematical expectation targeted global optimization model. In this paper, optimization method guides the direction of simulation to reduce the solution difficultly of the model.

Supplier selection and evaluation methods had been studied wicely, and it is assumed that the selection order of candidate supplier was provided: $\{S_i\}$ ($i = 1, 2, ..., K$).

3.1 Parameter Derivation

The objective function (1) includes some parameters must be derived from other parameters, because they are no real data source.

3.1.1 $\hat{\lambda}_L$

Set$\{N(t),\ t \geqslant 0\}$ is a *Poisson* process, and in the period of $[0,\ T]$, the events $S_1, S_2, \ldots, S_n$ is observed at t_1, t_2, ……, $t_n \leqslant T$, then The MLE is,

$L(t_1 .t_2,\cdots,t_n,\lambda) = \lambda^n e^{-\lambda T}$, let $\frac{dL}{d\lambda} = 0$, MLE of λ is $\hat{\lambda}_L = \frac{n}{T}$. Let $n=k$, $T=T_{\max}$. Therefore, MLE (Maximum Likelihood estimation) of λ :

$$\hat{\lambda} = \frac{k}{T_{\max}} \tag{2}$$

3.1.2 $E(R(t))\big|_{t=m}$

The outsourcing parts is stored in unit interval time sequentially, so the change estimation of outsourcing parts inventory in unit interval time is presented.

From above discussion, the process of outsourcing parts batch arriving is a *Poisson* process$\{N_k(t), t \geqslant 0, \lambda\}$, and the outsourcing parts will be assembled at t.

When *No. i* batch enter the inventory is R_i, so the sum of inventory time in period of [0, t] is

$R(t) = \sum_{i-1}^{N_k(t)} (t - R_i)$. According to the definition of *Poisson* process,

$$E(R(t) \mid N_k(t) = n) = E\{\sum_{i=1}^{N_k(t)} (t - R_i) \mid N_k(t) = n\} = nt - E[\sum_{i=1}^{n} R_i \mid N_k(t) = n]$$

According to the independent uniform distribution theorem of *Poisson* process, we get $E\{\sum_{i=1}^{n} R_i \mid N_k(t = n)\} = \frac{nt}{2}$.

Similarly, we get $E\{\sum_{i=1}^{n} (t - R_i) \mid N_k(t) = n\} = \frac{nt}{2}$.

Synthetically,

$$E[R(t)] = \sum_{n=0}^{\infty} \left(P\{N_k(t) = n\} E[\sum_{i=1}^{N_k(t)} (t - R_i) \mid N_k(t) = n] \right) = \frac{t}{2} E(N_k(t)) = \frac{\lambda t^2}{2}$$

For the average amount of each batch is $\frac{1}{k}\sum_{i\in G} p_i$, $E(R(t)) = \frac{\lambda}{2k} t^2 \sum_{i\in G} p_i$. Now, set $t=1$, we get the inventory Estimation form m-1 to m is $\frac{1}{2k}\lambda\sum_{i\in G} p_i$.

$$E(R(t))\big|_{t=m} = \frac{\lambda}{2k}\sum_{i\in G} p_i \tag{3}$$

3.2 The Simulation Data Generation (Poisson Process)

The interval of two adjacent events in Poisson process $X_n (n \geq 1)$ is exponential distribution duo to independent parameter $\lambda > 0$ [9], so we use the method as follow:

1. Produce independent uniform distribution series in [0, 1] $\{U_n, n \geq 1\}$
2. Let $X_k = -\lambda^{-1} \ln U_k$, get exponential distribution $\{X_n, n \geq 1\}$, which is independent and as same parameter as above
3. Set $D_0 = 0$ and $D_n = \sum_{k=1}^{n} X_k$
4. Produce Poisson process $N(t)$: if $0 \leq t < D_1$, then $N(t)=0$; if $D_n \leq t < D_{n+1}$, then $N(t)=n$

3.3 Simulation-Based Optimization

From introducing processes of optimization and detailing the main intermediary parameters, we describe the process of Simulation-Based Optimization in detail.

3.3.1 The Process of Simulation-Based Optimization

The basic idea of Simulation-Based Optimization is: The simulation system is core. Firstly, we get decision-making interval variable V_t and decision-making strategy variable u_t due to $\min E(\widehat{Z}_t)$; then the objective function is output in the inner loop by means of different condition of V_t and u_t; Finally V_t and u_t are changed and u^* and $\widehat{Z}^*$ is output in outside loop, as shown in fig. 2.

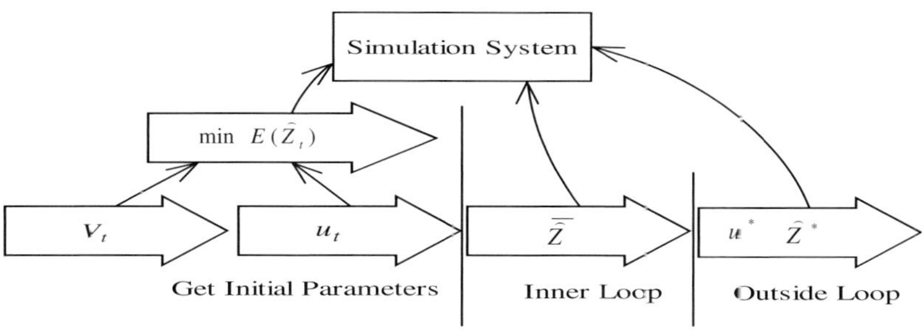

Fig. 2. Process of simulation-based optimization

3.3.2 $\min E(\widehat{Z}_t)$, V_t and u_t

$\min E(\widehat{Z}_t)$ is optimal function in feasible searched domain $\exists u_t$.

$$\begin{aligned} \min E(\widehat{Z}_t) &= \widehat{Z}_{t-1} + \min E(H_{t+1}) \\ \min E(H_{t+1}) &= \min_{u \in \exists u_t} f(H_t, u, \varpi_t) \end{aligned} \tag{4}$$

T is divided into different interval periods due to decision-making interval variable V_t and different replenishment strategy is adopted in different interval periods. Decision-making strategy variable u_t is the adjust method of decision variables (the amount of suppliers k and outsourcing parts batch $\{p_i\}$), which are divided into three types: the control variables μ_{Y1} means either to adjust the amount of suppliers firstly or to revise outsourcing parts batch firstly (1 means to adjust k, otherwise 0); the range of adjustment μ_{Y2} (S, M and L is the range of control variables adjustment); the sensitivity of adjustment μ_{Y3} (how much the objective function is changed, the control variables is adjusted).

Concretely, we employed an ES-(1+1) [11] approach to computer V_t and u_t. This algorithm operates with two individuals, one father and one descendent per generation, which competes with each other. If the descendent is better than the parent, the descendant replaces it as the father in the next generation.

3.3.3 The Process of Simulation-Based Optimization

The process of Simulation-Based Optimization is approached to output decision control variable and optimal objective function, as showed in fig. 3. By imported initial parameters, the system is looped as follow in total plan period T: After the different outsourcing parts demand ϖ_t is available, the next system status H_{t+1} and the next estimation of objective function $\hat{Z}_{t+1}$ is presented by decision-making strategy variable u_t and system status H_t. Then the feasible estimation of objective function $\min E(\hat{Z}_{t+1})$ (the direction of search) at t+1 is provided by means of $\hat{Z}_{t+1}$ and H_{t+1} in optimization unit, and then decision-making interval variable V_{t+1} and decision-making strategy variable u_{t+1} is derived from $\min E(\hat{Z}_{t+1})$. The next loop is executed. Therefore, a series of $\hat{Z}_t$ in period of T is presented, and then the numerical average of objective function $\bar{\hat{Z}}$ is approached duo to the system statuses H_t. The loop in period of T is over. Then, The new loop in period T begins when the counter of outside loop is increased. Finally, stop criteria is checked and the optimal objective function $\hat{Z}^*$ and optimal system decision u^* is output.

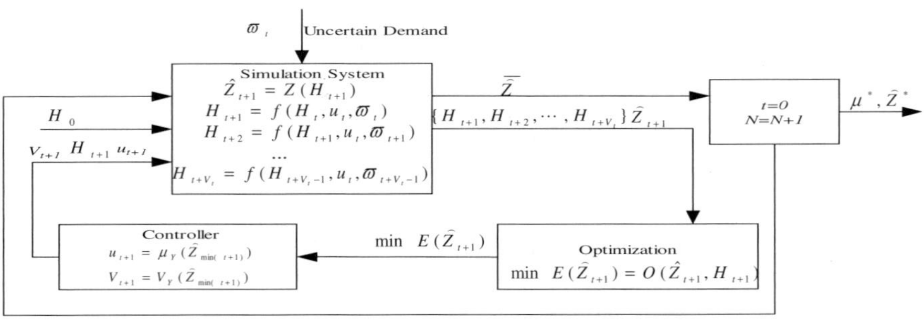

Fig. 3. Process of Simulation-Based Optimization

4 Example

The example as follow is based on the history demand of an aircraft model factory, and the results of optimization will be compared with the factory historical status.

Finally, by means of simulation based on three distinct typical demands, the stability of the optimization model is studied.

The values or ranges of main parameters is presented as follow firstly:

- The Amount of candidate suppliers K=6
- The selection order of candidate suppliers$\{S_i\}=\{S_4,\ S_2,\ S_1,\ S_5,\ S_6,\ S_3\}$
- The max capability of inventory W_{max}=500
- The Max batch of suppliers $\{P_i\}$={28, 27, 28, 25, 20, 23}
- The Initial amount of suppliers =4;
- The Initial set of suppliers $G=\{S_4,\ S_2,\ S_1,\ S_5\}$
- The Initial batch of supplier $\{p_i\}$={25, 27, 25, 18}

The order delivery plan of the factory in 2005 is showed in fig. 5, and there are 6 samplings in a month.

Because of controlling the ratio of out of outsourcing parts stock, the factory usually maintain a large number of outsourcing parts inventory. According to the 2005 annual statistics, the cost of outsourcing pats inventory is as much as 269,246 yuan. The inventory changes of outsourcing parts are shown in fig. 5, and the inventory smoothing effect is obvious. Compared with results of optimization are shown in table 2, the optimal cost of procurement savings than the factory historical status is 336,854-269,246=67,608 yuan.

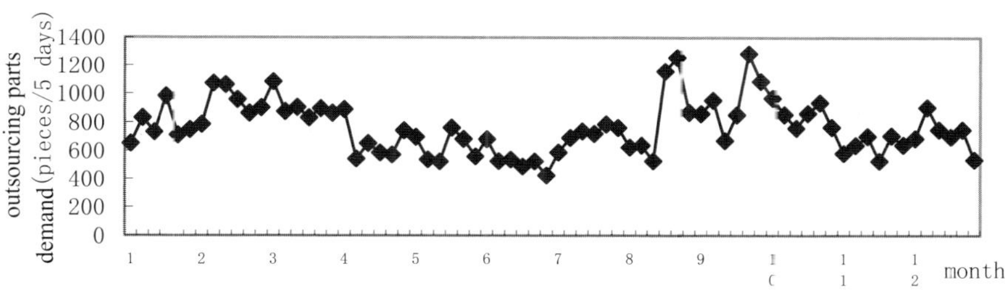

Fig. 4. Order delivery plan of 2005

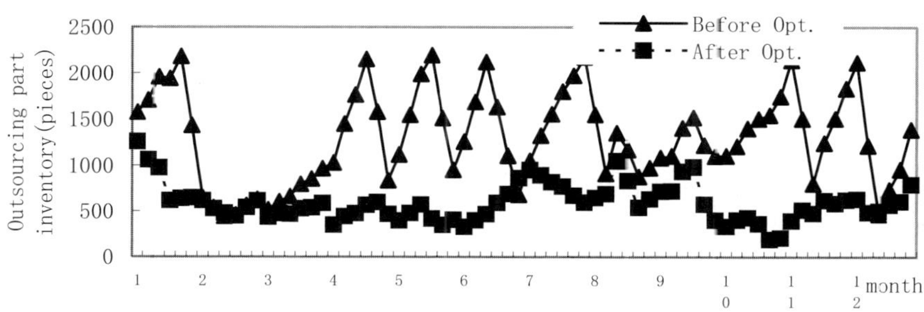

Fig. 5. Changes of inventory before Opt. and after Opt

Table 2. Result of optimization

Intervals of total plan time T	μ_{Y1}, μ_{Y2}, μ_{Y3}	Control point	Replenishment decision (G, $\{p_i\}$)	C_{total}
Jan. 1→Apr. 5	0,S,798	Jan. 25	$\{S_4, S_2, S_1, S_5\}\{25, 27, 21, 18\}$	59,236
		Feb. 20	$\{S_4, S_2, S_1, S_5\}\{25, 27, 28, 18\}$	
		Mar. 5	$\{S_4, S_2, S_1, S_5\}\{25, 27, 28, 6\}$	
Apr. 5 → Aug. 15	0,S,363	Apr. 5	$\{S_4, S_2, S_1, S_5\}\{25, 27, 28, 10\}$	94,885
		Apr. 25	$\{S_4, S_2, S_1, S_5\}\{25, 27, 28, 12\}$	
		May 15	$\{S_4, S_2, S_1, S_5\}\{25, 27, 28, 10\}$	
		June 25	$\{S_4, S_2, S_1, S_5\}\{25, 27, 28, 12\}$	
		July 5	$\{S_4, S_2, S_1, S_5\}\{25, 27, 28, 2\}$	
Aug. 15→Oct. 15	1,M,1250	Aug. 15	$\{S_4, S_2, S_1, S_5, S_6\}\{25, 27, 28, 10, 5\}$	57,692
		Aug. 25	$\{S_4, S_2, S_1\}\{25, 27, 28,\}$	
		Sep. 15	$\{S_4, S_2, S_1, S_5, S_6\}\{25, 27, 28, 10, 7\}$	
Oct. 15→Dec. 31	0,S,410	Oct. 15	$\{S_4, S_2, S_1, S_5, S_6\}\{25, 27, 28, 5, 5\}$	57,433
		Nov. 15	$\{S_4, S_2, S_1, S_5, S_6\}\{25, 27, 28, 5, 2\}$	
		Dec. 10	$\{S_4, S_2, S_1, S_5, S_6\}\{25, 27, 28, 8, 5\}$	
2005	Cost of procurement in 2005			269,246

5 Conclusions

The proposed program provides a useful tool for analysis of outsourcing procurement cost control challenges of MTO manufacturing firms, optimize the firms operation under different supply and demand configurations. The study developed a procurement cost optimization model based on simulation to achieve continuous effectively control of outsourcing supply, having integrated the evolutionary optimization. The simulation and analytical approaches is presented in this paper to ensure the validity and feasibility of the model. Note that the combination of optimization with simulation has shown to be a successful tool in this real industrial problem.

Further research efforts may be directed to ward the development of the model's shortcomings and deficiencies. For instance, there is the assumption of problem, i.e., supplier replenishment batch are fixed. In practice, if there would be an urgent demand of outsourcing parts, a not full batch is desired to enter the outsourcing inventory.

Acknowledgements

This paper is financed and supported by The National High Technology Research and Development Program (2006AA04Z157).

References

1. Chen, C.-L., Lee, W.-C.: Multi-objective optimization of multi-echelon supply chain networks with uncertain product demands and prices. Computer and Chemical Engineering 28, 1131–1144 (2004)

2. Weng, Z.K., MClurg, T.: Coordinated ordering decisions for short life cycle products with uncertainty in delivery time and demand. European Journal of Operation Research 151, 12–24 (2003)
3. Ross, S.M.: Introduction to Probability Models, 8th edn. Academic Press, New York (2003)
4. Fishman, G.S.: Monte Carlo, Concepts, Algorithms and Applications. Springer, New York (1996)
5. Azadivar, F.: A tutorial on simulation optimization. In: Proceedings of the 1992 Winter Simulation Conference, Arlington, VA, pp. 198–204 (1992)
6. Fu, M.C.: Optimization via simulation: A review. Annals of Operation Research 53, 199–247 (1994)
7. Alberto, I., Azcárate, C.: Optimization with simulation and multiobjective analysis in industrial decision-making: A case study. European Journal of Operational Research 140, 373–383 (2002)
8. Law, A.M., Kelton, W.D.: Simulation Modeling and Analysis, 3rd edn. McGraw-Hill, New York (2000)
9. Edward, Kac, P.C.: An Introduction to Stochastic Processes. Duxbury Press, Boston, MA (1997)
10. Fu, M.C., Healy, K.J.: Techniques for optimization via simulation: An experimental study on an (s, S) inventory system. IIE Transaction 29, 191–199 (1997)
11. Winter, G., Periaux, J.M., Cuesta, P.: Genetic Algorithms in Engineering and Computer Science. In: Proceedings of 1st Short Course EURO-GEN-95, pp. 127–140. Wiley, New York (1995)

Gene Optimization: Computational Intelligence from the Natures and Micro-mechanisms of Hard Computational Systems

Yu-Wang Chen and Yong-Zai Lu

Department of Automation, Shanghai Jiaotong University,
200240 Shanghai, P.R. China
{cywpeak,yzlu}@sjtu.edu.cn

Abstract. Research on evolutionary theory and statistic physics has provided computer scientists with powerful methods for designing intelligent computational algorithms, such as simulated annealing, genetic algorithm, extremal optimization, etc. These techniques have been successfully applied to a variety of scientific and engineering optimization problems. However, these methodologies only dwell on the macroscopic behaviors (i.e., the global fitness of solutions) and never unveil the microscopic mechanisms of hard computational systems. Inspired by Richard Dawkins's notion of the "selfish gene", the paper explores a novel evolutionary computational methodology for finding high-quality solutions to hard computational systems. This method, called gene optimization, successively eliminates extremely undesirable components of sub-optimal solutions based on the local fitness of genes. A near-optimal solution can be quickly obtained by the self-organized evolutionary processes of computational systems. Simulations and comparisons based on the typical NP-complete traveling salesman problem demonstrate the effectiveness and efficiency of the proposed intelligent computational method.

1 Introduction

Extremely hard computational systems are pervasive in fields ranging from scientific computation to engineering. A representative collection of these problems are called NP-complete problems: if someone finds a polynomial-time algorithm for one NP-complete problem, then this algorithm could be used as a subroutine in programs that would then solve all other NP-complete problems in polynomial time [1]. Examples of NP-complete problems include the traveling salesman problem (TSP) – a classic combinatorial optimization conundrum that requires calculating the shortest route among a selection of cities [2]. Generally, with the characteristics of combinatorial explosion, the solution time grows exponentially with the size of problem, making it impossible to solve very large instances in reasonable times.

In response to the difficulties, computer scientists have developed a variety of intelligent computational algorithms, which are inspired by evolutionary theory and statistic physics, such as simulated annealing [3], genetic algorithm [4], extremal

K. Li et al. (Eds.): LSMS 2007, LNCS 4688, pp. 182 – 190, 2007.

optimization [5], etc. These optimization methods, although not always guaranteed to find the optimal solution, often provide satisfactory solutions in a reasonable time. A variety of scientific and engineering problems have been successfully solved by related techniques. However, these methodologies only dwell on the macroscopic behaviors (i.e., the global fitness of solutions) and never unveil the microscopic mechanisms of computational systems.

Inspired by Richard Dawkins's theory of the "selfish gene" [6-7], which is a gene centered view of evolution [8], a framework broadly used in evolutionary theory, assumes that one can assign an effective fitness directly to each allele, this paper explores a novel evolutionary computation methodology for finding high-quality solutions to hard computational systems. This method, called gene optimization, successively eliminates extremely undesirable components of sub-optimal solutions based on the local fitness of genes. A near-optimal solution can be quickly obtained by the self-organized evolutionary processes of computational systems [9-10]. As conjectured by R. Dawkins [6], the evolution at the gene level is the most intrinsic mechanism of natural selection, the evolutionary computation methodology under study in this paper may affect the conceptual foundations of evolutionary computation, and can optimize computational systems from its natures and micro-mechanisms.

2 Anatomy of Hard Computational Systems

In this section, we anatomize hard computational problem from a microscopic and systematic viewpoint.

2.1 Systematic Insights on Hard Computational Problems

Mathematically, hard computational systems can be formulated as combinatorial optimization problems for finding an "optimal" configuration of a set of discrete variables to achieve some objectives. Generally, a combinatorial optimization problem $P = (S, F)$ is defined by:

– a set of variables $X = \{x_1, \ldots, x_n\}$, and variable domains $D_1, \ldots, D_n$;
– constraints among variables;
– an objective function f to be minimized, where $F: D_1\times\cdots\times D_n \rightarrow R$;

The set of all feasible solutions is: $S = \{s = \{x_1, \ldots, x_n\} \mid x_i \in D_i$, s satisfies all the constraints$\}$. S is usually called a solution space, as each element s of the set can be seen as a possible solution. Given a computational system, the aim is to find a microscopic configuration $s^* \in S$ which minimizes the objective function, i.e., $F(s^*) \leq F(s)$ for each solution s in the solution space S.

To illustrate the systematic and microscopic characteristics of computational systems, let us consider the typical NP-complete traveling salesman problem. It has wide applications in the disciplines of discrete mathematics, theoretical computer science, and computational biology. Due to the combinatorial complexity of the solutions and the strong non-convexity of the cost function in the hyper-dimensional solution space, the TSP has often served as a test-bed for many optimization algorithms. In the TSP we are given a set $\{c_1, c_2, \ldots, c_n\}$ of cities and for each pair $\{c_i, c_j\}$ of distinct cities a distance $d(c_i, c_j)$. The aim is to find an ordering π of the cities that minimizes the objective function:

$$F(s)=\sum_{i=1}^{n-1} d(c_{\pi(i)},c_{\pi(i+1)})+d(c_{\pi(n)},c_{\pi(1)}) \tag{1}$$

Based on the systematic picture of biological evolution wholly "from the point of view of the gene", each city in a directed Hamiltonian cycle containing all cities can be viewed as a gene in survival machines [11], and the microscopic configuration of solutions can be characterized by the states of all genes. As an illustration, the state variable of city i can be defined as $s_i = k$ $(1 \le k \le n-1)$ if it is connected to the k-th nearest neighbor. In the utopian society, all genes reside in the ideal environment and have the maximum probability of reproductive success, i.e., $s_i = 1$ for all gene i. However, the computational systems in practice are often "frustrated" by the interacting effects of the agents causing that the genes in any microscopic configurations have deterministic states, but not always stay on the fittest states. Consequently, a local fitness can be defined for evaluating the underlying impetus to be fittest for each gene. In a feasible TSP tour, let d_i be the length of the forward-directed edge starting from city i, the local fitness of city i is defined as:

$$f(c_i)=d_i-\min_{j\neq i} d(c_i,c_j) \tag{2}$$

Consequently, the global fitness for any possible solution s can be expressed as:

$$F(s)=\sum_{i=1}^{n} \min_{j\neq i} d(c_i,c_j)+\sum_{i=1}^{n} f(c_i) \tag{3}$$

Obviously, the first part of equation (3) is a constant for a given optimization instance. As a result, the optimal solution for optimization problems is equivalent to the microscopic configuration with the minimal sum of local fitness for all genes.

2.2 Global Fitness vs. Local Fitness

The evolutionary process of computational systems mainly correlates with fitness function and neighbor heuristic. Generally, neighbor heuristic, also called move class, is guided by fitness function. A computational system may evolve in different paths if it is guided by different fitness functions. For a specified computational system, different fitness function can also be constructed for searching the same optimization objective.

As a rule of thumb, a global fitness can be easily defined to evaluate how good a solution is for any computational systems. The evolution is driven by the iterative process of computing the global fitness of all new solutions and returning an optimized solution. It is a simple and natural idea to use global fitness to guide the evolution of computational systems to the optimal configuration.

Alternatively, local fitness can be defined to evaluate how fit a gene is. With no global view of computational systems, the evolution is guided by following procedures: compute the local fitness of all genes, and eliminate those inferior genes using some evolutionary strategies. As a biological system, some global emergent behaviors will arise from local improvements and interactions.

Since the local fitness is not as straightforward as global fitness in optimization problems, the following question should be considered: what is the relationship and differences between global fitness and local fitness for the evolution of computational systems? For answering this question, here we define the consistency and equivalence theory between global fitness and local fitness.

Definition 1. Suppose that solution s' is constructed from s ($\forall s \in S$) by altering the state of variable i using certain neighbor heuristic, local fitness is ***consistent*** with global fitness, if it is true that

$$\mathrm{sgn}(F(s')-F(s)) = \mathrm{sgn}(\sum\nolimits_{x\in X(s,s',i)} f_{s'}(x) - \sum\nolimits_{x\in X(s,s',i)} f_{s}(x)) \quad (4)$$

where sgn(x) is the symbolic function, and $X(s,s',i)$ is the variable set in which all variables' states are altered because of the interacting effects of altering the state of variable i.

If the move class is a reversible process, $X(s,s',i) = X(s',s,i)$, and the solution space will be defined as an undirected graph.

Definition 2. Local fitness is ***equivalent*** to global fitness if

(i) the local fitness is consistent with global fitness and

(ii) $\exists\alpha \in R^{+}$, $\beta \in R$ such that it is true that for all $s \in S$

$$F(s) = \alpha\sum\nolimits_{x\in N} f(x) + \beta \quad (5)$$

Obviously, equivalence is the subset concept of consistency. If local fitness is consistent with global fitness, a good transformation by altering one or some genes' states based on local fitness is also a good change to the whole computational system. The notion makes it easier for us to understand that local fitness can also guide the computational system to a near-optimal solution. Furthermore, if local fitness is consistent with global fitness, the neighbor heuristic derived from either global fitness or fitness can guide the system to global optimum.

3 Gene Optimization

This section seeks to present an effective evolutionary computation method at the level of genes, rather than at the usual level of individual solutions, and the optimization dynamics simulates the self-organized evolution process driven by extremal dynamics [12].

3.1 Self-organized Evolution

The concept of self-organization originates in the study of nonlinear physical and chemical systems, such as convection flows and chemical reactions that form waves [13]. In those self-organized systems, information and control are distributed among many interacting agents, and organization can seem to arise spontaneously from disorder. In biological systems, self-organization evolution can lead to surprisingly complex functional structures, and the local rules of interaction can be tuned by natural selection to produce larger-scale patterns that are adaptive.

The Bak-Sneppen (BS) evolution model [9] driven by extremal dynamics is perhaps the simplest prototype of self-organized evolution. In the "toy" model, species are located on the sites of a lattice, and have an associated "fitness" value between 0 and 1 (randomly sampling from a uniform distribution). At each update step, the extremal species, i.e., the one with the smallest value is selected. Then, that species and its interrelated species are updated with new random numbers. After a sufficient number of update steps, the system reaches a highly correlated self-organized criticality, and the fitness of almost all species has transcended a certain fitness threshold. However, the dynamical systems maintain punctuated equilibrium: those species with lower fitness can update their states and affect the interrelated neighbors. This co-evolutionary activity gives rise to chain reactions called "avalanches", large fluctuations that rearrange major parts of the system, potentially making any configuration accessible [14].

3.2 Gene Optimization

Base on the definition of local fitness and the self-organized evolution process discussed above, the gene optimization algorithm is presented as follows:

Step-1. Initialization

Define the mapping between the phenotype of a solution of the optimization problem with n variables x_i and the genotype $s = (x_1, \ldots, x_n)$ (i.e., chromosome),

Define the global fitness $F(s)$ for chromosomes and local fitness $f(x_i)$ for evaluating the local fitness contributions of each gene i,

Construct a initial solution s randomly, let s_{best} represent the best solution found so far, and set $s_{best} \leftarrow s$;

Step-2. For the "current" solution s:

(a) Evaluate $f(x_i)$ for all genes i $(1 \le i \le n)$,

(b) Find the gene sequence $\{\pi(1), \ldots, \pi(k), \ldots, \pi(n)\}$, in which $f(x_{\pi(1)}) \ge \ldots \ge f(x_{\pi(k)}) \ldots \ge f(x_{\pi(n)})$, i.e., the rank k is going from $k = 1$ for the gene with the worst fitness to $k = n$ for the gene with the best fitness,

(c) Obtain the gene $\pi(k)$ with the k-th "worse" fitness by the probability distribution $P_k \propto k^{-\alpha}$ $(1 \le k \le n)$, where α is an adjustable power parameter,

(d) Construct a neighborhood $N(s)$ of solution s by altering the allele of the gene $\pi(k)$,

(e) Sequence the solution in the neighborhood by the descending order of global fitness, choose a neighbor s' having the l-th "worse" global fitness by another power-law distribution $P_l \propto l^{-\beta}$ $(1 \leq l \leq |N(s)|)$, where β is another power parameter, and $|N(s)|$ is the cardinality of the neighbor $N(s)$;

(f) Accept $s \leftarrow s'$ unconditionally, independent of $F(s') - F(s)$, if $F(s') < F(s)$, set $s_{best} \leftarrow s'$;

Step-3. Repeat at step-2 until the termination criteria is satisfied, which is a certain number of iterations, or a predefined amount of *CPU* time;

Step-4. Return s_{best} and $F(s_{best})$.

As described above, gene optimization adopts the bottom-up approach. It puts emphasis on the level of genes, which is considered to be the intrinsic mechanisms of macroscopic behaviors at the usual level of individual organisms, species and eco-systems.

3.3 Co-evolution

It is worthy noting that co-evolution [15] plays an important role in the gene optimization. In the neighbor heuristic for constructing a neighborhood $N(s)$ from the current solution s, updating the states of those unfit genes has an impact on the fitness of the correlated genes. Consequently, co-evolution of interacting genes will give rise to a coupled local fitness landscapes and induce the emergent behaviors of computation systems. From the gene centered view of evolutionary computation, the fitness interaction of genes is a procedure of competitive and cooperative co-evolution. In the gene optimization, the co-evolution can be illustrated by Fig.1. We can see that competition between genes causes selection of genes with "worse" fitness, and collaboration between genes enables the updated solution to be competitive among the neighborhood.

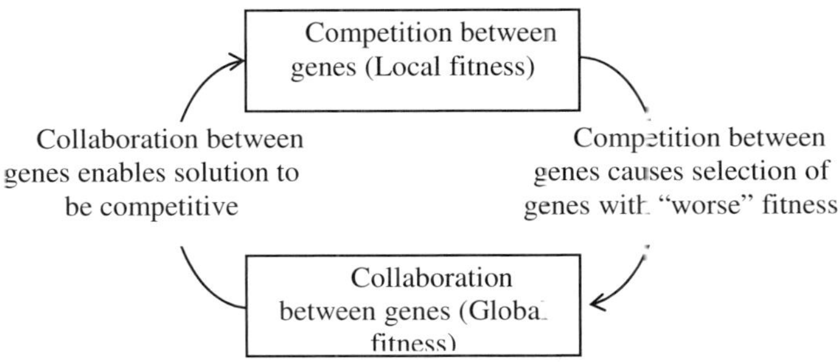

Fig. 1. Co-evolution of interacting genes in gene optimization

Consequently, a near-optimal solution can be quickly obtained by the self-organized co-evolutionary processes.

4 Simulations and Comparisons

To explicitly explain the self-organized evolutionary processes of gene optimization, here we simulate the proposed optimization techniques on the TSP. In our computational experiments, all the neighborhoods in the iterative evolutionary processes are constructed by the 2-*opt* move [16]: construct new Hamiltonian cycles by deleting two edges and reconnecting the two resulting paths in a different way. In other words, the paper puts emphasis on the analysis of the evolutionary processes for hard computational systems. By simulating the proposed optimization algorihtm on TSP instance (gr48) from the Reinelt's TSPLIB [17], a typical evolutionary dynamics of gene optimization ($\alpha = 1.45$, $\beta = 2.5$) is plotted in Fig.2.

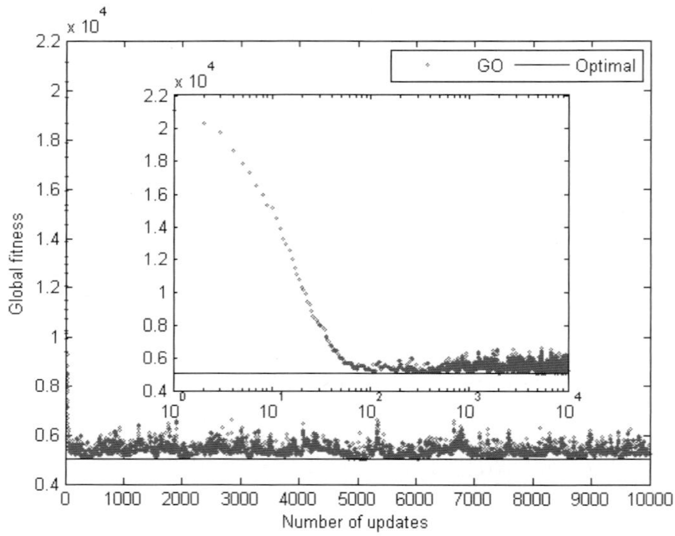

Fig. 2. Evolution of the global fitness in a typical run of gene optimization. The inset (log- lin plot) shows that gene optimization descends sufficiently fast to a near-optimal solution with enough fluctuations to pass through barriers of fitness landscape.

Comparatively, simulated annealing (SA) is a computational stochastic technique that simulates the thermodynamic process of annealing of molten metals to attain the lowest free energy. The primary mechanism of SA is to permit occasional updates with worsening system configuration according to Boltzmann probability $exp(-\Delta L/T)$, so that it can possibly help locate the neighborhood to the global minimum. For enforcing the system to reach the equilibrium dynamics, SA requires a carefully tuned temperature schedule. SA and its modifications as powerful optimization tools have been applied to a variety of disciplines. However, the major negative features of SA is that the technique only dwells on the macroscopic behavior of computational systems (i.e., global energy function), and does not investigate the microscopic states of solution configuration.

Extremal optimization (EO) is an algorithm that, starting from a random solution, searches for the optimal solution by successively replacing extremely undesirable elements of the sub-optimal solution with new random ones. To avoid getting stuck into a local optimum, a single parameter τ is introduced into EO and the new algorithm is called τ-EO [14]. Although EO and its variations attempt to optimize hard computational systems from the local rules, compared with gene optimization, the noticeable differences are the definition of local fitness and the parameter β for regulating the extent of avalanche-like fluctuations.

Genetic algorithm (GA) is an adaptive population-based search algorithm premised on the evolutionary ideas of natural selection and genetic. The basic concept of GA is designed to simulate processes in natural system necessary for evolution, specifically those that follow the principles first laid down by Charles Darwin of survival of the fittest. It mimics evolution on the solution level, keep track of entire "gene pools" of configurations and use many tunable parameters to select and "breed" an improved generation of solutions.

Comparatively, Table 1 reports the experimental results obtained by running SA [3], τ-EO [14], GA [18-19] and GO on a number of TSP benchmark problems, which are available the Reinelt's TSPLIB. All the algorithms are encoded in C++ and implemented on a Pentium IV 2.80GHz CPU. The computational results for each problem are averaged over 10 runs. The error percentage %opt. = $(F(s)-F(s^*))/F(s^*)\times 100\%$, and the computing time τ_{CPU} are used to evaluate the effectiveness and efficiency of different optimization methods respectively.

Table 1. Comparisons of SA, τ-EO, GA and GO applied to 6 TSP benchmark problems, in which the numbers of cities range from 24 to 1032

Problem	$F(s^*)$	SA		τ-EO		GA		GO	
		%opt.	τ_{CPU}	%opt.	τ_{CPU}	%opt.	τ_{CPU}	%opt.	τ_{CPU}
gr24	1272	0.00	1	0.00	1	0.00	4	0.00	1
gr48	5046	0.06	2	0.00	2	0.00	15	0.00	2
gr120	6942	0.61	11	0.26	12	0.53	74	0.07	9
si175	21407	0.92	25	0.35	25	0.67	121	0.08	21
si535	48450	0.75	210	0.55	197	0.56	1215	0.12	175
si1032	92650	0.91	450	0.52	391	0.76	3102	0.18	305

Experimental results show that the proposed GO algorithm always outperforms state-of-the-art SA, τ-EO and GA. It seems to be promising for solving a variety of hard computational systems.

5 Conclusions

In Summary, we have presented a novel evolutionary computation methodology for finding high-quality solutions to hard computational systems. The experimental results on the traveling salesman problem demonstrate that gene optimization performs very well and provides much better performance than existing stochastic search methods, such as simulated annealing, extremal optimization, and genetic algorithm. Since the evolutionary computation methodology under study in this paper takes aim at the natures and micro-mechanisms of hard computational systems, and guide the evolutionary optimization process at the level of genes, it may provide much better evolutionary algorithms and affect the conceptual foundations of evolutionary computation.

Acknowledgements. The authors would like to thank Bang-hua Yang and Ting Wu for their helps and comments on a draft of the paper.

References

1. Farhi, E., Goldstone, J., Gutmann, S., Lapan, J., Lundgren, A., Preda, D.: A Quantum Adiabatic Evolution Algorithm Applied to Random Instances of an NP-Complete Problem. Science 292, 472–476 (2001)
2. Gutin, G., Punnen, A.P. (eds.): The Traveling Salesman Problem and Its Variations. Kluwer Academic Publishers, Boston (2002)
3. Kirkpatrick, S., Gelatt Jr., C.D., Vecchi, M.P.: Optimization by simulated annealing. Science 220(4598), 671–680 (1983)
4. Forrest, S.: Genetic algorithms: principles of natural selection applied to computation. Science 261(5123), 872–878 (1993)
5. Boettcher, S.: Extremal optimization: heuristics via co-evolutionary avalanches. Computing in Science and Engineering 2, 75–82 (2000)
6. Dawkins, R.: The Selfish Gene, 2nd edn. Oxford University Press, Oxford (1989)
7. Heylighen, F.: Evolution, Selfishness and Cooperation. Journal of Ideas 2(4), 70–76 (1989)
8. Bar-Yam, Y.: Formalizing the gene centered view of evolution. Advances in Complex Systems 2, 277–281 (2000)
9. Bak, P., Sneppen, K.: Punctuated Equilibrium and Criticality in a Simple Model of Evolution. Physical Review Letters 71(24), 4083–4086 (1993)
10. Visscher, P.K.: How self-organization evolves. Nature 421, 799–800 (2003)
11. Sperber, D.: Evolution of the selfish gene. Nature 441, 151–152 (2006)
12. Gabrielli, A., Cafiero, R., Marsili, M., Pietronero, L.: Theory of self-organized criticality for problems with extremal dynamics. Europhysics Letters 38, 491–491 (1997)
13. Pepper, J.W., Hoelzer, G.: Unveiling Mechanisms of Collective Behavior. Science 294, 1466–1467 (2001)
14. Boettcher, S., Percus, A.G.: Nature's way of optimizing. Artificial Intelligence 119, 275–286 (2000)
15. Back, T., Fogel, D.B., Michalewicz, Z. (eds.): Handbook of Evolutionary Computation. IOP Publishing Ltd and Oxford University Press (1997)
16. Okano, H., Misono, S., Iwano, K.: New TSP Construction Heuristics and Their Relationships to the 2-Opt. Journal of Heuristics 5, 71–88 (1999)
17. Reinelt, G.: TSPLIB-a traveling salesman problem library. ORSA Journal on Computing 3(4), 376–384 (1991)
18. Michalewicz, Z.: Genetic Algorithms + Data Structures = Evolution Programs. Springer, London (1996)
19. Potvin, J.Y.: Genetic algorithms for the traveling salesman problem. Annals of Operations Research 63, 339–370 (1996)

Reinforcement Learning Algorithms Based on mGA and EA with Policy Iterations*

Changming Yin[1], Liyun Li[1], and Hanxing Wang[2]

[1]College of Computer and Communicational Engineering
Changsha University of Science and Technology
Changsha, Hunan, 410076, China
[2]College of Sciences, Shanghai University, Shanghai, 200444, China
yinchm@csust.edu.cn

Abstract. We contribute two new algorithms in this paper called PImGA and PIrlEA respectively in which we construct populations online in each iteration. Every iteration process in these two algorithms does not like the normal EA and GA in which they employ the inefficient value iteration method in general, instead of, in this paper, we employ the efficient policy iteration as the computation method for searching optimal control actions or policies. Meanwhile,these algorithms also do not like general EA and GA for selection operator to get a optimal policy, instead of we make the Agent learning a good or elite policy from its parents population. The resulted policy will be as one of elements of the next population. Because this policy is obtained by taking optimal reinforcement learning algorithm and greedy policy, the new population always can be constructed by applying better policies than its parents, that is to say, the child or offspring will inherit parents' good or elite abilities. Intuitively, for a finite problem, the resulted population from simulation will accommodate the near optimal policies after a number of iterations. Our experiments show that the algorithms can work well.

1 Introduction

Many literatures have described the conceptions about evolutionary computation. Here we would like to give a glancing description for some important definitions briefly. Roughly speaking, evolutionary techniques simulate Darwin's theory of the survival of the fittest during the evolution of species. Evolutionary methods have been used for many practical applications, such as designing patentable circuit boards, machine learning, reasoning, planing, optimal control and so on.

The central notion behind evolutionary computation is to treat programs as individuals in a population. In our learning control problem, a population is a group of policies, and the individuals , or genes, are these of policies in a population. An evolutionary algorithm(EA) iteratively updates a population of potential solutions to obtain more good or elite policies. The first problem

* This research was funded by NSFC No.10471088 and No.60572126.

K. Li et al. (Eds.): LSMS 2007, LNCS 4688, pp. 191–204, 2007.

in evolutionary computation had to solve was how to imitate the reproductive process whereby two individuals combine genetic material into their offspring. During each iteration, or at each generation, the EA evaluates solutions and generates offspring based on the fitness of each solution. A strictest way is the genetic algorithms(GA) to be employed. A less restrictive way of which is known as genetic programming enabling genetic material to crossover from the parents is to make the programs look like graphs[5,8].

The next problem to be addressed was how to simulate the survival of the fittest. To do this, some researchers must specify a fitness function for each application of evolutionary computation. Because large chunks of genetic material are swapped from both parents into the child, there is a good chance that the child program will inherit both parents' good abilities but slow progress.

Developing this slow progress situation in evolutionary computation applications is how to take thousands of generations to produce programs in the population which are adequate at the task in AI research. In combining of evolutionary computation and reinforcement learning methods, we can look at natural evolution try to overcome this problem through learning elite policy online for constructing new populations. The size of a population can be controlled by learning algorithm and the offspring producing can be dealt with using an easy way.

In paper [2],the authors, David E.M., Alan C.S. and John J.G., try to combine two methods of EA algorithm and reinforcement learning for solving optimal control problems. But their algorithm is fully based on the simplest EA framework and not more well than traditional methods. This idea told us we can use these existing EA algorithms to solve reinforcement learning problems. In literature [1], Chang H.S., Lee H-G., M.Fu, and Marcus S.I. discussed, using EA and policy iteration method, how to solve Markov Decision Processes(MDP) problem. This is a good idea to try to find a new way for dealing with the old topic. In their paper, the authors employed a kind of special EA frame to search the optimal or best policies, in which the algorithm constructs a series of populations and each population accommodates an elite policy from its parents as one of its individuals. This elite policy was calculated by greedy way. But a fatal disadvantage in this algorithm is that the policy searching speed still is not considered and only the convergence is proved theoretically. So, the application for a practical reinforcement learning problem is limited.

In this paper, we develop two new algorithms called PImGA and PIrlEA respectively in which we construct populations online in each iteration. They do not like general EA and GA for selection operator, instead of we make the Agent learning a good or elite policy from its parents population. The resulted policy will be as one of elements of the next population. Because this policy is obtained by taking optimal reinforcement learning algorithm and greedy policy, the new population always can be constructed by applying better policies than its parents, that is to say, the child will inherit parents' good or elite abilities. Meanwhile, the algorithms will employ the efficient policy iteration method instead of general value iteration. The main benefit for using policy iteration is that we can manipulate the policies directly, rather than finding it indirectly via the optimal value function. Intuitively, for a finite problem, the resulted population will accommodate the near optimal policies after a number of iterations.

This paper will arrange the context according to the following order. The next section spells out the reinforcement learning problem and reviews its model briefly. Section 3 and 4 are the main parts of this paper in which we shall propose two new algorithms respectively how to search optimal polices and solve reinforcement learning problems. In section 5, we show a simple but important experiment model to verify our new algorithms how good and efficient they are. The final section summarizes our new ideas and points out further research directions.

2 Reinforcement Learning Model and MDPs

In recent years, reinforcement learning (RL) has been an active research area not only in machine learning but also in control engineering, operations research, robotics, microeconomics and so on. It is a computational approach to understand and automate goal-directed learning and decision-making, without relying on exemplary supervision or complete models of the environment [11]. In RL, an Agent is placed in an initial unknown environment and only receives evaluative feedback from the environment. The feedback is called reward or cost, or both of them are called reinforcement signal. The ultimate goal of RL is to learn a strategy for selecting actions such that the expected sum of discounted rewards is maximized or costs is minimized. Since lots of problems in the real world are sequential decision processes with delayed evaluative feedback, the research in RL has been focused on theory and algorithms of learning to solve the optimal control problem of Markov decision processes (MDPs) which provide an elegant mathematical model for sequential decision-making. The common descriptions for reinforcement learning problem can be found in paper [11] in which the authors have a nice review for this research area. They also pointed its importance to machine learning study and the main research methods.

Below we shall review the reinforcement learning model briefly. Firstly, we represent the MDP model for an Agent as a 4-tuple $M = (X, A, P, r)$, where $X = \{s_1, s_2, \cdots, s_n\}$ is a finite set of states, $A = \{a_1, a_2, \cdots, a_m\}$ denotes a finite action set, $r : X \times A \longrightarrow \mathbf{R}$ is a reward function, and P gives the dynamics; we write $P(s, a, s')$ for the probability of reaching state s' when executing action a from state s. More of the descriptions of this model can be found in [7] et al. If $V \in \mathbf{R}^{|S|}$ is an arbitrary assignment of values to states, we define state-action values with respect to V by

$$Q_V(s,a) = r(s,a) + \gamma \sum_{s' \in X} P(s,a,s')V(s'), \tag{1}$$

where $\lambda \in (0, 1]$ is a discount factor. A deterministic or stationary policy is a function $\pi : X \rightarrow A$, where $\pi(s)$ is the action of that the agent takes at state s. The *state-action value function* $Q^\pi(s, a)$, defined over all possible combinations of states and actions, indicates the expected, discounted, total reward when taking action a in state s and following policy π thereafter. The exact Q-values for all state-action pairs can be found by solving the linear system of the Bellman's equations. For a policy π, we define the value function of π as the solution to the set of linear equations

$$V^{\pi}(s) = r(s, \pi(s)) + \gamma \textstyle\sum_{s' \in X} P(s, \pi(s), s')V^{\pi}(s'). \tag{2}$$

It is well-known that there exists an optimal value function V^*, and it follows the Bellman's equations [10]:

$$V^*(s) = \max_{a \in A} Q_{V^*}(s, a). \tag{3}$$

For every MDP, there exists a deterministic optimal policy, π^*, not necessarily unique, which maximizes the expected, discounted return or reward of every state. The state-action value function Q^{π^*} of an optimal policy is the fixed point of the non-linear Bellman's optimality equations:

$$Q^{\pi^*}(s, a) = r(s, a) + \gamma \max_{a'} \sum_{s' \in X} P(s, a, s')Q^{\pi^*}(s', a'). \tag{4}$$

Policy Iteration is a method of discovering an optimal policy by iterating through a sequence of monotonically improving policies. Each iteration consists of two phases: Value Evaluation or Determination computes the value function for a policy $\pi^{(t)}$ by solving these linear Bellman's equations, and Policy Improvement defines the next policy as

$$\pi^{(t+1)}(s) = \arg\max_{a} Q^{\pi^{(t)}}(s, a). \tag{5}$$

These steps are repeated until convergence to an optimal policy, often in a surprisingly small number of steps. Value Iteration is another method of approximating the Q^{π^*} values arbitrarily closely by iterating the equations above and it is similar to the Gauss iteration for linear systems.

For an arbitrary value function V,we define the (signed) Bellman error of V at state s by

$$\mathrm{be}_V(s) = V(s) - \max_{a \in A} Q_V(s, a). \tag{6}$$

The greedy policy with respect to V, greedy(V), is defined by

$$\mathrm{greedy}(V)(s) = \arg\max_{a \in A} Q_V(s, a). \tag{7}$$

In the absence of a model of the MDP, that is, when P and r are unknown, the decision maker or the Agent has to learn the optimal policy through interaction with the environment. Knowledge comes in the form of samples (s, a, r, s'), where s is a state of the process, a is the action taken in s, r is the reward received, and s' is the resulting state. Samples can be collected from actual (sequential) episodes or from queries to a generative model of the MDP. In the extreme case, they can be experiences of other Agents on the same MDP. The class of problems that fall under this framework is known as Reinforcement Learning (RL). So, according to this framework, the objective in RL for the Agent is how to learn the value function and get an optimal action or policy at each state,and then obtain an optimal policy. There are a series of existing methods can solve this traditional problem. Comparing with the value iteration method, policy iteration can work more efficient for a large scale learning control problem. In this paper, we assume that the MDP has an infinite horizon and that future rewards are discounted exponentially with a discount factor $\lambda \in [0, 1)$. Again, assuming that all policies are proper, i.e. that all episodes eventually terminate, our results generalize to the undiscounted case as well. We can learn more details about reinforcement learning in Sutton and Barto's book [10].

3 Messy Genetic Algorithms with Policy Iteration

In this section, we shall extend the mGA and propose a new algorithm using basic mGA and evolutionary policy iteration method [1,4] for solving reinforcement learning problem. We denote this kind of algorithm to be PImGA, it means Messy Genetic Algorithm with Policy Iteration. In generally, genetic algorithms are loosely modelled on processes that appear to be at work in biological evolution and the working of the immune systems [5]. A GA is a non-deterministic search and global optimization algorithm based on the ideas of genetics and try to find optimal solutions (control actions or polices) to problems instead of trying to solve them directly [5,8].

In [8], the authors developed a new learning method using a messy GA (mGA). This kind of learning method is for fuzzy control rule using messy genetic algorithms and provided a way for solving control problems based-on mGA. It is well known that mGA is more efficient than regular GA. The main differences between a mGA and a regular GA are that the mGA uses varying string lengths; the coding scheme considers both the allele positions and values; the crossover operator is replaced by two new operators called cut and splice; and it works in two phases which are called primordial and juxtapositional respectively for the evolutionary process. In our algorithm, it does not apply regular GA method for selection operator, instead the algorithm will construct a new population applying reinforcement learning method. The one of elements (individuals) in this population is a greedy policy inherited (calculated) from its parents.

During the primordial phase, in our PImGA method, the new population is constructed in accordance with the evolutionary policy iteration. This results the new population of building blocks whose combination will include optimal or near optimal solution which we call it elite policy from its father population. In the juxtapositional phase, we still employ the populations generated at primordial phase, and the GA invokes cut and splice operators acted on each generated population.

Given a fixed initial state probability distribution δ defined over states space X, we adopt the definition in [1] about the fitness value of a policy π for δ as following:

$$H_\delta(\pi) = \sum_{s \in X} V^\pi(s)\delta(s). \tag{8}$$

From [1], We can know that, for any $\pi \in \Pi$, an optimal policy π^* satisfies $H_\delta(\pi^*) \leq H_\delta(\pi)$, where Π is the set of all stationary policies $\pi : X \to A$.

Policy mutation can be done according to the following method: for each policy $\pi(S_i), i = 1, \cdots, n-1$, generate a globally mutated policy $\pi^m(S_i)$ w.p.P_m using P_g and μ ,or a locally mutated policy $\pi^m(S_i)$ w.p.$1-P_m$ using P_l and μ. Where P_m is the mutation selection probability, P_g the global mutation probability, P_l the local mutation probability and μ the given action selection distribution. The action to which it is changed would follow the given action selection distribution μ [1]. Generating $n-1$ random subsets $S_i, i = 1, \cdots, n$, of $P(k)$ is by selecting $m \in \{2, \cdots, n-1\}$ with the same probability, and selecting m policies in $P(k)$ with the same probability as the elements of set S_i [1,4].

The following is this algorithm's details.

Void PImGA { template = zeros;
select population size n;
$P(0) = \{\pi_1, \cdots, \pi_n\}$,where $\pi_i \in \Pi$; set $k = 0$;
$P_m, P_g, P_l \in (0, 1]$; the exploitation probability $q \in [0, 1]$.
for (*level*=1; *level* < *max-level*; *level* + +)
{
  //compare the elite policy with the last population.
  evaluate(template,$P(k)$);
  while (primordial-phase) **do** {
    //constructing populations.
    //calculating the value functions for each policy in $P(k)$.
    obtain V^π for each $\pi \in P(k)$:
      $V^\pi(s) = r(s, \pi(s)) + \gamma \sum_{s' \in X} P(s, \pi(s), s')V^\pi(s')$.
    //get a elite policy in $P(k)$.
    the elite policy of $P(k)$ and its fitness value defined as:
      $\pi^*(k)(x) \in \left\{ \arg \max_{\pi \in P(k)} (V^\pi(x))(x) \right\}, x \in X$;
      $H_\delta(\pi^*(k)) = \sum_{x \in X} V^{\pi^*(k)}(x)\delta(x)$;
    //an other stop condition.
    if $H_\delta(\pi^*(k)) = H_\delta(\pi^*(k-1))$
      then stop;
      else continue.
    //method for constructing individuals of a new population.
    generate $n-1$ random subsets S_i of $P(k), i = 1, \cdots, n-1$;
    //select a greedy policy in each subset S_i.
    generate $n-1$ policies $\pi(S_i)$ defined as

```
      π(S_i)(x) ∈ { arg max_{π∈S_i} (V^π(x))(x) }, x ∈ X;
      //policy mutation:
      for each policy π(S_i), i = 1, ···, n − 1, generate a mutated policy π^m(S_i);
      //construct a new population as following:
      P(k + 1) = {π*(k), π^m(S_i)}, i = 1, ···, n − 1;
      queue-in(P(k + 1)); //put the new population to a queue.
      k = k + 1; //to next population.
    } //end the first while.
    while (juxtaposition-phase) do {
      queue-out(P(k)); //take a population from the queue.
      //do cut and splice operation for resulted population.
      cut(P(k), P(k − 1));
      splice(P(k), P(k − 1));
      k = k + 1; //to the next population processing.
    } //end the second while.
    //get the elite policy from the population P(k + 1).
    template = elite(P(k + 1));
  } //end for.
} //end PImGA.
```

Algorithm 1: Pseudo-code of messy GA with evolutionary policy iteration.

Remark 1. In *while-do* section of the primordial phase, two stop conditions are intercalated for the algorithm. One is the *primordial-phase* condition, and the other is the Boolean variable which comes from the comparison in fitness formula $H_\delta(\pi^*(k)) = H_\delta(\pi^*(k-1))$. The second one means that if the fitness value of elite policy in father population is the same as in its child, then it can stop primordial phase. The reason is because the elite policy calculated according to greedy law in a population comes from its father population, this is to say, the elite policy is optimal or near optimal among all policies in father population. Obtaining the same fitness value of elite policy in father and child population demonstrates that the fitness value of elite policy in grandchild will be the same. So, at this time, the algorithm should stop this phase and continue other processing.

Remark 2. Setting the latest two populations ($P(k-1)$ and $P(k)$) as parents of the next population, this algorithm do *cut* and *splice* operations like a general mGA. The *cut* operator simply cuts the string (in our algorithm, the string is a group of policies) in two parts at an randomly chosen position. The *splice* operator concatenates two strings, which could have been previously cut, in a randomly chosen order. When the *cut* and *splice* operators are applied simultaneously to two parents strings they alt in a similar way to the ordinary *crossover* operator. In messy GAs the positions of cuts in strings, which are to be joined can be chosen independently, whereas in classical GAs the crossover points must coincide. Again, in this algorithm, there are two operators introduced which are called *queue-in* and *queue-out* respectively. The operator *queue-in* is to put a population to a queue and *queue-out* to take a population from this queue and each of them follows FIFO law. The obvious purpose for introducing these two operator is to remember the new populations at each step in primordial phase and then to reuse them in juxtapositional phase.

Remark 3. There are still two phases in this algorithm. In the primordial phase the initial population is created using a partial enumeration, and it is reduced using tournament selection. Next, in the juxtapositional phase partial solutions found in the primordial phase are mixed together. Since the primordial phase dominates the execution time, Merkle and Lamont tried several data distribution strategies to speed up this phase [6] extending previous work by Dymek [9]. The results indicate that there were no significant differences in the quality of the solutions for the different distribution strategies, but there were significant gains in the execution time of the algorithm. To evaluate the resulted policies, at each level, the elite policy in new population is stored in a variable *template* and the function *evaluate* compares this policy with the latest one. This function is Boolean. If it returns *ture*, then a final result in this level is obtained, otherwise continue.

4 Evolutionary Algorithms for Reinforcement Learning with Policy Iteration

We shall quote some basic knowledge about EA in [2]. Actually, we can learn the knowledge from many many literatures. Even though, we still would like to introduce some main ideas about evolutionary algorithm here. An EA iteratively updates a population of potential solutions, which are often encoded in structures called chromosomes. During each iteration, called a generation, the EA evaluates solutions and generates offspring based on the fitness of each solution in the task environment. Substructures, or genes, of the solutions are then modified through genetic operators such as mutation and recombination. The user must still decide on a (rather large) number of control parameters for the EA, including population size, mutation rates, recombination rates, parent selection rules, but there is an extensive literature of studies which suggest that EAs are relatively robust over a wide range of control parameter settings [5,8]. Thus, for many problems, EAs can be applied in a relatively straightforward manner [8].

In the case of RL, the user needs to make two major design decisions. First, how will the space of policies be represented by chromosomes in the EA? Second, how will the fitness of population elements be assessed? The answers to these questions depend on how the user chooses to bias the EA [2,5]. David E.M., Alan C.S. and John J.G. in their paper [2] developed an algorithm called EARL in which they simply combine EAs with reinforcement learning method. They also list some limitations of EARL in this paper. This section will present a new algorithm called PIrlEA that adopts the most straightforward set of design decisions. We try to overcome some disadvantages they mentioned. Specially, we apply the policy iteration methods to our new algorithm, and so, this algorithm will possess of advantages in policy iteration and EA for solving reinforcement learning problems.

The following shows details of this algorithm.

Function PIrlEA {
$k = 0$;
initialize population $P(0) = \{\pi_1, \cdots, \pi_n\}$;
evaluate policies in $P(k)$;
while (termination condition not satisfied) **do** {
$k = k + 1$;
alter structures in $P(k-1)$;
evaluate structures in $P(k-1)$;
//construct a population $P(k)$ from $P(k-1)$ as following;
obtain V^{π} for each $\pi \in P(k-1)$:
$V^{\pi}(s) = r\big(s, \pi(s)\big) + \gamma \sum_{s' \in X} P\big(s, \pi(s), s'\big) V^{\pi}(s')$.
generate the elite policy of $P(k-1)$ defined as

$$\pi^*(k)(x) \in \left\{ \arg \max_{\pi \in P(k-1)} \big(V^{\pi}(x)\big)(x) \right\}, x \in X.$$

generate $n-1$ random subsets $S_i, i = 1, \cdots, n-1$, of $P(k-1)$.
generate $n-1$ policies $\pi(S_i)$ defined as

$$\pi(S_i)(x) \in \left\{ \arg\max_{\pi \in S_i} \big(V^{\pi}(x)\big)(x) \right\}, x \in X.$$

//policy mutation for next population.
for each policy $\pi(S_i), i = 1, \cdots, n-1$, generate a mutated policy $\pi^m(S_i)$.
//constructe new population generation:
$P(k) = \{\pi^*(k), \pi^m(S_i)\}, i = 1, \cdots, n-1$.
record $\pi^*(k)$;
//to the next population.
} //end while.
return $\pi^*(k)$;
}. //end PIrlEA.

Algorithm 2: Pseudo-code of evolutionary algorithm with policy iteration.

Remark 4. In this algorithm, after constructing a new population, we still employ two operators which are alter and evaluate structures in this new population to adopt the advantages in EA algorithm. The main idea is the method for constructing a new population in which contains an elite policy inherits from its farther population. The new population always more elite than its parents or inherits its parents excellent abilities.

Remark 5. We would like to apply the Bellman errors as the termination condition in this algorithm. The specific formula is in the following:

$$\text{be}_{V^{\pi^*(k)}}(s) = V^{\pi^*(k)}(s) - \max_{a \in A} Q_{V^{\pi^*(k)}}(s, a). \qquad (9)$$

This is a easy way to calculate the values. We can set a gate threshold value ϵ and small enough. If the Bellman error $\text{be}_{V^{\pi^*(k)}}(s) < \epsilon$ at some stage, then the iteration will be stopped.

Remark 6. This algorithm kept away from the second problem above, that is, this algorithm does not need to consider how will the fitness of population individuals be assessed. And meanwhile, the first problem was solved through a easy way. So, intuitively, this algorithm can work well under the EA framework. The next section will demonstrate efficiency of it by a specific simulation task.

5 A Task for Solving Single Queue Problem

Our example is a model of a controlled single queueing system. This system have been used for studying the linear programming approach to approximate dynamic programming by D.P.De Farias and B.Van Roy [3], and for solving MDP problem with evolutionary policy iteration by J.Hu and Fu M.C. et al [4]. Here we will adopt the configurations in their experiment basically. We consider a Markov process with states $0, 1, 2, \cdots, N-1$, each representing a possible number of jobs in a queue, where N is the maximum queue length. In period t, a possible service completion is generated with probability $a(s_t)$, a cost of $c(s_t, a(s_t))$ is incurred, and resulting in a transition to state s_{t+1}. For any time period t, the system state $0 \leq s_t \leq N-1$ evolves according to

$$s_{t+1} = \begin{cases} s_t - 1, \text{with probability } a(s_t), \\ s_t + 1, \text{with probability } p, \\ s_t, \text{otherwise with probability } w. \end{cases} \tag{10}$$

Where s_t is the state variable which is the number of jobs in the system at the beginning of period t. Note that we use cost function instead of reward function in this experiment here. Actually, the negative cost function can be looked as reward function. So, these two type of functions have no difference essentially. From state 0, a transition to state 1 or 0 occurs with probabilities p or $1-p$, respectively. From state $N-1$, a transition to state $N-2$ or $N-1$ occurs with probabilities $a(N-2)$ or $1-a(N-2)$, respectively. The arrival probability p is the same for all states and we assume that $p < 1/2$ and $w \in [0, 1]$ randomly. The action to be chosen in each state s is the departure probability or service rate $a(s)$, which takes values in a discrete finite action set $A = \{a_i \in [0, 1], i = \text{action index}\}$. Moreover, in our algorithms, populations size is set to be $n = 20$ and 50 in two simulations respectively, search range or maximum level $max\text{-}level = 10$ in algorithm PImGA. The one-stage cost incurred in state s at any time if action a is taken is given by $c(s, a) = s + m(a)$, where m is a nonnegative and increasing function.

Here we only demonstrate a small scale problem to show our algorithms we have discussed how good and efficient are. In this experiment, the goal is to choose the optimal service completion probability for each state such that the total discounted cost

$$C(s_t, a_t) = E\big[\sum_{t=0}^{\infty} \gamma^t c(s_t, a(s_t))\big] \tag{11}$$

is minimized, or the objective of control is optimize the service rate a. We assume that jobs arrive at a queue with probability $p = 0.2$ in any unit of time(so no arrival with probability 0.8). Service rates/probabilities $a(s)$ are chosen from the action set A. The cost incurred at any time for being in state s and taking action a is given by $c(s, a) = s + 60a^3$. The number of states $N = 100$. Then, under the action $a(s)$ at some state s, the value function can be expressed as:

$$V^{a(s)}(s) = s + 60a^3(s) + \gamma \sum_{s_{ne}=0}^{99} P(s, a(s), s_{ne})V(s_{ne}), \tag{12}$$

and the Q-value function is:

$$Q_V(s, a(s)) = s + 60a^3(s) + \gamma \sum_{s_{ne}=0}^{99} P(s, a(s), s_{ne})V(s_{ne}), \tag{13}$$

where s_{ne} denotes the next state of s. The optimal Q-value function is

$$Q^*(s, a(s)) = s + 60a^3(s) + \gamma \min_{a'} \sum_{s_{ne}=0}^{99} P(s, a(s), s_{ne})Q^*(s_{ne}, a')), \tag{14}$$

and the optimal value function is

$$V^*(s) = \min_{a(s)\in A} Q^*(s, a(s)). \tag{15}$$

5.1 Simulation for PImGA Algorithm

In this simulation, we take the discount factor to be $\gamma = 0.98$, and the exploitation probability $q = 0.25, 0.5, 0.75$. We assume that the maximum queue length is $N = 100$,the start state $s_0 = 0$.We also assume that the action set $A = \{a_i \in [0, 1], i = 1, 2, \cdots, 20\}$ with 20 actions. The maximum number of generation is set to be 100.

Remark 7. The figure 1 above demonstrates that: (A) shows the plot of the value (object) function. This object function depicts the costs at per generation. The obvious characteristic of this plot is that the costs will decrease along with increasing the number of iterations in the algorithm. This feature is just what we expect. (B) gives the policy trend-line with 20 actions obtained randomly on all generations. The mean value in the last generation 100 is 0.5247. This value can be as the learning result of the Agent. Plot (C) gives 20 actions of all individuals. It also depicts the effect of each action for polices in a population.

5.2 Simulation for PIrlEA Algorithm

In this simulation, we still use the configurations as above.But we assume that the action set $A = \{a_i \in [0, 1], i = 1, 2, \cdots, 10\}$ only with 10 actions, the maximum number of generations is set to be 500, and, the population size is 50. The following figure shows the simulation results.

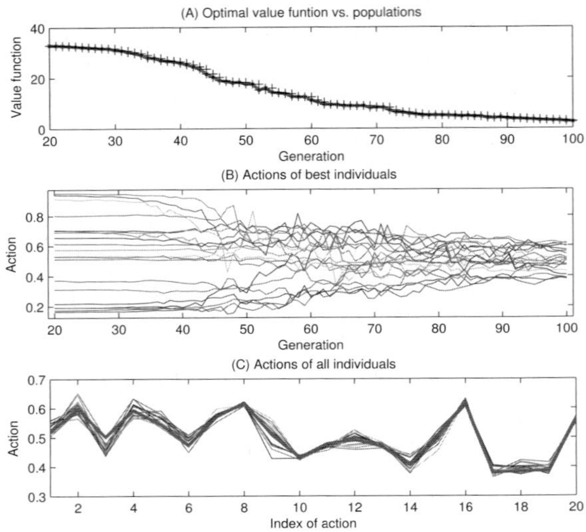

Fig. 1. The results of simulation for algorithm PImGA

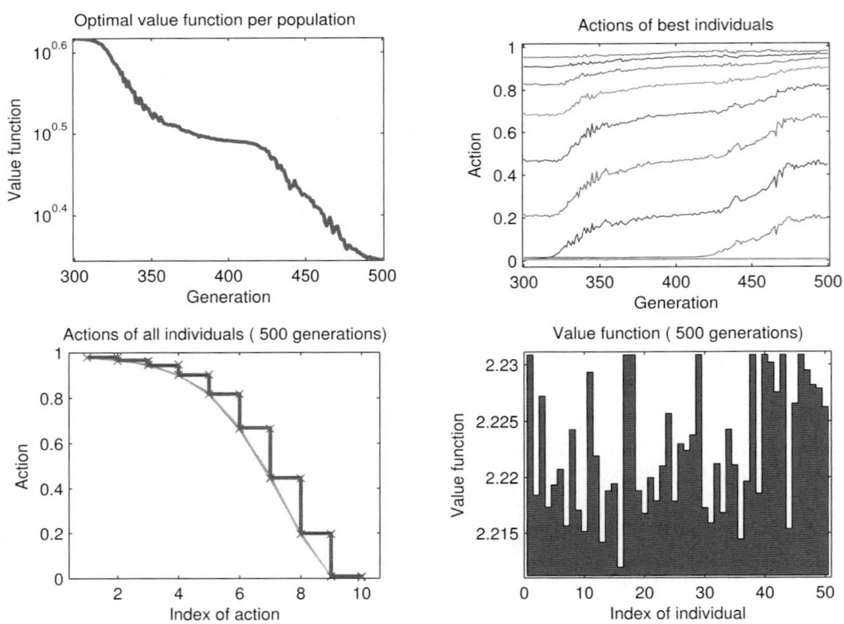

Fig. 2. The results of simulation for algorithm PIrlEA

Remark 8. From the figure above, we know that, the plot on up-left shows the value (object) function. As comments in the last subsection, the costs will also decrease along with increasing the number of iterations in this algorithm. The up-right plot demonstrates the relationship between 10 random actions and generations after 500 iterations. The plot on bottom-left shows actions of all individuals from 300 to 500 generation. The bottom-right plot gives the relationship of individuals and value function.

6 Conclusions

In this paper, we develop two new algorithms, PImGA and PIrlEA, in which we construct populations online in each iteration. These algorithms do not need selection operator as general EA and GA, instead the Agent learn a good or elite policy from its parents population. The resulted policy obtained by taking optimal reinforcement learning algorithm will be as one of elements (individuals) of the next population, and the new population always can be constructed by applying better policies than its parents. For a finite problem, the resulted population will accommodate the near optimal policies after a number of iterations (reproducing generations). The simulations demonstrate that both of the algorithms are good to work for solving learning control problems.

This paper does not provide the convergence of the algorithms and its proofs. We plan to continue our research for solving this problem in future work theoretically and mathematically.

Acknowledgments. This work was supported in part by Chinese National Science Foundation grants NSFC10471088 and NSFC60572126. We thank professor Anders Rantzer and professor Rolf Johansson in Lund University, Sweden, as advisors of the first author for providing insightful comments that led to the improvement of the paper. We thank specially for Dr. Hartmut Pohlheim, who provides a part of codes in his GEATbx evaluation version for our simulations. His GEATbx version 3.7 is really such a nice software for researching of GA and EA simulation.

References

1. Chang, H.S., Lee, H-G., Fu, M., Marcus, S.I.: Evolutionary Policy Iteration for Solving Markov Decision Processes. IEEE Trans. on Automatic Control 50(11), 1804–1808 (2005)
2. David, E.M., Alan, C.S., John, J.G.: Evolutionary Algorithms for Reinforcement Learning. Journal of Artificial Intelligence Research 11, 199–229 (1999)
3. De Farias, D.P., Van Roy, B.: The linear programming approach to approximate dynamic programming. Operations Research 51(6), 850–865 (2003)
4. Hu, J., Fu, M.C., Ramezani, V., Marcus, S.I.: An Evolutionary Random Search Algorithm for Solving Markov Decision Processes. TR2005-3 Institute for Systems Research, University of Maryland (2005)

5. Hoffmann, F., Pfister, G.: Learning of a Fuzzy Control Rule Base Using Messy Genetic Algorithms. In: Herrera, F., Verdegay, J.L. (eds.) Genetic Algorithms and Soft Computing, Physica-Verlag, Heidelberg (1996)
6. Merkle, L.D., Lamont, G.B.: Comparison of parallel messy genetic algorithm data distribution strategies. In: Forrest, S. (ed.) Proceedings of the Fifth International Conference on Genetic Algorithms, pp. 191–198. Morgan Kaufmann, San Mateo, CA (1993)
7. McMahan, H.B., Likhachev, M., Gordon, G.J.: Bounded Real-Time Dynamic Programming:RTDP with monotone upper bounds and performance guartees. In: Proceedings of the 22nd International Conference on Machine Learning, Bonn, Germany (2005)
8. Chowdhury, M.M.M., Li, Y.: Messy genetic algorithm-based new learning-method for structurally optimized neurofuzzy controllers. In: Proceedings of the IEEE International Conference on Industrial Technology, Shanghai, China, pp. 274–278. IEEE Computer Society Press, Los Alamitos (1996)
9. Dymek, A.: An examination of hypercube implementations of genetic algorithms. umt,Air Force Institute of Technology, Wright-Patterson Air Force Base, OH (1992)
10. Sutton, R.S., Barto, A.G.: Reinforcement Learning: An Introduction. MIT Press, Cambridge MA (1998)
11. Xu, X., He, H., Hu, D.: Efficient Reinforcement Learning Using Recursive Least-Squares Methods. Journal of Artificial Intelligence Research 16, 259–292 (2002)

It is obtained by the fact that the one-dimensional optimization problem (5)-(6) is equivalent to the optimization problem

$$\max_p \ R(n, Q_f, p),$$
$$\text{s.t. } p \text{ is a nonnegative integer.}$$

And since

$$1 = R(n, +\infty, 0) < R(n, +\infty, 1),$$
$$R(n, +\infty, 1) > R(n, +\infty, +\infty) = 0,$$

the optimization problem (13)-(14) also admits the finite solution $\bar{p}$.

Suppose $p_1 > 0$ is the solution to the optimization problem(5)-(6). Since

$$\frac{\partial w(n, Q_f, p)}{\partial p} = \frac{Q_I + 6Q_f}{2(p+1)^2}[3^{p+2}\ln(3^{p+1}/e) + 9 - 2\frac{Q_D + 6nQ_f}{Q_I + 6Q_f}]. \quad (15)$$

By $\partial w/\partial p = 0$, we can see that p_1 satisfies

$$3^{p_1+2}\ln(3^{p_1+1}/e) + 9 = 2(Q_D + 6nQ_f)/(Q_I + 6Q_f). \quad (16)$$

Carefully observe the two side of above equation, we can see that p_1 is increasing with respect to Q_f. In addition, the solution $p^* = p(n, Q_f)$ to (5)-(6) and the solution p_1 to (5)-(6) have the relationship

$$\lfloor p_1 \rfloor \leq p^* \leq \lceil p_1 \rceil, \quad (17)$$

where $\lfloor p_1 \rfloor$ and $\lceil p_1 \rceil$ denote the largest integer not larger than p_1 and smallest integer not smaller than p_1, respectively. We can get that $p(n, Q_f)$ is increasing with Q_f. Combine the existence of $p(n, 0)$ and $p(n, +\infty)$, we can get

$$\underline{p} \leq p^* \leq \bar{p}. \quad (18)$$

And by $w(n, 0, 0) > w(n, 0, 1)$, we have

$$\underline{p} \geq 1. \quad (19)$$

Therefore, the conclusion (10) is obtained.

Theorem 1 shows that $\underline{p}$ is the lower bound of p^*, and $\bar{p}$ is the upper bound of p^* . Table 1 lists the parameter value $\underline{p}$ when $0 \leq n \leq 10000$, and Table 2 lists the parameter value $\bar{p}$ when $0 \leq n \leq 10000$. From Table 1 and Table 2, we know that when $n = 500$, $2 \leq p^* \leq 3$; when $n = 5000$, $3 \leq p^* \leq 5$; $n = 10000$, $4 \leq p^* \leq 5, \cdots$.

$\underline{p}$ does not depend on Q_f, and it is the solution to the one dimensional optimization problem (11)-(12). Now we consider Algorithm NIMTH(p) with $p = \underline{p}$. To investigate the theoretical and practical efficiency of NIMTH($\underline{p}$). As a beginning, we give the following algorithm.

Algorithm EVNIMTH (Efficient Version of NIMTH)

EVNIMTH is Algorithm model NIMTH(p) when $p = \underline{p}$, where $\underline{p}$ is the solution to the optimization problem(11)-(12).

Table 1. The parameter $\underline{p}$ with respect to n

n	100 – 260	270 – 1230	1240 – 5120	5130 – 10000
$\underline{p}$	1	2	3	4

Table 2. The parameter $\bar{p}$ with respect to n

n	100 – 200	210 – 850	860 – 3280	3290 – 10000
$\bar{p}$	2	3	4	5

Theorem 2. *Suppose $n \geq 100$, $Q_f > 0$, $p*$ is the solution to the optimization problem (5)-(6), $\underline{p}$ is the solution to the optimization problem (11)-(12). Then, we have the following conclusions:*

(1) The efficiency ratio $R(n, Q_f, \underline{p})$ of EVNIMTH over IMTH satisfies

$$R(n, Q_f, p^*) \geq R(n, Q_f, \underline{p}) > R(n, 0, \underline{p}) > 1; \tag{20}$$

(2) $R(n,0,\underline{p})$ is strictly increasing with respect to n;
(3) When $n \to +\infty$,

$$R(n, Q_f, \underline{p}) > R(n, 0, \underline{p}) \sim \ln n / \ln 3.$$

Proof.
(1) Since $p^* = p(n, Q_f)$ is the solution to (5)-(6),

$$R(n, Q_f, p^*) \geq R(n, Q_f, \underline{p}). \tag{21}$$

Now, let's prove

$$R(n, Q_f, \underline{p}) > R(n, 0, \underline{p}). \tag{22}$$

First define

$$c(p) = p/\sigma(p), \tag{23}$$

where $\sigma(p) \overset{def}{=} \frac{1}{2}(3^{p+2} - 3^2)$. $\sigma(p)$ and $c(p)$ are functions with respect to continuous variable $p \geq 1$. Because

$$c'(p) = [\frac{p}{\frac{1}{2}(3^{p+2} - 3^2})]' = \frac{2(3^p - 1 - p3^p \ln 3)}{9(3^p - 1)^2} < 0,$$

by $p^* \geq 1$, we can get

$$c(p^*) = \frac{p^*}{\sigma^*} = \frac{p^*}{\sigma(p^*)} \leq \frac{1}{\sigma(1)} = \frac{1}{9}.$$

Therefore,

$$\frac{6 + 20c(p^*)}{6n + 20} \leq \frac{6 + 20/9}{6n + 20} < \frac{Q_I}{Q_D},$$

that is

$$\frac{6\sigma^* + 20p^*}{6n + 20} < \frac{\sigma^* Q_I}{Q_D}. \tag{24}$$

Then consider $R(n, Q_f, p)$ as a function with respect to Q_f, by (24) we have

$$\frac{(p+1)(6n+20)}{6n + 20 + 6\sigma + 20p} > \frac{(p+1)Q_D}{Q_D + \sigma Q_I}.$$

That is, in the right side of the following equation, the ratio of coefficients of Q_f is greater than the ratio of constants,

$$R(n, Q_f, p) = \frac{[(p+1)(6n+20)]Q_f + (p+1)Q_D}{(6n + 20p + 6\sigma + 20)Q_f + Q_D + \sigma Q_I}.$$

Therefore, when n and p are fixed, $R(n, Q_f, p)$ is increasing with respect to $Q_f \geq 0$. So, when $Q_f > 0$, (22) is establishable.

By Theorem 1, $\underline{p} \geq 1$,

$$R(n, 0, \underline{p}) > 1 \tag{25}$$

is obtained.

From (21), (22) and (25), the conclusion is proved.

The proof of Conclusions (2) and (3) is similar to the proof of Theorem 5.1 in [5], they are omitted.

Theorem 2 shows that the efficiency ratio $R(n, Q_f, \underline{p})$ of EVNIMTH over IMTH has a lower bound $R(n, 0, \underline{p})$ which is greater than 1 and $R(n, 0, \underline{p})$ is increasing with respect to n. In this case, with n increaseing, the efficiency ratio $R(n, Q_f, \underline{p})$ is showing a tendency to increase at the same rate of $\ln n / \ln 3$, therefore the improvement of EVNIMTH over IMTH will be more significant. moreover, the parameter $\underline{p}$ can be easily obtained instead of solving the optimization problem including Q_f.

Remark 1. *EVNIMTH has advantage over NIMTH for the following reasons:*

(1) The selection of p^* depends on Q_f, which is difficult to quantify exactly in practice, but $\underline{p}$ is easy to be obtained. So, EVNIMTH is concise and easy to be implemented in comparison with NIMTH.

(2) Those middle and large scale problems often have sparse structures, and the computational complexity Q_f is not very large in practice. Therefore, as usual, $\underline{p} = p^*$, more precisely, $R(n, Q_f, \underline{p}) = R(n, Q_f, p^*)$. This is also confirmed by the following numerical experiments.

3 Numerical Experiments

Both EVNIMTH and IMTH are tested by the standard unconstrained optimization problems. For the convenience of comparison, we choose the following four problems in [8]:

Table 3. Extended Rosenbrock function

n	p^*	p	r_{theo}	r_{prac}	I_E	I_I	g_{normE}	g_{normI}
100	1	1	1.46	1.55	7	6	5.50429e-007	1.41308e-012
200	1	1	1.64	1.65	7	6	7.78424e-007	1.99840e-012
300	2	2	1.85	2.07	7	6	9.53371e-007	7.51101e-007
400	2	2	2.03	2.13	7	7	0.00000e+000	8.67298e-007
500	2	2	2.16	2.14	7	7	0.00000e+000	9.69664e-007
600	2	2	2.25	2.45	8	7	0.00000e+000	3.26859e-011
700	2	2	2.33	2.46	8	7	0.00000e+000	3.53048e-011
800	2	2	2.39	2.48	8	7	0.00000e+000	3.77425e-011
900	2	2	2.44	2.49	8	7	0.00000e+000	4.00319e-011
1000	2	2	2.49	2.53	8	7	0.00000e+000	4.21974e-011

Table 4. Penalty function I

n	p^*	p	r_{theo}	r_{prac}	I_E	I_I	g_{normE}	g_{normI}
100	1	1	1.48	1.55	30	22	6.15203e-008	7.32700e-008
200	1	1	1.66	1.71	30	77	5.60580e-007	4.78073e-010
300	2	2	1.88	2.58	70	51	2.88054e-010	1.37944e-008
400	2	2	2.05	2.64	76	137	1.81885e-007	1.86497e-007
500	2	2	2.17	2.74	53	79	2.98782e-007	2.14062e-007
600	2	2	2.26	2.54	46	21	1.25965e-008	3.64552e-007
700	2	2	2.33	2.71	34	47	1.04853e-010	6.27596e-007
800	2	2	2.40	2.78	154	24	1.38151e-008	2.30061e-007
900	2	2	2.45	2.75	74	25	8.56669e-009	4.02877e-007
1000	2	2	2.49	2.55	120	47	4.14901e-011	5.74403e-007

Problem 1. The Extended Rosenbrock function
Problem 2. Penalty function I
Problem 3. The Extended Powell singular function
Problem 4. Discrete boundary value function

They are executed in C++ routines with double precision. The initial points of these problems are the standard start points. In addition, the condition

$$\|\nabla f(x_k)\| \leq 10^{-6} \tag{26}$$

is also used for termination test.

The main value we are interested in is the ratio

$$r_{prac} = \frac{\text{The CPU time by IMTH}/I_I}{\text{The CPU time by EVNIMTH}/I_E}. \tag{27}$$

The ratio shows the practical improvement of EVNIMTH over IMTH, where I_E and I_I are respectively the iteration numbers of EVNIMTH and IMTH.

For the above four problems with $n = 100, 200, \cdots, 1000$, they are all solved successfully by EVNIMTH and IMTH. The parameters $p = \underline{p}$ of EVNIMTH are

Table 5. Extended Powell singular function

n	p^*	$\underline{p}$	r_{theo}	r_{prac}	I_E	I_I	g_{normE}	g_{normI}
100	1	1	1.51	1.67	13	13	2.52066e-007	2.53305e-007
200	2	1	1.68	2.13	13	13	3.56475e-007	3.57780e-007
300	2	2	1.91	2.23	13	13	4.36591e-007	4.38166e-007
400	2	2	2.07	2.33	13	13	5.04131e-007	5.05964e-007
500	2	2	2.18	2.34	13	13	5.63636e-007	5.65685e-007
600	2	2	2.28	2.37	13	13	6.17432e-007	6.19677e-007
700	2	2	2.35	2.40	13	13	6.66903e-007	6.69328e-007
800	2	2	2.41	2.42	13	13	7.12949e-007	7.15542e-007
900	2	2	2.46	2.44	13	13	7.56197e-007	7.58947e-007
1000	2	2	2.50	2.43	13	13	7.97102e-007	8.00000e-007

Table 6. Discrete boundary value function

n	p^*	$\underline{p}$	r_{theo}	r_{prac}	I_E	I_I	g_{normE}	g_{normI}
100	1	1	1.58	1.31	9	8	3.66927e-007	2.04366e-008
200	2	1	1.81	1.72	10	9	3.38474e-008	3.18918e-008
300	2	2	2.00	1.88	10	9	1.65019e-007	4.92975e-008
400	2	2	2.14	1.97	10	9	2.58699e-007	6.70743e-008
500	2	2	2.24	2.13	10	9	3.44771e-007	7.97833e-008
600	2	2	2.33	2.18	10	9	4.22960e-007	9.01243e-008
700	2	2	2.39	2.32	10	9	4.94390e-007	9.88542e-008
800	2	2	2.44	2.38	10	9	5.60189e-007	1.06420e-007
900	2	2	2.49	2.48	10	9	6.21292e-007	1.13124e-007
1000	2	2	2.52	2.52	10	9	6.78443e-007	1.19172e-007

listed in column 3 of Table 3–6. For the convenience of comparison, these tables list the corresponding theoretical values

$$r_{theo} \stackrel{def}{=} R(n, Q_f, \underline{p}),$$

where Q_f is equal to $2n$, $3n$, $(7/4)n$ and $8n$ respectively.

In column 4, the practical values r_{prac} are listed in column 5. To get further comparison, the iteration numbers I_E and I_I are reported in column 6 and column 7 of the tables, and the termination gradient norms g_{normE} of EVNIMTH and g_{normI} of IMTH are reported in column 8 and column 9 of the tables. All of the optimization function values are the one reported in [8], here they are leaved out.

From Table 3-Table 6, all of the values r_{prac} are greater than 1. It shows that Algorithm EVNIMTH is much better than Algorithm IMTH. We can see that the practical values r_{prac} are approximately consistent with the theoretical values r_{theo}. This confirms the conclusion of Section 2.

For Problem 1-Problem 4, we can see $\underline{p} = p^*$ (except for Problem 3 and Problem 4 where $n = 200$), therefore, the efficiency of EVNIMTH is equal to that of NIMTH mostly, but the former is concise and easy to be implemented.

therefore, the efficiency of EVNIMTH is equal to that of NIMTH mostly, but the former is concise and easy to be implemented.

4 Conclusions

In this paper, we establish and implement Algorithm EVNIMTH by improving NIMTH and using AD algorithm. Further theoretical analysis and numerical experiments show that the version of the algorithm is competitive for middle and large scale unconstrained optimization problems.

References

1. Kalaba, R., Tischler, A.: A Generalized Newton Algorithm Using High Order Derivatives. Journal of Optimization Theory and Applications 39, 1–17 (1983)
2. Jackson, R., McCormick, G.: The Poliyadic Structure of Factorable Functions Tensors with Applications to High-Order Minimization Techniques. Journal of Optimization Theory and Applications 51(1), 63–94 (1986)
3. Sherman, A.H.: On Newton-iterative methods for the solution of systems of nonlinear equations. SIAM Journal Numerical Analysis 15, 755–771 (1978)
4. Toint, P.L.: Towards an Efficient Sparsity Exploiting Newton Method for Minimization. Sparse Matrices and Their Uses (I.S. Duff), pp. 57–88. Academic Press, London, England (1981)
5. Deng, N.Y., Zhang, H.B.: Theoretical Efficiency of a New Inexact Method of Tangent Hyperbolas. Optimization Methods and Software 19(3–4), 247–265 (2004)
6. Zhang, H.B., Deng, N.Y.: An improved inexact Newton method. Journal of Global Optimization (to appear), doi:10.1007/s10898-007-9134-4
7. Griewank, A.: Evaluating Derivatives Principles and Techniques of Algorithmic Differentiation, SIAM, Philadephia. Frontiers in Applied Mathematics, vol. 19 (2000)
8. Moré, J., Garbow, B., Hillstrom, K.: Testing unconstrained optimization software. ACM Transactions on Mathematical Software 7, 17–41 (1981)

Donor Recognition Synthesis Method Base on Simulate Anneal

Chen Dong and Yingfei Sun

School of Information Science and Engineering, Graduate University of Chinese Academy of Sciences, Beijing, 100080, P. R. of China
yfsun@gucas.ac.cn

Abstract. The recognition of splicing sites is an important step in gene recognition. We introduce a synthesis method for splicing sites prediction based on short sequence pattern and long sequence pattern. There are some weights in short sequence pattern and long sequence pattern. We regulate weights by simulated anneal. Applying the method to recognize both the true and false splicing sites, the result shows true positive rate and false positive rate and compared with GeneSplicer.

1 Introduction

In the past years, a lot of species have been sequenced. As a result, gene recognition is one of the most important problems in life science. For eukaryotic mRNA, prediction of splice sites is a key step of all gene structural prediction algorithms.

Pre-mRNA is spliced by spliceosome at the junctions of exon and intron. The spliceosome is a large dynamic complex assembled from RNA and protein components, including four small nuclear RNAs (snRNAs) and associated proteins that make up the small nuclear ribonucleoprotein particles (snRNPs)[1,2]. First, the spliceosome recognize donor site, acceptor site and branch point. Then the spliceosome excise intron and joint exons.

Existing methods fall into two groups with respect to the data they utilize. The first group consists of ab initio programs which use only the query genomic sequence as input. Examples are the programs GENSCAN [3], AUGUSTUS [4] and HMMGene [5] and GeneSplicer [6] and GENEID [7]. The second group of gene-finding methods, extrinsic methods, comprises all programs which use data other than the query genomic sequence. Some extrinsic methods use genomic sequences from other species. This approach is commonly referred to as comparative gene prediction. Examples are the programs SGP2 [8] and TWINSCAN [9] and N-SCAN [10] and SLAM [11] and DOUBLESCAN [12] and AGenDA [13]or methods based on evolutionary Hidden Markov Models [14,15,16]. Some programs integrated advantages of those two groups, such as GenomeScan and HMMGene. GenomeScan uses BLAST alignments with protein sequences [17]. HMMGene, has an extension that integrates into the HMM information from BLAST alignments of the query sequence with cDNA, EST and protein sequences [18]. Another program which uses extrinsic evidence is Gene Wise. It aligns a protein sequence to a genomic sequence [19].

K. Li et al. (Eds.): LSMS 2007, LNCS 4688, pp. 215–221, 2007.

In this paper we introduced a short sequences pattern(Local Property Method) and long sequences pattern(Probability Retrospect Method) in order to apply on two famous dataset and compared the result with GeneSplicer: http://www.cbcb.umd.edu/software/GeneSplicer/

2 Data Sets

We have two real donor set. First is Guigo's dataset [20]. (Download from http://genome.imim.es/datasets/genomics96/) The second real donor data comes from alternative splicing database (ASD), which is established by European Bioinformatics Institute (EBI) [21,22,23]. (http://www.ebi.ac.uk/asd/) The database has more than 180000 sequences in all. In the first dataset we choose 2000 sequences and in the second dataset we choose 135000 sequences base on two criteria: the donor must be GT, There are 70 bases at both sides of the donor junction. For example:

FSDE: -70 cgccactttgtcctcgtgtttgtcgtcttcttcat -36
-35 ctgctttggcctgaccatcttcgttgggatcagag -1
FSDI: 1 GTAAGGTTCGGTTTTACTCATTGAATCTTTTGCC 35
36 ATGCGCGGTGGCCCTGGCGTATTGCTTTCGGTCAG 70

FSDE has 70 bases before donor. We numbered them -70 to -1 from left to right. FSDI has 70 bases after donor. We numbered them 1 to number 70 from left to right. The number 1 must be G and number 2 must be T. In the first dataset, we use 1500 sequences as training set and 500 sequences as test set. In the second dataset we divide the selected sequences into 3 parts: Real donor set 1, Real donor set 2 and Real donor set 3. Every set has 135000/3=45000 sequences. There are three criteria to extract pseudo sequences. The first one and the second one are the same as real donor data we discussed above. The third one is that pseudo donor must be different from read donor. Corresponding to the first real data set, we extract 7000 pseudo donor sequences from http://genome.imim.es/datasets/genomics96/seqs/DNASequences.fasta. 6000 of them are training data set and 1000 of them are test set. Corresponding to the second real data set, we extract 135000 sequences from Human Genome 1.08 edition of NCBI and divide them into 3 parts. Pseudo donor set 1, Pseudo donor set 2 and Pseudo donor set 3. Every set has 135000/3=45000 sequences.

3 Local Property Method

From fig 1, we could see that the probabilities of the four bases are very different near real donor. This could help us to find pseudo donor. Now we focus on number -2,-1,3,4,5 bases. Combination of those 5 bases, there are 4X4X4X4X4=1024 situations. Then we make a statistic of the real donor set and the pseudo donor set, the result is shown in fig1.

By analyzing real donor statistical graph, we found two situations happened more than 1000 times per 10000 times. At the same time, more than 400 situations

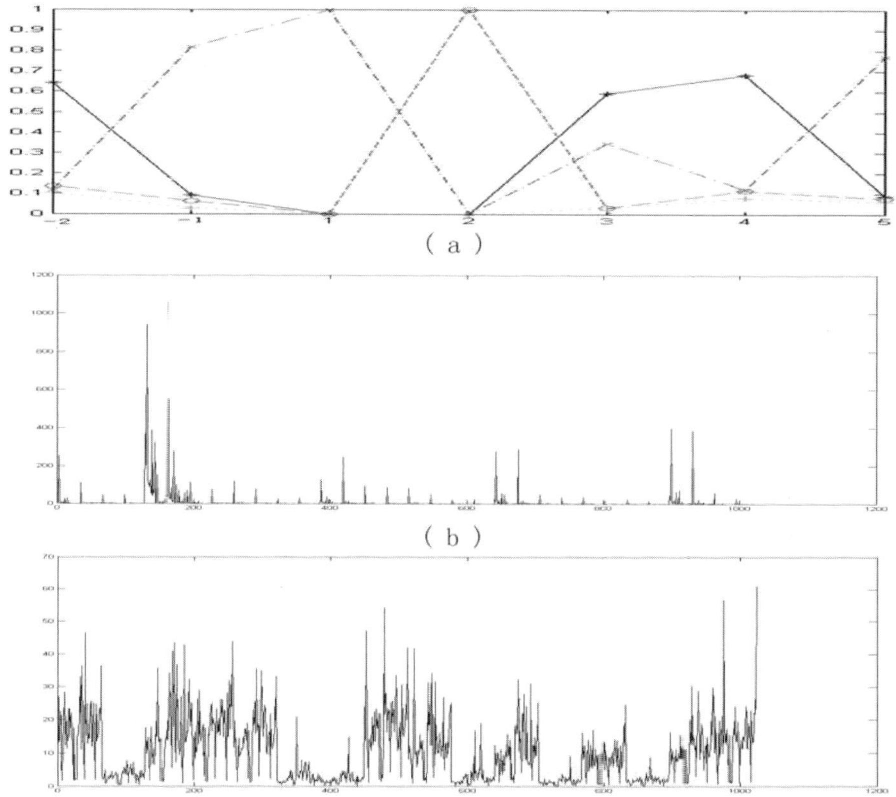

Fig. 1. (a) Bases frequency of real donor. X axis is the number of bases. Y axis is frequency. (b)Combination of number -2,-1,3,4,5 bases are 1024 situations.Frequency of 1024 situations of real donor. (c) Frequency of 1024 situations of pseudo donor. Y axis is frequency per 10000 times.

happened 0 time per 10000 times. By analyzing pseudo donor statistical graph, all situations happened less than 90 times per 10000 times. Based on this situation, we defined a Short Sequence Real/Pseudo Ratio (SSR/PR). For example: in real donor, the frequency that -2,-1,3,4,5 bases happened to be AAAAA is 9.8 times per 10000 times, and 37 times per 10000 times in pseudo donor. In this case, we calculate the SSR/PR(AAAAA) as 9.8/37=0.2649. If a sequence's SSR/PR is too large or too small, we can judge whether this sequence has a real donor.

4 Probability Retrospect Method (PRM)

By doing statistic, we know the probability of every base happened in every position. They constitute a probability matrix. When we dispose a sequence, we should give this sequence a score base on the probability matrix.

4.1 One Dimension Probability Retrospect Method(1DPRM)

If the length of a sequence is L. P(i,A) represents the times A happened in position i. We plus all P(i,N) as final score. N is A,C,G or T.

$$Score(S) = \sum_{i=1}^{L} P(i, N) \ . \tag{1}$$

Further more, we can give every base a weight so that it has more biological meaning.

$$Score(S) = \sum_{i=1}^{L} P(i, N)W(N) \ . \tag{2}$$

4.2 Two Dimension Probability Retrospect Method(2DPRM)

Two dimension means every two bases. The size of probability matrix is 139X16. The formula to calculate score is:

$$Score(S) = \sum_{i=1}^{L-1} P(i, NN)W(NN) \ . \tag{3}$$

4.3 Part Three Dimension Probability Retrospect Method(P3DPRM)

The combination of 3 bases is 4X4X4=64.If we consider all the situations, there will be as much as 64 parameters. Therefore, we only consider parts of the situations in case over study. We consider 18 situations because their contributions are different between real donor set and pseudo donor set.

5 Simulate Annealing

At the beginning, we made the weight W(i,NN) equal to the ratio of positive probability to negative probability. So there was a 139*16 weight matrix. We considered the number of sequences which were classified by error as the energy. We used e^{-c} to control temperature in order to find the minimum energy. The C is from 1 to 5. The weight matrix of Guigo's dataset is different from ASD's dataset.

6 Synthesis Method and Result

In our experiment, we use local property method in the first step. 2D Probability Retrospect Method in the second step, and then use 3D Probability Retrospect Method. At last, Short Sequence Real/Pseudo Ratio was used to regulate. Between each method, we use simulate anneal to regulate the weight of every method.

Table 1. The result of ASD's dataset

Training set	Test set	TP in 1st step	FN in 1st step	Undisposed in 1st step	TP in 2nd step	FN in 2nd step	TP in total	FN in total
RDS2,3	RDS1	0.01	0.37	44828	98.57	1.05	98.58	1.42
RDS1,3	RDS2	0.01	0.3	44859	98.25	1.44	98.26	1.74
RDS1,2	RDS3	0	0.3	44855	98.58	1.12	98.58	1.42
	RDSA	0.01	0.32	44847	98.46	1.21	98.47	1.53
Training set	**Test set**	**TN in 1st step**	**FP in 1st step**	**Undisposed in 1st step**	**TN in 2nd step**	**FP in 2nd step**	**TN in total**	**FP in total**
PDS2,3	PDS1	72.66	0	12300	25.92	1.65	98.58	1.65
PDS1,3	PDS2	71.67	0	12749	26.7	1.63	98.37	1.63
PDS1,2	PDS3	72.18	0	12520	26.19	1.63	98.37	1.63
	PDSA	72.17	0	12523	26.27	1.64	98.44	1.64
	Average Accuracy 98.41							

Table 2. The result of Guigo's dataset

Training	Test	Undisposed in first step	TP in total	FP in total	SP	SN
1500/6000	500/1000	487/267	89.8	2.9	93.54	89.8
	Same	Using GeneSplicer	93	75.6	65.6	93

In first step, we assume the number -2,-1,3,4,5 bases of a sequence is N1toN5. If SSR/PR(N1N2N3N4N5)>195, we judge donor in this sequence as real donor. If SSR/PR(N1N2N3N4N5)<0.042, we judge donor in this sequence as pseudo donor.

For ASD's datsset (Second Dataset) In the second step, we plus 2DPRM and P3DPRM and SSR/PR.

$$Score(s) = \sum_{i=1}^{L-1} P(i, NN)W(NN) + \sum_{i=1}^{L-2} P(i, NNN)W(NNN) \\ +W(SSR/PR)log(SSR/PR(NNNNN)) \,. \tag{4}$$

If the score is above 193, we judge donor in this sequence as real donor. If the score is below 193, we judge donor in this sequence as pseudo donor. The result showed in table1.The score of test set showed in fig2(a).The ROC curve of second step showed in fig2(b).

RDS=real donor set PDS=pseudo donor set
RDSA=real donor set on average PDSA=pseudo donor set on average
TP: real donor judge as real donor FN: real donor judge as pseudo donor TN: pseudo donor judge as pseudo donor FP: pseudo donor judge as real donor

For Guigo's dataset (First Dataset), the result compared with GeneSplicer shown in table2.

After first step,2DPRM was used. If the score is above 221, we judge donor in this sequence as real donor. If the score is below 221, we judge donor in this sequence as pseudo donor.

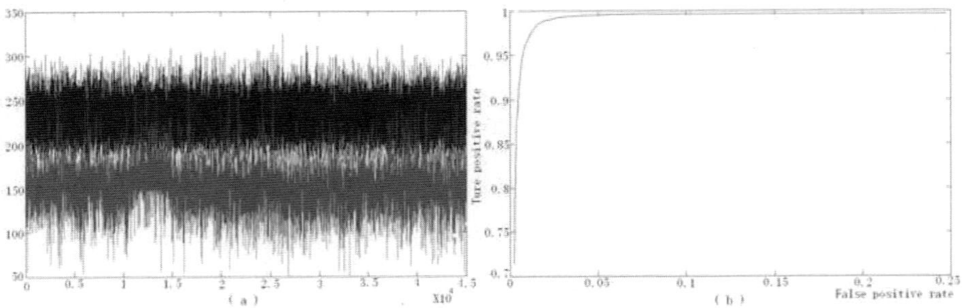

Fig. 2. The training set is RDS2 and 3, PDS2 and 3.The test set are RDS1 and PDS1. (a)Score of test set. Score of real donor is large than pseudo donor on average. (b)ROC curve of second step. X axis is FP. Y axis is TP.

7 Conclusion and Outlook

In this paper, we use short sequence pattern and long sequence pattern to recognize donor and obtain a good result. However, there is at least one property we didn't use. The distance between two GT in exon is different from in intron. If we consider that, the result is properly better. In addition, the Local property method and PRM we discussed could apply to acceptor identify.

If considering both donor recognition and acceptor recognition, there will be another important property: the length of exon. Because most exon are less than 300bp. It will help us find new gene.

Acknowledgment

This work is supported by a grant from the National Natural Science Foundation of China (No. 60472090).The authors would like to thank Zhao Shi, Wan Shuwu and Yang Hong for their continuous support and helpful discussions.

References

1. Sheth, N., Roca, X., Hastings, M.L., Roeder, T., Krainer, A.R., Sachidanandam, R.: Comprehensive splice-site analysis using comparative genomics. Nucleic Acids Res. 34, 3955–3967 (2006)
2. Will, C.L., Luhrmann, R.: Spliceosomal UsnRNP biogenesis, structure and function. Curr. Opin. Cell. Biol. 13, 290–301 (2001)
3. Burge, C.: Identification of Genes in Human Genomic DNA. PhD thesis. Stanford University (1997)
4. Stanke, M., Waack, S.: Gene prediction with a hidden Markov model and new intron submodel. Bioinformatics 19, ii215–ii225 (2003)
5. Krogh, A.: Two methods for improving performance of an HMM and their application for gene finding. In: Proc. Fifth Int Conf Intelligent Systems for Molecular Biology, pp. 179–186 (1997)

6. Pertea, M., Lin, X., Salzberg, S.L.: GeneSplicer: a new computational method for splice site prediction. Nucleic Acids Res. 29, 1185–1190 (2001)
7. Parra, G., Enrique, B., Guig, R.: GeneID in Drosophila. Genome Research 10, 511–515 (2000)
8. Parra, G., Agarwal, P., Abril, J., Wiehe, T., Fickett, J., Guig, R.: Comparative Gene Prediction in Human and Mouse. Genome Research 13 108–117 (2003)
9. Korf, I., Flicek, P., Duan, D., Brent, M.R.: Integrating Genomic Homology into Gene Structure Prediction. Bioinformatics 1, S1–S9 (2001)
10. Gross, S.S., Brent, M.R.: Using Multiple Alignments to Improve Gene Prediction. In: McLysaght, A., Huson, D.H. (eds.) RECOMB 2005. LNCS (LNBI), vol. 3678, pp. 374–388. Springer, Heidelberg (2005)
11. Alexandersson, M., Cawley, S., Pachter, L.: SLAM: Cross-Species Gene Finding and Alignment with a Generalized Pair Hidden Markov Model. Genome Research 13, 496–502 (2003)
12. Meyer, I.M., Durbin, R.: Comparative ab initio prediction of gene structures using pair HMMs. Bioinformatics 18, 1309–1318 (2002)
13. Taher, L., Rinner, O., Gargh, S., Sczyrba, A., Morgenstern, B.: AGenDA: gene prediction by cross-species sequence comparison. Nucleic Acids Research 32, W305–W308 (2004)
14. Pedersen, J.S., Hein, J.: Gene finding with a hidden Markov model of genome structure and evolution. Bioinformatics 19, 219–227 (2003)
15. Siepel, A., Haussler, D.: Computational identification of evolutionarily conserved exons. In: RECOMB 2004. Proceedings of the Eighth Annual International Conference on Research in Computational Molecular Biology, pp. 177–186 (2004)
16. Stanke, M., Schöffmann, O., Morgenstern, B., Waack, S.: Gene prediction in eukaryotes with a generalized hidden Markov model that uses hints from external sources. BMC Bioinformatics 7, 62 (2006)
17. Yeh, R.F., Lim, L.P., Burge, C.: Computational Inference of Homologous Gene Structures in the Human Genome. Genome Research 11, 803–816 (2001)
18. Krogh, A.: Using Database Matches with HMMGene for Automated Gene Detection in Drosophila. Genome Research 10, 523–528 (2000)
19. Birney, E., Clamp, M., Durbin, R.: Gene Wise and Genomewise. Genome Research 14, 988–995 (2004)
20. Burset, M., Guig, R.: Evaluation of gene structure prediction programs. Genomics 34, 353–357 (1996)
21. Stamm, S., Riethoven, J.-J.M., Le Texier, V., Gopalakrishnan, C., Kumanduri, V., Tang, Y., Barbosa-Morais, N.L., Thanaraj, T.A.: ASD: a bioinformatics resource on alternative splicing. Nucleic Acids Res. 34, D46–D55 (2006)
22. Thanaraj, T.A., Stamm, S., Clark, F., Riethoven, J.J.M., Le Texier, V., Muilu, J.: ASD: the Alternative Splicing Database. Nucl. Acids Res. 32, D64–D69 (2004)
23. Clark, F., Thanaraj, T.A.: Categorization and characterization of transcript-confirmed constitutively and alternatively spliced introns and exons from human. Human Molecular Genetics 11, 451–464 (2002)

An Improved Particle Swarm Optimization Algorithm Applied to Economic Dispatch

Yuzeng Li, Long Long*, and Shaohua Zhang

Key laboratory of Power Station Automation Technology,
Department of Automation, Shanghai University,
ZhaBei District, Shanghai, 200072, China
leolong1984@gmail.com

Abstract. This paper proposes a new practical optimization method applied to the economic dispatch (ED) in a power system. The proposed method is based on an improved particle swarm optimization algorithm and considers some restrict conditions of ED in a practical power system. By reinitializing them with some currently optimal values during every cycle of iteration, this proposed method can make some inactivity particles to be always within a very small area having an optimal solution. The proposed method can avoid effectively the "premature" of the classic particle swarm optimization (PSO) algorithm due to improve the cognized capacity of the classic PSO, thereby it is beneficial to obtain some optimal global solutions. The simulation results show that proposed method has some excellent characteristics of higher quality calculation precision and better computation efficiency, compared with some other PSO methods.

1 Introduction

In the rapid development of the power industry, it is very important for improving the stability, reliability and security of the power system so as to obtain its optimal benefits. The economic dispatch (ED) optimization problem is one of the fundamental issues in a practical power system [1].

At present, there is a great diversity of methods used for solving the ED problems, including the lambda-iteration method, the base point and participation factors method, the gradient method, the dynamic programming (DP) method and so on [2], [3-5]. However, they become seldom ideal and feasible for these conventional methods used to solve ED problems in a practical power system having some characteristics.

In fact, some global optimization techniques have been successfully used to solve ED problems in some power systems in the past decade. These optimization techniques are all well known, including, Genetic Algorithms (GA), Simulated Annealing (SA), Tabu Search (TS) and recently Particle Swarm Optimization (PSO),

* This project is support by National Natural Science Foundation of China (50377023) and the Development Foundation of Shanghai Municipal Education Commission(05AZ28).

K. Li et al. (Eds.): LSMS 2007, LNCS 4688, pp. 222 – 230, 2007.

as modern heuristic techniques, and they are now considered as a promising alternative to solve this type of the ED problem, since they have been proved robust and powerful tools that outperform the classical optimization for obtaining global or quasi-global optimums in a reasonable amount of time for many kinds of hard optimization problems in some power systems [6].

Among these modern optimization techniques, the PSO, since it is first introduced by Kennedy and Eberhart in 1995 [7], has received increasing attentions in many different areas because its simplicity, robustness and effectiveness in performing difficult optimization tasks. So, its features make it a real excellent competitor for the other prominent heuristic approaches such as GA, SA and TS in solving hard optimization problems.

This paper presents a novel optimization method based on an Improved Particle Swarm Optimization (IPSO) algorithm applied to solve ED in a practical power system. Under considering some nonlinear characteristics of a generator, such as the ramp rate limits, the prohibited operating zone and the cost functions, the proposed method can make some inactivity particles to be always within a very small area having an optimal solution by reinitializing them with some current optimal values during every cycle of the iteration. Therefore, the proposed method can effectively avoid "prematurity" of the classic PSO; thereby it is beneficial to obtain some optimal global solutions.

2 ED Problem Formulation

The ED of a power system is a nonlinear optimization problem, and its main objective is to achieve some optimizations of a practice system, namely, to simultaneously meet the load demand of the power system and minimize the generation cost rate while to satisfy various equality and inequality constraints. Therefore, ED problem must often be formulated so as to establish mathematics model. For a practical power system, the ED problem mainly relates to its objective function and some operation constraint conditions about this practical power system [8-10].

Objective Function:

$$\min\ F = \sum_{i=1}^{m} F_i(P_i) = \sum_{i=1}^{m} a_i + b_i P_i + c_i P_i^2 \tag{1}$$

where the generation cost function $F_i(P_i)$ is usually expressed as a quadratic polynomial; a_i, b_i and c_i are the cost coefficients of the ith generator, respectively; m is the number of generators committed to the operation system; P_i is the power output of the ith generator. The basic operation constraints are the power balance and power operating limits which can satisfy various operation states of the power system.

Power Balance:

$$\sum_{i=1}^{m} P_i = P_D + P_L, i = l, \ldots, m \tag{2}$$

where P_D and P_L are the total transmission network losses and load of the system, respectively.

Generator Operation Constraints:

$$P_i^{\min} \le P_i \le P_i^{\max}, i = 1,2,\ldots, m \tag{3}$$

where P_i^{max} and P_i^{min} are the maximum and minimum of the power output of the *i*th generator, respectively.

Of course it is also very important for solving ED problem in a practical power system to take into account of the ramp rate limits and prohibited operating zone in actual generator operation.

Ramp Rate Limits:

$$\max(P_i^{\min}, P_i^0 - DR_i) \le P_i \le \min(P_i^{\max}, P_i^0 + UR_i) \tag{4}$$

where P_i and P_i^0 are the current output power and the previous output power of the *i*th generator; DR_i and UR_i are the downramp limit and upramp limit of the *i*th generator (MW/time period), respectively.

Prohibited Operation Zone:

$$\begin{aligned} &P_i^{\min} \le P_i \le P_{i,1}^l \\ &P_{i,j-1}^u \le P_i \le P_{i,l}^l, \; j = 2,3,\ldots, n_i \\ &P_{i,n}^u \le P_i \le P_i^{\max} \end{aligned} \tag{5}$$

where j is the number of prohibited zones of the generator i, $P_{i,1}^l$ and $P_{i,n}^u$ are the lower and upper boundaries of prohibited operating zone j of generator i, respectively.

3 Particle Swarm Optimization Algorithm

The particle swarm optimization (PSO) is a population-based, self-adaptive, stochastic optimization technique which is first introduced by Kennedy and Eberhart in 1995 [8]. We will refer it as: Classic PSO, in which the search procedure is conducted parallel using a group of individuals to explore promising regions in the search space. Imagine that we have a population of particles looking around in a given search space for the global optimum. This particle movement mimics, in a way, the coordinated movement of flocks of birds, schools of fish or swarms of insects; this is a good image of a PSO optimization algorithm. In the context of PSO, the population is called swarm and the individuals are called particles. Each particle moves with an adaptable velocity within the search space, trying to find an even better position. The particle adjusts its trajectory toward its own best previously visited position and toward its own best position attained by any particle of its topological neighborhood.

This adoption is a stochastic process that depends on both the memory of each particle as well as the knowledge gained by the swarm.

Classic PSO is initialized with a group of random particles each of which has own position and velocity. The position and the velocity of the ith particle in the d-dimensional search space can be represented as xi= (xi1, xi2,..., xid) and vi= (vi1, vi2,..., vid), respectively. The best previous position of itch particle that gives the best fitness value of the defined objective function is represented by pi= (pi1, pi2,..., pad). The global best particle, which represents the fittest particle found so far in the entire swarm is denoted by pg=(pg1, pg2, ..., pgd). During every cycle of the iteration, each particle is updated by following the two best values (pi, pg). After finding the two best values, the particle can update its velocity and position according to the following equations:

$$v_{id}^{t+1} = \omega \cdot v_{id}^{t} + c_1 * rand() * \left(p_{id} - x_{id}^{t}\right) + c_2 * Rand() * \left(p_{gd} - x_{id}^{t}\right) \tag{6}$$

$$x_{id}^{t+1} = x_{id}^{t} + v_{id}^{t+1} \tag{7}$$

where $i = 1, 2, ..., n$ and $d = 1, 2, ..., m$; n and m are the number of particles in a group and the number of members in a particle, respectively; rand()and Rand () are uniform random values in the range [0,1]; c1 and c2 are two positive acceleration factors called the cognitive and social parameters, ω is the inertia weight factor, t is the pointer of iterations, v_i^t is the velocity of particle i at iteration t and $V_d^{\min} \le V_{id}^t \le V_d^{\max}$; and X_i^t is the current position of particle i at iteration t. Equation (6) is used to compute the particle's new velocity according to its previous velocity and the distance of its current position from its own best position and the group's best position. Then the particle moves from the current position to the next one according to equation (7). In equation (6), the first term indicates the current velocity of the particle, the second term represents the cognitive part of PSO where the particle changes its velocity based on its own thinking and memory. The third term represents the social part of PSO where the particle changes its velocity based on the social-psychological adaptation of knowledge [12].

The use of linearly decreasing inertia weight factor ω has provided improved performance in all the applications. Its value is decreased linearly from 0.9 to 0.4 during a run. Suitable selection of the inertia weight can provide a balance between global and local exploration and exploitation, and results in a fewer iteration on average to find a sufficiently optimal solution. Its value is set according to the following equation:

$$\omega = \omega_{\max} - \frac{\omega_{\max} - \omega_{\min}}{t_{\max}} * t \tag{8}$$

where ω_{max} and ω_{min} are both random numbers called initial weight and final weight, respectively. t_{max} and t are the maximum iteration number and the current iteration number, respectively.

4 Improved Particle Swarm Optimization Algorithm

The PSO algorithm could be utilized mainly in the power system in order to determine its optimal generation power at the specific period, thus minimizing the total system generation cost.

However, when PSO applied to ED problems with some nonlinear constrains, the constrained optimization problem must be commonly converted firstly into a non-constrained optimization problem by using penalty function, which is a general way to handle constrains, and then PSO could be applied to the converted ED problem [11]. Therefore, the fitness function of the particle is generally defined as following equation:

$$f = F + \sigma \left| \sum_{i=1}^{N} P_i - p_D - P_L \right| \tag{9}$$

where σ is the penalty factor. In generally, the choice of σ is depended upon the accuracy and speed of convergence, and the greater value of σ, the more fast its speed and convergence rate is. It is obvious that when the total generator cost and power balance error become lower, the value of this particle's fitness function is lower, and then the solution to find is also more approached to the optimal one.

Except for the fitness function is used to guide search direction, some constraints in (3), (4) and (5) can be used to check the feasibility of particles. When updating the memories (P_{id} and P_{gd}), all the particles only keep feasible solution in their memory. Infeasible particles can not guide particle flying in the next iteration. To ensure better global search performance, all initial particles can be kept in the feasible space by adjusting some constraints of the particles which are not satisfying the equality constraint and inequality constraints. There are some adjusting strategies used to handle constraints [12], [13].

However, the initial values of power are generated randomly within power limits of a generator. As particles explore the searching space, starting from initial values, which are generated randomly within the predefined power limit, they do encounter case whereby the power generated exceeds the margin boundary and therefore violate some constraints. It is possible for a swarm of particle to loss its own diversity at its evolutional procedure, and this particle lost global or local search capability is called as an "inactivity" one. The particle will lose its global search capability when it is closed to the position of P_g and its velocity also approach zero, of course, this particle could only fly within a very small space, and it therefore cannot affect its fitness value at the evolutional procedure. When the particle P_g approach the local extremum and all particles have become inactivity particles, the classic PSO will emerge a phenomena called as "prematurity". This paper will be considered as an "inactivity" one, once the position difference between the particle P_i and P_g have been kept within a given area at *Tc* times.

For making them always have an optimal global searching capacity so as to obtain an optimal global solution, this proposed algorithm will handle some inactivity particles by modify them with its currently optimal value during every cycle of iteration according to the difference between the fitness values of the particle P_i and P_g, so as to improve PSO calculation precision.

Firstly, the difference between the fitness values of the particle P_i and P_g, namely, ΔF_i may be calculated and used to judge whether the particle is in feasible space or not. ΔF_i can be represented as follow:

$$\Delta F_i = \frac{F_i - F_g}{\min \{|F_i|, |F_g|\}} \tag{10}$$

where Fi and Fi are the fitness values of the particle Pi and Pg, respectively.

Secondly, when the resulting ΔF i of this particle is less than the threshold value, which is predetermined according to calculation iteration times and calculation precision of the PSO, this particle could then be considered as an inactivity one and should be modified using its current optimal value at each iteration.

The value of the particle, which is an inactivity one, can be modified by using its current optimal value at each iteration. This modified process is not an only simple reinitialized one, but is a modified one by using the cognized capacity of the PSO. So, this process could be taken not only within a very small area to provide an optimal solution, but also at the anaphase of the particle's iteration.

Then, the modified inertia is set according to the following equation:

$$x_{id}(i) = a * rand() + P_{gd} \tag{11}$$

where a is set to 0.5.

Thus, the calculation precision of the PSO algorithm can be improved due to improving some cognized capacities of the PSO, which is obtained by using the current optimal value to modify particle's value and thereby avoid the local prematurity for the classic PSO algorithm.

5 Numerical Examples and Results

To verify the feasibility of the proposed method based on improved PSO to solve the ED problem, some power systems were tested by using the software written in Matlab

7.0 platform. The results obtained for solving ED problem are compared with other methods.

The example tested here has six generator units, and its system load is 1263MW. Some parameters used for this generator unit are seen in Table 1 and Table 2, and also include: population size: 20 and iterations/generations: 1000. The calculation of PL is cited from reference [1]. The simulation results are shown in Table 3 and Table 4. These simulation results show that the method proposed here has excellent calculation precision and computation efficiency, compared with some other PSO methods.

Table 1. Generator parameters for six-generator system

Generator number	a_i	b_i	c_i	$\overline{P}_i$ / MW	$\underline{P}_i$ / MW
1	0.0070	7.0	240	500	100
2	0.0095	10.0	200	200	50
3	0.0090	8.5	220	300	80
4	0.0090	11.0	200	150	50
5	0.0080	10.5	220	200	50
6	*0.0075*	12.0	190	120	50

Table 2. Ramp constraints and output prohibited operation zones of unit

Generator number	$P_{i,o}$/MW	$R_{u,i,}$/(MW/h)	$R_{ud,}$/(MW/h)	Output prohibited operation zones/MW
1	440	80	120	[210, 240] [350,380]
2	170	50	90	[90, 110] [140,160]
3	200	65	100	[150, 170] [210, 240]
4	150	50	90	[80, 90] [110, 120]
5	190	50	90	[90, 110] [140, 150]
6	110	50	90	[75, 85] [100,105]

Table 3. Simulation results of the proposed method and some other methods

Generator output	Proposed method	PSO[13]	PSO[1]
P_1/MW	448.8646	448.3	447.50
P_2/MW	173.2821	173.55	173.32
P_3/MW	264.7999	264.55	263.47
P_4/MW	136.4987	126.95	139.06
P_5/MW	165.7965	175.98	165.48
P_6/MW	86.2264	86.67	87.13
Total Output/MW	1275.47	1276.00	1276.01
P_L/MW	12.47	12.94	12.96
Total cost/($/h)	15442.8	15448	15450

Table 4. Computation results of the proposed method and some other methods

algorithm	Total cost / ($/h)		
	maximum	minimum	average
Proposed method	15489	15442	15450
PSO [13]	15492	15448	15454
PSO[1]	15492	15459	15469

It is obvious that this proposed method based on improved PSO algorithm has a better calculation precision, as shown from above simulation results, compared with other methods.

6 Conclusions

The improved PSO method applied to solve ED of the power system is demonstrated in this paper. In this proposed method, not only some inactivity particles could be handled, but also these inactivity particles are re-initialized in every iterations so as to make them always lie within a very small area which has an optimal solution. This makes that the proposed method can effectively avoid "prematurity" of the classic PSO, thereby it is beneficial to obtain some optimal global solution. The test and simulation results show that proposed method has some excellent characteristics, such as a high-quality calculation precision and a good computation efficiency, compared with some other PSO methods.

References

1. Gaing, Z.-l.: Particle swarm optimization to solving the economic dispatch considering the generator constraints [J]. IEEE Trans. on Power System 18, 1187–1195 (2003), Bakirtzis, A., Petridis, V., Kazarlis, S.: Genetic algorithm solution to the economic dispatch problem [J]. Proc. Inst. Elect. Eng-Gen, Transm. Dist., 141(4), 377–382 (1994)
2. Lee, F.N., Breipohl, A.M.: Reserve constrained economic dispatch with prohibited operating zones [J]. IEEE Trans. on Power System 8, 246–254 (1993)
3. Su, C.-T., Chiou, G-J.: A fast-computation Hopfield method to economic dispatch of power system [J]. IEEE Trans. Power syst. 12, 1759–1764 (1997)
4. Zi-Xiong, L., Duncan Glover, J.: A zoom feature for a dynamic programming solution to economic dispatch including transmission losses [J]. IEEE Trans. Power syst. 7, 817–827 (1997)
5. Samaan, N., Singh, C.: Using genetic algorithm for reliability calculations of complex power systems [C]. In: ISCAS 2005. International symposium on Circiut and system, May 23-26, 2005, vol. 5, pp. 4191–4195 (2005)
6. Kennedy, J., Eberhart, R.C.: Particle swarm optimization [C]. In: IEEE International conference on neural networks, Pert, Australia, IEEE Service Center Piscataway, NJ (1995)
7. Jeyakumar, D.N., Jayabarathi, T., Raghunathan, T.: Particle swarm optimization for various types of economic dispatch problems [J]. Electrical power and energy systems 28, 36–42 (2006)

8. Park, J.-B., Jeong, Y.-W., Lee, W.-N., Shin, J.-R.: An improved particle swarm optimization for economic dispatch problems with non-smooth cost functions [C]. IEEE Power Engineering society general meeting 18-22, 1–7 (2006)
9. Selvakumar, A.I., Thanushkodi, K.: A new particle swarm optimization solution to nonconvex economic dispatch problems [J]. IEEE trans.on power systems 22(1), 42–51 (2007)
10. Yang, B., Chen, Y., et al.: Solving optimal power flow problems with improved particle swarm optimization [C]. In: Proceeding of the 6th world congress on intelligent control and automation, Dalian, China, June 21-23, 2006, pp. 7457–7461 (2006)
11. Eberhart, R.C., Shi, Y.: Particle swarm optimization: developments, applications and resources [C]. In: Proc. Conf. Evol. Computation, pp. 81–86 (2001)
12. Jun-jie, Y., Jian-zhong, Z., et al.: Application of improved particle swarm optimization in economic dispatch [J]. Power system Technology 29(2), 1–4 (2005)

Immune Particle Swarm Optimization Based on Sharing Mechanism

Chunxia Hu, Jianchao Zeng, and Jing Jie

Division of System Simulation and Computer Application,
Taiyuan University of Science and Technology,
Taiyuan, 030024 China
huchunxia@126.com

Abstract. Sharing mechanism is introduced into particle swarm optimization. Fitness values of particles are updated to sharing fitness values. Particles with higher sharing fitness value are punished and particles with smaller sharing fitness value are remained as memory particles. Particles are updated with memory particles and clone selection when global best have not changed in some continuous generations. Population diversity is increased by this way. At the same time the particle with the best fitness value is saved. The modified algorithm can avoid the local optimization and has better search performance to multi-peak functions. The experimental results show the modified algorithm has better convergence performance than standard particle swarm optimization algorithm.

1 Introduction

Particle Swarm Optimization (PSO) is a new evolutionary computation technique. It was originally developed by Dr. Eberhart and Dr. Kennedy in 1995[1],[2]. The motivation for the development of this method was based on the simulation of simplified animal social behaviors such as fishing or bird flocking. The basic theory is finding the best value through collaboration and sharing information.

Particle Swarm Optimization has some merits that are other optimization algorithms have not. For example, its conception is simple, realization is easy, and search is fast etc. But as an evolutionary computation technique it also has some demerits. For example, standard PSO can appear premature convergence for some complex optimization problem. This can make algorithm get into local optimum [3],[4]. Efficiency of search becomes slow when it closes to local optimum. So it leads to not finding global optimum. As far as the premature problem is concerned, the main reason lies in that the diversity of population get worse and worse gradually with the search carrying on. In order to prevent PSO from premature converging, a feasible idea is to maintain appropriate diversity during the search and make it easy to escape from the local optimum. So, we introduce the sharing mechanism and immune operators into PSO to improve the diversity. In this paper sharing mechanism is introduced into the

K. Li et al. (Eds.): LSMS 2007, LNCS 4688, pp. 231 – 238, 2007.

modified algorithm to update fitness value of particles with sharing fitness value. The particles with higher sharing fitness value are punished and the particles with smaller fitness value are saved as memory particles. Particles are updated by memory particles and clone selection when global optimum don't change in some continuous evolution generations. Population diversity is increased by this way. At the same time the particle with the best fitness value is saved. So the modified algorithm can avoid the local optimization and has better search performance for multi-peak functions. At the same time it can improve convergence rate and precision. The modified PSO model is tested on the three standard test functions, and results are compared with that of standard PSO. The experimental results show the validity of the modified algorithm.

The remaining of the paper is organized as the follows. Section 2 provides some descriptions about the standard PSO version. Section 3 describes the techniques about SIPSO in details. Section 4 presents some experimental results and makes some comparison analysis. Finally, Section 5 concludes with some remarks.

2 Standard PSO

Particle swarm algorithm was first proposed by Eberhart and Kennedy. Instead of using evolutionary operators to manipulate the individuals, as in other evolutionary computational algorithms, each individual in PSO flies in the search space with a velocity which is dynamically adjusted according to its own flying experience and its companions' flying experience. Each individual is treated as a volume-less particle (a point) in the D-dimensional search space. The i-th particle is represented as $X_i = (x_{i1}, x_{i2}, \ldots, x_{iD})$. The best previous position (the position giving the best fitness value) of the i-th particle is recorded as $P_i = (p_{i1}, p_{i2}, \ldots, p_{iD})$ and represented by the symbol $Pbest$. The index of the best particle among all the particles in the population is represented by the symbol $Gbest$. The rate of the position change (velocity) for particle i is represented as $V_i = (v_{i1}, v_{i2}, \ldots, v_{iD})$. The update of the particles is accomplished according to the following equations [5]:

$$V_i(t+1) = wV_i(t) + c_1r_1(Pbest - X_i(t)) + c_2r_2(Gbest - X_i(t)) \tag{1}$$

$$X_i(t+1) = X_i(t) + V_i(t+1) \tag{2}$$

Where w is the inertia weight [6],[7]. The inertia weight has characteristics that are reminiscent of the temperature parameter in the simulated annealing [8]. A large inertia weight facilitates a global search. By linearly decreasing the inertia weight from a relatively large value to a small value through the course of the PSO run, the PSO tends to have more global search ability at the beginning of the run while having more local search ability near the end of the run. c_1 and c_2 are two coefficients (positive constants). r_1 and r_2 are two random numbers within the range [0,1]. There is also a V_{MAX}, which sets the upper and lower bound for velocity values.

3 Immune Particle Swarm Optimization Based on Sharing Mechanism (SIPSO)

3.1 Algorithm Principle

Sharing mechanism algorithm is a kind of niched realization method of sharing fitness value. It was proposed by Golerge in 1989 [9]. The algorithm was proposed because it is inspirited by the thought that " birds of a feather flock together. resources sharing ". The thought that " birds of a feather flock together " reflects each living thing survives under the specific environment in the nature evolution process. There exists outstanding individual among the homogeneous biology, and there exists mutual competence between various living things. There also exists exchange of information between unhomogeneous biology. "Resources sharing" manifests the homogeneous living things of co-existence in the specific environment share limited resources. These living things achieved evolution together through the coordination. There also exists a mechanism through which various living things mutually suppress. Under the premise the resources are insufficiently, the living things whose compatibility is weak are gradually eliminated. The population diversity of the particle swarm algorithm becomes worse and worse in later period of evolution, which makes efficiency of search reduce and be easy to fall into local optimum. In order to overcome these shortcomings, sharing function in sharing mechanism algorithm is introduced into when the population diversity becomes worse. The fitness values of original particles are updated to sharing fitness value according to the equation (3):

$$f_j' = f_j / \sum_{s=i}^{i+p_i-1} Shd_{js}, \; j = i, i+1, \cdots, i+p_i-1; \tag{3}$$

Among

$$Shd_{js} = \begin{cases} 1, & d_{js} = o; \\ 1-(d_{js}/\sigma)^{\lambda}, & d_{js} \prec \sigma; \\ 0, & else; \end{cases}$$

The procedure of sharing mechanism algorithm can be illustrated as follows:

Step 1: Set i =1;

Step 2: Compute distance of particles.

$$d_{ij} = \| x_i - x_k \|, \quad \text{i=1,2,...,N;k=i,i+1,...,N;}$$

Step 3: Make certain niched subgroup M_{P_i} base on $d_{ij} \prec \sigma_0$, i=1,2,…,N; k=i,i+1,…,N. p_i are numbers of element of subgroup. It is decided by the concrete problem (NFP)

Step 4: If the best fitness value of population is included in M_{P_i}, then save the particle that has the best fitness value. Else, the other particles update according to Step 5.

Step 5: Compute fitness value of individual in M_{P_i} after updating (namely sharing fitness).

Step 6: Use sharing fitness value f_j' and Penalty function punish the particle with higher sharing fitness value in subgroup. Namely compare sharing fitness value of two particles when $x_i, x_k \in M_{p_i}, \|x_i - x_k\| \prec L, L \prec \sigma$, then punish the particle with higher sharing fitness value.

$$f_{\max}(x_i, x_k) = Penalty, j = k = i, i+1, \cdots, i + p_i - 1;$$

Step 7: When $i + p_i \prec N$, set $i \leftarrow i + p_i$, return Step 2; or else, turn into next step.

We adjust fitness value of particles with sharing fitness value. And particles with higher density are suppressed. But the most superior individual (namely $Gbest$) is not influenced. This not only improves the population diversity but also preserves the best particle in the population and improves the search performance.

In this paper, two important immune operators that of clone selection and the memory cell are introduced into the particle swarm algorithm at the right moment. The clone selection may make the better candidate solution in the evolution community participate in evolution by definite selection, which provides an opportunity to mine better candidate solution. The evolution of memory cell not only provides an effective opportunity for the similar question solution but also provides the essential preparation for the local search of algorithm. So introduction of these two operators may effectively improve search rate and precision, and it can overcome the disadvantage that the search efficiency of the particle swarm algorithm becomes low in later period.

3.2 Algorithm Flow

The procedure of SIPSO can be illustrated as follows:

Step 1: Initialize a population which including N particles.

Step 2: Compute fitness value of each particle.

Step 3: For each particle, compare the evaluated fitness value of each particle with its $Pbest$ .If its current value is better than the $Pbest$. Then set the current location as the $Pbest$ location.

Step 4: For each particle, compare the evaluated fitness value of each particle with $Gbest$, if its current value is better than the $Gbest$, then reset $Gbest$ to the current location of particle.

Step 5: Chang the velocity and location of the particle according to the equation (1), (2), respectively.

Step 6: Stop evolution when global optimum have not changed in some continuous evolution generations and have not arrived ending conditions. Update particles as follows:

Step 7: Produce memory particles making use of sharing mechanism algorithm.

Step 8: Clone selection. Select particle (antibody) with smaller fitness value (namely higher affinity) according to selection ratio α. The number is N_0. Among, $N_0 \equiv round(\alpha, N)$, α is satisfied with $round(\alpha.N) + M = N$.

Step 9: The velocity and the position of updated particles continue evolve according to equation (1), (2).

Step 10: If have not searched particle with enough well fitness value or have not arrived the max evolution generations ($G_{\max}$) , return Step 2. Otherwise stop evolution.

4 Experiments

We have tested the modified PSO model on the three standard test functions, and compared result to standard PSO. All three are widely known as benchmark functions for testing the performance of different evolutionary optimization strategies such as evolutionary programming, simulate annealing, genetic algorithms and particle swarm optimization.

4.1 Benchmark Functions

F1. Griewank function. It gets the global Min 0 at $x_i = 0$.

$$f_1(x) = \sum_{i=1}^{n} x_i^2 / 4000 - \prod_{i-1}^{n} \cos(x_i / \sqrt{i}) + 1 \qquad -10 \le x_i \le 10$$

F2. Rosenbrock function. This is a classical optimization function. It gets the global Min 0 at $x_i = 1$.

$$f_2(x) = \sum_{i=1}^{n} (100(x_{i+1} - x_i^2)^2 + (x_i - 1)^2) \qquad -30 \le x_i \le 30$$

F3. Rastrigin function. It gets the global Min 0 at $x_i = 0$.

$$f_3(x) = \sum_{i=0}^{n} (x_i^2 - 10\cos(2\pi x_i) + 10) \qquad -5.12 \le x_i \le 5.12$$

Based on characteristic of each function, we set the number of evolution generation are F1=1000, F2=F3=5000. The errors are F1=0.01, F2=F3=0.1. There is also a V_{MAX}, which sets the upper and lower bound for velocity values. the other parameters are

$c_1 = c_2 = 1.8$, $\sigma_0 = \sigma = 1$, $\lambda = 2$, L=0.01. The inertia weight starts with a value close to 1 and linearly decreasing to 0.4. Ratio of clone selection is $\alpha = 0.3$. Selection ratio of memory particles is 0.7. Population size is 40. Dimension is 10.

4.2 Experimental Results

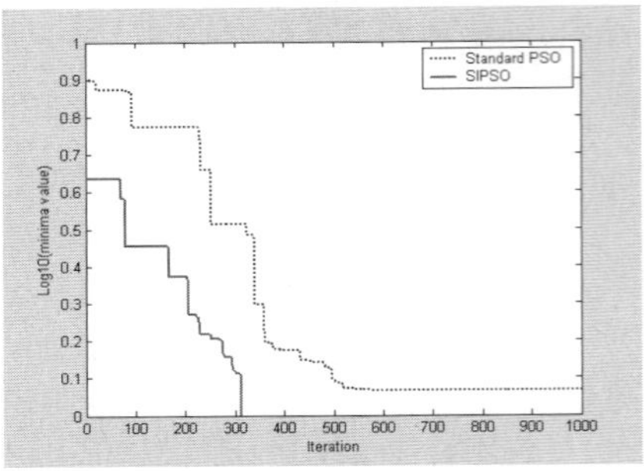

Fig. 1. Performance on Griewank

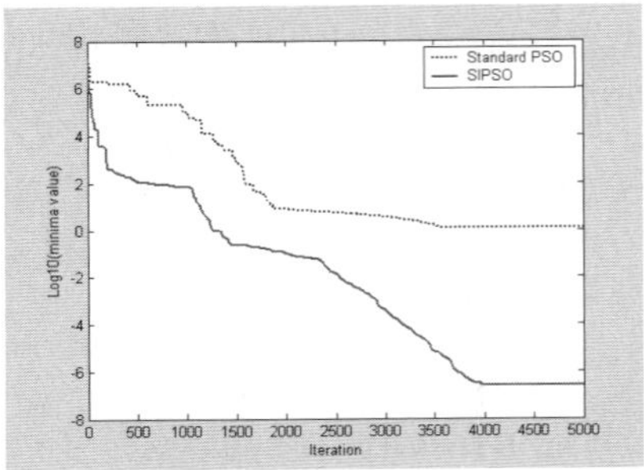

Fig. 2. Performance on Rosenbrock

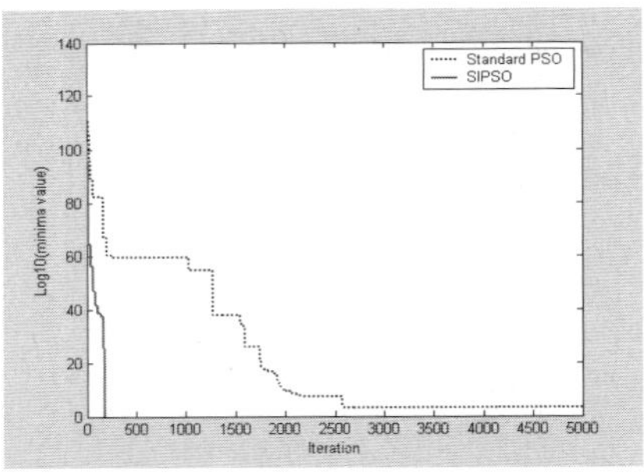

Fig. 3. Performance on Rastrigin

Table 1. Comparison among different simulation results (1) (ARC represents average convergence ratio, ACG represents average convergence generations and AVBF represents average value of the best fitness. Algorithm stop searching when arrive at enough precision)

Function	Algorithm	AC R	A C G	AVBF
F1	PSO	0/50	/	0.079366
	SIPSO	46/50	504	0.008756
F2	PSO	6/50	3466	1.627123
	SIPSO	50/50	1816	0.099306
F3	PSO	6/50	2629	1.837584
	SIPSO	50/50	604	0.001207

Table 2. Comparison among different simulation results (2) (TGBV represents times get the best value. Algorithm stop searching when arrive at the most generations).

Function	Algorithm	Generation	TGBV	AVBF
F1	PSO	1000	0	0.079366
	SIPSO	1000	46	0.008756
F2	PSO	5000	0	1.627123
	SIPSO	5000	50	0.099306
F3	PSO	5000	5	1.837584
	SIPSO	5000	50	0.001207

Fig.1-Fig.3 are graphs that the fitness value of function F1-F3 (namely the global best of each generation) changing with evolution. From the figures we can see that the convergence performance of modified algorithm is better than standard PSO. Especially for multi-peak functions F1, F2, It can search the global optimum in a short time. From the trend of development in figure we can get that the modified algorithm can converge onto the global optimum if have enough generations. This can be proved by Tab.2. So the modified algorithm is an algorithm that can ensure the global convergence.

5 Conclusions

Standard PSO is a simple, stochastic and global optimization technology. But its population diversity becomes worse and worse gradually with the search carrying on. So the algorithm is easy to fall into local optimum. The modified algorithm (SIPSO) introduces immune memory characteristic to update particles when population diversity becomes worse. This not only improves population diversity but also saves the particle with the best fitness value in population. The modified algorithm improves convergence rate and precision. At the same time it ensures the better convergence performance. Undoubted SIPSO is an effective algorithm. But the parameters that are introduced to algorithm have effect on search performance the certain degree. So it is an emphases and difficulty how to set the appropriate parameters according to concrete problem.

Acknowledgments. This work is supported by the National Natural Science Foundation of China under Grant No. 60674104, the Ministry of Education of China Key Project Science and Technology Foundation under Grant No.204018, the Natural Science Foundation of Shanxi province under Grant N0.2007011046.

References

1. Eberhart, R.C., Kennedy, J.: A new optimizer using particle swarm theory. In: Proceedings of the Sixth International Symposium on Micro Machine and Human Science, Nagoya, Japan, pp. 39–43 (1995)
2. Kennedy, J., Eberhart, R.C.: Particle swarm optimization. In: Proc. IEEE Int'l Conf. on Neura lNetworks, vol. IV, pp. 1942–1948. IEEE Press, Piscataway,NJ (1995)

3. Eberhart, R.C., Shi, Y.: Particle swarm optimization: developments, applications and resources[A]. In: Proc. 2001 Congress on Evolutionary Computation[C], pp. 81–86. Soul, South Korea (2001)
4. Parsopoulos, K.E., Vrahatis, M.N.: Recent approaches to global optimization problems through Particle Swarm Optimization[J]. Natural Computing: Netherlands, 235–306 (2002)
5. Kennedy, J., Eberhart, R., Shi, Y.H.: Swarm Intelligence [M], pp. 287–299. Morgan Kaufmann, San Francisco (2001)
6. Shi, Y.H., Eberhart, R.C.: Parameter selection in particle swarm optimization. In: 1998 Annual Conference on Evolutionary Programming, San Diego (March 1998)
7. Shi, Y.H., Eberhart, R.C.: A modified particle swarm optimizer. In: IEEE International Conference on Evolutionary Computation, Anchorage, Alaska, pp. 4–9 (May 1998)
8. Eberhart, R.C., Shi, Y.H.: Comparison between genetic algorithm and particle swarm optimization. In: Annual Conference on Evolutionary Programming, San Diego (1998)
9. Zhou, M., Sun, S.D.: Genetic algorithm and application [M], p. 202. National Defense Industry Press, Beijing (2000)
10. Huang, X.Y., Zhang, Z.H., He, C.J., et al.: Modern Intelligent Algorithm Theory & Application [M]. Science Press, Beijing (2005)

A Diversity-Guided Particle Swarm Optimizer for Dynamic Environments

Jing Hu, Jianchao Zeng, and Ying Tan

School of Computer Science and Technology,
Taiyuan University of Science and Technology,
Taiyuan 030024,
hujingtaiyuan@sina.com.cn

Abstract. For many real-world changeable problems over time, the goal of optimization is not only to acquire an optimal solution, but also to track its progression through the search space as closely as possible. In this paper, an improved detection technique at the particle level is designed. Then, a new method of response, learning from the changing global optimum for new environments guided by population diversity, is designed. It defines response condition as well as part of particles to be reset and flying direction after a change. Then, the parabolic benchmark functions with various severities are used to test, compared with the Eberhart-PSO and APSO, and the results show the modified strategies are effective in tracking changes.

1 Introduction

Particle Swarm Optimization (PSO) is an evolutionary computation technique motivated by the simulation of birds social behavior and was developed by Kennedy and Eberhart in 1995[1]. PSO is used to socio-organizing replacement of natural selection in evolutionary algorithm, which searches the optimum by individual cooperation. The PSO algorithm is already successfully applied in the optimization of static optimization realms, such as multi-objective optimization, decision-making and simulation, etc. However, many real world problems are dynamic and stochastically change over time, so that the optimum changes correspondingly over time. The dynamic problems require systems that are adaptable to change over time. In contrast to static optimization, the goal in dynamic optimization is not only to detect the changes automatically but also to respond a variety of changes so that the moving optimum can still be tracked as closely as possible. Hu and Eberhart proposed a method to re-evaluate "gbest" and "second-gbest" before each iteration to detect the change[2].Carlisle and Dozier also proposed a similar algorithm by monitoring randomly "sentry" particle which is randomly chosen at each iteration[3]. These methods are based on the assumption the corresponding fitness values of particles also change after dynamic environment has changed. However, in some cases, the corresponding fitness values do not change after a change. For response techniques, Hu

K. Li et al. (Eds.): LSMS 2007, LNCS 4688, pp. 239–247, 2007.

and Eberhart proposed to re-randomization of a certain number of particles, making them away from the direction to the previous goal[2]. It brings some question: firstly, the randomization implies loss of "memory" velocity and position information gathered during the searching so far, especially the randomization does not guarantee quick convergence the optimum on the beginning of the search. Secondly, it is no necessary to randomize entire or part of swarm if area of search space is still more extensive, where it is likely to search the optimum with current situation using memory information.

The paper is organized as follows: Section 2 gives a brief introduction about on PSO. Section 3 provides details of modified strategies. Finally, several simulation results are analyzed.

2 Standard PSO

In PSO, each particle represents a possible solution to the problem at hand and is treated as a point in j-th dimensional search space. Each particle adjusts its flight over the search space based on its own flight experience and that of its neighbors. Each particle is characterized by its position and current velocity. Furthermore, a particle has a memory where it stores the best position that it has found so far. Particles adjust their flight trajectories using the following equations:

$$v_{ij}(t+1) = wv_{ij}(t) + c_1 r_{1j}(t)(p_{ij}(t) - x_{ij}(t)) + c_2 r_{2j}(t)(p_{gj}(t) - x_{ij}(t)) \quad (1)$$

$$x_{ij}(t+1) = x_{ij}(t) + v_{ij}(t+1) \quad (2)$$

Where w is the inertia factor, whose goal is to control the impact of the history of the velocities of a particle over the current velocity, by influencing the local and global exploration abilities of the algorithm. v_{ij} is the velocity of the i-th particle in the j-th dimension, c_1 and c_2 are the cognitive and social learning rates, respectively. The two rates control the relative influence of the memory of the neighborhood to the memory of individual. $r_{1j}(t)$ and $r_{2j}(t)$ are random values (with a uniform distribution) with the interval [0, 1].

3 Diversity-Guided PSO (DGPSO) Model

3.1 Environment Detection Method

This method is invalid when the fitness value of re-evaluating the "gbest" or "sentry" is equal to its previous fitness after environment has changed.

For a particle, individual fitness value may be changeable when environment has changed. If $|F(X_i(t) - F(X_i(t-1)|\rangle \ell$, where is $|X_i(t) - X_i(t-1)|\langle \eta$, there is a conclusion that environment has changed.(ℓ is a predefined threshold, η is small positive values).

$$G_i(t) = \frac{F(X_i(t)) - F(X_i(t-1))}{R_i(t)} \tag{3}$$

$$R_i(t) = \sqrt{\sum_{j=1}^{n} (x_{ij}(t) - x_{ij}(t-1))^2} \tag{4}$$

$$M_i(t) = \frac{G_i(t)}{G_i(t-1)} \tag{5}$$

Where t is iteration, $X_i(t)$ is the position of i-th particle. $F(X_i(t))$ is the fitness value of i-th particle. $R_i(t)$ is space-distance of i-th individual within iteration t-1 and t. $G_i(t)$ is the ratio of $F(X_i(t)) - F(X_i(t-1))$ and $R_i(t)$. $M_i(t)$ is the ratio of $G_i(t)$ and $G_i(t-1)$. General considering that $M_i(t)$ is closed to 1 if environment has not changed and $M_i(t)$ is springing if environment has changed. Based on above analysis, the global optimum is reevaluated before iteration and compared with its previous fitness and there is a change if the compared result is not equal. Otherwise, there is a change if the number of particles whose $G_i(t)$ is more than a predefined threshold and $M_i(t)$ is not closed to 1 exceeds the certain scale. Contrarily, there is no change.

This method detects environmental changes at the particle level and reduces the optimization cost of triggering "sentry" and re-evaluating the "gbest".

3.2 Environment Response Method

The next goal for optimization method is to provide an effective mechanism to respond to these changes. The information stored in the memory, i.e., individual local best location and the corresponding fitness value, may no longer be true and may actually be misguiding the search. Hu and Eberhart proposed re-randomize of entire or part of the swarm[2].Susana used a dynamic macro-mutation operator hybridized with evolutionary algorithm concepts[4]. The problems of these methods in this category are that randomization implies loss of information gathered so far and it seems difficult to determine the right amount of randomization. Too much will resemble restart, while too little find the changeable optimum with difficulty. And these methods do not make the particles randomize around the area of the new environment so that it is a random response during iterations. It is necessary to design more adaptive response method to react to a change.

The standard PSO is a random search algorithm charactered by its fast convergence, the whole swarm has already stuck to a small area and as a result it seems not to be easy to dump out to keep tracking the changes. The whole swarm clusters so excessively that search space contracts and make it impossible to find the optimum in the contracted and "outdated" area of search space. It needs to reset the memory of the swarm with aim so that the swarm spread out (disperse) and make it possible to find the optimum in the expanded area of search space. Otherwise, the whole swarm distribute so uniformly in the search space that there is no necessary to reset the memory of particles. Concluded, it is direct relation between particles to be reset and population distribution.

Population diversity describes population distribution. It is with respect and independence of the population size, the dimensionality of problem and search range of each of the variables. A measure for N-dimensional numerical problems is the "distance-to-average-point" measure defined as:

$$diversity(S) = \frac{1}{|S| \cdot |L|} \cdot \sum_{i=1}^{|S|} \sqrt{\sum_{j=1}^{N} (p_{ij} - \bar{p}_j)^2} \tag{6}$$

where S is the population, |S| is the population size, |L| is the length of the longest diagonal in the search place, N is dimensionality of the problem, p_{ij} is the *j-th* value of the *i-th* individual and $\bar{p}_j$ is the *j-th* value of the average point.

Population diversity is above a certain threshold d_{high}, the swarm distributes uniformly. There is more chance to relocate the optimum for new environment in the search space and it is no necessary to reset. Otherwise, the swarm distributes clustered, simply because the population covers a small part of the search space, it is necessary to expand area of search space and improves population distribution by reset.

The DGPSO applies diversity-increasing operators by resetting the current position of individual when population diversity drop a certain threshold d_{low} .The DGPSO switches to diversity-decreasing operators until a diversity of d_{high} is reached.

Zhang proposed individual current position $X_i(t)$ and local best position $P_i(t)$ is possible to the optimum position after a change[5], thereby, $X_i(t)$ and $P_i(t)$ are reevaluated after a change, corresponding fitness value marked as $f'(X_i(t))$ and $f'(P_i(t))$ .Then, $f'(X_i(t))$ and $f'(P_i(t))$ are compared and choose the better as $P_i'(t)$ from them (corresponding fitness value marked as $f'(P_i'(t))$. Finally, the best of $P_i'(t)$ is selected as $P_g'(t)$ (corresponding fitness value marked as $f'(P_g'(t))$.

When population diversity is above a certain threshold d_{high} , it is no necessary to reset the memory of swarm. Local exploration with exploring the optimum for current situation is necessary to enhance. The velocity-update formulas of the particles are as follows:

$$v_{ij}(t+1) = wv_{ij}(t) + c_1 r_{1j}(t)(p_{ij}'(t) - x_{ij}(t)) + c_2 r_{2j}(t)(p_{gj}'(t) - x_{ij}(t)) \tag{7}$$

$$x_{ij}(t+1) = x_{ij}(t) + v_{ij}(t+1) \tag{8}$$

When population diversity drops below a certain threshold (d_{low}), resetting is necessary until d_{high} is reached.

Continuous resetting increases population diversity and expands areas of search space of swarm. The course of resetting enhances global exploitation and makes it possible to find the optimum for current situation. How to reset the particles? "Social cognition" of standard PSO model guarantees particle moving to the best location in the entire swarm.

Motivated by it, a new response method, learning from the global optimum for new environments, is designed. Particles escape from their flight trajectories and learn from $P_g'(t)$, namely, flying towards $P_g'(t)$ named the global optimum for new environments. The courses of resetting are guided by following equations:

$$x_{ij}'(t) = x_{ij}(t) + k(t)\left|p_{gj}(t) - p_{gj}'(t)\right| \tag{9}$$

$$k(t) = \frac{f(P_g(t)) - f'(P_g'(t))}{\sqrt{\sum_{j=1}^{n}(p_{gj}(t) - p_{gj}'(t))^2}} \tag{10}$$

$x_{ij}(t)$ is *i-th* individual position in *j-th* dimension, $x_{ij}(t)$ is *i-th* individual position in *j-th* dimension after reset, $k(t)$ is the influence of environmental changes on the global optimum, smaller the $k(t)$ is, smaller the influence is, bigger the $k(t)$ is, bigger the influence is. $k(t)=0$ denotes that change has not happened. $\left|p_{gj}(t)-p_{gj}(t)\right|$ is the offset of global optimum in *j-th* dimension because of environmental change. Equation (9) denotes *i-th* individual learning from $p_g(t)$.

In order to avoid abruptly cleaning the memories of all particles simultaneously, particles to be reset are defined. The equation $D_i(t) \le \delta(t)$ defines particles to be reset. $D_i(t)$, which is the distance between the current position $x_{ij}(t)$ and global best position $p_{gj}(t)$, is defined:

$$D_i(t) = \sqrt{\sum_{j=1}^{n}(x_{ij}(t) - p_{gj}(t))^2} \tag{11}$$

δ(t),which is the mean distance between all of particles and global best position $P_{gj}(t)$,is defined:

$$\delta(t) = \frac{1}{n}\sum_{i=1}^{n}\sqrt{\sum_{j=1}^{n}(x_{ij}(t) - p_{gj}(t))^2} \tag{12}$$

By above equations, parts of particles are repeated to reset until the diversity in excess of a certain threshold d_{high} .The velocity-update formulas of particles are as follows:

$$v_{ij}(t+1) = wv_{ij}(t) + c_1 r_{1j}(t)(p_{ij}'(t) - x_{ij}'(t)) + c_2 r_{2j}(t)(p_{gj}'(t) - x_{ij}'(t)) \tag{13}$$

$$x_{ij}(t+1) = x_{ij}'(t) + v_{ij}(t+1) \tag{14}$$

4 Experiments and Evaluation

As the parabolic function is easy to control the dynamic changes, it is chosen as the test function described as formula (15). A severity factor called offset in the function changes in various ways.

$$f(x)=\sum_{i=1}^{n}(X_i - Offset)^2,\ X_i \in [-50,50] \tag{15}$$

In the experiments, the number of particles is set as 30 and the dimension of each particle is 10. The dynamic range for the parabolic function was set to [-50, 50] on each dimension. The maximum velocity v_{max} is equal to the dynamic range on each dimension. The inertia weight was $0.5+(Rnd/2.0)$. We call this modification to the standard PSO for dynamic environments as Diversity-Guided PSO (DGPSO).

4.1 The Influence of Different *d* on DGPSO

The relation between threshold of diversity named d and d_{low} and d_{high} is as follows:

$$diversity(S(t))=\begin{cases} d_{low} & if \quad diversity(S(t)) \le d \times diversity(S(0)) \\ d_{high} & if \quad diveristy(S(t)) > d \times diversity(S(0)) \end{cases} \tag{16}$$

The influence of different d is tested on DGPSO. The severity factor called *offset* takes on values of 0.01, 0.1 and 1.0.The frequency of change is 10, 50, and 100 iterations. The first time denotes the number of iterations needed to reach the required error level. The second time denotes the number of iteration to re-reach the required error level after changes are applied to the function.

From the Table1 shown, The average number of iterations of finding the optimum reduces gradually until $diversity(S(t))$ is less than the 50 percent of $diversitiy(S(0))$ .Whereas, the results are closed to above-mentioned when $diversity(S(t))$ is less than 10 percent of $diversitiy(S(0))$ .It indicates that the algorithm is not enough improved through the proportion of resetting particles increases gradually. According to Table1 results, $d=50\%$ is fitter and set in all following experiments.

Table 1. The influence of different *d* on DGPSO

d(%)	*Offset*=0.01 Δg=10		*Offset*=0.1 Δg=50		*Offset*=1.0 Δg=100	
	first time	second time	first time	second time	first time	second time
100	168.3	1150.2	170.5	163.2	169.5	342.2
80	166.6	321.6	174.2	141.1	168.3	265.5
50	166.1	139.9	165.2	122.7	165.1	248.5
10	169.2	124.1	163.1	123.7	166.6	260.3
1	164.9	123.4	164.0	126.5	165.9	236.6

4.2 Convergence Performance Comparison

The severity factor *offset* takes on values from 0.00001 to 10. The frequency of change can be 1 (*offset* is not more than 1) and 100 (*offset* is equal to10) iterations. The two algorithms are repeated for 100 run.

Table 2. The averaged number of iterations tracking the moving optimum

Offset	Eberhart- PSO first time	Eberhart- PSO second time	DGPSO first time	DGPSO second time
0.00001	222.79	0.06*	169.75	0
0.0001	221.68*	0.77*	175..0	0
0.001	221.48*	14.86*	174.4	0.25
0.01	220.55*	54.86*	177.9	12.40
0.1	223.35*	108*	167.40	54.70
1.0	220.20*	86.74*	162.30	104.95
10	220.46*	213.97*	167.10	159.75

*The data come form reference [1]

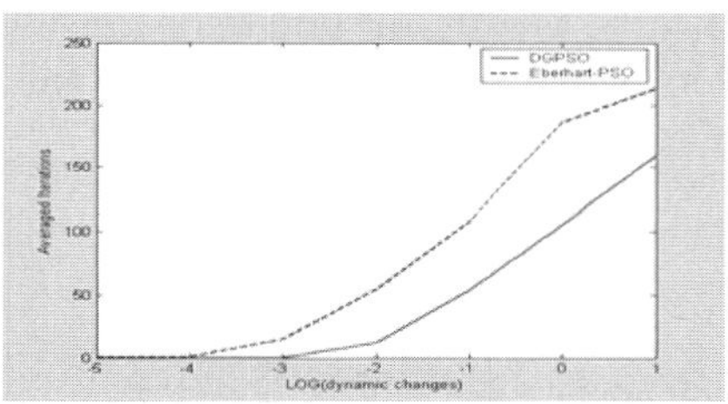

Fig. 1. Comparison of the convergence performance

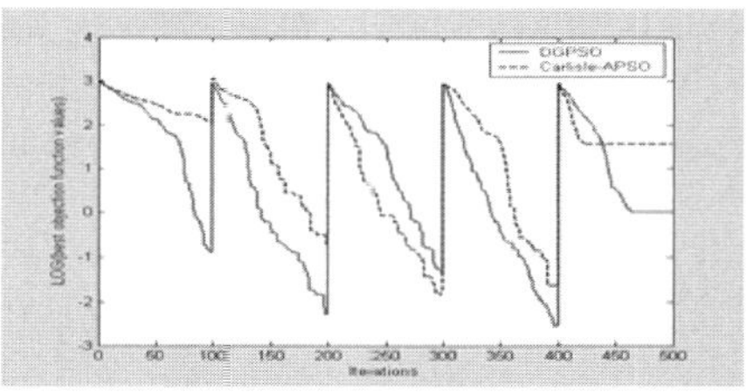

Fig. 2. Comparison of the tracking performance

Table.2 and Fig.1 show the comparison of DGPSO and Eberhart-PSO[2]. The Eberhart-PSO responds these changes by randomizing a scale of particles[2], While for the DGPSO, part of particles is reset only if $diversity(S(t))$ is less than 50% percent of $diversity(S(0))$ .From the table2, the DGPSO tracks the optimum without any iterations when *offset* is less than 0.001.The average number of iterations of the DGPSO is superior to the Eberhart-PSO when *offset* is more than 0.01.From the Fig.1 shown, the DGPSO outperforms the Eberhart-PSO.

4.3 Performance Tracking Comparison

In order to test the performance of tracking the moving optimum, The DGPSO was compared with the Carlisle-APSO[6]. The severity factor *offset* takes on values 10.The frequency of change is 100 iterations. Two algorithms are repeated for 20 runs. The max iteration is 500.

As Fig.2.shown, two algorithms adapt to changes and track the changing optimum. On the beginning, the DGPSO exhibits better performance in convergent speed and accuracy. Due to not escaping from the "outdated" areas of search space, the performance of DGPSO is inferior to Carlisle-APSO in the following changes. The DGPSO outperforms Carlisle-APSO in the next change.

5 Conclusions and Future Research

In this paper, a new modified PSO for dynamic environments, combined with population diversity and learned from the global optimum for new environments, is proposed. The reasons as well as necessity of putting forward population diversity and relation with reset are analyzed. The proposed method not only detects the various changes for dynamic environments exactly but also balances between exploitation and exploration under the guide of population diversity. These experiments show the modified algorithm convergent the optimum with respect to the current situation. Here only a simple parabolic function is tested and further investigation is needed for more complex problems.

Acknowledgments. This work is supported by the National Natural Science Foundation of China under Grant No. 60674104, the Ministry of Education of China Key Project Science and Technology Foundation under Grant No.204018, the Natural Science Foundation of Shanxi province under Grant N0.2007011046.

References

1. Kennedy, J., Eberhart, R.C.: Particle Swarm Optimization. In: Proc. of the IEEE International Conf.on Neural Networks, vol. IV, pp. 1942–1948. IEEE Press, Piscataway, ND (1995)
2. Xiaohui, H., Eberhart, R.C.: Adaptive Particle Swarm Optimization: Detection and Response to Dynamic Systems. In: Proceedings of the IEEE Congress on Evolutionary Computation, Honolulu, Hawaii, USA, pp. 1666–1670. IEEE Computer Society Press, Los Alamitos (2002)
3. Carlisle, A., Dozier, G.: Adapting Particle Swarm Optimization to Dynamic Environments. In: Proceeding, ICAI, Las Vegas,NV, vol. 1, pp. 429–434 (2000)
4. Esquivel, S.C., Coello Coello, C.A.: Particle Swarm Optimization in Non-stationary Environments. In: Lemaître, C., Reyes, C.A., González, J.A. (eds.) IBERAMIA 2004. LNCS (LNAI), vol. 3315, pp. 757–766. Springer, Heidelberg (2004)
5. Zhang, X., Du, Y., Qin, Z., Qin, G., Lu, D.: A Modified Particle Swarm Optimizer for Tracking Dynamic Systems. Advances in Natural Computation, 592–601 (2005)
6. Carlisle, A., Dozier, G.: Tracking Changing Extrema with Adaptive Particle Swarm Optimizer. In: ISSCI, 2002. World Automation Congress, Orlando, FL, USA (2002)
7. Blackwell, T., Branke, D.: Multi-swarm Optimization in Dynamic Environments. In: Raidl, G.R., Cagnoni, S., Branke, J., Corne, D.W., Drechsler, R., Jin, Y., Johnson, C.G., Machado, P., Marchiori, E., Rothlauf, F., Smith, G.D., Squillero, G. (eds.) EvoWorkshops 2004. LNCS, vol. 3005, pp. 489–500. Springer, Heidelberg (2004)

8. Cui, C.X., Hardin, T.: Tracking non-stationary Optimal Solution by Particle Swarm Optimizer. In: Proceedings of the Sixth International Conference on Software Engineering
9. Li, X., Dam, K.H.: Comparing Particle Swarms for Tracking Extrema in Dynamic Environments. Congress on Evolutionary Computation 3, 1772–1779 (2003)
10. Hu, X., Eberhart, R.C.: Tracking dynamic systems with PSO: where's the cheese? In: Proceedings of the workshop on Particle Swarm Optimization (2001)

Colony Algorithm for Wireless Sensor Networks Adaptive Data Aggregation Routing Schema

Ning Ye[1,2], Jie Shao[4], Ruchuan Wang[2,3], and Zhili Wang[1]

[1] Dept .of Information Science, Nanjing College for Population Programme Management, 210042Nanjing, China
[2] Department of Computer Science and Technology, NJUPT, Nanjing, China, 210003
[3] State Key Laboratory for Novel Software Technology, Nanjing University, 210093 Nanjing, China
[4] College of Information Science and Technology, Nanjing University of Aeronautics and Astronautics, 210016 Nanjing, China
Cathery163@163.com

Abstract. Wireless sensor network should decrease the power costs of redundancy information and delay time. The technology of data aggregation can be adopted. A routing algorithm for data aggregation based on ant colony algorithm (ACAR) is presented. The main idea of this algorithm is optimization of data aggregation route by some cooperation agents called ants using the three heuristic factors about energy, distant and aggregation gain. For realizing data aggregation by positive feedback of the ants, the nodes of wireless sensor networks should not maintain the global information. The algorithm is a distributed routing algorithm and realizes data aggregation trade-off in energy and delay. The analysis and the experimental results show that the algorithm is efficient.

1 Introduction

Recent advances in technologies allowed the development of small devices with capability of sensing, processing and transmission. These sensors have the ability to be integrated into wireless sensor network (WSNs). Many applications have been emerged to use WSNs especially for environmental monitoring, health monitoring, vehicle tracking system, military surveillance and earthquake observation [1]. The most distinguishing attribute of WSNs is its limited power supply, thus, energy efficiency is a very important criterion for any existing and new applications.

In sensor networks, data collected by many sensors are based on common phenomena. Normally, there is a large amount of data that can be collected from these WSNs, which make it possible to eliminate redundancy, minimize the number of transmissions [2]. It is because the dominant energy consumer within a sensor node is the radio transceiver [3]. The effort to reduce the number of packet transmission with the in-network processing is called data aggregation.

The main contribution of this paper is to propose an ant colony algorithm based routing scheme for wireless sensor network (ACAR). Inspired by the intelligent entity and the social behaviors of natural ants in finding the sources of food. We assign the exploring task of data aggregation routing to multi mobile agents which have ant-like

K. Li et al. (Eds.): LSMS 2007, LNCS 4688, pp. 248–257, 2007.

logically analyses ability. These intelligent agents are coordination in adaptive information decision-making by pheromone renewal. The method also produces an optimize data aggregation routing based on energy efficient. The simulation results which compared with three other protocols show that it decrease energy consume.

This paper is organized as follows. Section II outlines the relate work about data aggregation routing in WSNs. Section III exhibits the details of ACAR and Section IV evaluates the performance of ACAR. Finally, Section V gives the conclusions and further work.

2 Relate Work and Problem Formulation

Data aggregation techniques have been investigated recently as efficient approaches to achieve significant energy savings in the WSNs [4,5]. How to optimize the data aggregation routing that means select the best data aggregation nodes have being drawn much attention from the research community from the very beginning.

Directed Diffusion (DD) [6] is an important milestone in the data-centric routing research of sensor networks. The idea aims at diffusing data through sensor nodes by using a naming scheme for the data. A sink through its neighbors broadcasts the interest. Each node receiving the interest can do caching for later use. The nodes have the ability to do in-network data aggregation. In addition, DD is highly energy efficient since it is on demand and there is no need for maintaining global network topology. However, DD cannot be applied to all sensor network applications since it is based on a query-driven data delivery model.

Heinzelman et al. have proposed a LEACH algorithm for WSNs in which the sensors elect themselves as cluster heads with some probability and broadcast their decisions [7]. Once the data from each sensor node are received, the cluster head aggregates and transfers them directly to the base station.

A chain-based protocol, PEGASIS, was proposed [8]. In PEGASIS, a chain is created so that each node receives aggregate information and forwards it to a nearby neighbor. It presents mechanisms that can support the variation of the energy parameters.

Qi et al. [9] proposes the concept of mobile agent-based distributed sensor networks (MADSNs). Mobile agents are special programs that can be dispatched from a source node to be executed at remote nodes. Where in a mobile agent selectively visits the sensors and incrementally fuses the appropriate measurement data. For a particular multi-resolution data integration application, it is shown that a mobile-agent implementation saves up to 90 percent of the data transfer time due to savings in avoiding the raw data transfers [10].

Inspired by the society behave of ant colony, Dorigo M et al. proposes the Ant Colony Optimization Algorithm (ACO) in solving the TSP problem [11,12]. ACO algorithms are proposed as approximate algorithms targeting on the hard combinatorial optimization problems. Nowadays, ACO algorithms have successfully applied in many different applications. [13, 14, 15].

In this paper, we propose the idea of ant-like agent, which combine the mobile agent technology with the ACO algorithms for network control in adaptive route selection.

3 Ant Colony Algorithm Based Data Aggregation Routing Scheme

3.1 Data Aggregation Model

To simplify the network model, we adopt a few reasonable assumptions as follows:

- N sensors are uniformly dispersed within a square field.
- All sensors and base station are stationary after deployment.
- The communication is based on the single-hop.
- Communication is symmetric and a sensor can compute the approximate distance based on the received signal strength if the transmission power is given.

We use a simplified model shown in [9] for the data aggregation as follows. Each node in WSNs is equipped with an ant-like agent context where mobile agent is operating. Information analysis programs are embedded into mobile.

Each ant-like agent may only be a small signaling packet transporting state information and builds a path from its source nodes to its destination by sharing their knowledge among neighbors. During data aggregation cycle, each ant-like agent gets quantitative information about the path cost and qualitative information about the amount of traffic in the WSNs.

For a network of N sensor nodes, each node has an attribute vector λ . $\lambda = (\mathrm{ID}, \mathrm{E}, \psi, \tau, \mathrm{p}_{\mathrm{re}}, \mathrm{p}_{\mathrm{tr}}, \mathrm{Tag})$, ID indicates the identification number of node. E is the residual energy of node. $\Psi$is a distance vector table which is in the format of 3-tuple(i, j, dis), where indicates the distance hop between the node i and node j which initialize by broadcasting the message. τ is pheromone trails table. Assuming to be known a priori, the number of sensors in the neighborhood of node i, $\tau_{i,j}$ indicates the pheromone strength number of the path between two nodes. P_{re} and P_{tr} indicate the number of data packets received /transmitted by current node respectively. Tag indicates the instance of being visited by agent. Tag=1 indicates that the current node has been visited by ant-like agent, otherwise, Tag=0.

Considering bi-direct data transmission between source node and base station, we makes use of two kinds of ant-like mobile agent (AMA).

Active AMA (AAMA)**:** travels from the source node to the destination in charge of exploring new paths and gathering information.

Passive AMA (PAMA)**:** is dispensed by the base station and travel back to the source node to update the information in each sensor node as they move.

In this paper, we adopt the attribute definition of mobile agent provided in Qi [8]. Being an entity, the AMA has five attributes (Identification, Itinerary, Data, Method, Interface).

- Identification is used to uniquely identify the AMA by the format of 2-tuple (i, j). For $AAMA_{i,j}$ or $PAMA_{i,j}$, i indicates the ID of its dispatcher and j indicates the ID of its receiver.
- **Itinerary** includes information about migration route assigned by processing node (base station/source node) before dispatched.

- **Data** is AMA's private data buffer, which mainly carries code information or integration results.
- **Method** is the prior implementation algorithm for optimization routing. Here, we use ACO algorithm.
- **Interface** provides interface functions for processing node to access AMA's information as well as the communication between AMA and processing node.

3.2 Adaptive Routing Schema Based on Ant Colony Algorithm

In previous model of data aggregation, one AMA is placed initially at each of the node before the nodes have been deployed in WSNs. Once received the routing task, the AMA tries to find a path in the network, providing minimum cost. The new location determined of AMA is implemented by defining a probability for next neighbor node to be selected, with a higher probability being allotted to more promising connection. Similar to ACO algorithm, the AMA's position may be updated using the following equation [15].

$$P_{i,j}^{k} = \begin{cases} \dfrac{[\tau_{i,j}]^{\alpha}[\eta_{i,k}]^{\beta}}{\sum\limits_{s\in \mathrm{allowed}_K}[\tau_{i,s}]^{\alpha}[\eta_{i,s}]^{\beta}}, & \forall j \in \mathrm{allowed}_K \\ 0, & otherwise \end{cases} \tag{1}$$

Where the quantity $P_{i,j}^{k}$ is the migrate probability from node i to node j for AMA k. $\mathrm{allowed}_k$ indicates the next node which be allowed selecting by AMA k. $\tau_{i,j}$ is the value of pheromone. Pheromone values are stored in node's memory. Each node has information about amount of pheromone on the paths to their neighbor nodes. In ACO algorithm, a negative feedback, AMA cooperating together to migrate along the path related to the total cost of the routing. The trail updating is performed at the end of each trial. The trail updating follows the simple rules.

$$\tau_{i,j} = (1-\rho)\tau_{i,j} + \rho\Delta\tau_{i,j} \tag{2}$$

Where ρ is a control coefficient called the trail evaporation rate. $\rho \in (0,1)$ is used to measure how rapidly the trails evolve. We adapt Ant-Quantity model [14] that the general form of the incremental update $\Delta\tau_{i,j}$, is given by,

$$\Delta\tau_{i,j}^{k}(t) = \begin{cases} \dfrac{Q}{d_{i,j}} & \text{if the AMA K pass } path(i,j) \\ 0 & otherwise \end{cases} \tag{3}$$

Where Q represents the intensity of the pheromone. It affects the rate converging of this algorithm. In equation (1), the quantities α and β are two parameters that are maintained constant throughout the algorithm. $\eta_{i,k}$ is the value of heuristic.

For one thing, minimizing energy consumption is the major constraint to of the nodes involved in a routing task. For another thing, the ability of aggregation processing of nodes is not identical due to different memory. So we consider two facts constraints $\eta_{i,k}$.

Aggregation Gain G: The aggregation gain in WSNs is a measure of the benefits of applying aggregation to the system in terms of communication traffic reduction. The following expression define the aggregation gain:

$$G = \frac{p_{tr}}{p_{re}} \tag{4}$$

Where P_{re} is the number of transmissions that are necessary to perform a given application level task. P_{tr} is the number of transmissions to perform the same task when aggregation is applied. Then the heuristic value for the node k $\eta_{i,k}$ can be expressed as

$$\eta_{i,k} = \frac{e_k}{d_{i,k} G_k} = \frac{e_k p_{re}^{(k)}}{d_{i,k} p_{tr}^{(k)}} \qquad i \in \text{allowed}_K \tag{5}$$

Where e_k is the residual energy level of the node k. G_k is the data aggregation gain of the node k. When an AMA makes a decision, it is mean that the node has higher probability to be chosen if it has higher residual energy and lower gain of data aggregation. In above algorithm, pheromone values are stored in node's memory.

Given our choice of naming scheme, a data aggregation of WSNs is formation in the following manner during a single Ant-like agent life cycle:

Data Request Stage: The base station periodically broadcasts an interest message to the data sense region using flooding method [6]. This initial interest contains the specified recto and duration attributes. After receiving the interest, the node update neighbor tableΨ. In a task cycle, the sense nodes that sampled data match to the interest will be active from idle state and become source nodes.

AAMA Forward Stage: Every source node generates an AAMA. The total number of AAMA is equal to the number of source nodes. Each AAMA builds a path form its locality to the base station. Each such agent maintains a route table of nodes already visited. The route table is initialized to the source sensor where the AAMA is located. Each node updates its data aggregation gain using equation (4). While an AAMA builds a path, it gets quantitative information about the path cost from attribute vector λ of nodes and computes the heuristic value η using equation (5). Then The AAMA migrates to next node, which is chosen based on probability $P_{i,j}^{k}$ using equation (1). The new location node j is pushed into route table. The tag of node j is examined. If $tag_j = 0$, indicating that node j is previously unvisited then set $tag_j = 1$.

PAMA Feedback Stage: It is assumed that the base station at the beginning of the computation knows the total number of source nodes. PAMA are generated at the base station when all AAMA have arrived at the base station. Each PAMA carry a copy of the base station variable cost, which is used to update the pheromones. When PAMA moves from node j to node i according $tag = 1$, it updates the pheromones of node i using equation (2), (3).

Data Transmit Stage: As the migrate probability reaching a predefined threshold ,the data packets transmit according to the optimum probability of migration routing.

4 Simulation and Experiment Results

In this section, we evaluate the performance of data aggregation routing schema based on ACO. We use the network simulator for WSNs developed by NRL (Naval Research Laboratory) [14], which is an extended simulation environment based on NS2 [17, 18]. For simplicity, we assume the probability of signal collision and interference in the wireless channel is ignorable. And we adapt the same MAC protocols and energy mode in AMA as in DD.

The simulation area was 100×100m and, the sensor nodes were distributed randomly. The base station is within the sensor area (50, 50). During the simulation time, sensor nodes were responsible to report to the base station the phenomenon activities (simulated carbon monoxide). The parameters of simulations are listed in Table 1.

Table 1. Parameters of simulations

Parameters	Value
Area	100m×100m
Phenomenon (the creator of environment event)	2 (static)
Base station	1 (static)
Sensor node	50,100,150,200,250,300
MAC	802.11
Queue type	DropTail/PriQueue
Antenna type	OmniAntenna
Packet size	64bytes
Initial energy	0.2J
Transmitting/ p Receiving power	660mw/395mw
Simulation time	50 seconds

In order to get sufficiently good results from the limited runs, all the parameters of the algorithm were empirically fine-tuned. The exponents α and β in equation (1) are set to 1 and 5, respectively. The trail evaporation rate, ρ , is set to a value of 0.005[19]. The coefficients for the local and global trail updating were fixed at the following values, Q = 0.1. The updating of the trail was always done in a local as well in a global sense. In our simulation experiments, the threshold of migrate probability could have the values 0.7 by default as experimentally found to be good in the literature [20].

Unless otherwise specified, every simulation result shown below is the average of 20 independent experiments where each experiment uses a different randomly generated uniform topology of sensor nodes.

The consumption of energy is the biggest limitation in this type of networks. First we compared the impact on energy efficiency in different algorithm. The Fig.1 (a) illustrates the average dissipated energy under different network size. By the comparison, we can say, that the ACAR algorithm has better efficiency in energy than DD in larger network size. The Fig.1 (b) illustrates the average residual energy of nodes. As simulation time varies from 1 sec to 7 sec, the energy that consume in ACAR is decrease faster than DD. That means the AMA spends more cost in computation for making decision and migration in initial stages. So the energy consumption is more than DD in post stages.

The Fig.2 illustrates the percentage of total remaining energy in different simulation time. By the comparison, we can say, that the ACAR algorithm has better efficiency in energy than LEACH. But as simulation time varies from 10 sec to 20 sec, for ACAR algorithm the energy consumes faster than PEGASIS. That means the AMA spends more cost in computation for making decision and migration in initial stages. With the pheromone increasing, the probability of selecting routing goes to stabilization. So the energy consumption is less than PEGASIS in post stages.

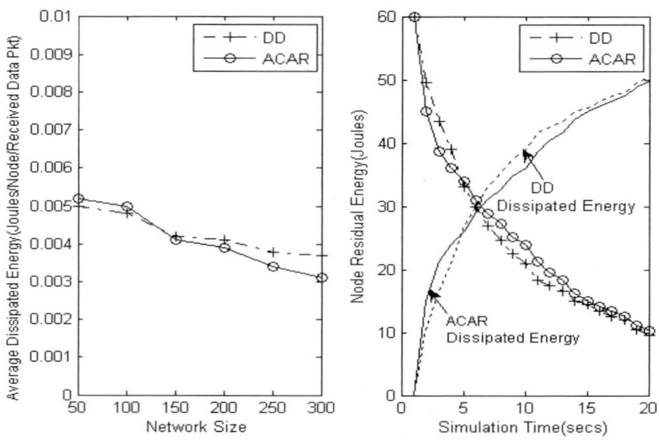

Fig. 1. Percentage of Residual Energy vs. Simulation Time

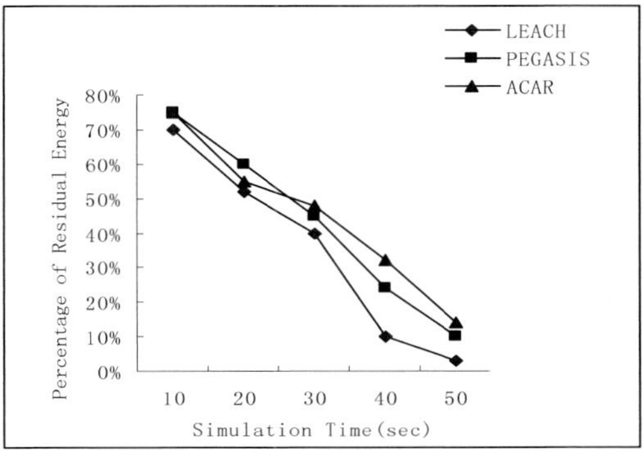

Fig. 2. Percentage of Residual Energy vs. Simulation Time

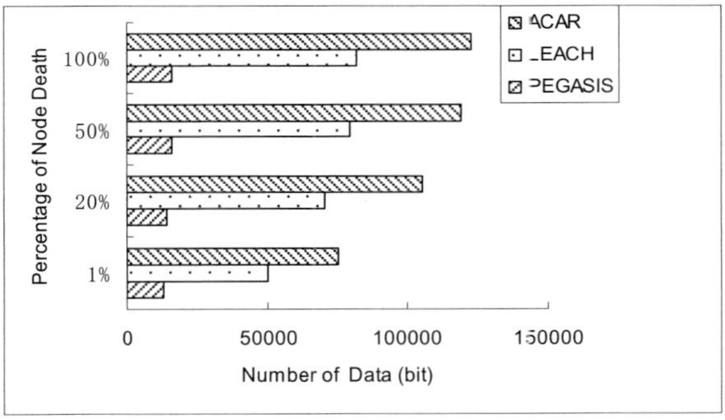

Fig. 3. Data amount when 1%, 20%, 50%, 80% and 100% nodes die

Due to the main objective of energy-efficiency is to send biggest amount of data with lowest energy cost, we showed the effect of data amount that related to the different dead node percentage. The percentage of dead node varies from 1% to 100% in experiment. In Fig.3, we observed that ACAR performs about more 1.5 times better than PEGASIS, and more that 7 times better than LEACH in terms of energy efficiency.

5 Conclusions and Further Work

In this paper, we explored the application of mobile agent based on ant colony algorithms to the data aggregation routing in WSNs. This problem involves establishing paths from multiple data sources in a sensor network to a base station. Data is aggregated at intermediate stages in the paths for optimal dissemination. We presented an evaluation ACO algorithm with data aggregation gain. Furthermore, we evaluated the impact of energy efficiency compared with DD, PEGASIS and LEACH. Our simulation results show that ACAR performs well than above protocols.

We will extend our research in the impact of large scale and multi mobile base station for WSNs. We expect that ACAR will outperform the other protocols in terms of system lifetime and the quality of the network.

Acknowledgments. Supported by the National Natural Science Foundation (60573141), Science Foundation of Nanjing College for Population Programme Management(2006C14, 2006B03)

References

1. Sankarasubramaniam, Y., Akyildiz, I.F., Su, W., Cayirci, E.: Wireless Sensor Networks: A Survey. Computer Networks, 393–422 (2002)
2. Krishnamachari, B., et al.: The Impact of Data Aggregation In Wireless Sensor Networks. In: The 22nd International Conference on Distributed Computing Systems Workshops (ICDCSW'02), Los Alamito, pp. 1–11 (2002)
3. Hill, J., Szewczyk, R., Woo, A., et al.: System Architecture Directions For Networked Sensors. In: 9th International Conference on Architectural Support for Programming Languages and Operating Systems (ASPLOS-IX), New York, NY, USA, pp. 93–104 (2000)
4. Heinzelman, W.R., Chandrakasan, A., Balakrishnan, H.: Energy-Efficient Communication Protocol for Wireless Microsensor Networks. In: Proceedings of IEEE HICSS(2000), pp. 3005–3015. IEEE Computer Society Press, Los Alamitos (2000)
5. Manjeshwar, A., Agrawal, D.P.: TEEN: A Routing Protocol For Enhanced Efficiency in Wireless Sensor Networks. In: Proc. 15th Int'l Parallel and Distributed Processing Symp. (IPDPS'01), SanFrancisco, CA, pp. 2009–2015 (2001)
6. Intanagonwiwat, C., Govindan, R., Estrin, D.: Directed Diffusion: A Scalable and Robust Communication Paradigm For Sensor Networks. In: ACM / IEEE International Conference on Mobile Computing and Net2 works (MobiCom2000), Boston, Massachusetts, pp. 56–67 (2000)
7. Handy, M.J., Haase, M., Timmermann, D.: Low Energy Adaptive Clustering Hierarchy with Deterministic Cluster-Head Selection. In: Proc. of the 4th IEEE Conf. on Mobile and Wireless Communications Networks, pp. 368–372. IEEE Communications Society, Stockholm (2002)
8. Lindsey, S., Raghavendra, C.S.: Pegasis: Power-Efficient Gathering in Sensor Information Systems. In: Proc. IEEE Aerospace Conference, pp. 1125–1130. IEEE Computer Society Press, Los Alamitos (2002)
9. Qi, H., Iyengar, S.S., Chakrabarty, K.: Multi-Resolution Data Integration Using Mobile Agents in Distributed Sensor Networks. IEEE Trans. Systems, Man, and Cybernetics Part C: Applications and Rev., 383–391 (2001)
10. Lange, D.B., Oshima, M.: Seven Good Reasons for Mobile Agents. Communications of the ACM, 88-89 (1999)
11. Colorni, A., Dorigo, M., Maniezzo, V., et al.: Distributed Optimization By Ant Colonies. In: Proceeding of the 1st European Conference on Artificial Life, pp. 134–142 (1991)
12. Dorigo, M.: Optimization, Learning and Natural Algorithm. Ph.D. Thesis, Department of Electronics, Politecnico diMilano, Italy (1992)
13. Dorigo, M., Maniezzo, V., Colorni, A.: The Ant System: An Auto catalytic Optimizing Process. Technical Report No. 91-016 Revised, Politecnico di Milano, Italy (1991)

14. Stützle, T., Dorigo, M.: ACO Algorithms for the Traveling Salesman Problem. In: Miettinen, K., Makela, M., Neittaanmaki, P., Periaux, J. (eds.) Evolutionary Algorithms in Engineering and Computer Science, pp. 163–183. Wiley, Chichester (1999)
15. Stützle, T., Grün, A., Linke, S., Rüttger, M.: A Comparison of Nature Inspired Heuristics on The Traveling Salesman Problem. In: Deb, K., Rudolph, G., Lutton, E., Merelo, J.J., Schoenauer, M., Schwefel, H.-P., Yao, X. (eds.) PPSN VI. LNCS, vol. 1917, pp. 661–670. Springer, Heidelberg (2000)
16. NRL's Sensor Network Extension to Ns-2: http://nrlsensorsim.pf.itd.nrl.navy.mil
17. The Network Simulator - ns-2: http://www.isi.edu/nsnam/ns/
18. Ant-like Mobile Agents NS2 Patch: http://www.item.ntnu.no/ wittner/ns/index.html
19. Ye, Z.W., Zheng, Z.B.: Research on The Configuration of Parameter α, β, ρ in Ant Algorithm Exemplified by TSP. In: Proceedings of the International Conference on Machine Learning and Cybernetics, pp. 2106–2111 (2003)
20. Gambardella, L.M., Dorigo, M.: Ant-Q: A Reinforcement Learning Approach To The Traveling Salesman Problem. In: Proceedings of the 12th International Conference on Machine Learning, pp. 252–260 (1995)

The Limited Mutation Particle Swarm Optimizer

Chunhe Song, Hai Zhao, Wei Cai, Haohua Zhang, and Ming Zhao

College of Information Science and Engineering, Northeastern University,
110006 Shenyang, China
{songchh,zhhai,caiv,zhhh,zhm}@neuera.com

Abstract. Similar with other swarm algorithms, the PSO algorithm also suffers from premature convergence. Mutation is a widely used strategy in the PSO algorithm to overcome the premature convergence. This paper discusses some induction patterns of mutation (IPM) and typical algorithms, and then presents a new PSO algorithm – the Limited Mutation PSO algorithm. Basing on a special PSO model depicted as "social-only", the LMPSO adopts a new mutation strategy – limited mutation. When the distance between one particle and the global best location is less than a threshold predefined, some dimensions of the particles will mutate under specific rules. The LMPSO is compared to other five different types of PSO with mutation strategy, and the experiment results show that the new algorithm performances better on a four-function test suite with different dimensions.

1 Introduction

Eberhart and Kennedy (1995) proposed the Particle Swarm Optimizer (PSO) algorithm [1], a simple but effective evolutionary algorithm, is motivated from the simulation of birds' social behavior. With many advantages of computing with real number, few parameters to be adjusted, the PSO algorithm is applied in many fields such as NN-training, Optimization, and Fussy Control etc.

One major problem with PSO algorithms in multi-model optimization is premature convergence, which results in great performance loss. Some optimization strategies such as sub swarm[14], global-local model [3,17], neighborhood[15,16] and some other strategies [4,5,6,7,12] were introduced to overcome it. In this paper, a certain type of optimization strategy in the PSO algorithm - mutation is discussed, and some typical PSO algorithms with different types of strategy are mentioned, then a new PSO algorithm - the limited mutation PSO is presented.

The remaining of the paper is organized as follows: in section 2 the Basic PSO is presented. In the next section, a discussion is given on different types of mutation and their typical algorithms. While in section 4 gives the new PSO algorithm -LMPSO. In Section 5 the experiment results will be presented. Finally, we give some concluding remarks and future research consideration in section 6.

2 Basic PSO Algorithm

PSO is an optimization algorithm generally employed to find a global best point. The basic PSO (BPSO) algorithm begins by scattering a number of "particles" in the

K. Li et al. (Eds.): LSMS 2007, LNCS 4688, pp. 258 – 266, 2007.

function domain space. Each particle is essentially a data structure that keeps track of its current position and its current velocity, and the 'good' or 'bad' of the position is judged by a problem-dependent function. Additionally, each particle remembers the "best" position it has obtained in the past, denoted by p_i, fitness of which is denoted by f_i. The best of these values among all particles is denoted by p_g, and the fitness is denoted by f_g .At each time step, a particle updates its position and velocity using the following equations:

$$v_{ij}(t+1) = w * v_{ij}(t) + c_1 r_{1j}(t)(p_{ij}(t) - x_{ij}(t)) + c_2 r_{2j}(t)(p_{gj}(t) - x_{ij}(t)) \tag{1}$$

$$x_{ij}(t+1) = x_{ij}(t) + v_{ij}(t+1) \tag{2}$$

where $j \in \{1,2,....Dn\}$; $i \in \{1,2,...,n\}$; n is the size of the population and Dn is the dimension of the space searched; w is the inertia weight; c_1 and c_2 are two positive constants; r_1 and r_2 are two random values into the range .

3 Different Induction Patterns of Mutation in PSO Algorithms

Similar with other evolutionary algorithms, the PSO algorithm also suffers from premature convergence. So far, many optimization strategies have been adopted to solve it, and the mutation is a widely used one. As the sake of convenience, the PSO algorithm with mutation strategy is described as PSOM in the rest of this paper.

Mutation process can be depicted as follows: when specific conditions are met, some particles' velocity and position will be updated according to particular equations (PE) rather than the normal ones. These conditions could be described as Induction Patterns of Mutation (IPM). In this paper some types of IPMs and their typical algorithms will be discussed. In the section 5, these algorithms will be compared to the new algorithm – LMPSO.

3.1 Random Number as the IPM

The Dissipative Particle Swarm Optimization (DPSO)[5] is a kind of PSOM which uses a random number as the IPM. In this algorithm, each particle's velocity and position will be re-initialized in a certain range according to a small probability in every evolution generation; the PE can be depicted as follows.

$$\begin{aligned}&\text{if } rand(0,1) < c_v \text{ then } v_{id} = rand(-1,1) * v_{\max,d} \\ &\text{if } rand(0,1) < c_l \text{ then } x_{id} = rand(l_d, u_d)\end{aligned} \tag{3}$$

where c_v and c_l are chaotic factors that in the range[0, 1] , $rand(l_d, u_d)$ is a random value between l_d and u_d .

3.2 Swarm's Fitness Variance as the IPM

The AMPSO[9] is the typical algorithm of this type. In this algorithm, when swarm's fitness variance is less than a threshold, the current global best particle will swing randomly according to a certain probability if it is not in the theoretical best position. The PE can be depicted as follows:

$$\begin{aligned}&\text{if } \sigma^2 < \sigma_d^2 \text{ and } f_g > f_g^{'} \text{ then } p_m = k; \text{else } p_m = 0\\&\text{if } rand(0,1) < p_m \text{ then } p_{gk} = p_{gk} * (1 + 0.5 * \eta)\end{aligned} \tag{4}$$

where η is subject to range[0, 1] in the Gaussian distribution and is the $f_g^{'}$ theoretical global best fitness.

3.3 Distances as the IPM

If the distance (Euclidean distance) between one particle and is smaller than a threshold, which means that this particle is searching the space some other particles has already searched, mutation need implement. In the GPSO [4], the speed inertia and the local best position are ignored, the PE is shown below:

$$\begin{aligned}&\text{if } \|xi(t-1) - g(t-1)\| \le \varepsilon \text{ then } \forall j\ v_{ij}(t) = rand(-v_{\max}, v_{\max})\\&\text{else } \forall j\ v_{ij}(t) = \gamma * rand(-1,1) * (x_{ij}(t-1) - g_j(t-1))\\&\text{if} (f(g(t)) < f(g(t-1))) \text{ then } \gamma = \max(\gamma - \delta, \gamma_{\min})\\&\text{else } \gamma = \min(\gamma + \delta, \gamma_{\max})\end{aligned} \tag{5}$$

4 The Limited Mutation Particle Swarm Optimizer (LMPSO)

Kennedy presented a simplified PSO version called 'social-only' model [1], in which the particles are only guided by the social knowledge. In the 'social-only' model, velocities of dimensions are updated as equ.6.

$$v_{ij}(t) = \beta * rand * (g_j(t-1) - x_{ij}(t-1)) \tag{6}$$

In the LMPSO, the 'social-only' model is also employed, and a new mutation strategy – limited mutation (LM) will be adopted. Especially each dimension of particles has a weight m , m_{ij} is the weight of the dimension of the i-th particle. The LM strategy means that when mutation happens, not all dimensions of the chosen particles join in this process; only some dimensions meeting special rules will be updated using the PE. In the LMPSO, three issues - the IPM, the electing principle of dimensions and the value of β are the factors to be considered.

4.1 The IPM of the Limited Mutation Particle Swarm Optimizer

In the LMPSO, particles move only towards the current global best position. Mutation happens when the distance between one particles and the global best position is less than a threshold α. The distance is defined as follows:

$$\text{distance}_i(t) = \sum_{j=1}^{n} m_{ij} * (g_j(t) - x_{ij}(t))^2 / Dn * \sum_{j=1}^{n} m_{ij} \tag{7}$$

When the distance between i-th particle and the current global best position is less than a threshold α, some dimensions of this particle will mutate.

4.2 Electing Principle of the Dimensions in the LMPSO

In order to ensure the diversity of mutation, the roulette algorithm is employed in the LMPSO. $flag_{ij}$ is employed to indicate one dimension is whether elected to mutate or not. The electing process can be depicted as follows: at the beginning, all $flag_{ij}$ is set at 0. Then if the j-th dimension' weight of the i-th particle meets the equ.12, this dimension is elected to mutate and $flag_{ij}$ is set at 1.

$$\sum_{p=1}^{j-1} m_{ip} / \sum_{p=1}^{n} m_{ip} \le rand(0,1) < \sum_{p=1}^{j} m_{ip} / \sum_{p=1}^{n} m_{ip},\ flag_{ij} = 1 \tag{8}$$

In fact more than one dimension of a particle need mutate, so the process above will be repeated several times. When one particle's mutation processing is over, the dimensions' weight will be updated as follows, and finally all dimensions' flag of this particle are set to zero.

$$\forall\ j, w_{ij} = \begin{cases} w_{ij} + 1, & flag_{ij} = 0; \\ w_{ij}, & flag_{ij} = 1; \end{cases} ; flag_{ij} = 0 \tag{9}$$

4.3 The Value of β

The factor β determines the step size of each particle in the direction of the global best position. Large values of will accelerate the search process but may make the current particle go past p_g, while small values will result in small step sizes which will lead to slow convergence to the p_g. In the LMPSO, bounded by the limits $[\beta_{\min}, \beta_{\max}]$, β is self-adjusted as follows:

$$f_{avg} = \frac{1}{n}\sum_{i=1}^{n} f(x_i),\ f'_{avg} = \frac{1}{m}\sum_{i=1}^{m} f(x_i),\ \forall f(x_i) < f_{avg}$$
$$\beta = \beta_{\max} - (\beta_{\max} - \beta_{\min}) * (f'_{avg} - f_g)/(f_{avg} - f_g) \quad (10)$$

4.4 The LMPSO Algorithm

The LMPSO algorithm is shown in the fig.1. MP is the proportion of the mutated dimensions. N is the size of the swarm. The integer part of the MP*N is denoted by [MP*N], which devotes the number of mutated dimensions.

initialize $x_{ij}, v_{ij}, f(x_i)$, p_i, p_g
$\forall i, j\ flag_{ij} = 0;$
for k=1 to maxStep
  for i=1 to n
    update distance$_i(t)$ using equ.7;
    update β using equ.10
    if distance$_i(t-1) \le \alpha$ then
      $k = 0;\ KT = [n * MP];$
      $while(k < KT)$
        elect one mutated dimension using equ.8
        suppose the $j-th$ dimension is elected
        if $flag_{ij} == 0$ then $v_{ij}(t) = rand(-V\max, V\max);\ flag_{ij} = 1; k = k + 1;$
      $\forall j,\ flag_{ij} = 0;$ update p_i
    else
      $for\ j = 1 : Dn$
        update v_{ij} using equ.6
        update x_{ij} using equ.2
      update p_i
  update p_g

Fig. 1. The LMPSO Algorithm

5 The Limited Mutation Particle Swarm Optimizer (LMPSO)

Two different sets of experiments were conducted in our experiments. In the first set of experiments, the performances of LMPSO with different MP on the benchmark functions have been measured. In the second experiment, the proposed algorithm has been compared with other five PSO algorithms- the BPSO, APSO, ARPSO, AMPSO and the GPSO.

5.1 Benchmark Functions

We have chosen four standard multi-model objective functions as the benchmark functions, which are presented in the table 1. In the benchmark functions, factor n represents the number of dimensions; f1, f2 and f4 have a global minimum at {0,0,…,0} , while f3 has a global minimum at {1,1,…,1}. All functions have a best fitness value of 0.

Table 1. Benchmark Functions

f1 Criewank	$f_1(x)=\frac{1}{4000}\sum_{i=1}^{n}x_i^2-\prod_{i=1}^{n}\cos(\frac{x_i}{\sqrt{i}})+1, \quad -600\le x_i\le 600$
f2 AckeyF1	$f_2(x)=e+20-20*\exp(-0.2*\sqrt{\frac{1}{n}\sum_{i=1}^{n}x_i^2})-\exp(\frac{1}{n}\sum_{i=1}^{n}\cos(2\pi*x_i)),$ $-30\le x_i\le 30$
f3 Rosenbrock	$f_3(x)=\sum_{i=1}^{n-1}100*(x_{i+1}-x_i^2)^2+(x_i-1)^2 \;;-100\le x_i\le 100$
f4 Rastrigin	$f_4(x)=\sum_{i=1}^{n}x_i^2+10-10*\cos(2\pi x_i) \;;-5.12\le x_i\le 5.12$

5.2 Setting of the Algorithms

In the first type of experiments, the maxStep parameter is set at 500 in the f1 and f3 while in the f1 and f3, the maxStep is set at 2000. In all the benchmark functions, Dn=20, $\beta_{max}=4$, $\beta_{min}=1.5$, $\alpha=1E-7$. The MPs are set at 0, 0.2, 0.4,0.6, 0.8 and 1. In the second type of experiments, the maxStep is set at 2000, Dn=50,100,200. In the LMPSO, $\beta_{max}=4$, $\beta_{min}=0.5$. And other five algorithms settings are the same as depicted in the [4,5,6,7,8].

5.3 Simulation Results and Discussion

Figures 2-5 show the performances of the LMPSO with different MPs. Figures 6-9 show the performances of the LMPSO and other five PSO algorithms on the four benchmark functions. Furthermore, the results from the second experiment are shown in the table 3. All results presented in this paper are the averages of fifty repeated tests.

From the fig.2~fig.5, we can see that the LMPSO performance worst when the MP is equal to 0, which means those particles will never mutate. And suitable values of MP can get better results. It is also clearly that in different benchmark functions the best MPs are different, which can be seen in the table 2.

On the Griewank benchmark function, although the LMPSO has the quickest convergence speed and could get a better result at the maximal evolution step, the premature convergence is still serious. While on the other three benchmark functions, the LMPSO can overcome premature convergence effectively.

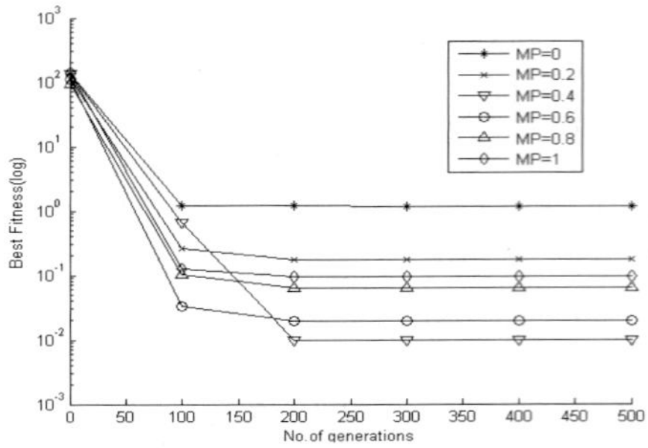

Fig. 2. LMPSO performance on Criewank

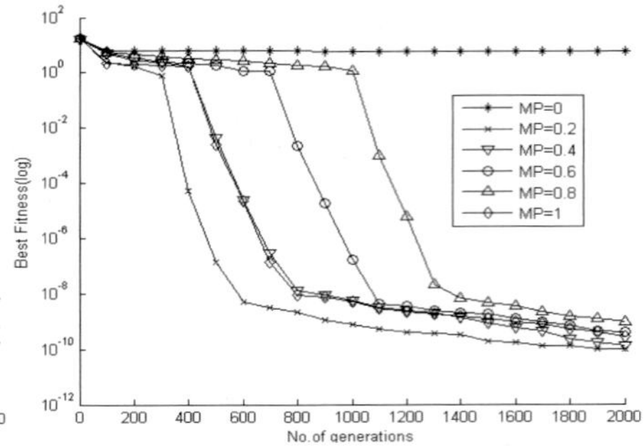

Fig. 3. LMPSO performance on AckeyF1

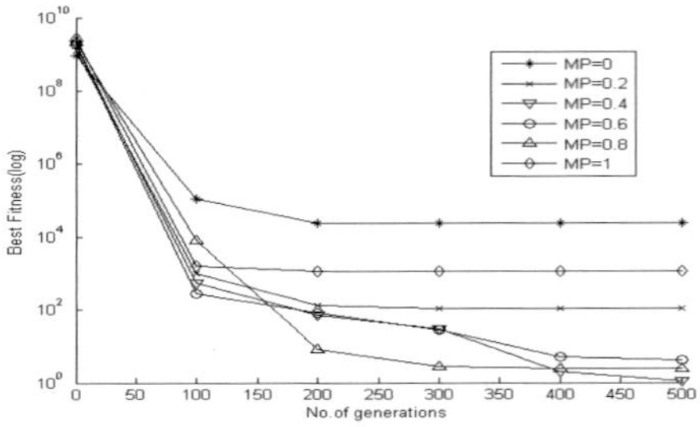

Fig. 4. LMPSO performance on Rosenbrock

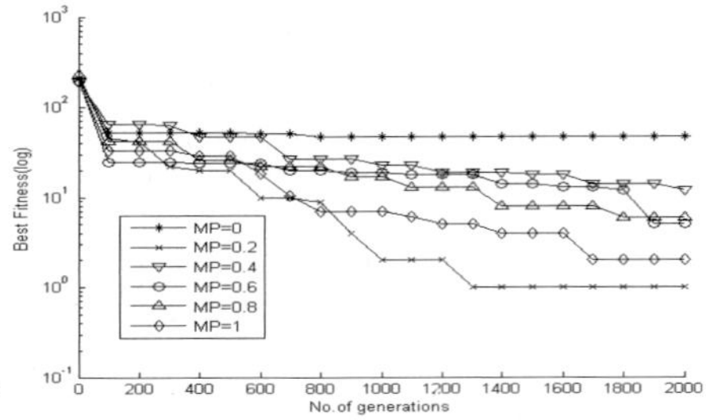

Fig. 5. LMPSO performance on Rastrigin

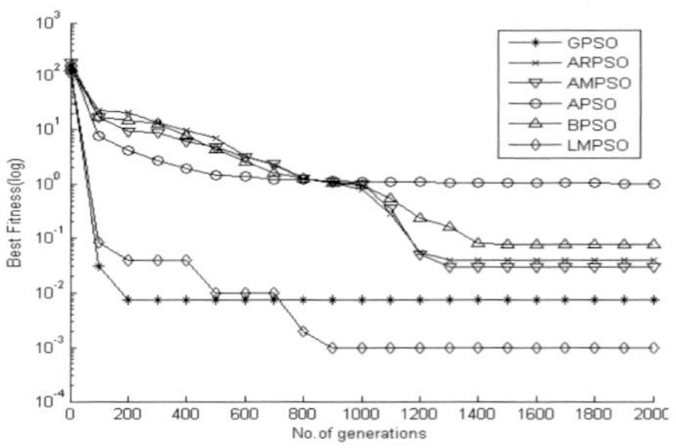

Fig. 6. Experimental results on Criewank

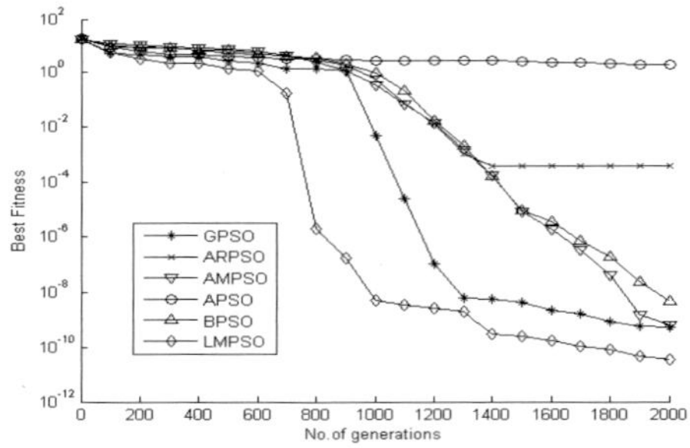

Fig. 7. Experimental results on AckeyF1

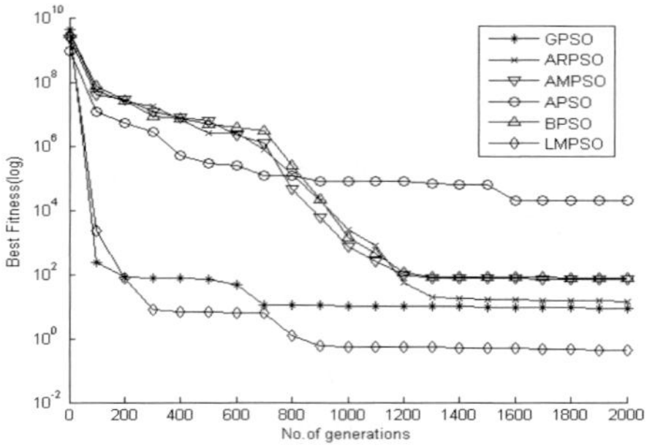

Fig. 8. Experimental results on Rosenbrock

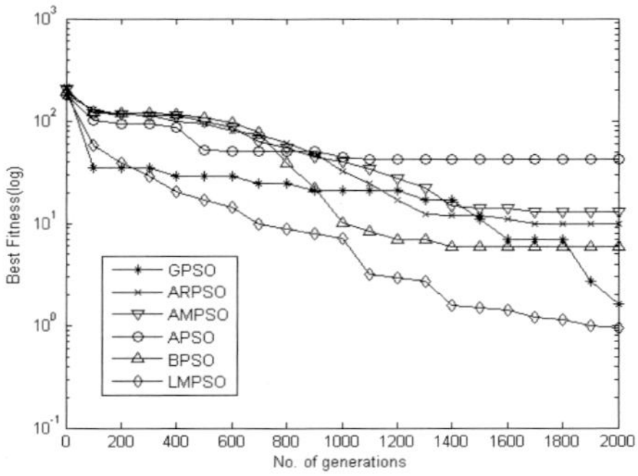

Fig. 9. Experimental results on Rastrigin

Table 2. Average best fitnesses on the benchmark Functions .The results in the rows LMPSO* are obtained by running the LMPSO until fitness stagnation in 1000000 evaluations.

Dim	50	100	200
Performance on Griewank			
GPSO	3.4290e-8	5.0426e-5	0.0099
ARPSO	2.0462e-6	3.4290e-4	0.5157
AMPSO	3.0148e-6	3.2575e-6	0.9607
APSO	1.1214e-4	1.3650e-2	17.6225
BPSO	1.0613e-7	8.5093e-4	0.0880
LMPSO	1.2587e-8	6.9775e-7	7.3217e-3
LMPSO*	0	4.2230e-10	6.0335e-6
Performance on AckeyF1			
GPSO	2.0952e-010	1.2432e-6	1.5842e-4
ARPSO	3.9790e-004	1.2035e-5	9.7665e-3
AMPSO	6.2172e-10	1.7926e-5	2.2163e-3
APSO	2.8911e-6	8.1833e-2	1.6631e-1
BPSO	2.7534e-10	1.1367e-4	1.5584e-2
LMPSO	7.1281e-11	5.5972e-6	1.5393e-3
LMPSO*	4.1868e-12	3.2348e-8	4.2238e-4
Performance on Rosenbrock			
GPSO	0.1637	11.4528	52.6854
ARPSO	1.0846	15.7789	109.4583
AMPSO	12.309	11.2445	189.5304
APSO	7.0683	191.45	604.2833
BPSO	1.07048	103.2393	165.5743
LMPSO	0.0528	1.4525	11.4892
LMPSO*	5.4782e-3	0.3367	3.2785
Performance on Rastrigin			
GPSO	0.9950	92.5310	315.4010
ARPSO	8.9546	102.5523	296.3705
AMPSO	6.9647	78.6028	416.6549
APSO	80.5668	276.8857	951.6185
BPSO	5.9698	102.7591	297.9143
LMPSO	0.1059	14.3914	23.4615
LMPSO*	0.08302	1.5981	5.3518

6 Conclusions

Basing on the 'social-only' model and the LM strategy, the LMPSO algorithm gets better performances in our tests. The inspiration of LM strategy comes from the experimental results that in a multi-model optimization process, when premature convergence happens, some dimensions' position of the particles have already been close enough to the theoretic global best position. If all dimensions of the mutated particles are re-initialized, this 'good' information would be lost, so choosing some special dimensions to mutate seems a better idea. Although the LMPSO is based on the 'social-only' model, we can see that the LM strategy does not conflict to other IPMs and PSO models. In this view point, the LM can be served as the accessorial strategy in all PSO with mutation strategy.

References

1. Kennedy, J., Eberhart, R.C.: Particle Swarm Optimization. In: Proc. IEEE Conf.on Neural Networks, vol. IV, pp. 1942–1948. IEEE Service Center, Piscataway, NJ (1995)
2. Shi, Y., Eberhart, R.C.: A Modified Particle Swarm Optimizer. In: Proceedings of the IEEE International Conference on Evolutionary Computation, pp. 69–73. IEEE Press, Piscataway, NJ (1998)
3. Kennedy, J.: The particle swarm: Social adaptation of knowledge. In: IEEE International Conference on Evolutionary Computation, pp. 303–308. IEEE Computer Society Press, Los Alamitos (1997)
4. Pasupuleti, S.: The Gregarious Particle Swarm Optimizer(G-PSO). In: GECCO'06, July 8-12, 2006, Seattle, Washington, USA (2006)
5. Xie, X.-F.: A Dissipative Particle Swarm Optimization. In: Congress on Evolutionary Computation (CEC), Hawaii, USA, pp. 1456–1461 (2002)
6. Riget, J.: A Diversity-Guided Particle Swarm Optimizer – the ARPSO
7. Ran, H.: An Improved Particle Swarm Optimization Based on Self-Adaptive Escape Velocity. Journal of Software (2005)
8. Jiao-Chao, Z.: A Guaranteed Global Convergence Particle Swarm Optimizer. Journal of Computer Research and Developemnt, 41 (2004)
9. Zhen-su, L.: Particle Swarm Optimization with Adaptive Mutation 32(3) (March 2004)
10. Jiang-hong, H.: Adaptive Particle Swarm Optimization Algorithm and Simulation. Journal of System Simulation 18(10) (October 2006)
11. Hao-yang, W., Chang-chun, Z.: Adaptive Genetic Algorithm to Improve Group Premature Convergence. Journal of Xi'an Jiaotong University 33 (1999)
12. Ratnaweera, A., Halgamuge, S., Watson, H.: Self-organizing hierarchical particle swarm optimizer with time-varying acceleration coefficients. IEEE Transactions on Evolutionary Computation 8, 240–255 (2004)
13. Angeline, P.J.: Using Selection to Improve Particle Swarm Optimization. In: IEEE International Conference on Evolutionary Computation, Anchorage, Alaska, USA, pp. 84–89. IEEE Computer Society Press, Los Alamitos (1998)
14. Lovbjerg, M., Rasmussen, T.K., Krink, T.: Hybrid Particle Swarm Optimizer with Breeding and Subpopulations. In: Proceeding of the third Genetic and Evolutionary Computation Conference (2001)
15. Suganthan, P.N.: Particle Swarm Optimizer with Neighborhood Operator. In: Proceedings of the 1999 Congress on Evolutionary Computation, pp. 1958–1962. IEEE Service Center, Piscataway, NJ (1999)
16. Kennedy, J.: Small worlds and Mega-minds: effects of neighborhood topology on particle swarm performance. In: Proc. Congress on Evolutionary Computation, 1931-1938, IEEE Service Center, Piscataway, NJ (1999)
17. Van den Bergh: A New Locally Convergent Particle Swarm Optimizer. In: 2002 IEEE International Conference on Systems, Man and Cybernetics, IEEE Computer Society Press, Los Alamitos (2002)

Research on Coaxiality Errors Evaluation Based on Ant Colony Optimization Algorithm

Ke Zhang

School of Mechanical and Automation Engineering
Shanghai Institute of Technology
200235 Shanghai, China
zkwy@hotmail.com

Abstract. Based on the analysis of existent evaluation methods for coaxiality errors, an intelligent evaluation method is provided in this paper. The evolutional optimum model and the calculation process are introduced in detail. According to characteristics of coaxiality error evaluation, ant colony optimization (ACO) algorithm is proposed to evaluate the minimum zone error. Compared with conventional optimum evaluation methods such as simplex search and Powell method, it can find the global optimal solution and the precision of calculating result is very good. Then, the objective function calculation approaches for using the ACO algorithm to evaluate minimum zone error are formulated. Finally, the control experiment results evaluated by different method such as the least square, simplex search, Powell optimum methods and GA, indicate that the proposed method does provide better accuracy on coaxiality error evaluation, and it has fast convergent speed as well as using computer expediently and popularizing application easily.

1 Introduction

Coaxiality error is the elementary and important one of position error evaluation. The evaluation methods of coaxiality error are mainly the least-squares and minimum zone method. Although the least-squares method, because of its simplicity in computation and uniqueness of the solution provided, is most widely used in industry for determining form and position error, it provides only an approximate solution that does not guarantee the minimum zone value [1], and does not satisfy requirement of modern exact measurement. The results of minimum zone method not only verge on ideal error value, but also accord with ISO standard. Therefore, much research has been devoted to finding the minimum zone solutions for flatness error and other form errors using a variety of methods. Some researchers applied the numerical methods of linear programming [2,3], such as the Monte Carlo method, the simplex search, spiral search, and the minimax approximation algorithm etc. Another approach has been to find the enclosing polygon for the minimum zone solution, such as the eigen-polyhedral method, the convex polygon method, and the convex hall theory etc. The methods mentioned above generally proceed initially with the random selection of

K. Li et al. (Eds.): LSMS 2007, LNCS 4688, pp. 267 – 276, 2007.

data points and then follow with an iterative data exchange procedure. Some of them are not prone to perform in a computer, and a longer computation time is naturally required by this kind of approach in order to reach the final minimum zone condition. Thus, it is necessary to study the more effective, faster and simpler the algorithm for coaxiality evaluation.

The optimization algorithms are commonly used to approach the minima of coaxiality error objective function through iteration when a microcomputer is applied to assess coaxiality errors by minimum zone method. The essential prerequisite for convergence of any optimization algorithm is that the objective function to be solved has only one minimum in its definition domain, that is, it is a single valley one. If an objective function has more local minima in its definition domain, it solution searched for by an optimization algorithms may not be its global minimum which is the wanted coaxiality error. Therefore, the mathematical models and algorithms for coaxiality error evaluation may be influenced in their solutions' reliability and practical value.

The traditional optimization methods are employed to refine the least-squares solution further. However, these traditional optimization methods have drawbacks in finding the global optimal solution, because it is so easy for these traditional methods to trap in local minimum points [4].

The ant colony optimization (ACO) algorithm is a novel simulated ecosystem evolutionary algorithm. It takes inspiration from the observations of ant colonies foraging behavior with which ants can find the shortest paths from food sources to their nest [5]. Preliminary study has shown that the ant colony algorithm is very robust and has great abilities in searching better solutions. ACO algorithm has been successfully used to solve many problems of practical significance including the Quadratic Assignment Problem [6], Traveling Salesman Problem [7], Single Machine Total Weighted Tardiness Problem [8], and many more [9-11].

In this paper, according to characteristics of coaxiality error evaluation, a novel algorithm ACO is presented to find the value of coaxiality error in terms of the minimum zone criterion. The rest of this paper is organized as follows: The coaxiality errors analysis will be given in Section 2. The ant colony optimization is described summarily in Section 3. In Section 4, based on ACO algorithm, coaxiality errors evaluation and an example to test its validity are presented therein. Finally, conclusions are provided in Section 5.

2 Coaxiality Errors Analysis

The minimum zone method is defined by using axes of ideal cylinder containing actual datum axes with minimum diameter as datum axes to evaluate coaxiality error. Here, a part in Cartesian coordinates is shown in Fig.1. Let intersection point between datum axes and XOY coordinate plane be $P_0(x_0, y_0, 0)$, and a set of direction number of datum axes be $(u, v, 1)$.

The distances R_j $(j = 1, 2, \cdots, m)$ between the centre O_j (a_j, b_j, z_j) of a least-squares circle of datum factor sampling section profile and datum axes L are as follows.

$$R_j = \frac{\left|(\vec{O}_j - \vec{P}_0) \times \vec{S}\right|}{\left|\vec{S}\right|} \quad (j = 1,2,\cdots,m) \tag{1}$$

where $\vec{O}_j = \{a_j, b_j, z_j\}$, $\vec{P}_0 = \{x_0, y_0, 0\}$, $\vec{S} = \{u, v, 1\}$, m is the number of datum sampling section.

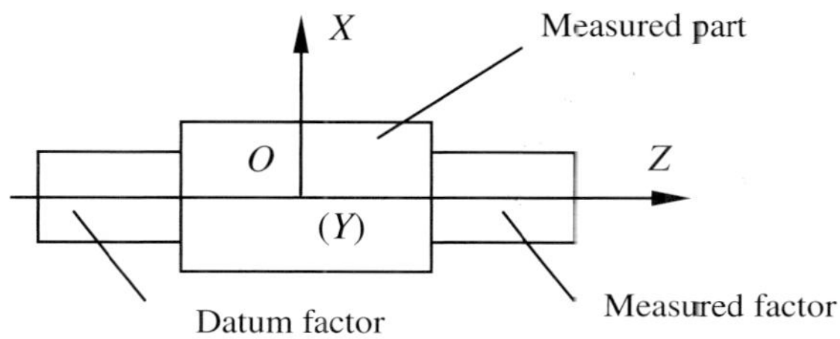

Fig. 1. Part in Cartesian Coordinates

Define $R_{jmax} = max\{R_j\}$, where $R_{j\max}$ is radius of ideal cylinder containing actual datum axes. We may regard R_j as the function of the four variables (x_0, y_0, u, v), minimum zone datum axes may be expressed as unconstraint optimization problem as follows:

$$\min f(x_0, y_0, u, v) \tag{2}$$

where objective function is $f(x_0, y_0, u, v) = R_{j\max}$. It is a function of variables (x_0, y_0, u, v). Consequently, evaluating the minimum zone coaxiality error is translated into searching the values of variables (x_0, y_0, u, v), so that the objective function f (x_0, y_0, u, v) is minimum.

We assume optimal solution $V = \{x_0^*, y_0^*, u^*, v^*\}$, the distances R_J^* $(J = 1,2,\cdots,M)$ between the centre O_J (a_J, b_J, z_J) of a least-squares circle of measured factor sampling section profile and datum axes are

$$R_J^* = \frac{\left|(\vec{O}_J - \vec{P}_0^*) \times \vec{S}^*\right|}{\left|\vec{S}^*\right|} \quad (J = 1,2,\cdots,M) \tag{3}$$

where $\vec{O}_J = \{a_J, b_J, z_J\}$, $\vec{P}_0^* = \{x_0^*, y_0^*, 0\}$, $\vec{S}^* = \{u^*, v^*, 1\}$, M is the number of measured sampling section. Thus, coaxiality error according with minimum zone is

$$f^* = 2\max\{R_J^*\} \quad (J = 1,2,\cdots,M) \tag{4}$$

3 Ant Colony Optimization

Although not guaranteed, global search methods have a much better chance of converging to the global minimum. Global search methods are structured to search the entire design space and find regions of low cost function values where global minimum is likely to exist. Ant colony optimization (ACO) algorithm, as a kind global search algorithm, has been successfully used to solve many problems of practical significance. According to characteristics of coaxiality error evaluation, ACO algorithm is proposed to evaluate the minimum zone coaxiality error in the paper.

3.1 Development Background

ACO algorithm, originally introduced by M. Dorige in the early 1990s, is a novel nature-inspired metaheuristic. ACO algorithm is a stochastic approximate method for tackling combinatorial optimization problems. The central component of an ACO algorithm is the pheromone model, which is used to probabilistically sample the search space.

Ants are social insects which live in highly organized swarms. The way ants organize their lives in colonies through sophisticated behavioral patterns which include ant foraging, proper division of labor, brood sorting, and cooperative transport have fascinated naturalists, as they try to discover and explain their organizational patterns and rules. Ethologists have discovered that one of the most critical skills ants possess, even the ones belonging to blind species, is their ability to find the shortest path between the food source and the nest.

Ants and other animal species, especially insects, secret a chemical substance called pheromone which alters the environment and provides a medium for what is known as stigmergic communication. The pheromone trail influences the behavior and development of others of the same species. The pheromone deposited during ant foraging along a trail helps ants find the shortest path between their nest and food source by the enactment of a natural autocatalytic or positive feedback process. Higher amounts of pheromone are deposited over short paths because they require less time to cross. Thus, over time, more pheromone is deposited over a short path than the longer ones, eventually attracting other ants to follow the same path. Pheromones concentration guides an ant to continuously correct its path over time and eventually finds the most efficient one. Inspired by this type of behavior, engineers have developed a class of algorithms based on foraging of artificial ants collectively known as Ant Colony Optimization in which the management of pheromone deposition and evaporation during a search play a central role.

3.2 Ant Colony Optimization

Starting with a random initial solution, artificial ants involved in a search start exploring promising paths. During the search, the higher pheromones concentration on the most frequently visited paths will have a higher probability of being visited in the next iteration. This probability is represented mathematically by the random proportional transition rule as follows [9]:

$$p_{ij}^{k}(t)=\frac{\left[\tau_{ij}(t)\right]^{\alpha}\left[\eta_{ij}\right]^{\beta}}{\sum_{N_i^k}\left[\tau_{ij}(t)\right]^{\alpha}\left[\eta_{ij}\right]^{\beta}} \tag{5}$$

where i is a decision point (node) for ant k to decide which of the ij possible branches or edges to pursue and t represents time. N_i^k is the neighborhood of all possible branches ant k can pursue while at decision node i. τ_{ij} is the trail intensity, which is the amount of pheromone deposited on each ij edge. η_{ij} is the visibility factor which is problem-specific reflecting a certain description of the specific problem being optimized. η_{ij} is constant for each edge and does not change during iterations. α and β in equation (5) are two adjustable parameters that control the relative weight of trail intensity and visibility. The trail intensity τ_{ij} is updated after each iteration step according to the rule [9]

$$\tau_{ij}(t+1)=(1-\rho)\tau_{ij}(t)+\Delta\tau_{ij}(t) \tag{6}$$

$\Delta\tau_{ij}(t)$ represents the total amount of updated pheromones by all involved ants and is given by

$$\Delta\tau_{ij}(t)=\sum_{k=1}^{m}\Delta\tau_{ij}^{k}(t) \tag{7}$$

where m is the total number of ants , and $\Delta\tau_{ij}^{k}(t)$ is the pheromone of ij edge left by ant k in this iteration.

The evaporation rate ρ $(1<\rho<1)$ in equation (6) is an important parameter which is used to ensure that pheromones on unvisited paths evaporate over time and that those paths are affixed low visiting probabilities in the future. A judicious compromise between pheromone evaporation and deposition is necessary.

4 Evaluation and Example

4.1 Optimization Model

Ant colony optimization is a kind of discrete algorithm. There for the optimization problem must be converted in to a discrete one before can be solved by an ant colony algorithm. According to parameter values of datum axes calculated by the least-square method, we can get the range of the variable x_i .

(1) Discretization of Variables

First discretize each variable in its feasible region, i. e., divide the variables to be optimized in its value range. The number of the division relates with the range of the variable and the complexity of the problem, etc.

Suppose the variable to be optimized is divided into q_i nodes, denote the node value as x_{ij} where i is variable number ($i=1,2,\dots n$) and j is node number of x_i ($j=1,2,\dots q_{ij}$).

After the division, a network which composed of variables and their division is constructed as depicted in figure 1. where "•"is the node of each variable division. The number of the division of each variable can be unequal and the division can be linear or nonlinear.

(2) Optimization with Ant Colony Algorithm

Suppose using m ants to optimize n variables, for ant k at time $t=0$, set the ant at origin O. Then the ant will start searching according to the maximum transition probability. The searching starts from x_1 and goes on in sequence. Each ant can only select one path in its division to do transition. The ant k transit from x_i to x_{i+1} is called a step and is denoted as t. The ant k starts searching from the first variable x_1 to the last variable x_n is called an iteration and is denoted as NC. An iteration includes n steps. Figure 2 shows an iteration path of an ant.

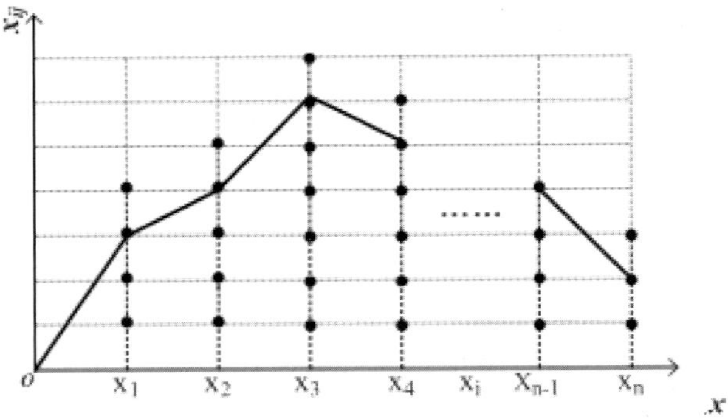

Fig. 2. Variable division and ant search path

4.2 Transition Probability

At time t, the transition probability of ant k transit from node h of x_i to node l of x_{i+1} is defined as equation (8)

$$p_{hl}^k(t)=\frac{[\tau_{hl}(t)]^\alpha[\eta_{hl}]^\beta}{\sum_{j=1}^{q_{i+1}}[\tau_{hj}(t)]^\alpha[\eta_{hj}]^\beta} \tag{8}$$

where $\tau_{hl}(t)$ is the pheromone level of node l of x_{i+1} at time t. $\eta_{hl}(NC)$ is the visibility of node l of x_{i+1} in NC iteration. The denominator represents the sum of pheromone level multiple visibility of all node of x_{i+1} at time t. Given $\alpha, \beta > 0$ are heuristic factors, they will control the relative importance of pheromone level $\tau_{hl}(t)$ and visibility $\eta_{hl}(NC)$ respectively. While searching, the ant will select the node which has the maximum transition probability node to transit.

4.3 The Definition of Visibility

The visibility of node j of variable x_i in iteration *NC* is defined as equation (9)

$$\eta_{ij}(NC) = \frac{(r_{i_{\text{upper}}} - r_{i_{\text{lower}}}) - \left|x_{ij} - x^*_{ij}(NC-1)\right|}{r_{i_{\text{upper}}} - r_{i_{\text{lower}}}} \tag{9}$$

where i is variable number ($i = 1,2,\dots n$) and j is node number of x_i ($j = 1,2,\dots q_{ij}$). $r_{i_{\text{upper}}}, r_{i_{\text{lower}}}$ are the upper and lower limit value of range of variable x_i respectively. $x^*_{ij}(NC-1)$ is the optimal value of the variables (parameter values of datum axes) in iteration $(NC-1)$. In the first iteration, $x^*_{ij}(NC-1)$ can be obtained by using the least-square method. After that ($NC > 1$), $x^*_{ij}(NC-1)$ will be the node value of the variable relevant the optimal path produced in last iteration.

4.4 The Update of Pheromone Level

At the initial time $t = 0$, the pheromone levels of each node are equal, i. e., $\tau_{ij}(0) = c$ and all of the ants are at origin *O*. After n time units, all of the ants will crawl from start node to terminal node and the pheromone levels of each node can be adjusted according following formation.

$$\tau_{ij}(t+n) = \rho\tau_{ij}(t) + \Delta\tau_{ij} \tag{10}$$

where i is variable number ($i = 1,2,\dots n$) and j is node number of variable x_i ($j = 1,2,\dots q_{ij}$). $(1-\rho)$ represents the decay of pheromone level of node x_{ij} from t to time $t+n$.

$$\Delta\tau_{ij} = \sum_{k=1}^{m} \Delta\tau^k_{ij} \tag{11}$$

where $\Delta\tau_{ij}^k(t)$ is the pheromone of node x_{ij} left by ant k in this iteration. It can be calculated according to equation (12)

$$\Delta\tau_{ij}^k = \begin{cases} Q/f_k, & \text{if kth ant pass through this node in this iteration} \\ 0, & \text{else} \end{cases} \tag{12}$$

where f_k is the value of objective function of the ant k in this iteration. It is calculated with equation (2). Q is a positive constant.

4.5 Evaluation Procedure of the Ant Colony Optimization Algorithm

The procedure of the ant colony optimization is divided into ten steps as follows.

Step1: Create the objective function of the optimization problem. Decide the number of variable and discrete the variables by dividing them. Each variable is divided into q_i nodes which constitute the number vector $\mathbf{q}$. Node value is x_{ij}.

Step2: Given the ant number m and define a one dimensional vector $\mathbf{path}_k$ for ant k which has n elements. The one dimensional vector $\mathbf{path}_k$ is called path table and used to save the path that the ant k passed through.

Step3: Initialization. Let time counter $t=0$ and the maximum iteration number is NC_{max}. Let the pheromone level of each node $\tau_{ij}(0)=c$, $\Delta\tau_{ij}=0$. Set the m ants at origin O. According equation (9), calculate the visibility $\eta_{ij}(1)$ used in the first iteration.

Step4: Let variable counter $i=1$, variable division number counter $q=1$.

Step5: Use equation (8) to calculate the transition probability of ant k transit from node h of variable x_i to node l of variable x_{i+1} and select the maximum probability to do transition. Then save node l of x_{i+1} to the ith element of $\mathbf{path}_k$. Transit all ants from node of variable x_i to node of variable x_{i+1}.

Step6: Let $i=i+1$, if $i \le n$, then go to Step 5, else, go to Step 7.

Step7: Based on the path that ant k walked, i. e., $\mathbf{path}_k$, calculate the objective function f_k of ant k according to equation (2). Save the node values relevant optimal path in this iteration.

Step8: Let $t \leftarrow t+1$, $NC \leftarrow NC+1$. Update the pheromone level of each node according to equations (10), (11), (12) and clear all elements in $\mathbf{path}_k$ ($k=1 \sim m$). Calculate visibility of each node according to equation (9).

Step9: If $NC < NC_{max}$ and the whole ant colony have not converged to the same path yet, then set all of the ants at the origin O again and go to Step 4; If $NC < NC_{max}$ and the whole ant colony have converged to the same path then the iteration complete and the optimal path is the optimal solution.

Step10: By using the above optimal solution $V = \{x_0^*, y_0^*, u^*, v^*\}$, calculate coaxiality error according to equations (3) and (4).

4.6 Examples

To test the validity of the proposed procedure and its ability to provide exact values of coaxiality error, an example problem was solved. A set of data points measured is shown in Table 1 and 2. The procedures were programmed in the Matlab programming language. The result of calculation using the ant colony optimization algorithm is $\phi f^* = 13.4732$ μm . The results of coaxiality error from different methods are provided in Table 3. The comparison shows that the global optimum solution of coaxiality evaluation problem using the proposed procedure can be obtained and accord with the fact of measured profile.

Table 1. Data Measured of Datum Profile

x (mm)	y (mm)	z (mm)	x (mm)	y (mm)	z (mm)
32.221	0.017	-91.226	32.183	-0.023	-62.327
-32.175	-0.023	-91.223	-32.143	0.027	-62.326
-0.024	32.186	-91.225	-0.027	32.178	-62.329
0.016	-32.188	-91.225	0.015	-32.173	-62.328
32.167	0.013	-79.973	32.194	0.014	-50.105
-32.161	0.016	-79.975	-32.201	-0.020	-50.101
0.029	32.227	-79.973	0.013	32.169	-50.104
-0.023	-32.223	-79.972	0.027	-32.167	-50.104

Table 2. Data Measured of Measured Profile

x (mm)	y (mm)	z (mm)	x (mm)	y (mm)	z (mm)
32.183	0.017	90.254	32.164	0.012	60.097
-32.201	-0.023	90.254	-32.156	-0.021	60.099
0.016	32.154	90.252	-0.027	32.206	60.094
0.025	-32.161	90.255	0.021	-32.221	60.096
32.153	-0.019	79.232	32.179	0.015	49.314
-32.143	0.026	79.233	-32.163	-0.021	49.313
-0.025	32.161	79.230	-0.023	32.175	49.318
0.018	-32.169	79.233	0.011	-32.187	49.315

Table 3. Coaxiality Results

Calculation method	Coaxiality error (μm)
The least-square	14.8930
Simplex search	14.4515
Powell	14.6644
Standard GA	13.7929
Ant Colony Optimization	13.4732

5 Conclusions

In this paper, an intelligent optimization approach to evaluate coaxiality error was presented. The coaxiality evaluation problem was formulated as unconstraint optimization problems. An ant colony solution procedure was developed to solve the optimization problem. The techniques were compared to some existing techniques. It is shown through example that the procedure of this paper provides exact values of coaxiality error. The result also shows that the proposed procedure converges to the global optimum more rapidly than conventional methods. The evaluation method is the same with the others form and position errors evaluation.

References

1. Kanad, T., Suzuki, S.: Evaluation of Minimum Zone Flatness by Means of Nonlinear Optimization Techniques and Its Verification. Precision Engineering 15, 93–99 (1993)
2. Huang, S.T., Fan, K.C., Wu, J.H.: A New Minimum Zone Method for Evaluating Flatness Error. Precision Engineering 15, 25–32 (1993)
3. Cheraghi, S.H., Lim, H.S., Motavalli, S.: Straightness and Flatness Tolerance Evaluation: an Optimization Approach. Precision Engineering 18, 30–37 (1996)
4. Singiresu, S.R.: Engineering Optimization. John Wiley & Sons, New York (1996)
5. Colorni, A., Dorigo, M., Maniezzo, V.: Distributed Optimization by Ant Colonies. In: Proc. First European Conf. Artificial Life, pp. 134–142 (1991)
6. Gambardella, L.M., Taillard, E.D., Dorigo, M.: Ant Colonies for the Quadratic Assignment Problem. J. Oper. Res. Soc. 50, 167–176 (1999)
7. Gambardella, L.M., Bianchi, L., Dorigo, M.: An Ant Colony Optimization Approach to the Probabilistic Traveling Salesman Problem. In: Guervós, J.J.M., Adamidis, P.A., Beyer, H.-G., Fernández-Villacañas, J.-L., Schwefel, H.-P. (eds.) PPSN VII. LNCS, vol. 2439, Springer, Heidelberg (2002)
8. Besten, M., Stutzle, T., Dorigo, M.: An Ant Colony Optimization Application to the Single Machine Total Weighted Tardiness Problem. In: Deb, K., Rudolph, G., Lutton, E., Merelo, J.J., Schoenauer, M., Schwefel, H.-P., Yao, X. (eds.) PPSN VI. LNCS, vol. 1917, Springer, Heidelberg (2000)
9. Bonabeau, E., Dorigo, M., Theraulaz, G.: Swarm Intelligence: From Natural to Artificial Systems. Oxford University Press, New York (1999)
10. Botee, H.M., Bonabeau, E.: Evolving Ant Colony Optimization. Adv. Complex Syst. 1, 149–159 (1998)
11. Dorigo, M., Stuzle, T.: Ant Colony Optimization. MIT Press, Cambridge (2004)

A Novel Quantum Ant Colony Optimization Algorithm

Ling Wang, Qun Niu, and Minrui Fei

Shanghai Key Laboratory of Power Station Automation Technology, School of Mechatronics and Automation, Shanghai University, Shanghai 200072, China
wangling@shu.edu.cn

Abstract. Ant colony optimization (ACO) is a techniqule for mainly optimizing the discrete optimization problem. Based on transforming the discrete binary optimization problem as a "best path" problem solved using the ant colony metaphor, a novel quantum ant colony optimization (QACO) algorithm is proposed to tackle it. Different from other ACO algorithms, Q-bit and quantum rotation gate adopted in quantum-inspired evolutionary algorithm (QEA) are introduced into QACO to represent and update the pheromone respectively. Considering the traditional rotation angle updating strategy used in QEA is improper for QACO as their updating mechanisms are different, we propose a new strategy to determine the rotation angle of QACO. The experimental results demonstrate that the proposed QACO is valid and outperforms the discrete binary particle swarm optimization algorithm and QEA in terms of the optimization ability.

1 Introduction

Ant colony optimization (ACO) was firstly developed by M. Dorigo as a novel nature-inspired metaheuristic for solving the discrete combinatorial optimization problems in the early 1990s [1], [2], [3]. It is biologically inspired from the foraging behavior of real ants. When searching food, as soon as an ant finds the food, it evaluates quantity and quality of the food and deposits a chemical pheromone trail on the ground which will guide other ants to the food source. At last, the indirect communication between the ants via the pheromone trails allows them to find the shortest paths to the food source by swarm intelligence. This characteristic of real ant colonies is exploited in artificial ant colonies to tackle optimization problems. Due to the excellent optimization performance of ACO, it has drawn more and more attention of researchers, and has been widely used in varieties applications, such as traveling salesman problem [4], vehicle routing problem [5], scheduling problem [6], allocation problem [7] and assignment problem [8]. Now ACO algorithm has been a hotspot in the research of intelligent optimization algorithms, and more details about ACO can be found in [9], [10].

Quantum computing was proposed by Benioff and Feynman in the early 1980s. As quantum computing can solve some specialized problem more effectively than the classical computing, there has been a great interest in the application of the quantum computing. In the late 1990s, researches have been conducted on merging evolutionary

K. Li et al. (Eds.): LSMS 2007, LNCS 4688, pp. 277 – 286, 2007.

computing and quantum computing. These works can be classified into two fields. One concentrates on generating new quantum algorithm using evolutionary computing technique; the other concentrates on quantum-inspired evolutionary computing for the classical optimization algorithm [11-13], that is, the evolutionary algorithm characterized by certain principles of quantum mechanics. One of the most important works is the quantum-inspired evolutionary algorithm proposed by Han [14-15], which is based on the Q-bit representation and quantum gate. And the previous researches have reported that QEA outperform the classical genetic algorithm in some applications.

General speaking, binary representation is the most efficient and economical data representation, and two-value logic is the easiest way to simulate and implement in nature. And it is seen that an optimizer which operates on two-valued functions might be advantageous as any problem, discrete or continuous, can be expressed in a binary notation [16]. So our interest in this paper is to research and propose a new quantum ant colony optimization algorithm for solving discrete binary optimization problem, which is based on concepts and principles of ACO and QEA.

The reminder of the paper is organized as follows. Section 2 introduces the original ACO and QEA. Section 3 presents the proposed QACO algorithm. The experimental results of the proposed QACO are offered in Section 4. The comparisons with the classical QEA and discrete binary particle swarm optimization, known as another famous swarm intelligence optimization algorithm, are also given in this section. Finally, concluding remarks follow in Section 5.

2 Original ACO and QEA

2.1 Ant Colony Optimization Algorithm

ACO algorithm inspired by colonies of real ants has been successful employed to solve various optimization problems. It is an evolutionary approach where several generations of artificial agents in a cooperative way search for good solutions. Agents are initially randomly generated on nodes, and then stochastically move from a start node to feasible neighbor nodes. Agents collect and store information in pheromone trails during the process of finding feasible solutions. Agents can online release pheromones while building solutions. In addition, the pheromones will be evaporated in the search process to avoid local convergence and to explore more search spaces. Thereafter, additional pheromone is deposited to update pheromone trail offline so as to bias the search process in favor of the currently path. The pseudo code of classical ACO algorithm can be described as:

```
Procedure: Ant colony optimization
Begin
   While (ACO has not been stopped) do
      Agents_generation_and_activity();
      Pheromone_evaporation();
      Daemon_actions();
    End;
End;
```

In ACO [7], agents find solutions starting from an initial node and moving to feasible neighbor nodes in the process of Agents_generation_and_activity. During the process, information collected by agents is stored in the so-call pheromone trails. In this process, agents can release pheromone while building the solution (online step-by-step) or while solution is built (online delayed). An agent-decision rule, made up of the pheromone and heuristic information, guides agents' search toward neighbor nodes stochastically. The *k-th* ant at time t positioned on node r move to the next node s with the rule governed by

$$s = \begin{cases} \arg\left\{ \max\limits_{u=allowed_k(t)} [\tau_{ru}(t)^{\alpha} \eta_{ru}^{\beta}] \right\} & when\ q \le q_0 \\ S & \text{otherwise} \end{cases} . \tag{1}$$

where τ_{ru} *(t)* is the pheromone trail at time t, η_{ru} is the problem-specific heuristic information, α is a parameter representing the importance of heuristic information, q is a random number uniformly distributed in [0,1], q_0 is a pre-specified parameter ($0 \le q_0 \le 1$), $allowed_k(t)$ is the set of feasible nodes currently not assigned by ant k at time t, and S is an index of node selected from $allowed_k(t)$ according to the probability distribution given by

$$P_{rs}^{k}(t) = \begin{cases} \dfrac{\tau_{rs}(t)^{\alpha} \eta_{rs}^{\beta}}{\sum\limits_{u \in allowed_k(t)} \tau_{rs}(t)^{\alpha} \eta_{ru}^{\beta}} & \text{if } s \in allowed_k(t) \\ 0 & \text{otherwise} \end{cases} . \tag{2}$$

Pheromone_evaporation is a process of decreasing the intensities of pheromone trails over time. This process is used to avoid locally convergence and to explore more search space. Daemon_actions are optional for ACO, and they are often used to collect useful global information by depositing additional pheromone.

2.2 Quantum-Inspired Evolutionary Algorithm

Quantum computing takes its origins from the foundations of the quantum physics. The parallelism that the quantum computing provides reduces obviously the algorithmic complexity. Such an ability of parallel processing can be used to solve the problems which require the exploration of large solutions spaces. However, the quantum machines that these algorithms require to be efficiently executed are not available yet. So some researches on combining those algorithms to conventional methods, for instance QEA, have been engaged and successfully applied to solve optimization problems such as knapsack problem, filter design and parameter estimation.

Although QEA is based on the concept and principles of quantum computing, such as quantum bit, superposition of state and quantum gate, however, QEA is not a quantum algorithm, but an evolutionary algorithm. In QEA, the smallest information unit is called Q-bit, which is defined as $[\alpha, \beta]^T$. α and β are complex numbers that specify the probability amplitudes of the Q-bit corresponding states. $|\alpha|^2$ gives the probability that the Q-bit will be "0" while $|\beta|^2$ presents the probability that the Q-bit will be "1". The individual xi of QEA with m-bit is defined as:

$$x_i = \left[\begin{array}{c|c|c|c} \alpha_1 & \alpha_2 & \ldots & \alpha_m \\ \beta_1 & \beta_2 & \ldots & \beta_m \end{array}\right]. \quad (3)$$

where $|\alpha_i|^2 + |\beta_i|^2 = 1$, i =1,2,…,m.

For instance, for a three Q-bits system Q with three pairs of amplitudes as

$$Q = \left[\begin{array}{c|c|c} \frac{\sqrt{2}}{2} & \frac{\sqrt{2}}{2} & \frac{1}{2} \\ \frac{\sqrt{2}}{2} & -\frac{\sqrt{2}}{2} & \frac{\sqrt{3}}{2} \end{array}\right]. \quad (4)$$

its states can be described as :

$$\frac{1}{4}|000\rangle + \frac{\sqrt{3}}{4}|001\rangle - \frac{1}{4}|010\rangle - \frac{\sqrt{3}}{4}|011\rangle + \frac{1}{4}|100\rangle + \frac{\sqrt{3}}{4}|101\rangle - \frac{1}{4}|110\rangle - \frac{\sqrt{3}}{4}|111\rangle,$$

which means that the probabilities of being each state are 1/16, 3/16, 1/16, 3/16, 1/16, 3/16, 1/16, and 3/16, respectively. By consequence, the above three Q-bits system contain the information of eight states.

In QEA, to evaluate each individual's performance for guiding updating of the algorithm and tackle the optimization problems, the corresponding solutions represented in the conventional form are necessary. A conventional binary solution can be constructed by observing the Q-bits. That is to say, for a bit ri of the binary individual r, a random number η between [0, 1] is generated and compared with α_i of the Q-bit individual p. If α_i satisfies $|\alpha_i|^2 > \eta$, then set r_i="0", otherwise set r_i = "1", i.e.

$$\begin{cases} r_i = 0 \ \ if \ \ |\alpha_i|^2 > \eta \\ r_i = 1 \ \ if \ \ |\alpha_i|^2 \le \eta \end{cases}. \quad (5)$$

Following this steps, the whole binary solutions can be built by observing the states of the current Q-bit solutions.

After the corresponding conventional solutions are generated, the fitness of each individual is evaluated. And then, a quantum rotation gate $R(\theta)$ is employed to update the Q-bit individual as follows:

$$\begin{bmatrix} \alpha_{id}' \\ \beta_{id}' \end{bmatrix} = R(\theta_{id}) \begin{bmatrix} \alpha_{id} \\ \beta_{id} \end{bmatrix} = \begin{bmatrix} \cos(\theta_{id}) & -\sin(\theta_{id}) \\ \sin(\theta_{id}) & \cos(\theta_{id}) \end{bmatrix} \cdot \begin{bmatrix} \alpha_{id} \\ \beta_{id} \end{bmatrix}. \quad (6)$$

where θ_{id} is the rotation angle. The quantum rotation gate updating makes the Q-bit individual converge to the fitter states. The best solution among the individuals is selected, and if the solution is better than the stored best solution, the best solution will be changed by it.

As quantum rotation gate updates the individuals, rotation angle θ_{id} is a very important parameter. However, it has not had the theoretical basis on the value of θ_{id}. Usually, θ_{id} is defined as:

$$\theta_{id} = s(\alpha_{id}, \beta_{id}) \cdot \Delta\theta_{id} \,. \quad (7)$$

where $s(\alpha_{id}, \beta_{id})$ is the sign of θ_{id} that determines the direction; and $\Delta\theta_{id}$ is the magnitude of rotation angle. The values of $s(\alpha_{id}, \beta_{id})$ and $\Delta\theta_{id}$ are determined by the lookup table as shown in Table 1. In Table 1, b_d is the *d-th* bit of the best solution b and r_{id} is the *d-th* bit of the *i-th* individual of current solution

Table 1. The lookup table of rotation angle of QEA

x_{id}	b_d	$f(x_i)>f(b)$	$\Delta\theta_{id}$	$S(\alpha_i,\beta_i)$			
				$\alpha_{id}\beta_{id}>0$	$\alpha_{id}\beta_{id}<0$	$\alpha_{id}=0$	$\beta_{id}=0$
0	0	False	0	0	0	0	0
0	0	True	0	0	0	0	0
0	1	False	0	0	0	0	0
0	1	True	0.05 π	+1	−1	0	±1
1	0	False	0.01 π	+1	−1	0	±1
1	0	True	0.025 π	−1	+1	±1	0
1	1	False	0.005 π	−1	+1	±1	0
1	1	True	0.025 π	−1	+1	±1	0

3 Quantum Ant Colony Optimization Algorithm

In QEA, the individual in Q-bit representation has the advantage as it is able to represent a linear superposition of states probabilities. Only one individual is enough to represent all the solutions. So QEA can treat the balance between exploration and exploitation more easily than conventional evolutionary algorithm. In this paper, the main motivation that we introduce the Q-bit representation and quantum rotation gate into ACO is to develop a discrete binary ACO algorithm and implement the hyper-cube framework [17] as

1) The rotation angle of quantum rotation gate is independent on the fitness value or the scale of the problem;
2) The range of values that the pheromone trail parameters can assume is limited to the interval [0, 1] using Q-bit representation and quantum rotation gate.

In the last decade, ACO algorithm has been recognized that not only routing problems as the traveling salesperson problem, but also any other type of combinatorial optimization problems can be encoded as "best path" problems and solved using the ant colony metaphor [9]. To tackle the discrete binary optimization problems by using QACO, we transform the binary optimization problem as a "best path" problem figured as Fig1, i.e., a number of ants cooperate together to search the best solution in the binary solution domain by walking sequentially from "ant nest" to "food ".

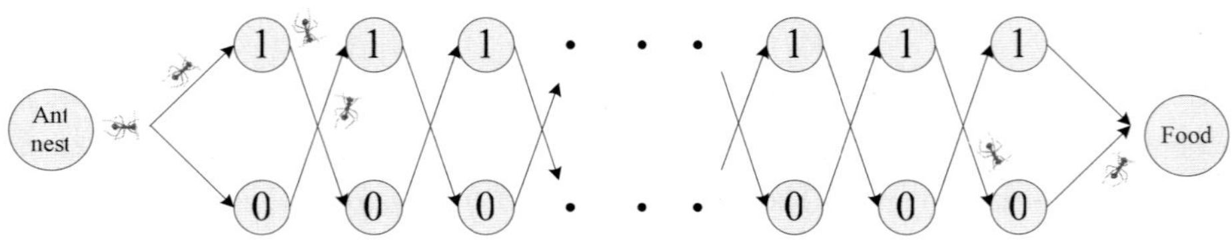

Fig. 1. Binary ant colony optimization algorithm

In the proposed QACO algorithm, the pheromone is represented with Q-bit. A pair of node "0" and "1" in Fig.1 is represented by a Q-bit, that is, α and β of Q-bit represents node "0" and "1", respectively. And $|\alpha|^2$ and $|\beta|^2$ means the probability that ants choose the path to node "0" and node "1", respectively. By observing the whole Q-bit pheromone, a valid solution will be generated and displayed in the binary form by an ant walking from the nest to the food.

The main steps of QACO can be described as follows:

Step1: Initializing QACO, setting population of ant colony, the number of generations and the initial pheromone τ with $\alpha = \beta = 1/\sqrt{2}$ i.e.,

$$\tau = \left[\begin{array}{c|c|c|c} \tau_{1\alpha} & \tau_{2\alpha} & \cdots & \tau_{m\alpha} \\ \tau_{1\beta} & \tau_{2\beta} & \cdots & \tau_{m\beta} \end{array}\right] = \left[\begin{array}{c|c|c|c} 1/\sqrt{2} & 1/\sqrt{2} & \cdots & 1/\sqrt{2} \\ 1/\sqrt{2} & 1/\sqrt{2} & \cdots & 1/\sqrt{2} \end{array}\right]. \tag{8}$$

which means that all solutions are generated with the same probability at the beginning.

Step2: Each ant traveling from the nest to the food to build the solution completely by observing the Q-bit pheromone.

Firstly, a random number p is generated and compared with the exploiting probability parameter pe. If the p is less than or equal to the parameter pe, the solution of ant i of j-th bit is determined as Eq(9)

$$Solution_{i,j} = \begin{cases} 0 \text{ if } \tau_{j,\beta} \le \tau_{j,\alpha} \\ 1 \text{ if } \tau_{j,\beta} > \tau_{j,\alpha} \end{cases} \tag{9}$$

If p is greater than pe, *the* solution of ant i of j-*th* bit is determined by the following threshold function:

$$\eta_c(x) = \begin{cases} 0 & c < \eta_0 \\ 1 & c \ge \eta_0 \end{cases} \tag{10}$$

where c is a random number and η_0 is a constant.

Step3: Calculating the fitness of each ant after all ant agents completing building solutions.

Step4: Terminating if termination criteria are reached, else going on the next step.

Step5: Updating the pheromone intensity by using quantum rotation gate, i.e.:

$$\begin{bmatrix} \tau_{i\alpha}' \\ \tau_{i\beta}' \end{bmatrix} = R(\theta_i) \begin{bmatrix} \tau_{i\alpha} \\ \tau_{i\beta} \end{bmatrix} = \begin{bmatrix} \cos(\theta_i) & -\sin(\theta_i) \\ \sin(\theta_i) & \cos(\theta_i) \end{bmatrix} \cdot \begin{bmatrix} \tau_{i\alpha} \\ \tau_{i\beta} \end{bmatrix} \tag{11}$$

where θ_i is the rotation angle, i=1,2,…,m.

Step6: Going to the step 2.

As the pheromone updating mechanism of QACO is different from the individual updating mechanism of QEA, it is obviously that the quantum rotation angle updating strategy of QEA is improper for the pheromone updating in QACO. To tackle this problem, a novel quantum rotation angle strategy is developed and given in Table 2.

Table 2. The lookup table of rotation angle of QACO

x_i	b_i	$f(x)>f(b)$	$\Delta\theta_i$	$S(\alpha_i,\beta_i)$			
				$\alpha_i\beta_i > 0$	$\alpha_i\beta_i < 0$	$\alpha_i = 0$	$\beta_i = 0$
0	0	False	0.01 π	−1	+1	±1	±1
0	0	True	0.01 π	−1	−1	±1	±1
0	1	False	0.025 π	−1	−1	±1	±1
0	1	True	0.025 π	+1	−1	±1	±1
1	0	False	0.025 π	+1	−1	±1	±1
1	0	True	0.025 π	−1	−1	±1	±1
1	1	False	0.01 π	+1	−1	±1	±1
1	1	True	0.01 π	+1	−1	±1	±1

The exploiting probability *pe* is another important parameter in QACO. A big *pe* can speed convergence of ACO, while a small *pe* maybe make QACO algorithm escape from local optimal more effectively.

4 Numerical Experiment

To test the proposed QACO, a set of 5 optimization functions was employed to evaluate the optimization ability of QACO. The adopted benchmark functions are listed as follows:

$$F_1 = 100(x_1^2 - x_2)^2 + (1 - x_1)^2 \qquad -2.048 \le x_i \le 2.048 . \tag{12}$$

$$\begin{aligned} F_2 = &[1 + x_1 + x_2 + 1)^2 \cdot (19 - 14x_1 + 3x_1^2 - 14x_2 + 6x_1x_2 + 3x_2^2)] \\ &\cdot [30 + (2x_1 - 3x_2)^2 \cdot (18 - 32x_1 + 12x_1^2 + 48x_2 - 36x_1x_2 + 27x_2^2)] \quad -2 \le x_i \le 2 \end{aligned} . \tag{13}$$

$$F_3 = (x_1^2 + x_2^2)^{0.25}[\sin^2(50(x_1^2 + x_2^2)^{0.1}) + 1.0] \qquad -100 < x_i < 100 . \tag{14}$$

$$F_4 = 0.5 - \frac{\sin^2 \sqrt{x_1^2 + x_2^2} - 0.5}{(1 + 0.001(x_1^2 + x_2^2))^4} \qquad -100 < x_i < 100 . \tag{15}$$

$$F_5 = (x_1^2 + x_2 - 11)^2 + (x_1 + x_2^2 - 7)^2 \qquad -10 \le x_i \le 10 . \tag{16}$$

where F_4 has the global maximum, others have the global minimum.

4.1 The Setting of Parameter *pe*

As the value of *pe* is directly related with the searching ability and convergence speed of QACO, an analysis of *pe* based on numerical experiment was conducted firstly. We used QACO to optimize the benchmark function F_2 and F_5 with different *pe*. The population number of QACO was 20, and the length of Q-bit pheromone was 30. All experiments were repeated for 100 runs, and the fixed maximum generation was 2000. The experimental results, i.e., the percent of finding optimal value ($r_{opt}\%$), the average optimizing value (Av), the minimum step of finding the global optimum (Sm) and the average step of finding the global optimum (Sa) are listed in Table 3.

Table 3. The test results of QACO with different *pe*

pe	F_2				F_5			
	$r_{opt}\%$	Av	Sm	Sa	$r_{opt}\%$	Av	Sm	Sa
0.0	0	3.0896	/	/	0	0.1403	/	/
0.1	0	3.0672	/	/	0	0.1108	/	/
0.3	0	3.0649	/	/	0	0.0884	/	/
0.5	1	3.0330	72	72	0	0.0653	/	/
0.7	74	3.0212	15	195	75	0.0223	29	616
0.9	95	3.0113	5	55	89	0.1462	23	746
1.0	0	107.01	/	/	0	39.29	/	/

From Table 3, we can conclude that the value of *pe* seriously affects the optimization ability of QACO. A small *pe* makes QACO work like the random search algorithm, which results in a very poor optimization performance. With the increment of *pe*,

QACO performs better as ants can utilize the swarm intelligence to guide their search, which effectively improve the optimization ability of QACO. But the over big value of *pe*, for instance *pe*=1, can badly spoil the performance of QACO as QACO lack of enough exploring ability and can not escape from the local optima.

4.2 Experimental Results

Based on the simulation results and analysis on parameter *pe*, an adaptive strategy of *pe* as Eq.(18) is adopted in the proposed QACO to improve its searching ability.

$$p_{ex} = 0.93 - 0.15 \cdot (I + G) / G \tag{17}$$

where I represents the current generation and G is the maximum generation of the algorithm. The other parameters of QACO are set at the same values used in section 4.1. For a comparison, QEA [14] and DBPSO [16] with the same settings were used to optimize the above 6 benchmark functions. The experimental results are given in Table 4.

Table 4. The optimization experiment results of QEA, DBPSO and QACO

	Optimal Value	QEA		DBPSO		QACO	
		$r_{opt}\%$	Av	$r_{opt}\%$	Av	$r_{opt}\%$	Av
F_1	0	77	0.0111	56	2.62E-7	100	0
F_2	3	42	29.87	30	3+1.15E-3	100	3
F_3	0	100	0	100	0	100	0
F_4	1	100	1	93	0.9994	100	1
F_5	0	40	0.0761	39	1.15E-4	100	0

From the results given in the Table 4, it is obvious that QACO is a valid and effective optimization algorithm, and outperforms the QEA and DBPSO algorithm for all the 5 test functions.

5 Conclusions

This paper presents a novel quantum ant colony optimization algorithm to tackle the discrete binary combinational problems. In QACO, Q-bit and quantum rotation gate are introduced to represent and update the pheromone. And due to the characteristic of Q-bit and quantum rotation gate, QACO implements the hyper-cube framework which makes the algorithm more robust [17]. A set of 5 benchmark functions has been used to test the proposed QACO algorithm. The experimental results prove that QACO is an effective optimization algorithm and outperforms the traditional QEA and DBPSO.

Acknowledgements. This work is supported by Shanghai Leading Academic Disciplines under Grant No.T0103, Shanghai Key Laboratory of Power Station Automation Technology, and Key Project of Science & Technology Commission of Shanghai Municipality under Grant 06DZ22011.

References

1. Colorni, A., Dorigo, M., Maniezzo, V.: Distributed optimization by ant colonies. In: Proceedings of ECAL91- European Conf. on Artificial Life, vol. 1, pp. 134–142 (1991)
2. Dorigo, M., Maniezzo, V., Colorni, A.: Positive feedback as a search strategy. Tech. Report 91-016, Dipartimentodi Elettronica, Politecnico di Milano, Italy (1991)
3. Dorigo, M., Maniezzo, V., Colorni, A.: Ant system: optimization by a colony of cooperating agents. IEEE Trans. on Systems, Man and Cybernetics-part B 26, 29–41 (1996)
4. Dorigo, M., Gambardella, L.M.: Ant colony system: A cooperative learning approach to the traveling salesman problem. IEEE Trans. Evolutionary Comput. 1, 53–66 (1997)
5. Reimann, M., Doerner, K., Hartl, R.F.: D-ants: savings based ants divide and conquer the vehicle routing problems. Comput. Oper. Res. 31, 563–591 (2004)
6. Liao, C.J., Juan, H.C.: An ant colony optimization for single-machine tardiness scheduling with sequence-dependent setups. Computers & Operations Research 34, 1899–1909 (2007)
7. Lee, Z.J., Lee, C.Y.: A hybrid search algorithm with heuristics for resource allocation problem. Information Sciences 173, 155–167 (2005)
8. Stutzle, T., Hoos, H.H.: MAX-MIN ant system. Future Gen. Comput. Systems 16, 889–914 (2000)
9. Dorigo, M., Blum, C.: Ant colony optimization theory: A survey. Theoretical Computer Science 344, 243–278 (2005)
10. Blum, C.: Ant colony optimization: Introduction and recent trends. Physics of Life Reviews 2, 353–373 (2005)
11. Narayanan, A., Moore, M.: Quantum Inspired Genetic Algorithms. In: Proc. of the 1996 IEEE Intl. Conf. on Evolutionary Computation (ICEC96), vol. 1, pp. 212–221. IEEE Computer Society Press, Los Alamitos (1996)
12. Han, K.H.: Genetic Quantum Algorithm and its Application to Combinatorial Optimization Problem. In: IEEE Proc. Of the 2000 Congress on Evolutionary Computation, pp. 1354–1360. IEEE Computer Society Press, Los Alamitos (2000)
13. Wang, Y., Feng, X.Y., Huang, Y.X.: A novel quantum swarm evolutionary algorithm and its applications. Neurocomputing 70, 633–640 (2007)
14. Han, K.H., Kim, J.H.: Quantum-inspired evolutionary algorithm for a class of combinatorial optimization. IEEE Transactions on Evolutionary Computation 6, 580–593 (2002)
15. Han, K.H., Kim, J.H.: Quantum-Inspired Evolutionary Algorithms With a New Termination Criterion,Hε Gate and Two-Phase Scheme. IEEE Transaction on Evolutionary Computation 8, 156–169 (2004)
16. Kennedy, J., Eberhart, R.C.: A discrete binary version of the particle swarm optimization algorithm. In: Proc. of the 1997 Conf. on Systems, Man, and Cybernetics, pp. 4104–4108 (1997)
17. Blum, C., Dorigo, M.: The hyper-cube framework for ant colony optimization. IEEE Transactions on Systems, Man and Cybernetics, Part B 34, 1161–1172 (2004)

Genetic Particle Swarm Optimization Based on Estimation of Distribution

Jiahai Wang

Department of Computer Science, Sun Yat-sen University,
No.135, Xingang West Road, Guangzhou 510275, P.R.China
wjiahai@hotmail.com

Abstract. Estimation of distribution algorithms sample new solutions from a probability model which characterizes the distribution of promising solutions in the search space at each generation. In this paper, a modified genetic particle swarm optimization algorithm based on estimation of distribution is proposed for combinatorial optimization problems. The proposed algorithm incorporates the global statistical information collected from local best solution of all particles into the genetic particle swarm optimization. To demonstrate its performance, experiments are carried out on the knapsack problem, which is a well-known combinatorial optimization problem. The results show that the proposed algorithm has superior performance to other discrete particle swarm algorithms.

1 Introduction

The PSO is inspired by observing the bird flocking or fish school [1]. A large number of birds/fishes flock synchronously, change direction suddenly, and scatter and regroup together. Each individual, called a particle, benefits from the experience of its own and that of the other members of the swarm during the search for food. Comparing with genetic algorithm, the advantages of PSO lie on its simple concept, easy implementation and quick convergence. The PSO has been applied successfully to continuous nonlinear function [1], neural network [2], nonlinear constrained optimization problems [3], etc. Most of the applications have been concentrated on solving continuous optimization problems [4].

To solve discrete (combinatorial) optimization problems, Kennedy and Eberhart [5] also developed a discrete version of PSO (DPSO), which however has seldom been utilized. DPSO essentially differs from the original (or continuous) PSO in two characteristics. First, the particle is composed of the binary variable. Second, the velocity must be transformed into the change of probability, which is the chance of the binary variable taking the value one. Furthermore, the relationships between the DPSO parameters differ from normal continuous PSO algorithms [6] [7]. Although Kennedy and Eberhart [5] have tested the robustness of the discrete binary version through function optimization benchmark, few applications for combinatorial optimization have ever been developed based on their work.

K. Li et al. (Eds.): LSMS 2007, LNCS 4688, pp. 287–296, 2007.

Though it has been proved the DPSO can also be used in discrete optimization as a common optimization method, it is not as effective as in continuous optimization. When dealing with integer variables, PSO sometimes are easily trapped into local minima [5]. Therefore, Yang et al. [8] proposed a quantum particle swarm optimization (QPSO) for discrete optimization in 2004. Their simulation results showed that the performance of the QPSO was better than DPSO and genetic algorithm. Recently, Yin [9] proposed a genetic particle swarm optimization (GPSO) with genetic reproduction mechanisms, namely crossover and mutation to facilitate the applicability of PSO to combinatorial optimization problem, and the results showed that the GPSO outperformed the DPSO for combinatorial optimization problems.

In the last decade, more and more researchers tried to overcome the drawbacks of usual recombination operators of evolutionary computation algorithms. Therefore, estimation of distribution algorithms (EDAs) [10] have been developed. These algorithms, which have a theoretical foundation in probability theory, are also based on populations that evolve as the search progresses. EDAs use probabilistic modeling of promising solutions to estimate a distribution over the search space, which is then used to produce the next generation by sampling the search space according to the estimated distribution. After every iteration, the distribution is re-estimated.

In this paper, a modified genetic particle swarm optimization algorithm based on estimation of distribution is proposed for combinatorial optimization problems. The proposed algorithm incorporates the global statistical information collected from local best solution of all particles into the genetic particle swarm optimiza-tion. To demonstrate its performance, experiments are carried out on the knapsack problem, which is a well-known combinatorial optimization problem. The results show that the proposed algorithm has superior performance to other discrete particle swarm algorithms.

The paper is organized as following: In Section 2, Section 3 and Section 4, the basic ideas of PSO, GPSO, and EDA are briefly described respectively. Then, in Section 5, the modified GPSO based on EDA is proposed. Section 6 presents simulated results and Section 7 concludes the paper.

2 Particle Swarm Optimization

PSO is initialized with a group of random particles (solutions) and then searches for optima by updating each generation. In every iteration, each particle is updated by following two best values. The first one is the local best solution (fitness) a particle has obtained so far. This value is called personal best solutions. Another best value is that the whole swarm has obtained so far. This value is called global best solution. The philosophy behind the original PSO is to learn from individual's own experience (personal best solution) and best individual experience (global best solution) in the whole swarm, which can be described by Fig.1.

Denote by N particle number in the swarm. Let $X_i(t)=(x_{i1}(t), \cdots, x_{id}(t), \cdots, x_{iD}(t))$, be particle i with D bits at iteration t, where is treated as a potential solution. Denote the velocity as $V_i(t) = (v_{i1}(t), \cdots, v_{id}(t), \cdots, v_{iD}(t))$, $v_{id}(t) \in R$. Let $PBest_i(t) = (pbest_{i1}(t), \cdots, pbest_{id}(t), \cdots, pbest_{iD}(t))$ be the best solution that particle i has obtained until iteration t, and $GBest(t) = (gbest_1(t), \cdots, gbest_d(t), \cdots, gbest_D(t))$ be the best solution obtained from $PBest_i(t)$ in the whole swarm at iteration t. Each particle adjusts its velocity according to previous velocity of the particle, the cognition part and the social part. The algorithm is described as follows [1]:

$$v_{id}(t+1) = v_{id}(t) + c_1 \cdot r_1 \cdot (pbest_{id}(t) - x_{id}(t)) + c_2 \cdot r_2 \cdot (gbest_d(t) - x_{id}(t)), \quad (1)$$

$$x_{id}(t+1) = x_{id}(t) + v_{id}(t+1), \quad (2)$$

where

c_1 is the cognition learning factor and c_2 is the social learning factor;

r_1 and r_2 are two random numbers uniformly distributed in [0,1].

Most of the applications have been concentrated on solving continuous optimization problems. To solve discrete (combinatorial) optimization problems, Kennedy and Eberhart [5] also developed a discrete version of PSO (DPSO), which however has seldom been utilized. DPSO essentially differs from the original (or continuous) PSO in two characteristics. First, the particle is composed of the binary variable. Second, the velocity must be transformed into the change of probability, which is the chance of the binary variable taking the value one. The velocity value is constrained to the interval [0, 1] using the following sigmoid function:

$$s(v_{id}) = \frac{1}{1 + \exp(-v_{id})}, \quad (3)$$

where $s(v_{id})$ denotes the probability of bit x_{id} taking 1. Then the particle changes its bit value by

$$x_{id} = \begin{cases} 1 & \text{if } \mathrm{rand}() \le s(v_{id}) \\ 0 & \text{otherwise} \end{cases}, \quad (4)$$

where rand() is a random number selected from a uniform distribution in [0,1]. To avoid $s(v_{id})$ approaching 1 or 0, a constant $V_{\max}$ as a maximum velocity is used to limit the range of v_{id}, that is, $v_{id} \in [-V_{\max}, V_{\max}]$. The basic flowchart of PSOs (including continuous PSO and discrete PSO) is shown by Fig.1.

3 Genetic Particle Swarm Optimization (GPSO)

To facilitate the applicability of PSO to combinatorial optimization problem, Yin [14] proposed a genetic particle swarm optimization (GPSO) with genetic reproduction mechanisms, namely crossover and mutation. Denote by N the number of particles in the swarm. The GPSO with genetic recombination for the d-th bit of particle i is described as follows:

$$\begin{aligned} x_{id}(t+1) = {} & w(0, w_1) \cdot rand(x_{id}(t)) + w(w_1, w_2) \cdot rand(pbest_{id}(t)) \\ & + w(w_2, 1) \cdot rand(gbest_d(t)), \end{aligned} \quad (5)$$

where $0 < w_1 < w_2 < 1$, $w()$ and rand( ) are the threshold function and the probabilistic bit flipping function, respectively, and they are defined as follows:

$$w(a,b) = \begin{cases} 1 & \text{if } a \le r_1 < b \\ 0 & \text{otherwise} \end{cases}, \tag{6}$$

$$rand(y) = \begin{cases} 1-y & \text{if } r_2 < p_m \\ y & \text{otherwise} \end{cases}, \tag{7}$$

where r_1 and r_2 are the random numbers uniformly distributed in [0,1]. Thus, only one of the three terms on right hand side of Eq. (5) will remain dependent on the value r_1, and $rand(y)$ mutates the binary bit y with a small mutation probability p_m. The updating rule of the genetic PSO is analogue to the genetic algorithm in two aspects. First, the particle derives its single bit from the particle x_{id}, $pbest_{id}$ and $gbest_d$. This operation corresponds to a 3-way uniform crossover among X_i, $PBest_i$ and $GBest_g$, such that the particle can exchange building blocks (segments of ordering or partial selections of elements) with personal and global experiences. Second, each bit attained in this way will be flipped with a small probability p_m, corresponding to the binary mutation performed in genetic algorithms. As such, genetic reproduction , in particular, crossover and mutation, have been added to the DPSO, and this new genetic version, named GPSO, is very likely more suitable for solving combinatorial optimization problems than the original one.

4 Estimation of Distribution Algorithms (EADs)

Evolutionary Algorithms that use information obtained during the optimization process to build probabilistic models of the distribution of good regions in the search space and that use these models to generate new solutions are called estimation of distribution algorithms (EDAs) [11]. These algorithms, which have a theoretical foundation in probability theory, are also based on populations that evolve as the search progresses. EDAs use probabilistic modeling of promising solutions to estimate a distribution over the search space, which is then used to produce the next generation by sampling the search space according to the estimated distribution. After every iteration, the distribution is re-estimated.

An algorithmic framework of most EDAs can be described as:

InitializePopulation( ) /*Initialization*/
While Stopping criteria are not satisfied do /*Main Loop*/
P_{sel}=Select(P)/*Selection*/
$P(x) = P(x|P_{sel})$=EstimateProbabilityDistribution( ) /*Estimation*/
P=SampleProbabilityDistribution( ) /*Sample*/
EndWhile

An EDA starts with a solution population P and a solution distribution model $P(x)$. The main loop consists of three principal stages. The first stage is to select the best individuals (according to some fitness criteria) from the population.

These individuals are used in a second stage in which the solution distribution model $P(x)$ is updated or recreated. The third stage consists of sampling the updated solution distribution model to generate new solutions offspring. EDAs are based on probabilistic modelling of promising solutions to guide the exploration of the search space instead of using crossover and mutation like in the well-known genetic algorithms (GAs).

There has been a growing interest for EDAs in the last years. More comprehensive presentation of the EDA field can be found in Refs. [12] [13].

5 Modified GPSO Based on EDA

In this section, we describe the proposed modified GPSO algorithm which uses global statistic information gathered from the local best solutions of all particles during the optimization process to guide the search.

In the second term of original GPSO's updating rule, Eq. (5), a particle can only learn from oneself best experience. In the proposed algorithm, we incorporate the idea of EDA into the original GPSO by modifying the second term of Eq. (5), and therefore a particle can learn from the global statistical information collected by the personal best experiences of all the particles in the proposed algorithm.

Several different probability models have been introduced in EDAs for modeling the distribution of promising solutions. The univariate marginal distribution (UMD) model is the simplest one and has been used in univariate marginal distri-bution algorithm [14], population-based incremental learning (PBIL) [15], com-pact GA [16].

In the proposed algorithm, as defined in the previous section, denote by N the number of particles in the swarm. Let $X_i(t) = (x_{i1}(t), \cdots, x_{id}(t), \cdots, x_{iD}(t))$, $x_{id}(t) \in \{0, 1\}$, be particle i with D bits at iteration t, where $X_i(t)$ being treated as a potential solution. Firstly, all the local best solutions are selected; then, the UMD model is adopted to estimate the distribution of good regions over the search space based on the selected local best solutions. The UMD uses a probability vector $P = (p_1, \cdots, p_d, \cdots, p_D)$ to characterize the distribution of promising solutions in the search space, where p_d is the probability that the value of the d-th position of a promising solution is 1. New offspring solutions are thus generated by sampling the updated solution distribution model. The probability vector $P = (p_1, \cdots, p_d, \cdots, p_D)$ guides a particle to search in binary 0-1 solution space in the following way:

$$\begin{aligned} x_{id}(t+1) = w(0, w_1) \cdot rand(x_{id}(t)) + w(w_1, w_2) \cdot rand(EDA_{id}(t)) \\ +w(w_2, 1) \cdot rand(gbest_d(t)), \end{aligned} \tag{8}$$

where

$$EDA_{id} = \begin{cases} 1 & \text{if rand()} < p_d \\ 0 & \text{otherwise} \end{cases}, \tag{9}$$

In the sample process above, a bit is sampled from the probability vector P randomly. The probability vector P is initialized by the following rule:

$$p_d = \frac{\sum_{i=1}^{N} pbest_{id}}{N}, \tag{10}$$

p_d is the percentage of the binary strings with the value of the d-th element being 1. P can also be regarded as the center of the personal best solutions of all the particles.

The probability vector in the proposed algorithm can be learned and updated at each iteration for modeling the distribution of promising solutions. Since some elements of the offspring are sampled from the probability vector P, it can be expected that should fall in or close to a promising area. The sampling mechanism can also provide diversity for the search afterwards. At each iteration t in the proposed algorithm, the personal best solutions of all the particles are selected and used for updating the probability vector P. Therefore, the probability vector P can be updated in the same way as in the PBIL algorithm [15]:

$$p_d = (1-\lambda)p_d + \lambda\frac{\sum_{i=1}^{N} pbest_{id}}{N}, \tag{11}$$

where $\lambda \in (0, 1]$ is the learning rate. As in PBIL [15], the probability vector P is used to generate the next set of sample points; the learning rate also affects which portions of the problem space will be explored. The setting of the learning rate has a direct impact on the trade-off between exploration of the problem space and exploitation of the exploration already conducted. For example, if the learning rate is 0, there is no exploitation of the information gained through search. As the learning rate is increased, the amount of exploitation increases, and the ability to search large portions of the problem space diminishes.

The proposed algorithm is different from the pure EDAs or GPSO in several aspects. Pure EDAs extract global statistical information from the previous search and then represent it as a probability model, which characterizes the distribution of promising solutions in the search space. New solutions are generated by sampling from this model. However, the location information of the locally optimal solutions found so far, for example, the global best solution in the population, has not been directly used in the pure EDAs. In contrast to that a particle can only learn from oneself best experience in the original GPSO, a particle can learn from the global statistical information collected by the personal best experiences of all the particles in the proposed algorithm. That is, all particles can potentially contribute to a particle's search via the probability vector P, which can be seen as a kind of comprehensive learning ability.

6 Simulation Results

To demonstrate the performance of the proposed algorithm, experiments are carried out on the knapsack problem, which is a well-known combinatorial optimization problem. The classical knapsack problem is defined as follows: We are given a set of n items, each item i having an integer profit p_i and an integer weight w_i. The problem is to choose a subset of the items such that their total profit is maximized, while the total weight does not exceed a given capacity C. We may formulate the problem to maximize the total profit $f(X)$ as the following [17]:

$$f(X) = \sum_{i=1}^{n} p_i x_i,$$

subject to

$$\sum_{i=1}^{n} w_i x_i \leq C,$$

where the binary decision variables x_i are used to indicate whether item i is included in the knapsack or not. Without loss of generality it may be assumed that all profits and weights are positive, that all weights are smaller than the capacity C, and that the total weight of the items exceeds C.

In all experiments, strongly correlated sets of data were considered:

$$w_i = \text{uniformly random}[1, R],$$

$$p_i = w_i + R/10,$$

and the following average knapsack capacity was used:

$$C = \frac{1}{2} \sum_{i=1}^{n} w_i.$$

The traditional test instances with small data range are too easy to draw any meaningful conclusions; therefore we test the proposed algorithm on a class of difficult instances with large coefficients [17]. That is, the weights are uniformly distributed in a large data rang $R = 10^6$. This makes the dynamic programming algorithms run slower, but also the upper bounds get weakened since the gap to the optimal solution is scaled and cannot be closed by simply rounding down the upper bound to the nearest smaller integer [17]. Five knapsack problems with 100, 500, 1000, 5000, and 10000 items were considered.

For comparison, QPSO, DPSO and GPSO for this problem were also implemented in C on a DELL-PC (Pentium4 2.80 GHz). the parameters $\alpha = 0.1$, $\beta = 0.9$, $c_1 = c_2 = 0.1$, and $c_3 = 0.8$ are used. In the DPSO, the parameters, two acceleration coefficients $c_1 = c_2 = 1.2$, and velocity limit $V_{max} = 4$ were used in the simulations.

In the GPSO, the standard parameters are adopted from Ref. [9]: the value of w_1 is dynamically tuned from 0.9 to 0.4 according to the number of generations such that more exploration search is pursued during the early generations and the exploitation search is emphasized afterward. The value of w_2 determines the relative importance of $PBest$ and $GBest$, therefore $w_2 = 0.2w_1 + 0.8$ is set. The bit mutation probability p_m is set to 0.001. The proposed algorithm has power global exploration search mechanism because of the global statistic information

Table 1. Simulation results of 5 test problems

Algorithm		$n = 100$	$n = 500$	$n = 1000$	$n = 5000$	$n = 10000$
QPSO	Best	34885926	155155453	323632482	1601064252	3208126993
	Av.	34855727.7	154988339.05	323187543.7	1598475266.6	3200372726.95
GPSO	Best	34885971	**155431535**	325342751	1607304363	3214945549
	Av.	34885853.7	155431513.1	325342740.2	1607145495	3214588236.15
DPSO	Best	34885545	155431349	325142736	1607159843	3214745648
	Av.	34885545	155431349	325142736	1606874334.95	3214032772.4
Proposed algorithm	Best	**34885973**	**155431535**	**325342752**	**1607459852**	**3216238062**
	Av.	**34885919.35**	**155431528.7**	**325342745.9**	**1607459823.9**	**3216052194.15**

Table 2. Computation time of the algorithm of QPSO, DPSO, GPSO and the proposed algorithm (seconds)

Algorithm	$n = 100$	$n = 500$	$n = 1000$	$n = 5000$	$n = 10000$
QPSO	6.32	13.81	28.57	144.9	293.17
GPSO	2.58	12.51	24.99	125.86	255.68
DPSO	3.48	28.5	56.48	287.47	565.73
Proposed algorithm	3.68	16.67	35.21	175	338.29

gathered from the local best solutions of all particles, and exploitation search should be emphasized in the proposed algorithm. Therefore, in the proposed algorithm, the value of w_1 is dynamically tuned from 0.4 to 0 according to the number of generations. The parameters, $w_2 = 0.2w_1 + 0.8$ and $\lambda = 0.1$ are also be set. In all algorithms, the population size and maximum iteration number are set to 40 and 1500, respectively. In all algorithms, the greedy repair is adopted to handle the constraint of knapsack problems [18].

Table 1 shows simulation results. The best total profit ("Best") and average total profit ("Av.") produced by QPSO, DPSO, GPSO and the proposed algorithm respectively within 20 simulation runs are shown. Simulation results show that the proposed algorithm can obtain better solutions than the other particle swarm optimization algorithm. Furthermore, all the average solutions of the proposed algorithm are better than the best solutions of QPSO and DPSO algorithms. The better average performance of the proposed algorithm shows that the proposed algorithm is of a certain robustness for the initial solutions. Bold figures indicate the best results among the four algorithms.

Table 2 shows the comparison of computation time which is the average of 20 simulations. The proposed algorithm requires a little more CPU computational time than QPSO and GPSO because the proposed algorithm spends time in computing and updating the estimation of distribution at each iteration. DPSO is the slowest algorithm because it spends a lot time in computing the nonlinear sigmoid function at each iteration. Therefore, we can conclude that the proposed algorithm can search better solution within reasonable time.

7 Conclusions

In this paper, a modified genetic particle swarm optimization algorithm based on estimation of distribution is proposed for combinatorial optimization problems. The proposed algorithm incorporates the global statistical information collected from local best solution information of all particles into the genetic particle swarm optimization. To demonstrate its performance, experiments are carried out on the knapsack problem, which is a well-known combinatorial optimization problem. The results show that the proposed algorithm has superior performance to other discrete particle swarm algorithms. The future work is to use the high order probability models to estimation the distribution of promising solutions and their applications to different kinds of hard combinatorial optimization problems.

Acknowledgments. The Project was supported by the Scientific Research Foundation for Outstanding Young Teachers, Sun Yat-sen University.

References

1. Kennedy, J., Eberhart, R.C.: Particle Swarm Optimization. In: Proceedings of IEEE International Conference on Neural Networks. pp. 1942–1948. Piscataway, NJ (1995)
2. Van den Bergh, F., Engelbrecht, A.P.: Cooperative Learning in Neural Network Using Particle Swarm Optimizers. South African Computer Journal 26, 84–90 (2000)
3. El-Galland, A.I., El-Hawary, M.E., Sallam, A.A.: Swarming of Intelligent Particles for Solving the Nonlinear Constrained Optimization Problem. Engineering Intelligent Systems for Electrical Engineering and Communications. 9, 155–163 (2001)
4. Parsopoulos, K.E., Vrahatis, M.N.: Recent approaches to global optimization problems through Particle Swarm Optimization. Natural Computing 1(2-3), 235–306 (2002)
5. Kennedy, J., Eberhart, R.C.: A Discrete Binary Version of the Particle Swarm Algorithm. In: Proceedings of the World Multiconference on Systemics, Cybernetics and Informatics, pp. 4104–4109. Piscatawary, NJ (1997)
6. Franken, N., Engelbrecht, A.P.: Investigating Binary PSO Parameter Influence on the Knights Cover Problem. IEEE Congress on Evolutionary Computation 1, 282–289 (2005)
7. Huang, Y.-X., Zhou, C.-G., Zou, S.-X., Wang, Y.: A Hybrid Algorithm on Class Cover Problems. Journal of Software (in Chinese) 16(4), 513–522 (2005)
8. Yang, S.Y., Wang, M., Jiao, L.C.: A Quantum Particle Swarm Optimization. Proceeding of the 2004 IEEE Congress on Evolutionary Computation 1, 320–324 (2004)
9. Yin, P.Y.: Genetic Particle Swarm Optimization for Polygonal Approximation of Digital Curves. Pattern Recognition and Image Analysis 16(2), 223–233 (2006)
10. Mühlenbein, H., Paaβ, G.: From Recombination of Genes to the Estimation of Distributions. In: Ebeling, W., Rechenberg, I., Voigt, H.-M., Schwefel, H.-P. (eds.) PPSN IV. LNCS, vol. 1141, pp. 178–187. Springer, Heidelberg (1996)
11. Pelikan, M., Goldberg, D.E., Lobo, F.: A Survey Of Optimization by Building and Using Probabilistic Models. Computational Optimization and Applications 21(1), 5–20 (2002)

12. Kern, S., Muller, S.D., Hansen, N., Buche, D., Ocenasek, J., Koumoutsakos, P.: Learning Probability Distributions in Continuous Evolutionary Algorithms-A Comparative Review. Natural Computing 3(1), 77–112 (2004)
13. Larrañaga, P., Lozano, J.: Estimation of Distribution Algorithms: A New Tool for Evolutionary Computation. In: Genetic Algorithms and Evolutionary Computation, vol. 2, Springer, Heidelberg (2001)
14. Müehlenbein, H.: The Equation for Response to Selection and Its Use for Prediction. Evol. Comput. 5(3), 303–346 (1997)
15. Baluja, S.: Population-Based Incremental Learning: A Method for Integrating Genetic Search Based Function Optimization and Competitive Learning. School of Comput. Sci., Carnegie Mellon Univ., Pittsburgh, PA, Tech. Rep. CMU-CS-94-163 (1994)
16. Harik, G.R., Lobo, F.G., Goldberg, D.E.: The Compact Genetic Algorithm. IEEE Trans. Evol. Comput. 3(4), 287–297 (1999)
17. Pisinger, D.: Where Are the Hard Knapsack Problem? Computer & Operations Research 32, 271–2284 (2006)
18. Michalewicz, Z.: Genetic Algorithm+ Data Structure=Evolution Programs, pp. 59–65. Science Press, Beijng (2000)

A Memetic Algorithm with Genetic Particle Swarm Optimization and Neural Network for Maximum Cut Problems

Jiahai Wang

Department of Computer Science, Sun Yat-sen University,
No.135, Xingang West Road, Guangzhou 510275, P.R.China
wjiahai@hotmail.com

Abstract. In this paper, we incorporate a chaotic discrete Hopfield neural network (CDHNN), as a local search scheme, into a genetic particle swarm optimization (GPSO) and develop a memetic algorithm GPSO-CDHNN for the maximum cut problem. The proposed algorithm not only performs exploration by using the population-based evolutionary search ability of the GPSO, but also performs exploitation by using the CDHNN. Simulation results show that the proposed algorithm has superior ability for maximum cut problems.

1 Introduction

One of the most important combinatorial optimization problems in graph theory is the maximum cut problem [1]. Given a weighted, undirected graph $G = (V, E)$, the goal of this problem is to find a partition of vertices of the graph G into two disjoint sets, such that the total weight of the edges from one to the other is as large as possible. Many optimization problems can be formulated to find a maximum cut in a network or a graph. Besides its theoretical importance, the maximum cut problem has applications in the design of VLSI circuits, the design of communication networks, circuit layout design and statistical physics [2] [3]. This problem is one of Karp's original NP-complete problems [1].

Since the maximum cut problem is NP-complete for which the exact solution is difficult to obtain, different heuristic or approximation algorithms have been proposed. In 1998, Bertoni et al [4] proposed a simple and effective LORENA algorithm. Recently, algorithms based on the SDP relaxation proposed by Goemans et al. [5], Helmberg et al. [6], Benson et al. [7] and Choi et al. [8], and heuristic algorithm proposed by Burer et al. [9] for maximum cut problem produced very good results. Galán-Marín et al. [10] [11] proposed a optimal competitive Hopfield model (OCHOM), which guarantees and maximizes the descent of any Lyapunov energy function as the groups of neurons are updated. By introducing a stochastic dynamics into the OCHOM, we proposed a new algorithm that permits temporary energy increases which help the OCHOM escape from local minima [12]. The results produced by the SOCHOM for maximum cut problem were better than several other existing methods [12]. Wu et al. [13]

K. Li et al. (Eds.): LSMS 2007, LNCS 4688, pp. 297–306, 2007.

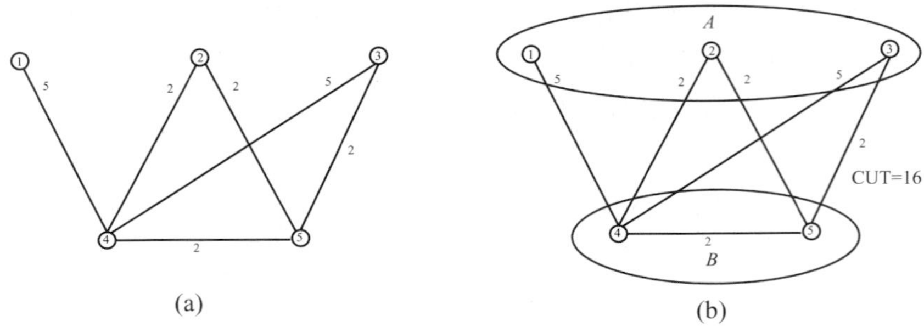

Fig. 1. (a) A simple undirected graph composed of five vertices and six edges, (b) one of its maximum cut graphs

gave new convergence conditions for discrete Hopfield neural networks (DHNN) [14] recently and proposed a DHNN with negative diagonal weight matrix for the maximum cut problem, which improved the solution quality dramatically. In order to prevent the DHNN to be trapped in a stable state corresponding to a local minimum, we proposed chaotic DHNN (CDHNN), which helps the DHNN escape from local minima [15].

During the past decade, a novel evolutionary computation technique, particle swarm optimization (PSO), was proposed by Kennedy and Eberhart [16] and it has attracted much attention. Most of the applications have been concentrated on solving continuous optimization problems. To solve discrete (combinatorial) optimization problems, Kennedy and Eberhart [17] also developed a discrete version of PSO (DPSO), which however has seldom been utilized. Yin [18] proposed a genetic particle swarm optimization (GPSO) with genetic reproduction mechanisms, namely crossover and mutation to facilitate the applicability of PSO to combinatorial optimization problem, and the results showed that the GPSO outperformed the DPSO for combinatorial optimization problems.

In this paper, the CDHNN, as a local search scheme, is incorporated into the proposed GPSO to improve its performance and a memetic algorithm, called GPSO-CDHNN is proposed for the maximum cut problem. A number of instances have been simulated, and the simulation results show that the proposed algorithm can get better results than other algorithms.

2 Memetic Algorithm for Maximum Cut Problems

In this Section, we incorporate the chaotic discrete Hopfield neural network (CDHNN) [15], as a local search scheme, into the genetic particle swarm optimization (GPSO) and develop a memetic algorithm GPSO-CDHNN for the maximum cut problem.

2.1 Maximum Cut Problem

Let $G = (V, E)$ be an edge-weighted undirected graph, where V is the set of vertices and E is the set of edges. The edge from vertex i to vertex j is represented by $e_{ij} \in E$. $e_{ij} = e_{ji}$ also can be seen as weights on edges whose endpoints are vertex i to vertex j. The maximum cut problem is to find a partition of V into two nonempty, disjoin sets A and B, such that $A \cup B = V$ and $A \cap B = \phi$, $\sum_{i\in A, j\in B} e_{ij}$ is maximum. The maximum condition can also be represented as $\sum_{ij\in A} e_{ij} + \sum_{ij\in B} e_{ij}$ is minimum. Figure 1(b) shows one of the solutions of Fig.1(a) [12][15].

2.2 Genetic Particle Swarm Optimization (GPSO)

The PSO is inspired by observing the bird flocking or fish school [16]. A large number of birds/fishes flock synchronously, change direction suddenly, and scatter and regroup together. Each individual, called a particle, benefits from the experience of its own and that of the other members of the swarm during the search for food. Due to the simple concept, easy implementation and quick convergence, the PSO has been applied successfully to continuous nonlinear function, neural network, nonlinear constrained optimization problems [19], etc. Most of the applications have been concentrated on solving continuous optimization problems [19].

To solve discrete (combinatorial) optimization problems, Kennedy and Eberhart [17] also developed a discrete version of PSO (DPSO), which however has seldom been utilized. Yin [18] proposed a genetic particle swarm optimization (GPSO) with genetic reproduction mechanisms, namely crossover and mutation to facilitate the applicability of PSO to combinatorial optimization problem, and the results showed that the GPSO outperformed the DPSO for combinatorial optimization problems.

Denote by N the number of particles in the swarm. Let $X_i(t) = (x_{i1}(t), \cdots, x_{id}(t), \cdots, x_{iD}(t))$, $x_{id}(t) \in \{-1, 1\}$, be particle i with D bits at iteration t, where $X_i(t)$ is treated as a potential solution. Let $P_i(t) = (p_{i1}(t), \cdots, p_{id}(t), \cdots, p_{iD}(t))$ be the best solution that particle i has obtained until iteration t, and $P_g(t) = (p_{g1}(t), \cdots, p_{gd}(t), \cdots, p_{gD}(t))$ be the best solution obtained from $P_i(t)$ in the whole swarm at iteration t. The GPSO with genetic recombination for the d-th bit of particle i is described as follows:

$$\begin{aligned} x_{id}(t+1) = {} & w(0, w_1) \cdot rand(x_{id}(t)) + w(w_1, w_2) \cdot rand(p_{id}(t)) \\ & + w(w_2, 1) \cdot rand(p_{gd}(t)), \end{aligned} \tag{1}$$

where $0 < w_1 < w_2 < 1$, $w()$ and rand( ) are the threshold function and the probabilistic bit flipping function, respectively, and they are defined as follows:

$$w(a, b) = \begin{cases} 1 & \text{if } a \le r_1 < b \\ 0 & \text{otherwise} \end{cases}, \tag{2}$$

$$rand(y) = \begin{cases} -y & \text{if } r_2 < p_m \\ y & \text{otherwise} \end{cases}, \tag{3}$$

where r_1 and r_2 are two random numbers uniformly distributed in [0,1]. Thus, only one of the three terms on right hand side of Eq. (1) will remain dependent on the value r_1, and $rand(y)$ mutates the binary bit y with a small mutation probability p_m. The updating rule of the genetic PSO is analogue to the genetic algorithm in two aspects. First, the particle derives its single bit from the particle x_{id}, p_{id} and p_{gd}. This operation corresponds to a 3-way uniform crossover among X_i, P_i and P_g, such that the particle can exchange building blocks (segments of ordering or partial selections of elements) with personal and global experiences. Second, each bit attained in this way will be flipped with a small probability p_m, corresponding to the binary mutation performed in genetic algorithms. As such, genetic reproduction , in particular, crossover and mutation, have been added to the DPSO, and this new genetic version, named GPSO, is very likely more suitable for solving combinatorial optimization problems than the original one.

2.3 Chaotic Discrete Hopfield Neural Network (CDHNN)

Let M be a discrete Hopfield neural network (DHNN) with n neurons, where each neuron is connected to all the other neurons. The input, output state and threshold of neuron i is denoted by u_i, v_i and t_i, respectively, for $i = 1, \ldots, n$; $w_{i,j}$ is the interconnection weight between neurons i and j, where symmetric weights are considered, for $i, j = 1, \ldots, n$. Therefore, $W = (w_{ij})$ is an $n \times n$ weight matrix and $T = (t_1, \ldots, t_n)^T$ is a threshold vector. The Lyapunov function of the DHNN is given by [8] [10]:

$$E = -\frac{1}{2} \sum_{i=1}^{n} \sum_{j=1}^{n} w_{ij} v_i v_j + \sum_{x=1}^{n} t_i v_i. \tag{4}$$

The inputs of the neurons are computed by

$$u_i(t+1) = \sum_{j=1}^{n} w_{ij} v_j + t_i, \tag{5}$$

and the output states of the neurons are computed by

$$v_i(t+1) = \mathrm{sgn}(u_i(t+1)), \tag{6}$$

where sgn is the signum function (or bipolar binary function) defined by

$$\mathrm{sgn}(x) = \begin{cases} 1 & \text{if } x \geq 0 \\ -1 & \text{otherwise} \end{cases}. \tag{7}$$

Given initial input values for $u_i(0)(t = 0)$, the DHNN, which runs in the asynchronous mode, will always converge to a stable state if its diagonal elements of weight matrix W are nonnegative (that is $w_{ii} > 0$) [8].

Wu et al. [8] gave new convergence conditions for the DHNN. They proved that the DHNN can converge to a stable state or a stable cycle by rearranging the weight matrix W to even negative diagonal element. That is: Let $d_i = \min\{\sum_{j \neq i}^{n} w_{ij} v_j + t_i | \sum_{j \neq i}^{n} w_{ij} v_j + t_i \neq 0, v_i \in \{-1, 1\}\}$, if $w_{ii} > -d_i, i = 1, \ldots, n$, then the DHNN can converge to a stable state or a stable cycle. Furthermore, they pointed out that smaller diagonal elements may result in fewer stable states of a network, then reduce the possibility to be trapped in a state corresponding to a local minimum or poor quality solution and may approach a better solution.

In order to prevent the network to be trapped in a stable state corresponding to a local minimum, we introduce a nonlinear self-feedback term to the motion equation (6) of the DHNN as defined below [10]:

$$u_i(t+1) = \sum_{j=1}^{n} w_{ij} v_j + t_i + g(u_i(t) - u_i(t-1)). \tag{8}$$

The self-feedback term in Eq.(8) is switched on at $t = 1$, so that $u_i(0)$ is sufficient to initialize the system. In the Eq.(8), g is a nonlinear function as defined below:

$$g(x) = p_1 x \exp(-p_2|x|), \tag{9}$$

where p_1 and p_2 are adjustable parameters, which determine the chaotic dynamics of the function $g(x)$. When p_2 is fixed, the network can transfer from one minimum to another in a chaotic fashion at large p_1. Generally, chaotic transition among local minima can be obtained provided that p_1 is not too small and p_2 not too large [15][20].

For the nonlinear function disturbance, we apply a simulated annealing strategy to the parameter p_1 and the effect of the disturbance is gradually eliminated with the annealing process and finally the nonlinear function disturbance does not affect the evolution of DHNN, and then the neural network will converge to a stable state and be expected to get the optimal solution. The simulated annealing strategy is defined below:

$$p_1(t+1) = p_1(t) \times (1 - \beta) \tag{10}$$

where β is damping factor for p_1.

DHNN with the nonlinear self-feedback is called chaotic DHNN (CDHNN). The n vertices maximum cu problem can be mapped onto the CDHNN with n neuron representation. The output $v_i = 1$ means vertex i is assigned to one set, the $v_i = -1$ means vertex i is assigned to the other set. The energy function for the maximum cu problem is given by

$$E_{CUT} = -\frac{1}{2} \sum_{i=1}^{n} \sum_{j=1}^{n} e_{ij} \frac{1 - v_i v_j}{2}, \tag{11}$$

where e_{ij} is weight on the edge whose endpoints are vertex i and j. A stable state of the CDHNN corresponds to a good solution of the maximum cut problem.

2.4 Procedure of Memetic Algorithm (GPSO-CDHNN)

In this section, we attempt to incorporate the CDHNN, as a local search scheme into the GPSO algorithm (called GPSO-CDHNN) to improve its performance. The basic idea is as follows. Given the current solution S_1, the mechanism of the

GPSO leads the solution to an intermediate solution S_2. Then a CDHNN is applied to S_2 to reach a solution S_3. We represent a particle as $X = (v_1, v_2, \cdots, v_n)$, where $v_i = 1$ means that vertex i is assigned to one set, the $v_i = -1$ means vertex i is assigned to the other set. Therefore, we propose a reasonable procedure for GPSO-CDHNN as follows:

1. Initialize.
  1.1 Generate 1 particle using the CDHNN, and generate other $N - 1$ particles randomly.
2. Repeat until a given maximal number of iterations ($MaxIter$) is achieved.
  2.1 Evaluate the fitness of each particle using $F = -E_{CUT}$.
  2.2 Determine the best solution obtained so far by each particle.
  2.3 Determine the best solution obtained so far by the whole swarm.
  2.4 Update the potential solution using Eq.(1).
  2.5 Improve the solution quality of each particle using the CDHNN for every fixed number (say 50) of iterations.

At the beginning of the algorithm, we incorporate the CDHNN into the random initialization of the GPSO. Therefore the proposed algorithm can guarantee to obtain a solution no worse that that of the CDHNN because the CDHNN can generate a suboptimal solution rapidly. Moreover, the diversity of the initial population can be maintained to a certain extent because the other solutions are still generated randomly. At the same time, the solutions generated randomly can share information with the suboptimal solution generated by the CDHNN. In order to allow each particle to arrive at an intermediate solution, the local search should be implemented occasionally. Therefore, in the loop of the DPSO procedure, the CDHNN is carried out once for every fixed number (say 50) of iterations. In the memetic algorithm, the GPSO is applied to perform global exploration and the CDHNN is employed to perform locally oriented search (exploitation) for the solutions resulted by the GPSO. Furthermore, The CDHNN has strong ability to escape from local minima.

When the algorithm is terminated, the best solution obtained by the whole swarm, and the corresponding fitness value are output and considered as the maximum cut problem solution.

3 Simulation Results

The proposed algorithm was implemented in C on a DELL-PC (Pentium4 2.80 GHz). The parameters, $N = 10$, and $MaxIter = 200$ were used in the simulations. In the GPSO, the standard parameters are adopted from Ref. [18]: the value of w_1 is dynamically tuned from 0.9 to 0.4 according to the number of generations and $w_2 = 0.2w_1 + 0.8$. The bit mutation probability p_m is set to 0.001. The CDHNN is applied to all particles once for every 50 iterations.

In order to verify the proposed algorithm, we used a machine independent graph generator, **rudy**, created by G. Rinaldi [1] [8] [13]. We tested the maximum cut problem on the G set benchmark graph used by [5] [7] [13]. This set of

Table 1. Description of 15 maximum cut test problems

Graph	Size	
	$(\|V\|, \|E\|)$	Spars(%)
G11	(800,1600)	0.63
G12	(800,1600)	0.63
G13	(800,1600)	0.63
G14	(800,4694)	1.59
G15	(800,4661)	1.58
G20	(800,4672)	1.59
G21	(800,4667)	1.58
G22	(2000,19990)	0.15
G23	(2000,19990)	0.15
G24	(2000,19990)	0.15
G30	(2000,19990)	0.15
G31	(2000,19990)	0.15
G32	(2000,4000)	0.25
G33	(2000,4000)	0.25
G34	(2000,4000)	0.25

Table 2. Simulation results of 15 maximum cut problems

Graph	DSDP	Circuit	DHNN	SOCHOM		CDHNN		GPSO-CDHNN	
				Best	Av.	Best	Av.	Best	Av.
G11	542	554	560	534	518.64	**564**	561.58	**564**	**562.56**
G12	540	552	548	524	510.92	554	552.14	**556**	**554.4**
G13	564	572	574	548	530.86	**580**	577.52	**580**	**579.92**
G14	2922	3053	3024	3043	3030.63	3055	3042.72	**3058**	**3057.43**
G15	2938	3039	3013	3028	3011.92	3039	3023.99	**3047**	**3046.67**
G20	838	939	895	920	869.4	939	931.69	**940**	**939.58**
G21	841	921	881	893	855.69	921	908.12	**928**	**926.65**
G22	12960	13331	13167	13275	13202.4	13344	13325.8	**13346**	**13344.4**
G23	13006	13269	13157	13256	13203.6	13320	13293.3	**13323**	**13321.3**
G24	12933	13287	13140	13262	13204.3	13319	13303.2	**13329**	**13321.8**
G30	3038	3377	3200	3293	3215.63	3398	3377.63	**3405**	**3394.62**
G31	2851	3255	3111	3192	3132.64	3270	3256.52	**3293**	**3290.73**
G32	1338	1380	1378	1310	1286.34	1390	1377.02	**1392**	**1391.86**
G33	1330	1352	1354	1286	1258.16	1362	1351.54	**1368**	**1367.58**
G34	1334	1358	1356	1298	1267.38	1368	1354.8	**1370**	**1367.57**

problems, given by Table 1, becomes a standard test set for graph optimization. The sizes of the graphs are given as $(|V|, |E|)$ where $|V|$ denotes the number of vertices and $|E|$ denotes the number non-zero weights.

To evaluate the performance of the proposed algorithm, the results of proposed algorithm are compared with the results of DSDP method [8], one of the latest algorithms based on SDP relaxation, CirCuit method [9], a new heuristic algorithm for the maximum cut problem, the DHNN with negative diagonal [13],

Table 3. Computation time of the algorithm of DHNN, SOCHOM and the proposed algorithm (seconds)

Graph	DHNN	SOCHOM	CDHNN	GPSO-CDHNN
G11	1.31	7.24	2.06	58.46
G12	1.31	7.24	2.06	58.2
G13	1.31	7.24	2.06	57.4
G14	1.31	7.24	2.06	58.23
G15	1.31	7.24	2.06	60.88
G20	1.31	7.24	2.06	60.89
G21	1.31	7.24	2.06	60.23
G22	8.14	42.72	12.55	348.72
G23	8.14	42.72	12.55	361.03
G24	8.14	42.72	12.55	361.15
G30	8.14	42.72	12.55	349.95
G31	8.14	42.72	12.55	360.70
G32	8.09	42.61	12.28	347.59
G33	8.09	42.61	12.28	359.23
G34	8.09	42.61	12.28	360.24

the SOCHOM algorithm [12] and the CDHNN [15]. These algorithms are the most efficient algorithms for the maximum cut problem at present [13]. In these test problems, the weights of edges are all integers, therefore the values of the diagonal elements in the DHNN and CDHNN are reassigned to -0.4 [13]. The parameter, $\lambda = 50$, was used in the SOCHOM algorithm [12]. In the CDHNN, The parameters, $p_1(1) = 15$ (the self-feedback term is switched on at $t = 1$), $p_2 = 5$, and $\beta = 0.02$ were used.

Table 2 shows the simulation results for the G set graphs by six algorithms. The results in Table 2 are the best values obtained in 100 tests with random initial points. At the same time, we also listed the average results of SOCHOM, CDHNN and the proposed algorithm. From Table 2, we can find that the best results produced by the proposed algorithm are better than those produced by other algorithms. Furthermore, average results of the proposed algorithm are even better than the best results of other algorithms. In the proposed algorithm, all the average solutions are close or even equal to the best solutions, which shows that the proposed algorithm is of a certain robustness for the initial solutions. The reasons why the proposed algorithm is better than other algorithms are: (1) the stochastic nature of the GPSO enables the proposed algorithm to escape from local minima, (2) the local search algorithm, CDHNN with chaotic search dynamics, also has ability of escaping from local minima, and especially, (3) the proposed algorithm combines the local search method CDHNN into the global search method GPSO; therefore the proposed algorithm has good performance of exploration and exploitation. Moreover, the equivalence between quadratic zero-one optimization problem and maximum cut problem has been pointed out by Hammer [21], therefore, the proposed algorithm can be extended to other combinatorial optimization problems, for example, maximum clique problem [11].

Table 3 shows the real running time of the algorithm of DHNN, SOCHOM, CDHNN and the proposed algorithm (seconds) which all were implemented in C on a DELL-PC (Pentium4 2.80 GHz). The computation time of CDHNN is longer than that of DHNN because the CDHNN spent much time in chaotic searching for global or near-global minimum. The proposed algorithm requires more CPU computational time to perform evolutionary computations in GPSO phase and chaotic searching in CDHNN scheme. Although the proposed algorithm requires much more computational time than other algorithms, the other algorithms cannot achieve such a quality even with the same computational time. We think the moderate increase in computational time should not discourage practitioners from considering the method because it is possible to carry out the computations using high-speed computers.

4 Conclusions

In this paper, we incorporate a chaotic discrete Hopfield neural network (CDHNN), as a local search scheme, into the genetic particle swarm optimization (GPSO) and develop a memetic algorithm GPSO-CDHNN for the maximum cut problem. The proposed algorithm not only performs exploration by using the population-based evolutionary search ability of the GPSO, but also performs exploitation by using the CDHNN. Simulation results show that the proposed algorithm has superior ability for maximum cut problem. Furthermore, the proposed algorithm can be extended to other combinatorial optimization problems.

Acknowledgments. The Project was supported by the Scientific Research Foundation for Outstanding Young Teachers, Sun Yat-sen University.

References

1. Karp, R.M.: Reducibility Among Combinatorial Problems. Complexity of Computer Computations, Plenum, New York, pp. 85–104 (1972)
2. Barahona, F., Grotschel, M., Junger, M., Reinelt, G.: A Application of Combinatorial Optimization to Statistical Physics and Circuit Layout Design. Operations Research 36(3), 493–513 (1988)
3. Chang, K., Du, D.: Efficient algorithm for the Layer Assignment Problem. IEEE Transactions on Computer-Aided Design of Integrated Circuits and Systems CAD-6(1), 67–78 (1987)
4. Bertoni, A., Campadelli, P., Grossi, G.: An Approximation Algorithm for the Maximum Cut Problem and Its Experimental Analysis. In: Proceedings of algorithms and experiments, Trento, Italy, pp. 137–143 (1998)
5. Goemans, M.X., Williamson, D.P.: Improved Approximation Algorithms for Maximum Cut and Satisfiability Problems Using Semidefinite Programming. Journal of ACM 42(6), 1115–1145 (1995)
6. Helmberg, C., Rendl, F.: A Spectral Bundle Method for Semidefinite Programming. SIAM Journal on Optimization 10(3), 673–696 (2000)

7. Benson, S., Ye, Y., Zhang, X.: Solving Large Scale Sparse Semidefinite Programs for Combinatorial Optimization. SIAM Journal on Optimization 10(2), 443–461 (2000)
8. Choi, C., Ye, Y.: Solving Sparse Semidefinite Programs Using the Dual Scaling Algorithm with an Iterative Solver. Working paper, Department of Management Science, University of Iowa, Iowa (2000)
9. Burer, S., Monteiro, R.D.C., Zhang, Y.: Rand-Two Relaxation Heuristics for Maximum Cut and Other Binary Quadratic Programs. Technical Report TR00-33, Department of Computational and Applied Mathematics, Rice University, Texas (2000)
10. Galán-Marín, G., Muñoz-Pérez, J.: Design and Analysis of Maximum Hopfield Networks. IEEE Trans. Neural Networks 12(2), 329–339 (2001)
11. Galán-Marín, G., Mérida-Casermeiro, E., Muñoz-Pérez, J.: Modelling competitive Hopfield networks for the maximum clique problem. Computer & Operations Research 30(4), 603–624 (2003)
12. Wang, J., Tang, Z., Cao, Q., Wang, R.: Optimal Competitive Hopfield Network with Stochastic Dynamics for Maximum Cut Problem. International Journal of Neural Systems 14(4), 257–265 (2004)
13. Wu, L.Y., Zhang, X.S., Zhang, J.L.: Application of Discrete Hopfield-Type Neural Network for Max-Cut Problem. In: Proceedings of 8th International Conference on Neural Information Processing, pp. 1439–1444. Fudan University Press, Shanghai (2001)
14. Hopfield, J.J., Tank, D.W.: Neural Networks and Physical Systems with Emergent Collective Computational Abilities. Proceedings of the National Academy of Sciences, USA 79(8) (1982)
15. Wang, J., Tang, Z.: An Improved Optimal Competitive Hopfield Network for Bipartite Subgraph Problems. Neurocomputing 61C, 413–419 (2004)
16. Kennedy, J., Eberhart, R.C.: Particle Swarm Optimization. In: Proceedings of IEEE International Conference on Neural Networks, pp. 1942–1948. Piscataway, NJ (1995)
17. Kennedy, J., Eberhart, R.C.: A Discrete Binary Version of the Particle Swarm Algorithm. In: Proceedings of the World Multiconference on Systemics, Cybernetics and Informatics, pp. 4104–4109. Piscatawary, NJ (1997)
18. Yin, P.Y.: Genetic Particle Swarm Optimization for Polygonal Approximation of Digital Curves. Pattern Recognition and Image Analysis 16(2), 223–233 (2006)
19. Parsopoulos, K.E., Vrahatis, M.N.: Recent Approaches to Global Optimization Problems Through Particle Swarm Optimization. Natural Computing 1(2–3), 235–306 (2002)
20. Zhou, C.S., Chen, T.L.: Chaotic Annealing for Optimization. Physical Review E 55(3), 2580–2587 (1997)
21. Hammer, P.L.: Some Network Flow Problems Solved with Pseudo-Boolean Programming. Operation research 13, 388–399 (1965)

A Novel Watermarking Scheme Based on PSO Algorithm

Ziqiang Wang, Xia Sun, and Dexian Zhang

School of Information Science and Engineering, Henan University of Technology, Zheng Zhou 450052,China
wzqagent@126.com

Abstract. In this paper, a novel blind watermark extracting scheme using the Discrete Wavelet Transform (DWT) and Particle Swarm Optimization(PSO) algorithm is introduced. The watermark is embedded to the discrete multiwavelet transform (DMT) coefficients larger than some threshold values, and watermark extraction is efficiently performed via particle swarm optimization algorithm.The experimental results show that the proposed watermarking scheme results in an almost invisible difference between the watermarked image and the original image, and is robust to common image processing operations and JPEG lossy compression.

1 Introduction

The success of the Internet allows for the prevalent distribution of multimedia data in an effortless manner. Due to the open environment of Internet downloading, copyright protection introduces a new set of challenging problems regarding security and illegal distribution of privately owned images. One solution to these problems is digital watermarking, i.e., the insertion of information into the image data in such a way that the added information is not visible and yet resistant to image alterations. A variety of techniques has already been proposed; an overview of the subject can be found in[1].

A wide variety of image watermarking scheme has been proposed in recent years.Such techniques can be broadly classified in two categories: spatial-domain and transform-domain based.First generation watermarking techniques typically embed a secret message or label bit string into an image via characteristic pseudorandom noise patterns. These noise patterns can be generated and added in the either spatial,Fourier, DCT, or wavelet domain.Commonly the amplitudes of the noise patterns are made dependent on the local image content as so trade-off the perceptual image degradation due to the noise and the robustness of the embedded information against image processing-based attacks. Other approaches use a particular order of discrete cosine transform (DCT) coefficients to embed the watermark.More recent approaches use salient geometric image properties in fractal compression algorithm to embed the watermark[2].

It is possible to state that the most important features a watermarking technique to be used for intellectual property rights protection should exhibit are

K. Li et al. (Eds.): LSMS 2007, LNCS 4688, pp. 307–314, 2007.

unobtrusiveness and robustness: in practice, it is required that a signal is accurately hidden into image data in such a way to result very difficult to be perceived but also very difficult to be removed. Another important characteristic is blindness, i.e., the watermark decoder must not require the original nonwatermarked image for extracting the embedded code.

It is today widely accepted that robust image watermarking techniques should largely exploit the characteristics of the HVS[3], for more effectively hiding a robust watermark,A widely used technique exhibiting a strong similarity to the way the HVS processes images is the Discrete Wavelet Transform (DWT). As a matter of fact, the next-generation image coding standard JPEG2000 will strongly rely on DWT for obtaining good quality images at low coding rates.In addition, the particle swarm optimization (PSO) algorithm[4], which has emerged recently as a new meta-heuristic derived from nature, has attracted many researchers' interests. The algorithm has been successfully applied to several minimization optimization problems and neural network training. Nevertheless, the use of the algorithm for image watermarking is still a research area where few people have tried to explore.

In this paper, a novel blind watermark extracting scheme using the Discrete Wavelet Transform (DWT) and Particle Swarm Optimization(PSO) algorithm is introduced. The watermark is embedded to the discrete multiwavelet transform (DMT) coefficients larger than some threshold values, and watermark extraction is efficiently performed via particle swarm optimization algorithm.The experimental results show that the proposed watermarking scheme results in an almost invisible difference between the watermarked image and the original image, and is robust to common image processing operations and JPEG lossy compression.

The remainder of this paper is organized as follows. In the next section, we present the basic idea and key techniques of the PSO algorithm. In section 3,The watermark embedding algorithm in the wavelet domain is described, In section 4,the PSO-based watermarking extraction method is proposed. Experimental results are given in Section 5. Finally, the paper ends with conclusions and future research directions.

2 The Particle Swarm Optimization(PSO) Algorithm

The PSO is a population-based stochastic optimization method first proposed by Kennedy and Eberhart [4]. PSO technique finds the optimal solution using a population of particles. Each particle represents a candidate solution to the problem. PSO is basically developed through simulation of bird flocking in two-dimensional space. Some of the attractive features of the PSO include ease of implementation and the fact that no gradient information is required. It can be used to solve a wide array of different optimization problems; some example applications include neural network training and function minimization.

The PSO is a population based optimization technique, where the population is called a swarm. A simple explanation of the PSO's operation is as follows. Each particle represents a possible solution to the optimization task.During each iteration each particle accelerates in the direction of its own personal best solution found so far, as well as in the direction of the global best position discovered so far by any of the particles in the swarm. This means that if a particle discovers a promising new solution, all the other particles will move closer to it, exploring the region more thoroughly in the process.

Let s denote the swarm size.Each individual particle $i(1 \leq i \leq s)$ has the following properties: a current position x_i in search space, a current velocity v_i, and a personal best position p_i in the search space, and the global best position p_{gb} among all the p_i.During each iteration, each particle in the swarm is updated using the following equation .

$$v_i(t+1) = k[w_i v_i(t) + c_1 r_1 (p_i - x_i(t)) + c_2 r_2 (p_{gt} - x_i(t))] \quad (1)$$

$$x_i(t+1) = x_i(t) + v_i(t+1) \quad (2)$$

where c_1 and c_2 denote the acceleration coefficients, and r_1 and r_2 are random numbers uniformly distributed within [0,1].

The value of each dimension of every velocity vector v_i can be clamped to the range $[-v_{max}, v_{max}]$ to reduce the likelihood of particles leaving the search space. The value of v_{max} chosen to be $k \times x_{max}$(where $0.1 \leq k \leq 1$).Note that this does not restrict the values of x_i to the range $[-v_{max}, v_{max}]$.Rather than that, it merely limits the maximum distance that a particle will move.

Acceleration coefficients c_1 and c_2 control how far a particle will move in a single iteration. Typically, these are both set to a value of 2.0, although assigning different values to c_1 and c_2 sometimes leads to improved performance.The inertia weight w in Equation (6) is also used to control the convergence behavior of the PSO.Typical implementations of the PSO adapt the value of w linearly decreasing it from 1.0 to near 0 over the execution. In general, the inertia weight w is set according to the following equation[5]:

$$w_i = w_{max} - \frac{w_{max} - w_{min}}{iter_{max}} \cdot iter \quad (3)$$

where $iter_{max}$ is the maximum number of iterations, and $iter$ is the current number of iterations.

In order to guarantee the convergence of the PSO algorithm, the constriction factor k is defined as follows:

$$k = \frac{2}{|2 - \varphi - \sqrt{\varphi^2 - 4\varphi}|} \quad (4)$$

where $\varphi = c_1 + c_2$ and $\varphi > 4$.

3 The Watermark Embedding Algorithm

The watermark insertion is performed in the discrete multiwavelet transform (DMT) domain by applying the three-level multiwavelet decomposition based on the well-known DGHM multiwavelet from [6] and optimal prefilters from [7].The reason is that DGHM multiwavelets simultaneously possess orthogonality, compact support, an approximation order of 2, and symmetry. Since the approximation subband contains the high-energy components of the image, we do not embed the watermark in this subband to avoid visible degradation ofwatermarked image. Furthermore, thewatermark is not embedded in the subbands of the finest scale due to the low-energy components to increase the robustness of the watermark. In other subbands, we choose all DMT coefficients that are greater than the embedding threshold T . These coefficients are named V_i and applied to the following equation:

$$V_i^w = V_i + \beta(f_1, f_2)|V_i|x_i \tag{5}$$

where i runs over all DWT coefficients $> T$, V_i denotes the corresponding DWT coefficients of the original image and V_i^w denotes the coefficients of the watermarked image.The variable x_i denotes the watermark signal which is generated from a Gaussian distribution with zero mean and unit variance, and $\beta(f_1, f_2)$ is the embedding strength used to control the watermark energy to be inserted.The watermarking algorithm is adaptive by making use of human visual system (HVS) characteristics,which increase robustness and invisibility at the same time. The HVS $\beta(f_1, f_2)$ can be represented by [8]:

$$\beta(f_1, f_2) = 5.05e^{-0.178(f_1+f_2)}(e^{0.1(f_1+f_2)} - 1) \tag{6}$$

where f_1 and f_2 are the spatial frequencies (cycles/visual angle). However,the watermark will be spatially localized at high-resolution levels of the host image. By this,the watermark will be more robust. At the end,the inverse DWT is applied to form the watermarked image.

4 The PSO-Based Watermark Detection Algorithm

The PSO-based watermark detection algorithm consists of two steps. First,the suspected image is decomposed into three levels using the DMT. We choose all the DMT coefficients greater than the detection threshold T from all subbands.Then,the watermark is extracted in the above selected DMT coefficients. The PSO realizes efficient watermark extraction by selecting the best particle and updating other particle's velocity and position to generate optimal solutions.

The PSO algorithm process is initialized with a group of random particles (solutions), N. The ith particle is represented by its position as a point in a S-dimensional space, where S is the number of variables. Throughout the process, each particle i monitors three values: its current position $X_i = (x_{i1}, x_{i2}, ..., x_{iS})$; the best position it reached in previous cycles $P_i = (p_{i1}, p_{i2}, ..., p_{iS})$; its flying velocity $V_i = (v_{i1}, v_{i2}, ..., v_{iS})$. In each time interval (cycle), the position (P_g) of the best particle (g) is calculated as the best fitness of all particles. Accordingly, each particle updates its velocity V_i to catch up with the best particle g, as follows:

$$V_i(t+1) = \omega \times V_i(t) + c_1 r_1(P_i - X_i(t)) + c_2 r_2(P_g - X_i(t))] \tag{7}$$

$$X_i(t+1) = X_i(t) + V_i(t); V_{max} \geq V_i \geq -V_{max} \tag{8}$$

where c_1 and c_2 are two positive constants named learning factors (usually $c_1 = c_2 = 2$); r_1 and r_2 are two random functions in the range [0, 1], V_{max} is an upper limit on the maximum change of particle velocity , and ω is an inertia weight employed as an improvement proposed by Shi and Eberhart [9] to control the impact of the previous history of velocities on the current velocity. The operator ω plays the role of balancing the global search and the local search; and was proposed to decrease linearly with time from a value of $1.4 - 0.5$ [9]. As such, global search starts with a large weight and then decreases with time to favor local search over global search.

In order to evaluate the feasibility of extracted watermark, normalized correlation (NC) is used. The number of mismatched data between the inserted and the extracted watermarks is used to represent the similarity of watermarks. NC for valid watermarks, which represents the characteristics of the extracted watermark, is defined as

$$NC = \frac{\sum w_{x,y} w'_{x,y}}{\sum w_{x,y}^2} \tag{9}$$

where w represents the inserted watermark, w' the extracted watermark.The NC value can be anywhere between 0 and 1. The closer the NC value is to 1, the higher the accuracy is for the recovered watermark.The NC for random noise is about 0.5 and possibility of distinguishing extracted logo is higher than $0.7 \sim 0.8$ NC.

The steps of the PSO-based watermark detection algorithm are described as follows.

Step1:Generate random population of N solutions (particles);

Step2:Watermark-inserted populations are transformed and applied the 3-level DWT to the restored image.

Step3:For each individual $i \in N$, calculate its fitness (i) according to Eq.(9);

Step4:Initialize the value of the weight factor ω;

Step5:For each particle;

Step5.1:Set *pBest* as the best position of particle i;

Step5.2:If fitness (i) is better than *pBest*;

Step5.3:$pBest(i) = fitness(i)$;

Step5.4: End;

Step6:Set *gBest* as the best fitness of all particles;

Step7:For each particle;

Step7.1:Calculate particle velocity according to Eq.(7);

Step7.2:Update particle position according to Eq.(8);

Step7.3:End;

Step8:Update the value of the weight factor ω according to Eq.(3);

Step9:Check if NC is larger than a predetermined value, the extracted watermark is decided to be feasible, and the algorithm terminates. Otherwise go back to step3.

5 Experimental Results

The algorithm has been extensively tested on standard image 'Lena' and attempting different kinds of attacks: in this section some of the most significant results will be shown. For the experiments presented in the following, the Daubechies-6 filtering kernel has been used for computing the DWT. To estimate the quality of our method, we used the peak signal to noise ratio (PSNR) to evaluate the distortion of the watermarked image.

$$PSNR = 10log_{10}\frac{255^2}{MSE}(db) \tag{10}$$

$$MSE = \frac{1}{m \times n}\sum_{i=0}^{m-1}\sum_{j=0}^{n-1}(X_{ij} - X_{ij}^{'})^2 \tag{11}$$

where X_{ij} and $X_{ij}^{'}$ represent the pixel value of original image and the attacked image respectively.

The accuracy ratio (AR) is used to evaluate the similarity between the original watermark and the extracted one [10,11], which is defined as follows:

$$AR = \frac{CB}{NB} \tag{12}$$

where NB is the number of the original watermark bits, and CB is the number of correct bits between the original watermark and the extracted one. The more close to 1 the AR value is, the more like the original watermark the extracted one is.

First, watermark invisibility is evaluated: in Fig. 1 , the original "Lena" image is presented, while in Fig. 2, the watermarked copy is shown: the images are evidently undistinguishable, thus proving the effectiveness of PSO-based DWT watermarking scheme.

In addition,a good watermark technique should be robust to withstand different kind of attacks. In the following experiments, several common attacks are used to measure the robustness of the proposed scheme,such as JPEG compression, sharpening, blurring and cropping,etc.The detailed experimental results are shown in Table 1.we can see that the experimental results are acceptable.

Fig. 1. Original image "Lena"

Fig. 2. The watermarked image "Lena"

Table 1. The experimental results under different attacks

Attack–Scheme	PSNR Values	AR Values
JPEG attack	30.4	1
Scaling attack	22.16	1
noise adding attack	11.39	1
blurring attack	28.16	1
sharpening attack	18.25	1

Table 2. The comparison results with several existing schemes

Attack–Scheme	PSO-Based Scheme	Ref.[10] Scheme	Ref.[11] Scheme
JPEG (quality factor)	1 (QF=10)	0.939 (QF=75)	0.91 (QF=80)
Scaling (reduced 1/4)	1	0.923	0.883
Sharpening	1	0.950	0.895
Blurring	1	0.908	0.887
Cropping (cropped 10%)	1	0.975	0.959

Finally, the proposed scheme is compared with several existing schemes. Table 2 shows the experimental results. It is clear that the proposed scheme has higher robustness compared with Hsieh–Huang's [10] scheme and Chang–Chung's [11] scheme.

6 Conclusions

The PSO algorithm,new to the image watermarking,is a robust stochastic evolutionary algorithm based on the movement and intelligence of swarms.In this paper, a PSO-based algorithm for image watermarking is presented. The experimental results show that the proposed watermarking scheme results in an almost invisible difference between the watermarked image and the original image, and is robust to common image processing operations and JPEG lossy compression.

References

1. Hartung, F., Kutter, M.: Multimedia watermarking techniques. Proceedings of the IEEE 87, 1079–1107 (1999)

2. Rongen, P.M.J., Maes, M.J., van Overveld, K.W.: Digital image watermarking by salient point modification: practical results. In: Proceedings of SPIE Security and Watermarking of Multimedia Contents, vol. 3657, pp. 273–282 (1999)
3. Jayant, N., Johnston, J., Safranek, R.: Signal compression based on models of human perception. Proceedings of the IEEE 81, 1385–1422 (1993)
4. Eberhart, R., Kennedy, J.: A new optimizer using particle swarm theory. In: Proceedings of the Sixth International Symposium on Micro Machine and Human Science, Nagoya, Japan, pp. 39–43 (1995)
5. Kennedy, J.: The particle swarm: social adaptation of knowledge. In: Proceedings of 1997 IEEE International Conference on Evolutionary Computation, Indianapolis, IN, USA, pp. 303–308. IEEE Computer Society Press, Los Alamitos (1997)
6. Geronimo, J.S., Hardin, D.P., Massopust, P.R.: Fractal functions and wavelet expansions based on several scaling functions. Journal of Approximation Theory 78, 373–401 (1994)
7. Attakitmongcol, K., Hardin, D.P., Wilkes, D.M.: Multiwavelet prefilters II: optimal orthogonal prefilters. IEEE Transactions on Image Processing 10, 1476–1487 (2001)
8. Clark, R.: An introduction to JPEG 2000 and watermarking. In: IEE Seminar on Secure Images and Image Authentication, London, UK, pp. 3–6 (2000)
9. Shi, Y., Eberhart, R.: A modified particle swarm optimizer. In: Proceedings of the 1998 IEEE International Conference on Evolutionary Computation, Anchorage, AK, USA, pp. 69–73. IEEE Computer Society Press, Los Alamitos (1998)
10. Hsieh, S.-L., Huang, B.-Y.: A copyright protection scheme for gray-level images based on image secret sharing and wavelet transformation. In: Proceedings of International Computer Symposium, pp. 661–666 (2004)
11. Chang, C.-C., Chung, J.-C.: An image intellectual property protection scheme for gray-level images using visual secret sharing strategy. Pattern Recognition Letters 23, 931–941 (2002)

On the Problem of Group Decision Making Based on Intuitionistic Fuzzy Judgment Matrices

Zaiwu Gong

College of Economics and Management and China Institute for Manufacture Developing, Nanjing University of Information Science and Technology, Nanjing 210044, China
zwgong26@163.com

Abstract. The problem of group decision making based on intuitionistic fuzzy judgment matrix is investigated. Approaches to intuitionistic fuzzy group decision making are proposed from three different preference views. Using the operations of intuitionistic fuzzy values, the ranking method of intuitionistic fuzzy judgment matrix is given. It is illustrated by a numerical example that the approaches proposed are in accord with the rules of group decision making.

1 Introduction

Group decision making can be defined as aggregating all the individual preferences to collective preference and ranking all the alternatives from best to worst. How to settle conflict of different individual preferences, and how to synthesize the different individual preferences to an unanimous approval, is one of the most important problems that must be given research in group decision making [1,2,3]. There exist plenty of imprecise and uncertainty problems in multi-attribute decision makings. These kinds of problems can be dealt with by the fuzzy set theory introduced by Zadeh [4]. Up to now, the fuzzy set theory has developed a comprehensive theory system and system method in dealing with fuzzy information, and has been widely used in decision making research. In 1980's, Atanassov [5] extended and developed the classical fuzzy set theory into intuitionistic fuzzy set theory which has also been applied to the fields of relationship operating, logic reasoning and pattern recognizing, etc [6,7,8]. And moreover, it has taken more serious to the study of intuitionistic fuzzy set theory in the field of decision making. For instances, in literature [9], the method to measure the degree of consensus of collective preference by means of intuitionistic fuzzy group decision theory is put forward; the similarity approach using intuitionistic fuzzy group decision theory to judge whether the opinions of collective are unanimous is also proposed [10]. But how to aggregate the different individual preferences into the collective preference, and how to rank all the alternatives from best to worst, has been given little research in these literatures.

In this text, approaches to aggregate the collective preferences in intuitionistic fuzzy group decision making are proposed from three different preference views.

K. Li et al. (Eds.): LSMS 2007, LNCS 4688, pp. 315 – 325, 2007.

Using the operations of intuitionistic fuzzy numbers, the ranking method of intuitionistic fuzzy judgment matrix is put forward. At last, a numerical example is given.

2 Some Concepts

Definition 1. Let $X = \{x_1, x_2, \dots, x_n\}$ be an ordinary finite nonempty set. An intuitionistic fuzzy set in X is an expression given by $A = \{< x, \mu_A(x), v_A(x) > | x \in X\}$, where $\mu_A : x \mapsto [0,1], v_A : x \mapsto [0,1]$ denote the membership degree and non-membership degree of the element x in A, respectively, and for all x in X, $0 \le \mu_A(x) + v_A(x) \le 1$.

For each finite intuitionistic fuzzy set in X, $\pi_A(x) = 1 - \mu_A(x) - v_A(x)$ is called a hesitation margin (or an intuitionistic fuzzy index) of $x \in A$, and it is a hesitation degree of whether x belongs to A or not. It is obvious that $0 \le \pi_A(x) \le 1$ for each $x \in A$. If $\pi_A(x) = 0$, then $\mu_A(x) + v_A(x) = 1$, which denotes that the intuitionistic fuzzy set A degenerated to the classic fuzzy set $A = \{< x, \mu_A(x) > | x \in X\}$. In consequence, the intuitionistic fuzzy set is the generalization and development of the classic fuzzy set.

A physics depiction of the intuitionistic fuzzy set: In a voting, there are 10 voters, of which 5 persons vote for, 2 persons vote against and the rest abstain or give invalid votes. We can express this by an intuitionistic fuzzy set $A = \{< x, 0.5, 0.2 > | x \in X\}$, where X is the set of voters.

By definition 1, we can see that the degree of element x in set A is bounded to a subinterval $[\mu_A(x), 1 - v_A(x)] \subseteq [0,1]$, which is called the intuitionistic fuzzy value in literature [12], meanwhile, some operations of the intuitionistic fuzzy values are developed:

Let $\tilde{a}_i = [\mu_i, 1 - v_i], i \in N = \{1, 2, \dots, n\}$ be any n intuitionistic fuzzy values, then

$$\tilde{a}_i \oplus \tilde{a}_j = [\mu_i + \mu_j - \mu_i \mu_j, 1 - v_i v_j] \tag{1}$$

$$\lambda \tilde{a}_i = [1 - (1 - \mu_i)^\lambda, 1 - v_i^\lambda] \tag{2}$$

$$\tilde{a}_i \le \tilde{a}_j \Leftrightarrow \mu_i \le \mu_j, v_i \ge v_j; \tilde{a}_i \ge \tilde{a}_j \Leftrightarrow \mu_i \ge \mu_j, v_i \le v_j \tag{3}$$

By (1) and (2), we can easily get the following equation:

$$\lambda(\tilde{a}_1 \oplus \cdots \oplus \tilde{a}_n) = \lambda \tilde{a}_1 \oplus \cdots \oplus \lambda \tilde{a}_n = \left[1 - \prod_{i=1}^{n} (1 - \mu_i)^\lambda,\ 1 - \prod_{i=1}^{n} v_i^\lambda\right] \tag{4}$$

where $\lambda > 0$, and if $\lambda = \frac{1}{n}$, the equation (4) is called the arithmetic average of the intuitionistic fuzzy values.

For the convenience of discussing, we also call $(\mu_A(x), v_A(x), \pi_A(x))$ an intuitionistic fuzzy value in this text. So $\tilde{a}_i = [\mu_i, 1 - v_i]$ can be also denoted by $\tilde{a}_i = (\mu_i, v_i, \pi_i)$, $i \in N$, where $\pi_i = 1 - v_i - \mu_i$. The equivalent forms of equation (1-4) are as follows:

$$\tilde{a}_i \oplus \tilde{a}_j = (\mu_i + \mu_j - \mu_i\mu_j, v_i v_j, 1 - \mu_i - \mu_j + \mu_i\mu_j - v_i v_j) \quad (5)$$

$$\lambda \tilde{a}_i = (1 - (1 - \mu_i)^\lambda, v_i^\lambda, (1 - \mu_i)^\lambda - v_i^\lambda) \quad (6)$$

$$\tilde{a}_i \le \tilde{a}_j \Leftrightarrow \mu_i \le \mu_j, v_i \ge v_j; \tilde{a}_i \ge \tilde{a}_j \Leftrightarrow \mu_i \ge \mu_j, v_i \le v_j \quad (7)$$

$$\lambda(\tilde{a}_1 \oplus \cdots \oplus \tilde{a}_n) = (1 - \prod_{i=1}^{n}(1 - \mu_i)^\lambda, \prod_{i=1}^{n} v_i^\lambda, \prod_{i=1}^{n}(1 - \mu_i)^\lambda - \prod_{i=1}^{n} v_i^\lambda) \quad (8)$$

If $\lambda = \frac{1}{n}$, we also called the equation (8) the arithmetic average of the intuitionistic fuzzy values.

Let $S = \{s_1, \ldots, s_n\}$ be the alternative set, an intuitionistic fuzzy preference relation in S can be defined as $R' = \{< (s_i, s_j), \mu_{R'}(s_i, s_j), v_{R'}(s_i, s_j) > | (s_i, s_j) \in S \times S\}$, where $\mu_{R'} : S \times S \to [0,1]$, $v_{R'} : S \times S \to [0,1]$, $\mu_{R'}(s_i, s_j)$ is the degree to which the alternative s_i is preferred to s_j, and $v_{R'}(s_i, s_j)$ is the degree to which the alternative s_i is not preferred to s_j [9]. Moreover, the inequality $0 \le \mu_{R'}(s_i, s_j) + v_{R'}(s_i, s_j) \le 1$ holds for every $(s_i, s_j) \in S \times S$.

Based on the definition of fuzzy judgment matrix (or fuzzy preference relation), the definition of intuitionistic fuzzy judgment matrix (it is also called intuitionistic fuzzy preference relation [10,11], but only to distinguish the concept given above) is given as follows:

Definition 2. Let R' be the intuitionistic fuzzy preference relation in S, for any $i, j \in N$, $r_{ij} = \mu_{R'}(s_i, s_j)$, $t_{ij} = v_{R'}(s_i, s_j)$, if the following conditions

$$r_{ii} = 0.5, t_{ii} = 0.5; r_{ij} = t_{ji}, t_{ij} = r_{ji}; 0 \le r_{ij} + t_{ij} \le 1.$$

hold, then $R = (r_{ij})_{n \times n}$ and $T = (t_{ij})_{n \times n}$ are called intuitionistic fuzzy judgment matrices (intuitionistic preference relations).

The matrix below is also called an intuitionistic fuzzy judgment matrix for the convenience of discussing.

Definition 3. For any $i, j \in N$, if the judgment matrix $M = (m_{ij})_{n \times n} = (r_{ij}, t_{ij}, \pi_{ij})_{n \times n}$ satisfies $m_{ii} = (0.5, 0.5, 0); r_{ij} = t_{ji}, t_{ij} = r_{ji}, \pi_{ij} = \pi_{ji}; r_{ij} + t_{ij} + \pi_{ij} = 1,$ where $0 \le r_{ij} \le 1$, $0 \le t_{ij} \le 1$, $0 \le \pi_{ij} \le 1$, then the matrix is called an intuitionistic fuzzy judgment matrix. ($\pi_{ij} = \pi_{ji}$ implies that the hesitation degree to which the alternative s_i being preferred to s_j and the hesitation degree to which the alternative s_j being preferred to s_i is equivalent.)

Definition 4. If we consider $\tilde{a}_i = (\mu_i, v_i, \pi_i), \pi_i = 1 - v_i - \mu_i, i \in N$ as vectors of 3-dimension (In the rest of this text we also regard the intuitionistic fuzzy values as vectors which makes it easier for us to discuss the problems), then cosine of the angle between the vector $\tilde{a}_i$ and the vector $\tilde{a}_j$

$$C(\tilde{a}_i, \tilde{a}_j) = \frac{\mu_i \mu_j + v_i v_j + \pi_i \pi_j}{(\mu_i^2 + v_i^2 + \pi_i^2)^{\frac{1}{2}} (\mu_j^2 + v_j^2 + \pi_j^2)^{\frac{1}{2}}}$$

can be defined as the correlation degree between the intuitionistic fuzzy values $\tilde{a}_i$ and $\tilde{a}_j$. It is obvious that $0 \le C(\tilde{a}_i, \tilde{a}_j) \le 1$, and the greater $C(\tilde{a}_i, \tilde{a}_j)$, the larger the correlation degree between the two values. (Note: We can also use $C^2(\tilde{a}_i, \tilde{a}_j) = C(\tilde{a}_i, \tilde{a}_j) \times C(\tilde{a}_i, \tilde{a}_j)$ to denote the correlation degree between $\tilde{a}_i$ and $\tilde{a}_j$.)

Definition 5. If we transform the vector $\tilde{a}_i, \tilde{a}_j$ to unit vectors, that is

$$\tilde{a}_i' = \frac{(\mu_i, v_i, \pi_i)}{(\mu_i^2 + v_i^2 + \pi_i^2)^{\frac{1}{2}}}; \tilde{a}_j' = \frac{(\mu_j, v_j, \pi_j)}{(\mu_j + v_j + \pi_j)^{\frac{1}{2}}}.$$

then the equations $C(\tilde{a}_i, \tilde{a}_j) = C(\tilde{a}_i', \tilde{a}_j')$ and $C^2(\tilde{a}_i, \tilde{a}_j) = C^2(\tilde{a}_i', \tilde{a}_j')$ also hold.

Let $D(\tilde{a}_i, \tilde{a}_j) = [(\mu_i - \mu_j)^2 + (v_i - v_j)^2 + (\pi_i - \pi_j)]^{\frac{1}{2}}$ denotes the Euclidean distance between $\tilde{a}_i$ and $\tilde{a}_j$, then $S(\tilde{a}_i, \tilde{a}_j) = 1 - \frac{1}{3} D^2(\tilde{a}_i, \tilde{a}_j)$ denotes the similarity degree between $\tilde{a}_i$ and $\tilde{a}_j$. Obviously, the greater $S(\tilde{a}_i, \tilde{a}_j)$, the larger similarity degree between $\tilde{a}_i$ and $\tilde{a}_j$.

The correlation degree and the similarity degree both can be regarded as a measure of the consistent degree between any two intuitionistic fuzzy values.

3 Group Decision Making Model of Intuitionistic Fuzzy Judgment Matrices

Generally speaking, due to the conflict of opinion or interest of each decision maker (DM), the aim of group decision is to reach a compromise or a consensus. The compromise or the consensus can be determined by a standard that all the DMs agree with. In fact, this standard is the common preference of the DMs. In the following, we will show how to aggregate the different preferences in three different kinds of formats.

- Intuitionistic fuzzy arithmetic average.
- From the point of correlation, maximize the consistency of the DMs' preferences.
- From the point of similarity, maximize the consistency of the DMs' preferences.

Let X be a decision-making problem, $S=\{s_1,s_2,\cdots,s_n\}$ be an alternative set, and $D=\{d_1,d_2,\cdots,d_m\}$ be the DMs set. Under a given criterion, the intuitionistic fuzzy values given by $d_i, i\in M=\{1,2,\ldots,m\}$ are $R_i=(\mu_i,v_i,\pi_i)$, where

$$\mu_i+v_i+\pi_i=1, 0\le\mu_i\le 1, 0\le v_i\le 1, 0\le\pi_i\le 1.$$

3.1 The Arithmetic Average Aggregation Method

Let $R_*=f(R_1,R_2,\cdots,R_m)$ be the unanimous opinion, where $R_i, i\in M$ is the intuitionistic fuzzy estimation value of each DM. Let function f denotes the arithmetic average of m DMs' preferences, then the result of aggregation is as follows:

$$R_*=f(R_1,R_2,\cdots,R_m)=\frac{1}{m}R_1\oplus\frac{1}{m}R_2\oplus\cdots\oplus\frac{1}{m}R_m=\frac{1}{m}(R_1\oplus R_2\oplus\cdots\oplus R_m)$$

$$=(1-\prod_{i=1}^{m}(1-\mu_i)^{\frac{1}{m}},\prod_{i=1}^{m}v_i^{\frac{1}{m}},\prod_{i=1}^{m}(1-\mu_i)^{\frac{1}{m}}-\prod_{i=1}^{m}v_i^{\frac{1}{m}}).$$

3.2 The Optimal Aggregation Method Based on Correlation

As we have mentioned in previous, the essence of decision-making process is to reach a compromise or a consensus. Therefore, we must find an "ideal" estimation value presented by an "ideal" DM: the consistent degree between the "ideal" estimation value and the estimation value of each DM is larger than the consistent degree between any two estimation values. This consistent degree can be measured by the correlation method.

Supposed that the "ideal" estimate value $R_*=(\mu_*,v_*,\pi_*)$ is given by the "ideal" DM. Obviously, the following equation holds.

$$max\frac{1}{m}\sum_{i=1}^{m}C^2(R,R_i)=\frac{1}{m}\sum_{i=1}^{m}C^2(R_*,R_i) \tag{9}$$

where $R=(\mu,v,\pi)$ is an intuitionistic fuzzy value.

In order to get the optimal value $R_*=(\mu_*,v_*,\pi_*)$, we will introduce two lemmas [13] firstly.

Lemma 6. Let $A\in M_n$, if $A=A^T$, then $max\ r(x)=max\frac{x^TAx}{x^Tx}=\lambda_{max}$, where λ_{max} is the largest eigenvalue of A, the expression $r(x)=\frac{x^TAx}{x^Tx}$ is known as Rayleigh-Ritz ratio, and $x\neq 0$ is a column vector of n-dimension.

Lemma 7. Let $R=(r_{ij})_{n\times n}\in M_n$ be irreducible and nonnegative, then

(1) R has one and only one maximum eigenvalue λ_{max};

(2) All the elements of eigenvector corresponding to eigenvalue λ_{max} must be positive, and the only differences among all the eigenvectors are the ratio factors.

Let $b=(b_1,b_2,b_3)=(\mu',v',\pi')$, $x_i=(\mu_i',v_i',\pi_i')$, and $b_*=(\mu_*',v_*',\pi_*')$ be the unit vector of $R=(\mu,v,\pi)$, $R_i=(\mu_i,v_i,\pi_i)$, and $R_*=(\mu_*,v_*,\pi_*)$, respectively, where $i\in M$. The equation (9) can be denoted as:

$$max\frac{1}{m}\sum_{i=1}^{m}[(\mu',v',\pi')(\mu_i',v_i',\pi_i')^T]^2=\frac{1}{m}\sum_{i=1}^{m}[(\mu_*',v_*',\pi_*')(\mu_i',v_i',\pi_i')^T]^2 \tag{10}$$

Let

$$f(b)=\sum_{i=1}^{m}[(\mu',v',\pi')(\mu_i',v_i',\pi_i')^T]^2=\sum_{i=1}^{m}(bx_i^Tbx_i^T)=b(\sum_{i=1}^{m}x_i^Tx_i)b^T=bX^TXb^T \tag{11}$$

where $X=\begin{pmatrix}\mu_1' & v_1' & \pi_1'\\ \mu_2' & v_2' & \pi_2'\\ \cdots & \cdots & \cdots\\ \mu_m' & v_m' & \pi_m'\end{pmatrix}$. Obviously, in equation (11), X^TX is a symmetric matrix, and $bb^T=1$. By lemma 6, we have $max\ f(b)=max\frac{bX^TXb^T}{bb^T}=\lambda_{max}$, where λ_{max} is the maximum eigenvalue of X^TX. In the following, we will prove that b_* is the unique eigenvector of X^TX corresponding to eigenvalue λ_{max}.

Construct a Lagrange function

$$g(b,\lambda) = f(b) - \lambda(bb^T - 1) \tag{12}$$

Let $\frac{\partial g(b,\lambda)}{\partial b_i} = 0, i = 1,2,3$, we have

$$X^T Xb = \lambda b \tag{13}$$

It is obvious that $X^T X$ is irreducible. By lemma 7, we have that $X^T X$ must have the only one maximum eigenvalue λ_{max}, and the elements of eigenvector b_* of $X^T X$ corresponding to eigenvalue λ_{max} are all positive. Because b_* is a unit vector, then b_* must be the unique vector corresponding to eigenvalue λ_{max}. By equation(13), if we can get the maximum value of f, then λ_{max} is the eigenvalue of $X^T X$, and b_* is the eigenvector of $X^T X$ corresponding to λ_{max}.

Therefore, we get the following theorem.

Theorem 8. For any $b = (b_1, b_2, b_3) = (\mu', v', \pi')$, we have $max\, bX^T Xb^T = b_* X^T X b_*^T = \lambda_{max}$, where λ_{max} is the maximum eigenvalue of $X^T X$, b_* is the unique positive eigenvector corresponding to λ_{max}, and $b_* b_*^T = 1$.

According to theorem 8, if we have got $b_* = (\mu_*', v_*', \pi_*')$, by the equation $\mu_* + v_* + \pi_* = 1$ and $(\mu_*', v_*', \pi_*') = (\mu_*, v_*, \pi_*)(\mu_*^2 + v_*^2 + \pi_*^2)^{-\frac{1}{2}}$, it is easily to get the value of $R_* = (\mu_*, v_*, \pi_*)$.

3.3 The Aggregation Method Based on Similarity

In the decision-making problem, the consistent degree between the estimation value of the "ideal" DM and the estimation value of each DM is larger than the consistent degree between any two estimation values. This consistency can be measured by similarity method. That is, R_* should satisfy the following equation.

$$max\,\{1 - \frac{1}{3m}\sum_{i=1}^{m}[(\mu - \mu_i)^2 + (v - v_i)^2 + (\pi - \pi)^2]\}$$

$$= 1 - \frac{1}{3m}\sum_{i=1}^{m}[(\mu_* - \mu_i)^2 + (v_* - v_i)^2 + (\pi_* - \pi_i)^2] \tag{14}$$

The optimal solution to the nonlinear optimization model (14) is

$$\mu_* = \frac{1}{m}\sum_{i=1}^{m}\mu_i,\ \nu_* = \frac{1}{m}\sum_{i=1}^{m}\nu_i, \pi_* = \frac{1}{m}\sum_{i=1}^{m}\pi_i.$$

This optimization problem is very simple, and the proof process can be found in many nonlinear programming books, so we omit it.

3.4 The Relation Between the Three Models

Suppose that estimation values of the group decision-making derived by intuitionistic fuzzy arithmetic average method, correlation method and similarity method are as follows:

$$R_*^1 = (\mu_*^1, \nu_*^1, \pi_*^1), R_*^2 = (\mu_*^2, \nu_*^2, \pi_*^2), R_*^3 = (\mu_*^3, \nu_*^3, \pi_*^3),$$

where $R_i, i = 1, 2, \ldots, m$ is the estimation value of each DM.

Making use of the relation between the arithmetic average and geometric average, we have

$$1 - \prod_{i=1}^{m}(1-\mu_i)^{\frac{1}{m}} \geq \frac{1}{m}\sum_{i=1}^{m}\mu_i \tag{15}$$

$$\prod_{i=1}^{m}\nu_i^{\frac{1}{m}} \leq \frac{1}{m}\sum_{i=1}^{m}\nu_i \tag{16}$$

Utilizing the comparison method of intuitionistic fuzzy values, we get

$$\begin{aligned} R_*^1 &= (1 - \prod_{i=1}^{m}(1-\mu_i)^{\frac{1}{m}}, \prod_{i=1}^{m}\nu_i^{\frac{1}{m}}, \prod_{i=1}^{m}(1-\mu_i)^{\frac{1}{m}} - \prod_{i=1}^{m}\nu_i^{\frac{1}{m}}) \\ &\geq (\frac{1}{m}\sum_{i=1}^{m}\mu_i, \frac{1}{m}\sum_{i=1}^{m}\nu_i, \frac{1}{m}\sum_{i=1}^{m}\pi_i) = R_*^2 \end{aligned} \tag{17}$$

Thus we have:

Property 1. The estimation value of the group decision-making derived by intuitionistic fuzzy arithmetic average method satisfies $R_*^1 \geq R_*^2$.

By equation (9) and (14), it is easily to get the properties below:

Property 2. The ideal estimation value of the group decision-making derived by correlation method satisfies $\frac{1}{m}\sum_{i=1}^{m} C^2(R_*^2, R_i) \geq \frac{1}{m}\sum_{i=1}^{m} C^2(R_*^t, R_i), t = 1, 3.$

Property 3. The ideal estimation value of the group decision-making derived by similarity method satisfies $1-\frac{1}{3m}\sum_{i=1}^{m}[(\mu_*^3-\mu_i)^2+(v_*^3-v_i)^2+(\pi_*^3-\pi_i)^2]\geq 1-\frac{1}{3m}\sum_{i=1}^{m}[(\mu_*^t-\mu_i)^2+(v_*^t-v_i)^2+(\pi_*^t-\pi_i)^2], t=1,2.$

From these results, we can get the following conclusions:

(1) In fact, the consensus of group decision-making derived through the correlation method and similarity method is the arithmetic average of each element in the same position of every intuitionistic fuzzy value estimated by the DMs (The similarity model is actually a least square optimal model, while the correlation model is actually a variational least square optimal model. Therefore, the difference between the two "ideal" estimation values derived through these two method is very small.)

(2) By equation (15-17), the group preference derived through the intuitionistic fuzzy arithmetic average is more inclinable to the "positive" attitude of a certain problem, while the group preference derived through correlation method or similarity method is more inclinable to the "negative" attitude of a certain problem.

4 Numerical Example

Consider a group decision-making problem with a given alternative set $S=\{s_1,s_2,s_3\}$ using the opinions of three DMs k_1,k_2,k_3. Suppose that the intuitionistic fuzzy judgment matrices are given as follows:

$$M^1=\begin{pmatrix}(0.5,0.5,0) & (0.2,0.6,0.2) & (0.4,0.5,0.1)\\ (0.6,0.2,0.2) & (0.5,0.5,0) & (0.6,0.3,0.1)\\ (0.5,0.4,0.1) & (0.3,0.6,0.1) & (0.5,0.5,0)\end{pmatrix};$$

$$M^2=\begin{pmatrix}(0.5,0.5,0) & (0.3,0.6,0.1) & (0.6,0.1,0.3)\\ (0.6,0.3,0.1) & (0.5,0.5,0) & (0.5,0.3,0.2)\\ (0.1,0.6,0.3) & (0.3,0.5,0.2) & (0.5,0.5,0)\end{pmatrix};$$

$$M^3=\begin{pmatrix}(0.5,0.5,0) & (0.2,0.6,0.2) & (0.7,0.2,0.1)\\ (0.6,0.2,0.2) & (0.5,0.5,0) & (0.6,0.4,0)\\ (0.2,0.7,0.1) & (0.4,0.6,0) & (0.5,0.5,0)\end{pmatrix}.$$

Aggregated every element in the same position of these three matrices by the different methods mentioned in section 3, the corresponding group judgment matrices can be got as follows:

a) The group decision-making matrix derived through the intuitionistic fuzzy arithmetic average method is

$$M_*^1 = \begin{pmatrix} (0.5,0.5,0) & (0.2348,0.6000,0.1652) & (0.5840,0.2154,0.2006) \\ (0.6000,0.2348,0.1652) & (0.5,0.5,0) & (0.5691,0.3302,0.1007) \\ (0.2154,0.5840,0.2006) & (0.3302,0.5691,0.1007) & (0.5,0.5,0) \end{pmatrix}.$$

b) The group decision-making matrix derived through correlation method is

$$M_*^2 = \begin{pmatrix} (0.5,0.5,0) & (0.2337,0.6000,0.1663) & (0.5651,0.2670,0.1679) \\ (0.6000,0.2337,0.1663) & (0.5,0.5,0) & (0.5640,0.3308,0.1052) \\ (0.2670,0.5651,0.1679) & (0.3357,0.5696,0.0946) & (0.5,0.5,0) \end{pmatrix}.$$

c) The group decision-making matrix derived through similarity method is

$$M_*^3 = \begin{pmatrix} (0.5,0.5,0) & (0.2333,0.6000,0.1667) & (0.5667,0.2667,0.1667) \\ (0.6000,0.2333,0.1667) & (0.5,0.5,0) & (0.5667,0.3333,0.1000) \\ (0.2667,0.5667,0.1667) & (0.3333,0.5667,0.1000) & (0.5,0.5,0) \end{pmatrix}.$$

Calculating the arithmetic average of the elements in the same row (not including the element of the principal diagonal of the matrix) of each group decision-making matrix, we obtain the whole degree matrix, we obtain the whole degree $t_i, i = 1,2,3$ to which the alternative $s_i, i = 1,2,3$ is preferred to all the other alternatives.

For M_*^1, we have $t_1 = (0.4358, 0.3595, 0.2047), t_2 = (0.5848, 0.2784, 0.1368), t_3 =$ $(0.2751, 0.5765, 0.1484)$. Thus the optimal ranking order of the alternatives is given by $s_2 \succ s_1 \succ s_3$.

For M_*^2, we have $t_1 = (0.4227, 0.4002, 0.1771), t_2 = (0.5824, 0.2780, 0.1396), t_3 =$ $(0.3022, 0.5673, 0.1305)$. Thus the optimal ranking order of the alternatives is given by $s_2 \succ s_1 \succ s_3$.

For M_*^3, we have $t_1 = (0.4236, 0.4000, 0.1764), t_2 = (0.5837, 0.2789, 0.1374), t_3 =$ $(0.3008, 0.5667, 0.1325)$. Thus the optimal ranking order of the alternatives is given by $s_2 \succ s_1 \succ s_3$.

5 Notes and Comments

The degree of membership, the degree of nonmembership and the hesitation margin of intuitionistic fuzzy set are corresponding to the position, negation and uncertainty of the attitude to the evidence or opinion that people holds in behavior model, so intuitionistic fuzzy set can reflect the cognitive processes of behavior in group decision-making. Therefore, the research on the group decision-making of intuitionistic fuzzy set is more significance than that of the common fuzzy set. At present, for lack of intensive study, there are still lots of problems to be solved in the field of intuitionistic fuzzy judgment matrix, in which the properties of intuitionistic fuzzy judgment matrix need to be developed further research.

References

1. Qiu, W.H.: Management Decisions and Entropy in Appliaction. Machine Press, Beijing (2002)
2. Chen, Y., Fan, Z.P.: Study on the adverse judgment problem for group decision making based on linguistic judgment matrices. Journal of Systems Engineering, 211–215 (2005)
3. Tan, C.Q., Zhang, Q.: Aggregation of opinion in group decision making based on intuitionistic fuzzy distances. Mathematics in practice and theory, 119–124 (2006)
4. Zadeh, L.A.: Fuzzy sets, Information and Control, 338–353 (1965)
5. Atanassov, K.: Intuitionistic fuzzy sets. Fuzzy Sets and Systems, 87–96 (1986)
6. Lei, Y.J., Wang, B.S., Miao, G.Q.: On the intuitionistic fuzzy relations with compositional operations. Systems Engineering-Theory and Practice, 113–118 (2006)
7. Li, D.F.: Multiattribute decision making models and methods using intuitionistic fuzzy sets. Journal of Computer and System Sciences, 73–85 (2005)
8. Dimitrov, D.: The Paretian liberal with intuitionistic fuzzy preferences: A result, Soc. Choice Welfare, 149–156 (2004)
9. CSzmidt, E., Kacprzyk, J.: A new concept of a similarity measure for intuitionistic fuzzy sets and its use in group decision making. Lecture Notes in Computer Science, pp. 272–282 (2005)
10. Szmidt, E., Kacprzyk, J.: A consensus-reaching process under intuitionistic fuzzy preference relations. International Journal of Intelligent Systems, 837–852 (2003)
11. Pankowska, A., Wygralak, M.: General IF-sets with triangular norms and their applications to group decision making. Information Sciences, 2713–2754 (2006)
12. Xu, Z.S., Yager, R.R.: Some geometric aggregation operators based on intuitionistic. International Journal of General Systems, 417–433 (2006)
13. Horn, R.A., Johnson, C.R.: Matrix Analysis. Cambridge University Press, Cambridge (1990)

Optimal Design of TS Fuzzy Control System Based on DNA-GA and Its Application

Guangning Xu and Jinshou Yu

Research Institute of Automation, East China University of Science and Technology, Shanghai, 200237, China
niatxgn@126.com

Abstract. A TS fuzzy model modeling method is presented in this paper. The input parameters of the TS fuzzy model are identified via fuzzy c-means clustering method and the output parameters are optimized via DNA genetic algorithm. Finally, the proposed method is applied to build the soft sensing model for the yield of acrylonitrile. Examining results demonstrate the effectiveness of this method.

1 Introduction

Fuzzy controllers are inherently nonlinear controllers, and hence fuzzy control technology can be viewed as a new, cost effective and practical way of developing nonlinear controllers. The major advantage of this technology over the traditional control technology is its capability of capturing and utilizing qualitative human experience and knowledge in a quantitative manner through the use of fuzzy sets, fuzzy rules and fuzzy logic. However, carrying out analytical analysis and design of fuzzy control systems is difficult not only because the explicit structure of fuzzy controllers is generally unknown, but also due to their inherent nonlinear and time-varying nature.

There exist two different types of fuzzy controllers: the Mamdani type and the Takagi-Sugeno (TS, for short) type. They mainly differ in the fuzzy rule consequent: a Mamdani fuzzy controller utilizes fuzzy sets as the consequent whereas a TS fuzzy controller employs linear functions of input variables. By using TS fuzzy model modeling approach [1], a complex non-linear system can be represented by a set of fuzzy rules of which the consequent parts are linear state equations. Then the complex nonlinear plant can be described as a weighted sum of these linear state equations. This TS fuzzy model is widely accepted as a powerful modeling tool. But the fuzzy rules' parameters of the TS fuzzy model are more and it is difficult to optimize these parameters. The authors of [2] suggested an identification method of TS fuzzy model for nonlinear systems via Fuzzy Neural Networks (FNN). It is considerable effective in describing systems. A new method to generate weighted fuzzy rules from relational database systems for estimating null values using genetic algorithms was given in reference [3].

K. Li et al. (Eds.): LSMS 2007, LNCS 4688, pp. 326 – 334, 2007.

Although the GA algorithm provides a way to possibly obtain the global optimization solution, it has some limitations. In a GA, an optimal solution is achieved through the manipulation of a population of string structures known as chromosomes. Each chromosome is a simple coding of a potential solution to the problem to be solved. With successive generations of population by reproduction and recombination operators, such as crossover and mutation, the overall quality of the population assessed by the fitness function can be improved. But, some biological operations at the gene level cannot be effectively adopted in the existing GAs.

Deoxyribonucleic acid (DNA) is the major genetic material for life, and it encodes plentiful genetic information. Since Watson and Crick's discovery, many ways to manipulate DNA have been developed. DNA was first used for computations in 1994 [4]. In his ground breaking Science article, Adleman described that his experiment solved a seven-node instance of the Hamiltonian path problem from graph theory, which is a NP-complete problem. He devised a code for the edges of a graph based on the encodings of their nodes. Since then, many new ideas about DNA computing have appeared. Moreover, research has been directed to the soft computing aspect of the DNA computing, that is, the integration of DNA computing with intelligent technologies [5], eg., evolutionary computation and neural networks. In order to overcome the limitations of the GAs, a few of GAs based on the mechanism of the biological DNA, such as double stranded DNA and DNA encoding method, have been developed. In this paper, a TS fuzzy model modeling method is presented. The input parameters of the TS fuzzy model are identified via fuzzy c-means clustering method and the output parameters are optimized via DNA genetic algorithm. Finally, the proposed method is applied to build the soft sensing model for the yield of acrylonitrile.

2 Fuzzy C-Means(FCM) Clustering Method

FCM clustering method is formulated as minimization of the following functional:

$$\begin{aligned} &\text{minimize} \sum_{i=1}^{c}\sum_{j=1}^{l} u_{ij}^{\xi} \left\| x_j - v_i \right\|^2 \\ &\text{subject to: } \sum_{i=1}^{c} u_{ij} = 1, 0 < \sum_{j=1}^{l} u_{ij} < l \end{aligned} \tag{1}$$

where c is the number of cluster, l is the number of data points(samples), v_i is the center of ith cluster; x_j is the jth data point, $\|\bullet\|$ is the Euclidean distance, $u_{ij} \in [0,1]$ is the degree of membership of the jth data point to the ith cluster and the so-called "fuzzifier" ξ is chosen in advance and influences the fuzziness of the final partition (common values for ξ are within 1.5 and 4, 2 is most frequently used and also in this paper). The solution of the optimization problem of Eq.(1) is calculated by iteration, namely, in every iteration step, minimization with respect to u_{ij} and v_i is done separately according to the following update equations, Eq.(2) and Eq.(3), respectively.

$$u_{ij} = \frac{(1/\|x_j - v_i\|^2)^{1/\xi-1}}{\sum_{k=1}^{c}(1/\|x_j - v_k\|^2)^{1/\xi-1}} \tag{2}$$

$$v_i = \frac{1}{\sum_{j=1}^{l}(u_{ij})^{\xi}}\sum_{j=1}^{l}(u_{ij})^{\xi}x_j \tag{3}$$

FCM clustering method converges either at $\left|u_{ij}^{t} - u_{ij}^{t-1}\right| \le \delta$ or at $\sum_{i=1}^{c}\left\|v_i^t - v_i^{t-1}\right\| < \delta$, where δ is a positive constant. More detailed description on the FCM method is available in [6].

3 TS Fuzzy Model

The TS fuzzy model used in this paper is written as the followings:
ith rule:

$$\text{if } x_{j1} \text{ is } A_{i1}^{j} \text{ and } x_{j2} \text{ is } A_{i2}^{j} \text{ and } \cdots \text{ and } x_{jm} \text{ is } A_{im}^{j}, \text{ then } y_i^j = a_i \mathrm{X} + b_i$$

$$\mathrm{X}=\begin{bmatrix} x_{j1} & x_{j2} & \cdots & x_{jm}\end{bmatrix}^T;\ a_i = [\, a_{i1}\ a_{i2}\ \cdots\ a_{im}\,];\ i = 1,2,\cdots k$$

where k is the number of rules in the TS fuzzy model, $x_{j1}, x_{j2}, \cdots, x_{jm}$ is the jth input sample in the ith rule and y_i^j is the output of the ith rule for the jth input sample. m is the number of input variables. For the ith rule, $A_{i1}^j, A_{i2}^j, \cdots, A_{im}^j$ is the fuzzy set of the corresponding input variables. a_i and b_i are the coefficient of the consequent. Once the model input variables and its fuzzy set are available, the output of TS model can be calculated by the weighted-average of each y_i^j:

$$y^j(X) = \frac{\sum_{i=1}^{k}\left[\gamma_i^j(X)\, y_i^j(X)\right]}{\sum_{i=1}^{k}\gamma_i^j(X)} \tag{4}$$

where y_i^j is determined by consequent equation of the ith rule. γ_i is defined as

$$\gamma_i^j(X) = \prod_{h=1}^{m} \mu_{ih}^j(X) \tag{5}$$

and μ_{ih}^j is the membership function of hth fuzzy set in the ith rule for the jth input sample.

Now defining

$$w_i^j(X) = \frac{\gamma_i^j(X)}{\sum_{i=1}^{k} \gamma_i^j(X)} \tag{6}$$

we can write (4) as

$$y^j(X) = \sum_{i=1}^{k} w_i(X)(a_i X + b_i) \tag{7}$$

It is assumed that

$$\gamma_i \geq 0,\ \sum_{i=1}^{k} \gamma_i(X) > 0 \tag{8}$$

therefore,

$$0 \leq w_i \leq 1 \tag{9}$$

$$\sum_{i=1}^{k} w_i = 1 \tag{10}$$

In this paper, the number of cluster c is equal to the number of rules k. So in the Eq.(1) and the Eq.(2), if $x_j = \left[x_{j1}\ x_{j2}\ \cdots\ x_{jm} \right]^T$, then $u_{ij} = \left[u_{ij}^1\ u_{ij}^2\ \cdots\ u_{ij}^m \right]^T$. Now the μ_{ih}^j can be calculated as:

$$\begin{aligned} &\mu_{ih}^j = u_{ij}^h \\ &i = 1, 2, \cdots, k \\ &j = 1, 2, \cdots, l \\ &h = 1, 2, \cdots, m \end{aligned} \tag{11}$$

where k is the number of rules in the TS fuzzy model, l is the whole number of input samples, m is the number of input variables. a_i and b_i can be identified and be optimized by DNA-GA.

4 DNA Genetic Algorithm (DNA-GA)

As we known, the basic elements of biological DNA are nucleotides. Due to their different chemical structure, nucleotides can be classified as four bases: Adenine (A), Guanine (G), Cytosine (C) and Thymine (T). A chromosome consists of combinations of the four bases. So there are 64 possible triplet codes (see Table 1). The range of the design parameters can be adjusted with respective to [0, 63] according to different design problems. For a particular application, one may transfer a value in the range of [0, 63], into the proper range of the design parameters. After the translation, the fuzzy controller with these design parameters can be used and the fitness function can be computed.

Table 1. Translation the codons into the numbers

First	Second Base				Third
Base	T	C	A	G	Base
T	0	4	8	12	T
	1	5	9	13	C
	2	6	10	14	A
	3	7	11	15	G
C	16	20	24	28	T
	17	21	25	29	C
	18	22	26	30	A
	19	23	27	31	G
A	32	36	40	44	T
	33	37	41	45	C
	34	38	42	46	A
	35	39	43	47	G
G	48	52	56	60	T
	49	53	57	61	C
	50	54	58	62	A
	51	55	59	63	G

4.1 Coding

A single strand of DNA can be considered as a string consisting of the combination of four different symbols, A, G, C, T. Mathematically, this means we have a four-letter alphabet $\sum\{$ A, G, C, T $\}$ to encode information, which is more than enough, considering that an electronic computer needs only two digits, 0 and 1, for the same purpose. In the artificial DNA model, the design parameters of a problem to be solved are encoded by four-letter alphabet $\sum\{$ A, G, C, T $\}$ to form a chromosome. Based on the DNA model, we can introduce features of the biological DNA into the GA and develop the DNA-GA.

The DNA-GA coding is shown in Fig. 1.

4.2 Fitness Function

The fitness function we adopt in this paper is chosen as the following:

$$f_{fit} = C - \frac{1}{l}\sum_{j=1}^{l}(t^j - y^j)^2 \tag{12}$$

where C is a constant, t^j is the real value of the output variable for the *j*th input samples, y^j is the output of TS model.

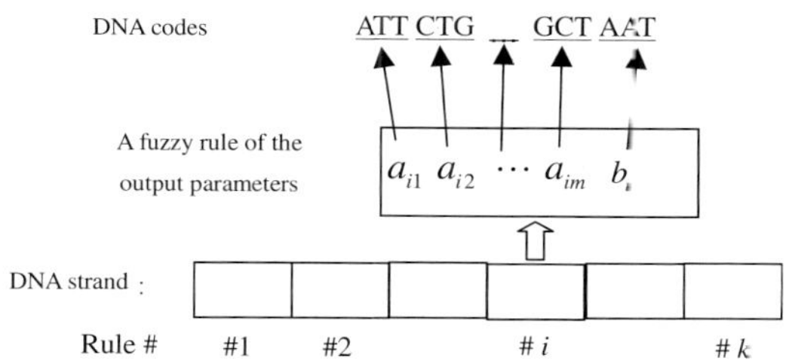

Fig. 1. The DNA coding

4.3 DNA-GA Operations

1. Selection

The first DNA-GA operation is to select individuals with the probability p_i, the value of p_i can be obtained as the following equation:

$$p_i = f_i / (\sum_{j=1}^{m} f_i) > 0 \tag{13}$$

In practice, the interval I=[0,1] is divided into m subintervals so that each individual corresponds to each subinterval. The *i*th individual is selected as the next generation if $p_0+p_1+\ldots+p_{i-1}<\lambda\leqslant p_0+p_1+\ldots+p_i$, $\lambda\in[0,1]$ is generated randomly. And the larger the fitness value of an individual, the more the chance to be selected. We also accept the idea that the best individual (which has the largest fitness value) in the father generation replaces directly the worst individual (which has the smallest fitness value) in the child generation.

2. Crossover and Mutation

Here, two-point crossover method is adopted. In DNA-GA, two mutation transition methods are employed. One is the mutual transition between purines and pyrimidines, eg. A to T. The other is purines to purines or pyrimidines to pyrimidines, eg. A to G, T to C.

3. Cut and Sticker
The cut and sticker operation is shown in Fig. 2.

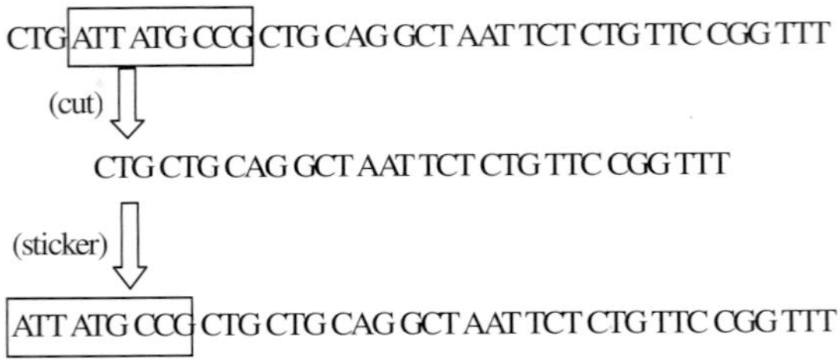

Fig. 2. The cut and sticker operation

5 DNA-GA Application in the Soft-Sensing Modeling

5.1 Introduction of Engineering Background

Acrylonitrile is the major raw material of polyacrylonitrile production, and usually produced by adapting the method of the propylene ammonia oxidation. Its practical industrial process is complex, mainly involving the process of reaction, refrigeration, assimilation, recycle, separation and so on [7]. First, liquid ammonia and liquid propylene are respectively transferred into different evaporator, and then enter into a reactor after being exchanged heat, absterged and overheated. Meanwhile, air is transferred into the same reactor after being compressed and preheated. Thus, with the help of the activator, oxidation reaction occurs and acrylonitrile is produced, with the by-products. Finally, after a series of the process of refrigeration, assimilation, recycle, and separation, we can obtain the pure acrylonitrile.

During the process of acrylonitrile processing, acrylonitrile yield is a very important product indicator. Obtaining the acrylonitrile yield on-line and precisely is the key to control the process of acrylonitrile processing. However, special online analysis instrument is expensive and has a large lag, which will decrease the control quality so severely that it can't meet the demand. Therefore, it's significant for refineries to predict the acrylonitrile yield on-line by using the soft-sensing technology.

5.2 Soft-Sensing Modeling of Acrylonitrile Yield Based on TS Fuzzy Model and DNA-GA

According to process mechanism and practical experience, acrylonitrile yield is related mainly to the following seven variables: reaction pressure, temperature, the amount of pure propylene, the amount of ammonia, the amount of air, activator's density, and reaction speed. The relationship between acrylonitrile yield and the above-mentioned seven variables is complex nonlinear relationship. To estimate the acrylonitrile yield, we must find the relationship between them. In this section, the TS fuzzy model, which has seven input variables corresponding to the above seven variables and an output variable that is acrylonitrile yield, is constructed to find the relationship between the output variable and the input variables.

The objective function of the soft-sensing model can be expressed as follow:

$$\min E = \sum_{j=1}^{l} (t^j - y^j)^2 \tag{14}$$

Where, t is the real value of acrylonitrile yield; y is the predictive value of acrylonitrile yield.

To evaluate the performance of soft-sensing model conveniently, the mean square error (*RMSE*) and the maximal absolute error (*MAXE*) are defined as follow:

$$RMSE = \frac{1}{l}\sum_{j=1}^{l} (t^j - y^j)^2 \tag{15}$$

$$MAXE = \max_{1\le j\le l}\left(\left|t^j - y^j\right|\right) \tag{16}$$

5.3 Experiment and Discussion

There are 328 sets of sample data that consist of seven operating variables in different operating states and one output variable, the real value of acrylonitrile yield. After being processed by error-detected, smoothed, filtered and standardized there are still 317 sets of data remained. At first, all samples are normalized to [-1,1], and then the first 240 samples are chosen as the learning samples and the other 77 as testing samples. For the 240 samples, they are classified to 8 subsets by FCM clustering method. The parameters setting of DNA-GA are as following: the maximal loop number MAX_LOOPS=500, crossover rate P_c=0.9, mutation rate P_m=0.05, cut and sticker rate P_{cs}=0.05, the number of the individuals of the initial population m=30. After learning and optimizing, statistical results show that the mean square error of the learning samples is 0.046 and the maximal absolute error of the learning samples is 0.476 in soft-sensing based on TS fuzzy model and DNA-GA. Then, another 77 samples are used as on-line examining data sets. The mean square error of the examining samples is 0.272, and the maximal generalization absolute error of the examining samples is

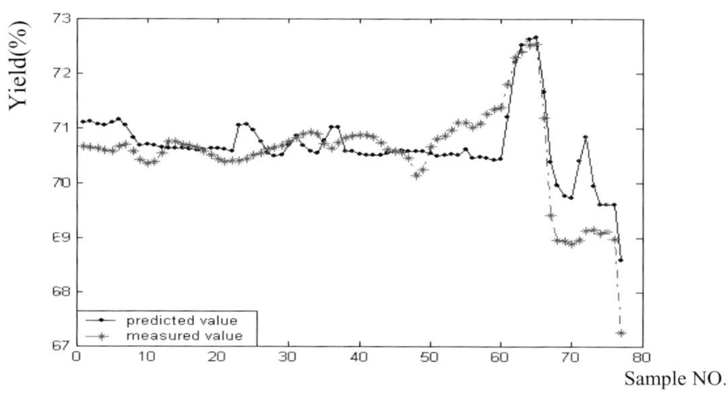

Fig. 3. Comparison of examining results

1.704 in soft-sensing based on TS fuzzy model and DNA-GA. Moreover, the generalization errors of 96.1 percent of all the examining samples are less than 1. This result indicates that this model's predicting values are very closed to the real values, the model can predict precisely the acrylonitrile yield. Examining results is showed as the following figure 3.

In order to compare the result of soft-sensor based on TS fuzzy model and DNA-GA(TS-DNA-GA), we also build a model based on principal component analysis(PCA). The performances of the two models are listed in table 2. Table 2 shows that the examining result of soft-sensing modeling based on TS-DNA-GA is better than the examining result of soft-sensing modeling based on PCA and has the better generalization property.

Table 2. Comparison of the two soft-sensing model

Model type	Learning error		Generalization error		Absolute error less than 1
	RMSE	MAXE	RMSE	MAXE	
TS-DNA-GA	0.046	0.476	0.272	1.704	74/77
PCA	0.228	1.351	0.329	1.364	61/77

6 Conclusion

In this paper, a TS fuzzy model modeling method is presented. The input parameters of the TS fuzzy model are identified via fuzzy c-means clustering method and the output parameters are optimized via DNA genetic algorithm. Finally, the proposed method is applied to build the soft sensing model for the yield of acrylonitrile. Examining results demonstrate the effectiveness of this method.

References

1. Takagi, T., Sugeno, M.: Fuzzy identification of systems and its applications to modelling and control. IEEE Trans. Syst. Man Cybernet. 15, 116–132 (1985)
2. Keming, X., Jianwei, Z.: A linear fuzzy model identification method based on fuzzy neural networks. In: Proceedings of the 2nd Worldwide Chinese Intelligence Control and Intelligence Automation Conference (1997)
3. Chen, S.M., Huang, C.M.: Generating weighted fuzzy rules from relational database systems for estimating null values using genetic algorithms. IEEE Transactions on Fuzzy Systems 11, 495–506 (2003)
4. Adleman, L.M.: Molecular computation of solutions to combination problems. Science 266, 1021–1023 (1994)
5. Ding, Y.S., Ren, L.H., Shao, S.H.: Automatic design of Takagi-Sugeno fuzzy controllers by a new DNA-based evolutionary algorithm. Acta Automatica Sinica 27, 510–520 (2001)
6. Lozano, S., Dobado, D., Larrañeta, J., Onieva, L.: Modified fuzzy C-means algorithm for cellular manufacturing. Fuzzy Sets and Systems 126, 23–32 (2002)
7. Nie, D., Zhang, Q., Chen, Z.: The Research and Application of Acrylonitrile. Chemical Industry and Engineering Technology 26, 35–36 (2005

An Interactive Fuzzy Multi-Objective Optimization Approach for Crop Planning and Water Resources Allocation

Huicheng Zhou*, Hui Peng, and Chi Zhang

School of Civil and Hydraulic Engineering, Dalian University of Technology, 116024 Dalian,China
penghui_1981@yahoo.com.cn

Abstract. A fuzzy multi-objective crop planning model is formulated for determining the optimal crop pattern and irrigation water resources allocation in the irrigated agriculture, in which the individual farmers and irrigation administrators put their views on three conflicting objectives: maximum of net returns, maximum of expected grain yield, and maximum of environmental returns. An interactive fuzzy multi-objective optimization (IFMOO) approach is applied to develop the sustainable crop planning for solving problems of vagueness and imprecision information related to data, model formulation and decision maker's preferences involving in multi-objective linear programming (MOLP). The methodology is illustrated in a case study of crop pattern alternatives in Xingkaihu Lake Irrigation District of northeastern China. The incorporation of these socio-economic and environmental objectives demonstrates the capability of IFMOO approach and also works suitably in water resources management by trade-off procedures.

1 Introduction

The growing population and intense economic development have been placing heavy demands on the increasing food supplies in China. Water has been considered one of the most important factors limiting the crop yield. The crop planning is a problem of irrigation water management, in which how much water should be allocated to each crop is the key decision; consequently, under the constraints of availability of land and water resources, the optimal water resources allocation along with corresponding crop pattern has been called for.

Mathematical models based on the linear programming have been used extensively in the analysis and planning of water resources management [1]. However, the decision making issues like water resources management involve multiple, conflicting and non-commensurable criteria. They are called multiple criteria decision making (MCDM) problems, where the decision maker (DM) generally desires a satisfying solution rather than the maximization of objectives[2]. Numerous multi-objective crop planning models were developed and well described in the literature[3,4,5,6,7].

* The authors gratefully acknowledge the financial support from the Dalian University of Technology in China (893222).

K. Li et al. (Eds.): LSMS 2007, LNCS 4688, pp. 335 – 346, 2007.

Nomenclature	
X decision variable vector	c_j costs per unit area from production of jth crop,(dollars/ha)
R constraints sets	e_j environmental returns per unit area of jth crop,(dollars/ha)
$f_i(X)$ objective function i	y_j yield per unit area of jth crop,(tons/ha)
$f_i(X)^m$ individual minimum of $f_i(X)$	Q_s available surface water allocated to crops,(m^3)
$f_i(X)^M$ individual maximum of $f_i(X)$	Q_g available groundwater allocated to crops, (m^3)
λ_i weight vector of ith objective	q_j volume of water requirements of per unit area of jth crop, (m^3/ha)
$\mu(f_i(X))$ membership function of ith objective	A total cultivable area available in planning unit,(ha)
$\mu_D(\mu(f(X)))$ fuzzy decision of $f(X)=\left[f_1(X), f_2(X), \cdots f_p(X)\right]$	x_j area of jth crop, $j=1,2,3,4$ (ha)
b_j net returns per unit area from production jth crop,(dollars/ha)	$\min x_j$ minimum area available of jth crop, (ha)
	$\max x_j$ maximum area available of jth crop, (ha)

However, for multi-objective water resources management issues, a decision maker (DM) might not be able to express his or her preferences precisely, or the real world decision making problems can rarely be defined precisely in mathematical terms because the basic input information, constraints, and consequences of possible actions might not be incompletely available. Therefore, water resources administrators and modelers are bound to deal with the imprecision information—mostly due to insufficient data and imperfect knowledge in the decision making. Hence, it is more realistic to consider fuzzy goals and constraints of the mathematical model[8]. The fuzzy multi-objective linear programming (FMOLP) approach was first developed by Zimmermann [9]. Bogardi et al. applied FMOLP approach to plan an aquifer and mineral extraction under water hazard, incorporating fuzzy economic and environmental objectives[10]. Guillermo A. Mendoza et al. applied a FMOLP approach to the forest planning[11]. A.P. Gupta et al. developed a FMOLP model for crop planning in Narmada river basin[2]. L.S.Vamvakeridou-Lyroudia et al. applied FMOLP approach to design the optimal water distribution networks[12]. Elcin Kentel and Mustafa M. Aral applied the FMOLP approach to select the optimal groundwater management strategy[13].

One of the most effective approaches for solving MOLP problems in the fuzzy environment is to use an interactive approach[14], since the interactive technique can yield a single preferred solution by the DM according to his or her preferences. In the last two decades, the interactive fuzzy multiple criteria decision making (IFMCDM) techniques have experienced spectacular growth. Sakawa and Seo perhaps first proposed an IFMCDM technique—fuzzy sequential proxy optimization approach, for the industrial pollution control problems in Osaka[15]. Sakawa introduced and applied an interactive fuzzy goal programming approach for the case study of the Bow River Valley[16]. Slowinski developed an interactive fuzzy linear programming approachfor the long-term planning of a water supply and wastewater system[17]. Lee and Chang extended some previous works and applied IFMOO approach to develop a water quality management planning as sustainable water quality management strategy in a river system[18].

In this paper, a multi-objective interactive approach in terms of fuzzy set theory [19,20], is applied to identify the optimal (satisfying) crop pattern alternative and conjunctive use of surface and groundwater. The term 'interactive' in a fuzzy multi-objective optimization problem is defined as adjustment of weights of the fuzzy goals according to the DM's preferences. The specific objectives of the study are: (1) to present the interactive fuzzy multi-objective optimization(IFMOO) approach; (2) to develop a multi-objective crop planning model of three fuzzy objectives: maximum of net returns, maximum of expected grain yield, and maximum of environmental returns; (3) to apply the IFMOO approach to determine the 'satisfying' crop planning in Xingkaihu Lake Irrigation District of northeastern China.

2 Interactive Fuzzy Multi-Objective Optimization Approach

Fuzzy Multi-Objective Linear Programming Problem. A classical model of MOLP problem is as follows:

$$\begin{array}{ll} \max\limits_{X\in R} & f(X)=\left[f_1(X),f_2(X),\cdots f_p(X)\right] \\ s.t. & R=\left\{X\left|G(X)\le 0, X\ge 0\right.\right\} \end{array} \tag{1}$$

The solution of MOLP model employs the concept of Pareto optimal, also is known as the non-inferior solution or satisfying solution. However, considering the imprecise nature inherent in human judgments in MOLP problems, the concept of Membership-Pareto (M-Pareto) optimal solution is introduced[19], which is defined in term of membership function instead of objective function. Detailed discussions of fuzzy set mathematics and their operations can be found in[19,20,21,22].

The fuzzy goal can be quantified by eliciting a corresponding membership function. To elicit a membership function $\mu(f_i(X))$ for each of the objective functions $f_i(X), i=1,2,\cdots p$, a pay-off table (Table 1), is helpful in calculating the individual minimum $f_i(X)^m$ and maximum $f_i(X)^M$ of $f_i(X)$ under original constraints.

A linear fuzzy-max membership function $\mu(f_i(X))$ can be defined as:

$$\mu(f_i(X))=\begin{cases} 0 & f_i(X)\le f_i(X)^m \\ \dfrac{f_i(X)-f_i(X)^m}{f_i(X)^M-f_i(X)^m} & f_i(X)^m< f_i(X)< f_i(X)^M \\ 1 & f_i(X)\ge f_i(X)^M \end{cases} \tag{2}$$

A linear fuzzy-min membership function $\mu(f_i(X))$ can be defined as:

$$\mu(f_i(X))=\begin{cases} 0 & f_i(X)\ge f_i(X)^M \\ \dfrac{f_i(X)^M-f_i(X)}{f_i(X)^M-f_i(X)^m} & f_i(X)^m< f_i(X)< f_i(X)^M \\ 1 & f_i(X)\le f_i(X)^m \end{cases} . \tag{3}$$

Table 1. A pay-off table with $p-$objective optimization

Solution	Objective Value			
	$f_1(X^k)$	$f_2(X^k)$	$\cdots$	$f_p(X^k)$
X^1	$f_1(X^1)$	$f_2(X^1)$	$\cdots$	$f_p(X^1)$
X^2	$f_1(X^2)$	$f_2(X^2)$	$\cdots$	$f_p(X^2)$
$\vdots$	$\vdots$	$\vdots$	$\cdots$	$\vdots$
X^p	$f_1(X^p)$	$f_2(X^p)$	$\cdots$	$f_p(X^p)$

After the membership functions $\mu(f_i(X)), i=1,2,\cdots p$ have been elicited, the FMOLP can be defined as:

$$\max \quad \left[\mu(f_1(X), \cdots \mu(f_i(X), \cdots \mu(f_p(X)\right]. \tag{4}$$

By introducing a general aggregation function:

$$\mu_D(\mu(f_i(X))) = \min\left[\mu(f_1(X), \cdots \mu(f_i(X), \cdots \mu(f_p(X)\right]. \tag{5}$$

The general fuzzy multi-objective decision making problem can be defined as:

$$\max \quad \mu_D(\mu(f(X))). \tag{6}$$

Or equivalently

$$\begin{aligned} \max \quad & v \\ s.t. \quad & v \le \mu(f_i(X)) \\ & X \in R' \end{aligned} \tag{7}$$

Weights of Fuzzy Objectives(Goals). For FMOLP problems, the DM must specify his or her preferences to the fuzzy goals. To capture the knowledge and preferences of the DM, weights are used to indicate the importance of competing objectives. Chen proposed a pair-wise comparison fuzzy decision making analysis (PCFDMA) approach, for quantifying the weights of fuzzy goals in mathematical terms [20].

Definition 1. (Qualitative Ordering Scale Matrix).Let $G=\{g_i\}, i=1,2,\cdots p$ denotes a collection of fuzzy goals, make pair-wise comparison between g_k and g_l regarding the fuzzy concept of 'importance' from the DM in terms of the complementary rule. The complementary rule can be summarized as follows:

(1) If g_k is more important than g_l, then ordering scale $e_{kl}=1, e_{lk}=0$;
(2) If g_l is more important than g_k, then ordering scale $e_{lk}=1, e_{kl}=0$; (8)
(3) If g_k is as important as g_l, then ordering scale $e_{kl}=e_{lk}=0.5$.

Suppose the pair-wise comparison matrix

$$E=\begin{bmatrix} e_{11} & e_{12} & \cdots & e_{1p} \\ e_{21} & e_{22} & \cdots & e_{2p} \\ \vdots & \vdots & \vdots & \vdots \\ e_{p1} & e_{p2} & \cdots & e_{pp} \end{bmatrix}=(e_{kl}) \tag{9}$$

subject to complementary rule, then E is called the qualitative ordering scale matrix regarding the fuzzy concept of 'importance' of objective collection G.

The PCFDMA approach for quantifying weights of fuzzy objectives is generally summarized as follows: Firstly, elicit the qualitative ordering scale matrix E from the DM, add elements of matrix E line by line and make sort ascending, then obtain the qualitative sort of g_i; Secondly, according to the relationship between mood operator and weight, shown in Table 2, elicit the non-normalized weight vector $\lambda_i'=(\lambda_1',\lambda_2',\cdots\lambda_p')$; Thirdly, normalize λ_i', attain the quantitative weight vector $\lambda_i=(\lambda_1,\lambda_2,\cdots\lambda_p)$.

After the membership functions $\mu(f_i(X))$ and weight vector $\lambda_i, i=1,2,\cdots p$ have been elicited from the DM for objective functions, we can attain the corresponding M-

Table 2. Relationship between mood operator and weight of the fuzzy goal

Mood operator	Equally	Slightly	Somewhat	Rather	Obviously	Remarkably
Weight	1.0	0.818	0.667	0.538	0.429	0.333
Mood operator	Very	Extra	Exceedingly	Extremely	Incomparably	
Weight	0.250	0.176	0.111	0.053	0	

Pareto optimal solution by solving the following problem:

$$\min \quad \max \quad \left[\lambda_i^r-\mu(f_i(X))\right]. \tag{10}$$

Or equivalently:

$$\begin{aligned} &\min && v \\ &s.t. && \lambda_i^r-\mu(f_i(X))\le v. \\ & && X\in R' \end{aligned} \tag{11}$$

The Interactive Fuzzy Multi-Objective Optimization Algorithm. The IFMOO algorithm used to derive the optimal (satisfying) solution for the DM from the M-Pareto optimal solution set, is briefly indicated as follows[18,23]:

Step1. Attain the individual optimal solution X^p for each objective function under given constraints, by solving the single-objective linear programming using the MATLAB Optimization Toolbox[24].

Step2. Construct a pay-off table to obtain the individual minimum $f_i(X)^m$ and maximum $f_i(X)^M$ of $f_i(X)$ under the given constraints.

Step3. Elicit the membership function $\mu(f_i(X))$ for each objective function, for example, the linear fuzzy-max membership function can be defined as Eq.(2), the linear fuzzy-min membership function can be defined as Eq.(3).

Step4. Set initial weight vector $\lambda_i^r = (0.334;0.333,0.333)$. Set iteration index $r = 1$.

Step5. Set $\lambda_i^* = \lambda_i^r$ and then solve the corresponding problem of Eq.(11) by using the MATLAB Optimization Toolbox, to obtain a (local) M-Pareto optimal solution $X^r, f_i(X^r)$.

Step6. Provide the information obtained in Step 5 to the DM. If DM is satisfied with $f_i(X^r)$, then terminate the solution procedures. The current M-Pareto optimal solution X^r is a satisfying solution to the DM. Otherwise, update the current weight vector of $\lambda_i^r, i = 1,2,\cdots p$ to new weight vector $\lambda_i^{r+1}, i = 1,2,\cdots p$, by using PCFDMA approach from the DM. Then set $r = r+1$ and return to Step 5. If DM finds nothing will satisfy him or her in achieving the aspiration level of each objective, this multi-objective model will not be a feasible solution.

3 Case Study

Research Area. The research area, Xingkaihu Lake Irrigation District, is located in Heilongjiang Province of northeastern China, and is bounded on the east by Songacha River, on the west by Xiaohei River, on the south by Xingkaihu Lake and on the north by Wusulijiang River. It has a population of 0.0327 million, the cultivated land acreage of 136400 ha, the available volume of groundwater resources is 159.01 million m^3, and the available volume of surface water resources is 251.79 million m^3.

Besides a crucial grain supply area of China, the research area is also a vital national natural conservation. However, in recent decades, the extensive cultivation has resulted in the availability of rapid economic growth, simultaneously serious damage of environment. For example, the volume exceeding capacity of existing groundwater abstraction has reached to approximately 130 million m^3; and lower of groundwater level has brought on the sharp shrinking of wetland areas. From the viewpoint of a socio-economic and environmental balance with sustainable use of water resources, the issue of optimal conjunction use of groundwater and surface water along with scientific planning of crop pattern has been urgently called for.

Model Building. The multi-objective crop planning model consists of three objectives and their related constraints. The objectives and their constraints are as follows:

(1) Maximum of net returns: the net returns maximization is commonly aspired by every decision maker in the planning process. However, farmers are more biased to such objectives and always prefer a crop planning which can provide them with more financial returns. Hence, it can be formulated as:

$$\max \quad f_1(X) = \sum_{j=1}^{n} (b_j - c_j) x_j \,. \tag{12}$$

(2) Maximum of the expected crop yield: considering the government's policy of providing adequate food supply, the suggested crop planning should ascertain the maximal grain supply. Mathematically, it can be expressed as:

$$\max \quad f_2(X) = \sum_{j=1}^{n} y_j x_j \,. \tag{13}$$

(3) Maximum of environmental returns: besides the value of food production, the crop has value of climate regulation, water resources conservation, soil protection, etc. Chen and Zhang used the parameters of [25] to estimate the ecosystem services value and natural capital of China[26](Table 3.). The irrigation administrators prefer to a crop pattern that can provide more environmental returns for a sustainable irrigation planning. Thus, mathematically:

$$\max \quad f_3(X) = \sum_{j=1}^{n} e_j x_j \,. \tag{14}$$

These objective functions subject to the following constraints:

(1) Water allocation constraint

$$\sum_{j=1}^{n} q_j \le Q_g + Q_s \,\cdot \tag{15}$$

(2) Land areas constraints

$$\sum_{j=1}^{n} x_j \le A \,\cdot \tag{16}$$

$$\min x_j \le x_j \le \max x_j \,. \tag{17}$$

(3) Non-negativity constraints

$$x_j \ge 0 \,. \tag{18}$$

4 Results

Formulation of Membership Functions. Take the areas of rice, soybean, spring wheat and summer maize as decision variables. The multi-objective fuzzy model for optimal crop planning is as follows:

$$\max \quad f_1(X) = 426.24x_1 + 190.45x_2 + 32.65x_3 + 181.38x_4 \,. \tag{19}$$

$$\max \quad f_2(X) = 6.75x_1 + 2.25x_2 + 3.00x_3 + 6.00x_4 \,. \tag{20}$$

$$\max \quad f_3(X) = (6709.70x_1 + 397.70x_2 + 117.00x_3 + 224.60x_4) \,. \tag{21}$$

After a pay-off table of Xingkaihu Lake Irrigation District case is constructed (Table 4.), The membership functions of these three objectives are listed as follows:

$$\mu(f_1(X)) = \begin{cases} 0 & f_1(X) \le 38.27 \\ f_1(X)/0.74\times10^6 - 51.72 & 38.27 < f_1(X) < 39.01 \\ 1 & f_1(X) \ge 39.01 \end{cases} \,. \tag{22}$$

$$\mu(f_2(X)) = \begin{cases} 0 & f_2(X) \le 0.6 \\ f_2(X)/0.02\times10^6 - 30 & 0.6 < f_2(X) < 0.62 \\ 1 & f_2(X) \ge 0.62 \end{cases} \,. \tag{23}$$

$$\mu(f_3(X)) = \begin{cases} 0 & f_3(X) \le 443.77 \\ f_3(X)/5.33\times10^6 - 83.26 & 443.77 < f_3(X) < 449.10 \\ 1 & f_3(X) \ge 449.10 \end{cases} \,. \tag{24}$$

Trade-Off Process and DM's Preferences. Set iteration index $r = 1$, weight vector $\lambda^1 = (0.334; 0.333, 0.333)$; solve corresponding problem of Eq.(11), attain the solution;

Table 3. Environmental services value per unit area of ecosystem of China(dollars/yr · ha)

Ecosystem type	Gas regulation	Climate regulation	Water conservation	Soil protection	Waste disposal
Wetland	192.59	1829.61	1658.43	182.96	1945.18
Farmland	53.49	95.22	64.20	155.22	175.48
Ecosystem type	**Biomass protection**	**Food production**	**Material production**	**Recreation**	**Total**
Wetland	267.50	32.10	7.48	593.82	6709.7
Farmland	75.96	107.00	10.70	1.06	739.3

Table 4. A pay-off table of Xingkaihu Lake Irrigation District case

Solution	Objective Value		
	$f_1(X)$	$f_2(X)$	$f_3(X)$
X^1=(62336.75,58842.74,10106.12,5075.87)	39.01	0.61	443.77
X^2=(62825.73,52932.12,15527.76,5057.87)	38.27	0.62	445.33
X^3=(63283.56,57550.16,15527.76,0)	38.42	0.60	449.10
$f_i(X)^m$	38.27	0.60	443.77
$f_i(X)^M$	39.01	0.62	449.10

provide the information to DM, the DM is unsatisfied with the solution. Continue the trade-off procedures, set iteration index $r=2$, elicit the pair-wise comparison consistent scale matrix regarding the fuzzy concept 'importance' is as follows:

$$
{}_2E = \begin{bmatrix} 0.5 & 1 & 1 \\ 0 & 0.5 & 1 \\ 0 & 0 & 0.5 \end{bmatrix} \begin{matrix} \textit{sort} & \\ (1) & \textit{economic} \\ (2) & \textit{social} \\ (3) & \textit{environmental} \end{matrix} . \tag{15}
$$

attain the sort of economic objective, social objective and ecological objective is (1)(2)(3); since the sort order of economic objective is (1), regarded it as comparison standard, considering economic objective is 'slightly' important than social objective, and is 'somewhat' important than environmental objective, according to Table 2, obtain the non-normalized weight vector of $\lambda_2' = (1.0, 0.818, 0.667)$; normalize $\lambda^{2'}$ attain the normalized weight vector $\lambda_2 = (0.402; 0.329; 0.269)$; solve the problem of Eq.(11), the DM is unsatisfied with the solution. Continue the procedures, until iteration $r=4$, DM is satisfied with the solution, then terminate the procedures. The trade-off procedures are generated and displayed in Table 5.

Table 5. Trade-off procedures of Xingkaihu Lake Irrigation District case

r	Solution							
	λ_p	x_1	x_2	x_3	x_4	$f_1(X)$	$f_2(X)$	$f_3(X)$
1	(0.334,0.333,0.333)	55702.77	58842.74	15527.76	6288.21	36.58	0.59	400.18
2	(0.402,0.329, 0.269)	62053.14	58842.74	10391.86	5075.87	38.90	0.61	441.90
3	(0.290,0.378, 0.332)	62825.73	52932.12	15527.76	5075.87	38.27	0.62	442.33
4	(0.301,0.332, 0.367)	62551.26	58842.74	9891.61	5075.87	38.79	0.61	445.18

The satisfying crop planning of the case study is shown in Table 6.

The comparison of existing crop pattern as well as conjunctive use of groundwater and surface water with the optimum crop pattern as well as conjunctive water use is presented in Table 7.

The integrated comparison of existing crop planning with optimum crop planning is presented in Table 8.

Table 6. The satisfying crop planning of Xingkaihu Lake Irrigation District case

Crops	Rice	Soybean	Wheat	Maize	Total
Area (ha)	62551.26	58842.74	9891.61	5075.87	136361.48
Ground water volume (million m^3)	105.85	44.11	2.97	6.09	159.01
Surface water volume (million m^3)	251.79	0.00	0.00	0.00	251.79
Net returns (million dollars)	26.65	11.20	0.32	0.92	39.09
Grain yield (million tons)	0.42	0.13	0.03	0.03	0.61
Environmental returns (million dollars)	419.49	23.39	1.16	1.14	445.18

Table 7. Comparison of exiting crop pattern and water allocation with the optimum crop pattern and water allocation of Xingkaihu Lake Irrigation District case

Crops	Existing			Optimum		
	Area (ha)	Groundwater (million m^3)	Surface water (million m^3)	Area (ha)	Groundwater (million m^3)	Surface water (million m^3)
Rice	83401.68	243.75	234.65	62551.26	105.85	251.79
Soybean	39226.27	29.41	0.00	58842.74	44.11	0.00
Wheat	10351.84	3.10	0.00	9891.61	2.97	0.00
Maize	3381.69	4.06	0.00	5075.87	6.09	0.00
Total	136361.48	280.32	234.65	136361.48	159.01	251.79

Table 8. Integrated comparison of existing crop planning with optimum crop planning of Xingkaihu Lake Irrigation District case

	Existing	Optimum
Volume of groundwater (million m^3)	280.32	159..01
Volume of surface water (million m^3)	234.65	251.79
Net returns (million dollars)	43.95	39.09
Crop yield (million tons)	0.70	0.61
Environmental returns (million dollars)	291.9	445.18

5 Conclusions

Generally, the multi-objective linear programming problems require much information that is difficult to obtain in real-word setting, sometimes the complexity of mathematical formulations alienates the potential users. In this paper, a simple, yet potential fuzzy multi-objective crop planning model with three conflicting and fuzzy objectives involving socio-economic and environmental issues is formulated. An explicit trade-off among economic (represented as net returns), social(considered crop yield) and environmental(defined as environmental returns) can be accomplished by using an interactive fuzzy multi-objective optimization approach. The application to determine optimal irrigation water allocation with corresponding crop pattern in Xingkaihu Lake Irrigation District has demonstrated feasibility and efficiency of the algorithm.

References

1. Loucks, D.P., Stedginger, J.R., Haith, D.A.: Water resources systems planning and analysis. Prentice Hall, Englewood Cliffs, NJ (1981)
2. Gupta, A.P., Harboe, R., Tabucanon, M.T.: Fuzzy multiple-criteria decision making for crop area planning in Narmada river basin. Agr. Syst. 63, 1–18 (2000)
3. Onta, P.R., Das Gupta, A., Paudyal, G.N.: Integrated irrigation development planning by multiobjective optimization. Int. J. Water Resourc. Dev. 7, 185–193 (1991)
4. Mohammed, M., Ashim, D.G., Pushpa, R.O.: Optimal crop planning model for an existing groundwater irrigation project in Thailand. Agr. Water Manage. 33, 43–62 (1997)
5. de Wit, C.T., van Keulen, H., Seligman, N.G., Spharim, I.: Application of interactive multiple goal programming techniques for analysis and planning of regional agricultural development. Agr. Syst. 26, 211–230 (1988)
6. Rabbinge, R.: Long-term options for land use in the european community. Agr. Syst. 40, 195–210 (1992)
7. Lu, J.B., Wang, Z.Q., F.W.T.: Application of interactive multiple goal programming for red soil watershed development: A case study for Qingshishan watershed. Agr. Syst. 73, 313–324 (2002)
8. Bellman, R.E., Zadeh, L.A.: Decision making in a fuzzy environment. Management Science (application series) 4, B141–164 (1970)
9. Zimmermann, H.J.: Fuzzy programming and linear programming with several objective functions. Fuzzy Sets Syst. 1, 45–56 (1978)
10. Bogardi, I., Bardossy, A.: Regional management of an aquifer for mining under fuzzy environmental objectives. Water Resour. Res. 6, 1394–1402 (1983)
11. Guillermo, A., Mendoza, B., Bare, B., Zhou, Z.: A fuzzy multiple objective linear programming approach to forest planning under uncertainty. Agr.Syst. 3, 257–274 (1993)
12. Vamvakeridou-Lyroudia, L.S., Walters, G.A., Savic, D.A.: Fuzzy Multiobjective Optimization of Water Distribution Networks. J. Water Resour. Plann. Manage. ASCE 6, 467–476 (2005)
13. Kentel, E., Mustafa, M.A.: Fuzzy Multiobjective Decision-Making Approach for Groundwater Resources Management. J. Water Resour. Plann. Manage. ASCE 2, 206–217 (2007)
14. Mohan, C., Nguyen, H.T.: Reference direction interactive method for solving multiobjective fuzzy programming problems. Eur. J. Oper. Res. 107, 599–613 (1998)

15. Sakawa, M., Seo, F.: Interactive multiobjective decision- making in environmental systems using the fuzzy sequential proxy optimization technique. Large Scale Syst. 4, 223–243 (1983)
16. Sakawa, M.: Interactive fuzzy goal programming for multi-objective nonlinear programming problems and its applications to water quality management. Control Cybern. 13, 217–228 (1984)
17. Slowinski, R.: A multicriteria fuzzy linear programming method for water supply system development planning. Fuzzy Sets Syst. 19, 217–237 (1986)
18. Lee, C.S., Chang, S.P.: Interactive fuzzy optimization for an economic and environmental balance in a river system. Water Res. 39, 221–231 (2005)
19. Sakawa, M., Inuiguchi, M., Kato, K., Ikeda, T.: A fuzzy satisficing method for multiobjective linear optimal control problems. Fuzzy Sets System 78, 223–229 (1993)
20. Chen, S.Y.: Engineering Fuzzy Set Theory and Application [M] (in Chinese). National Defense and Industry Press, Peking (1998)
21. Jairaj, P.G., Vedula, S.: Multireservoir System Optimization using Fuzzy Mathematical Programming. Water Resour. Manage 14, 457–472 (2000)
22. Carlsson, C., Fullér, R.: Fuzzy multiple criteria decision making: Recent developments. Fuzzy Sets Syst. 78, 139–153 (1996)
23. Shang, S.H.: System analysis of water resources: methods and application (in Chinese). Tsinghua University Press, Beijing (2006)
24. Su, L.M., Wang, Y.L.: The enchiridion of MATLAB 7.0 (in Chinese). Electronic Industries Press (2004)
25. Costanza, R., Arge, R., deGroot, R., et al.: The value of the world's ecosystem services and natural capital. Nature 387, 253–260 (1997)
26. Chen, Z.X., Zhang, X.S.: The value of Chinese ecosystem services (in Chinese). Chinese Science Bulletin 1, 17–22 (2000)

A P2P Trust Model Based on Multi-Dimensional Trust Evaluation

Xinsheng Wang, Peng Liang, Huidong Ma, Dan Xing, and Baozong Wang

College of Information Science and Engineering, Yanshan University,
Qinhuangdao, Hebei, China
Liangpeng1981@126.com

Abstract. The current P2P trust models suffer much from dishonest feedbacks and the trust metrics depending on subjective judgment excessively results in rough trust value. So this paper proposes a versatile trust model based on multi-dimensional trust evaluation, MDTrust. The concept of multi-dimensional trust is introduced and the approach of evaluation which uses the joint of subjective and objective evaluation is presented. We use the discrete degree of the set of evaluation similarity between peers' evaluation to measure the feedback credibility, in order to resolve the problem of collusion and bad-mouthing. Consequently, simulation and analysis shows that MDTrust can calculate the trust value effectively and reasonably and discard the malicious peers from P2P system.

1 Introduction

The open and anonymous nature of P2P networks exposes a number of security vulnerabilities, so peers are not responsible for their behaviors, and what's more, lots of malicious nodes provide faked files. As shown by existing work, such as research [1, 3, 4, 5], establishing a trust model in P2P system can successfully mitigate this risk by finding reliable trustworthy communication partners.

Oram's research shows that trust-based P2P networks have great similarity with interpersonal networks [2]. So recurring to the trust model of society network, P2P trust model can divide into global reputation-based trust model and local reputation-based trust model. Research [1] and EigenRep [3] are all global reputation-based trust models. In these systems, every participant in the distributed applications has a global reputation, which is calculated by iterating the mutual satisfaction degree among neighbors. But, the global reputation-based trust model is restrained and not suitable for complex environments for lacking of differentiation of different trust classes.

Contrarily, many researchers consider that the trust value of a node is different for different observers. PeerTrust [4] is a trust model based on local reputation. In this model, a peer's reputation is evaluated by collecting the feedback information of its history transactions.

Those reputation-based trust models are brought forward to prevent peers abusing of network resources and restrain the malicious behaviors of peers. But there are some vulnerabilities of these representative trust models:

K. Li et al. (Eds.): LSMS 2007, LNCS 4688, pp. 347 – 356, 2007.

1) The concept of trust and the reasonability of trust measure should to be further expatiated. The trust computation metrics employed by the existing systems depend on the subjective view excessively.
2) Another widely recognized vulnerability is the shilling attack [6], where malicious peers submit dishonest feedback and collude with each other to boost their own ratings or bad-mouthing non-malicious peers.
3) The existing trust models could not be used in different application situation and the trust calculation mechanism of current models is lack of flexible manner.

Recurring to sociology research, we propose a versatile P2P trust model based on multi-dimensional trust evaluation, MDTrust, in which peers calculate potential partners' trustworthiness based on their behaviors along several dimensions. We use the discrete degree of peers' the set of evaluation similarity to prevent collusion and bad-mouthing attacks. Consequently, theoretic analysis and simulation show that this model is effective and has the ability to discard the malicious peers from P2P system.

2 Related Concepts

Trust model is to use the method of mathematics to obtain peers' trust intention and reliability information based on the definition of trust [5]. In the sociology, the concept of trust is a judgment based on multi-dimensions that has been widely accepted by sociologists. Moreover, in the computer research field, Maximilien and Singh [7] have acknowledged the need for multi-dimensionality in trust.

In addition, research [5] shows that the metric of trust can be divided into two categories: subjective measure and objective measure. Dou considers that peer's trustworthiness in p2p system is essential a synthesis of peer's subjective and objective characteristics. In the evaluation of peer's trust value, subjective and objective measures are complementary and interdependent.

A general problem with the only dependent of subjective measures is that it is difficult to protect against unfair ratings. Another potential problem is that the act of referring negative subjective trust in an entity can lead to accusations of slander. An advantage with objective measures is that the correctness of ratings can be verified by others, or automatically generated based on automated monitoring of events. Contrarily, it is difficult to find a peer's "participation strategy" based on objective measures solely. An effective solution is that a peer to integrate subjective measures to detect another peer whether it has malicious behaviors after transaction.

Building on their previous work, we view that, firstly, dimensions are the fundamental factor affecting a peer's trust assessment. Secondly, the setting of trust dimensions should reflect peers' subjective participant attitudes and objective behaviors of the specific application domain. So we propose a P2P trust model based on multi-dimensional trust evaluation, MDTrust. We divide trust evaluating dimensions into two kinds: subjective dimension and objective dimension.

Definition 1. Subjective Dimension, denoted by SD, is reflected a peer's subjective judgment about another peer's transaction behaviors. Subjective dimension reflects the target peer's participant strategy.

The number of SD is restricted only one in MDTrust. But SD has different meaning for different application. For example, in P2P file-sharing systems, when a peer A completes download from another peer B, A will check the integrity and authenticity of the download file from B. And then A makes subjective judgment for B on SD dimension according to whether B exists malicious behaviors or is a good service provider.

Definition 2. Objective Dimension, denoted by OD, is reflected a service-provide peer's practical performance on some objective attributes. Objective dimension is the basic rational factor.

OD could reduce the side-effect of excessive use of subjective evaluation and make peers' trust value more effectively and reasonably. MDTrust is not restricted to specific prescribed objective dimensions, and different P2P application systems can model trust along any number of dimensions. Moreover, different applications have different objective dimensions' setting and metric. For example, in electronic commerce: the amount of the transaction, transaction time; in file sharing application: file size, the rate of successful download and so on.

When we set objective dimensions of specific application, we should note that: due to a peer have different expectations on different objective dimensions, there are some distinctions among objective dimensions' calculation methods. But these calculation methods should be a monadic and congeneric function.

The different setting of trust dimensions constitutes different multi-dimensional trust type. Finally, we give the formalization definition of multi-dimensional trust type as follow:

Definition 3. Multi-dimensional Trust Type is defined to be a set, denoted by {(Id, T), (SD, w_s), (OD_1, w_1), (OD_2, w_2)… (OD_n, w_n)}, and $w_s+w_1+w_2\ldots+w_n=1$, where Id denotes the identification of application type (service type).T denotes service provider's trust value. w_s denotes the weight of subjective dimension SD. The weight of objective dimension OD_k denoted by w_k.

According to the definition (3) of multi-dimensional trust type, we propose a versatile trust metric which satisfies different application's requirements through the different setting of trust dimensions.

3 Multi-Dimensional Trust Model

Trust in P2P networks is composed of direct trust and recommendation trust [1]. The direct trust A to B is based on the past behaviors of direct trades between A and B. If there is no trade history between them, recommendation from other peers who have interacted with B is the only way to take, which is called recommendation trust.

Let peer *i* denote the service requestor in P2P networks, the transaction service provider is denoted by peer *j*, and let *DTij* denote the direct trust value between node *i* and *j*. The recommendation trust of *i* to *j*, denoted by RT_{ij}, can be computed according to the transaction feedback information of peers who have interacted with *j*. Finally, we denote the trust value of peer *j* in the view of peer *i* as T_{ij}. T_{ij} is defined in formula (1):

$$T_{ij} = \lambda DT_{ij} + (1-\lambda)RT_{ij} \quad 0 \le \lambda \le 1 . \tag{1}$$

Where the trust value T_{ij} is defined in [-1, 1] and λ denotes the "trust-confidence factor", which means that how a peer is confident to its evaluation of direct trust value.

3.1 Direct Trust Metric

Let E_{ij}^{s} denotes peer *j*'s evaluation scores in the view of peer *i* on subjective dimension, SD. Let E_{ij}^{k} denotes peer *j*'s evaluation in view of peer *i* on objective dimension *k*, OD_k. According to the definition (3) of multi-dimensional trust type, DT_{ij} is defined in formula (2):

$$DT_{ij} = w_s E_{ij}^{s} + w_1 E_{ij}^{1} + w_2 E_{ij}^{2} + \cdots + w_n E_{ij}^{n} = w_s E_{ij}^{s} + \sum_{k=1}^{n} w_k E_{ij}^{k} \quad n = 1, 2 \cdots \cdot \tag{2}$$

The value of E_{ij}^{s} can be calculated in Table1 according to peer *j*'s transaction results or service quality in the detection of peer *i*. The calculation of E_{ij}^{k} should be different from different the setting of P2P application domains. Malicious behaviors or interrupting transaction will lead to low scores of service provider. So if peer *j* wants to get higher evaluation scores, it must provide good services and not terminate the service arbitrarily. Therefore, subjective dimension evaluation has the function of rewards and punishment.

Table 1. The value of E_{ij}^{k}

Value	Value Explanation
1	Very successful transaction, good service quality of j.
0.5	Successful transaction, general service quality of j
0	Uncertain(default situation, i does not estimate)
-0.5	Failed transaction, low service quality of j or j terminates service arbitrarily.
-1	Worst transaction, j exists malicious behaviors.

Let $V_{ij}^{s} = w_s E_{ij}^{s}$, $V_{ij}^{k} = w_k E_{ij}^{k}$ denotes peer *i*'s evaluation value to peer *j* on subjective dimension and objective dimension *k* respectively. So DT_{ij} can be also calculated in formula (3):

$$DT_{ij} = V_{ij}^{s} + V_{ij}^{1} + V_{ij}^{2} + \cdots + V_{ij}^{n} = V_{ij}^{s} + \sum_{k=1}^{n} V_{ij}^{n} \ . \tag{3}$$

3.2 Recommendation Trust Metric

Recommendation trust can acquire more comprehensive trust information and correct the deviation of direct experience, and to improve the reasonability and veracity of trust metric.

However, for the open and self-determination of the P2P networks, the peers may take collusion and bad-mouthing attacks to the reputation-based trust model to maximize their own benefits [8]. For example, in e-commerce application systems, a peer may collude with some peers in order to bad-mouthing other peers that it wants to drive out of the market. In such a situation, the conspiring peers provide unfairly negative ratings to the targeted peers, thus lowering their trustworthiness.

To resist these two malicious behaviors, PeerTrust proposes a personalized similarity measure (PSM) to rate recommendation peer's feedback credibility [4], and a set of experiments that show the feasibility and benefit of this approach. Therefore, in this paper, the credibility of feedback is evaluated based on the discrete degree of "evaluation similarity set". Firstly, the evaluation similarity between peer i and recommendation peer r, S_{ir}, is defined as follows:

Definition 4. S_{ir} reflects the similarity of feedback evaluation information between rating peer i and recommendation peer r. We use the cosine of the angle to measure S_{ir}.

If the cosine of the angle is greater than some threshold value, the two vectors are considered to be similar.Let *IS(i)* denote the set of peers that have interacted with peer i, the common set of peers that have interacted with both peer i and r, denoted by $CS(i,r)$, $CS(i,r)=IS(i)\cap IS(r)$. Service provider j which have transacted with peer i and r, is a peer in $CS(i,r)$. The evaluation values in the view of i and r to j on every trust dimensions constitute into two vectors respectively, and denoted by: $\overrightarrow{IJ}=(V_{ij}^{s},V_{ij}^{1},V_{ij}^{2},\cdots V_{ij}^{n})$ and $\overrightarrow{RJ}=(V_{ij}^{s},V_{ij}^{1},V_{ij}^{2},\cdots V_{ij}^{n})$.

So the evaluation similarity between i and r to service provider peer j, $S_{ir}(j)$, can be measured by the cosine of the angle θ between the $\overrightarrow{IJ}$ and $\overrightarrow{RJ}$ in formula (4):

$$S_{ir}(j)=\cos(\theta)=\overrightarrow{IJ}\bullet\overrightarrow{RJ}=\frac{V_{ij}^{s}V_{rj}^{s}+\sum_{k=1}^{n}V_{ij}^{k}V_{rj}^{k}}{\sqrt{(V_{ij}^{s})^{2}+\sum_{k=1}^{n}(V_{ij}^{k})^{2}}\sqrt{(V_{rj}^{s})^{2}+\sum_{k=1}^{n}(V_{rj}^{k})^{2}}}\,. \tag{4}$$

Let $CS(i,r)=\{c_1,c_2,\ldots c_n\}$ denote the common set of peers between peer i and the recommendation peer r. The evaluation similarity between i and r to each peer C_i in $CS(i,r)$ have formed the set of evaluation similarity $S=\{S_{ir}(c_1), S_{ir}(c_2), \ldots, S_{ir}(c_n)\}$. The feedback credibility of r in view of i, C_{ir}, can be measured by the standard deviation's reciprocal of the set of evaluation similarity S in formula (5):

$$C_{ir}=\left(\sqrt{\frac{\sum_{k\in CS(i,r)}(S_{ir}(k)-\overline{S_{ir}})}{|CS(i,r)|}}\right)^{-1}\,. \tag{5}$$

Where $\overline{S_{ir}}$ denotes the mean value of the evaluation similarity set *S*. If the feedback credibility C_{ir} is greater, the evaluation views between peer *i* and recommendation peer *r* are more consistent and peer *r*'s feedback evaluation information is more credible. So the recommendation trust value from *i* to *r* is defined in formula (6):

$$RT_{ij} = \sum_{r \in IS(j)} \frac{C_{ir} \times DT_{rj}}{\sum_{r \in IS(j)} C_{ir}}. \tag{6}$$

Where *IS*(*j*) denotes the set of recommendation peers, namely, the peers interacted with *j* directly except peer *i*. Finally, according to formula (1), the trust value of peer *j* in the view of peer *I*, T_{ij} is defined in formula (7):

$$T_{ij} = \lambda\left(V_{ij}^{s} + \sum_{k=1}^{n} V_{ij}^{k}\right) + (1-\lambda)\left(\sum_{r \in IS(j)} \frac{C_{ir} \times DT_{rj}}{\sum_{r \in IS(j)} C_{ir}}\right). \tag{7}$$

4 Experiments and Analyses

In order to analyse our trust model, we conduct simulation experiments for file sharing in P2P system. According to successful downloads and authentic files, we judge whether the trade is successful. In MDTrust, the trust information can be preserved through distribution hash table (DHT) mechanism. We use CAN to implement this mechanism [9]. Because of the single computer simulation environment, the DHT mechanism has been simplified, that is, every peer's trust and transaction information is stored in the designated peer arbitrarily.

4.1 Simulation Parameters

We simulate a P2P network with 300 peers and every peer is both requesters and service providers. 3000 sharing files (including fake files) are distributed in nodes randomly and the queries for sharing files are also random. The peers randomly request a certain file every time window, 5 seconds. After downloads finish, replicas of files will be stored in local nodes. Every user will download queried files from peers with the highest trust value. The initial trust value is 0. The scale of virtuous peers is 50%~ 100% .The scale of malicious peers is 0%~50%. Let trust-confidence factor λ is 0.5.

We simulate two kinds of peers, virtuous peers and malicious peers. Virtuous peers have good reputation and provide honest service and recommendation feedback information. However, malicious peers can be divided into two kinds, simple malicious peers and complex malicious peers, denoted by SM and CM peers. SM peers provide fake files or false service, but truthfully report feedback information. CM peers not only provide fake files or false service but also have collusion and badmouthing malicious behaviors, providing false feedback information to other querying peers.

We set four trust dimensions in MDTrust for P2P file sharing system. There are subjective dimension and three objective dimensions, success ratio dimension, cost time dimension and file size dimension, denoted by S_{od}, T_{od}, F_{od} respectively. The evaluation value of subjective dimension E_{ij}^{s} can be calculated by Table1. The metrics of S_{od}, T_{od}, F_{od} dimensions are defined as follow:

$$S_{od} = \frac{S_{count}(i,j) - F_{count}(i,j)}{S_{count}(i,j) + F_{count}(i,j)}. \tag{8}$$

Where $S_{count}(i,j)$ denotes the number of successful downloads between peer i and j. $F_{count}(i,j)$ represents the number of failed transactions. Peer i update the transaction success information of j according to j's acquired evaluation on subjective dimension. If the evaluation scores on subjective dimension $E_{ij}^{s}>0$, then $S_{count}(i,j)+1$. If $E_{ij}^{s}<0$, then $F_{count}(i,j)+1$.

$$C_{od} = 1 - \frac{C_{ij}}{C_{i}}. \tag{9}$$

Where C_{ij} represents the transaction time between peer i and j in evaluated transaction. C_i denotes the transaction time between peer i and all peers who have transacted with i.

$$F_{od} = \frac{F_{ij}}{F_{i}}. \tag{10}$$

Where F_{ij} represents the size of download file which is peer i download from j in evaluated transaction. F_i represents the size of all download files which are peer i have transacted with all peers.

The weight of every dimension is different to reflect the different preference of peers. We use the method of Analytical Hierarchy Process (AHP) which presented by Satty to set the weight of every dimension [10]. In this paper we design three kinds of peers which have different weights in our experiment. The weight of each dimension of peers is defined in Table 2:

Table 2. Weights of every dimension

Peer	w_s	w_{sod}	w_{cod}	w_{fod}
Peer I	0.5	0.25	0.125	0.125
Peer II	0.5	0.125	0.25	0.125
Peer III	0.5	0.125	0.125	0.25

4.2 Evaluation and Comparison

Firstly, we evaluate the effectiveness of MDTrust. We suppose 30% peers are malicious peers, where 1/3 malicious peer is CM peers. We calculate the changes of successful downloads ratio with increasing simulation time windows. The result of simulation is showed in Fig.1. With the simulation time increasing, when the simulation time window is 3500, MDTrust model is effective and the successful downloads ratio is very high (at 82%).

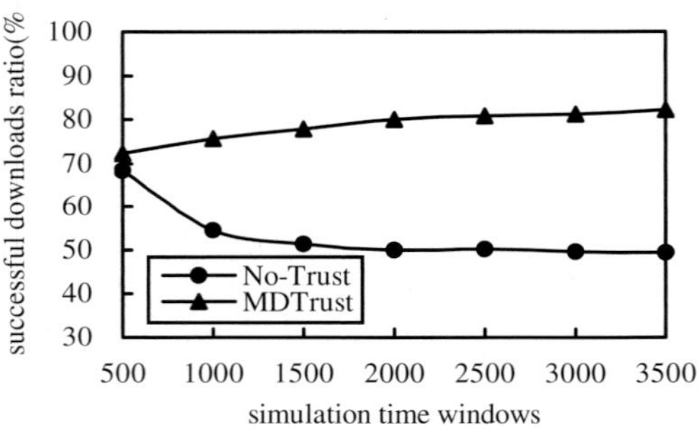

Fig. 1. Successful downloads ratio changes with simulation time windows

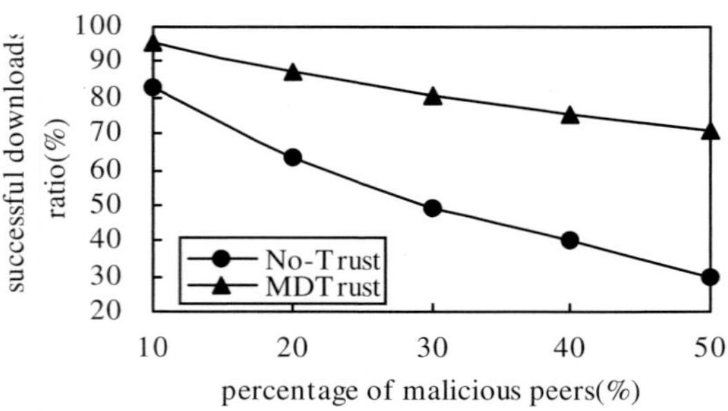

Fig. 2. Successful downloads ratio in different-scale malicious nodes

Secondly, we suppose different percentage of malicious peers provide fake service. Every scenario runs 2000 simulation time windows. From the Fig.2 we can conclude, with the ratio of malicious peers increasing, our trust model can maintain the successful downloads ratio in a high level, even when malicious peers take 50% of the whole peers, successful downloads ratio is still above 70%.

Thirdly, we evaluate the ability of the trust model based on multi-dimensional trust evaluation to resist collusion and bad-mouthing attacks. The result of this simulation experiment is showed in Fig.3 and Fig.4.

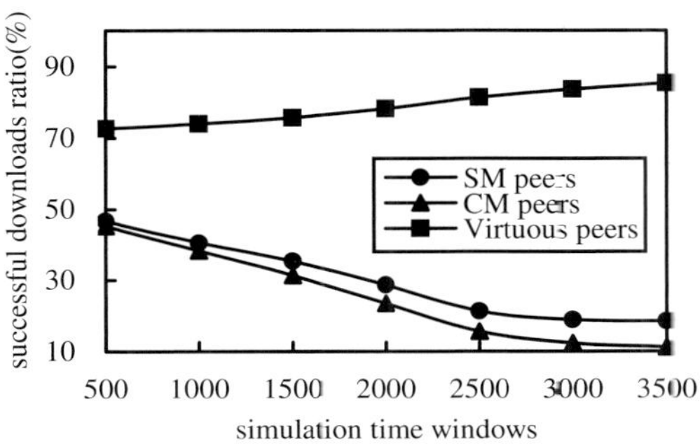

Fig. 3. Successful downloads ratio of three types peer changes with simulation time windows

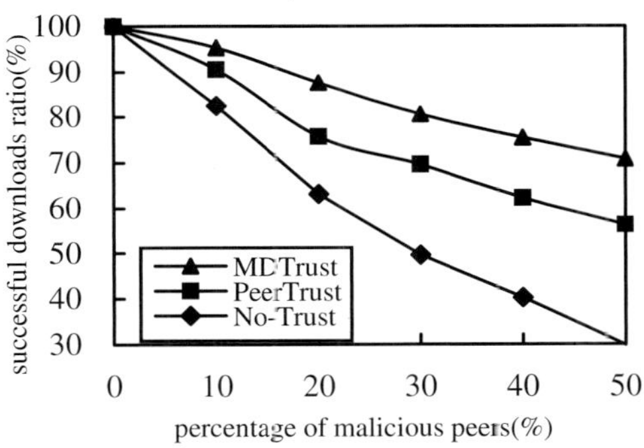

Fig. 4. Successful downloads ration with different scale malicious peers in three situations

In Fig.3 we can conclude that, CM malicious peers' collusion and bad-mouthing behaviors hardly disturb virtuous peers' judgment. However, the collusive behaviors of CM malicious peers have greatly influence on it selves. Because the malicious peers' trust value is very low, they acquire download service very hardly, so their successful download ratio is not high. From the Fig.4, we conclude that the metric of feedback credibility has enhanced the veracity of the ability of aggregating recommendation information. MDTrust improve the successful downloads ratio when existing a lot of false recommendation information and slanders, and it works better than PeerTrust in collusive setting.

5 Conclusions

In this paper, we propose a trust model based on multi-dimensional trust evaluation in P2P system, which introduces the concept of dimensional trust and takes consideration for subjective and objective evaluation. In this model, the metric of feedback credibility of recommendation peer is provided to resist collusion and bad-mouthing attacks. The results of simulation show that this model can quickly build up the trust relationships among peer, and effectively improve the successful transaction ratio.

References

1. Dou, W., Wang, H.M., Jia, Y., Zou, P.: A recommendation-based peer-to-peer trust model (in Chinese with English abstract). Journal of Software 15(4), 571–583 (2004)
2. Oram, A.: Peer-to-Peer: Harnessing the Power of Disruptive Technology, pp. 222–238. O'Reilly Press, Sebastopol (2001)
3. Kamvar, S.: EigenRep: Reputation management in P2P networks. Stanford University, Tech. Rep: SCCM-02-16 (2002)
4. Xiong, L., Liu, L.: PeerTrust: Supporting reputation-based trust in peer-to-peer communities. IEEE Transactions on Data and Knowledge Engineering, Special Issue on Peer-to-Peer Based Data Management 16(7), 843–857 (2004)
5. Dou, W.: The research on trust-aware P2P topologies and constructing technologies [Ph.D. Thesis] (in Chinese with English abstract). National University of Defense Technology, Changsha (2003)
6. Lam, S.K., Riedl, J.: Shilling recommender systems for fun and profit. In: Proceedings of the 13th World Wide Web Conference (2004)
7. Maximilien, E.M., Singh, M.P.: Agent-Based Trust Model Involving Multiple Qualities. In: Proceedings of International Conference on Autonomous Agents and Multi-Agent Systems(AAMAS 05), The Netherlands, pp. 519–526 (2005)
8. Dellarocas, C.: Immunizing Online Reputation Reporting Systems Against Unfair Ratings and Discriminatory Behavior. In: ACM Conference on Electronic Commerce, pp. 150–157. ACM Press, New York (2000)
9. Ratnasamy, S.P.: A Scalable Content Addressable Network. A dissertation submitted in partial satisfaction of the requirements for the degree of Doctor of Philosophy in Computer Science in the Graduate Division of the University of California at Berkeley [D] (2002)
10. Satty, T.L.: Inner and outer Dependence in the Analytic Hierarchy Process: The Supermatrix and Super-hierarchy [A]. In: Proceeding of the 2nd ISAHP[C], Pittsburgh (1991)

A Fuzzy Mapping from Image Texture to Affective Thesaurus

Haifang Li[1], Jin Li[1], Jiancheng Song[2], and Junjie Chen[1]

[1] College of Computer and Software,
[2] College of Electrical and Power Engineering
Taiyuan University of Technology, Taiyuan, Shanxi, China, 030024
sxlhf123@163.com, lenven.li@gmail.com

Abstract. Most of the image analysis methods focus on affective information by using color. In this paper a new indexing scheme, called FRD (Fuzzy Recognize Degree)-Clustering is proposed for affection-assisted semantic image analysis on the basis of the texture feature. A set of perceptual relevant features, used for indexing is hence introduced: directionality, contrast and coarseness. This FRD scheme allows us to retrieve images based on high-level affective concepts. Experiments with images of the nature and landscape domain demonstrate the performance of the proposed approach.

1 Introduction

As feature for the human visual system, texture reveals scene depth and surface orientation; moreover, it describes properties like the smoothness, coarseness, and regularity of a region [1] [2] [3].

Texture is not as well defined as color features; some systems do not use texture features. However, texture provides important information in image classification as it describes the content of many real-world images such as fruit skin, clouds, trees, bricks, and fabric. Hence, texture is an important feature in defining high-level semantics for image retrieval purpose [4].

Textures have attracted considerable attention both from the Computer Vision community and the Computer Graphics community. Indeed to understand the processes behind the human perception of an image as a "pattern" or a "texture" involves deep questions about human brain organizations that are relevant for those that try to simulate the human vision mechanism. The first and oldest approach to texture classification is based on statistics. Textures are described using mathematical models and a set of purely statistics features of a texture can be computed in order to achieve both supervised and unsupervised classification.

In order to bridge this semantic gap, the usage of appropriate prior knowledge is very important [5]. Ontology, user preference profiles, and relevance feedback techniques were developed to utilize such knowledge. However, such methods require an enormous effort and consequently can only be applied in a limited domain [6]. We address the semantic gap from another angle, since we aim at developing techniques

K. Li et al. (Eds.): LSMS 2007, LNCS 4688, pp. 357 – 367, 2007.

that are human-based and may lead to generic methods that were applied in an unlimited domain.

In this paper a new clustering scheme, called FRD-Clustering is proposed to guarantee higher searching quality. By following this scheme, images can be searched by affective concepts, which are derived from Tamura texture space. Affective concepts, which are kind of semantic interpretation on images felt by human, are automatically extracted and represented as FRD (Fuzzy Recognize Degree) within image databases. Using these FRD cluster degrees, the proposed system can support image searching by affective concepts. For example, the affective queries, such as "find cool images" and "find lovely images," can be directly supported based on the semantic features derived from texture information of images. We propose an algorithm to generate FRD cluster degrees from texture information in Tamura texture degrees. We also show how these FRD cluster degrees can be used for grouping and searching images by affective concepts.

2 Texture Space Understanding

The term "texture", although it has been used widely, does not have a commonly accepted definition. Haindl states that: "Texture is generally a visual property of a surface, representing the spatial information contained in object surfaces." Corresponding to the many definitions of texture, there are several approaches to analyzing it.

Texture provides important information in image classification as it describes the content of many real-world images such as fruit skin, clouds, trees, bricks, and fabric. Hence, texture is an important feature in defining high-level semantics for image retrieval purpose.

The description of texture by perceptual relevant features has involved many researchers both in computer vision and visual perception. The seminal work in this research line is due to Tamura et alii that described and provided computational measures for six perceptual features validating their proposal with subjective experiments.

In the rest of these section three perceptual features, directionality, contrast and coarseness will be described and computationally defined. These three indices represent an "affective-sign" used to discriminate textures in a database in a coherent way to the human perceptual system.

2.1 Directionality

Our measure of directionality is based on co-occurrence matrices, first introduced by Haralick.

Differently than where the preferred direction (vs. isotropy) is derived analyzing the distribution of the angles through the gradient values, we focused on the computation of the confidence degree for a given orientation of the texture. In other words instead of computing a global value of "directionality", we provide a "degree of confidence of relevance" for a set of orientations.

Let T a texture of size *n x m x* c colors and *v=(x, y)* an offset vector, the co-occurrence matrix *C (T, v)* is a c x c matrix defined in each point by:

$$C(T,v)_{i,j} = |\{ (p,q) \, in \, T \, x T : q = p + v \, and \, L(p) = i \, and \, L(q) = j\}| \quad (1)$$

where *L(p)* is the luminance value of the pixel *p*.

Then a point *(i, j)* in *C* contains the number of pixels pairs in *T* that have respectively gray level *i* and *j* and with displacement vector *v*.

Co-occurrences matrices collect second-order statistics of a texture that, according to Julesz's conjecture, are used for discrimination; they have been largely used also for classification and retrieval trough a set of well-known measures.

The proposed measure is based on a simple idea: the plot of the main diagonal of a co-occurrence matrix with offset v is as closer to the histogram of the image as v is relative to a relevant direction.

Starting from this observation we suggest to quantitatively estimating the relevance of a direction computing how much the diagonal of the co-occurrence matrix relative to such a direction deviates from the histogram of the image.

More precisely, given a co-occurrence matrix *C (T, v)*, let $d_C(t) = C(T,v)_{t,t}$, $t = 0 \ldots c$ and *H (t)* the histogram of *T*. We propose to compute the area between the plot of the histogram and d_C as a measure of the deviation of the co-occurrence matrix with respect to the histogram.

The discrepancy of matrix C is defined by the expression:

$$d(C) = \sum_{t=2}^{c-1} [H(t) - d_c(t)] + \frac{1}{2} \times [H(1) - d_c(1)) + H(c) - d_c(c)] \quad (2)$$

The expression above is similar to trapezium method formulae for the numerical integration.

Alternatively the co-occurrences matrices are often described in literature to emphasize the distance from pixel and direction. This representation uses distance (radius) d and angle θ instead of vector v. In this notation for example *C (T, v (1, 0))* becomes *C (T, d=1, θ=0)*. The advantage of this representation lies on the immediate interpretation of the spatial relationship of pixel in the co-occurrence matrix. Thus, in order to observe the behavior of the directionality of the texture, the discrepancy values must be computed for a set S of vectors with increasing distance *d=1, 2..., D*.

To obtain the relevance of a direction from the discrepancy measure defined above over a finite family *S* of offsets we suggest the following: let D_S be the matrix of the discrepancy values of the co-occurrence matrices relative to the offsets in S and d_s be the set of the normalized values (levels *d=1, 2, ..., D*) of D_S in *[0,1]*. We claim that the set $R_s = 1 - d_s$ provides a set of relevance measure of the directions in the family S for the texture under examination.

Fig.1 shows the 2 textures in the VisTex database on which the proposed measure with respect to the right diagonal direction gives the two extreme values.

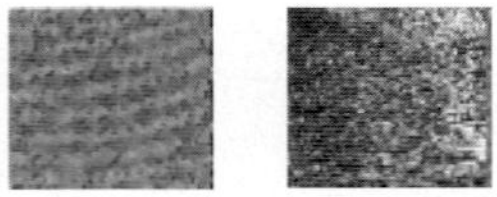

Fabric0 Fabric4

Fig.1. The less and the most directional (as respect to the right diagonal direction) texture on view

2.2 Contrast

Local contrast is commonly defined for each pixel as an estimate of the local variation in a neighborhood. More precisely, given a pixel *p= (i,j)* and neighbor mask W . W of the pixel, local contrast is computed as:

$$local\_contrast\,(i,j) = \frac{max_{p\in W\times W}(p) - min_{p\in W\times W}(p)}{max_{p\in W\times W}(p) + min_{p\in W\times W}(p)} \tag{3}$$

We propose to measure the global contrast as the global arithmetic mean of all the local contrast values over the image:

$$contrast = \frac{1}{m*n} * \sum_{i=1}^{n}\sum_{j=1}^{m} local\_contrast\,(i,j) \tag{4}$$

where n, m are the dimensions of the image.

We have computed global contrast using different windows sizes (3x3, 5x5 …). As the window size changes the contrast measure changes of a small factor that is approximately the same for all the images in our test databases. For this reason we decided to take into account windows of fixed size (i.e. 9x9).

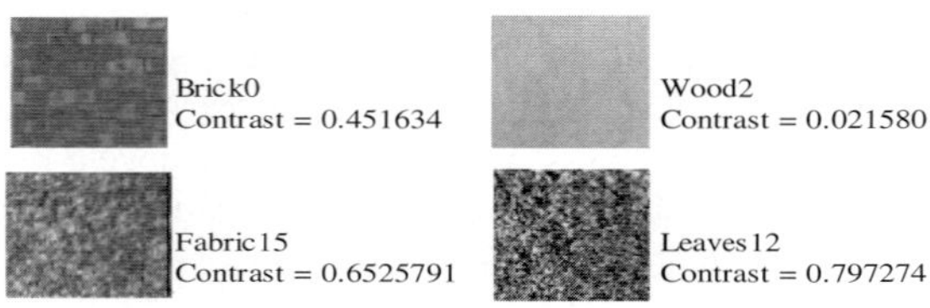

Fig. 2. The less and the most contrasted texture on view

2.3 Coarseness

Coarseness is probably the most essential perceptual features, since in many cases the word "coarseness" is used instead of "texture". If two textures differ only in scale, then the magnified one is coarser.

According to WordNet definition coarseness is: "the quality of being composed of relatively large particles [syn: graininess, granularity]".

The computational definition used in this paper, derived from Rosenfeld, consists of the computation of the size of the structural element of texture using various sized operators.

Their method is a multi-step algorithm that is here briefly described:

Step 1: Build K images in which each element is the average of intensities in a neighboring $2^k \times 2^k$, $k=1, 2... K$:

$$A_k(x,y) = \frac{1}{2^{2k}} * \sum_{i=x-2^{k-1}}^{x+2^{k-1}-1} \sum_{j=y-2^{k-1}}^{y+2^{k-1}-1} T(i,j) \tag{5}$$

Step 2: Take differences between pairs of average that correspond to non- overlapping neighborhood both in vertical and horizontal orientation:

$$E_{k,horizontal}(x,y) = | A_k(x+2^{k-1}, y) - A_k(x-2^{k-1}, y) | \tag{6}$$

$$E_{k,vertital}(x,y) = | A_k(x, y+2^{k-1}) - A_k(x, y-2^{k-1}) | \tag{7}$$

Step 3: For each point, compute the best size of the neighbor mask (that is the structural element) as the maximum difference among adjacent regions. $S_{best}(x,y) = 2^k$, where k maximizes the differences: $E_k = max(E_{k,horizontal}, E_{k,vertical})$ In our experiments we prefer to take $S_{best}(x,y) = k$.

Step 4: Compute a global coarseness value as the average of the S_{best}:

$$coarseness = \frac{1}{m * n} \sum_{i=1}^{n} \sum_{j=1}^{m} S_{best}(i,j) \tag{8}$$

In order to obtain a degree of coarseness, we normalized S_{best} values in the range [0, 1].

3 FRD-Clustering Technologies for Texture Image

In this section, the overall process for FRD-cluster based texture image retrieval is explained. At first, the Tamura texture features are extracted from images and are stored into databases. These Tamura texture features are converted into FRD-cluster.

FRD-cluster is stored in an affective categories table. To search an image by affective concept, FRD-cluster degrees are utilized. The detailed procedure is summarized in Fig.4.

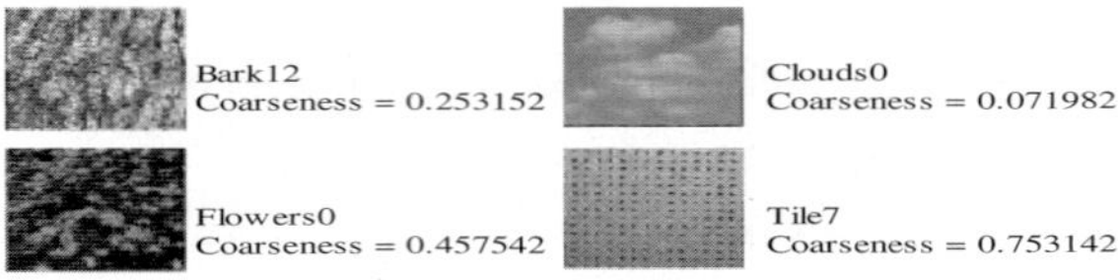

Fig. 3. The coarsest and the finest texture on view

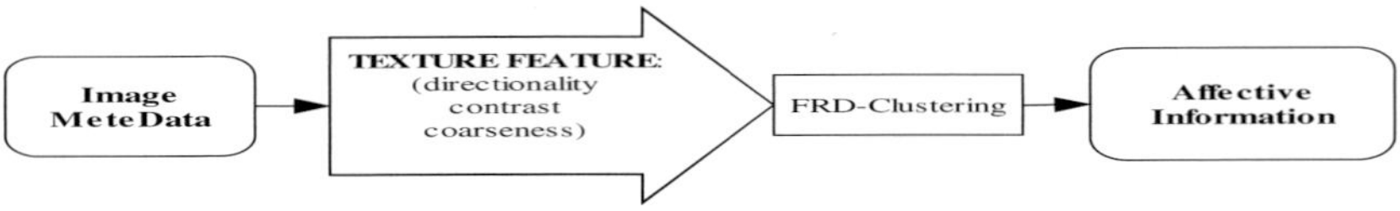

Fig. 4. Procedure Diagram and Framework for FRD-Clustering

3.1 A High-Level Semantics of Textures

FRD-cluster represents a conceptual distance between affective terms as fuzzy recognize functions and estimates texture semantics based on Tamura (directionality, contrast and coarseness) degree. Table 1 shows twelve classes about affective terms by Tamura texture features. According to directionality, an image can give regular or irregular feeling. And according to contrast, an image can give soft or hard feeling. In addition, an image can give dynamic or static feeling by coarseness.

Table 1. Texture Semantics

Class	Semantics
1	homogeneous / non homogeneous
2	geometrical / non geometrical
3	soft / rough
4	fine / coarse
5	with lines / without lines
6	flat / non flat
7	regular / irregular
8	symmetrical / non symmetrical
9	clear / dark
10	simple / complex
11	defined / diffuse
12	natural / artificial

Affective terms, which are felt by humans after looking at images, are very diverse and ambiguous. But some textures that make their distinction very clear make classification easier. In case of directionality, "regularly" or "smooth" feeling could be made. In this way, contrast makes a feeling such as "real" or "active." Therefore, recognize degrees of textures and affective terms can be measured by the following formula (9):

$$Frd(f,a) = Max\{Min[Frd(f,t_i), Frd(t_i,a)]\}, 0 \leq i \leq 11 \quad (9)$$

Such affective term space for a texture represents the term recognizes degree as fuzzy measures and stores it into databases. Table 2 shows a graph, which represents affective term, recognize degrees as fuzzy recognize degree, FRD.

Table 2. Fuzzy Recognize Degree for Sensation of Texture

Affective terms	Softness	Clearness	Regularity
active	0.98	0.90	0.50
strong	0.90	0.70	0.48
romantic	0.70	0.90	0.8
rhythmic	0.80	0.88	0.65
beautiful	0.75	0.90	0.65
fair	0.78	0.70	0.60
handsome	0.64	0.75	0.65
pretty	0.70	0.90	0.80
lovely	0.65	0.85	0.78

If F is related to affective terms about texture, the possibility of "a" being included in F, is calculated. For instance, let's assume that a fuzzy set of affective terms about "softness" and "clearness" are clearness = { (0.90, active), (0.70, strong), (0.9, romantic), (0.90, beautiful)}, (0.9,pretty)}, softness = { (0.98, active), (0.9, strong), (0.7, romantic), (0.75, beautiful), (0.7, pretty)} and a fuzzy set of affective term "dynamic" is dynamic = { (0.98, softness), (0.95, clearness), (0.8, regularity), (0.5, fineness), (0.1, flatness), (0.9, geometrics)}. Then, recognize degree of affective terms such as "active" and "dynamic" can be calculated like this:

$$\begin{aligned} Frd(dynamic, active) &= Max\{Min[Frd(dynamic, softness), Frd(softness, active)], \\ &\quad Min[Frd(dynamic, clearness), Frd(clearness, active)]\} \\ &= Max\{Min(0.95, 0.98), Min(0.90, 0.98)\} \\ &= 0.95 \end{aligned}$$

Therefore, an image with a "softness" or "clearness" texture holds "dynamic" feelings and relationship probability, whereas an affective term such as "active" included in the term "dynamic" can reach to 0.95. As a result, an image having "softness" or "clearness" texture could be an "active" image.

3.2 FRD-Cluster Generation

Texture semantics of images are grouped together according to a distribution of directionality. FRD-clusters for classified images are stored into databases. Tamura degree is composed of 360 degrees on a directionality basis. In this paper, basic texture, used to distinguish texture by human's eyesight, is classified into twelve categories, as shown in Table 1.

In case of softness, it has a degree of range (15>=d>= 0 and, 360>= d>= 345). Twelve textures become more dim or deep in accordance with contrast and more soft or hark according to coarseness. For example, if (Direction, Contrast, Coarseness) degree of image I_n is (5, 250, 30), degree of directionality exists between the range 15>= d >= 0, 360>= d>= 345 and concludes a texture of the image I_n as softness category. But, an important point is that the depth of softness texture can be less according to a degree of contrast, even if the degree of directionality lies in the range 15>= d >= 0, 360>= d >= 345. If the degree of contrast lies in the range (20>= con >= 0), it is impossible to recognize a softness texture.

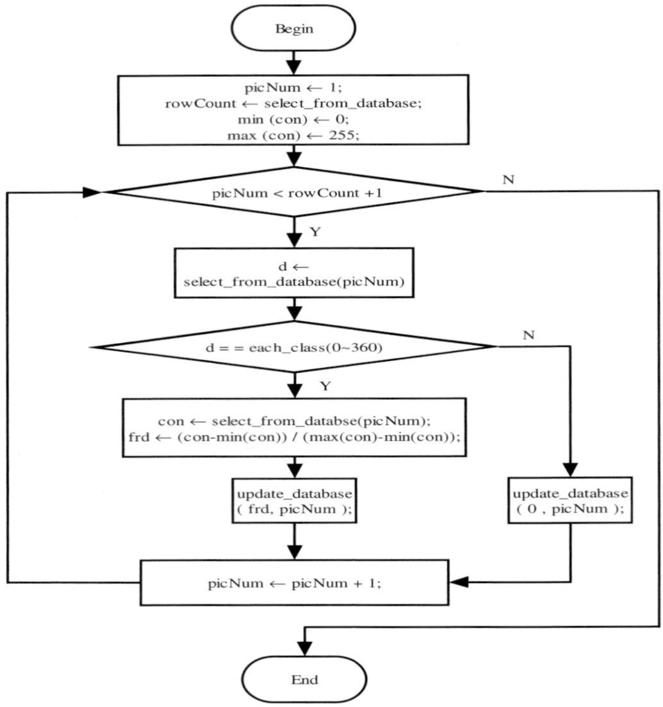

Fig. 5. Algorithm of FRD-Clustering Generation

In this respect, it can be inferred that the degree of contrast plays a key role for human to feel emotions by seeing textures. In addition, a feeling about texture could be confused when the degree of directionality varies. For example, if a degree of directionality becomes 0, distinction of a texture is clear because a degree of contrast is an axis of texture. If a degree of directionality lies between 15 and 20, distinction of a texture could be ambiguous. For that reason, when texture varies, a range of error tolerance is ±5 based on twelve textures. Fig 5 shows an algorithm that FRD-cluster is automatically produced using Tamura (directionality, contrast and coarseness) degree stored in a database.

3.3 Image Indexing and Searching by FRD Cluster

For each texture image, FRD cluster degrees are produced by FRD cluster algorithm and its category is determined according to the predefined twelve affective grouping classes. Fig 6 shows the process that stores image data into affective grouping class. Each affective grouping class is built up of an affective term thesaurus to support affective adjectives and search images with a terminology dictionary related to affective adjectives. In case of a query, "find active images," affective adjectives are first scanned. If there is no proper adjective, a resembling word is used as an affective adjective. That affective adjective is used to calculate fuzzy recognize, which is used to search FRD cluster in an affective grouping table. At this moment, a weight (α) can be assigned by affiliated degree.

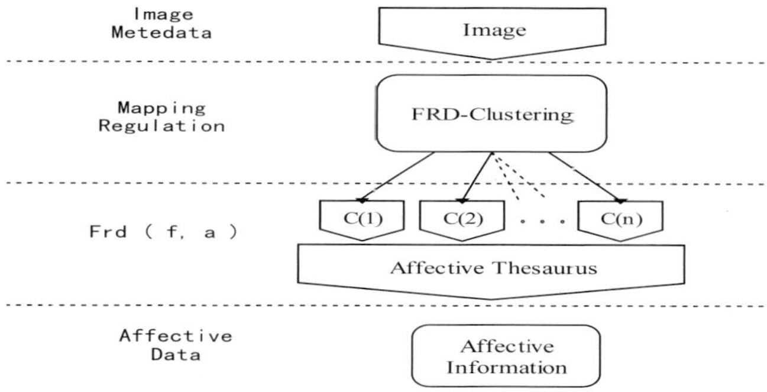

Fig. 6. Classification and Searching Images by Affective Concepts

4 Experimental Results

We used several databases in order to evaluate the FRD-Clustering performances, to study more about the FRD and to compare the FRD performances with other traditional methods. The experiments were conducted on the MIT Vision Texture (VisTex) database.

Different color texture images are used for the experiments described in this Section. Randomly selected 200 images such as scenery, animal and flower images were used in the experiment.

Image searching techniques with affective concepts uses similarity-based matching technique instead of exact-matching technique. Recall signifies the percentage of relevant images in the database that are retrieved in response to the query. Precision is the proportion of the retrieved images that is relevant to the query:

$$P_r = \frac{number\ retrieved\ that\ are\ relevant}{total\ number\ retrieved}$$

$$R_t = \frac{number\ retrieved\ that\ are\ relevant}{total\ number\ relevant}$$

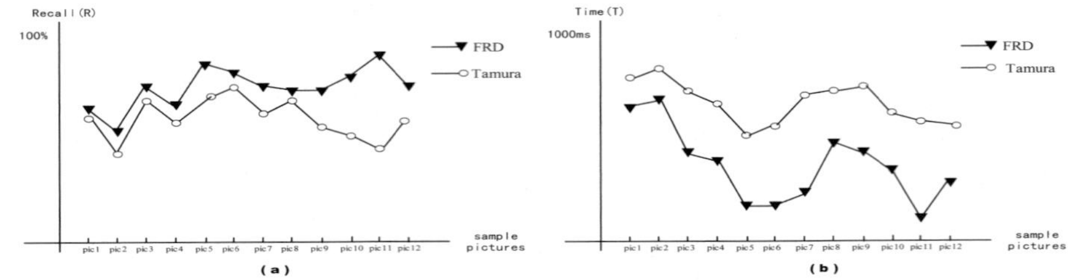

Fig. 7. Performance comparison of Tamura and FRD in Recall (a) and Time (b)

5 Conclusions

In this paper, FRD clustering techniques on the basis of texture features of image, with related grouping and searching algorithms are proposed to allow image searching by affective concepts. An efficient method using some perceptual features such as directionality, contrast and coarseness is proposed. The affective adjectives applied to images and fuzzy degrees are automatically obtained based on human visual features. The proposed FRD cluster-searching scheme does support semantic-based retrieval that depends upon human sensation and emotion, as well as traditional texture-based retrieval. Affective concepts are classified into twelve classes according to affective expression and images are classified into these categories as well. Image searching speed can be improved by assigning FRD cluster degree to each class. As a result, more user-friendly and accurate searching, with affective expression, can be realized. The efficiency of the proposed techniques is compared with FRD and Tamura texture methods.

We believe that there should be further research on how to finely categorize texture information so that searching efficiency based on FRD cluster can be further improved. To guarantee much higher searching quality, research on how to adopt other image features combination together, such as color and shape, within FRD- Clustering scheme are also required.

Acknowledgements. The work is supported in part by the Key Project of The Natural Science Foundation of Shanxi Province (No. 200601130 and No.2007011050).

References

1. Battiato, S., Gallo, G., Nicotra, S.: Perceptive visual texture classification and retrieval. In: Ferretti, M., Albanesi, M.G. (eds.) Proceedings of the 12th International Conference on Image Analysis and Processing, Mantova, Italy, September 17-19, 2003, pp. 524–529 (2003)
2. Broek, E., Kisters, P.M.F., Vuurpijl, L.G.: Content-based image retrieval benchmarking: Utilizing texture categories and texture distributions. Journal of Imaging Science and Technology 49 (2005)
3. Rousson, M., Brox, T., Deriche, R.: Active unsupervised texture segmentation on a diffusion based feature space. In: Proceedings of the 2003 IEEE Conference on Computer Vision and

Pattern Recognition, Madison, Wisconsin, vol. 2, pp. 699–704. IEEE Computer Society Press, Los Alamitos (2003)

4. Liu, Y., Zhang, D., Lu, G., Ma, W.-Y.: A survey of content-based image retrieval with high-level semantics. Pattern Recognition 40(1), 262–282 (2007)
5. Landy, M.S., Graham, N.: Visual perception of texture. Chalupa, L.M., Werner, J.S. (eds.), MIT Press, Cambridge, MA (Under review for publication) (2002)
6. Wang, J.Z., Li, J., Wiederhold, G.: SIMPLIcity: Semantics-Sensitive Integrated Matching for Picture Libraries. IEEE TKDE 23(9), 947–963 (2001)

A New Grey Prediction Based Fuzzy Controller for Networked Control Systems

Lisheng Wei[1], Minrui Fei[1], Taicheng Yang[2], and Huosheng Hu[3]

[1] School of Mechatronics and Automation, Shanghai University, Shanghai 200072, China
lshwei_11@163.com
[2] Department of Engineering and Design, University of Sussex, Brighton, BN1 9QT, UK
[3] Department of computer science, University of Essex, Colchester CO4 3SQ, UK

Abstract. This paper proposes a novel fuzzy control method based on the research of grey theory and switching algorithm in networked control systems (NCSs). A grey prediction structure is used to obtain more important information from remote sensors, and an on-line rule switching mechanism is constructed to provide an appropriate forecasting step size to the grey predictor. The overall mathematical model of this grey prediction based fuzzy controller is established. By using this method to get an appropriate positive or negative forecasting step size in NCSs, the random characteristic of some non-stationary time series from the sensors can be reduced so as to improve control performance. Experiments on a nonlinear plant via communication network show its precision and robustness is better than other traditional control methods.

1 Introduction

Traditional Fuzzy Control (TFC) theory and method has been widely used in industry for its easy realization. Much work has been done on the analysis of control rules and membership function parameters. However, TFC strategies adopt the previous state information; mostly belong to “delay control”[1]. In many circumstances, these control strategies are feasible and practicable, but due to the limitations of our cognitive abilities and the information we can reach about the system, is always uncertain and limited in scope. So they can’t attain true real-time, accuracy and adaptability. And it is difficult to increase the quality of the control system further, especially in NCSs.

In 1980s, Chinese professor Deng J. L. put forward grey prediction theory[2,3]. Because grey model is built by only a few given data, it looks for the development rule of the system according to characteristic value of actions occurred, and predicts actions in the future, then confirm corresponding control strategies from the tendency. So it has better self-adapting, versatility, real-time, accuracy and good foreground. However, the size of the predicted distance could affect the variations of the output performance.

So a novel grey prediction based fuzzy controller (GPBFC) is proposed. A grey theory algorithm is introduced into the TFC to predict the future output error of the system and the error change, rather than the current output error of the system and the

K. Li et al. (Eds.): LSMS 2007, LNCS 4688, pp. 368 – 377, 2007.

current error change, as input variables of the TFC, and adopts a dynamic forecasting-step to predict and control a plant in different system response regions which is divided in terms of error. This design can not only reduce the system overshoot efficiently but also can maintain the characteristic of the shorter rise time of system.

This paper is organized as follows. In section 2, the structure and the mathematical model of grey prediction based fuzzy controller is constructed. In section 3, simulation results with the proposed control scheme are obtained. Conclusion remarks are given in section 4.

2 The Structure of Grey Prediction Based Fuzzy Controller

The traditional grey controller (TGC) structure uses a fixed forecasting step size to reduce or prevent the overshot, but it lengthens the rise time of the system response. We can conclude that it is difficult to improve the rise time and overshoot in chorus when we use a fixed step to control the system. In this paper we propose an on-line rule switching mechanism to regulate a suitable forecasting step size for each control action during the system response. When the error is big, we adopt negative prediction step. On the contrary, we adopt big positive prediction step. So the short rise time and the small overshot can be taken about at same time.

The block diagram of the on-line rule tuning GPBFC system is shown in Figure.1, in which the on-line rule switching mechanism is used to regulate the appropriate forecasting step size of the grey predictor. The grey predictor user a suitable forecasting step size to forecast the output of the plant and sends the forecasting information to the fuzzy controller.

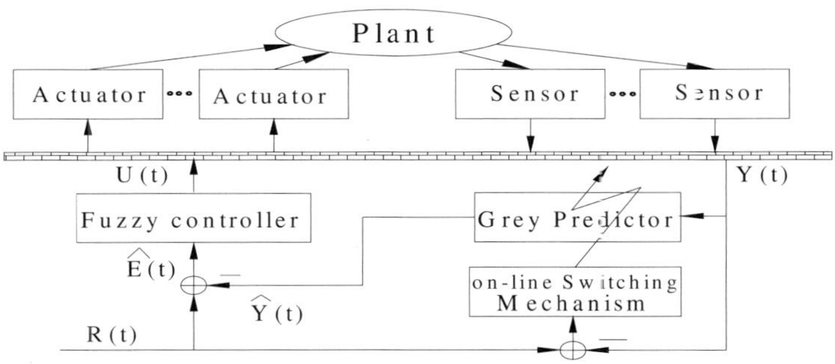

Fig. 1. The Structure of Grey Prediction Based Fuzzy Controller in NCSs

Where R(t) is the set value, Y(t) is the sensor output value, $\hat{Y}(t)$ is the grey prediction value, $\hat{E}(t)$ is the deviation: $\hat{E}(t) = R(t) - \hat{Y}(t)$, and U(t) is the output of the fuzzy controller. In Figure.1, the system response is divided into three regions, such that an algorithm to switch and adjust the size of the prediction step is developed. So three forecasting modes are obtained, which are one small positive-step prediction mode, one small positive-step prediction mode and one negative-step prediction mode. When the system error is large, the negative-step mode is used to increase the upward momentum of the output curve. When the system error is small, the positive-step mode is used to prevent the overshooting. The last condition is the middle error occurred. The whole grey prediction fuzzy controlling process is described as follow.

2.1 Traditional Grey Prediction Model

Different from the existing statistic methods for prediction, grey theory uses data generation method, such as ratio checking (RC) and accumulated generating operation (AGO) to reduce the stochastic of raw delay datum and obtain more regular sequence from the existing information. The regular sequence is used to estimate the parameters of the differential equation. The solution of the differential equation is then used to predict the moving trend of the target. The general form of a grey differential model is GM (n, m), where n is the order of the ordinary differential equation of grey model and m the number of grey variables. The variables n and m define the order of AGO, and the computing time increases exponentially as n and m increases. So we will use the traditional grey prediction model GM (1, 1) in this paper. The whole modeling procedure of optimization grey prediction will be introduced next.

In general, the traditional grey modeling differential equation can be derived by the follow basic steps: Ratio checking (RC), accumulated generation operations (AGO) and build a grey prediction model (GPM)[4-6].

First, check the plant output sequences $Y^{(0)}=(y^{(0)}(1), y^{(0)}(2), \cdots, y^{(0)}(n))$ ratio and transform the original sequences into a new sequences with AGO, We get

$$\text{RC:} \qquad \sigma^{(0)}(k)=\frac{y^{(0)}(k-1)}{y^{(0)}(k)} \tag{1}$$

$$\text{AGO:} \qquad y^{(1)}(k)=\sum_{m=1}^{k} y^{(0)}(m) \tag{2}$$

Where $k=2,\cdots,n$ is the sequences number; the new ratio sequences overlay area is$(e^{-2/(n+1)}, e^{2/(n+1)})$. When the ratio sequences are in overlay area, we can transform the original sequences into new more smooth sequences with AGO. Otherwise, we should take some pretreatment.

Second, build a grey model GM (1, 1). The GM (1, 1) model can be constructed by establishing a first order differential equation. Set up this equation as follows:

$$y^{(0)}(k)+az^{(1)}(k)=b \tag{3}$$

Where a and b are estimation parameters. $z^{(1)}$ is affected by $Y^{(1)}$,i.e.

$$z^{(1)}(k) = 0.5y^{(1)}(k) + 0.5y^{(1)}(k-1) \tag{4}$$

Here, we appoint

$$Y_N = [y_1^{(0)}(2) \quad y_1^{(0)}(3) \quad \cdots \quad y_1^{(0)}(n)]^T$$

$$B = \begin{bmatrix} -z^{(1)}(2) & -z^{(1)}(3) & \cdots & -z^{(1)}(n) \\ 1 & 1 & \cdots & 1 \end{bmatrix}^T; \qquad M = \begin{bmatrix} a \\ b \end{bmatrix}$$

So the Eq.(3) can be substituted as:

$$Y_N = B \bullet M \tag{5}$$

Then the optimal parameters M can be obtained by the minimum least square estimation algorithm:

$$M = (B^T B)^{-1} B^T Y_N \tag{6}$$

According to first order differential equation, the grey model GM (1, 1) can be derived in Eq.(7).

$$y^{(0)}(k) + 0.5 \cdot a[y^{(1)}(k) + y^{(1)}(k-1)] = b \tag{7}$$

That is,

$$(1+0.5a)y^{(0)}(k) + ay^{(1)}(k-1) = b$$

Based on the above description, we use inverse accumulated generation operations to obtain the grey prediction model as follow.

$$\hat{y}^{(0)}(n+p) = \left(\frac{1-0.5a}{1+0.5a}\right)^{(n+p-2)} \frac{b - ay^{(0)}(1)}{1+0.5a} \tag{8}$$

Where p is the prediction step; $\hat{y}^{(0)}(n+p)$ is the prediction value which is sent to the fuzzy controller.

2.2 Design of the Fuzzy Controller

TFC achieves many practical successes which guarantee the very basic requirements of global stability and acceptable performance. Two-input FC is easy to implement and receive great performance responses from simulations. And it is the most commonly researched in real systems. The framework of fuzzy control system is shown in Figure.2.

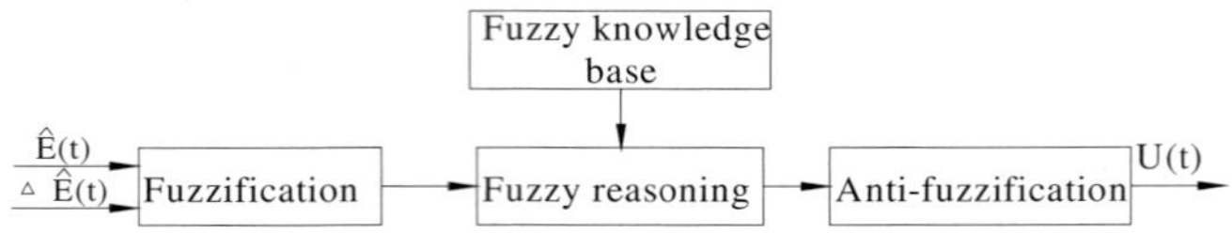

Fig. 2. Framework of fuzzy control system

Where $\hat{E}(t)$ is the deviation: $\hat{E}(t) = R(t) - \hat{Y}(t)$, $\Delta\hat{E}(t)$ is the change ratio of voltage deviation: $\Delta\hat{E}(t) = d\hat{E}(t)/dt$, and U(t) is the output of the fuzzy controller. The design follows the steps below[7,8].

First, Fuzzification

Fuzzification is to translate the exact data of the input variables into language variables (fuzzy inputs) in a proper scale, that is, to fix on the varying range of the input and the output as well as the quantification coefficients K_E , $K_{\Delta E}$, K_U , and get the elements of their corresponding language variables. The membership functions of input and output are shown in Figure.3 and Figure.4 respectively.

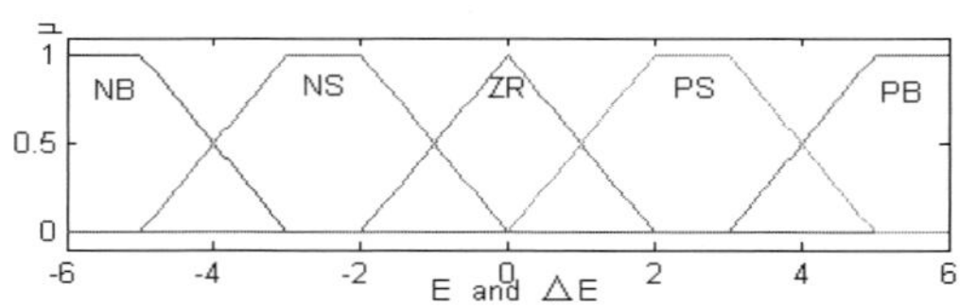

Fig. 3. Membership function of input (E and △E)

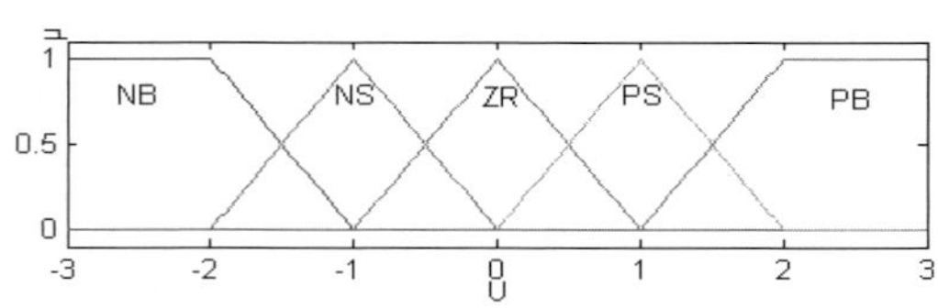

Fig. 4. Membership function of output (U)

Second, Fuzzy Reasoning

Fuzzy reasoning is the process of getting fuzzy output from fuzzy inputs by some reasoning rules and on the basis of the repository. Given the proper rules, we input the variables and the control variable in five language data: NB, NS, ZR, PS and PB. According to the working characters of the system in this paper, we conclude the reasoning rule in the form of "IF THEN". All the control rules are shown in Table 1.

Table 1. Fuzzy reasoning rule

U (rows: △E, columns: E)	NB	NS	ZR	PS	PB
NB	PB	PB	PS	PS	ZR
NS	PB	PS	PS	ZR	ZR
ZR	PS	PS	ZR	ZR	NS
PS	PS	ZR	ZR	NS	NS
PB	ZR	ZR	NS	NS	NB

By Zedder's fuzzy rule, using "the maximum and minimum" method and barycenter principle, we get the control variable U.

$$U = \frac{\sum_{i=1}^{l} U^i[\mu_\beta^i(u^i)]}{\sum_{i=1}^{l} \mu_\beta^i(u^i)} \tag{9}$$

Where $\mu_\beta^i(u^i)$ is membership value, l is the number of relevant fuzzy rule.

Last, Anti-fuzzification and Fuzzy Control of the System

In this step, we collect grey prediction outputs $\hat{y}^{(0)}(n+p)$ from every control cycle and calculate the actual deviation $\hat{E}(t)$ and the deviation variance $\Delta\hat{E}(t)$, then multiply them with their quantification coefficients K_E , and $K_{\Delta E}$ relatively, to get input1, input2. Consult the fuzzy reasoning rule table with input1, input2 and we will find U, which is within the discussing field of outputs. Multiply the fuzzy value of the corresponding drift in the fuzzy control table with a ratio coefficient K_U , namely, u= $K_U \times$ U and we will get the actual control variable u.

3 Simulation and Experiment of GPBFC in NCSs

With the development of network technology, computer science and control theory, the concept of NCSs was proposed. NCSs are feedback control loops closed through a real time network. Compared with traditional "point to point" control system, NCSs can realize resource sharing and coordinate manipulation, reduce cost of installation and offer ease of maintenance [9,10]. Thus, NCSs have great potential in applications in manufacturing plants, vehicles, aircrafts, and spacecrafts.

Despite of the great advantages and wide applications, communication networks in the control loops make the analysis and design of a networked control system complicated. One main issue is the network-induced delay (sensor-to-controller and controller-to-actuator)[11-13]. It is well known that the network-induced delay is brought into the control systems along with the inserted communication network, which not only prevents us from applying some conventional theorem to NCSs, but brings about so many unstable factors that degrade the stability and control performance. So we introduce the fuzzy controller integrating grey prediction model to control the plant and avoid the effect of time delay in NCSs. The structure of grey prediction based fuzzy controller in NCSs is shown in Figure.5.

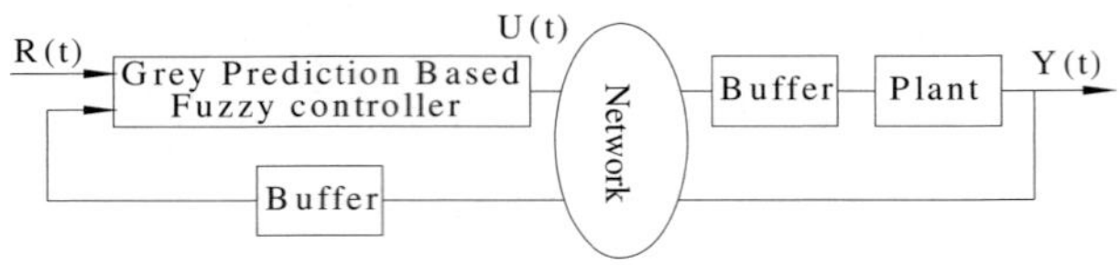

Fig. 5. Grey Prediction Based Fuzzy Controller in Networked Control Systems

The buffers are used to reorder the transmitted data. And the size of each buffer should be equal to the length of the data of the signal from the zero-step to maximum-step network delay. In this way, the random network delay can be treated as a constant delay after the buffer. This implies that if the transmission delay of the data on the network is less than the maximum delay, they have to stay in the buffer until the maximum delay is reached. For the convenience of investigation, we make the following assumptions[14,15]:

1) The upper bounds of the communication delays in the forward and backward channels are k and f of sampling period T, respectively;
2) The data transmitted through the communication network are with a time stamp in order that the packets can't arrive at the grey prediction node in a wrong order;
3) There are no data packets lost;
4) The output of the plant is transmitted with one data packet. That is, a single packet is enough to transmit the output of plant at every sampling period.

For showing the efficiency, we apply the proposed method to control a second-order non-linear system with saturation area of 0.7 and the dead zone of 0.07. The transfer function of the plant is:

$$G(s) = \frac{20}{1.6s^2 + 4.4s + 1}$$

The system sampling period T is 0.01sec. From the extrapolation in section 2, the dynamic prediction mode performs the big positive-step $P_3 = 10$, small positive-step $P_2 = 2$ and negative-step $P_1 = -2$ mode based on the deviation between the set value R(t) and the sensor output value Y(t). In order to reduce the rise time and minish the overshoot in the meantime, the switching mechanism is defined by

$$p = \begin{cases} p_1 = -2 & \text{if } E(t) = R(t) - Y(t) > 0.6 \\ p_2 = 2 & \text{if } 0.1 < E(t) = R(t) - Y(t) < 0.6 \\ p_3 = 10 & \text{if } E(t) = R(t) - Y(t) < 0.1 \end{cases}$$

By using the Eq.(8), we get the grey prediction values and send them to fuzzy controller in order to control the nonlinear plant via communication network. And in the fuzzy controller part, we set $K_E = 60$, $K_{\Delta E} = 2.5$ and $K_U = 0.8$ .The unit step response simulation results, when the networked control systems have no time delays, the upper bounds of the communication time delays in the forward and backward channels are equal to the system sampling period (that is k=f=1×T), and the upper bounds of the communication time delays in the forward and backward channels are three times of system sampling period (that is k=f=3×T), are shown in Figure.6, Figure.7 and Figure.8, respectively.

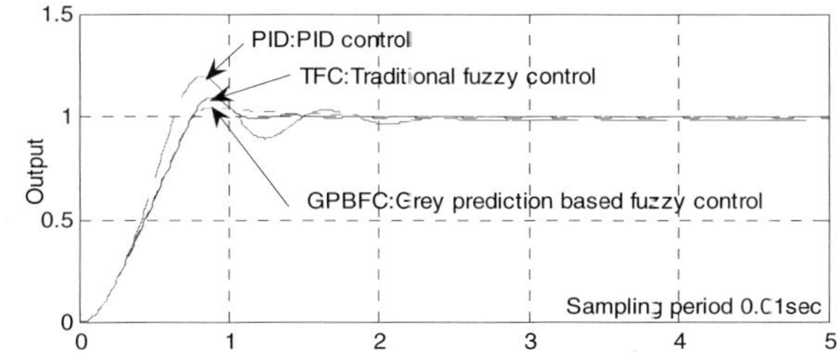

Fig. 6. Simulation results with no time-delays in NCSs

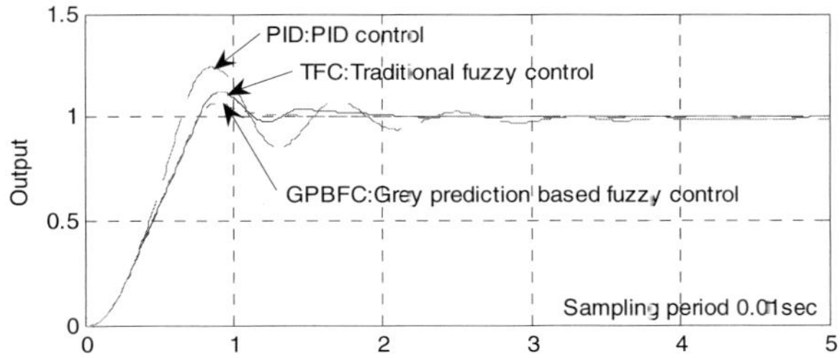

Fig. 7. Simulation results with the upper bounds of forward and backward channels k=f=T in NCSs

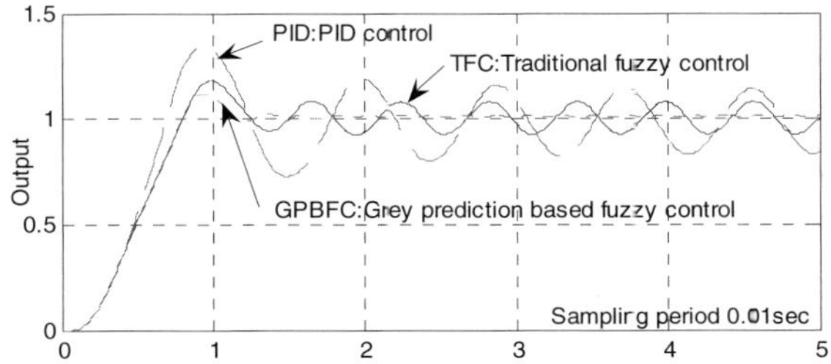

Fig. 8. Simulation results with the upper bounds of forward and backward channels k=f=3T in NCSs

In these figures, the dotted line is the simulation curve of grey prediction based fuzzy controller (GPBFC), the actual line is the simulation curve of traditional fuzzy control (TFC), and the imaginal line is the simulation curve of traditional PID control with Kp=5, ki=0.1, kd=0.01.The performance indices of NCSs with these three different methods are shown as follow:

Table 2. The performance indices of NCSs with three different methods

Performance		Rise time	Overshoot(%)
No time-delays	**PID**	1.7	19.58
	TFC	1.0	9.33
	GPBFC	0.7	4.39
The upper bounds of forward and backward channels k=f=T	**PID**	2.3	24.1
	TFC	1.1	12.03
	GPBFC	0.9	6.45
The upper bounds of forward and backward channels k=f=3T	**PID**	instability	34.26
	TFC	instability	18.25
	GPBFC	1.2	11.93

From the observations on the results, we use three experiments of different time delays in forward and backward channels to implement the GPBFC design. By simulating a non-linear plant, the achievements are successful to improve the system performances about the rise time and the max overshoot. It is better than the traditional fuzzy and PID control strategies.

4 Conclusions

This paper combines the advantages of grey and fuzzy theory and the technology of on-line state evaluation switching algorithm to design a novel grey prediction based fuzzy controller. Using the on-line rule switching mechanism, the system response is divided into three regions. And then we get three forecasting modes, which are one small positive-step forecasting mode, one small positive-step forecasting mode and one negative-step mode. When the system error is large, the negative-step mode is used to increase the upward momentum of the output curve. When the system error is small, the positive-step mode is used to prevent the overshooting. And the last condition is the middle error occurred. Simulation results indicate that the GPBFC method's precision and robustness is better than other traditional control methods.

Acknowledgments. This work was supported by Program for New Century Excellent Talents in University under grant NCET-04-0433, Research Project for Doctoral Disciplines in University under grant 20040280017, Key Project of Science & Technology Commission of Shanghai Municipality under Grant 061107031 and 061111008, Sunlight Plan Following Project of Shanghai Municipal Education Commission, and Shanghai Leading Academic Disciplines under grant T0103.

References

1. Chen, J.H.: Fuzzy Controller and Grey Prediction Model Applied to the Motion Control of a Robot Arm under the Windows. Master Thesis, NCU (2001)
2. Deng, J.: Introduction to grey system theory. The Journal of Grey System 1, 1–24 (1989)
3. Julong, D.: Grey predicting & grey decision-making[M]. Huazhong university of science & technology press, Wuhan (2002)
4. Wenbin, H., Ben, H., Changzhi, Y.: Building thermal process analysis with grey system method[J]. Building and Environment 37(6) (2002)
5. Huang, S.-J., Huang, C.-L.: Control of an Inverted Pendulum Using Grey Prediction Model. IEEE Transactions on Industry Applications 36(2) (March/April 2000)
6. Liting, Z., Guorui, Z., Shijian, Z., Tieding, L.: Grey model experimentation for time series[J]. Wuhan: engineering journal of wuhan university 38(1) (2005)
7. Lisheng, W., Minrui, F., Haikuan, W.: New Grey Modeling Method of GM(1,N)based on self-adaptive data fusion. In: 1st International Symposium on Test Automation and Instrumentation, Beijing, China, September 13-16, 2006, vol. 3, pp. 1791–1795 (2006)
8. Ji-fu, G., Yun-peng, W., Hua, H., Liang, G.: The Fuzzy Control for the Semi-active Suspension Vehicle[J]. Journal of System Simulation 19(5), 1030–1033 (2007)
9. Shousong, H., Qixin, Z.: Stochastic Optimal Control and Analysis of Stability of Networked Control Systems with Long Delay[J]. Automatica 39(11), 1877–1884 (2003)
10. Lei, Z., Hristu-Varsakelis, D.: Communication and Control Co-design for Networked Control Systems[J]. Automatica 42(6), 953–958 (2006)
11. Zhang, W., Branicky, M.S., Phillips, S.M.: Stability of networked control system. IEEE Control Systems Magazine 21(2), 84–99 (2001)
12. Walsh, G.C., Ye, H., Bushnell, L.G.: Stability analysis of networked control systems. IEEE Transactions on Control Systems Technology 10(3), 438–446 (2002)
13. Xu, S.: Analysis of GM(1,N) forecasting model and its applications. Grey Systems, pp. 180-194, China Ocean press, Beijing (1988)
14. Li, H., Sun, Z., Chen, B.: Modeling and Control of Networked Control Systems. In: Proceedings of the 25th Chinese Control Conference, Harbin, Heilongjiang, August 7-11, 2006, pp. 2018–2023 (2006)
15. Lisheng, W., Minrui, F.: A Real-time Optimization Grey Prediction Method For Delay Estimation in NCS. In: The Sixth IEEE International Conference on Control and Automation, Guangzhou, China, May 30-June 1, 2007, pp. 514–517. IEEE Computer Society Press, Los Alamitos (2007)

Guaranteed Cost Robust Filter for Time Delayed T-S Fuzzy Systems with Uncertain Nonlinearities*

Fan Zhou[1], Li Xie[2], and Yaowu Chen[1]

[1] Department of Instrument Sciences and Engineering, Zhejiang University, 310027 Hangzhou, P.R.China
[2] Department of Information and Electronic Engineering, Zhejiang University, 310027 Hangzhou, P.R.China
Xiehan@Zju.edu.cn

Abstract. The problem of guaranteed cost robust fuzzy filtering design for a class of time delayed systems with uncertain nonlinearities have been investigated. The nonlinear uncertain time delayed system is represented by state-space Takagi-Sugeno fuzzy model. In terms of linear matrix inequalities (LMIs), the stability of the filter error systems are analyzed by basis-dependent Lyapunov function approach. Then the sufficient conditions for the existence of the fuzzy filter are given in the form of convex optimization. The optimal upper bound of the guaranteed cost value of the filtering error systems can be solved when the LMIs are feasible.

1 Introduction

In recent years, there are many applications of fuzzy systems theory in various fields, such as control systems, communication systems, and signal processing [1-3]. It has been recognized that fuzzy logic method is one of the most useful techniques for systems analysis and design by utilizing the qualitative knowledge. Among various fuzzy logic approaches, a global method proposed by Takagi and Sugeno gives an effective way to analysis the complex nonlinear systems [4]. The main procedures of this approach are as follows: firstly, the considered nonlinear system is represented by a Takagi-Sugeno (TS) fuzzy model, in which each local dynamic in different state-space region is represented by a linear local model. Secondly, by blending of all these fuzzy models, the overall fuzzy system model is obtained. Then, the analysis and design methods can be carried out via a parallel distributed compensation (PDC) scheme. In the applications of filter design, the main idea of this approach is that for each local linear model, a linear filter is designed. Then, by fuzzy blending of each linear individual linear filter, the nonlinear overall filter is achieved.

It is well known that time delays often occur in many dynamical systems, such as chemical processes, biological systems, long transmission lines in electrical networks. The existence of the delays usually becomes the source of the instability and poor performance of these systems. Many results have been proposed by extending the

* This work was supported by Chinese Nature Science Foundation (60473129).

K. Li et al. (Eds.): LSMS 2007, LNCS 4688, pp. 378–388, 2007.

conventional procedure to T-S model with time delays [5]. Motivated by the idea of basis-dependent Lyapunov methods [6], the guaranteed cost robust filter design issue is investigated in this paper for time delayed systems with uncertain nonlinearities within T-S fuzzy delayed model framework. The time-varying uncertainties of the system parameters are assumed to norm bounded. The nonlinear approximation errors are assumed to satisfy Lipschitz conditions. The problem we address in the paper is to design a stable filter such that the filter error dynamical system is stable and to achieve the minimization of the guaranteed cost value of the filter error system. The main contribution of this paper is the sufficient condition for the existence of fuzzy guaranteed cost filters in terms of linear matrix inequalities (LMIs).

The paper is organized as follows. Section 2 provides preliminaries and the formulation of the problem. In Section 3, the design scheme is proposed for robust fuzzy guaranteed cost filter for uncertain nonlinear systems with time delay. It is shown that the problem of filter design with the minimization of guaranteed cost value of the filter error system can be solved by solving a set of certain LMIs. Some concluding remarks are given in Section 4.

2 Systems Descriptions

A class of continuous T-S fuzzy dynamical model with parameter uncertainties and time varying delays can be described by the following IF THEN rules:

Plant Rule i : IF $\theta_1(t)$ is M_{i1} and ... and $\theta_k(t)$ is M_{iz} THEN

$$\dot{x}(t) = A_i(t)x(t) + A_{di}(t)x(t-d(t)) + B_i(t)f(x(t)) \tag{1}$$

$$y(t) = C_i(t)x(t) + C_{di}(t)x(t-d(t)) + D_i(t)g(x(t)) \tag{2}$$

$$z(t) = L_i(t)x(t) \tag{3}$$

$$x(t) = \varphi(t), \quad \forall t \in [-\tau, 0] . \tag{4}$$

where M_{ij} is a fuzzy set, $\theta_j(t)$ is the premise variables, $j = 1,\ldots,k$, r is the number of IF-THEN rules, $x(t) \in R^n$ is the state vector, $y(t) \in R^m$ is the measurement output vector, $z(t) \in R^p$ is the estimation output vector, $\varphi(t)$ is the initial value function. The time-varying delay $d(t)$ is a positive real number satisfying:

$$0 \le d(t) \le \tau, \ \dot{d}(t) \le \eta < 1 . \tag{5}$$

$A_i(t)$, $A_{di}(t)$, $B_i(t)$, $C_i(t)$, $C_{di}(t)$, $D_i(t)$ and $L_i(t)$ are appropriately dimensioned real valued system matrices:

$$A_i(t) = A_i + \Delta A_i(t), \ A_{di}(t) = A_{di} + \Delta A_{di}(t), \ B_i(t) = B_i + \Delta B_i(t) \tag{6}$$

$$C_i(t) = C_i + \Delta C_i(t), \ C_{di}(t) = C_{di} + \Delta C_{di}(t), \ D_i(t) = D_i + \Delta D_i(t) \tag{7}$$

$$L_i(t) = L_i + \Delta L_i(t) . \tag{8}$$

where A_i , A_{di} , B_i , C_i , C_{di} , D_i and L_i are known constant matrices that describe the nominal system, ΔA_i , ΔA_{di} , ΔB_i , ΔC_i , ΔC_{di} , ΔD_i and ΔL_i represent time-varying parameter uncertainties that are assumed to be of the form

$$\begin{bmatrix} \Delta A_i(t) & \Delta A_{di}(t) & \Delta B_i(t) \\ \Delta C_i(t) & \Delta C_{di}(t) & \Delta D_i(t) \end{bmatrix} = \begin{bmatrix} E_{1i} \\ E_{2i} \end{bmatrix} F_i(t) \begin{bmatrix} H_{1i}(t) & H_{2i}(t) & H_{3i}(t) \end{bmatrix} \tag{9}$$

$$\Delta L_i(t) = E_{3i} F_i(t) H_{1i} \,. \tag{10}$$

where E_{1i}, E_{2i}, E_{3i}, H_{1i}, H_{2i} and H_{3i} are known constant matrices with compatible dimension, and $F_i(.): R^n \to R^{s\times q}$ are uncertain real-valued time-varying matrix functions satisfying

$$F_i^T(t) F_i(t) \le I \,. \tag{11}$$

It is assumed that the elements of $F_i(t)$ are Lesbesgue measurable.

Thoughout the paper, we assume that the nonlinear functions $f(x(t))$ and $g(x(t))$ in the system (1) satisfy the following conditions:

(A1). $\|f(x_1) - f(x_2)\|^2 \le \|W(x_1 - x_2)\|^2$; $f(0) = 0$. (12)

(A2). $\|g(x_1) - g(x_2)\|^2 \le \|V(x_1 - x_2)\|^2$; $g(0) = 0$. (13)

By the standard fuzzy inference method, the global dynamical of the T-S fuzzy system (1)-(4) is described by

$$\dot{x}(t) = \sum_{i=1}^{r} \mu_i(\theta(t))[A_i(t)x(t) + A_{di}(t)x(t-d(t)) + B_i(t)f(x(t))] \tag{14}$$

$$y(t) = \sum_{i=1}^{r} \mu_i(\theta(t))[C_i(t)x(t) + C_{di}(t)x(t-d(t)) + D_i(t)g(x(t))] \tag{15}$$

$$z(t) = \sum_{i=1}^{r} \mu_i(\theta(t)) L_i(t) x(t) \tag{16}$$

$$x(t) = \varphi(t) \,, \quad \forall t \in [-\tau, 0] \,. \tag{17}$$

where $\mu_i(\theta(t))$ is the normalized membership function of the inferred fuzzy set $h_i(\theta(t))$, that is

$$\mu_i(\theta(t)) = h_i(\theta(t)) \Big/ \sum_{i=1}^{r} h_i(\theta(t)) \,, \; h_i(\theta(t)) = \prod_{j=1}^{g} M_{ij}(\theta_j(t)) \tag{18}$$

$$\theta(t) = \begin{bmatrix} \theta_1(t) & \theta_2(t) & \ldots & \theta_g(t) \end{bmatrix}. \tag{19}$$

in which $M_{ij}(\theta_j(t))$ is the grade of membership of $\theta_j(t)$ in M_{ij} . It is assumed that

$$h_i(\theta(t)) \ge 0 \,, \; i = 1, 2, \ldots, r \,, \; \sum_{j=1}^{r} h_j(\theta(t)) > 0 \,. \tag{20}$$

for all t . Then we can obtain the following conditions:

$$\mu_i(\theta(t)) \ge 0 \,, \; i = 1, 2, \ldots, r \,, \; \sum_{j=1}^{r} \mu_j(\theta(t)) = 1 \,. \tag{21}$$

for all t . Associated with the plant, consider the following stable robust fuzzy filter for the estimate of $z(t)$:

Filter Rule i: IF $\theta_1(t)$ is M_{i1} and … and $\theta_g(t)$ is M_{ig} THEN

$$\dot{\tilde{x}}(t) = G_i\tilde{x}(t) + K_i y(t) \tag{22}$$

$$\tilde{z}(t) = L_i\tilde{x}(t),\ i = 1,2,\ldots,r. \tag{23}$$

where $\tilde{x}(t)\in R^n$ is the state vector of the filter, $\tilde{z}(t)\in R^p$ is the estimation vector, G_i, K_i are matrices to be determined. Here, the initial state of the filter is assumed to be zero. Let

$$\xi(t) = \begin{bmatrix} x^T(t) & \tilde{x}^T(t)\end{bmatrix}^T,\ \bar{z}(t) = z(t) - \tilde{z}(t). \tag{24}$$

Then the filtering error dynamical systems can be described by

$$\dot{\xi}(t) = \sum_{i=1}^{r}\sum_{j=1}^{r}\mu_i(\theta(t))\mu_j(\theta(t))[\bar{A}_{ij}^{\xi}\xi(t) + \bar{A}_{dij}^{\xi}\xi(t-d(t)) + \bar{B}_{ij}^{\xi}\hat{f}(\xi(t))] \tag{25}$$

$$\bar{z}(t) = \sum_{i=1}^{r}\sum_{j=1}^{r}\mu_i(\theta(t))\mu_j(\theta(t))\bar{L}_{ij}^{\xi}\xi(t) \tag{26}$$

$$\xi(t) = \varphi^{\xi}(t) = \begin{bmatrix}\varphi^T(t) & 0\end{bmatrix}^T,\ \forall t\in[-\tau,0] \tag{27}$$

where

$$\bar{A}_{ij}^{\xi} = A_{ij}^{\xi} + \Delta A_{ij}^{\xi}(t),\ \bar{A}_{dij}^{\xi} = [A_{dij}^{\xi} + \Delta A_{dij}^{\xi}(t)]H,\ \bar{B}_{ij}^{\xi} = B_{ij}^{\xi} + \Delta B_{ij}^{\xi}(t), \tag{28}$$

$$\bar{L}_{ij}^{\xi} = L_{ij}^{\xi} + \Delta L_{ij}^{\xi}(t),\ \hat{f}(\xi(t)) = \begin{bmatrix} f^T(x(t)) & g^T(x(t))\end{bmatrix}^T, \tag{29}$$

$$A_{ij}^{\xi} = \begin{bmatrix} A_j & 0 \\ K_iC_j & G_i\end{bmatrix},\ A_{dij}^{\xi} = \begin{bmatrix} A_{dj} \\ K_iC_{dj}\end{bmatrix},\ B_{ij}^{\xi} = \begin{bmatrix} B_j \\ K_iD_j\end{bmatrix}, \tag{30}$$

$$\Delta A_{ij}^{\xi} = \begin{bmatrix} \Delta A_j(t) & 0 \\ K_i\Delta C_j(t) & 0\end{bmatrix},\ \Delta A_{dij}^{\xi} = \begin{bmatrix} \Delta A_{dj}(t) \\ K_i\Delta C_{dj}(t)\end{bmatrix},\ \Delta B_{ij}^{\xi} = \begin{bmatrix} \Delta B_j(t) \\ K_i\Delta D_j(t)\end{bmatrix}, \tag{31}$$

$$H = \begin{bmatrix} I & 0\end{bmatrix},\ L_{ij}^{\xi} = \begin{bmatrix} L_j & -L_i\end{bmatrix},\ \Delta L_{ij}^{\xi}(t) = \begin{bmatrix}\Delta L_j & 0\end{bmatrix}. \tag{32}$$

Consider the assumption (A1) and (A2), we have

$$\left\|\hat{f}(\xi(t))\right\|^2 = \left\|\begin{bmatrix} f(x(t)) \\ g(x(t))\end{bmatrix}\right\|^2 \le \left\|W^{\xi}H\xi(t)\right\|^2,\ W^{\xi} = \begin{bmatrix} W^T & V^T\end{bmatrix}^T. \tag{33}$$

Introduce the guaranteed cost function as following:

$$J = \int_0^{\infty}\bar{z}^T(s)\bar{z}(s)ds. \tag{34}$$

The objective of the paper is formulated as following: Determine a stable filter in the form of (22)-(23) that achieve the minimization of the guaranteed cost value in the filtering error dynamical systems (25)-(27).

3 Main Results

In this section, the robust guaranteed cost filter design methods are presented for time delayed T-S fuzzy systems with uncertain nonlinearities.

Theorem 1. The stable robust fuzzy filter (22)-(23) is a guaranteed cost filter for TS fuzzy uncertain nonlinear time-delayed systems (1)-(4) if there exist symmetric matrices $P_l > 0$ and $R > 0$ such that the following LMIs are satisfied:

$$\text{(a).}\ \Omega_{iil} = \begin{bmatrix} \Omega^1_{iil} & P_l \overline{A}^{\xi}_{dii} \\ \overline{A}^{\xi\,T}_{dii} P_l & -(1-\eta)R \end{bmatrix} < 0,\ i, l = 1, 2, \ldots, r. \tag{35}$$

where

$$\Omega^1_{iil} = \overline{A}^{\xi T}_{ii} P_l + P_l \overline{A}^{\xi}_{ii} + R + 2\alpha^{-1} P_l \overline{B}^{\xi}_{ii} \overline{B}^{\xi T}_{ii} P_l + \frac{\alpha H^T W^{\xi T} W^{\xi} H}{2} + \overline{L}^{\xi T}_{ii} \overline{L}^{\xi}_{ii} \tag{36}$$

$$\text{(b).}\ \Omega_{ijl} = \begin{bmatrix} \Omega^1_{ijl} & P(\overline{A}^{\xi}_{dij} + \overline{A}^{\xi}_{dji}) \\ (\overline{A}^{\xi}_{dij} + \overline{A}^{\xi}_{dji})^T P & -2(1-\eta)R \end{bmatrix} < 0,\ 1 \le i < j \le r,\ \ l = 1, 2, \ldots, r. \tag{37}$$

where

$$\begin{aligned} \Omega^1_{ijl} = {} & (\overline{A}^{\xi}_{ij} + \overline{A}^{\xi}_{ji})^T P_l + P_l(\overline{A}^{\xi}_{ij} + \overline{A}^{\xi}_{ji}) + \alpha^{-1} P_l (\overline{B}^{\xi}_{ij} + \overline{B}^{\xi}_{ji})(\overline{B}^{\xi}_{ij} + \overline{B}^{\xi}_{ji})^T P_l \\ & + \alpha H^T W^{\xi T} W^{\xi} H + 2R + (\overline{L}^{\xi}_{ij} + \overline{L}^{\xi}_{ji})^T (\overline{L}^{\xi}_{ij} + \overline{L}^{\xi}_{ji})/2 \end{aligned} \tag{38}$$

Proof. Introduce a Lyapunov-Krasovskii functional as

$$V(\xi(t), t) = \xi^T(t) P(\mu) \xi(t) + \int_{t-d(t)}^{t} \xi^T(s) R \xi(s) ds. \tag{39}$$

where $P = P^T = \sum_{l=1}^{r} \mu_l(\theta(t)) P_l > 0$, $R = R^T > 0$. For $q > 1$, it is easy to verify that

$$0 \le \lambda_{\min}(P(\mu)) \|\xi(t)\|^2 \le V(\xi(t), t) \le [\lambda_{\max}(P(\mu)) + 2\tau q \lambda_{\max}(R)] \|\xi(t)\|^2. \tag{40}$$

The derivative of the Lyapunov function along the solution of system (25) is

$$\begin{aligned} \dot{V}(\xi(t), t) = {} & \dot{\xi}^T(t) P(\mu) \xi(t) + \xi^T(t) P(\mu) \dot{\xi}(t) \\ = {} & \sum_{i=1}^{r} \sum_{j=1}^{r} \sum_{l=1}^{r} \mu_i(\theta(t)) \mu_j(\theta(t)) \mu_l(\theta(t)) \{ [\xi^T(t) \overline{A}^{\xi T}_{ij} + \xi^T(t-d(t)) \overline{A}^{\xi\,T}_{dij} \\ & + \hat{f}^T(\xi(t)) \overline{B}^{\xi T}_{ij}] P_l \xi(t) + \xi^T(t) P_l [\overline{A}^{\xi}_{ij} \xi(t) + \overline{A}^{\xi}_{dij} \xi(t-d(t)) \\ & + \overline{B}^{\xi}_{ij} \hat{f}(\xi(t))] \} + \xi^T(t) R \xi(t) - (1-\dot{d}(t)) \xi^T(t-d(t)) R \xi(t-d(t)) \\ = {} & \frac{1}{2} \sum_{i=1}^{r} \sum_{j=1}^{r} \sum_{l=1}^{r} \mu_i(\theta(t)) \mu_j(\theta(t)) \mu_l(\theta(t)) \{ [\xi^T(t) (\overline{A}^{\xi}_{ij} + \overline{A}^{\xi}_{ji})^T \\ & + \xi^T(t-d(t)) (\overline{A}^{\xi}_{dij} + \overline{A}^{\xi}_{dji})^T + \hat{f}^T(\xi(t)) (\overline{B}^{\xi}_{ij} + \overline{B}^{\xi}_{ji})^T] P_l \xi(t) \\ & + \xi^T(t) P_l [(\overline{A}^{\xi}_{ij} + \overline{A}^{\xi}_{ji}) \xi(t) + (\overline{A}^{\xi}_{dij} + \overline{A}^{\xi}_{dji}) \xi(t-d(t)) \\ & + (\overline{B}^{\xi}_{ij} + \overline{B}^{\xi}_{ji}) \hat{f}(\xi(t))] \} + \xi^T(t) R \xi(t) \\ & - (1-\dot{d}(t)) \xi^T(t-d(t)) R \xi(t-d(t)) \end{aligned} \tag{41}$$

By applying matrix inequalities knowledge [3], we can deduce that for any $\alpha > 0$

$$2\xi^T(t)P_l(\overline{B}_{ij}^{\xi} + \overline{B}_{ji}^{\xi})\hat{f}(\xi(t)) \le \alpha\xi^T(t)H^T W^{\xi T} W^{\xi} H\xi(t) + \alpha^{-1}\xi^T(t)P_l(\overline{B}_{ij}^{\xi} + \overline{B}_{ji}^{\xi})(\overline{B}_{ij}^{\xi} + \overline{B}_{ji}^{\xi})^T P_l\xi(t) \tag{42}$$

$$\sum_{i=1}^{r}\sum_{j=1}^{r}\sum_{k=1}^{r}\sum_{l=1}^{r}\mu_i(\theta(t))\mu_j(\theta(t))\mu_k(\theta(t))\mu_l(\theta(t))\xi^T(t)\overline{L}_{ij}^{\xi T}\overline{L}_{kl}^{\xi}\xi(t) \le \frac{1}{4}\sum_{i=1}^{r}\sum_{j=1}^{r}\mu_i(\theta(t))\mu_j(\theta(t))\xi^T(t)(\overline{L}_{ij}^{\xi} + \overline{L}_{ji}^{\xi})^T(\overline{L}_{ij}^{\xi} + \overline{L}_{ji}^{\xi}) \tag{43}$$

Let $\Gamma = \left[\xi^T(t) \quad \xi^T(t-d(t))\right]^T$, condition (35)-(38) implies that

$$\dot{V}(\xi(t),t) \le \Gamma^T\{\sum_{i=1}^{r}\sum_{l=1}^{r}\mu_i^2(\theta(t))\mu_l(\theta(t))\Omega_{iil} + \sum_{i,j=1,i<j}^{r}\sum_{l=1}^{r}\mu_i(\theta(t))\Omega_{ijl}\}\Gamma - \overline{z}^T(t)\overline{z}(t) < -\overline{z}^T(t)\overline{z}(t) < 0 \tag{44}$$

which ensures the asymptotic stability of the filtering error systems (25)-(27). Furthermore, by integrating both sides of the inequality (44) from 0 to T_f, we have

$$-\int_0^{T_f}\overline{z}^T(t)\overline{z}(t)dt > \xi^T(T_f)P(\mu)\xi(T_f) - \xi^T(0)P(\mu)\xi(0) + \int_{T_f-d(t)}^{T_f}\xi^T(s)R\xi(s)ds - \int_{0-d(0)}^{0}\xi^T(s)R\xi(s)ds \tag{45}$$

As the filtering error systems (25)-(27) is asymptotic stable, when $T_f \to \infty$,

$$\lim_{T_f\to\infty}\xi^T(T_f)P(\mu)\xi(T_f) = 0,\ \lim_{T_f\to\infty}\int_{T_f-d(t)}^{T_f}\xi^T(s)R\xi(s)ds = 0 \tag{46}$$

and using the initial condition (27), we get

$$\int_0^{\infty}\overline{z}^T(t)\overline{z}(t)dt < \xi^T(0)P(\mu)\xi(0) + \int_{-d(0)}^{0}\xi^T(t)R\xi(t)dt = J^* \tag{47}$$

Here, J^* is an upper bound of the guaranteed cost value of the filter error dynamical system (25)-(27). This completes the proof.

Remark 1. In Theorem 1, a sufficient condition is presented for the existence of robust guaranteed cost filter for the TS fuzzy uncertain nonlinear systems (1)-(4). In the following theorem, the main result of this section, shows that the feasibility of a set of LMIs is sufficient for the existence of robust fuzzy guaranteed cost filter.

Theorem 2. The stable robust fuzzy filter (22)-(23) is a guaranteed cost filter for T-S fuzzy uncertain nonlinear time-delayed systems (1)-(4) if there exist symmetric matrices P_{l1}, P_{l2}, R_1, R_2, S_1, $S_2 > 0$ such that the following LMIs are satisfied:

$$Min\{\sigma + Tr(M_1)\} \tag{48}$$

S.t.

(a). $\begin{bmatrix} -\sigma & \varphi^T(0) \\ \varphi(0) & -Q_{l1} \end{bmatrix} < 0$, $\begin{bmatrix} P_{l1} & I \\ I & Q_{l1} \end{bmatrix} \le 0$, $l = 1, 2, \ldots, r$,

$$\begin{bmatrix} -M_1 & N^T \\ N & -S_1 \end{bmatrix} < 0, \begin{bmatrix} R_1 & I \\ I & S_1 \end{bmatrix} \le 0. \tag{49}$$

where

$$\int_{-d(0)}^{0} \varphi^T(t)\varphi(t)dt = NN^T \tag{50}$$

$$\text{(b).} \quad \begin{bmatrix} \Phi_{iil}^{11} & \Phi_{iil}^{12} & \Phi_{iil}^{13} & \Phi_{iil}^{14} & \Phi_{iil}^{15} \\ \Phi_{iil}^{21} & \Phi_{iil}^{22} & \Phi_{iil}^{23} & \Phi_{iil}^{24} & \Phi_{iil}^{25} \\ \Phi_{iil}^{31} & \Phi_{iil}^{32} & \Phi_{iil}^{33} & 0 & 0 \\ \Phi_{iil}^{41} & \Phi_{iil}^{42} & 0 & \Phi_{iil}^{44} & 0 \\ \Phi_{iil}^{51} & \Phi_{iil}^{52} & 0 & 0 & \Phi_{iil}^{55} \end{bmatrix} < 0, \; i, l = 1, 2, \ldots, r. \tag{51}$$

where

$$\begin{aligned}
\Phi_{iil}^{11} &= P_{l1}A_i + A_i^T P_{l1} + R_1 + \frac{\alpha}{2}(W^T W + V^T V) + \beta H_{1i}^T H_{1i} \\
\Phi_{iil}^{12} &= \left(\Phi_{iil}^{21}\right)^T = C_i^T F_{1i}^T \\
\Phi_{iil}^{13} &= \left(\Phi_{iil}^{31}\right)^T = P_{l1}A_{di} + \beta H_{1i}^T H_{2i} \\
\Phi_{iil}^{14} &= \left(\Phi_{iil}^{41}\right)^T = [P_{l1}E_{1i} \quad P_{l1}B_i] \\
\Phi_{iil}^{15} &= \left(\Phi_{iil}^{51}\right)^T = \left[P_{l1}E_{1i} \quad P_{l1}L_i^T \quad P_{l1}H_{1i}^T\right] \\
\Phi_{iil}^{22} &= F_{2i} + F_{2i}^T + R_2 \\
\Phi_{iil}^{23} &= \left(\Phi_{iil}^{32}\right)^T = F_{1i}C_{di} \\
\Phi_{iil}^{24} &= \left(\Phi_{iil}^{42}\right)^T = [F_{1i}E_{2i} \quad F_{1i}D_i] \\
\Phi_{iil}^{25} &= \left(\Phi_{iil}^{52}\right)^T = \left[F_{1i}E_{2i} \quad -P_2L_i^T \quad 0\right] \\
\Phi_{iil}^{33} &= -(1-\eta)R_1 + \beta H_{2i}^T H_{2i} \\
\Phi_{iil}^{44} &= diag\{-\beta I, \varepsilon H_{3i}^T H_{3i} - \alpha I\} \\
\Phi_{iil}^{55} &= diag\{-\varepsilon I, \delta E_{3i}E_{3i}^T - I, -\delta I\}
\end{aligned} \tag{52}$$

$$\text{(c).} \quad \begin{bmatrix} \Phi_{ijl}^{11} & \Phi_{ijl}^{12} & \Phi_{ijl}^{13} & \Phi_{ijl}^{14} & \Phi_{ijl}^{15} \\ \Phi_{ijl}^{21} & \Phi_{ijl}^{22} & \Phi_{ijl}^{23} & \Phi_{ijl}^{24} & \Phi_{ijl}^{25} \\ \Phi_{ijl}^{31} & \Phi_{ijl}^{32} & \Phi_{ijl}^{33} & 0 & 0 \\ \Phi_{ijl}^{41} & \Phi_{ijl}^{42} & 0 & \Phi_{ijl}^{44} & 0 \\ \Phi_{ijl}^{51} & \Phi_{ijl}^{52} & 0 & 0 & \Phi_{ijl}^{55} \end{bmatrix} < 0, \; 1 \le i < j \le r, \; l = 1, 2, \ldots, r. \tag{53}$$

where

$$\Phi_{ijl}^{11} = P_{l1}A_i + A_i^T P_{l1} + P_{l1}A_j + A_j^T P_{l1} + 2R_1 + \frac{\alpha}{2}(W^T W + V^T V)$$
$$+\beta(H_{1i}^T H_{1i} + H_{1j}^T H_{1j})$$
$$\Phi_{ijl}^{12} = \left(\Phi_{ijl}^{21}\right)^T = C_j^T F_{1i}^T + C_i^T F_{1j}^T$$
$$\Phi_{ijl}^{13} = \left(\Phi_{ijl}^{31}\right)^T = P_{l1}(A_{di} + A_{dj}) + \beta(H_{1i}^T H_{2i} + H_{1j}^T H_{2j})$$
$$\Phi_{ijl}^{14} = \left(\Phi_{ijl}^{41}\right)^T = \begin{bmatrix} P_{l1}E_{1j} & P_{l1}E_{1i} & P_{l1}(B_i + B_j) \end{bmatrix}$$
$$\Phi_{ijl}^{15} = \left(\Phi_{ijl}^{51}\right)^T = \begin{bmatrix} P_{l1}E_{1j} & P_{l1}E_{1i} & P_{l1}(L_j^T + L_i^T) & P_{l1}H_{1j}^T & P_{l1}H_{1i}^T \end{bmatrix}$$
$$\Phi_{ijl}^{22} = F_{2i} + F_{2i}^T + F_{2j} + F_{2j}^T + 2R_2$$
$$\Phi_{ijl}^{23} = \left(\Phi_{ijl}^{32}\right)^T = F_{1i}C_{dj} + F_{1j}C_{di}$$
$$\Phi_{ijl}^{24} = \left(\Phi_{ijl}^{42}\right)^T = \begin{bmatrix} F_{1i}E_{2j} & F_{1j}E_{2i} & P_{l1}(B_i + B_j) \end{bmatrix}$$
$$\Phi_{ijl}^{25} = \left(\Phi_{ijl}^{52}\right)^T = \begin{bmatrix} F_{1i}E_{2j} & F_{1j}E_{2i} & -P_{l2}(L_j^T + L_i^T) & 0 & 0 \end{bmatrix}$$
$$\Phi_{ijl}^{33} = -2(1-\eta)R_1 + \beta(H_{2i}^T H_{2i} + H_{2j}^T H_{2j})$$
$$\Phi_{ijl}^{44} = diag\{-\beta I, -\beta I, \varepsilon(H_{3i}^T H_{3i} + H_{3j}^T H_{3j}) - \alpha I\}$$
$$\Phi_{ijl}^{55} = diag\{-\varepsilon I, -\varepsilon I, \delta(H_{3i}^T H_{3i} + H_{3j}^T H_{3j}) - I, -\delta I, -\delta I\} \tag{54}$$

then $J^* = Min\{\sigma + Tr(M_1)\}$ is the upper bound of the guaranteed cost value of the filtering error systems (25)-(27), and the fuzzy guaranteed cost filter gains are $G_i = \sum_{l=1}^{r} \mu_l(\theta(t)) P_{l2}^{-1} F_{2i}$, $K_i = \sum_{l=1}^{r} \mu_l(\theta(t)) P_{l2}^{-1} F_{1i}$, $i = 1, \ldots, r$.

Proof. Firstly, taking into account the condition (51), we have

$$\alpha I - \varepsilon H_{3i}^T H_{3i} > 0, \quad I - \delta E_{3i} E_{3i}^T > 0 \tag{55}$$

Then, it can be obtained that

$$\begin{bmatrix} P_l \Delta A_{ii}^{\xi} + \Delta A_{ii}^{\xi T} P & P_l \Delta A_{dii}^{\xi} H \\ H^T \Delta A_{dii}^{\xi\,T} P_l & 0 \end{bmatrix} \le \beta \begin{bmatrix} H^T H_{1i}^T \\ H^T H_{2i}^T \end{bmatrix} \begin{bmatrix} H_{1i} H & H_{2i} H \end{bmatrix}$$
$$+\beta^{-1} \begin{bmatrix} P_l \begin{bmatrix} E_{1i} \\ K_i E_{2i} \end{bmatrix} \\ 0 \end{bmatrix} \begin{bmatrix} \begin{bmatrix} E_{1i}^T & E_{2i}^T K^T \end{bmatrix} P_l & 0 \end{bmatrix} \tag{56}$$

$$\alpha^{-1} \overline{B}_{ii}^{\xi} \overline{B}_{ii}^{\xi T} \le B_{ii}^{\xi} (\alpha I - \varepsilon H_{3i}^T H_{3i})^{-1} B_{ii}^{\xi T} + \varepsilon^{-1} \begin{bmatrix} E_{1i} \\ K_i E_{2i} \end{bmatrix} \begin{bmatrix} E_{1i}^T & E_{2i}^T K_i^T \end{bmatrix} \tag{57}$$

$$\overline{L}_{ii}^{\xi T} \overline{L}_{ii}^{\xi} \le L_{ii}^{\xi T} (I - \delta E_{3i} E_{3i}^T)^{-1} L_{ii}^{\xi} + \delta^{-1} H^T H_{1i}^T H_{1i} H \tag{58}$$

By using Schur complement, it is obvious that the inequality (35) hold if the following LMI is satisfied

$$\begin{bmatrix} \Theta'_{ii} & P_l A^{\xi}_{dii} + \begin{bmatrix} \beta H^T_{1i} H_{2i} & 0 \\ 0 & 0 \end{bmatrix} & P_l \begin{bmatrix} E_{1i} \\ K_i E_{2i} \end{bmatrix} & P_l B^{\xi}_{ii} & P_l \begin{bmatrix} E_{1i} \\ K_i E_{2i} \end{bmatrix} & P_l L^{\xi T}_{1i} & P_l \begin{bmatrix} H^T_{1i} \\ 0 \end{bmatrix} \\ A^{\xi T}_{dii} P_l + \begin{bmatrix} \beta H^T_{2i} H_{1i} & 0 \\ 0 & 0 \end{bmatrix} & \Theta''_{ii} & 0 & 0 & 0 & 0 & 0 \\ \begin{bmatrix} E^T_{1i} & E^T_{2i} K^T_i \end{bmatrix} P_l & 0 & -\beta I & 0 & 0 & 0 & 0 \\ B^{\xi T}_{ii} P_l & 0 & 0 & \varepsilon H^T_{3i} H_{3i} - \alpha I & 0 & 0 & 0 \\ \begin{bmatrix} E^T_{1i} & E^T_{2i} K^T_i \end{bmatrix} P_l & 0 & 0 & 0 & -\varepsilon I & 0 & 0 \\ L^{\xi}_{1i} P_l & 0 & 0 & 0 & 0 & \delta E_{3i} E^T_{3i} - I & 0 \\ \begin{bmatrix} H_{1i} & 0 \end{bmatrix} P_l & 0 & 0 & 0 & 0 & 0 & -\delta I \end{bmatrix} < 0 \quad (59)$$

where

$$\Theta'_{ii} = A^{\xi T}_{ii} P_l + P_l A^{\xi}_{ii} + R + (\alpha H^T W^{\xi T} W^{\xi} H)/2 + \beta H^T H^T_{1i} H_{1i} H^T \quad (60)$$

$$\Theta''_{ii} = -(1-\eta) R + \beta H^T H^T_{2i} H_{2i} H \quad (61)$$

Let $P_l = diag\{P_{l1}, P_{l2}\}$, $R = diag\{R_1, R_2\}$, it is easy to conclude that the condition (51) is equivalent to (35).

Secondly, from (53), it is obvious that for $0 \le i < j \le r$, we have

$$\alpha I - \varepsilon \begin{bmatrix} H^T_{3j} & H^T_{3i} \end{bmatrix} \begin{bmatrix} H_{3j} \\ H_{3i} \end{bmatrix} > 0,\ I - \delta \begin{bmatrix} E_{3j} & E_{3i} \end{bmatrix} \begin{bmatrix} E^T_{3j} \\ E^T_{3i} \end{bmatrix} > 0 \quad (62)$$

then, it can be obtained that

$$\begin{bmatrix} P_l(\Delta A^{\xi}_{ij} + \Delta A^{\xi}_{ji}) + (\Delta A^{\xi}_{ij} + \Delta A^{\xi}_{ji})^T P_l & P_l(\Delta A^{\xi}_{dij} + \Delta A^{\xi}_{dji}) \\ (\Delta A^{\xi}_{dij} + \Delta A^{\xi}_{dji})^T P_l & 0 \end{bmatrix}$$

$$\le \beta \begin{bmatrix} \begin{bmatrix} H^T H^T_{1j} & H^T H^T_{1i} \end{bmatrix} \\ \begin{bmatrix} H^T H^T_{2j} & H^T H^T_{2i} \end{bmatrix} \end{bmatrix} \begin{bmatrix} \begin{bmatrix} H_{1j} H \\ H_{1i} H \end{bmatrix} & \begin{bmatrix} H_{2j} H \\ H_{2i} H \end{bmatrix} \end{bmatrix}$$

$$+\beta^{-1} \begin{bmatrix} P_l \begin{bmatrix} \begin{bmatrix} E_{1j} \\ K_i E_{2j} \end{bmatrix} & \begin{bmatrix} E_{1i} \\ K_j E_{2i} \end{bmatrix} \end{bmatrix} \\ 0 \end{bmatrix} \begin{bmatrix} \begin{bmatrix} \begin{bmatrix} E^T_{1j} & E^T_{2j} K^T_i \end{bmatrix} \\ \begin{bmatrix} E^T_{1i} & E^T_{2i} K^T_j \end{bmatrix} \end{bmatrix} P_l & 0 \end{bmatrix} \quad (63)$$

$$\alpha^{-1}(\overline{B}^{\xi}_{ij} + \overline{B}^{\xi}_{ji})(\overline{B}^{\xi}_{ij} + \overline{B}^{\xi}_{ji})^T$$

$$\le (B^{\xi}_{ij} + B^{\xi}_{ji})(\alpha I - \varepsilon \begin{bmatrix} H^T_{3j} & H^T_{3i} \end{bmatrix} \begin{bmatrix} H_{3j} \\ H_{3i} \end{bmatrix})^{-1} (B^{\xi}_{ij} + B^{\xi}_{ji})^T$$

$$+\varepsilon^{-1} \begin{bmatrix} \begin{bmatrix} E_{1j} \\ K_i E_{2j} \end{bmatrix} & \begin{bmatrix} E_{1i} \\ K_j E_{2i} \end{bmatrix} \end{bmatrix} \begin{bmatrix} \begin{bmatrix} E^T_{1j} & E^T_{2j} K^T_i \end{bmatrix} \\ \begin{bmatrix} E^T_{1i} & E^T_{2i} K^T_j \end{bmatrix} \end{bmatrix} \quad (64)$$

$$\begin{aligned}&(\overline{L}_{ij}^{\xi}+\overline{L}_{ji}^{\xi})^{T}(\overline{L}_{ij}^{\xi}+\overline{L}_{ji}^{\xi})\\&\leq(L_{ij}^{\xi}+L_{ji}^{\xi})^{T}(I-\delta\begin{bmatrix}E_{3j}&E_{3i}\end{bmatrix}\begin{bmatrix}E_{3j}^{T}\\E_{3i}^{T}\end{bmatrix})^{-1}(L_{ij}^{\xi}+L_{ji}^{\xi})\\&+\delta^{-1}\begin{bmatrix}H^{T}H_{1j}^{T}&H^{T}H_{1i}^{T}\end{bmatrix}\begin{bmatrix}H_{1j}H\\H_{1i}H\end{bmatrix}\end{aligned}\tag{65}$$

Similar to the proof of (51), by using Schur complement, and let $P_l = diag\{P_{l1}, P_{l2}\}$, $R = diag\{R_1, R_2\}$, we obtain that the inequality of (37) hold if the LMI (53) hold. That is to say, if the condition of LMI (53) is satisfied for $1 \le i < j \le r$ and the condition of LMI (51) hold for $i = 1, \ldots, r$, then the filtering error systems (25)-(27) is robust stable.

Since

$$\begin{aligned}&\begin{bmatrix}-\sigma&\varphi^{T}(0)\\\varphi(0)&-Q_{1l}\end{bmatrix}<0,\ \begin{bmatrix}P_{1l}&I\\I&Q_{1l}\end{bmatrix}\leq 0\\&\Leftrightarrow\xi^{T}(0)P_{l}\xi(0)=\varphi^{T}(0)P_{1l}\varphi(0)\leq\varphi^{T}(0)Q_{1l}^{-1}\varphi(0)<\sigma\end{aligned}\tag{66}$$

$$\begin{aligned}&\begin{bmatrix}-M_{1}&N^{T}\\N&-S_{1}\end{bmatrix}<0,\ \begin{bmatrix}R_{1}&I\\I&S_{1}\end{bmatrix}\leq 0,\ \int_{-d(0)}^{0}\varphi^{T}(t)\varphi(t)dt=NN^{T}\\&\Leftrightarrow\int_{-d(0)}^{0}\xi^{T}(t)R\xi(t)dt=\int_{-d(0)}^{0}\varphi^{T}(t)R_{1}\varphi(t)dt=trace\{N^{T}R_{1}N\}\\&\qquad\leq-trace\{N^{T}S_{1}N\}=trace\{M_{1}\}\end{aligned}\tag{67}$$

Form (47), we know that $J^* = Min\{\sigma + Tr(M_1)\}$ is the upper bound of the guaranteed cost value of the filtering error dynamical systems (25)-(27). Therefore, the robust filter (22)-(23) designed by Theorem 2 is the robust fuzzy guaranteed cost filter which minimized the guaranteed cost value of the filtering error dynamical systems (25)-(27) and $Min\{\sigma + Tr(M_1)\}$ is the guaranteed cost value.

This completes the proof. □

Remark 2. If the LMIs (48)-(58) are feasible, there exists the fuzzy guaranteed cost filter for the consider systems. Furthermore, the filtering error systems are asymptotical stable with guaranteed cost performance. By using the software LMI toolbox in Matlab, the fuzzy guaranteed cost filter can be easily obtain based on Theorem 2. Due to the limited space, the numerical examples are omitted here.

4 Conclusions

In this paper, the problem of robust guaranteed cost fuzzy filtering design for uncertain nonlinear system with time delay has been investigated. A sufficient condition for the existence of a robust stable filter, which ensures the filtering error system is asymptotical stable with guaranteed cost performance, has been derived. All the results in the paper are presented in terms of linear matrix inequalities. The proposed approach is flexible in computation by using software such as LMI toolbox in Matlab.

References

1. Lee, J.H., Park, J.B., Chen, G.R.: Robust Fuzzy Control of Nonlinear Systems with Parametric Uncertainties. IEEE Trans. on Fuzzy Systems 9, 366–379 (2001)
2. Tseng, C.S., Chen, B.S.: H∞ Fuzzy Estimation for a Class of Nonlinear Discrete-Time Dynamic Systems. IEEE Trans. on SP 49, 2605–2619 (2001)
3. Xu, S.Y., Lam, J.: Exponential H∞ Filter Design for Uncertain Takagi-Sugeno Fuzzy Systems with Time Delay. Eng. App. of AI 17, 645–659 (2004)
4. Takagi, T., Sugeno, M.: Fuzzy Identification of Systems and Its Applications to Modeling and Control. IEEE Trans. on SMC 15, 116–132 (1985)
5. Xie, L., Liu, J.L., Lu, G.D.: Robust Guaranteed Cost Fuzzy Filtering for Nonlinear Uncertain Systems with Time Delay. Dyn. Con. Dis. and Impl. Sys. A. Suppl. 13, 488–495 (2006)
6. Zhou, S.S., Lam, J., Xue, A.K.: H∞ filtering of discrete-time fuzzy systems via basis-dependent Lyapunov function approach. Fuzzy Sets and Systems 16, 180–193 (2007)

Satellite Cloud Image De-Noising and Enhancement by Fuzzy Wavelet Neural Network and Genetic Algorithm in Curvelet Domain

Xingcai Zhang[1] and Changjiang Zhang[2]

[1] Satellite Sensing Center, Zhejiang Normal University, Postcode 321004 Jinhua, China
jhqxzxc@zjnu.cn

[2] College of Mathematics, Physics and Information Engineering, Zhejiang Normal University, Postcode 321004 Jinhua, China
zcj74922@zjnu.cn

Abstract. A satellite cloud image is decomposed by discrete curvelet transform (DCT). In-complete Beta transform (IBT) is used to obtain non-linear gray transform curve so as to enhance the coefficients in the coarse scale in the DCT domain. GA determines optimal gray transform parameters. Information entropy is used as fitness function of GA. In order to calculate IBT in the coarse scale, fuzzy wavelet neural network (FWNN) is used to approximate the IBT. Hard-threshold method is used to reduce the noise in the high frequency subbands of each decomposition level respectively in the DCT domain. Inverse DCT is conducted to obtain final de-noising and enhanced image. Experimental results show that proposed algorithm can efficiently reduce the noise in the satellite cloud image while well enhancing the contrast. In performance index and visual quality, the proposed algorithm is better than traditional histogram equalization and unsharpened mask method.

1 Introduction

There are kinds of noise in a satellite cloud image. If they cannot be reduced efficiently, this will affect the image quality so as not to extract some important information. Thus it is important to reduce efficiently the noise in the satellite cloud image. Over the last decade, there has been abundant interest in wavelet methods for noise removal in signals and images. Also, many investigators have experimented with variations on the basic schemes—modifications of thresholding functions, level-dependent thresholding, block thresholding, adaptive choice of threshold, Bayesian conditional expectation nonlinearities, and so on [1, 2]. Recently, curvelet transform proposed by Donoho has been widely applied to image processing [3-6], such as image de-noising, image enhancement, image retrial, image compression and image fusion, etc. It is considered to be a very useful new technique in image processing. Curvelet transform is proposed based on the wavelet transform. It overcomes the disadvantage of wavelet transform. In addition, it is necessary to expand the range of gray levels in the satellite cloud image. Traditional image enhancement algorithms are as following: point operators, space operators, transform operators and pseu-color enhancement. Existing many contrast enhancement algorithms' intelligence and

K. Li et al. (Eds.): LSMS 2007, LNCS 4688, pp. 389–395, 2007.

adaptability are bad and much artificial interference is required. Employing DCT, an efficient de-noising and enhancing algorithm for a satellite cloud image is proposed. DCT is implemented to an image. Noise is reduced in the high frequency sub-bands of each decomposition level respectively so that the maximum signal-noise-ratio can be obtained in the high frequency sub-bands respectively. The contrast is enhanced by incomplete Beta transform with genetic algorithm and fuzzy wavelet neural network in the low frequency sub-band in the DCT domain. Experimental results show that the new algorithm can reduce efficiently the GWN in the satellite cloud image while the contrast can be well enhanced. In visual quality, the proposed algorithm is better than traditional histogram equalization (HE) and unsharpened mask (USM) method.

2 De-Noising Based on DCT

The discrete curvelet transform of a continuum function $f\left(x_1,x_2\right)$ makes use of a dyadic sequence of scales, and a bank of filters $\left(P_0 f,\Delta_1 f,\Delta_2 f,\cdots\right)$ with the property that the passband filter Δ_s is concentrated near the frequencies $\left[2^{2s},2^{2s+2}\right]$, e.g., $\Delta_s=\Psi_{2s}*f$, $\hat{\Psi}_{2s}\left(\xi\right)=\hat{\Psi}\left(2^{-2s}\xi\right)$. The sub-bands used in the DCT of continuum functions have the nonstandard form $\left[2^{2s},2^{2s+2}\right]$. This is nonstandard feature of the DCT well worth remembering. We now employ a sketch of the DCT algorithm [7]:

(1) Apply the a trous algorithm with J scales;

(2) Set $B_1=B_{\min}$;

(3) For $j=1,\cdots,J$ do

  a) Partition the subband w_j with a block size B_j and apply the digital ridgelet transform to each block;

  b) If j modulo 2=1 then $B_{j+1}=2B_j$;

  c) Else $B_{j+1}=B_j$

The sidelength of the localizing windows is doubled at every other dyadic subband, hence maintaining the fundamental property of the curvelet transform, which says that elements of length about $2^{-j/2}$ serve for the analysis and synthesis of the jth subband $\left[2^j,2^{j+1}\right]$. Note also that the coarse description of the image c_J is not processed. This implementation of the curvelet transform is also redundant. The redundancy factor is equal to $16J+1$ whenever J scales are employed. Finally, the method enjoys exact reconstruction and stability, because this invertibility holds for each element of the processing chain. We will use Ref. [3] method to reduce the noise in the satellite cloud image. We consider discrete image model as follows:

$$x_{i,j} = f(i,j) + \sigma z_{i,j} \tag{1}$$

Where f is the image to be recovered and z is white noise. Discrete redgelet (resp. curvelet) transform is not norm-preserving and, therefore, the variance of the noisy ridgelet (resp. curvelet) coefficients will depend on the ridgelet (resp. curvelet) index λ. For instance, letting F denote the discrete curvelet transform matrix, we have $Fz \overset{i.i.d}{\rightarrow} N\left(0, FF^T\right)$. Because the computation of FF^T is prohibitively expensive, we calculated an approximate value $\tilde{\sigma}_\lambda^2$ of the individual variances using MonteCarlo simulations where the diagonal elements of FF^T are simply estimated by evaluating the curvelet transforms of a few standard white noise images.

Let $\hat{y}_\lambda$ be the noisy curvelet coefficients ($y = Fx$). We use the following hard-thresholding rule for estimating the unknown curvelet coefficients

$$\hat{y}_\lambda = y_\lambda, \qquad if \frac{|y_\lambda|}{\sigma} \geq k\tilde{\sigma}_\lambda \tag{2}$$

$$\hat{y}_\lambda = 0, \qquad if \frac{|y_\lambda|}{\sigma} \prec k\tilde{\sigma}_\lambda \tag{3}$$

In our experiments, we actually chose a scale-dependent value for k ; we have $k = 4$ for the first scale ($j = 1$) while $k = 3$ for the other ($j \succ 1$).

3 Enhancement by GA and FWNN in the DCT Domain

We will use incomplete Beta transform to obtain the optimal non-linear parameters so as to enhance global contrast of the remote sensing image. The incomplete Beta function can be written as following:

$$F(u) = B^{-1}(\alpha, \beta) \times \int_0^u t^{\alpha-1}(1-t)^{\beta-1}dt, \quad 0 < \alpha, \beta < 10 \tag{4}$$

All the gray levels of original image have to be unitary before implementing IBT. All the gray levels of enhanced image have to be inverse-unitary after implementing IBT. We will employ the GA to optimize transform parameters [11]. Let $\mathbf{x} = (\alpha, \beta)$, $F(\mathbf{x})$ is the fitness function for GA, where $a_i < \alpha, \beta < b_i$ （$i = 1,2$）, a_i and b_i （$i = 1,2$） can be determined by Ref.[8]. Before running GA, several issues must be considered as follows. In this study, the population size is set to 20 and the initial population will contain 20 chromosomes (binary bit strings), which are randomly selected. The maximum number of iterations (generations) of GA is set as 60 in our experiment. In this paper, Shannon's entropy function is used to quantify the gray-level histogram complexity [9].

$$E_{ntr} = -\sum_{i=1}^{n} \left(p_i \cdot \log p_i \right) \tag{5}$$

Where $p_i \cdot \log p_i = 0$ by definition for $p_i = 0$. Since P is a probability distribution, the histogram should be normalized before applying the entropy function. In this study, the fitness function is formed by Equation (6):

$$F_{ctr} = -E_{ntr} \tag{6}$$

Less F_{ctr} is, better contrast of enhanced image is. In this study, a multi-point crossover is used. Mutation probability of applying mutation, P_m, is set as 0.001. The GA will be iteratively performed on an input degraded image until a stopping criterion is satisfied. The stopping criterion is the number of iterations is larger than another threshold (here it is set as 50). Then the chromosome (the solution) with the smallest fitness function value, i.e., the optimal set of IBT for the input degraded image, is determined. Using the optimal set of IBT enhances the degraded image. Totally there are two parameters in the set of IBT (α and β). The two parameters will form a solution space for finding the optimal set of IBT for image enhancement. Applying GA, the total two parameters will form a chromosome (solution) represented as a binary bit string, in which each parameter is described by 20 bits. We will employ the GA to optimize continuous variables [10]. If the algorithm is used directly to enhance image contrast, it will result in large computation cost.

IBT is calculated pixel-to-pixel. Operation burden is very large when pixels in original image are large. Different IBT have to be calculated to different α and β. Different IBT need to be calculated once in every iteration during optimization. To improve operation speed during the whole optimization, we will employ fuzzy wavelet neural network (FWNN) to approximate IBT [11]. The architecture of FWNN is shown in Fig.1 [11]. With a four-layered architecture of FWNN, we fuzzify the wavelet-transformed inputs.

The four layers are wavelet transformation, fuzzification, inference, and defuzzification, respectively. Once we determine the numbers of inputs, rules and fuzzy sets of for each input, the structure of FWNN is determined. In this, each input is presented to the wavelet block so as to exploit the compression and denoising properties leading to the architecture FWNN. This wavelet-transformed input is then fuzzified and presented to the neural network to get the actual output. The activation function is fuzzification layer is characterized by the Gaussian membership function. Detail information about FWNN can be found in Ref. [11].

IBT can be calculated by the above FWNN. Parameters α, β, g are input to trained FWNN and output g' for IBT is obtained directly. 100000 points are selected as sample sets. Parameter α and β, which are between 1 and 10, are divided into 10 parts at the same interval. Parameter x, which is between 0 and 1, is divided into 1000 parts at the same interval.

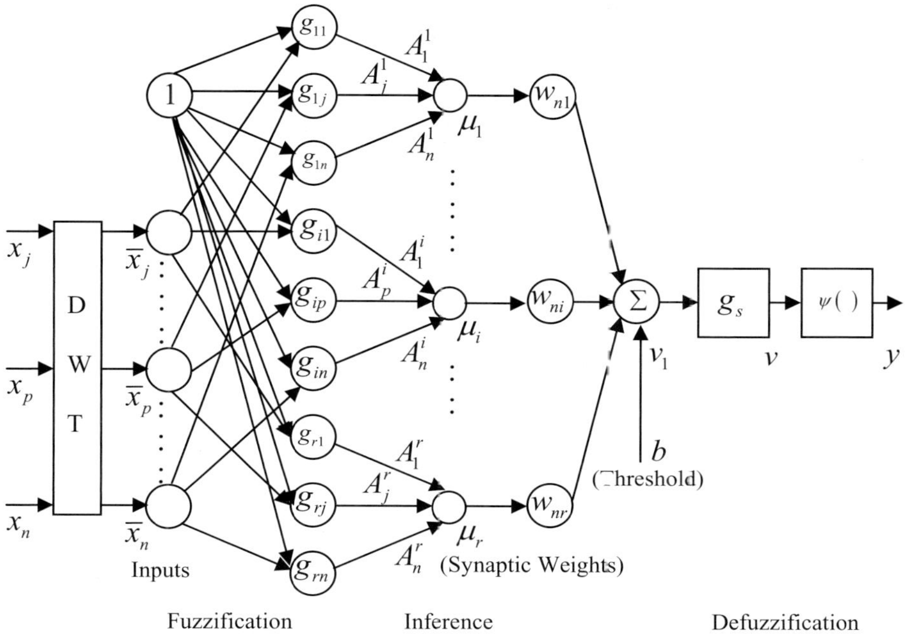

Fig.1. Structure of FWNN

4 Experimental Results

In experiments, a satellite cloud image added corrupted by additive gauss white noise (GWN) is used to prove the efficiency of our algorithm. Figure 2 indicates incompleteBeta curve, which is obtained by GA. In order to prove the efficiency of the new

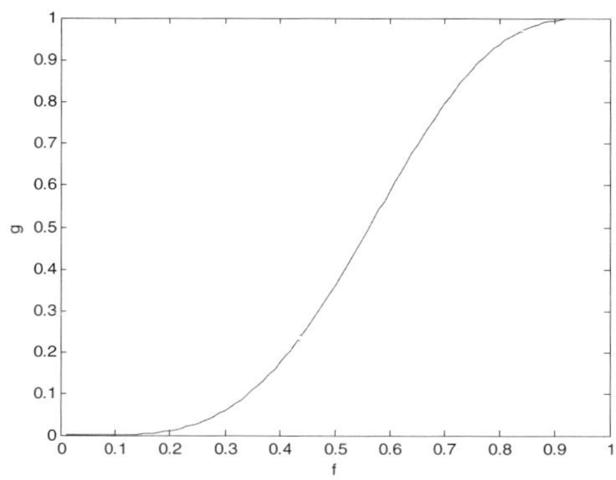

Fig.2. Incomplete Beta transform curve ($\alpha = 4.8184$, $\beta = 3.8258$)

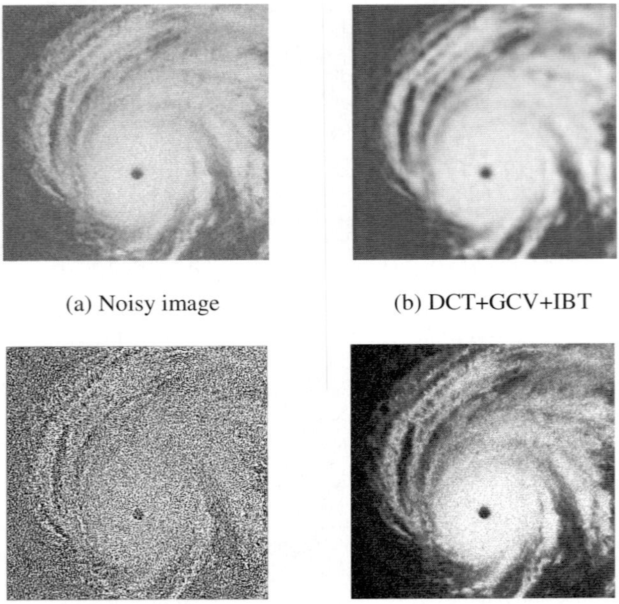

(a) Noisy image (b) DCT+GCV+IBT

(c) USM (d) HE

Fig.3. Processed results by five methods

algorithm, we will compare the performance between our algorithm (DCT+ GCV+ IBT), USM and HE. Figure 3(a) represents a noisy satellite cloud image (standard variance of noise is 10). Figure 3 (b)–(d) respectively represent the results of DCT+GCV+IBT, USM and HE. Noises are enhanced when we use USM to enhance the contrast of Fig.3 (a). Some important information is emerged into background and noise, this is obvious in Fig.3 (c). Although the whole contrast is well when HE is used to enhance the contrast of Fig.3 (a), the noise in Fig.3 (a) is also greatly enhanced. This can be seen in Fig.3 (d). Compared with USM and HE, the noise in the satellite cloud image is well reduced while the contrast of the image is also well enhanced. This can be seen in Fig.3 (b).

5 Conclusions

In this paper, we propose a method to reduce the noise and enhance contrast in the satellite cloud image by the GCV, GA, FWNN and IBT in DCT domain. The global contrast is enhanced in the low frequency coefficients in the DCT domain. Compared with wavelet transform, USM and HE, USM and HE do not consider the noise in the image, we consider both noise and contrast.

Acknowledgments. This work was supported by the Science and Technology Department of Zhejiang Province of P.R. China (2005C23070).

References

1. Donoho, D.L., Johnstone, I.M.: Ideal spatial adaptation via wavelet shrinkage. Biometrika 81, 425–455 (1994)
2. Chang, S.G., Yu, B., Vetterli, M.: Spatially adaptive wavelet thresholding with context modeling for image denoising. IEEE Trans. on Image Processing 9, 1522–1531 (2000)
3. Starck, J.L., Candes, E.J., Donoho, D.L.: The curvelet transform for image de-noising. IEEE Trans. on Image Processing 11, 670–684 (2002)
4. Starck, J.L., Candès, E.J., et al.: Very high quality image restoration by combining wavelets and Curvelets. In: Proc. SPIE, pp. 9–19. SPIE Press, San Jose, CA (2001)
5. Ulfarsson, M.O., et al.: Speckle reduction of SAR images in the curvelet domain. In: Proc. IGARSS' 02, pp. 315–317. IEEE Computer Press, Oakland (2002)
6. Saevarsson, B., et al.: Combined wavelet and curvelet denoising of SAR images. In: Proc. IGARSS'04, pp. 4235–4238. IEEE Computer Press, Oakland (2004)
7. Johnstone, I.M., Silverman, B.W.: Wavelet threshold estimators for data with correlated noise. Journal of the Royal Statistical Society, Series B 59, 319–351 (1997)
8. Tubbs, J.D.: A note on parametric image enhancement. Pattern Recognition 30, 616–621 (1997)
9. Fu, J.C., Lien, H.C., Wong, S.T.C.: Wavelet-based histogram equalization enhancement of gastric sonogram images. Computerized medical imaging and graphics 24, 59–68 (2000)
10. Shyu, M.-S., Leou, J.-J.: A genetic algorithm approach to color image enhancement. Pattern Recognition 31, 871–880 (1998)
11. Srivastava, S., Singh, M., Hanmandlu, M., Jha, A.N.: New wavelet neural networks for system identification and control. Applied Softing Computing 6, 1–17 (2005)

The Research of the Sensor Fusion Model Based on Fuzzy Comprehensive Theory

Xiaodan Zhang, Zhendong Niu, Xiaomei Xu, Kun Zhao, and Yunjuan Cao

School of Computer Science and Technology, Beijing Institute of Technology University, Beijing, P.R.China
Xiaodanzhang75@126.com

Abstract. A new decision sensor fusion model based on the fuzzy theory, which introduces fuzzy comprehensive assessment into traditional decision sensor fusion technology, is proposed in this paper. Through compare the difference between the architecture of hard decision and soft decision, the soft decision architecture had been applied. At the fusion center, the process of fusion is composed of the comprehensive operation and the global decision, and the global decision of the concerned object could be obtained by fusing the local decision of multiple sensors. In the practical application, the model has been successfully applied in the temperature fault detection and diagnosis system of Jilin Fengman Hydroelectric Simulation System. In the analyses of factual data, the performance of the system precedes that of the traditional diagnosis method.

1 Introduction

Sensor fusion or information fusion is a new information process technology for the alliance of data obtained from multiple sources, such as sensors, database, knowledge base and so on. It aims at obtaining coherent explanation and description of the concerned object and environment, through making use of the multi-sensor resource, combining the redundant, complement information that each sensor has obtained, by rationally employ each sensor and its sensor data. So from this view, information fusion is a kind of comprehensive, multiple angles, multiple layers analyses process to the concerned object [1][2].

Information fusion could be classified into three levels according to the abstract level of data, that is, pixel level fusion, feature level fusion and decision level fusion [3]. Decision fusion is a kind of high-level fusion process, and its result is often utilized as the basis for the system decision. Because the process of the decision level fusion often concerns all kinds of factors, besides the data that obtained by sensor, database, etc. further more the evidence using in decision fusion process is often uncertain and fuzzy, it is very difficult to construct the accurate mathematic model that has high reliability for a certain problem. But in practical application, the decision level fusion can bring some especial benefit, such as high robustness, processing different class information, and so on, so it has been paid attention to by scientist and engineer, and become an important subject in the study of information fusion theory and its application.

K. Li et al. (Eds.): LSMS 2007, LNCS 4688, pp. 396–402, 2007.

In this paper, a new decision level fusion model, in which the fuzzy property of the decision level fusion is considered and the fuzzy decision architecture of information fusion is adopted, is researched. The model introduces fuzzy comprehensive assessment of fuzzy theory into decision assessment at the process of the fusion. In the practical application, the model has been successfully applied in the temperature fault detection and diagnosis system of Jilin Fengman Hydroelectric Simulation System. In the analyses of data, the performance of the model precedes that of the traditional diagnosis method.

2 Model of Fuzzy Comprehensive Assessment

Fuzzy comprehensive assessment is comprehensive assessment method to the object and phenomena that is influenced by multiple factors using fuzzy set theory [4]. The method has been successfully applied into the process of industry, evaluation of product; supervise of quality and so on.

In the process of fuzzy comprehensive assessment, it is denoted that (U,V,R) is assessment model of fuzzy comprehensive assessment, and the Factor Set U consists of all elements that relates to the assessment, it can be represented as $U=\{u_1,u_2,\cdots,u_m\}$, In general, and every factor u_i has its different weight a_i. The weight set A is a fuzzy set, which is represented by a fuzzy vector, $A=(a_1,a_2,\cdots,a_m)$, where a_i is the value of the membership function of the factor u_i relating A, that is, it represents the degree of every factor in the comprehensive assessment. In general, it satisfies $\sum_{i=1}^{m} a_i = 1 \quad a_i > 0$.

The set V is the Assessment Set, which is the set that consists of the assessment degree of the object. It can be represented by $V=\{v_1,v_2,\cdots,v_n\}$, where v_i is the assessment degree for this assessment.

The matrix $R=(r_{ij})_{m\times n}$ is a fuzzy mapping from ***U to V***, where r_{ij} express the possibility degree of jth assessment when considering the ith factor, that is the membership degree of from u_i to v_j.

In the process of fuzzy comprehensive assessment, let $A=(a_1,a_2,\cdots,a_m)$ be the fuzzy set on the Factor Set U, in which a_i is the weight of u_i. $B=(b_1,b_2,\cdots,b_n)$ is the fuzzy set on the assessment set V, the comprehensive assessment can be represent as following,

$$B = A \circ R = \{b_1, b_2, \cdots, b_n\} \tag{1}$$

In formula (1) the operator $\circ$ is often defined as the assessment arithmetic operator $(\wedge^*, \vee^*)$, so formula 1 can be written as

$$\forall b_i \in B,\ \ b_i = (a_1 \wedge^* r_{i1}) \vee^* (a_2 \wedge^* r_{i2}) \vee^* \cdots \vee^* (a_m \wedge^* r_{in}) \tag{2}$$

In general, the assessment arithmetic operator can be defined according to the practical application, such as common matrix operation (multiplication and addition) or zadeh fuzzy operation (min and max) and so on.

Following the comprehensive process, the synthetic evaluation of $\{b_1, b_2, \cdots, b_n\}$ a defuzzification process of making a fuzzy value to a precise value. The methods, such as max membership principle, centroid method, weighted average method etc, can be applied. In general, max-membership principle is also known as the height method, which is limited to peaked output. The centroid method is also called the center area or center of gravity, it is the most prevalent and physically appealing of all the defuzzification methods. Weighted average method is only valid for symmetrical output membership functions, but is simple and convenient. In practical application the exact method of synthetic evaluation is usually depend upon the application.

3 The Architecture of the Decision Information Fusion

The objects of decision information fusion is usually the local decisions of the sensors, that is, the process of decision information fusion is that of global decision under the basis of local decisions made by the multiple sensors. The method or architecture of the decision information fusion is usually classified into either the "hard" decision or the "soft" decision according to the results of local decision of the sensor [1].

In the "hard" decision, the local decision of the sensor is usually the problem of binary hypothesis, the result of hypothesis test is either zero or one according to the threshold level. So the local decision of the sensor that is directly sent to the fusion center is either zero or one. The Figure 1 is the architecure of "hard" decision fusion.

In the "soft" decision, the whole region of sensor decision is usually divided into multiple regions, and the result of the sensor includes not only the region of decision but also the reliability value belonging to the region, so the information that is sent to the fusion center in "soft" decision is the possibility of each hypothesis. The Figure 2 is the architecure of "soft" decision fusion.

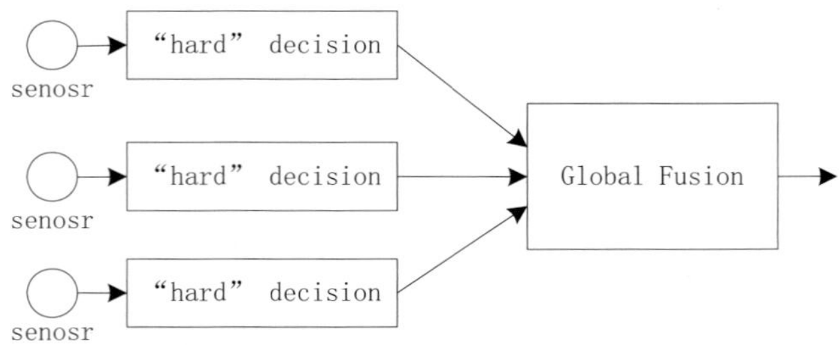

Fig. 1. Architecure of "hard" decision fusion

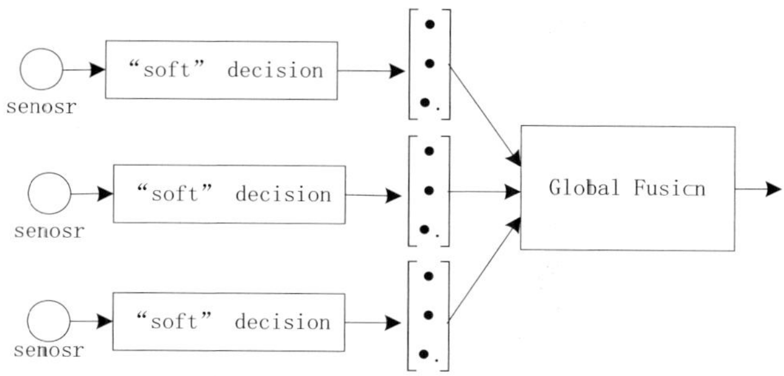

Fig. 2. Architecure of "soft" decision fusion

In the process of "hard" decision, the sensor couldn't provide any information the is lower or higher than the threshold level, so the information that is lower or higher than the threshold level is ignored in the process of fusion at the fusion center. Compared with the process of "hard" decision, the process of the "soft" decision provides not only the region of decision, but also the reliability value of the region. In the fusion center, the object including the region and the reliability value of the region can be utilized for the process of the fusion.

The architecture of the process of the "soft" decision under the fuzzy comprehensive assessment is that, the global fusion is composed by comprehensive operation and global decision. In comprehensive operation, local decisions of multisensor are integrated, that is the map from multi-dimension to one- dimension. The result of comprehensive operation is sent to the global decision module, the global decision.

4 The Decision Fusion Algorithm Based on the Fuzzy Comprehensive Theory

From the figure 3, the architecture of the "soft" decision is applied in the model of decision level information fusion based on the fuzzy comprehensive assessment. In the algorithm, we consider an information fusion system consisted of *M* sensors that observe the same phenomenon, and the sensors could be homogeneous or heterogeneous. Each sensor makes its local decision based on its observation, the local decision that include the decision region and its reliability value is sent to the fusion center, the global decision based on the local decisions of *M* sensors is obtained at the fusion center.

It is denoted that the set S is the sensor set, that is $S = \{s_1, s_2, \cdots, s_M\}$, the result of the fusion center is classified into *N* regions, is called as the assessment set Y, that is $Y = \{y_1, y_2, \cdots, y_N\}$. In the process of the "soft" decision of each sensor, the result of each sensor is the value of possibility on the assessment *Y*, for the ith sensor, the result of local decision can be described as the vector $r_i = (r_{i1}, r_{i2}, \cdots, r_{in})$, through the process of normalization, the input of the fusion center for the ith sensor is the vector $\overline{r_i} = \left(\overline{r_{i1}}, \overline{r_{i2}}, \cdots, \overline{r_{in}}\right)$.

For the $\forall s_i \in S$, the vector $\overline{r_i}$ consist of the $m \times n$ matrix *R*, that is called as the Fusion Matrix of the fusion center, can be described as following.

$$R = \begin{bmatrix} \overline{r_{11}} & \overline{r_{12}} & \cdots & \overline{r_{1n}} \\ \overline{r_{21}} & \overline{r_{22}} & \cdots & \overline{r_{2n}} \\ \vdots & \vdots & \vdots & \vdots \\ \overline{r_{m1}} & \overline{r_{m2}} & \cdots & \overline{r_{mn}} \end{bmatrix} \tag{3}$$

For each sensor in the fusion system, the effect of each sensor is always different, it is denoted that *A* is the sensors' the vector of weight power, it is a fuzzy set on the sensor set S, and described as the normalized fuzzy vector $A' = (a_1, a_2, \cdots, a_M)$ and $\alpha_i = \mu(\alpha_i), i = 1,2,..., M$.

In the comprehensive operation of the algorithm, the comprehensive result of the sensor weigh vector and the fusion matrix is the fuzzy set of the assessment set. The result can be described as following,

$$B = A \circ R = \{b_1, b_2, \cdots, b_n\} \tag{4}$$

For the comprehensive operator, the algorithm adopted the comprehensive operator $\left(\wedge^*, \vee^*\right)$ in the fuzzy comprehensive assessment.

In the process of the global decision at the fusion center, the input is the vector $\{b_1, b_2, \cdots, b_n\}$ result from the comprehensive operation, in this research, the max membership principle is adopted, that is if $\exists i \in \{1,2,\cdots,m\}$, stasficify b_i = max $\{b_1$, $b2,\ldots,b_m\}$, so the result of global decision of the fusion center is b_i.

5 Analysis

In the Hydroelectric Simulation System of Jilin Fengman, the generator system is the important component of the system, its working condition has great influence to the whole system, and so real-time fault detection and diagnosis is necessary to the hydro turbine. In the detection system in the Hydro power plant, the senor-threshold level method is often applied for fault detection and diagnosis, that is, in the system of fault detection and diagnosis of the running equipment, primary parameter of the equipment is supervised by a sensor, the data which is obtained and preprocessed from the sensor system is sent to the detection center of the system. In the detection center, threshold level of the parameter is usually set in advance, when the data that is obtained exceed the threshold level, touch off the corresponding fault alarm. So the sensitivity of the whole detection system is dependent upon the threshold level. But in the practical application, the threshold level is set artificially according to the experience. If the value of the threshold level is too high, it is possible to fail to detect the fault, otherwise if the value is too lower, it is possible to cause the system alarm when the equipment is in order.

Aimed to the disadvantage of the traditional detection and diagnosis system, the information fusion technology can be applied into fault detection and diagnosis system. In the practical diagnosis system, multiple sensors have been embedded into the running equipment, gathered the current data of circumstance, and sent to the fusion center of the system. At the fusion center, redundant and complemented data have been made full use of, so precise estimation about equipment status can be achieved, belief quantity of diagnosis system can be enhanced and fuzzy level of status is decreased. So the application of information fusion improves detection performance by making full use of resource of multiple sensors [5][6]. In the experiment of the simulation system, we have applied the new decision information fusion model into the temperature fault detection and diagnosis of the hydro turbine.

In the experiment, three embedded temperature sensors have been embedded into the hydro turbine, and the temperature of equipment has been periodically gathered and sent to the fusion center [7] [8] . The set of sensors can be defined as S={s_1, s_2, s_3}.

In the practical application of the system, it has been found that the reason of temperature alarm of the hydro turbine can be classified into the fault of Cycle Water Equipment, Cooling Water Equipment and misplay of operator, etc. So the assessment set in the temperature fault diagnosis system could be defined as follows.

$Y=\{y_1, y_2, y_3, y_4, y_5\}$ ={Circulation water valve shutdown by error, low pressure of circulation, cooling water valve shutdown by error, cooling water pump lose pressure and backup pump not switched, other undefined reason}.

The effect of the three sensors is different in the diagnosis system because of its position, precision and so on, so in the practical application, the weigh power vector has been allotted according to the experience of the hydroelectric experts. That is

$$A=(a_1, a2, a3) = (0.3400,0.2000,0.2400) \tag{5}$$

The three embedded sensors gather the data of the hydro turbine and make their local decision about the current state, the local decisions which are the reliability value of the fault have been send to the fusion center, the process of the diagnosis in the fusion center as following. In the fusion center, Firstly, the local decision of the sensor has been normalized the fusion matrix is constructed by the results of normalization of each sensor. Secondly, comprehensive operation is made between the sensor weigh power vector and the matrix of decision. At last, global decision about the fault is made according to the result of comprehensive operation under the max membership principle.

In the global decision, according to the max membership principle, the decision about the fault is made as the cooling water pump lose pressure and backup pump not switched.

6 Conclusion

In this paper, a new decision information fusion model is proposed. The soft decision architecture had been applied in the model, and the process of fusion is composed of the comprehensive operation and the global decision under the circumstance of multi-sensor system. In the comprehensive operation, local decisions of the sensors are integrated, that is the map from multi-dimension to one- dimension had been finished. In the global decision module, the global decision could be made according to the decision principle. In the practical application, the model had been successfully applied in the temperature fault detection and diagnosis system of Jilin Fengman

Hydroelectric Simulation System, and the performance of the diagnosis system precedes that of the traditional diagnosis method.

References

1. Liu, T.M., Xia, Z.X., Xie, H.C.: Data Fusion Techniques and its Applications. National Defense Industry Press, Beijing (1999)
2. He, Y., Wang, G.H.: Multisensor Information Fusion with applications. Publishing House of Electronics Industry, Beijing (2000)
3. Linas, W.E.: Multisensor data fusion. Artech House Inc., Norwood (1991)
4. Hall, D.: Mathematical Techniques in Multisensor Data Fusion. Artech House Inc., Norwood (1992)
5. Jlinals, J.: Assessing the Performance of Multisensor Fusion System. SPIE, p. 1661 (1991)
6. Goebel, K.F.: Conflict Resolution using Strengthening and Weakening Operations in Decision Fusion. In: Proceedings of The 4th International Conference on Information Fusion, vol. 1 (2001)
7. Xu, L.Y., Du, D.Q., Zhao, H.: Study on information fusion methods of embeded power plant fault prediction system. Journal of Northeastern University (Natural Science) 1, 8–11 (2000)
8. Du, Q., Zhao, H.: D-S Evidence Theory Applied to Fault Diagnosis of Generator Based on Embedded Sensors. In: Proceedings of The 3th International Conference on Information Fusion, vol. 1 (2000)
9. Goebel, K., Krok, M., Sutherland, H.: Diagnostic Information Fusion: Requirements Flowdown and Interface Issues. In: Proceedings of the IEEE Aerosense Conference, p. 1103. IEEE Computer Society Press, Los Alamitos (2000)

Real Time Object Recognition Using K-Nearest Neighbor in Parametric Eigenspace*

MyungA Kang[1] and JongMin Kim[2]

[1] Dept. Computer Science and Engineering, KwangJu University, Korea
makang@gwangju.ac.kr
[2] Computer Science and Statistic Graduate School, Chosun University, Korea
mrjjoung@chosun.ac.kr

Abstract. Object recognition technologies using PCA(principal component analysis) recognize objects by deciding representative features of objects in the model image, extracting feature vectors from objects in a image and measuring the distance between them and object representation. Given frequent recognition problems associated with the use of point-to-point distance approach, this study adopted the k-nearest neighbor technique(class-to-class) in which a group of object models of the same class is used as recognition unit for the images inputted on a continual input image. However, the robustness of recognition strategies using PCA depends on several factors, including illumination. When scene constancy is not secured due to varying illumination conditions, the learning performance the feature detector can be compromised, undermining the recognition quality. This paper proposes a new PCA recognition in which database of objects can be detected under different illuminations between input images and the model images.

1 Introduction

Object recognition is one of the most actively researched areas in computer vision [1]. An object recognition system can be described in various ways, but it simply finds objects in a given image that match models of known objects in the database [2][3]. Under this definition, the object recognition system responds differently in the presence of clutter: If there is almost no amounts of clutter in the image, a single object exists, and the system will find out whether the detected image matches an object representation in the database. In high clutter environments, the system will not only find out whether detected object matches the models in the database but also identifies characteristics features of objects such as area in the image.

In this study, an image containing a single object in absence of background clutter was tested. The image of objects can be acquired in either two-dimensional (2D) image or three-dimensional (3D) image. A 2D shape recognition approach maintains images of objects in one stable position, making it useful for flat objects, whereas a 3D shape recognition approach obtains images of objects from different viewpoints, making it useful for every kind of object.

* This Study was conducted by research funds from Gwangju University in 2007.

K. Li et al. (Eds.): LSMS 2007, LNCS 4688, pp. 403–411, 2007.

Because objects appear differently from the point of view, 3D images demand more complex object recognition systems than 2D images.

In this study, a collection of images was developed by rotating a 3D object 5 degrees at a time and making a full turn. The performance of recognition systems using principal component analysis is very sensitive to rotation, translation, scale and illumination [4][5]. It is therefore necessary to create many images of objects to be tested and normalize the size of images to keep the performance of recognition system stable. Because of frequent recognition errors involved in the use of point-to-point approach, this study used K-Nearest Neighbor approach (class-to-class), in which a group of object models of the same class is used as recognition unit for images inputted on a continual basis, to improve the recognition quality. Normalization and histogram equalization were also performed in object images to ensure a stable recognition rate under varying illumination conditions.

2 Object Recognition Algorithm

Normalization was applied to visual images obtained on a real time basis, and recognition was performed using principal component analysis. A local eigenspace was designed to apply the K-Nearest Neighbor decision rule rather than simple distance measures. The proposed algorithm is presented in Fig 1.

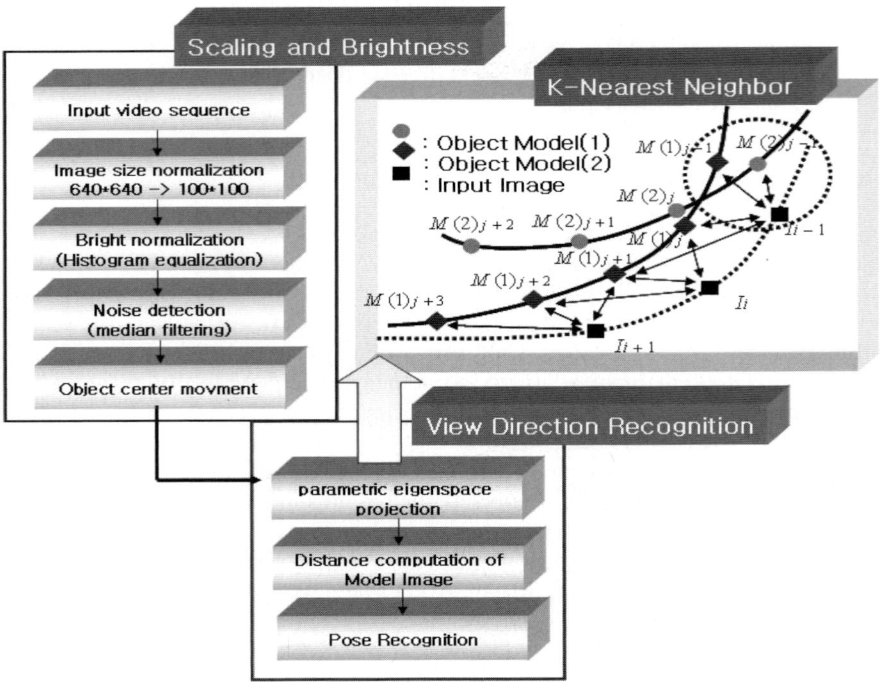

Fig. 1. The Whole structure of proposed algorithm

3 Normalization

Histogram equalization was performed to normalize varying illumination conditions, and median filters were also used to clean up noise. The outcome of histogram equalization is presented in Fig. 2.

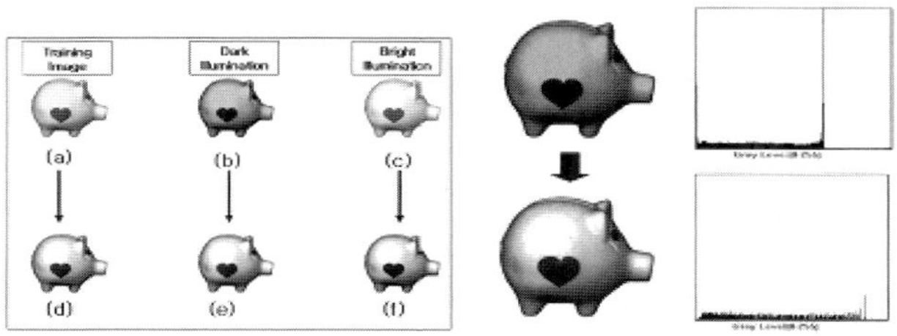

Fig. 2. Object images under variable illumination and result after histogram equalization

To eliminate the noise added to the images as a result of the process of histogram equalization, the following equation was employed for median filtering.

$$med(x_i) = \begin{cases} x_{v+1}, & n = 2v+1 \\ \frac{1}{2}(x_v + x_{v+1}), & n = 2v \end{cases} \tag{1}$$

4 Object Recognition Using PCA

It is much difficult to show images of object that is rotating continually using a camera. This study proposes a solution based on principal component analysis to obtain images of object that is rotating a certain degree at a time and making a full turn.

The solution involved creating a low dimensional vector space to model the overall appearance of object using the equalized output. This space was created on the basis of principal component analysis, which reconstructs the values of eigenvectors having

(a) (b)

Fig. 3. (a)Object Set (b) Image set obtained by rotating object 5°

high probability in the order of descending eigenvalues after computing the distribution of spatial data representing the images of objects. Thus this statistical method is designed to yield eigenvectors and eigenvalues and an average model of object. The images of objects used for this study is presented in Fig. 3(a) and 72 images of an object that rotated 5° degrees at a time is presented in Fig. 3(b).

4.1 Eigenspace Calculation with the PCA Method

In order to calculate eigenvectors, the average image was calculated as the center value and then the difference from this center to each image was calculated using the following formulas (2) and (3):

$$C = (1 - N)\sum_{i=1}^{N} x_i \tag{2}$$

$$X \stackrel{\Delta}{=} \{x_1^{(1)} - c, x_2^{(2)} - c, \ldots\ldots, x_R^{(p)} - c\} \tag{3}$$

Where C is average image and X is a set of images.

The image matrix X is $N \times M$, where M is the total number of images in the universal set, and N is the number of pixels in each image.

Next, we define the covariance matrix **:**

$$, Q \stackrel{\Delta}{=} XX^T \tag{4}$$

That is, eigenvalue λ and eigenvector e of the covariance matrix Q of the images were calculated according to the following equations:

$$\lambda_i e_i = Qe_i \tag{5}$$

Among the studied matrixes, the matrix used as eigenvector is U as it size equals to matrix X Eigenvectors yielded from the process of SVD(Singular Value Decomposition) can be reconstructed in the order of descending with eigenvalues. An eigenvector's eigenvalue reflects the importance of eigenvector and is calculated using the formula(6). At this time, it is possible not to include every eigenvector into eigenspace and select only main eigenvectors representing features of objects.

$$\frac{\sum_{i=1}^{k} \lambda_i}{\sum_{i=1}^{N} \lambda_i} \geq T_1 \tag{6}$$

Where T_1 is the threshold value used for determining the number of vectors.

K was 3 for the low dimensional space used for learning and pose evaluation.

4.2 Correlation and Distance in Eigenspace

Once object models are decided by normalized images in the eigenspace, the next step needed for recognition is very simple. After the difference between the input mage X

and the average image C was calculated and reflected in the eigenspace according to the following formula:

$$f_j = [e_1, e_2, e_3, \ldots\ldots\ldots\ldots, e_k]^T (x_n - c) \tag{7}$$

The new image f_j was represented as a point in the eigenspace. When reflected, a set of scattered points in the eigenspace correspond to all the objects. Vectors that have a similar feature or value were closely clustered in the eigenspace, meaning the images of two identical objects have a similar value and are situated in the same area.

The distribution of feature vectors for a set of images of a rotating object is presented in Fig. 4, and the distribution of feature vectors for images of all the rotating objects used for the study is presented in Fig. 5.

The closer distance between the fixed point to a point in the eigenspace is interpreted as a higher correlation between the new image and the previously stored image.

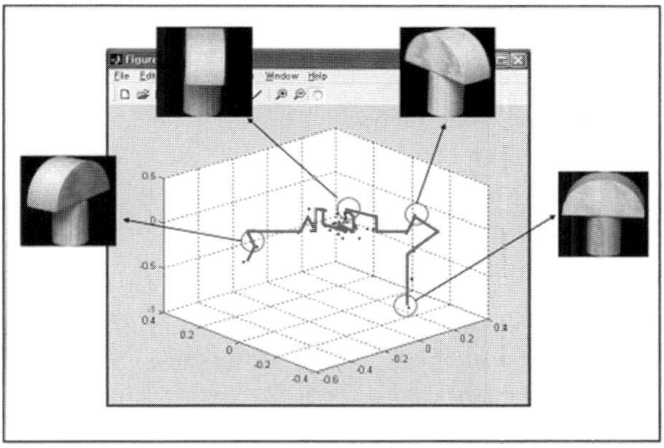

Fig. 4. The distribution of cat images in the eigenspace

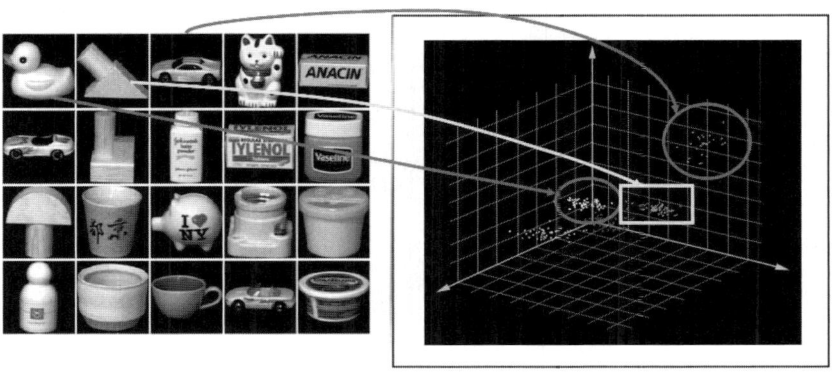

Fig. 5. The distribution of images of all objects in eigenspace

4.3 Improved K-Nearest Neighbor Algorithm

The use of point-to-point approach in pattern classification frequently resulted in a recognition error even if a new image was matched to the previously stored image because of features that were unnecessarily detected in the new image.

To solve this recognition problem associated with this individual feature matching technique, features can be analyzed by using a group of object models of the same class as recognition unit for images inputted on a continual basis (Class to Class).

$$w = \frac{(\arg S(M_j) - Min(\arg S(M_j)}{d(k-1)} \tag{8}$$

Where $\arg S(M_j) = j$ is an operator meaning the number of object model. K - Nearest Neighbor matching algorithm was used as presented in formulas (8) and (9).

$$\frac{\sum\sum w(I_j - M_j)}{k} \tag{9}$$

Where $K = 3$.

Recognition of model image and the input image is decided by the value obtained from the formula (9). Based on these formulas, input images were matched to model images as illustrated in the eigenspace(Fig. 6). It was found in the same eigenspace that the input image can be identified as a different object despite its proximity to the model image.

To solve this recognition problem resulted from the distance measures between points, this study used K -Nearest Neighbor matching algorithm in which features are analyzed for matching by using a group of object models of the same class as recognition unit for images inputted on a continual basis. As a result, Recognition quality improved.

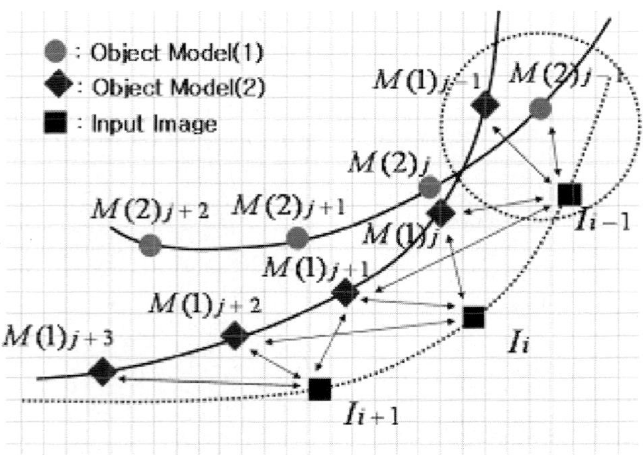

Fig. 6. Recognition based on K -Nearest Neighbor matching algorithm

5 Experiment for Object Recognition

5.1 Results from K-Nearest Neighbor Algorithm

The images were taken while each object was rotating 5° at a time and making a full turn. A set of these images is called the image of the object. The images in the size of 640*480 pixels were normalized to the size of 100*100 pixels.

Eigenvectors were calculated from a set of object images, and the five-dimensional vectors showing high probabilities were defined as a feature space. As a result, 1000 dimensional images (100*100) were compressed to 5-dimensional images, which were suitable for real-time object recognition. Matching rates of point-to-point approach and edited K -Nearest Neighbor rule are compared in Table 1. As shown in the table, the matching rates were high with improved K-Nearest Neighbor rule. A larger number of mismatches were corrected when using the k-nearest neighbor rule.

Table 1. Comparison of matching rates between two classification techniques

Matching	input image	A failure to find a match	Mismatching	Matching rate
Distance measure (Point to Point)	With object models	12.5 %	11 %	76.5 %
	Without object models	15.8 %	22.2 %	64 %
Improved K-Nearest Neighbor (Class to Class)	With object models	5.1 %	3.7 %	91.2 %
	Without object models	11.2 %	16.8 %	72 %

5.2 Effects of Illumination on Object Recognition

Effects of varying illumination conditions on recognition rates are presents in Table2. Illumination conditions were defined as the unaltered illumination condition (simple principal component analysis), normalized brightness and the condition after histogram equalization.

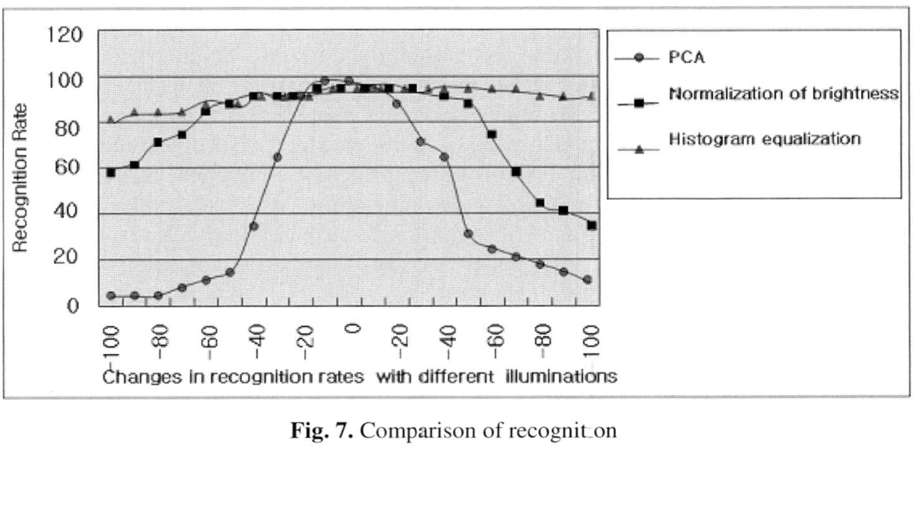

Fig. 7. Comparison of recognition

Table 2. Comparison of recognition rate in the varialbe illumination

Changes in recognition rates with different illuminations	PCA	Normalization of brightness	Histogram equalization
-100	3.33%	57.00%	81.23%
-90	3.33%	61.33%	85.00%
-80	3.34%	70.67%	86.00%
-70	5.00%	76.00%	86.33%
-60	7.67%	83.67%	88.67%
-50	11.33%	88.67%	92.00%
-40	35.00%	91.33%	92.67%
-30	65.66%	91.33%	93.33%
-20	88.20%	91.67%	93.33%
-10	97.00%	92.33%	92.00%
0	96.25%	93.33%	94.67%
10	95.33%	91.33%	92.67%
20	90.67%	90.67%	92.67%
30	68.30%	90.00%	92.67%
40	62.67%	89.67%	92.67%
50	30.33%	84.33%	92.67%
60	25.00%	73.00%	91.00%
70	20.33%	57.66%	91.33%
80	17.00%	44.00%	91.00%
90	14.33%	38.00%	91.00%
100	12.67%	33.67%	90.77%
Total	40.61%	75.70%	90.65%

6 Conclusions

It was found that matching rates were high when features were classified with k-nearest neighbor rule, compared with those obtained from the use of point-to-point measures. The study findings provided the evidence that k-nearest neighbor was more effective for simple and stable recognition than other techniques using geometric information or stereo images. The images of objects were taken by rotating objects 5° at a time to recognize 3D objects from 2D images, and features of objects of the same class were grouped to be used as recognition unit for 3D object recognition.

It is known that recognition systems using principal component analysis undermines recognition quality because of their vulnerability to changes in illumination conditions. The improved K-Nearest Neighbor rule proposed in this study maintained recognition rate of more than 90% under varying illumination conditions caused by histogram equalization, and the recognition rate was higher than that obtained from the illumination condition in which brightness was normalized.

However, mismatches often occurred during the process of 3D object recognition with objects rotated by 90°. In addition, it was difficult to separate characteristics features of objects such as area from cluttered background. It is therefore important to develope a more stable algorithm for 3D object recognition by solving these problems.

References

1. Weng, J., Ahuja, N., Huang, T.S.: Learning recognition and segmentation of 3-D object from 2-D images. In: Proc. of Fourth Int'l Conf. on Computer Vision, pp. 121–128 (1993)
2. Viola, P., Jones, M.: Robust real-time object detection. In: International Conference on Computer Vision (2001)
3. Murase, H., Nayar, S.K.: Visual Learning and Recogntion 3-Dobject from appearance. International journal of Computer Vision 14 (1995)
4. Arita, D., Yonemoto, S., Taniguchi, R.-i.: Real-time Computer Vision on PC-cluster and Its Application to Real-time Motion Capture (2000)
5. Yang, J., Zhang, D.: Two-Dimensional PCA: A New Approach to Appearance-Based Face Representation and Recognition. IEEE Transactions on Pattern analysis and Machine Intelligence 26 (2004)
6. Bourel, F., Chibelushi, C.C., Low, A.A.: Robust facial expression recognition using a state-based model of spatially localised facial dynamics. In: Proceedings of Fifth IEEE International Conference on Automatic Face andGesture Recognition, pp. 106–111. IEEE Computer Society Press, Los Alamitos (2002)
7. Georghiades, A.S., Belhumeur, P.N., Kriegman, D.J.: From Few to Many: Illumination Cone Models for Face Recognition under Variable Lighting and Pose. IEEE Trans. Pattern Analysis and Machine Intelligence 23, 643–660 (2001)
8. Segen, J., Kumar, S.: Shadow Gestures: 3D Hand Pose Estimation Using a Single Camera. In: CVPR99, Fort Collins, Colorado, vol. 1, pp. 479–485 (1999)
9. Yang, H.-S., Kim, J.-M., Park, S.-K.: Three Dimensional Gesture Recognition Using Modified Matching Algorithm. In: Wang, L., Chen, K., Ong, Y.S. (eds.) ICNC 2005. LNCS, vol. 3611, pp. 224–233. Springer, Heidelberg (2005)
10. Belhumeur, P.N., Hepanha, J.P., Kriegman, D.J.: Eigenfaces vs. Fisherfaces: Recognition Using Class Specific Linear Projection. IEEE Trans. Pattern Analysis and Machine Intelligence 19(7), 711–720 (1997)
11. Yang, J., Yang, J.Y.: From Image Vector to Matrix: A Straightforward Image Projection Technique-IMPCA vs. PCA. PCA Pattern Recognition 35(9), 1997–1999 (2002)
12. De Ritter Dick, Tax, D., Duin, R.P.W.: An Experimental Comparison of One-Class Class ification Methods. In: Proceedings of the Fourth Annual Conference of the Advanced School for Computing and Imaging, ASCI, Delft, June 1998 (1998)
13. Martinez, A.M., Benavente, R.: The AR Face Database. CVC Technical Report, no.24 (June 1998)
14. Martinez, A.M.: Recognizing Imprecisely Localized, Partially Occluded, and Expression Variant Faces from a Single Sample per Class. IEEE Trans. Pattern Analysis and Machine Intelligence 24(6), 748–763 (2002)
15. Park, B.G., Lee, K.M., Lee, S.U.: A Novel Face Recognition Technique Using Face-ARG Matching. In: Proceedings of the 6th Asian Conference on Computer Vision, January 2004, pp. 252–257 (2004)
16. Kim, J.-M., Yang, H.-S., Park, S.-K.: Network-Based Face Recognition System using Multiple Images. In: Shi, Z.-Z., Sadananda, R. (eds.) PRIMA 2006. LNCS (LNAI), vol. 4088, pp. 626–631. Springer, Heidelberg (2006)

Fuzzy Neural Network-Based Adaptive Single Neuron Controller

Li Jia[1], Pengye Tao[1], and MinSen Chiu[2]

[1] Shanghai Key Laboratory of Power Station Automation Technology
College of Machatronics Engineering and Automation, Shanghai University
Shanghai 200072, China
jiali@staff.shu.edu.cn
[2] Department of Chemical and Biomolecular Engineering
National University of Singapore
Singapore, 119260

Abstract. To circumvent the drawbacks in nonlinear controller designing of chemical processes, an adaptive single neuron control scheme is proposed in this paper. A class of nonlinear processes is approximated by a fuzzy neural network-based model. The key of this work is, an adaptive single neuron controller, which mimics PID controller, is considered in the proposed control scheme. Applying this result and Lyapunov stability theory, a novel-updating algorithm to adjust the parameters of the single neuron controller is presented. Simulation results illustrate the effectiveness of the proposed adaptive single neuron control scheme.

1 Introduction

The most widely used controllers in industrial chemical processes are based on conventional linear control technologies, such as PID controllers because of their simple structure and being best understood by process operators. However, demands for tighter environment regulation, better energy utilization, higher product quality and more production flexibility have made process operations more complex within a larger operating range. Consequently, conventional linear control technologies cannot give satisfactory control performance [1-3]. Recently, the focus has been shifting toward designing advanced control schemes based on nonlinear models. The nonlinear techniques using neural network and fuzzy system have received a great deal of attention for modeling chemical process, due to their inherent ability to approximate an arbitrary nonlinear function [4-9].

In this paper, an adaptive single neural control scheme based on fuzzy neural network model is proposed. A class of nonlinear processes is approximated by a fuzzy neural network-based model. The key of this work is, an adaptive single neuron controller, which mimics PID controller, is considered in the proposed control scheme. Applying this result and Lyapunov stability theory, a novel-updating algorithm to adjust the parameters of the single neuron controller is presented. Consequently the

K. Li et al. (Eds.): LSMS 2007, LNCS 4688, pp. 412 – 423, 2007.

proposed design method is able to guarantee the stability of the closed-loop system and the convergence of the predictive tracking error. Examples are used to illustrate the effectiveness of the proposed design method.

This paper is organized as follows. Section 2 introduces the fuzzy neural network used in this paper. The proposed adaptive single neural control scheme is described clearly in Section 3. Simulation results are presented in Section 4 to illustrate the effectiveness of the proposed control scheme. Finally, concluding results are given in Section 5.

2 Fuzzy Neural Network

Fuzzy neural networks (FNN) is recently developed neural network-based fuzzy logic control and decision system, and is more suitable for online nonlinear systems identification and control. The FNN is a multilayer feedforward network which integrates the TSK-type fuzzy logic system and RBF neural network into a connection structure. Considering of computational efficiency and easiness of adaptation, the zero-order TSK-type fuzzy rule is chosen is this paper, which is formulated as:

$$R^l : IF\ x_1\ is\ F_1^l\ AND\ x_2\ is\ F_2^l\ \cdots x_M\ is\ F_M^l,\quad THEN\ y = h_l \tag{1}$$

Where R^l denote the lth fuzzy rule, N is the total number of fuzzy rules, $x \in R^M$ is the input variables of the system, $y \in R$ is the single output variable of the system. $F_i^l (l = 1, 2, \cdots, N, i = 1, 2, \cdots, M)$ are fuzzy sets defined on the corresponding universe [0, 1], and h_l is the lth weight or consequence of fuzzy system.

The FNN contains five layers. The first layer is called input layer. The nodes in this layer just transmit the input variables to next layer, expressed as:

$$\begin{aligned} &Input : I_i^{(1)} = x_i, i = 1, 2, \cdots, M \\ &Output : O_i^{(1)} = I_i^{(1)}, i = 1, 2, \cdots, M \end{aligned} \tag{2}$$

The second layer is composed of N groups (N represents the number of fuzzy IF-THEN rules), each of which is with M neurons (M represents the number of rule antecedent, namely the dimension of the input variables). Neurons of the lth group in the second layer receive inputs from every neurons of the first layer. And every neuron in lth group acts as a membership function. Here Gaussian membership function is chosen, the function of lth group in this layer can then be expressed as:

$$\begin{aligned} &Input : I_i^{(2)} = x_i \\ &Output : O_{li}^{(2)} = \exp(-\frac{(I_i^{(2)} - c_{li})^2}{\sigma_l^2}) \\ &i = 1, 2, \cdots, M ; l = 1, 2, \cdots, N \end{aligned} \tag{3}$$

The third layer consists of N neurons, which compute the fired strength of a rule. The lth neuron receives only inputs from neurons of lth group of second layer. The input and output of every neuron is represented as follows:

$$\begin{aligned}&Input: I_i^{(3)} = O_l^{(2)}\\&Output: O_l^{(3)} = \exp\Big(-\sum_{i=1}^{m}\frac{(I_i^{(2)}-c_{li})^2}{\sigma_l^2}\Big)\\&l = 1,2,\cdots,N\end{aligned} \tag{4}$$

There are two neurons in fourth layer. One of them connects with all neurons of the third layer through unity weight and another one connects with all neurons of the third layer through h weights, described as following:

$$\begin{aligned}Input: I_1^{(4)} &= [O_1^{(3)}, O_2^{(3)}, \cdots, O_N^{(3)}]\\I_2^{(4)} &= [O_1^{(3)}, O_2^{(3)}, \cdots, O_N^{(3)}]\\Output: O_1^{(4)} &= \sum_{l=1}^{N} O_l^{(3)}\\O_2^{(4)} &= \sum_{l=1}^{N} h_l O_l^{(3)}\end{aligned} \tag{5}$$

The last layer has a single neuron to compute y. It is connected with two neurons of the fourth layer through unity weights. The integral function and activation function of the node can be expressed as:

$$\begin{aligned}Input: I^{(5)} &= [O_1^{(4)}, O_2^{(4)}]\\Output: O^{(5)} &= \frac{O_2^{(4)}}{O_1^{(4)}}\end{aligned} \tag{6}$$

The output of the whole FNN is:

$$y = O^{(5)} = \sum_{l=1}^{N}\Phi_l h_l \tag{7}$$

Where

$$\Phi_l = \frac{\exp\Big(-\sum_{i=1}^{M}\frac{(x_i-c_{li})^2}{\sigma_l^2}\Big)}{\sum_{l=1}^{N}\exp\Big(-\sum_{i=1}^{M}\frac{(x_i-c_{li})^2}{\sigma_l^2}\Big)}, (l=1,2,\cdots,N) \tag{8}$$

In this paper, the adjustable parameters of FNN are c_{li} , σ_l and h_l (l= 1,2, ⋯, N; i=1, 2, ⋯, M).

3 Adaptive Single Neuron Control Scheme

In this section, the proposed adaptive single neuron control scheme will be described clearly and systematically. Firstly, a class of nonlinear processes is approximated by a fuzzy neural network model (FNNM). Next, motivated by the conventional PID control design technique in process industries, an adaptive single neuron controller, which mimics a PID controller, is employed in the proposed control scheme. Applying this result and Lyapunov stability theory, a novel-updating algorithm to adjust the parameters of the single neuron controller is presented. The proposed adaptive single neuron control system is designed as Fig. 1, where P is the controlled nonlinear process, ASNC represents the proposed adaptive single neuron controller and FNNM is the nonlinear model. The variables u , r , y represent, respectively, the input, the setpoint and the output of the controlled nonlinear process.

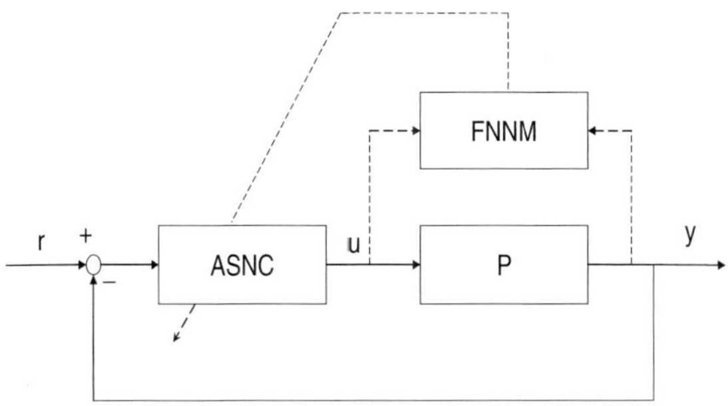

Fig. 1. The structure of adaptive single neuron control system

3.1 Nonlinear Process Modeling

The SISO nonlinear process can be represented by following discrete nonlinear function

$$y(k+1) = f(x(k)) \tag{9}$$

where

$$x(k) = (y(k), y(k-1), \cdots, y(k-n_y), u(k-n_d)^T, u(k-n_d-1)^T, \cdots, u(k-n_d-n_u)^T)^T \tag{10}$$

where u and y are the input and output of the system, respectively.

In this research, the FNN is employed to estimate the nonlinear process (denoted as FNNM) due to its capability of uniformly approximating any nonlinear function to any degree of accuracy, namely,

$$\hat{y}(k+1) = FNNM(x(k)) \tag{11}$$

The method proposed for the identification of FNNM can be summarized as follows:

(1) Define the similarity parameter β. Choose the input data point $x(1)$ as the first cluster (fuzzy rule) and set its cluster center as $c_1 = x(1)$, the number of input data point belonging to the first cluster N_1 and the number of fuzzy clusters N at this time are respectively $N_1 = 1$ and $N = 1$;
(2) For the kth training data point $x(k)$, compute the similarity of the kth training data points to every cluster centers $c_l (l = 1, 2, \cdots, N)$ by similarity criteria and find the biggest one. This is equivalent to finding a cluster (fuzzy rule) that $x(k)$ could belong to. The chosen cluster is denoted by L, where

$$S_L = \max_{1 \le l \le N} (\sqrt{e^{-\|x(k)-c_l\|_2^2}}) \tag{12}$$

(3) Decide whether a new cluster a new cluster (fuzzy rule) should be added or not according to the following criteria:
•If $S_L < \beta$, the kth training data point doesn't belong to all the existing cluster and a new cluster and a new cluster will be established with its center located at $c_{N+1} = x(k)$, and then let $N = N+1$ and $N_N = 1$ (represent the number of training data points belong to the Nth cluster), and keep other cluster unmodified;
•If $S_L \ge \beta$, the kth training data point belong to the Lth cluster and the Lth cluster will be adjusted as follows

$$\begin{aligned} c_L &= c_L + \lambda(x(k) - c_L) \\ \lambda &= \lambda_0 / (N_L + 1), \lambda_0 \in [0,1] \\ N_L &= N_L + 1 \end{aligned} \tag{13}$$

(4) Let $k = k+1$, then go to (2) until all training data points are clustered to corresponding cluster. After finishing the first three steps, the number of cluster (fuzzy rule) is decided as N, and the width of fuzzy can be calculated as:

$$\sigma_l = \min_{\substack{j=1,2,\cdots,N \\ j \ne l}} \frac{|c_l - c_j|}{\rho} \tag{14}$$

where ρ is overlap parameter, usually $1 \le \rho \le 2$.

(5) Estimate the consequent parameters $h_l, l = 1, 2, \cdots, N$,

$$h_l = AVE\_Y_l \tag{15}$$

where AVE_Y_l represents the average output of lth cluster.

3.2 FNN- Based Adaptive Single Neuron Controller Design

Based on above-mentioned FNN model, the proposed adaptive single neuron controller (ASNC) is designed as Fig. 2.

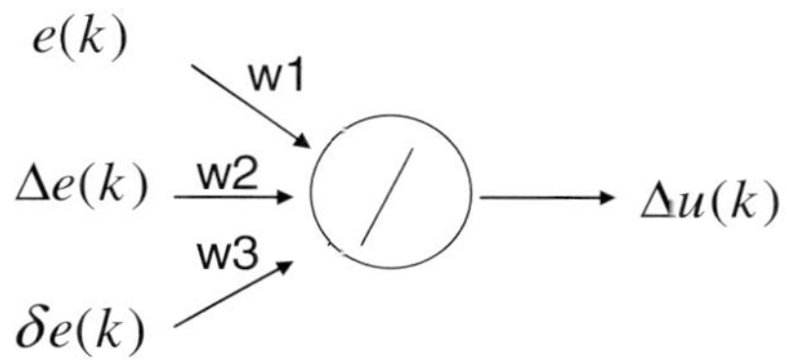

Fig. 2. The structure of ASNC

Where $e(k)$ is the error between the *k-th* setpoint $r(k)$ and the *k-th* process output $y(k)$, $\Delta e(k) = e(k) - e(k-1)$ is the difference between the current and previous error, $\delta e(k) = \Delta e(k) - \Delta e(k-1)$, and w_1, w_2 and w_3 are the neuron weights.

It is shown that the proposed controller ASNC is made by one neuron and has same controller structure as PID controller, so it is much easier to be used in practice. The control law of the proposed control system is obtained as follows:

$$\begin{aligned} u(k) &= u(k-1) + \Delta u(k) \\ \Delta u(k) &= w_1(k)e(k) + w_2(k)\Delta e(k) + w_3(k)\delta e(k) \end{aligned} \tag{16}$$

where $w_1(k)$, $w_2(k)$ and $w_3(k)$ are the ASNC controller parameters obtained at the kth sampling instant.

In the following section, Lyapunove stability theory will be developed to adjust the parameters of the single neuron controller at each sampling instant.

Denote $e(k) = r(k) - y(k)$ to be the prediction error at the sampling instant k, and the following auxiliary error $e_r(k+1)$ is introduced:

$$e_r(k+1) = r(k+1) - \hat{y}(k+1) \tag{17}$$

The following mapping function is introduced to guarantee the parameters w_1, w_2 and w_3 to be positive or negative [11]:

$$w_i(k) = \begin{cases} e^{\varsigma_i(k)}, & \text{if} \quad w_i(k) \geq 0 \\ -e^{\varsigma_i(k)}, & \text{if} \quad w_i(k) < 0 \end{cases}, \quad i = 1 \sim 3 \tag{18}$$

where $\varsigma_i(k)$ is adjustable.

Before proceeding the derivation, define

$$e_u(k) = [e(k) \quad \Delta e(k) \quad \delta e(k)] \tag{19}$$

$$w(k) = [w_1(k) \quad w_2(k) \quad w_3(k)]^T \tag{20}$$

$$\varsigma(k) = [\varsigma_1(k) \quad \varsigma_2(k) \quad \varsigma_3(k)]^T \tag{21}$$

where $\varsigma(k)$ is adjustable vector.

Lyapunov method is used to derive the parameter updating equations as follows:

$$\begin{aligned} &\varsigma(k+1) = \varsigma(k) + \Delta\varsigma(k) \\ &\Delta\varsigma(k) = \frac{1}{\dfrac{\partial \hat{y}(k+1)}{\partial u(k)}} \cdot \begin{bmatrix} \eta_1 & 0 & 0 \\ 0 & \eta_2 & 0 \\ 0 & 0 & \eta_3 \end{bmatrix} \cdot \left[\frac{\partial w(k)}{\partial \varsigma(k)}\right]^{-1} \cdot \frac{e_u(k)^T e_r(k)}{e_u(k) e_u(k)^T} \\ &\frac{\partial w(k)}{\partial \varsigma(k)} = \begin{bmatrix} w_1(k) & 0 & 0 \\ 0 & w_2(k) & 0 \\ 0 & 0 & w_3(k) \end{bmatrix} \end{aligned} \tag{22}$$

where η_1, η_2 and η_3 are learning rate to adjust the self-learning speed of the ASNC controller.

The following theorem shows the tracking ability of the proposed adaptive single neuron controller and the stability of the closed-loop system.

Theorem 1. Considering nonlinear processes (9) controlled by the adaptive single neuron controller (16) with the updating law (22), if the Lyapunov function candidate is chosen as

$$v(k) = \gamma e_r^2(k) \tag{23}$$

where γ is a positive constant and the parameter $\eta_i \le 2 (i = 1, 2, 3)$, then the $\Delta v(k) < 0$ always holds. Thus, the prediction error is guaranteed to converge to zero.

Proof. Define

$$e_r(k+1) = e_r(k) + \Delta e_r(k+1) \tag{24}$$

Considering Equations (23) and (24), we have

$$\Delta v(k) = v(k+1) - v(k) = 2\gamma e_r(k)\Delta e_r(k+1) + \gamma \Delta e_r^2(k+1) \tag{25}$$

By using Equation (17), $\Delta e_r(k+1)$ can be expressed as

$$\Delta e_r(k+1) = \frac{\partial e_r(k+1)}{\partial k} = -\frac{\partial \hat{y}(k+1)}{\partial u(k)} \cdot \frac{\partial u(k)}{\partial w(k)} \cdot \frac{\partial w(k)}{\partial \varsigma(k)} \cdot \Delta \varsigma(k) \tag{26}$$

Based on Equation (26), Equation (25) is rearranged as

$$\begin{aligned} \Delta v(k) = & -2\gamma e_r(k) \cdot \frac{\partial \hat{y}(k+1)}{\partial u(k)} \cdot \frac{\partial u(k)}{\partial w(k)} \cdot \frac{\partial w(k)}{\partial \varsigma(k)} \cdot \Delta \varsigma(k) \\ & + \gamma \left(-\frac{\partial \hat{y}(k+1)}{\partial u(k)} \cdot \frac{\partial u(k)}{\partial w(k)} \cdot \frac{\partial w(k)}{\partial \varsigma(k)} \cdot \Delta \varsigma(k) \right)^2 \end{aligned} \tag{27}$$

From fuzzy neural network model, we have

$$\frac{\partial \hat{y}(k+1)}{\partial u(k)} = \sum_{l=1}^{N} 2h_l t_l \cdot \frac{\sum_{l=1}^{N} \frac{(u(k) - c_{lm})t_l}{\sigma_l^2} - \frac{(u(k) - c_{lm})}{\sigma_l^2} \sum_{l=1}^{N} t_l}{(\sum_{l=1}^{N} t_l)^2} \tag{28}$$

where $t_l = \exp[-\sum_{i=1}^{M} \frac{(x_i - c_{li})^2}{\sigma_l^2}]$, and $u(k)$ is the mth variable of $x(k)$.

From Equation (16) and Equation (19), we have

$$\frac{\partial u(k)}{\partial w(k)} = e_u(k) \tag{29}$$

According to Equations (22), (28) and (29), Equation (27) is expressed as

$$\begin{aligned}\Delta v(k) &= -2\gamma e_r(k)\cdot\frac{\partial\hat{y}(k+1)}{\partial u(k)}\cdot e_u(k)\cdot\frac{\partial w(k)}{\partial\varsigma(k)}\cdot\frac{1}{\frac{\partial\hat{y}(k+1)}{\partial u(k)}}\cdot\begin{bmatrix}\eta_1 & 0 & 0\\ 0 & \eta_2 & 0\\ 0 & 0 & \eta_3\end{bmatrix}\\ &\cdot\left[\frac{\partial w(k)}{\partial\varsigma(k)}\right]^{-1}\frac{e_u(k)^T e_r(k)}{e_u(k)e_u(k)^T}+\gamma(-\frac{\partial\hat{y}(k+1)}{\partial u(k)}\cdot e_u(k)\cdot\frac{\partial w(k)}{\partial\varsigma(k)}\cdot\\ &\frac{1}{\frac{\partial\hat{y}(k+1)}{\partial u(k)}}\cdot\begin{bmatrix}\eta_1 & 0 & 0\\ 0 & \eta_2 & 0\\ 0 & 0 & \eta_3\end{bmatrix}\cdot\left[\frac{\partial w(k)}{\partial\varsigma(k)}\right]^{-1}\frac{e_u(k)^T e_r(k)}{e_u(k)e_u(k)^T})\\ &=\gamma e_r^{\,2}(k)\cdot\frac{\eta_1 e^2(k)+\eta_2\Delta e^2(k)+\eta_3\delta e^2(k)}{e^2(k)+\Delta e^2(k)+\delta e^2(k)}\cdot\\ &\left(\frac{(\eta_1-2)e^2(k)+(\eta_2-2)\Delta e^2(k)+(\eta_3-2)\delta e^2(k)}{e^2(k)+\Delta e^2(k)+\delta e^2(k)}\right)\end{aligned}\tag{30}$$

If it is chosen as $0<\eta_i<2$, $i=1,2,3$, we have

$$\frac{\eta_1 e^2(k)+\eta_2\Delta e^2(k)+\eta_3\delta e^2(k)}{e^2(k)+\Delta e^2(k)+\delta e^2(k)}>0\tag{31}$$

then

$$\frac{(\eta_1-2)e^2(k)+(\eta_2-2)\Delta e^2(k)+(\eta_3-2)\delta e^2(k)}{e^2(k)+\Delta e^2(k)+\delta e^2(k)}<0\tag{32}$$

Thus $\Delta v(k)<0$.

It is evident that equation (32) shows that the $\Delta v(k)<0$ always holds, if $\eta_i<2$ will be chosen to adjust the self-learning speed of the ASNC controlled. According to the Lyapunov theorem, the close-loop dynamic system is stable, indicating that e, Δe and δe will asymptotically converge to zero, respectively. This completes the proof. Q.E.D.

4 Example

To demonstrate the effectiveness of the proposed adaptive single neuron control, a distillation process is considered in this example, in which the output variable is the top column composition, y, and the input variable is the reflux flow rate u. This process can be defined by the following equations [10]:

$$\begin{aligned} y(k) &= 0.757y(k-1) + 0.243g(u(k-1)) \\ g(x) &= 1.04x - 14.11x^2 - 16.72x^3 + 562.7x^4 \end{aligned} \tag{33}$$

where the input and output variables are both defined as deviations from nominal values.

Independent random signal with uniform distribution between [-0.1 0.1] is used to simulate 100 input-output data points. To proceed with the proposed algorithm, $x(k)$ is chosen to be $x(k)=[y(k),u(k)]^T$. Then generated input-output data points are used to identify the proposed FNN model.

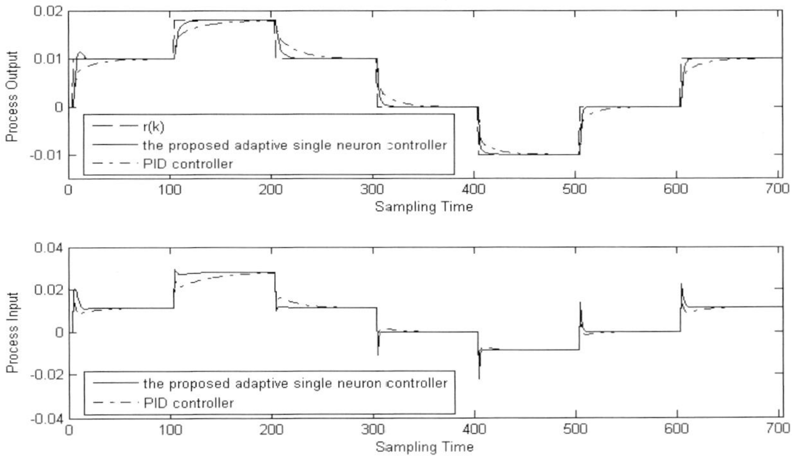

Fig. 3. Response comparison between the proposed single neuron controller and PID controller

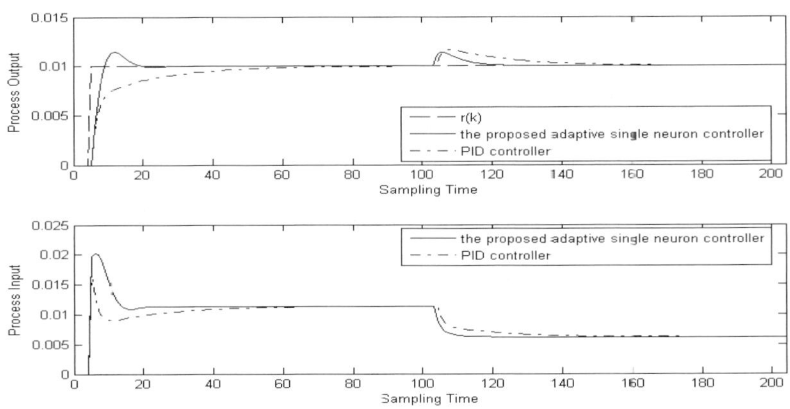

Fig. 4. Response comparison between the proposed single neuron controller and PID controller

Based on the obtained model, the adaptive single neuron control can be designed next. Simulation parameters are chosen as follows: $\eta_1 = 0.000028$, $\eta_2 = 0.00003$, $\eta_3 = 0.0005$.

The simulation compares the performance of the proposed adaptive single neuron control with the IMC based PID controller. Fig. 3 shows that the response of proposed control system is quicker.

To evaluate the disturbance rejection capability of the proposed adaptive single neuron controller, 10% step disturbance is assumed to occur in the top column composition y . It is evident from Fig. 4 that the proposed controller achieves better performance than the PID controller.

5 Conclusions

An adaptive single neuron controller, which mimics PID controller, is considered in the proposed control scheme. And a fuzzy neural network-based model is employed to approximate the controlled nonlinear process. Applying this result and Lyapunov stability theory, a novel-updating algorithm to adjust the parameters of the single neuron controller is presented. The proposed controller has same controller structure as PID controller, so it is much easier to be used in practice. Furthermore, the parameters of the controller are tuned to guarantee the stability of the closed-loop system and the convergence of the tracking error. Simulation results illustrate the performance and applicability of the proposed adaptive single neuron control scheme.

Acknowledgements. This work is supported by Shanghai Leading Academic Discipline Project, Project Number T0103, and supported by Shanghai Municipal Education Commission Project, Project Number: 05AZ22.

References

1. Kim, S., Lee, M., Park, S.: A neural linearizing control scheme for nonlinear chemical processes. Computer Chem. Engng. 21, 187–200 (1997)
2. Gao, F., Wang, F., Li, M.: An analytical predictive control law for a class of nonlinear process. Ind. Eng. Chem. Res. 39, 2029–2034 (2000)
3. Zhang, J.: A nonlinear gain scheduling control strategy based on neuron-fuzzy networks. Ind. Eng. Chem. Res. 40, 3164–3170 (2001)
4. Aswin, N., Ravindra, D.: Fuzzy segregation- based identification and control of nonlinear dynamic system. Ind. Eng. Chem. Res. 41, 538–552 (2002)
5. Narendra, K., Parthasarathy, K.: Identification and control of dynamic systems using neural networks. IEEE Trans. Neural Networks 1, 4–27 (1990)
6. Wang, L., Langari, R.: Complex system modeling via fuzzy logic. IEEE Trans. System, Man and Cybernetics-Part B: Cybernetics 26, 100–106 (1996)
7. Wang, Y., Rong, G.: A self-organizing neuralnetwork-based fuzzy system. Fuzzy Sets and Systems 103, 1–11 (1999)
8. Sugeno, M., Tanaka, K.: Successive identification of a fuzzy model and its application to prediction of a complex system. Fuzzy Sets and Systems 42, 315–334 (1991)

9. Cao, S., Rees, N., Feng, G.: Analysis and design for a class of complex control systems part 1: Fuzzy modeling and identification. Automatica 33, 1017–1028 (1997)
10. Eskinat, E., Johoson, E., Luyben, W.: Use of Hammerstein models in identification of nonlinear systems. AIChE J. 37, 255–268 (1991)
11. Cheng, C., Chiu, M.-S.: Aaptive Single-Neuron Controller Design For Nonlinear Process Control (to be published)

Fuzzy Sliding Mode Tracking Control for a Class of Uncertain Nonlinear Systems

Jinping Wang, Yanhua Wang , Xiqin He, and Shengjuan Huang

Institute of Applied Mathematics, University of Science and Technology Liaoning, 114051 Liaoning, China
wjp65@tom.com

Abstract. A class of T-S fuzzy models with parameter perturbations, input and external disturbances is studied in this paper. Both the case of matched uncertainties and the case of mismatched uncertain- ties are investigated using variable structure control technique and LMI method. Firstly, we study the problem of stabilization of T-S fuzzy systems by using a variable structure controller. Secondly, for the given reference model, we design another variable structure controller such that the system output tracks the reference system output in a finite time interval. Finally, the simulation example is employed to illustrate the validity and effectiveness of the proposed method.

1 Introduction

Fuzzy control has been widely and successfully used in nonlinear system modeling and control[1]. In many of the model-based fuzzy control approaches, the well-known Takagi-Surgeon (T-S) fuzzy model[2] is a popular and convenient tool. Recently, Takagi-Surgeon (T-S) type fuzzy controllers have been successfully applied to the stabilization controller design of nonlinear systems[2]-[6]. However, there are very few studies concerning with tracking control design based on the T-S fuzzy model, especially for continuous time systems. The author in [7] obtained the output tracking by using H_∞ model reference tracking performance, but he didn't consider parameter perturbations of the T-S model, which would weaken the robustness of systems. The author in [8] studied output tracking control of a class of multi-input multi-output(MIMO) fuzzy non-linear systems by employing variable structure control （VSC） technique, however, all the uncertainties and disturbances in the T-S fuzzy model must satisfy matching condition[8]. Chih-Lyang[9] developed a stable adaptive fuzzy sliding-mode control for affine nonlinear systems with application to four-bar linkage systems. In that paper, the external part of a transformed system via feedback linearized control evolves a linear dynamic system with uncertainties, and only single-input-single-output dynamic systems was considered. Chih-Lyang studied a nonlinear discrete-time system by using fuzzy linear pulse-transfer function-based sliding-mode control in [10]. Chi-Ying Liang[11] derived a new approach to the design of a fuzzy sliding mode controller, and shown that his control scheme reduced the ultimate bound of tracking error at least by one-half.

K. Li et al. (Eds.): LSMS 2007, LNCS 4688, pp. 424 – 433, 2007.

In this paper, we study a T-S fuzzy model in the existence of parameter perturbations, input and external disturbances, and discuss two cases of matched uncertainties and mismatched uncertainties in the state variable. For the given reference model, we design variable structure controller such that the output of systems tracks the reference system output in a finite time interval by employing the designed controller.

This paper is organized as follows. Section II gives the problem formulation. Section III presents the stabilization results for the considered fuzzy systems. Based on variable structure control, the output tracking problem of T-S fuzzy model is studied in Section IV. Illustrative examples are presented to demonstrate the whole design procedure in Section V.

2 Problem Formulation

In this paper, we consider a nonlinear system which could be described by the following T-S fuzzy model:

Plant rule i:

$$IF\ \theta_1\ is\ \mu_{i1},\ and,\ \cdots,\ and\ \theta_p\ is\ \mu_{ip}\ ,THEN$$

$$\dot{x}(t) = (A_i + \Delta A_i(x))x(t) + B_i(I + \Delta_i(t))u(t) + d_i(x,t) \tag{1}$$

where $x \in R^n$ and $u \in R^m$ are the state and control input, respectively; A_i and B_i are constant real matrices with appropriate dimensions; θ_j and μ_{ij} are respective the premise variables (which are the functions of state variable) and the fuzzy sets. It is assumed that the premise variable are independent of the input variable; r is the number of the IF-THEN rules; p is the number of the premise variables; ΔA_i represents the parameter perturbations in A_i and $d_i(x,t) \in R^n$ denotes external disturbances; $\Delta_i(t) \in R^{m\times m}$ satisfying $\|\Delta_i(t)\| \le \delta_{B_i} < 1$ is the uncertainty in the input.

The overall fuzzy model is as follows:

$$\dot{x}(t) = \sum_{i=1}^{r} h_i(\theta)[(A_i + \Delta A_i(x))x(t) + (B_i(I + \Delta_i(t))u(t) + d_i(x,t)] \tag{2}$$

Let $\delta_B = \max\{\delta_{B_i}, i = 1,2,\cdots r\}$ .In the following, we make the following assumptions.

Assumption 1: In the model (2), assume that $B_i = B$ ($i = 1,\cdots,r$), and B has full-column rank.

Assumption 2: Assume that the disturbances $d_i(x,t)$ in the model (2) satisfy the following matching conditions: There exist $\tilde{d}_i(x,t) \in R^m$, with, where $s_i(t) \ge 0$ ($i = 1,2,\cdots r$)are uniformly bounded continuous functions of t , such that

$$d_i(x,t) = B\tilde{d}_i(x,t), i = 1,2,\cdots,r \tag{3}$$

Consider a reference model as follows:

$$\dot{x}_r(t) = A_r x_r(t) + r(t) \tag{*}$$

where, $x_r(t)$ is the reference state; A_r is the specific asymptotically stable matrix; $r(t)$ is the bounded reference input.

Assumption 3: There exist $\Delta\bar{A}_i(x)$, with $\|\Delta\bar{A}_i(x)\| \le \rho_i$ ($i = 1, \cdots, r$), where ρ_i is a nonnegative scalar, such that the following hold:

$$\Delta A_i(x) = B\Delta\bar{A}_i(x) \tag{4}$$

Assumption 4: There exist nonnegative ρ_i, such that the following hold:

$$\|\Delta A_i(x)\| \le \rho_i (i = 1, \cdots, r) \tag{5}$$

The following lemma is needed in the proofs of the results in this paper.

Lemma 1. [13]. For any $z, y \in R^n$ and any positive definite matrix $X \in R^{n\times n}$,

$$2z^T y \le z^T X^{-1} z + y^T X y .$$

3 Stabilization of T-S Fuzzy Models Based on VSC Approach

In this section, we prove that variable structure controller stabilizes T-S fuzzy system (2) [15].

A. Matched Uncertainty Case:

Theorem 1. Consider uncertain system (2) and assume Assumptions 1-3 hold. If there exit matrices $Q > 0$, and Y_i ($i = 1, \cdots, r$), satisfying the LMIs

$$A_i Q + Q A_i^T + B Y_i + Y_i^T B^T < 0 \tag{6}$$

then the variable structure controller

$$u = \begin{cases} -\sum_{i=1}^{r} h_i(\theta) H A_i x(t) - \dfrac{1}{1-\delta_B}(\varphi_1 + \varepsilon)\dfrac{\sigma}{\|\sigma\|}, \sigma \ne 0 \\ -\sum_{i=1}^{r} h_i(\theta) H A_i x(t), \sigma = 0 \end{cases} \tag{7}$$

stabilizes system (2), where $\varepsilon > 0$ is a user defined scalar, and

$$\begin{cases} P = Q^{-1} \\ H = (B^T P B)^{-1} B^T P \in R^{m\times n} \\ \sigma(t) = Hx(t) \\ \varphi_1 = \sum_{i=1}^{r} h_i(\theta)[(\delta_B \|HA_i\| + \rho_i)\|x(t)\| + s_i(t)] \end{cases} \tag{8}$$

Proof. First, we show that the trajectory of system (2) under (7) will reach the sliding surface $S := \{x \in R^n : Hx = 0\}$ in a finite time and will stay on it thereafter. Then, we show that on the sliding surface S, the motion of (2) is asymptotically stable.

Let us prove the first part. According to Assumptions 1-3, and from u in (7), the derivative of $\sigma(t)$ along the trajectory of the closed-loop system given by (2) and (7) is

$$\dot{\sigma}(t) = \sum_{i=1}^{r} h_i(\theta)[\Delta\bar{A}_i x(t) + \tilde{d}_i] - \sum_{i=1}^{r} h_i \Delta_i(t)(\sum_{j=1}^{r} h_j(\theta) HA_i x(t))$$

$$-\frac{1}{1-\delta_B}(\varphi_1 + \varepsilon)\frac{\sigma}{\|\sigma\|} - \sum_{i=1}^{r} h_i \Delta_i(t)\frac{1}{1-\delta_B}(\varphi_1 + \varepsilon)\frac{\sigma}{\|\sigma\|}$$

Noticing that $\varphi_1 \geq 0$, $\varepsilon > 0$ and by Assumptions 2, 3 the above gives

$$\sigma^T \dot{\sigma} \leq \sum_{i=1}^{r} h_i(\theta)((\delta_B \cdot \|HA_i\| + \rho_i)\|x(t)\| + s_i)\cdot\|\sigma\| - (\varphi_1 + \varepsilon)\|\sigma\| = -\varepsilon\|\sigma\| < 0$$

which means that the reach-ability condition $\sigma^T \dot{\sigma} < 0$ for $\sigma \neq 0$ is satisfied.

We now prove the second part. Let the Lyapunov function be $V(x(t)) = x^T Px$ and from Assumptions 1-3, and on the switching surface $S := \{x \in R^n : Hx = 0\}$, $x^T(t)PB = 0$, then the derivative of V along the trajectory of (2) is

$$\dot{V}(t) \leq \sum_{i=1}^{r} h_i(\theta) x^T(t) P[(A_i Q + BY_i) + (QA_i + BY_i)^T]Px$$

and from (6), we have $\dot{V}(x) < 0$. Due to Lyapunov stability theorem, system (2) is asymptotically stable. Thus, the proof of Theorem 1 is completed.

B. Mismatched Uncertainty case:

Theorem 2. Consider uncertain system (2) and assume Assumtions1,2, and 4 hold. If there exist matrices $Q > 0$, Y_i and scalars $\varepsilon_i > 0$, $i = 1, \cdots, r$, satisfying the following LMIs:

$$\begin{bmatrix} A_i Q + QA_i^T + BY_i + Y_i^T B^T + \varepsilon_i & \rho_i^{1/2} Q \\ \rho_i^{1/2} Q & -\varepsilon_i \end{bmatrix} < 0 \tag{9}$$

Then the following variable structure controller

$$u = \begin{cases} -\sum_{i=1}^{r} h_i(\theta) HA_i x(t) - \dfrac{1}{1-\delta_B}(\varphi_2 + \varepsilon)\dfrac{\sigma}{\|\sigma\|}, \sigma \neq 0 \\ -\sum_{i=1}^{r} h_i(\theta) HA_i x(t), \sigma = 0 \end{cases} \tag{10}$$

stabilizes system (2), where $\varepsilon > 0$ is a user-defined scalar, and

$$\begin{cases} P = Q^{-1} \\ H = (B^T PB)^{-1} B^T P \in R^{m\times n} \\ \sigma(t) = Hx(t) \\ \varphi_2 = \sum_{i=1}^{r} h_i(\theta)[(\delta_B \|HA_i\| + \rho_i \|H\|)\|x(t)\| + s_i(t)] \end{cases} \tag{11}$$

Its proof is as that of theorem 1.

4 Output Tracking of T-S Fuzzy Models Based on VSC Approach

Definition 1 [8]. The state $x(t)$ of system (2) is said to track a desired reference trajectory $x_r(t)$ if the following two claims are true:i) $\lim_{x\to\infty}(x(t)-x_r(t))=0$;ii) the state $x(t)$ of (2) will be globally bounded.

Consider the general nonlinear system:

$$\dot{x} = f(t,x,u) \tag{**}$$

Definition 2 [14]. System (**) is input-state-stability if there exists a KL-function β and a K-function γ such that, for bounded input $u(t)$ and each $x(t_0)$, the inequality $\|x(t)\| \le \beta(\|x(t_0)\|, t-t_0) + \gamma(\sup_{t_0\le\tau\le t}\|u(\tau)\|)$ holds for each $t \ge t_0$ in the above, $x(t;x_0,u)$ denotes the trajectory of (**)started from x_0 and with input u .

Lemma 2[14]. System (**) is input-state-stability (ISS) if and only if there is a smooth function $V:[0,\infty)\times R^n \to R$ such that there exist K_∞-functions α_1, α_2 and K-function ρ, $W_3(x)$ is a continuous positive- definite function on R^n, such that

$$\alpha_1(\|x\|) \le V(x,t) \le \alpha_2(\|x\|), \frac{\partial V}{\partial t} + \frac{\partial V}{\partial x} f(t,x,u) \le -W_3(x), \forall \|x\| \ge \rho(\|u\|) > 0$$

$$\forall (t,x,u) \in [0,\infty)\times R^n \times R^m, \ \gamma = \alpha_1^{-1}\circ\alpha_2\circ\rho .$$

In the sequel, we discuss output tracking in both uncertainty satisfying matching condition and uncertainty satisfying mismatching condition.

A. Output Tracking in the Case of Matched Uncertainty

Theorem 3. Consider uncertain model (2) satisfying Assumptions 1-3.Suppose that there are a positive-definite matrix $Q > 0$ and a matrix Y_i ($i = 1,\cdots,r$), such that the following LMIs hold:

$$A_iQ + QA_i^T + BY_i + Y_i^T B^T < 0 , \quad i = 1,\cdots,r \tag{12}$$

$$\begin{cases} H=(B^TPB)^{-1}B^TP\in R^{m\times n}, \sigma(t)=Hx(t) \\ \varphi_3=\sum_{i=1}^{r}h_i(\theta)[(\delta_B\|HA_i\|+\rho_i)\|x(t)\|+\delta_B\|H\|\|\dot{x}_r(t)\|+s_i(t)] \end{cases} \tag{13}$$

$$u(t)=\begin{cases} -\sum_{i=1}^{r}h_i(\theta)HA_ix(t)-\frac{1}{1-\delta_B}(\varphi_3+\varepsilon)\frac{e}{\|e\|}+H\dot{x}_r, \sigma\neq 0 \\ -\sum_{i=1}^{r}h_i(\theta)HA_ix(t)+H\dot{x}_r, \sigma=0 \end{cases} \tag{14}$$

Where $P=Q^{-1}$, $e(t)=H(x(t)-x_r(t))$, then the above variable structure controller (14) will make the closed-loop system given by (2) and (14) achieve output tracking in the sense of Definition 1.

Proof. We prove the first part: According to Assumptions 1-3 and (14), the derivative of $e(t)$ along the trajectory of closed-loop system (2), (14) is

$$\dot{e}(t)=\sum_{i=1}^{r}h_i(\theta)[\Delta\bar{A}_ix(t)+\tilde{d}_i]-\sum_{i=1}^{r}h_i\Delta_i(t)\ (\sum_{j=1}^{r}h_j(\theta)HA_ix(t)-H\dot{x}_r)-\frac{1}{1-\delta_B}(\varphi_3+\varepsilon)\frac{\sigma}{\|\sigma\|}$$

$$-\sum_{i=1}^{r}h_i\Delta_i(t)\frac{1}{1-\delta_B}(\varphi_3+\varepsilon)\frac{\sigma}{\|\sigma\|}$$

Noticing that $\varphi_3\geq 0$ and $\varepsilon>0$ and according to Assumptions 2, 3, the above gives

$$e^T\dot{e}\leq-\varepsilon\|e\|<0$$

Thus, we are able to realize the first term in definition 1, that is to say, $\lim_{x\to\infty}(x(t)-x_r(t))=0$, denote by t_r the instant when e hits zero.

Now, we prove the second part. We view $x_{rr}=[x_r^T,\dot{x}_r^T]^T$ as the new input. Choose $V(x(t))=x^TPx$ as an ISS-Lyapunov function candidate. The time derivative of V along the trajectory of (2) is given by

$$\dot{V}(x)=2x^T(t)P\dot{x}(t)$$

$$\leq\sum_{i=1}^{r}h_i(\theta)x^T(t)(P(A_i+BY_iP)+(A_i+BY_iP)^TP)x$$

$$+2x^T(t)PB\sum_{i=1}^{r}h_i(\theta)\times(\rho_ix(t)+u+\delta_Bu(t)+s_i-Y_iPx]$$

Let

$$V_1=2x^T(t)PB\sum_{i=1}^{r}h_i(\theta)(\rho_ix(t)+(-\sum_{j=1}^{r}h_j(\theta)HA_jx(t))$$

$$-\frac{1}{1-\delta_B}(\varphi_3+\varepsilon)\frac{e}{\|e\|}+H\dot{x}_r\ )\times(1+\delta_B)+s_i-Y_iPx]$$

Because $x^T(t)PB$ is sliding surface, so $x(t)=x_r(t)$, when $t\geq t_r$, noticing that $\|x_r\|\leq\|x_{rr}\|$, $\|\dot{x}_r\|\leq\|x_{rr}\|$, we have

$$V_1\leq\Gamma_1\|x(t)\|\bullet\|x_{rr}(t)\|+\Gamma_2\|x_{rr}(t)\|+\Gamma_3\|x_{rr}(t)\|^2$$

Let $W_i=P(A_iQ+QA_i^T+BY_i+Y_i^TB^T)P$, $i=1,\cdots,r$ and $\lambda=\min\{\lambda_{\min}(-W_i),i=1,\cdots r\}$, where $\lambda_{\min}$ denotes the minimum eigenvalue of the concerned matrix. We obtain:

$$\dot{V}(t)\leq-(1-\eta)\lambda\|x\|^2-\eta\lambda\|x\|^2+\Gamma_1\|x(t)\|\bullet\|x_{rr}(t)\|+\Gamma_2\|x_{rr}(t)\|+\Gamma_3\|x_{rr}(t)\|^2$$

where $0<\eta<1$, and let $-\eta\lambda\|x\|^2+\Gamma_1\|x(t)\|\bullet\|x_{rr}(t)\|+\Gamma_2\|x_{rr}(t)\|+\Gamma_3\|x_{rr}(t)\|^2<0$

Define $\rho(\|x_{rr}(t)\|)=\dfrac{\Gamma_1\|x_{rr}(t)\|+\sqrt{\Gamma_1^2\|x_{rr}(t)\|^2+4\eta\lambda(\Gamma_2\|x_{rr}(t)\|+\Gamma_3\|x_{rr}(t)\|^2)}}{2\eta\lambda}$

$\|x(t)\|\geq\ \rho(\|x_{rr}(t)\|)$. Clearly, $\rho\in K$, for $\forall\|x(t)\|\geq\rho(\|x_{rr}(t)\|)$, we have $\dot{V}(x)<-(1-\eta)\lambda\|x\|^2$.

Therefore, from lemma 2, it is seen that the closed-loop systems (2), (14) is ISS when $t\geq t_r$. On the other hand, noting that when $0\leq t\leq t_r$, the function describing the differential equation of closed-loop system (2), (14) is globally Lipschitz. The above two points show that the state trajectory of (2), (14) is globally bounded.

B. State Tracking in the Case of Mismatched Uncertainty

Theorem 4. Consider uncertain system (2) and assume Assumptions 1, 2, and 4 hold. If there exist matrices $Q>0$, Y_i and scalars $\varepsilon_i>0$, $i=1,\cdots,r$, satisfying following LMIs:

$$\begin{bmatrix} A_iQ+QA_i^T+BY_i+Y_i^TB^T+\varepsilon_i & \rho_i^{1/2}Q \\ \rho_i^{1/2}Q & -\varepsilon_i \end{bmatrix}<0 \tag{15}$$

$$\begin{cases} H=(B^TPB)^{-1}B^TP\in R^{m\times n} \\ \sigma(t)=Hx(t) \\ \varphi_4= \\ \sum_{i=1}^{r}h_i(\theta)[(\delta_B\|HA_i\|+\rho_i\|H\|)\|x(t)\|+\delta_B\|H\|\|\dot{x}_r(t)\|+s_i(t)] \end{cases} \tag{16}$$

$$u(t)=\begin{cases}-\sum_{i=1}^{r}h_i(\theta)HA_ix(t)-\dfrac{1}{1-\delta_B}(\varphi_4+\varepsilon)\dfrac{e}{\|e\|}+H\dot{x}_r,\sigma\neq 0\\ -\sum_{i=1}^{r}h_i(\theta)HA_ix(t)+H\dot{x}_r,\sigma=0\end{cases} \tag{17}$$

will make the closed-loop system given by (2) and (17) achieve output tracking in the sense of Definition 1. Where $P=Q^{-1}$, $e(t)=H(x(t)-x_r(t))$ .Its proof is as that of theorem 3.

5 Illustrative Examples

To illustrate the use of our results, we consider an example of a continuous stirred tank reactor (CSTR). We depart directly from the following model, which is taken from [8]:

$$\begin{cases}\dfrac{dx_1}{d\tau}=-x_1+Da(1-x_1)\exp(-\dfrac{1}{x_2+\gamma})\\ \dfrac{dx_2}{d\tau}=-(1+\beta)x_2+HDa(1-x_1)\exp(-\dfrac{1}{x_2+\gamma})+\beta(u+\tilde{d}(\tau))\end{cases} \tag{18}$$

The parameters as $Da=0.072, H=8, \gamma=20, \beta=0.3$, and let $u_e=0$. $x_e=\begin{bmatrix}0.0642\\0.3948\end{bmatrix}$. Define new state variable as $\delta x_1=x_1-x_e(1)$, $\delta x_2=x_2-x_e(2)$.

Then, the dynamical model of (18) can be rewritten as shown in (19):

$$\begin{cases}\dfrac{d\delta x_1}{d\tau}=-\delta x_1-x_e(1)+Da(1-\delta x_1-x_e(1))\exp(-\dfrac{1}{\delta x_2+x_e(2)+\gamma}):=f_1(\delta x)\\ \dfrac{d\delta x_2}{d\tau}=-(1+\beta)(\delta x_2+x_e(2))+HDa(1-\delta x_1-x_e(1))\exp(-\dfrac{1}{\delta x_2+x_e(2)+\gamma})\\ \qquad +\beta(u+\tilde{d}(\tau)):=f_2(\delta x)+\beta(u+\tilde{d}(\tau))\end{cases} \tag{19}$$

Our objective is to make the temperature T (namely, state x_2) follows a desired signal. Hence, the system output equation is $y=Cx=C(\delta x+x_e)$. with $C=[0,1]$.

The local fuzzy models of the system, taking parameter, input and external disturbance into account are obtained through the linearization of (19) around the points $\delta x^1=[0,-1]^T$, $\delta x^2=[0,0]^T$, $\delta x^3=[0,1]^T$ respectively. In terms of $IF-THEN$ rule, the fuzzy models admit the following form.

*Rule*1 : *IF* δx_2 is small (i.e., δx_2 is about -1), *THEN*

$$\frac{d\delta x(\tau)}{d\tau}=(A_1+\Delta A_1)\delta x(\tau)+B(I+\Delta_1(t))u+B\tilde{d}$$

*Rule*2 : *IF* δx_2 is middle (i.e., δx_2 is about 0), *THEN*

$$\frac{d\delta x(\tau)}{d\tau}=(A_2+\Delta A_2)\delta x(\tau)+B(I+\Delta_2(t))u+B\tilde{d}$$

$$Rule3: \ IF \quad \delta x_2 \text{ is large (i.e., } \delta x_2 \text{ is about } +1), THEN$$
$$\frac{d\delta x(\tau)}{d\tau} = (A_3 + \Delta A_3)\delta x(\tau) + B(I + \Delta_3(t))u + B\tilde{d}$$

where $B = [0, \beta]^T$ account for the parameter perturbation in states and $\tilde{d}$ represents the external disturbance,

$$A_1 = \begin{bmatrix} -1.0684 & 0.0002 \\ -0.5471 & -1.2987 \end{bmatrix}, A_2 = \begin{bmatrix} -1.0686 & 0.0002 \\ -0.5484 & -1.2988 \end{bmatrix}, A_3 = \begin{bmatrix} -1.0687 & 0.0001 \\ -0.5497 & -1.2988 \end{bmatrix}.$$

When all ΔA_i satisfy matching condition:

$$\Delta A_1 = \begin{bmatrix} 0 & 0 \\ -0.0274 & -0.0649 \end{bmatrix}, \Delta A_2 = \begin{bmatrix} 0 & 0 \\ -0.0274 & -0.0649 \end{bmatrix}, \Delta A_3 = \begin{bmatrix} 0 & 0 \\ -0.0275 & -0.0649 \end{bmatrix}.$$

When $\Delta A_3 = \begin{bmatrix} 0 & 0 \\ -0.0275 & -0.0649 \end{bmatrix}$ satisfy mismatching condition, we have $\Delta A_i = 0.1 * A_i$.

So, when all parameter uncertainties ΔA_i satisfy matching condition, $\rho(\Delta \bar{A}) = 0.0705$, when all param-eteruncertainties ΔA_i satisfy mismatching condition, $\rho(\Delta \bar{A}) = 0.1513$, suppose that $\tilde{d}(\tau) = 0.5\cos(10\pi\tau)$, $\Delta_1(t) = 0.1 * \sin(t)$, $\Delta_2(t) = 0.1 * \sin(t)$, $\Delta_3(t) = 0.1 * \cos(t)$, then we have that, $s(\tilde{d}) = 0.5$, $\delta_B = 0.1$.

In this example, we choose Gaussian functions as our membership functions:

$$\alpha_i = \frac{\beta_i}{\sum_{j=1}^{3} \beta_j}, i = 1,2,3. \text{ Where } \beta_1 = \exp(\frac{-(\theta+1)^2}{\sigma^2}), \beta_2 = \exp(\frac{-\theta^2}{\sigma^2}), \beta_3 = \exp(\frac{-(\theta-1)^2}{\sigma^2})$$

and $\theta = \delta x_2(\tau), \sigma = 0.8$. And from Theorem 1 and 2, we have

$$Q = \begin{bmatrix} 62.8217 & 0.0046 \\ 0.0046 & 111.4904 \end{bmatrix} \text{ and } Q = \begin{bmatrix} 64.5991 & -0.0008 \\ -0.0008 & 55.7527 \end{bmatrix}.$$

The desired model to be tracked by x is $A_r = \begin{bmatrix} -3 & 0 \\ 0 & -3 \end{bmatrix}$, $r(t) = \begin{bmatrix} 0 \\ 2 * \sin(3\pi t) \end{bmatrix}$.

The initial conditions are given as $\delta x = [-2 \quad 2]^T$, $x_r = [-1 \quad 1]^T$.

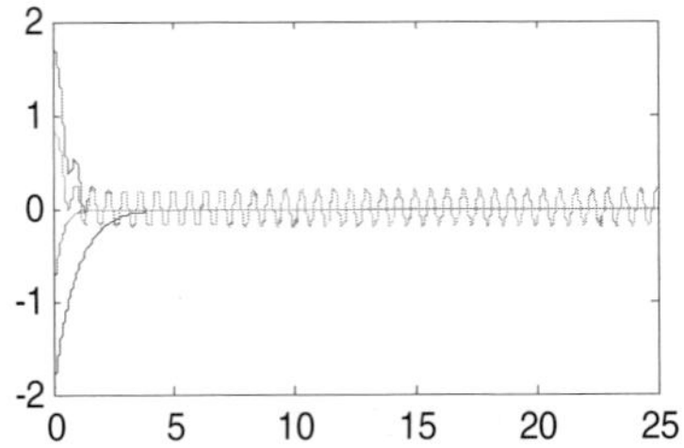

3

Fig. 1. Output tracking in theorem 3

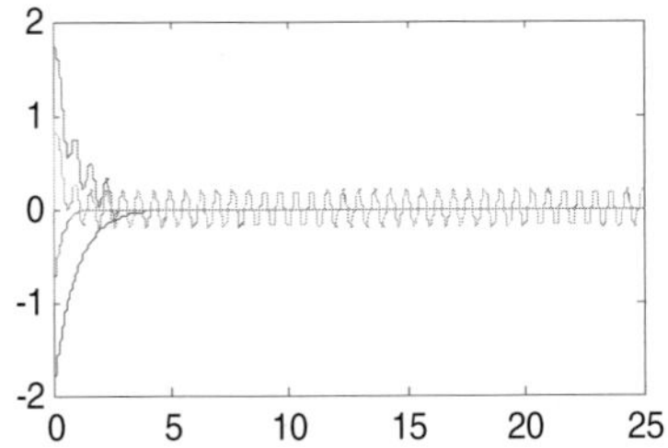

Fig. 2. Output tracking in theorem 4

6 Conclusions

Based on fuzzy T-S model in the existence of parameter perturbations, input and external disturbances, we first study the stabilization problem of nonlinear systems by using variable structure controller and linear matrix inequality (LMI) technique. Then we studied the output tracking problem such that the output of the system tracked the output of a given reference model. Simulation results show that the goal of output tracking can be achieved by the proposed controllers.

References

1. Wang, L.X.: Adaptive fuzzy systems and control: Design and stability analysis. Prentice-hall, Englewood Cliffs, NJ (1994)
2. Takagi, T., Sugeno, M.: Fuzzy identification of systems and its applications to modeling and control. IEEE Trans. Syst., Man, Cybern. 15, 116–132 (1985)
3. Wang, H.O., Tanaka, K., Griffin, M.F.: An approach to control of nonlinear systems:Stability and design issues. IEEE Trans. Fuzzy Syst. 4, 14–23 (1996)
4. Tanaka, K., Ikeda, T., Wang, H.O.: A unified approach to controlling chaos via an LMI-based fuzzy control system design. IEEE Trans. Circuits Syst. 45, 1021–1040 (1998)
5. Jadbabaie, A., Jamshidi, M., Titli, A.: Guaranteed-cost design of continuous-time Takagi-Sugeno fuzzy controllers via linear matrix inequalities. In: IEEE World Congress Computational Intelligence, Anchorge, AK, pp. 268–273. IEEE Computer Society Press, Los Alamitos (1998)
6. Chen, B.S., Tseng, C.S., Uang, H.J.: Robustness design of nonlinear dynamic systems via fuzzy linear control. IEEE Trans. Fuzzy Syst. 17, 571–585 (1999)
7. Tseng, C.S., Chen, B.S., Uang, H.J.: Fuzzy tracking control design for nonlinear dynamic systems via T-S fuzzy model[J]. IEEE Trans. on Fuzzy Systems 9, 381–392 (2001)
8. Zheng, F., Wang, Q.-G., Lee, T.H.: Output Tracking Control of MIMO Fuzzy Nonlinear Systems Using Variable Structure Control Approach. IEEE Transations on Fuzzy Systems 10, 686–697 (2002)
9. Hwang, C.L., Kuo, C.Y.: A stable adaptive fuzzy sliding-mode control for nonlinear systems with application to four-bar-linkage systems. IEEE Trans. Fuzzy Syst. 9, 238–250 (2001)
10. Hwang, C.-L.: Fuzzy linear pulse-transfer function-based sliding-mode control for nonlinear discretr-time systems. IEEE Trans. on Fuzzy Systems 10, 187–197 (2002)
11. Liang, C.-Y., Su, J.-P.: A new approach to the design of a fuzzy sliding mode controller. IEEE Trans. On Fuzzy sets and systems 139, 111–124 (2003)
12. Tong, S.C., Li, H.X.: Fuzzy adaptive sliding-mode control for MIMO nonlinear systems. IEEE Trans. on Fuzzy Systems 11, 354–359 (2003)
13. Cao, Y.Y., Sun, Y.X., Lam, J.: Delay-dependent robust H∞ control for uncertain systems with time-varying delays. IEE Proceeding online no (1998)
14. Khalil, H.K., Zhu, Y., Hui, D., Li, Z.: Nonlinear Systems, 3rd edn., pp. 125–126. Publishing House of Electronics Industry
15. Lin, C., Wang, Q.G., Lee, T.H.: Stabilization of uncertain fuzzy time-delay systems via variable structure control approach. IEEE Trans. on Fuzzy Systems 13 (2005)

Research on an Available Bandwidth Measurement Model Based on Regression Model

Fengjun Shang

College of Computer Science and Technology, Chongqing University of Posts and Telecommunications, Chongqing 400065, China
shangfj@cqupt.edu.cn

Abstract. Inferring the available bandwidth is of great importance for various network applications. In this paper, an available bandwidth measurement model is introduced based on regression model that can be used to high-speed IP network. The core of model has four parts as follows: 1) introducing the least absolute deviation method to compute regression coefficient. 2) deducing between delay and time interval. The advantages of the model are the lower cost and overhead, higher accurate result, independent clock synchronization. 3) deducing relation between sampling probability and background traffic. 4) introducing linear programming method to deal with the least absolute formula. The model is valid by simulation verification.

1 Introduction

Inferring the unused capacity or available bandwidth is of great importance for various network applications. Knowledge of the available bandwidth on an end-to-end path can improve rate-based streaming applications, end-to-end admission control, server selection, optimal route selection in overlay networks, congestion control as well as service level agreement verification. Obtaining useful estimates of the available bandwidth from routers is often not possible due to various technical and privacy issues or due to an insufficient level of measurement resolution or accuracy. Thus, it becomes necessary to infer the required information from the network edge via an active or passive probing scheme. Ideally, a probing scheme should provide an accurate estimate of a path's available bandwidth in as short a time as possible, preferably over only a few round trip times (RTTs), while imposing as light a load as possible on the network.

Characterizing end-to-end network available bandwidth is a problem that is both intellectually intriguing and of practical importance. However, the scale of the Internet, traffic volume and the diversity of network technologies make it a very challenging task. Furthermore, regular Internet users do not have access to network internals, adding to the complexity of understanding, characterizing, and modeling the performance of the Internet. While the problems of characterizing end-to-end latency and bottleneck link capacity have received a lot of attention, the question that is of most interest to applications is how much bandwidth is available to them along an end-to-end Internet path. What are good techniques for estimating available bandwidth and what

K. Li et al. (Eds.): LSMS 2007, LNCS 4688, pp. 434 – 444, 2007.

factors affect the measurement accuracy are still open questions. Those questions are the focus of this paper.

Network measurement techniques can be classified into two categories: passive measurement and active probing. Passive measurement tools use the trace history of existing data transmission. While potentially very efficient and accurate, their scope is limited to network paths that have recently carried user traffic. Active probing, on the other hand, can explore the entire network. The packet pair technique is one of the most popular active probing techniques. The basic idea of packet pairs is that the sender sends a pair of packets, which are echoed back by the destination. By measuring the changes in the packet spacing, the sender can estimate the bandwidth properties of the network path. While the packet pair mechanism is a reliable method for measuring the bottleneck link capacity of a network path, its use to measure the available bandwidth has had more mixed results [1].

The rest of the paper is organized as follows. In Section 2, we describe the related works. In Section 3, we describe available bandwidth. In Section 4, we present the regression model. In section 5, we introduce relation between sampling probability and background traffic. In section 6, we compute regression coefficient. In Section 7, we conclude.

2 Related Works

Current schemes for available bandwidth estimation fall into two broad classes. The first class of schemes is based on statistical cross-traffic models, such as Delphi and the methods proposed in [2]. While these schemes can potentially provide accurate estimates of cross-traffic, to date they have been designed for single hop (bottleneck) scenarios and as such may not be robust in multi-hop networks which are more common in practice.

The second class of schemes is based on the concept of self-induced congestion, which relies on a simple heuristic: If the probing rate exceeds the available bandwidth over the path, then the probe packets become queued at some router, resulting in an increased transfer time. On the other hand, if the probing rate is below the available bandwidth, the packets face no queuing delay. The available bandwidth can then be estimated as the probing rate at the onset of congestion. Such schemes are equally suited to single and multiple hop paths, since they rely only on whether the probe packets make it across the path with an unusual delay or not. Two examples of the self-induced congestion approach are pathload and TOPP.

Bottleneck link bandwidth is well understood in theory, and techniques to measure it are straightforward to validate in practice. Moreover, bottleneck link bandwidth measurement techniques have been shown to be accurate and fast in simulation. Furthermore, in some parts of the Internet, available bandwidth is frequently equal to bottleneck link bandwidth because either bottleneck link bandwidth is small or cross traffic is low. In addition to bottleneck link bandwidth's current utility, it can help the development of accurate and validated available bandwidth measurement techniques because of available bandwidth's dependence on bottleneck link bandwidth.

In [3], a regression model is proposed based on the least square fitting, but the least square fitting method is exposed to data fluctuating.

However, current tools to measure link bandwidth 1) measure all link bandwidths instead of just the bottleneck, 2) only measure the bandwidth in one direction, and/or 3) actively send probe packets. The tools pathchar, clink, pchar, and tailgater measure all of the link bandwidths along a path, which can be time-consuming and unnecessary for applications that only want to know the bottleneck bandwidth. Furthermore, these tools and bprobe can only measure bandwidth in one direction. These tools, tcpanaly, and pathrate actively send their own probe traffic, which can be more accurate than passively measuring existing traffic, but also results in higher overhead. The nettimer-sim tool only works in simulation [4].

3 Available Bandwidth Definition

Let us define the terms available bandwidth, bottleneck link, and tight link more precisely. Consider an end-to-end path that includes n links $L_1, L_2, ..., L_n$. Their capacities are $B_1, B_2, ..., B_n$, and the traffic loads on these links are $C_1, C_2, ..., C_n$. We define the bottleneck link as $L_b (1 \le b \le n)$, where

$$B_b = \min(B_1, B_2, ..., B_n)$$

The tight link is defined as $L_t (1 \le t \le n)$, where

$$B_t - C_t = \min(B_1 - C_1, B_2 - C_2, ..., B_n - C_n)$$

The unused bandwidth on the tight link, $B_t - C_t$, is also the available bandwidth of path.

4 Regression Formula

In [3], we acquire the simple expression for the mean message delay T as follows:

$$T = \frac{\lambda}{\mu C} \frac{1}{\mu C - \lambda} \tag{1}$$

From (1), we may conclude

$$\frac{1}{\lambda} = \frac{1}{\mu C} + \frac{1}{(\mu C)^2} \cdot \frac{1}{T} \tag{2}$$

If μC is constant, that is

$$M = a + bt \tag{3}$$

Where $M = \frac{1}{\lambda}, a = \frac{1}{\mu C}, b = \frac{1}{(\mu C)^2}, t = \frac{1}{T}$.

The formula (3) is called regression formula. If M and t is known, we may acquire parameter a and b. The initialization M may be acquired by means of SNMP. Initialization t may acquire by measuring time interval of the probe packet in active measurement. At last, after we have regression formula, we may acquire traffic by measuring t.

The packet transmitting delay t includes queuing time, transfers delay and transmission latency [5] shown in Figure 1. Given end to end path include n hops, in the i hop the queuing time of packet k is $t_{iq}^{(k)}$, transfers delay is $t_{id}^{(k)}$, transmission latency is $t_{il}^{(k)}$. The transfers delay includes the time of the packet transmitting link in router buffer. The transmission latency includes the time of the packet transmitting time by velocity of light in medium. The disposal time including lookup router table and others time is fixed toward the special instrument.

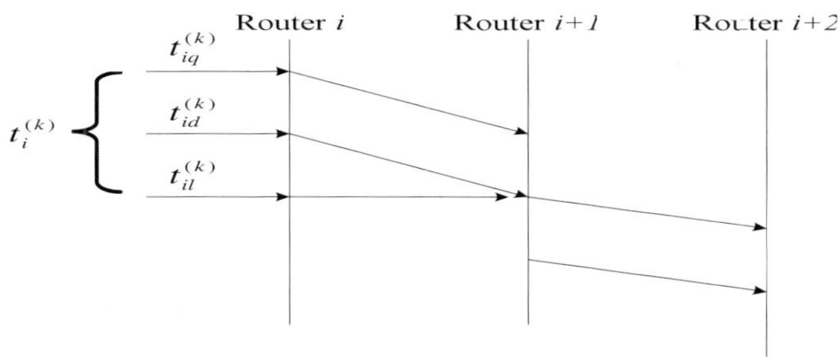

Fig. 1. The packet transmitting delay t in network

So the packet k delay is as follows in i hop:

$$t_i^{(k)} = t_{iq}^{(k)} + t_{id}^{(k)} + t_{il}^{(k)} \tag{4}$$

The packet k-1 delay is as follows in i hop:

$$t_i^{(k-1)} = t_{iq}^{(k-1)} + t_{id}^{(k-1)} + t_{il}^{(k-1)} \tag{5}$$

Where $t_{id}^{(k)} = t_{il}^{(k)}$, $t_{iq}^{(k)}$ is alterable.

When we uses(10)-(11), the result is

$$\begin{aligned}\Delta t_i^{(k,k-1)} &= t_i^{(k)} - t_i^{(k-1)} = (t_{iq}^{(k)} + t_{id}^{(k)} + t_{il}^{(k)}) - (t_{iq}^{(k-1)} + t_{id}^{(k-1)} + t_{il}^{(k-1})\\ &= t_{iq}^{(k)} - t_{iq}^{(k-1)}\end{aligned} \tag{6}$$

By formula (6), it is known that the time interval is time difference, and then this averts clock synchronization problem.

5 Relation Between Sampling Probability and Background Traffic

In order to reduce the probe packet affecting the link traffic, we introduce sampling method. The sampling measurement on high-speed IP traffic is aimed at selecting partly traffic to estimate the total traffic information. In order to know the number of probe packet, we analysis relation between sampling probability and background traffic.

Definition 1 Given network, the total traffic is n form all edge node to network, and total probing packet traffic is n_c .

If the probe probability is p,

$$p = \frac{n_c}{n} \tag{7}$$

The measurement error includes two part: 1) probing packet import error shown σ_p ; 2) route probing import error shown σ_m .

So, the total measurement error $\sigma = \sigma_p + \sigma_m$

Given k_{th} period, s send sd node pairs i route probe packet $n_i^{(k)}$, where transiting link j probing packet is $n_{ij}^{(k)}$, that is

$$p_{ij}^{(k)} = \frac{n_{ij}^{(k)}}{n_i^{(k)}} \tag{8}$$

Toward all route probe packet, p_{ij} follows Bernoulli distribution.

So the standard error of $p_{ij}^{(k)}$

$$p_{ij}^{(k)} = \sqrt{\frac{p_{ij}^{(k)}(1 - p_{ij}^{(k)})}{n_i^{(k)}}} \tag{9}$$

From formula (7), (8) and (9), we have

$$\sigma_m = \frac{1}{\sqrt{np}}\sqrt{\frac{n_i}{n}(1-\frac{n_i}{n})} \tag{10}$$

If the background traffic of link e_i is n_i and the probe probability is p, the measurement error σ_p is

$$\sigma_p = \frac{n \cdot p}{n_i} \tag{11}$$

In test network, n=500 (packet), n_i =250 (packet), we may know relation between σ_m and σ_p shown in Figure 2. From Figure 2, we know the point of optimization is about 0.05, so we probe 5 pairs packet toward the testing experiment because the bottl eneck link background traffic is about 200 packets.

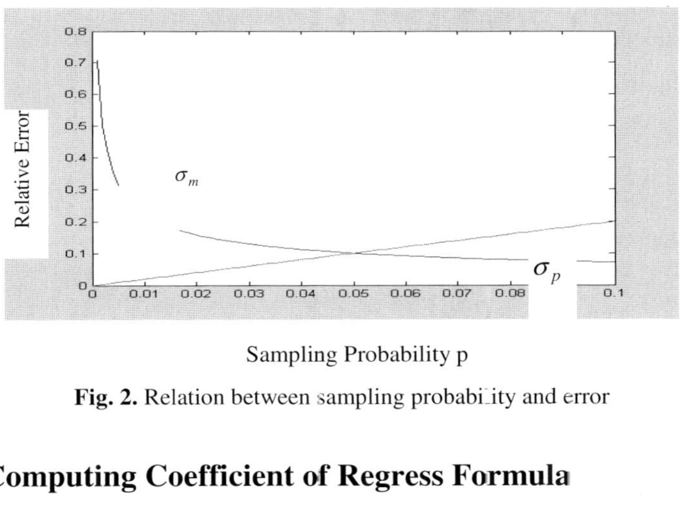

Fig. 2. Relation between sampling probability and error

6 The Computing Coefficient of Regress Formula

In order to decrease influence of burst traffic, we use linear regress formula to compute regress coefficient. At first, we acquire initial traffic value M by SNMP. Secondly, we measure time interval t by means of packet pairs method; At last, we may use linear regress method to compute regress coefficient

6.1 The Mathematical Model of Linear Regress

Linear regression has long been dominated by least squares techniques, mostly due to their elegant theoretical foundation and ease of implementation. The assumption in this method is that the model has normally distributed errors. In many applications, however, heavier-than-Gaussian tailed distributions may be encountered, where outliers in the measurements may easily ruin the estimates. To address this problem, robust regression methods have been developed so as to mitigate the influence of outliers. Among all the approaches to robust regression, the least absolute method is considered conceptually the simplest one since it does not require a "tuning" mechanism like most of other robust regression procedures. As a result, least absolute regression has drawn significant attentions in statistics, finance, engineering, and other applied sciences as detailed in a series of studies. The least absolute regression is based on the assumption that the model has Laplacian distributed errors. Unlike the least squares approach though, least absolute regression has no closed-form solution, hence numerical and iterative algorithms must be resorted to.

Toward two correlative variable， x denotes one variable, y denotes other variable. (x_1, y_1) (x_2, y_2) ,..... (x_n, y_n) is observation random variables.

In the paper, we assume that in the regression model

$$y_i = \alpha + \beta x_i + \sigma_i \qquad (i = 1,2,...,n)$$

Where x and y are observation random variable. σ_i is a sequence of non-observable independent random error.

This is basis of mathematical model of linear regress. We may estimate α， β and σ_i according to real observation values.

Assumed that in the regression formula

$$\hat{y}_i = a + bx_i$$

Where a is estimation value of α, b is estimation value β.

a、b make between estimation value and real value of y to be least absolute， that is,

$$\min Q = \sum \left| y_i - \hat{y}_i \right| = \sum \left| y_i - a - bx_i \right| \tag{12}$$

6.2 Computing Coefficient of Regression Formula

In order to compute (12), the absolute value must be unleashed. Given n pair data, we have

$$Q = y_i - a - bx_i \quad y_i \geq \hat{y}_i$$

$$Q = -y_i + a + bx_i \quad y_i < \hat{y}_i$$

Given n pair data, $Q \geq 0$, we have

$$\min Q_l = \sum y_i - na - b \sum x_i \tag{13}$$

In order to compute (13), we introduce linear programming method [6] and the algorithm is as follows:

Computing Coefficient of Regression Formula:

```
{
    for(l=0;l++;l<n)
    {
        Unleashing absolute value in formula (12);
        Computing Q_l using linear programming;
    }
    Select a,b value make Q least;
}
```

7 Conclusions

We select result in [3] to explain our idea. Table 1 is data in [3].

Table 1. Data in [3]

Real Traffic	Real Daley
129.2	4.15
229.32	10.71
77.81	2.82
287.52	34.86

According to (13), we have

$$\sum y_i = 0.02843, \sum x_i = 0.717631, n = 4$$

$$\min Q_l = 0.02843 - 4a - 0.717631b \tag{14}$$

The (14) is linear programming problem and constraint condition is as follows:

$a + 0.240964b \le 0.00774$

$a + 0.093371b \le 0.004361$

$a + 0.35461b \le 0.012852$

$a + 0.028686b \le 0.003478$

The result is $a = 0.0024, b = 0.0206$.

We totally compute 16 times and then select a, b value make Q least.

We may acquire the least absolute method by computing regression coefficient, that is $a = 0.0024$, $b = 0.0206$.

We may get the regression formula

$$M = 0.0024 + 0.0206t$$

$$\frac{1}{\lambda} = 0.0024 + \frac{0.0206}{T} \quad (1.5 < t < 35\ ,\ unit : ms)$$

In order to compare between the least absolute and the least square method, we also compute the coefficient of regression formula by means of the least square. Given n pairs data, toward the least square algorithm, the estimating formula is as follows:

$$Q = \sum (y - \hat{y})^2 = \sum (y - a - bx)^2 \tag{15}$$

Next step is how to compute a and b to make Q least. According to computing extremum method in calculous, we may get:

$$\frac{\partial Q}{\partial a} = -2\sum (y - a - bx) = 0$$

$$\frac{\partial Q}{\partial b} = -2\sum (y - a - bx)x = 0$$

We may acquire formula as follows

$$\begin{cases} an + b\sum x = \sum y \\ a\sum x + b\sum x^2 = \sum xy \end{cases}$$

We have

$$b = \frac{\sum xy - \frac{(\sum x)(\sum y)}{n}}{\sum x^2 - \frac{(\sum x)^2}{n}} = \frac{\sum (x - \overline{x})(y - \overline{y})}{\sum (x - \overline{x})^2} = \frac{SP_{xy}}{SS_x}$$

$$a = \overline{y} - b\overline{x}$$

Al last, the coefficient of regression formula is as follows:

$$a = 0.0007\ ,\ b = 0.0372$$

We may get the regression formula

$$M = 0.0007 + 0.0372t$$

$$\frac{1}{\lambda} = 0.0007 + \frac{0.0372}{T} \quad (1.5 < t < 35\ ,\ unit\ :\ ms)$$

Results of the two methods prove the least absolute method is better than the least square. The results are shown in figure 3.

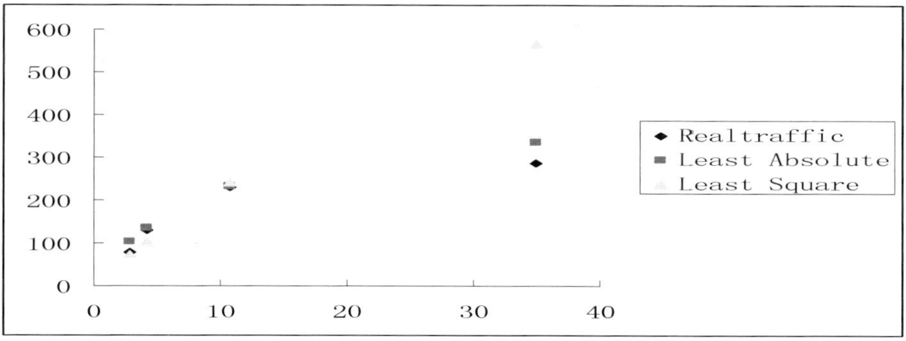

Fig. 3. Performance of model among Realtraffic, Least Absolute and Least Square, where x axis is time which unit is millisecond and y axis is traffic which unit is packets/second

Estimating the available bandwidth is of great importance for various network applications. The available bandwidth on an end-to-end path can improve rate-based streaming applications, end-to-end admission control, server selection, optimal route selection in overlay networks, congestion control as well as service level agreement verification. Thus, it becomes necessary to infer the required information from the network edge via an active or passive probing scheme. In this paper, an available bandwidth measurement model is used based on regression model that can be used to an end-to-end path. The core of model has four parts as follows. Firstly, the least absolute deviation method is introduced to compute regression coefficient. Secondly, relation is deduced between delay and time interval. The advantages of the model are the lower cost and overhead, higher accurate result, independent clock synchronization. Thirdly, relation is deduced between sampling probability and background traffic. Last but not least, the linear programming method is introduced to deal with the least absolute formula. The least absolute model is more valid than the least square by simulation verification.

Acknowledgments. The author would like to thank the Doctoral Research Fund of Chongqing University of Posts and Telecommunications (A2006-08). The author would also like to thank to Hao Fu for given the used data and the MATLAB software of the MathWorks Corpration.

References

1. Hu, N., Steenkiste, P.: Evaluation and characterization of available bandwidth probing techniques. IEEE JSAC Special Issue in Internet and WWW Measurement, Mapping, and Modeling 21(6) (August 2003)
2. He, G., Hou, J.C.: On exploiting long-range dependency of network traffic in measuring cross-traffic on an end-to-end basis. IEEE INFOCOM (2003)
3. Fu, H.: Research on IP network available bandwidth measurement based on regression model. A Dissertation for the Degree of Master (June 2004)

4. Ribeiro, V., Riedi, R., Baraniuk, R., Navratil, J., Cottrell, L.: pathChirp: Efficient Available Bandwidth Estimation for Network Paths. In: Proceedings of the 4th Passive and Active Measurements Workshop, La Jolla, CA (April 2003)
5. Cao-Gang, X., et al.: An Improved Method for Link Bandwidth Measurement. Journal of computer research and development 40(1), 68–74 (2003)
6. Xue, D., Chen, Y.: System simulation technology and application based on MATLAB/ Simulink[M], pp. 170–171. Tsinghua University Press, Beijing (2002)

Principle of 3-D Passive Localization Based on Vertical Linear Array and Precise Analysis

Yi Qu, Zhong Liu, and Hongning Hu

Electronics Eng. College, Naval University of Engineering,
430033 Wuhan, China
mailquyi@hotmail.com

Abstract. A system based on vertical linear array is brought forward. The localization model is presented, and the localization principle of vertical linear array is discussed. With range, elevation angle induced from time-delay and the measurement of azimuth angle, three dimensional equation of state and equation of measurement have been established. The localization precise according to length of baseline and precise of time-delay is simulated and analyzed. According to the observability principle of nonlinear system, the observability qualifications of the system are given.

1 Introduction

The passive localization is a technology, which detect the trajectory of a target solely through measurements of signals originating from the target. Since 1960s, the topic of isometric linear array of passive localization has been a subject of extensive studies [1]. Theoretically, we could get the position of the target as long as the three-point linear array receives the signals [1]. However, the design of a reliable and efficient algorithm is still a challenge, especially due to the fact that the problem is intrinsically nonlinear. In order to get a higher performance, some methods such as the increase of baseline and improvement of estimation error of time-delay are applied. Besides, the two-dimensional passive localization problem has been solved for a multitude of scenarios, thus giving a good insight into the three-dimensional problem has its significance. Therefore, with the elicitation of literature [1-2], a novel passive localization system based on vertical linear array is put forward. The model is presented, and the localization principle of vertical linear array is discussed. The simulation experiment results demonstrate that the new system has better performance.

2 The Problem Formation

In the ocean environment, supposing the target maneuver a trajectory of a constant velocity along a straight line. The geometry of the relative motion of the target and the vertical linear array is illustrated in Fig.1. The target's state vector is assumed to be $X_t = [x_t, y_t, z_t, v_{xt}, v_{yt}, v_{zt}]^T$. The vertical linear array is made up of three sensors,

K. Li et al. (Eds.): LSMS 2007, LNCS 4688, pp. 445–452, 2007.

$H_1 = [0,0,50]^T$ $H_2 = [0,0,0]^T$ and $H_3 = [0,0,-50]^T$. R is the distance between target and array. α is the azimuth angle, and γ is the elevation angle.

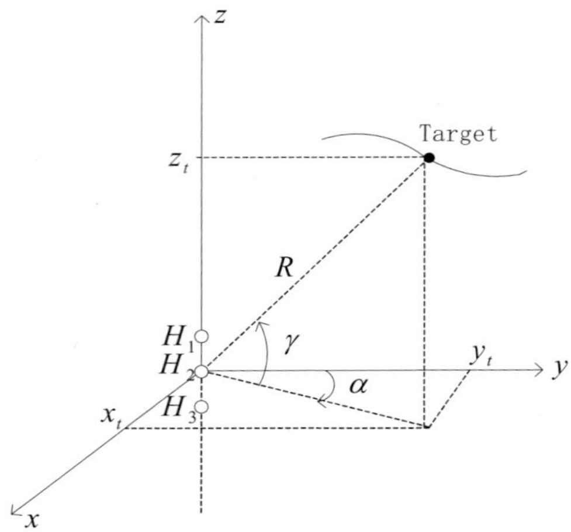

Fig. 1. Geometry of target and vertical linear array

The target motion is modeled as:

$$X(k+1) = \Phi(k+1,k)X(k) + W(k) \tag{1}$$

Where: $\Phi(k+1,k) = \begin{bmatrix} 1 & 0 & 0 & T & 0 & 0 \\ 0 & 1 & 0 & 0 & T & 0 \\ 0 & 0 & 1 & 0 & 0 & T \\ 0 & 0 & 0 & 1 & 0 & 0 \\ 0 & 0 & 0 & 0 & 1 & 0 \\ 0 & 0 & 0 & 0 & 0 & 1 \end{bmatrix}$ is the state transition matrix; T is the time interval between measurements; $W(k)$ is the process noise vector. The noise vector $W(k)$ is assumed to be a zero-mean white Gaussian noise vector with covariance matrix $Q(k)$. The initial state vector is assumed to be normally distributed with mean $X(0)$ and covariance $P(0)$.

Taking the plane consisted of target and z-axes, the 2-D geometry of the target and vertical linear array is illustrated in Fig. 2. $SH_1 = R_1, SH_2 = R_2 = R, SH_3 = R_3$. d is the length of baseline, and φ is the angle between SH_2 and u-axes.

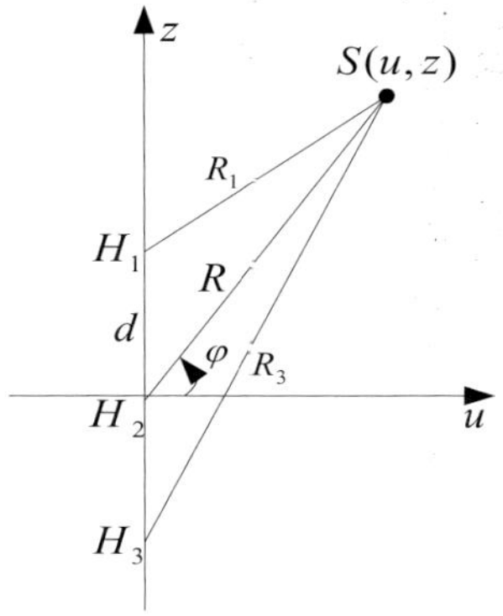

Fig. 2. 2-D geometry of vertical linear array

In ΔSH_2H_3:

$$[(u+d)^2+z^2]^{1/2}-(u^2+z^2)^{1/2}=c\tau_{23} \tag{2}$$

In ΔSH_1H_2:

$$(u^2+z^2)^{1/2}-[(u-d)^2+z^2]^{1/2}=c\tau_{12} \tag{3}$$

Because of $u^2+z^2=R^2$, we could get:

$$\begin{cases}(u+d)^2+z^2=c^2\tau_{23}^2+2c\tau_{23}R+R^2\\(u-d)^2+z^2=c^2\tau_{12}^2-2c\tau_{12}R+R^2\end{cases} \tag{4}$$

Combining equation (4), we could get:

$$R=\frac{2d^2-c^2(\tau_{12}^2+\tau_{23}^2)}{2c(\tau_{23}-\tau_{12})} \tag{5}$$

In ΔSH_1H_2:

$$\begin{aligned}c\tau_{12}&=R-R_1=R-[d^2+R^2-2Rd\cos(90-\varphi)]^{\frac{1}{2}}\\&=R\{1-[1+(\frac{d}{R})^2-2\frac{d}{R}\sin\varphi]^{\frac{1}{2}}\}\end{aligned} \tag{6}$$

Because of $(1+x)^{\frac{1}{2}}=1+\frac{1}{2}x-\frac{1}{8}x^2+\cdots$ and $\frac{d}{R}<<1$, we could get:

$$[1-2\frac{d}{R}\sin\varphi+(\frac{d}{R})^2]^{\frac{1}{2}}=1+\frac{1}{2}[-2\frac{d}{R}\sin\varphi+(\frac{d}{R})^2]-\frac{1}{8}[-2\frac{d}{R}\sin\varphi+(\frac{d}{R})^2]^2+\cdots$$
$$=1-\frac{d}{R}\sin\varphi+\frac{1}{2}(\frac{d}{R})^2-\frac{4}{8}\frac{d^2}{R^2}\sin^2\varphi+\cdots \tag{7}$$

Put equation (7) into equation (6):

$$c\tau_{12}\approx R\{1-[1-\frac{d}{R}\sin\varphi+\frac{1}{2}(\frac{d}{R})^2-\frac{1}{2}\frac{d^2}{R^2}\sin^2\varphi]\}=R[\frac{d}{R}\sin\varphi-\frac{1}{2}\frac{d^2}{R^2}\cos^2\varphi]$$
$$=d\sin\varphi-\frac{1}{2}\frac{d^2}{R}\cos^2\varphi \tag{8}$$

With the same method, we could get:

$$c\tau_{23}=d\sin\varphi+\frac{1}{2}\frac{d^2}{R}\cos^2\varphi \tag{9}$$

Combining equation (8) and equation (9), we could get : $2d\sin\varphi=c(\tau_{12}+\tau_{23})$.

So,

$$\varphi=\arcsin[\frac{c(\tau_{12}+\tau_{23})}{2d}] \tag{10}$$

Another expression could be [3] :

$$R=\frac{d^2\cos^2\varphi}{c(\tau_{23}-\tau_{12})} \tag{11}$$

Let distance measurement be R_m:

$$R_m = R + v_R \tag{12}$$

Let azimuth angle measurement be α_m:

$$\alpha_m = \arctan(\frac{x}{y}) + v_\alpha \tag{13}$$

Let azimuth angle measurement be γ_m [3]:

$$\gamma_m = \arcsin[\frac{c(\tau_{12} + \tau_{23})}{2d}] + v_\gamma \tag{14}$$

Where:

$$\begin{cases} \tau_{12} = [\sqrt{x^2 + y^2 + z^2} - \sqrt{x^2 + y^2 + (z-d)^2}]/c \\ \tau_{23} = [\sqrt{x^2 + y^2 + (z+d)^2} - \sqrt{x^2 + y^2 + z^2}]/c \end{cases}$$

The sound velocity in water is c. So the system measurement equation is:

$$Z(k) = \begin{pmatrix} R_m \\ \alpha_m \\ \gamma_m \end{pmatrix} = h(X(k)) + v(k) \tag{15}$$

$v(k) = \begin{pmatrix} v_R \\ v_\alpha \\ v_\gamma \end{pmatrix}$ is measurement noise vector. The mean of variables v_R、v_α、v_γ are zero and the variance are σ_R^2、σ_α^2、σ_γ^2.

3 Precise Analysis

According to equation (11), the variance of the estimator R is [4]:

$$Var(R) = c^2 Var(\tau_{23} - \tau_{12})(R / L_e)^4$$

Where , $L_e = d\cos\varphi$ is called equivalent half-length baseline. Taking $\varphi = 25^\circ$, $R = 12000m$, $c = 1500m/s$ [4] for example, the precise of time-delay is

$10\mu s$ 、 $50\mu s$ 、 $100\mu s$, the length of baseline is between $0 \sim 500m$, the precise of distance measurement is shown in Fig. 3.

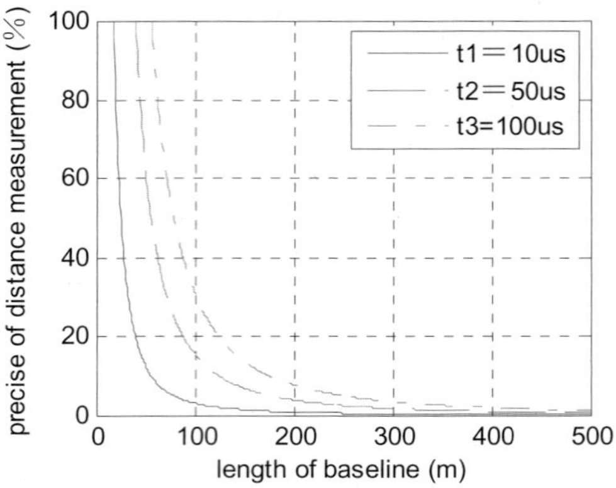

Fig. 3. Precise of distance measurement according to length of baseline

We can get some conclusions as follows:

1. As for the fixed time-delay precise, with the increase of the length of baseline, the precise of distance measurement is improved.
2. As for the fixed length of baseline, with the increase of time-delay precise, the precise of distance measurement is improved.

4 Observability Analysis

4.1 Observability Theorem of Nonlinear System

Assume the measurement equation is nonlinear, according to the principle of observability brought by Lee and Dum.

Theorem 1. As for nonlinear system:

$$\dot{x}(t) = f(x(t), t)\text{;} \quad x(t_0) = x_0 \tag{16}$$

$$y(t) = h(x(t), t) \tag{17}$$

We have:

$$M(x_0) = \int_{0} \Phi^T(\tau, t_0) H^T(\tau) H(\tau) \Phi(\tau, t_0) d\tau \tag{18}$$

Where, $H(t)=\dfrac{\partial h(x,t)}{\partial x}$, $\Phi(t,t_0)=\dfrac{\partial f(x,t)}{\partial x}$..

If equation (17) is positive-definite, the system is observable.

The equal conclusion of theorem above in the paper is:

$$\Gamma(k_0,k_0+N-1)=\begin{bmatrix} H_{k_0} \\ H_{k_0+1}\Phi_{k_0+1,k_0} \\ \vdots \\ H_{k_0+N-1}\Phi_{k_0+N-1,k_0} \end{bmatrix} \tag{19}$$

Where, $H_i=\left.\dfrac{\partial Z_i}{\partial X}\right|_{X=X_i}$.

Assume N is a positive integer, n is the dimension of state vector. If $rank\Gamma(k_0,k_0+N-1)=n$, system is observable.

4.2 Observability for Vertical Linear Array

As for the chosen relative state matrix's dimension is 6×6, and the measurement matrix's dimension is 3×6, we should measure twice($N=2$), get Z_k and Z_{k+1}.

$$\Gamma(k,k+1)=\begin{bmatrix} H_k \\ H_{k+1}\cdot\Phi_{k+1,k} \end{bmatrix}_{6\times 6} \tag{20}$$

According to the theorem mention above, if $\det(\Gamma(k,k+1)\neq 0$, the system is observable.

Assume $\det(\Gamma(k,k+1)=0$, we could get:

$$d(x_k^2+y_k^2)\left(\frac{z_k+d}{\sqrt{x_k^2+y_k^2+(z_k+d)^2}}+\frac{z_k-d}{\sqrt{x_k^2+y_k^2+(z_k-d)^2}}\right)=0 \tag{21}$$

According to (21), we could get some conclusions:

1. If $d=0$, vertical linear array degenerate to single observable platform, $\tau_{12}\equiv\tau_{23}\equiv 0$, system couldn't be observed.
2. If $x_k^2+y_k^2=0$, target takes a maneuvering along z-axis (except for $[-d,d]$), $\tau_{12}\equiv\tau_{23}$, system couldn't be observed.

3. If $z_k = 0$, target takes a maneuvering in xoy plane, $\tau_{12} \equiv \tau_{23}$, system couldn't be observed.

Form the analysis above, if $d = 0$ or target takes a maneuvering along z-axis or in xoy plane, system is not observability. In this circumstance, the time-delay doesn't apply any assistant information virtually. The vertical linear system degenerates to a bearings-only system, which is coincided with the conclusion in literature [5].

5 Conclusions

The system of vertical linear array is brought forward in this paper, and its localization problem is discussed. With range, elevation angle induced from time-delay and the measurement of azimuth angle, three dimensional equation of state and equation of measurement are established. The localization precise influenced by length of baseline and precise of time-delay is simulated and analyzed. According to the observability principle of nonlinear system, the observability qualifications of the system are given, and some conditions for the unobservable system are put forward. Theoretic basis for next step target tracking has been provided.

References

1. Hahn, W.R., Tretter, S.A.: Optimum Processing for Delay-Vector Estimation in Passive Signal Arrays. IEEE Transactions on Information Theory IT-19 (September 1973)
2. Van Cappel, D.: Target Motion Analysis using Time Delays Measured from a Nonlinear Array. In: Proceedings of the IEEE International Conference on Acoustics, Speech and Signal Processing (ICASSP'89), vol. 4, pp. 2724–2727. IEEE Computer Society Press, Los Alamitos (1989)
3. Zhi-rong, D.: The Necessary and Sufficient Condition of the Unique Solution of the Three-Point Isometric Linear Array's Target Positioning. Information Command Control System and Simulation Technology 27(6), 4–7 (2006)
4. Carter, G.C.: Passive sonar signal processing. In: Proc. of NATO ASI on Underwater Acoustics and Signal Processing, pp. 499–508 (1980)
5. Nadone, S.C., Lindgren, A.G., Gong, K.F.: Fundamental Properties and Performance of Conventional Bearings-Only Target Motion Analysis. IEEE Transaction on AC 29(9), 775–787 (1984)

The Study on Chaotic Anti-control of Heart Beat BVP System

Lü Ling, Chengye Zou, and Hongyan Zhao

College of Physics and Electronic Technology, Liaoning Normal University
116029, Dalian, China
luling1960@yahoo.com.cn

Abstract. The chaotic anti-control of the Bonhoeffer-Van der Pol system (BVP system) is investigated using the method of Lyapunov function, and the chaotifying of the BVP system is successfully realized. Taking the BVP system and the forced Brusselator system for example, the design project of controller is given, and the structure of error system is constructed between the state of the BVP system and that of the forced Brusselator system, leading the BVP system to track strictly the dynamical behavior of the forced Brusselator system.The control method also can be used to some systems of arbitrary dimensions.

1 Introduction

Chaos is a kind of random like motion which is produced in a certain system. It is often regarded as a disadvantage ingredient for its character of oddity, especially its highly sensitivity and instability to a small perturb in initial conditions. Therefore, chaos is often need to be controlled. Chaos control means the trace of a chaotic system is controlled to a certain periodic orbit or a balance point. Since the OGY method was initially proposed by Ott,Grebogi and Yorke, many technology and methods have been provided, and they have been validated in experiments and applications in laser physics, chemical reaction, electric circuit, artificial neural networks, biology system and so on [2-13]. But with further study, chaos is also found useful in many application fields, such as in biology, privacy communication and signal processing. Therefore, chaotic anti-control is made a new front topic in the research of chaos and chaos control. Chaotic anti-control means chaotifying a non-chaos system into a chaotic one according to the established objects,that is,to produce chaos behavior in a non-chaos system by adding a certain control.In biology,A person's or an animal's heart can be considered as a complicated oscillation system which can oscillate with its intrinsic rhythmicity without additional periodic stimulation. Bonhoeffer-Van der Pol system（BVP system）is originally used to simulate the oscillation in order to describe it.The chaotic anti-control of the BVP system is investigated using the method of Lyapunov function, and the chaotifying of the BVP system is successfully realized.Taking the BVP system and the forced Brusselator system for example, the design project of controller is given, and the structure of error system is constructed between the state of the BVP system and that

K. Li et al. (Eds.): LSMS 2007, LNCS 4688, pp. 453 – 461, 2007.

of the forced Brusselator system,leading the BVP system to track strictly the dynamical behavior of the forced Brusselator system.The control method also can be used to some systems of arbitrary dimensions.

2 Designing of the Controller

Assume the forced Brusselator system to be a drive system it can be discribed as [14]

$$\begin{cases} \dot{x}_1 = A + (B+1)x_1 + x_1^2 y_1 + d\cos\omega t \\ \dot{y}_1 = Bx_1 - x_1^2 y_1 \end{cases} \tag{1}$$

The form of the BVP system, proposed by FitzHugh is given as [15]

$$\begin{cases} \dot{x}_2 = x_2 - x_2^{\ 3}/3 - y_2 + u_1(t) \\ \dot{y}_2 = c(x_2 + a - by_2) + u_2(t) \end{cases} \tag{2}$$

where x is the membrane potential of myocardial cells, y is a variable related to the length of the restorative period, a, b, c are constants. $u_1(t)$, $u_2(t)$ are the controllers designed.

Suppose (x_0, y_0), which is the steady-state solution of Eq. (2), is given by $\frac{dx}{dt} = \frac{dy}{dt} = 0$, that is

$$\begin{cases} x_0 - x_0^3/3 - y_0 = 0 \\ x_0 + a - by_0 = 0 \end{cases} \tag{3}$$

A small deviation $u = x - x_0, v = y - y_0$ between (x, y) and (x_0, y_0) at the time t can be found by Eq. (2), and we obtain the linearized equations of the perturbation according to Eq.(2)

$$\frac{d}{dt}\begin{bmatrix} u \\ v \end{bmatrix} = L_c \begin{bmatrix} u \\ v \end{bmatrix} \tag{4}$$

Where L_c is the linearized operator.

$$L_c = \begin{bmatrix} 1 - x_{0c}^2 & -1 \\ c & -bc \end{bmatrix} \tag{5}$$

The eigenvalue equation is

$$\begin{vmatrix} 1-x_{0c}^2-\omega & -1 \\ c & -bc-\omega \end{vmatrix} = 0 \tag{6}$$

that is

$$\omega^2 + c_1\omega + c_0 = 0 \tag{7}$$

Where

$$\begin{cases} c_1 = bc - 1 + x_0^2 \\ c_0 = c - bc(1-x_0^2) \end{cases} \tag{8}$$

Suppose the system parameters $a = c = 0.1, b = 0.2$,we obtain

$$bc - 1 + x_0^2 < 0 \tag{9}$$

The trace of the BVP system is a single-period limited cycle, as is shown in Fig.2.

In the same way, when the system parameters are given as ω=0.85, d =0.04, according to the stability theory, the forced Brusselator system has a unstable saddle point, and the trace surrounding it is a chaotic attractor of a single helix, the phase map has the character of a single helix, as is shown in Fig.1. Accordingly, the states evolvement of the systems (1) and (2) with the lapses of times are shown in Fig.3 and Fig.4.

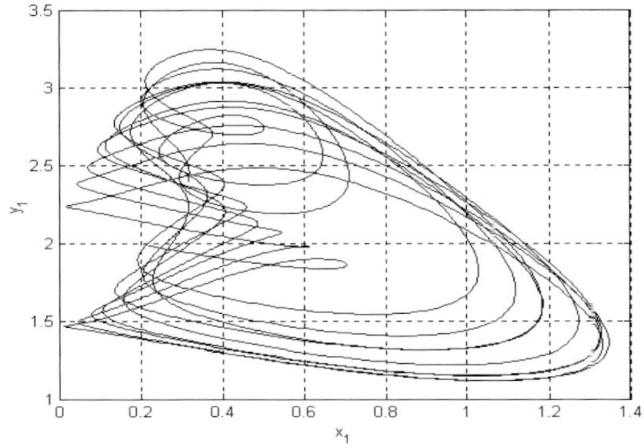

Fig. 1. The phase map of system (1)

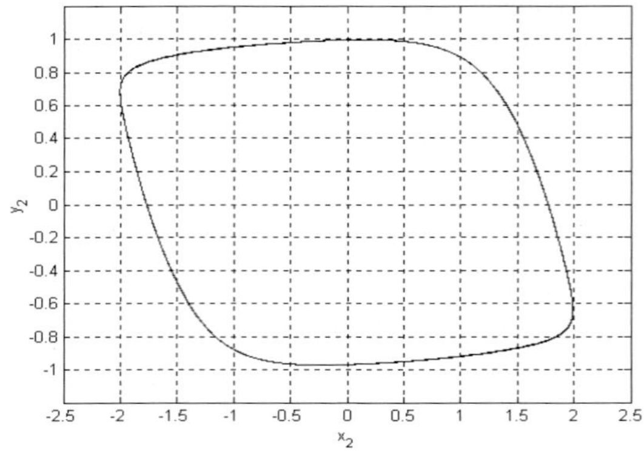

Fig. 2. The phase map of system (2)

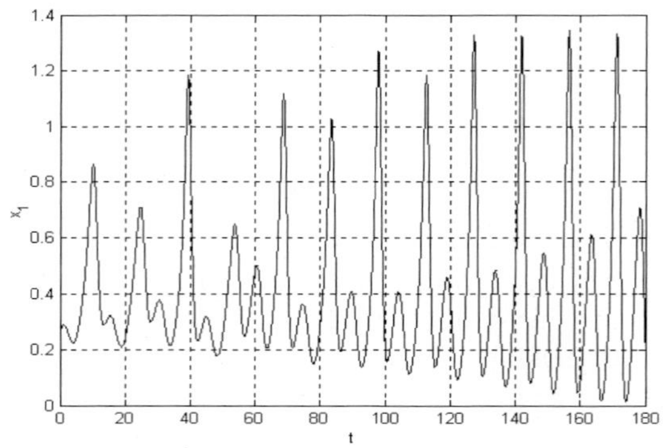

Fig. 3. State variables of system (1) vs time t

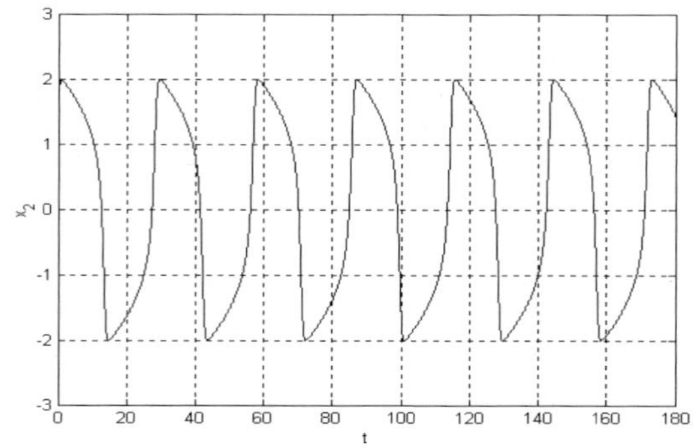

Fig. 4. State variables of system (2) vs time t

A controller is designed according to System (2), and the error variable between the drive and the respond systems is defined as follows

$$e_1 = x_2 - x_1 \tag{10}$$

$$e_1 = x_2 - x_1 \tag{11}$$

The Lyapunov function is constructed as

$$V = \frac{1}{2}e_1^{\ 2} + \frac{1}{2}e_2^2 \tag{12}$$

then

$$\begin{gathered} \dot{V} = e_1\dot{e}_1 + e_2\dot{e}_2 = -e_1^{\ 2} - e_2^{\ 2} + e_1(e_1 + x_2 - x_2^{\ 3}/3 - y_2 - A \\ -(B+1)x_1 - x_1^{\ 2}y_1 - d\cos\omega t + u_1(t)) + e_2(e_2 + c(x_2 + a - by_2) \\ -Bx_1 + x_1y_1^{\ 2} + u_2(t)) \end{gathered} \tag{13}$$

Surpose the structure of the controller is designed as follows

$$u_1(t) = -e_1 - x_2 + x_2^{\ 3}/3 + y_2 + A + (B+1)x_1 + x_1^2y_1 + d\cos\omega t \tag{14}$$

$$u_2(t) = -e_2 - c(x_2 + a - by_2) + Bx_1 - x_1^2y \tag{15}$$

then

$$\dot{V} = -e_1^{\ 2} - e_2^{\ 2} \tag{16}$$

According to Lyapunov stability theory [16], when

$$\dot{V} = -e_1^{\ 2} - e_2^{\ 2} \le 0 \tag{17}$$

the synchronization of the two systems is then achieved. and the error variables of the system approach zero with the lapses of times, that is

$$\lim_{t\to\infty} e_i(t) = 0 \qquad (i\text{=}1,2) \tag{18}$$

3 Simulation

Simulating is the chaotic trace of the two systems. The initial value of the two systems are respectly (x_1, y_1) = (0.25,2.5), (x_2, y_2) = (1.7, 2.8), the parameters are $a = c = 0.1$, $b = 0.2$, $\omega = 0.8$, $d = 0.04$, the calculus length is 0.01. The state viarables of the system are shown in Fig.5.1 and Fig 5.2, and the evolution of the error variables according to time are shown in Fig.6.1 and Fig 6.2.

,

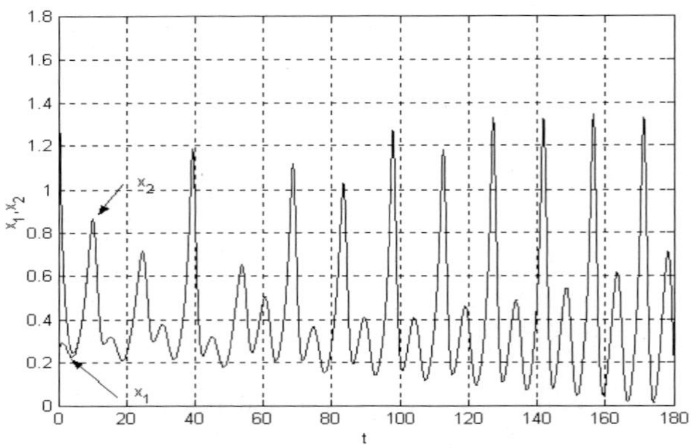

Fig .5.1. State variables of the systems vs time t

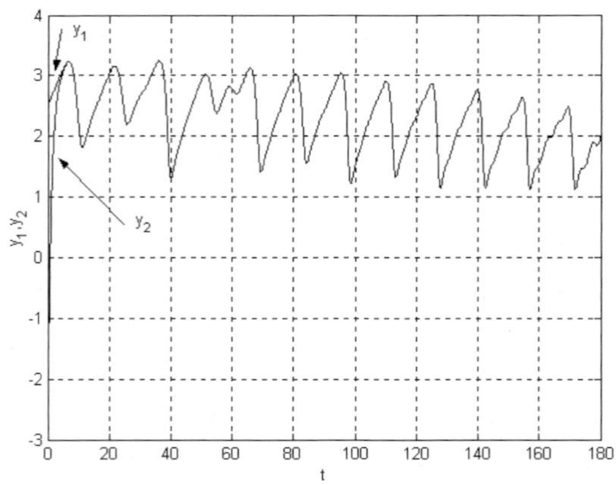

Fig .5.2. State variables of the systems vs time t

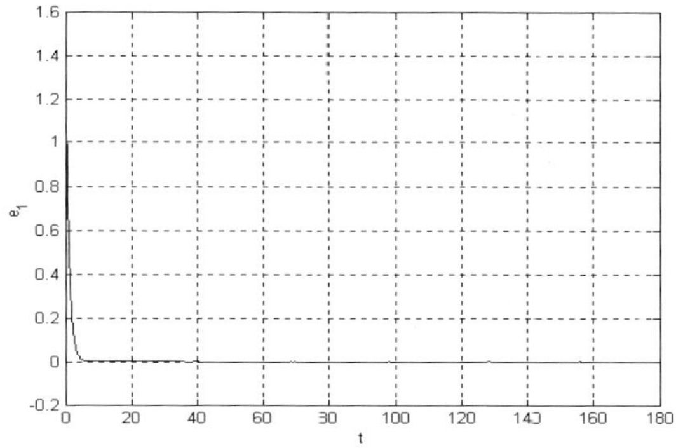

Fig .6.1. Error variable vs time t

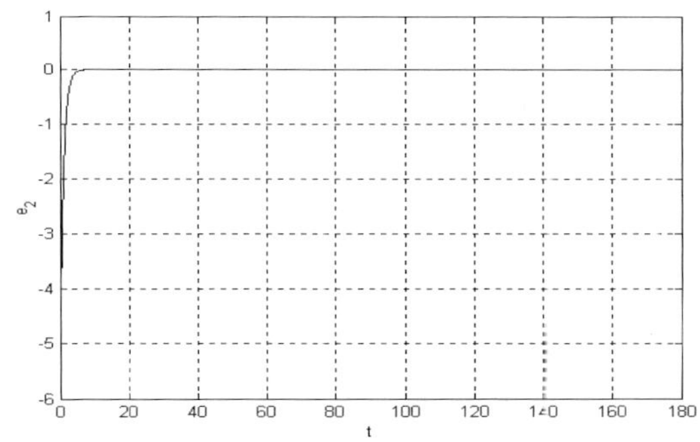

Fig. 6.2. Error variable vs time t

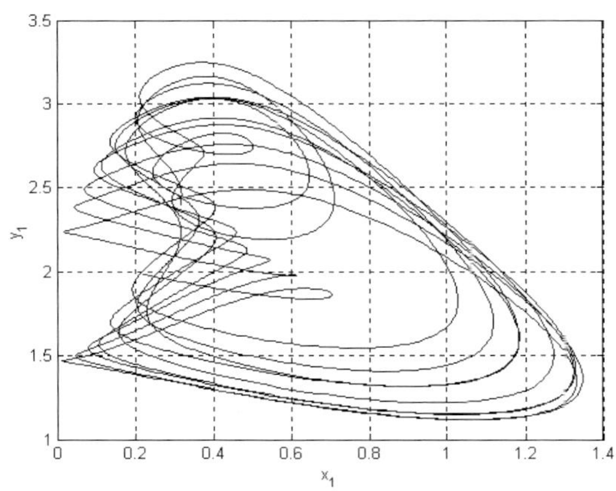

Fig. 7.1. Synchronization phase map of system (1) and (2)

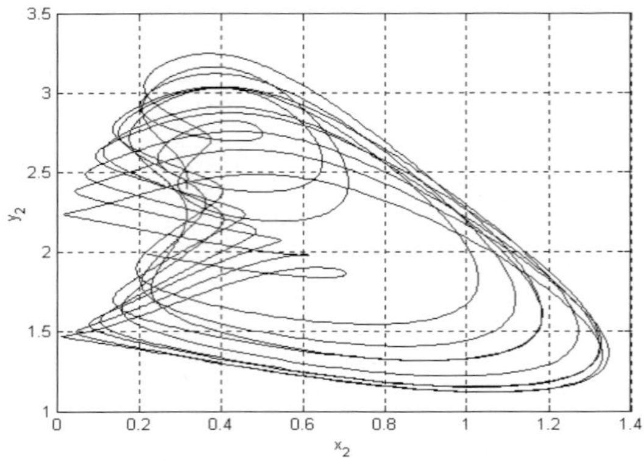

Fig. 7.2. Synchronization phase map of system (1) and (2)

As is seen in the figures, the state variables of BVP system track to strictly the chaotic orbits of the forced Brusselator system with the controller $u(t)$. The trace of the BVP system is altered from limited cycle to a chaotic attractor which is exactly the same as the trace of the Brusselator system, as is shown in Fig.7.1 and Fig 7.2.

4 Conclusions

The chaotic anti-control is investigated using the method of Lyapunov function. Taking the BVP system and the forced Brusselator system for example, the chaotifying of the BVP system is successfully realized. The design project of controller is given, and the structure of error system is constructed between the state of the BVP system and that of the forced Brusselator system , leading the BVP system to track strictly the dynamical behavior of the forced Brusselator system. One of the advantages of the project is that the control method can be used to some systems of arbitrary dimensions. The second, it can be used in both discrete systems and continuous systems.

Acknowledgements. Project supported by the National Natural Science Foundation of China (Grant No 20373021) and Natural Science Foundation of Liaoning Province,China(Grant No 20052151).

References

1. Ott, E., Grebogi, C., Yorke, J.A.: Cotrolling chaos [J]. Phys.Rev.Lett. 64, 1196–1199 (1990)
2. Lü, L., Guo, Z.A., Luan, L., et al.: The design and artificial realization of a controller of pulse coupling feedback [J]. Chin. Phys. 15, 958–962 (2006)

3. Lü, L., Luan, L., Du, Z., et al.: A new method for control chaos based on combination coefficient feedback of system variable rate [J]. Int. J. Infor. & Sys. Sci. 1, 237–243 (2005)
4. Trimper, S., Zabrocki, K.: Delay-controlled reactions [J]. Phys.Lett. (A) 321, 205–215 (2004)
5. Lü, L., Ou, Y.C.: Studies on Variable Rate Pulse Feedback of Chaos System [J]. Chin. J. Chem. Phys. 16, 459–462 (2003)
6. Alvarez-Ramirez, J., Espinosa-Paredes, G., Puebla, H.: Chaos control using small-amplitude damping signals [J]. Phys. Lett. (A) 316, 196–205 (2003)
7. Tian, Y.C., Tadé, M.O., Levy, D.: Constrained control of chaos [J]. Phys.Lett. (A) 296, 87–90 (2002)
8. Chen, L.Q.: An open-plus-closed-loop control for discrete chaos and hyperchaos [J]. Phys.Lett. (A) 281, 327–333 (2001)
9. Yang, L.F., Dolnik, M., Zhabotinsky, A.M., Epstein, I.R.: Oscillatory cluster in a model of the photosensitive Belousov-Zhabotinsky reaction system with global feedback [J]. Phys.Rev. (E) 62, 6414–6420 (2000)
10. Konishi, K., Kokame, H., Hirata, K.: Delayed-feedback control of spatial bifurcations and chaos in open-flow models [J]. Phys.Rev. (E) 62, 384–388 (2000)
11. Osipov, G.V., Kozlov, A.K., Shalfeev, V.D.: Impulse control of chaos in continuous systems [J]. Phys.Lett. (A) 247, 119–128 (1998)
12. Mirus, K.A., Sprott, J.C.: Controlling chaos in a high dimensional system with periodic parametric perturbations [J]. Phys.Lett. (A) 254, 275–278 (1999)
13. Meucci, R., Gadomski, W., Ciofini, M., et al.: Experimental control of chaos by means of weak parametric perturbations [J]. Phys.Rev. (E) 49, 2528–2531 (1994)
14. Bingzheng, L.: Nonlinear dynamics and chaos [M], pp. 343–347. Northeastern Normol University publishing house, Changchun (1994)
15. Bingzheng, L.: Chaos in the life science [M], pp. 141–144. Northeastern Normol University publishing house, Changchun (1999)
16. Lü Ling, L.: Nonlinear dynamics and chaos [M], pp. 130–133. Dalian publishing house, Dalian (2000)

A Modified Dynamic Model for Shear Stress Induced ATP Release from Vascular Endothelial Cells

Cheng Xiang[1], Lingling Cao[1], Kairong Qin[2,3], Zhe Xu[3], and Ben M. Chen[1]

[1] Department of Electrical and Computer Engineering, National University of Singapore, Singapore
[2] Institute of Mechanobiology and Medical Engineering, Shanghai Jiao Tong University, Shanghai 200240
[3] Institute of Biomechanics, Department of Mechanics and Engineering Science, Fudan University, Shanghai 200433
krqin@sjtu.edu.cn

Abstract. A modified dynamic model is proposed for shear stress induced adenosine triphosphate (ATP) release from endothelial cells (ECs) in order to incorporate the activation mechanism by time-varying shear stress. The dynamic behavior of ATP concentration at the endothelium-fluid interface by viscous shear flow is investigated via simulation studies. The numerical results demonstrate that the ATP concentration against time at the endothelium-fluid interface predicted by the modified dynamic ATP release model is more consistent with the experimental observations than that predicted by static ATP release model. The different behaviors of ATP concentration predicted by the original dynamic model and the new one are also compared and analyzed.

1 Introduction

The endothelium plays a critical role in vascular biology due to its unique location at the interface between vascular tissue and flowing blood. Endothelial cells (ECs) lining the inner surface of blood vessels are in direct contact with flowing blood, and are constantly affected by hemodynamic forces such as shear stress generated by flowing blood. It has been demonstrated by a number of recent studies that ECs recognize these hemodynamic forces and transmit signals to the interior of the cell, leading to cellular responses that involve changes in cell morphology, cell function, and gene expressions [1].Therefore, it is important to have a thorough investigation on the mechanisms underlying the shear stress induced cellular responses from ECs in order to better treat a variety of vascular diseases such as atherosclerosis, hypertension and aneurysm.

ATP, an important and ubiquitous extracellular signalling molecule, is often released under mechanical stimuli and plays an essential role in shear stress signal transduction [2-3]. For the modulation of extracellular ATP concentration the endothelium-fluid interface by fluid shear stress, much research has been conducted by a number of leading researchers. The research was first investigated in early 1990s by Nollert et al [4] and Shen et al [5]. It was suggested that convection-diffusion and

K. Li et al. (Eds.): LSMS 2007, LNCS 4688, pp. 462 – 472, 2007.

ATP hydrolysis alter the distributions of extracellular ATP concentrations in the perfusate and fluid shear stress indirectly modulates the extracellular ATP concentrations at the endothelium-fluid interface. In recent pursuit, further investigations revealed that, in addition to convection-diffusion effects and ATP hydrolysis, fluid shear stress also directly induces ATP release from ECs, and the endogenously released ATP mediates the ATP concentration at the endothelium-fluid interface [3], [6], [7].

A static model was proposed in [6] to describe the relationship between the shear stress and ATP release from ECs. It was assumed that ATP release rate is either a linear or a nonlinear function of the magnitude of the shear stress. Although the static models capture a number of important features of the shear stress induced ATP release process and have been widely applied in the modelling of calcium dynamics in endothelial cells [8]. They fail to characterize the dynamic relationship between the shear stress and ATP release. It is evident in more recent experimental studies [3] that ATP release rate should be a function of not only the magnitude of shear stress but also of the time.

In order to describe the time-dependency of shear stress induced ATP release rate, a dynamic model was first suggested in [9], which leads to better data fitting as well as predictions more consistent with experimental evidence [3]. However, the predicted ATP concentration under time-varying stimulus seems unreasonable due to lack of activation mechanism.

In this research, we aim to develop a modified dynamic model to incorporate the activation mechanism caused by time-varying shear stress on the basis of the original dynamic model in [9]. Different predictions of ATP concentration on ECs surface are demonstrated through simulation studies.

The rest of the paper is organized as follows. A modified dynamic model is developed in section 2. In section 3, the ATP concentrations at the surface of ECs are investigated by simulation studies, so as to compare the different dynamic behaviors predicted by the modified and original dynamic models and the static model. Discussion and conclusion are presented in section 4.

2 Model Development

2.1 Formulation of Governing Equation

A parallel-plate flow chamber is chosen in our studies as the apparatus to apply shear stress on ECs which are cultured on the bottom plate as shown in Figure 1. Prior to the onset of flow, the fluid within the chamber is assumed to contain ATP-free cell

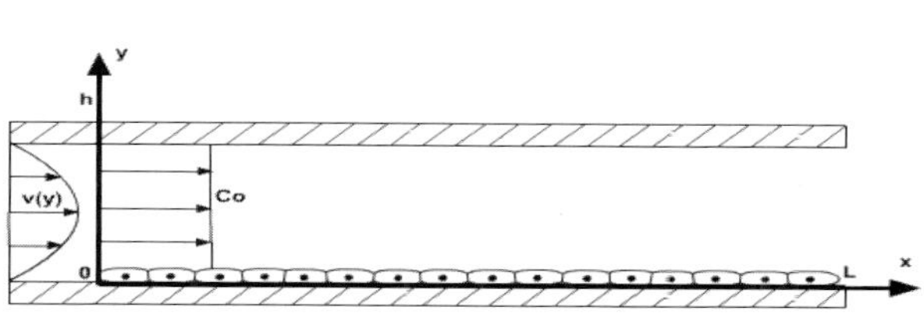

Fig. 1. Schematic diagram of a parallel-plate flow chamber

culture medium. The initial ATP concentration in the flow chamber is taken to be zero in all simulations. With the activation of flow, endogenously released ATP will convect and diffuse, which can be described by the standard convection-diffusion equation

$$\frac{\partial c}{\partial t} + v(y)\frac{\partial c}{\partial x} = D\left(\frac{\partial^2 c}{\partial x^2} + \frac{\partial^2 c}{\partial y^2}\right) \tag{1}$$

where c is the ATP concentration, D is the diffusion coefficient, and $v(y)$ is the flow velocity of the perfusate.

For steady flow, the velocity profile within the chamber can be obtained analytically and is expressed by *Poiseuille* formula as

$$v(y) = 6\bar{v}\frac{y}{h}\left(1 - \frac{y}{h}\right) \tag{2}$$

where $\bar{v}$ is the average velocity in the x direction. The shear stress to which the ECs are exposed in steady flow can then be determined directly from this profile as

$$\tau_w = \mu \left.\frac{\partial v}{\partial y}\right|_{y=0} = \frac{6\mu\bar{v}}{h} \tag{3}$$

where τ_w is the wall shear stress and μ is the dynamic viscosity of the fluid.

For the pulsatile flow, following the previous studies [6], [9], it is assumed that the flow is purely sinusoidal such that the velocity profile is given as

$$v(y) = 6\bar{v}\frac{y}{h}\left(1 - \frac{y}{h}\right)(1 + \sin\omega t) \tag{4}$$

where $\omega = 2\pi f$ is the angular frequency. Therefore the shear stress to which the ECs are exposed in pulsatile flow is that given in Equation (3) multiplied by the sinusoidal term $(1+\sin\omega t)$ in Equation (4).

It can be readily shown by order of magnitude analysis that $\partial^2 c/\partial x^2 \ll \partial^2 c/\partial y^2$, thus the term $\partial^2 c/\partial x^2$ can be ignored in Equation (1), and the diffusion is assumed to occur only in the y direction.

At the entrance of the flow chamber (x=0), the ATP concentration is assumed to be zero since the inflowing perfusate used in many of the experiments is usually fresh without ATP.

At the upper plate flow chamber ($y=h$), the flux of ATP is zero, i.e., the concentration gradient of ATP is zero, expressed as

$$\left.\frac{\partial c}{\partial y}\right|_{y=h} = 0 \tag{5}$$

At the bottom of the flow chamber ($y=0$), the net ATP mass flux is determined by the rate of ATP hydrolysis on the cell surface and the rate of shear stress induced ATP release by ECs. Similar to the previous studies [6], [9], it is assumed that the kinetics of ATP hydrolysis is described by an irreversible Michaelis-Menten formulation, while ATP release due to shear stress is included as a separate source term. Thus ATP flux at the ECs' surface is given as

$$D\left.\frac{\partial c}{\partial y}\right|_{y=0} = \left.\frac{V_{\max} c}{K_m + c}\right|_{y=0} - S(\tau_w, t) \equiv -S_{net,ATP} \tag{6}$$

where D is the diffusion coefficient for ATP in the cell culture medium, V_{max} is the maximum enzyme reaction velocity for ATP hydrolysis, K_m is the Michaelis constant for the enzyme, $S(\tau_w, t)$ is the source term for endothelial shear stress induced ATP release which depends upon not only the wall shear stress τ_w, but also the time t. The average net ATP release rate $S_{net,ATP}$ against time t under different wall shear stresses τ_w can be measured by *in vitro* cell experiments [3]. The mathematical details of the dynamic model will be given in the following subsection.

Given the initial and boundary conditions mentioned above, the convection-diffusion equation (1) can be solved numerically. The computer code developed for this purpose was based on a two-stage corrected Euler formulation with a central difference approximation in y direction and an upwind scheme in x direction.

2.2 Modified Dynamic Model of Shear Stress Induced ATP Release

In [9], a dynamic model is proposed in the form of

$$S(\tau_w, t) = p_1 p_2 \tag{7}$$

where the state variable p_1, summarizes both the effects of wall shear stress and the probability of the open states of all possible ATP release pathways. The state variable p_2 describes the activation levels of the various ATP release pathways. They satisfy the following equations

$$\frac{dp_1}{dt} = f(\tau_w) - \frac{p_1}{\tau_1} \tag{8}$$

$$\frac{dp_2}{dt} = -\frac{p_2}{\tau_2} + \lambda \frac{d\tau_w}{dt} \tag{9}$$

where τ_1 and τ_2 represent the time delay constants; λ is a positive coefficient; $f(\tau_w)$ is a function of the shear stress τ_w in the form of

$$f(\tau_w) = a_1 + \frac{a_2 \tau_w}{a_3 + \tau_w} \tag{10}$$

where a_1, a_2 and a_3 are positive constants. In the original dynamic model, a_1 is included to describe possible natural ATP release, which turns out to be very small. Since this term is negligible, it is assumed to be zero in the modified model for simplicity purpose.

In the original dynamic model, the state variable p_2 is intended to capture the characteristic of "receptor desensitization". Since p_2 is independent of shear stress in that model, it is always decreasing with time, which implies that the ATP release will decrease in the long run. However, it is well known that "receptor desensitization" usually happens only when the stimulus is a constant. When the stimulus is a time-varying signal, there could be other activation mechanism to balance this desensitization. In order to incorporate this activation mechanism, the dynamics of p_2 in this modified dynamic model is proposed to depend also on the change rate of shear stress as shown in Equation (9).

At t=0, the ATP release rate $S(\tau_w, t)$ is taken to be zero and initial conditions are expressed as follows

$$p_1(0) = 0 \tag{11}$$

$$p_2(0) = k \tag{12}$$

We set the initial value of p_2 to be a positive constant parameter k as it is difficult to determine the exact value of activation level of all the ATP pathways at the beginning of the experiment carried out in [3]. This parameter, together with a_1, a_2, a_3, τ_1, τ_2 and λ, is to be determined by experimental data [3], which will be discussed in the next subsection.

2.3 Identification of Model Parameters

Yamamoto and her co-workers published their experimental data [3] about the average net ATP release rate $S_{net,ATP}$, (see Equation (6)) against time using human pulmonary artery endothelial cells exposed to a stepwise increase shear stress (0-0.3-0.8-1.5 Pa). The model parameters are obtained by minimizing the difference E between experimental and corresponding model-predicted net ATP release rate, expressed by

$$E = \sum_{n=1}^{N} \left(S_{net,ATP}(n) - S_{net,ATP\_Exp}(n) \right)^2 \tag{13}$$

where N is the total number of experimental samples, $S_{net,ATP\_Exp}(n)$ is the observed value of the nth experimental sample point. $S_{net,ATP}(n)$ is the model predicted net ATP release rate. The calculation of $S_{net,ATP}$ should take the governing equation (1)-(3) and boundary and initial conditions into consideration. Since the derivatives of the cost function E with respect to parameters are not available, Nelder-Mead simplex algorithm [10], [11] is adopted to obtain the model parameters.

3 Simulation Results

3.1 Validation of Modified Dynamic Model for Shear Stress Induced ATP Release

Figure 2 shows the net ATP release rate against time under a stepwise increased shear stress. It is evident from Figure 2 that while the results fitted by linear static model show poor agreement with experimental data, the results fitted by our two dynamic models exhibit satisfactory agreements with the experimental results [3].

There is a sharp jump in the modified dynamic model-predicted ATP release rate during the periods when the shear stress suddenly increases from 0.3 to 0.8 Pa and from 0.8 to 1.5 Pa while the net ATP release rate curve fitted by the previous dynamic model is quite smooth. Such a difference is caused by the activation mechanism incorporated in the modified dynamic model since a sudden increase in shear stress will lead to a jump in p_2 and hence in $S_{net,ATP}$. However, it is still difficult to tell which of the two dynamic models gives a more accurate picture of net ATP release rate as the experimental data are sampled every 15 seconds in [3].

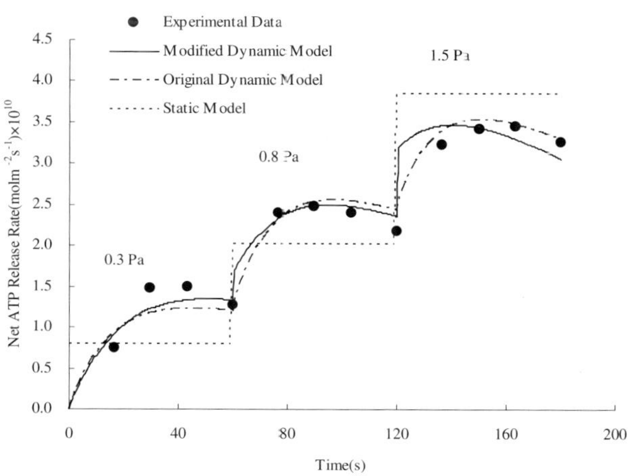

Fig. 2. Comparison between experimental and corresponding model-predicted average net ATP release rate $S_{net,ATP}$ against time t from the onset of steady fluid shear stress in a stepwise manner (0-0.3-0.8-1.5 Pa)

3.2 Comparison Between Dynamic and Static Model-Predicted Extracellular ATP Concentration at ECs Surface in Steady Flow in a Stepwise Manner

Figure 3 demonstrates the predicted dynamic behaviors of the average extracellular ATP concentration at ECs surface under stepwise manner shear stress stimulation. All the model parameters required by numerical simulations are listed in Table 1.

It can be readily seen from Figure 3 that the average ATP concentrations predicted by our two dynamic models are indeed dramatically different from that predicted by the static model. In particular, after ECs being activated for a long time (300-500s) by a step shear stress 1 Pa, the static model predicts a stable concentration while the dynamic models predict a gradually decreasing response. Unfortunately there is no direct experimental evidence for us to make judgment on which one is more reasonable. However, there are some indirect experimental evidences such as the experiment carried out on human umbilical vein ECs by Bodin and Burnstock [12]. It is clearly manifested in Figure 1 of [12] that being activated by a small step shear stress for one period and then a larger step shear stress for a much longer period, the ATP concentration initially increases to a maximum level and then gradually decreases, which agrees qualitatively with the predictions made by our dynamic models.

We intentionally make the shear stress decrease from 1 to 0.9 Pa at 500 second and then to 0.8 Pa at 600 second in order to see different behaviors of ATP concentration predicted by the modified and original dynamic models. The ATP concentration predicted by the original dynamic model does not have an obvious response to the decreased shear stress. However, in the modified dynamic model, there is a sudden drop of ATP concentration corresponding to the sudden decrease of shear stress. Direct experimental evidence is needed to help us judge on which of the two predictions is closer to the real case.

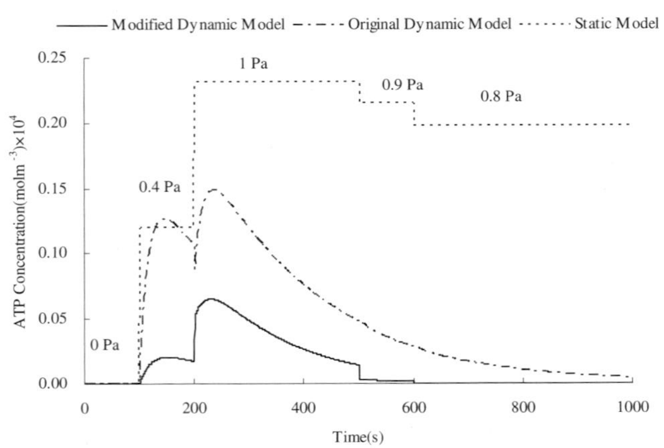

Fig. 3. Comparison between dynamic and static model-predicted extracellular ATP concentration in ECs surface against time from the onset of steady fluid shear stress in a stepwise manner (0-0.4-1-0.8-0.9 Pa)

3.3 Comparison Between Dynamic and Static Model-Predicted Extracellular ATP Concentration at ECs Surface in Pulsatile Flow

Despite the fact that none of the real experiments related to shear stress induced ATP release on ECs have been conducted for pulsatile flow, further numerical comparison

Table 1. Model parameter values for all the simulations

Parameter	Unit	Value	Source
L	m	0.025	
h	μm	200	[3]
μ	Nsm^{-2}	9.45×10^{-4}	
D	m^2s^{-1}	2.36×10^{-10}	
K_m	μM	475	[6]
V_{max}	$molm^{-2}s^{-1}$	0.8×10^{-3}	
a_1	$molm^{-2}s^{-2}$	0.65×10^{-3} *	
a_2	$molm^{-2}s^{-2}$	2.79×10^{-10}	
a_3	Pa	6.96	[9]
τ_1	s	17.4	
τ_2	s	218.9	
a_2	$molm^{-2}s^{-2}$	0.64×10^{-10}	
a_3	Pa	1.06	
τ_1	s	25.9	Modified dynamic model
τ_2	s	162.18	
k	dimensionless	0 **	
λ	Pa^{-1}	0.18	
S_{max}	$molm^{-2}s^{-1}$	3.0×10^{-10}	[6]

studies are carried out for ATP concentration prediction as this is more physiologically relevant to human ECs. Figure 4 and 5 display the dynamic behaviors of the extracellular average ATP concentration at ECs surface from the onset of pulsatile fluid shear stress.

It is noticed from Figure 4 that, during the initial period right after the onset of pulsatile flow, the dynamic behaviors of the average ATP concentration at endothelial surface predicted by our dynamic models are quite different from that predicted by the static model. However, after around 40 seconds, both the two dynamic models and static model predict a very similar characteristic of the ATP concentration at endothelial surface: an oscillation with the same period of 1 second as that of the pulsatile flow, even though the predicted magnitudes of oscillations are quite different.

It is also noticed that the magnitude of the oscillation predicted by the original dynamic model gradually decreases with time in the long run, due to the effect of "receptor desensitization", which usually occurs to a constant stimulus. Therefore the

* a_1 is set to be zero in all simulations as it is so small compared with other terms.

** k takes the value of 0.66 resulting from data fitting. We set it to be zero in all simulation studies in sections 3.2 and 3.3 assuming that all possible ATP pathways are closed at the beginning of the experiment.

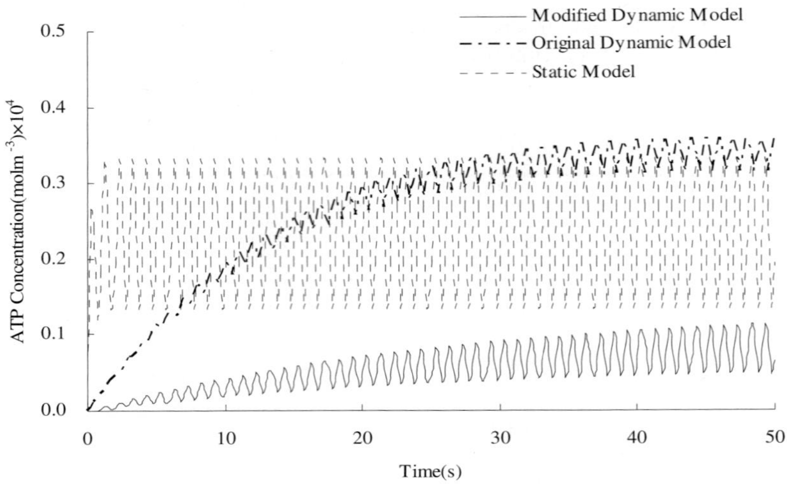

Fig. 4. Comparison between dynamic and static model-predicted extracellular ATP concent- ration at ECs surface against time from the onset of pulsatile fluid shear stress $\tau_w=1+sin2\pi\, t$; time course 0-50s

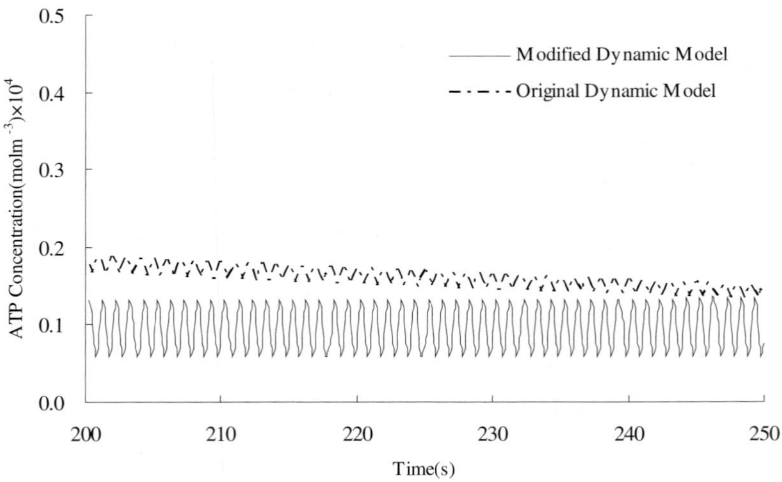

Fig. 5. Comparison between dynamic and static model-predicted extracellular ATP concent- ration in ECs surface against time from the onset of pulsatile fluid shear stress $\tau_w=1+sin2\pi\, t$, time course 200-250s

modified dynamic model-predicted ATP concentration seems to be more reasonable in face of a time-varying sinusoidal shear stress. As can be seen from Figure 5, the magnitude of its oscillation tends to stay within a fixed range. We omit the dynamic behaviors of ATP concentration for the static model in order for a clearer demonstration. It is not hard to imagine that during this period, the ATP concentration predicted by the static model tend to oscillate in the same manner as shown in Figure 4.

It is difficult to validate the predictions of the two dynamic models and the static one due to lack of experimental observations in the literature under the condition of pulsatile flow. It remains to be verified later by future experiments for pulsatile flow.

4 Discussion and Conclusions

Although a number of investigations by *in vitro* cell experiments have established that shear flow induces ATP release from endothelial cells, few quantitative analysis about the relationship between the shear stress and ATP release rate has been reported in the literature. The static model proposed by John and Barakat [6] captured a number of essential features of the shear stress induced ATP release process. A questionable assumption was made in their static model, that is, once a step shear stress acts on ECs, the ECs immediately release ATP, with release rate being a constant, which is independent of time, which apparently does not agree with the experimental evidences very well [3].

It is for the purpose of considering the time-dependency of the ATP release rate that a dynamic model was proposed in [9]. It was speculated that "receptor desensitization" occurs in the process of shear stress induced ATP release, i.e., all the ATP release pathways would be closed after being activated for a long time by a step shear stress. However such a situation usually happens only when the stimulus is a constant and hence it is not appropriate to describe ATP release under time-varying stimulus. Therefore, in this paper, we propose a modified dynamic model based on the previous work to incorporate the effect of time-varying stimulus. We currently assume a linear relationship between the stimulus change rate and the activation level of the ATP pathways. Further investigation into the relationship between shear stress variation and activation level of ATP pathways is necessary in order to obtain a more precise dynamic model.

Acknowledgements. The research reported here was supported by grant from National Natural Science Foundation of China, No. 10472027. The authors would like to thank Dr. Joji Ando and Dr. Kimiko Yamamoto at the University of Tokyo for their assistance in collecting experimental data of shear stress induced ATP release from ECs.

References

1. Davies, P.F.: Flow-mediated endothelial mechanotransduction. Physiol. Rev. 75, 519–560 (1995)
2. Ando, J., Ohtsuka, A., Korenaga, R., Kamiya, A.: Effect of extracellular ATP level on flow-induced Ca2+ response in cultured vascular endothelial cells. Biochem. Biophys. Res. Commun. 179, 1192–1199 (1991)

3. Yamamoto, K., Sokabe, T., Ohura, N., Nakatsuka, H., Kamiya, A., Ando, J.: Endogenously released ATP mediates shear stress-induced Ca2+ influx into pulmonary artery endothelial cells. Am. J. Physiol. Heart. Circ. Physiol. 285, H793–H803 (2003)
4. Nollert, M.U., Diamond, S.L., McIntire, L.V.: Hydrodynamic shear stress and mass transport modulation of endothelial cell metabolism. Biotechnol. Bioeng. 38, 588–602 (1991)
5. Shen, J., Gimbrone Jr., M.A., Luscinskas, F.W., Dewey Jr., C.F.: Regulation of adenine nucleotide concentration at endothelium-fluid interface by viscous shear flow. Biophys. J. 64, 1323–1330 (1993)
6. John, K., Barakat, A.I.: Modulation of ATP/ADP concentration at the endothelial surface by shear stress: effect of flow-induced ATP release. Ann. Biomed. Eng. 29, 740–751 (2001)
7. David, T.: Wall shear stress modulation of ATP/ADP concentration at the endothelium. Ann. Biomed. Eng. 31, 1231–1237 (2003)
8. Plank, M.J., Wall, D.J.N., David, T.: Atherosclerosis and calcium signaling in endothelial cells. Prog. Biophys. Mol. Biol. 91, 287–313 (2006)
9. Qin, K.R., Xiang, C., Xu, Z., Cao, L.L., Ge, S.S., Jiang, Z.L.: Dynamic Modeling for shear stress induced ATP release from vascular endothelial cells. Biomech. Model Mechanobiol. (in press), doi:10.1007/s10237-007-0088-8
10. Nelder, J.A., Mead, R.: A simplex method for function minimization. Computer Journal 7, 308–313 (1965)
11. Lagarias, J.C., Reeds, J.A., Wright, M.H., Wright, P.E.: Convergence properties of the nelder-mead simplex method in low dimensions. SIAM J. Optim. 9, 112–147 (1998)
12. Bodin, P., Burnstock, G.: Evidence that release of adenosine triphosphate from endothelial cells during increased shear stress is vesicular. J. Cadiovasc. Pharmacol. 38, 900–908 (2001)

Networked Control System: Survey and Directions

Xianming Tang and Jinshou Yu

Research Institute of Automation,
East China University of Science and Technology
Shanghai, 200237, China
mengfuqiao22@hotmail.com

Abstract. This paper extends a survey on Networked Control Systems (NCSs) published in previous papers. General frameworks and some fundamental issues for the study of NCSs are proposed. A significant emphasis has been on developing control methodologies to study the character of NCSs. This survey paper presents recent NCSs control methodologies, some promising research directions are also covered.

1 Introduction

Feedback control systems wherein the control loops are closed through a real-time network are called networked control systems (NCSs) [1]. One major trend in today's control community is the substantially increased research aimed at the integration of control, computer science, communications, and networking [2]. This includes principles, methodologies and tools for modeling and control of high level, networked, distributed systems, and rigorous techniques for reliable, embedded, real-time software. Roughly speaking, there are two main areas here: control of networks and control over networks. Control of networks is a quite large area, it studies control problems in networks, such as congestion control, routing control, data caching, and power management. Control over networks deals with control problems through data networks. The system here is called networked control system or network-based control system depending on different authors' preference. This paper gives a brief survey on NCSs.

The traditional communication architecture for control systems is point-to-point which could be traced back to the early 1970s when Distributed Control System (DCS) was introduced. New requirements, such as modularity, decentralization of control, integrated diagnostics, quick and easy maintenance and low cost, push the old architecture to its limits. Hence, the shared-data network architecture is introduced to meet the new requirements.

The insertion of the communication network in the feedback control loops makes the analysis and design of NCSs complex because this introduces different forms of time-delay uncertainty between sensors, actuators, and controllers. Conventional control theories with many ideal assumptions, such as synchronized control and non-delayed sensing and actuation, must be reevaluated before they can be applied to NCSs[1].

K. Li et al. (Eds.): LSMS 2007, LNCS 4688, pp. 473 – 481, 2007.

This article is organized as follow. First, in Section 2, general frameworks of NCSs are given before a survey on NCSs. Some fundamental issues of NCSs and research directions are then reviewed in Section 3. Finally, a summary of the paper is presented in Section 4. The main objective of the paper is to provide readers with an overview of the area, some future research is proposed at relevant places in the paper with references.

2 Overview of NCSs

2.1 General Framework of NCSs

There are two general frameworks for NCSs: direct structure and hierarchical structure [3]. The direct structure NCS is composed of one or several controllers and remote systems. The remote systems contain physical plants, sensors and actuators. The controllers and plants may physically locate at different locations and are directly linked by a data network in order to perform remote closed-loop control. There are two bottlenecks (shared resource) in this kind of structure: network bandwidth and controller CPU time. If several remote plants are controlled by one controller, there will be problem of scheduling the resource among plants which will be discussed later.

The control signal is encapsulated in a frame or a packet and sent to the plant via the network. The plant the returns the system output to the controller by putting the sensor measurement into a frame or a packet as well. Another issue is that plant outputs may be transmitted using multiple network packets (so-called multiple-packet transmission), due to the bandwidth and packet size constraints of the network. In a practical implementation, multiple controllers can be implemented in a single hardware unit to manage multiple NCS loops in the direct structure. Another general framework for NCSs is hierarchical structure. The hierarchical structure contains of a main controller and remote closed-loop control systems.

2.2 Delay in NCSs

Time delay is a very common issue in most control systems. Generally speaking, there are three kinds of delay in NCSs: plant delay, i.e. the delay within the plant itself, computational delay, i.e. the delay due to the control scheme adopted, and network-induced delay which is caused by the transfer protocol adopted. Plant delay has been considerably researched since the very beginning of control theory development. Computational delay has been studied since the digital processor was introduced to replace analogue controller [4]. Network-induced delays, due to their discrete and distributed nature, are quite different from the plant delays and computational delays that have been studied in the past.

In NCSs, the most significant delay is network-induced delay. With the rapid development in electronics, cheap and fast digital processors have become widely available, and the issue of computational delay in a digital controller has now become less important. However, in resource constrain situation such as embedded systems with limited computational ability, computational delay still should be handled carefully to prevent the system going unstable.

MAC protocols generally fall into two categories: random access and scheduling. Random access local area network such as CAN, DeviceNet, and Ethernet involve with more uncertain delays. The significant parts of random network delay are caused by queuing and frame collision on the networks. Other factors such as queuing time delays at a switch or a router and the propagation time delays from different network paths can increase the randomness of network delays. Therefore, CSMA networks are generally considered nondeterministic and the worst-case transmission time of packets is unbounded.

Scheduling network or cyclic service network has deterministic character. In local area network protocols with cyclic service such as IEEE 802.4, PROFIBUS, IEEE 802.5, and FIP, contention for the shared network medium is eliminated by allowing each node on the network to transmit according to a predetermined schedule. The time delays here can be both bounded and constant by transmitting packets periodically.

2.3 Packet Transmission

The second fundamental issue in NCSs is packet transmission. The network can be viewed as a web of unreliable transmission paths. Some packets not only suffer transmission delay but, even worse, can be lost during transmission. Thus, how such packet dropouts affect the performance of an NCS is an issue that must be considered. Another issue is that plant outputs may be transmitted using multiple network packets (so-called multiple-packet transmission), due to the bandwidth and packet size constraints of the network. Because of the arbitration of the network medium with other nodes on the network, chances are that all/part/none of the packets could arrive by the time of control calculation.

Current control system technology is based on simple communication architecture: all signals travel over synchronous dedicated links, with known (or worst-case bounded) delays, and no packet loss. Small dedicated communication networks can be configured to meet these demanding specifications for control systems, but it is worthwhile to study the packets dropout behavior and its influence on NCSs if the whole system is to be used in larger distributed style. Network packet drops occasionally happen on an NCS when there are node failures or message collisions. Although most network protocols are equipped with transmission-retry mechanisms, they can only retransmit for a limited time. After this time has expired, the packets are dropped. Furthermore, for real-time feedback control data such as sensor measurements and calculated control signals, it may be advantageous to discard the old, untransmitted message and transmit a new packet if it becomes available. In this way, the controller always receives fresh data for control calculation.

Normally, feedback-controlled plants can tolerate a certain amount of data loss, but it is valuable to determine whether the system is stable when only transmitting the packets at a certain rate and to compute acceptable lower bounds on the packet transmission rate.

2.4 Quality of Control (QoC)

From the control viewpoint, Quality of Control (QoC) is the most important aim of NCSs. In classical feedback control theory, several properties are used to evaluate the performance of closed-loop systems. The primary evaluation is mainly concerned with meeting the closed-loop system response characteristics (such as transient response and steady-state accuracy) and stability [4]. Beyond these requirements, controller designs attempt to minimize the system error for certain anticipated inputs or perturbations. The closed-loop system error is defined as the difference between the desired response and the actual response of the controlled system.

Traditional criteria such as IAE (Integral of the Absolute Error), ISE (Integral of Square Error) or ITSE (Integral of Time-weighted Square Error), provide quantitative measures of a control system response and are used to evaluate (and design) controllers. Some of them weight errors with time, penalizing steady-state errors and discounting the transient response errors.

In paper [5], QoC metric is defined in terms of the closed-loop system error and the IAE criterion are used because weighting all the errors equally is interested. Since the aim of controllers is to minimize the error (the deviation that the controlled system response is subject to, due to perturbation), better QoC will correspond to smaller errors (deviation). That is, there is an inverse relationship between the IAE index and the QoC.

3 Research Directions of NCSs

3.1 Periodic vs. Event-Based Control

A major problem in networked control is that of unknown and time-varying delays. The limited network bandwidth makes it hard to achieve short latencies and jitter-free delivery of measurement signals and control actions. The problem gets more pronounced as the network utilization becomes higher. One way to handle network delay in limited network bandwidth is to reduce the network utilization thus the whole NCSs efficiency will be improved. Preliminary results in event-based control theory promise a large reduction in the average network utilization with improved system performance [7]. Hence, periodic vs. event-based control is one research direction in NCSs that needs to be explored.

So far, there have been very little studies in event-based sampling and control theory. This is because the resulting system descriptions are both time-varying and nonlinear and hence very difficult to analyze.

Several research questions should be answered, they are:

A. When is it advantageous to use event-based rather than periodic control? One potential drawback of event-based control is that the communication jitter may increase. It is an open question how event-based schemes can be incorporated in network scheduling algorithms. Furthermore, it is not known how sensitive event-based controllers are to latency and jitter compared to ordinary periodic controllers.

B. How should network protocols for event-based control applications be designed?

C. How can multiple periodic or event-based control loops be incorporated into time-triggered, priority-based, or broadcast protocols?

3.2 NCSs Analysis and Design

Feedback control systems have been studied for at least 60 years. However, the insertion of a communication network into that structure presents some new challenges in NCS modeling, analysis and design. Techniques for modeling the plant/ sensor/ controller/actuator structure are well known, but techniques for modeling the main characteristics of network communication in an NCS, and the interactions between the network and the rest of the system are still not mature. Therefore, the fundamental importance in an analytical and/or a simulation study of NCS is to model and formulate the problem correctly. Since NCS is still a new area, there seems no need to follow certain pre-specified approaches at this stage. The exploration of diverse approaches should be encouraged, as long as the problem is realistically and correctly formulated.

Stability analysis is the first fundamental issue in NCSs research. One of the latest results on NCSs stability is [8]. The authors consider the stabilization problem for a class of networked control systems in the discrete-time domain with random delays. The sensor-to-controller and controller-to-actuator delays are modeled as two Markov chains, and the resulting closed-loop systems are jump linear systems with two modes. The necessary and sufficient conditions on the existence of stabilizing controllers are established. It is shown that state-feedback gains are mode-dependent. An iterative linear matrix inequality approach is employed to calculate the state-feedback gains. Other publications on NCSs stability will be [1, 9]. In paper [10], NCSs models with both network-induced delay and data packet dropout in transmission is considered.

Recently, there have been several efforts to apply middleware to assist networked control systems and teleoperation. Middleware is an implementation to seamlessly link applications and/or function calls together locally or over a network. In several implementations, middleware could also handle network resource allocation and reservation between two applications over a data network. A networked control system or a teleoperation system can utilize middleware to achieve certain network conditions such as guaranteed bandwidth, delay bound, or loss rate, by negotiating with the network counterpart for resource reservation. Replacing a widely used proportional-integral (PI) controller by a new networked controller for networked control capability can be costly and time-consuming. Paper[11,12] introduce novel methodology to apply middleware to enable an existing non-network-based controller so it can be used for networked control and teleoperation. The proposed methodology uses middleware to modify the controller output with respect to the current network traffic conditions in addition to utilize the generic features of middleware as mentioned. Controller output modification is performed based on gain scheduling. Since the controller does not need to be replaced, reinstalled, or redesigned, the proposed approach can be cost effective, and conveniently applied on existing systems.

3.3 Co-design and Co-simulation

In paper [13], the idea of NCSs control and scheduling co-design is discussed. Driving the research effort into NCSs is the point of view that the design of both the communication protocols and the interacting controlled system should not be treated as separate. In the co-design approach it is proposed that network issues such as bandwidth, quantization, survivability, reliability and message delay will be considered simultaneously with controlled system issues such as stability, performance, fault tolerance and adaptability. Thus, we could study network scheduling when a set of NCSs are connected to the network and arbitrating for network bandwidth.

Roughly speaking, the problem of network scheduling in NCSs is to assign a transmission schedule to each transmission entity (sensor, controller, actuator) on the network-based on a scheduling algorithm (a set of rules that, at any time, determines the order in which messages are transmitted).

In the past, control system design and CPU scheduling or network scheduling design have normally been separated. This separation has allowed the control community to focus on its own problem domain without worrying about how scheduling is being done; it has released the scheduling community from the need to understand what impact scheduling has on the stability and performance of the plant under control. However, when the two designs are combined together, many assumptions are not true. These will impact system performance.

However, network scheduling in NCSs is a relatively blank area that needs to be explored. Some pioneers should be [6, 13, 14]. One new method that could improve the idea of co-design is "feedback scheduling" which will be discussed in the next subsection.

3.4 Feedback Scheduling

Traditionally, the network bandwidth is allocated based on static schedule, which can be implemented using, for example, token rings, polling, or fixed priorities. In this way, the control community and the real-time scheduling community have normally been separated. This separation of concerns allows control engineers to focus on controllers design, leaving the bandwidth scheduling problem to be managed by network engineers. From the control perspective, the traditionally used static schedule is an open-loop solution. That is, once established at system set-up, the static schedule used to distribute the network bandwidth among control loops will not be adjusted at runtime. Given sufficient bandwidth, the static schedule can successfully guarantee real-time communications. However, when the network resource is scarce, e.g. in dynamic systems that feature both bandwidth constraints and workload fluctuations, scheduling the network with such an open-loop algorithm often causes pre-assigned network resources to be underutilized under light workload or yields unacceptable QoC due to resource unavailability under overload conditions[15].

To attack the above problems, a novel methodology is used, that is feedback scheduling [16]. The idea of feedback scheduling can be used in CPU scheduling as well as in network scheduling. Transmitting messages over a common-bus is comparable to executing tasks on one single CPU. Both cases involve allocating a shared resource to a set of concurrent tasks and frequent invocations of concurrent tasks, and both tasks have real-time constraints and have deadlines to be met. In NCSs, the shared resource becomes the control network bandwidth instead of CPU time, and the task execution is replaced by data packet transmission.

In [17], a novel feedback scheduler based on neural networks is suggested. With a modular architecture, the proposed feedback scheduler mainly consists of a monitor, a predictor, a regulator and an actuator. An online learning Elman neural network is employed to predict the network conditions, and then the control period is dynamically adjusted in response to estimated available network utilization. A fast algorithm for period regulation is employed. The idea of combining intelligent control with feedback scheduling method is very interesting and this could be a promising research direction for NCSs. For more knowledge of feedback scheduling and its research results, we can refer to the paper [15, 18, 19].

3.5 Bandwidth Management

NCSs have several challenges, which are due to the nondeterministic nature of communication networks. For example, communication delays, delay jitter, and packet losses adversely affect the performance of the system, and may even cause instability. Related research in this field has emerged into two distinct directions. The first direction aims at a control theoretical analysis while considering the network as a fixed parameter. The second direction targets the design of new communication network infrastructures, algorithms or protocols. Examples of the first approach include designing special controllers, and varying the sampling rate; and examples of the second include designing static and dynamic message scheduling algorithms [20].

Traditionally, bandwidth allocation and scheduling techniques for networked control systems have been based on static strategies that ensure average control performance at the expenses of permanently occupying the available bandwidth. However, static techniques may not be efficient when changing conditions occur at the control application level or at the network level, because pre-assigned resources may be underutilized. Ideally, these underutilized resources could be made available to other applications to provide new functionality or otherwise improve performance.

Network bandwidth arises as a crucial issue if distributed control is to be deployed ubiquitously over IP networks. This is true because when several processes compete for a finite resource, such as network bandwidth, and no proper coordination exists, congestion is a common consequence. In any bandwidth allocation scheme, a term that always accompanies efficiency is fairness.

In [21], an approach to dynamic bandwidth management in networked control systems that allows control loops to consume (or alternatively save) bandwidth according to the dynamics of the controlled process while attempting to optimize overall control performance is presented. This is achieved by augmenting the original state-space representation of each controlled process with a new state variable that describes the network dynamics.

Paper [22] proposes a scheme that efficiently allocates the network bandwidth among several control systems. The idea presented in this paper is that control systems vary their sampling periods (thus their bandwidth consumption) based on the congestion level fed back from the network.

Paper [23] presents a method by which controllers operating on real-time networks such as the Controller Area Network (CAN) can reduce their bandwidth requirements in response to periods of high network traffic from sporadic sources. The method derives from advances in the theory of dynamic systems on time scales.

The author of [24] presents an algorithm of jitter-dependent optimal bandwidth scheduling for the token-type based NCSs. The jitter caused by scheduling is considered only. In this algorithm, jitter is used as a connecting bridge between the bandwidth occupancy and the control performance. Through choosing the jitter, the permitted range of the sampling period of each control loop is constrained according to the requirement of the control performance. Choices of the sampling period and the jitter for each control loop not only affect the achievable control performance, but also affect the bandwidth utilization. Therefore a tradeoff between bandwidth occupancy and control performance could be determined.

4 Conclusions

This survey paper has introduced and reviewed the fundamental and recent control directions for NCSs. The study of networked control systems is an active research area with wide applications. With the rapid spreading of network applications to every place including homes, offices, and manufacturing plants, the research and reward in NCSs could be substantial in the near future.

References

1. Zhang, W., Phillips, M.: Stability of Networked Control Systems. IEEE Control Systems Magazine 21, 84–99 (2001)
2. Murray, R.M.: Future Directions in Control, Dynamics, and Systems: Overview, Grand Challenges, and New Courses. European Control Conference 9, 114–158 (2003)
3. Tipsuwan, Y., Chow, M.Y.: Control Methodologies in Networked Control Systems. Control Engineering Practice 11, 1099–1111 (2003)
4. Åström, K.J., Wittenmark, B.: Computer-Controlled Systems: Theory and Design. Prentice-Hall, Upper Saddle River, NJ, USA (1997)
5. Martí, P., Fohler, G., Ramamritham, K.: Improving Quality-of-Control Using Flexible Timing Constraints: Metric and Scheduling Issues. In: Proceedings of the 23rd IEEE Real-Time Systems Symposium, Washington, DC, USA, pp. 91–100. IEEE Computer Society Press, Los Alamitos (2002)
6. Martí, P., Yepez, J., Velasco, M., Fuertes, J.M.: Managing Quality-of-Control in Network-Based Control Systems by Controller and Message Scheduling Co-Design. IEEE Trans. on Industrial Electronics 51, 1159–1167 (2004)
7. Cervin, A.: Periodic and Event-Based Control over Networks (2005), http://www.control.lth.se/ anton/program.pdf
8. Zhang, L.Q., Shi, Y., Chen, T.W., Huang, B.: A New Method for Stabilization of Networked Control Systems With Random Delays. IEEE Trans. on Automatic Control 50, 1177–1181 (2005)
9. Nesic, D., Teel, A.R.: Input-Output Stability Properties of Networked Control Systems. IEEE Trans. on Automatic Control 49, 1650–1667 (2004)
10. Azimi-Sadjadi, B.: Stability of Networked Control Systems in the Presence of Packet Losses. In: 42nd IEEE Conference on Decision and Control, vol. 1, pp. 676–681. IEEE Computer Society Press, Los Alamitos (2003)
11. Zheng, L., Vanijjratikhan, R., Chow, M.Y., Viniotis, Y.: Comparison of Real-time Network Traffic Estimator Models in Gain Scheduler Middleware by Unmanned Ground Vehicle Network-Based Controller. Industrial Electronics Society, pp. 2433–2438 (2005)
12. Tipsuwan, Y., Chow, M.Y.: On the Gain Scheduling for Networked PI Controller on IP Network. IEEE/ASME Transactions on Mechatronics 9, 491–498 (2004)

13. Branicky, M.S., Philips, S.M., Zhang, W.: Scheduling and Feedback Co-Design for Networked Control Systems. In: Proceedings of the 41st IEEE Conference on Decision and Control, vol. 2, pp. 1211–1217. IEEE Computer Society Press, Los Alamitos (2002)
14. Branicky, M.S., Liberatore, V., Phillips, S.M.: Networked Control System Co-simulation for Co-design. American Control Conference 4, 3341–3346 (2003)
15. Xia, F., Li, S.B., Sun, Y.X.: Integrated Feedback Scheduling of Networked Control Systems. Dynamics of Continuous, Discrete and Impulsive Systems- Series B: Applications and Algorithms 4, 3274–3280 (2006)
16. Stankovic, J.A., Lu, C.Y., Son, S.H., Tao, G.: The Case for Feedback Control Real-Time Scheduling. In: Proceedings of the 11th Euromicro Conference on Real-Time Systems, York, UK, pp. 11–20 (1999)
17. Xia, F., Li, S.B., Sun, Y.X.: Neural Network-based Feedback Scheduler for Networked Control System with Flexible Workload. In: Wang, L., Chen, K., Ong, Y.S. (eds.) ICNC 2005. LNCS, vol. 3611, pp. 242–251. Springer, Heidelberg (2005)
18. Eker, J., Årzén, K.E.: A Feedback Scheduler for Real-time Controller Tasks. Control Engineering Practice 8, 1369–1378 (2000)
19. Montestruque, L.A., Antsaklis, P.J.: State and Output Feedback Control in Model-based Networked Control systems. In: Proceedings of the 41st IEEE Conference on Decision and Control, vol. 2, pp. 1620–1625. IEEE Computer Society Press, Los Alamitos (2002)
20. Al-Hammouri, A.T.: Optimization Congestion Control for Networked Control Systems. In: Parallel and Distributed Processing Symposium, pp. 25–29 (2006)
21. Velasco, M., Fuertes, J.M., Lin, C.X., Marti, P., Brandt, S.: A Control Approach to Bandwidth Management in Networked Control Systems. Industrial Electronics Society 3, 2343–2348 (2004)
22. Al-Hammouri, A.T., Liberatore, V., Phillips, S.M.: Decentralized and Dynamic Bandwidth Allocation in Networked Control Systems. In: Workshop on Parallel and Distributed Real-Time Systems, pp. 25–29 (2006)
23. Gravagne, I.A., Davis, J.M., Dacunha, J.J., Marks, R.J.: Bandwidth Reduction for Controller Area Networks Using Adaptive Sampling. In: IEEE International Conference on Robotics and Automation, vol. 5, pp. 5250–5255. IEEE Computer Society Press, Los Alamitos (2004)
24. Bai, T., Yang, G.K.: Optimal Bandwidth Scheduling of Networked Control Systems (NCSs) in Accordance with Jitter. Journal of Zhejiang University Science A 5, 535–542 (2005)

A Novel Fault Diagnosis of Analog Circuit Algorithm Based on Incomplete Wavelet Packet Transform and Improved Balanced Binary-Tree SVMs

Anna Wang[1], Junfang Liu[1], Hao Wang[2], and Ran Tao[3]

[1] School of Information Science and Engineering, Northeastern University, Shenyang, China
[2] School of Environment Science and Spatial Information, China University of Mining and Technology, China
[3] Northeast Yucai School
wanganna@mail.neu.edu.cn, junfang_0306@163.com

Abstract. Aiming at to the characteristics of fault diagnosis of analog circuit with tolerances, noise, circuit nonlinearities and small sample set, this paper presents a novel algorithm for the existent problems based on incomplete wavelet packet transform for feature extraction and improved balanced binary-tree support vector machines (BBSVMs) for multi-classification. Firstly, analyzing characters of the optimal wavelet packet transform and the incomplete wavelet packet transform, the conclusion that the latter is perfect for fault diagnosis of analog circuit is obtained. Secondly, in order to perform multi-class classification, binary-tree-based SVMs has been studied. By introducing the separability measure that based on the space distribution of pattern classes, an improved balanced binary-tree structure is constructed. The proposed algorithm for analog circuits with tolerances and noise is implemented, and simulation results show us that compared with several existent fault diagnosis methods, the current algorithm has highest classification speed and higher classification accuracy, which make fault diagnosis of analogy circuit on-line promising.

1 Introduction

Fault diagnosis and testing of electronic circuit and systems is a crucial and complex task in microelectronics and the semiconductor industry and has appealed to many researchers in the past two decades [1, 2]. Normally the techniques used in the field can be divided into two categories: the estimation methods and pattern recognition methods. The estimation methods require mathematical process models that represent the real process satisfactorily. However, no mathematical model of the process is required in the pattern recognition methods as the operation of the process is classified by matching the measurement data.

The process of fault diagnosis of analog circuit consists of two stages: feature extraction and pattern recognition. Automatic fault diagnosis and testing of analog circuits is subjected to many difficult problems, such as poor fault models, noise, nonlinearity and tolerance effects. The local of the signals caused by the faults can be

K. Li et al. (Eds.): LSMS 2007, LNCS 4688, pp. 482 – 493, 2007.

identified by analyzing the non-stationary components of the output signals. Generally, these output signals can be used to detect the incipient fault of the electronic components and reduced the possibility of catastrophic damage, through the on-line monitoring and diagnosis system, where the extracted features, such as time-frequency domain features of short-time Fourier transform, Wigner-Viller distribution, wavelet transform (WT), etc.. The WT method possesses perfect local property in both time and frequency space, and it is used widely in the region of machinery fault detection and identification. But the WT cannot split the high frequency band where the fault information of analog circuit always exists. The wavelet packet transform (WPT) can overcome the difficulty. However, there are still high noise content, irrelevant or redundant information in these extracted features in the WPT. Two improved methods of feature extraction that based on WPT are presented: optimal wavelet packet transform (OWPT) and incomplete wavelet packet transform (IWPT).

Some intelligent classification algorithms, such as artificial networks (ANNs) and support vector machines (SVMs) have been successfully applied to automated detection and diagnosis of machines conditions. The difference in risk minimization leads to a better generalization performance for SVMs than ANNs. SVMs are originally designed for two-class classification [3, 4]. Usually there are two schemes to obtain a multi-class pattern recognition system, one of which is to combine several binary classifiers, and another is to consider all the classes in one big optimization problem. Several common multi-classification methods based on above schemes are: one-against-all, one-against-one, decision directed acyclic graph (DDAG) SVMs and error correcting output codes (ECOC) [5]. However, for all these methods, too many binary tests are required to make a decision and there is a large nonseparable region. Tree architecture is always leveraged in decision theory. Support vector machines with binary tree architecture has been introduced to reduce the number of binary classifiers and to achieve a fast decision. By the defined separability measure method, a novel multi-classification based on improved balanced binary-tree SVMs is proposed in this paper.

The paper proceeds as follows. Second 2 presents the techniques used in this work, such as WPT, OWPT and IWPT. Second 3 introduces separability measure based on the space distribution of the training samples, defines three different binary-tree SVMs, analyzes briefly classification performance for four binary-tree SVMs. The proposed method is validated using signal from the circuit under test in Section 4. The conclusions are presented in Section 5.

2 Brief Introduction of the Techniques for WPT, OWPT and IWPT

2.1 WPT

Both WT and WPT have continuous and discrete formats. The discrete format of WPT is adopted in this work because it is more popularly used in engineering applications. To illustrate the underlying mathematical theory of WPT briefly, we denote $\{h_k\}_{k\in z}$ and $\{g_k\}_{k\in z}$ as the quadrature mirror filter banks. A signal can be decomposed at different scales on the basis functions with the form $2^{-j/2}u_n\left(2^{-j}t-k\right), j,k\in Z, n\in N$, where, Z denotes the integer and N denotes the non-negative integer. These functions are iterated as

$$u_{2n}(t)=\sqrt{2}\sum_{n\in z_+} h_k u_n(2t-k) \tag{1}$$

$$u_{2n+1}(t)=\sqrt{2}\sum_{n\in z_+} g_k u_n(2t-k) \tag{2}$$

where j is a scale parameter, k is a time localization parameter and n is an oscillation parameter. Thus, $u_0(t)$ is a scale function which corresponds to a low-pass filter. The filtered signal is an approximation of the analyzed signal. The function $u_1(t)$ is a wavelet function which corresponds to a high-pass filter. The filtered signal is a detail of the analyzed signal. The approximation and detail can be further sliced by dyadic decomposition using the dilated and translated scale functions and wavelet functions.

The wavelet packet function as well as the corresponding decomposition coefficients can be organized as a binary tree as shown in Fig. 1. Each node corresponds to a frequency band.

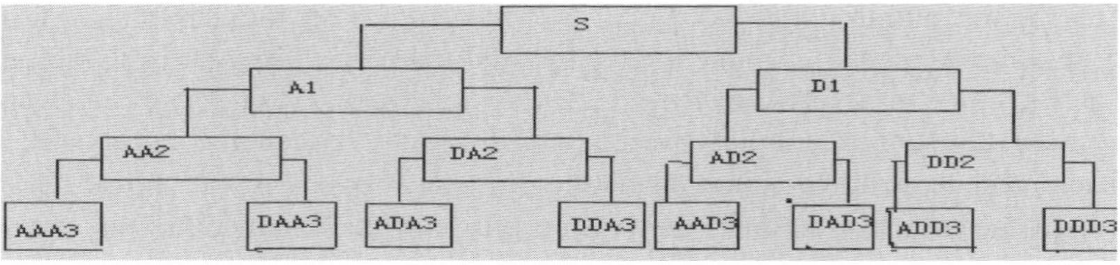

Fig. 1. A example of a three-level wavelet packet decomposition tree

The decomposition relationship is

$$S=\mathbf{AAA}_3+\mathbf{DAA}_3+\mathbf{ADA}_3+\mathbf{DDA}_3+\mathbf{AAD}_3+\mathbf{DAD}_3 \tag{3}$$

S is the discrete signal, A represents the approximation parts of low frequency, D represents the detail parts of high frequency, and the subscript is the layer number.

Suppose that the original signal is still $\{S_k, k\in Z\}$ and it is decomposed by a three-level wavelet packet decomposition, we get signal S_j , $S_j=\{s_{j,n,l}, j,n,l\in Z\}$. Where $s_{j,n,l}$ is the *l*th sample of the *n*th packet on level *j*. We define the relative energy for a packet as below

$$E_{j,n} = \sum_{l} s_{j,n,l}^2 / \sum_{k} s_k^2 \quad (4)$$

Where $\sum_{l} x_{j,n,l}^2$ denotes energy component of the nth packet on level j. $\sum_{k} s_k^2$ denotes the energy of the original signal and k is the number of frequency bands of he bottom level. The normalization feature vector is

$$T = \left[E_{30}, E_{31}, E_{32}, E_{33}, E_{34}, E_{35}, E_{36}, E_{37}\right] \quad (5)$$

2.2 OWPT

Wavelet packet bases bank consists of many different wavelet packet basis with special characters, which represent different signal features. The binary tree of bases can also be considered as 2D time-frequency plane. The information in the bases is redundant along two axes, i.e. information in child bases are overlapped with that in the parent basis. The best basis is preferably selected from the complete binary tree, so as to reduce the data analysis effort without losing information. For signal representation, best bases are defined in that they cover the complete tree and ensure no redundant information. The Shannon entropy-based criterion [6] is well suited to the selection of the best tree. When signals come from different classes and common best basis is required, a WPT-structure tree is used for the best basis selection. For pattern recognition, common best bases are selected in that they have the best classification capability. An example that the best basis is chosen is illustrated in Fig. 2.

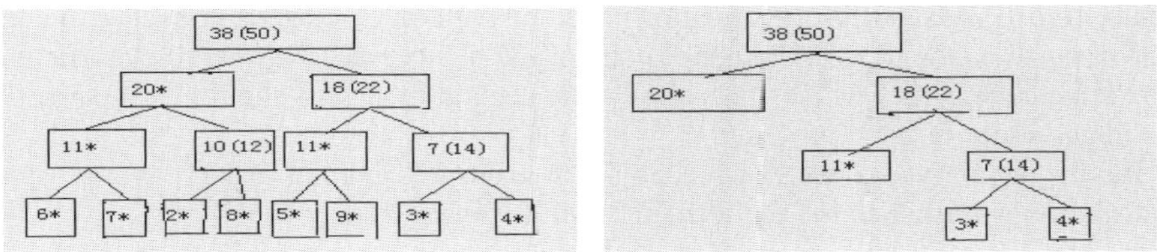

Fig. 2. (a) The wavelet packet decomposition tree with cost function values (b) The best wavelet packet decomposition tree

2.3 IWPT

Whether the WPT or the OWPT, aim to extract salient fault features to distinguish pattern classes best and decrease the computation complex and computer memory. According to the thought, a more directed preprocessor than the OWPT is presented, the incomplete wavelet packet transform. Because the fault information of a circuit always exists in the high frequency band of the output signal, that all the normalization energy values in WPT are computed is unnecessary. In other words, we can only compute energy values of frequency bands of the nodes A1, AAD3, DAD3, ADD3 and DDD3 in Fig. 1. In this way, nodes of high frequency bands are chosen to process, but low frequency aren't, which appears halfway wavelet packet transform, so we call it incomplete wavelet packet transform (IWPT). Corresponding to the WPT, the normalization feature vector of the IWPT is

$$T = [E_{10}, E_{34}, E_{35}, E_{36}, E_{37}] \tag{6}$$

3 Multi-class Classification Using Binary-Tree SVMs

The traditional SVM only provides two-class classification arithmetic and it is important to extend to multi-class classification. There are lots of multi-class classified methods that put forward before and comparatively mature. The multi-class classification problem is commonly solved by a decomposition to solve several two-class problems for which the standard SVM to be used. Generally we apply multi-classification modes as follows: one-against-all, o-against-one, DDAG and ECOC SVMs. But there are a large nonseparable region and lots of binary classifiers in the one-against-all and one-against-one modes, which lead to low classification speed and accuracy; there is without nonseparable region in DDAG mode, but different binary classifier choice of root node will give birth to different result for a decided class, which makes classification uncertain; depending on the encoding scheme, N times binary tests at least or N(N-1)/2 times at most are needed to make a decision in ECOC modes.

Tree architecture is always leveraged in decision theory. SVMs with binary tree architecture have been introduced to reduce the number of the binary classifiers and there is without nonseparable region. But there are lots of schemes to construct a binary-tree SVMs multi-classification with N pattern classes. If the classification performance is not good at the upper node of the binary tree, the overall classification performance becomes worse. Therefore more separable classes should be separated at the upper nodes of the binary tree [7].

Based on above reasons, a separability measure method based on the space distribution is introduced to the formation of the binary tree.

3.1 The Separability Measure Method [8]

The Euclidean distances or Mahalanobis distances between class centers are the commonly used separability measure, but the distance between the centers of the two classes can not always denote the separability between classes rightly if the distributions of the classes are not taken into consideration. Fig. 3 gives the comparison of the separability among classes.

The Euclidean distances among the centers of the three classes in Fig.3 are same, but it is obviously that class k can be classified more easily than that other classes. Therefore, the distribution of the classes is also an important factor of between-class separability measure.

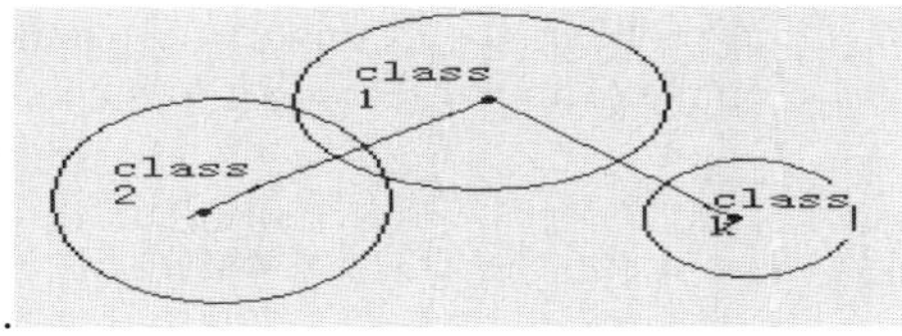

Fig. 3. The comparison of separability among classes with equal center distance

Based on this analysis, a new separability measure based on distributions of classes is proposed in reference [8]. Let $sm_{i,j}$ be the separability measure between class i $\{x_1, x_2, \cdots, x_{ni}\}$ and class j $\{x_1', x_2', \cdots, x_{nj}'\}$. Suppose that m_Φ is the class center and $m_\Phi = \frac{1}{n}\sum_{i=1}^{n}\Phi(x_i)$ in feature space, where n is the sample number, and Φ is the mapping. The distance between class centers is

$$d(m_\Phi^i, m_\Phi^j) = \sqrt{\frac{1}{n_i^2}\sum_{i=1}^{n_i}\sum_{j=1}^{n_i}k(x_i, x_i) - \frac{2}{n_i n_j}\sum_{i=1}^{n_i}\sum_{j=1}^{n_j}k(x_i, x_j') + \frac{1}{n_j^2}\sum_{j=1}^{n_j}\sum_{j=1}^{n_j}k(x_j', x_j')} \quad (7)$$

Let $d(x, m_\Phi)$ be the distance be the distance between training sample x and class center m_Φ in the feature space

$$d(x, m_\Phi) = \sqrt{k(x,x) - \frac{2}{n}\sum_{i=1}^{n}k(x, x_i) + \frac{1}{n^2}\sum_{i=1}^{n}\sum_{j=1}^{n}k(x_i, x_j)} \quad (8)$$

$\sigma = \frac{1}{n-1}\sum_{i=1}^{n} d(x_i, m_\Phi)$ is the class variance in feature space. The separability measure is defined as:

$$sm_{i,j} = \frac{d(m_\Phi^i, m_\Phi^j)}{(\sigma_i + \sigma_j)} \quad (9)$$

3.2 Three Improved Multi-classification Algorithm Based on Binary-Tree SVMs and the Separability Measure

Based on the idea that the most easily separated classes are separated firstly, and by introducing the between-class separability measure in feature space into the formation of the binary tree, an improved balanced binary-tree (BBT) SVMs is proposed. In order to verify that the binary-tree structure of the proposed method has higher generalization ability, we define another two kinds of binary-tree structures based on the separability measure: oblique binary tree (OBT) and unbalanced binary tree (UBT). Aiming to state expediently formations of different binary-tree structures, we take a 6-classes classification problem for example, and suppose that the matrix of separability measures based on the example according to the reference [8] is:

$$SM = \begin{bmatrix} \infty, 1.31, 0.82, 1.21, 0.84, 1.50 \\ 1.31, \infty, 1.10, 1.02, 1.10, 0.97 \\ 0.84, 1.10, \infty, 0.90, 0.87, 0.96 \\ 1.21, 1.02, 0.90, \infty, 1.01, 1.27 \\ 0.83, 1.10, 0.87, 1.01, \infty, 1.13 \\ 1.50, 0.97, 0.96, 1.27, 1,13, \infty \end{bmatrix}$$

The formation process of the UBT SVMs for multi-classifier:

Step 1: Calculate relative distances between two classes arbitrarily, which are based on the separability measure in feature space, such as the matrix SM above.

Step 2: Find out the two classes with the largest distance between them firstly (class 1 and class6 in the example), and then other classes belong to adaptively the node where one class nearer to them exists. By this clustering method, the left and right child nodes of the root node are chosen. Class1, 3, 4and class5 constitute the left node, and the class2 and 6 constitute the other. The process starts from the root node downwards, until all the classes at any node but leaf nodes are separated. The balanced binary structure of the example is illustrated in Fig. 4.

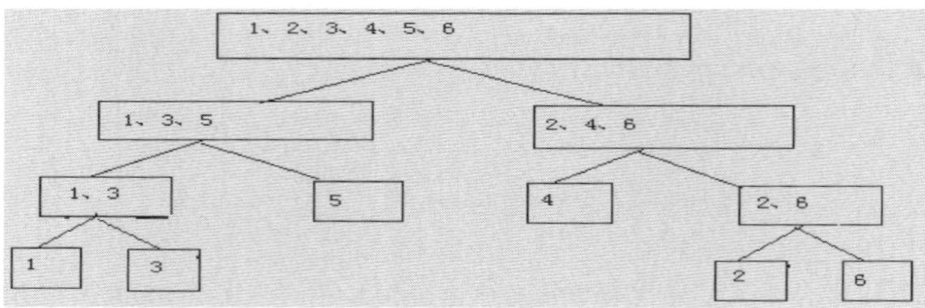

Fig. 4. The structure of the BBT about the example

Step 3: Based on the clustering results, all the classes at any node are nailed down. Every decision node is segmented accurately by SVM. All the nodes except leaf nodes correspond to a hyperplane separating some classes from the others.

The formation processes of the BBT SVMs are same to the UBT SVMs but the formation of the left and right nodes, so we state only different points in detail here. To construct the structure of the BBT SVMs, for an N-classes classification problem, N should be the power of 2, and if not, we can choose [N/2] and N-[N/2] approximatively instead as the class number of two child nodes. [N/2] represents the maximal integer which is not beyond N/2. For construction of the structure of the BBT SVMs, according to the clustering method of the UBT, all the classes are clustered adaptively into two sets for the quasi-left and quasi-right nodes firstly, and then if the class number of some node is beyond half [N/2], these redundant classes will be classified to the other node. At the same time, the real-left and real-right nodes are formed. In this way, some classes may be not classified to the nearer node from them to keep the balanced structure. For example, though the relative distance between class4 and class1 is nearer than the distance between class4 and class6, class4 is classified to the node where class6 exist to keep symmetrical structure, for the class number of the node where class1 exist arrives at half 3.

For construction of the structure of the OBT SVMs, the principle that the utmost class from the other classes is separated firstly is applied, which is discussed in detail in the literature [8] and will not be reviewed here. Fig.5 and Fig.6 illustrate the two structures based on the example above, respectively.

There are two important factors to affect classified accuracy and speed of SVM: one is distance between classes and the other is the number of training samples. The farther the distance is and the smaller the number of training samples is, the higher accuracy and speed of SVM obtains. Three multi-classification algorithms above are based on the largest distance between classes, but the average training times for a sample is different: $\log_2 N$ times and (N+1)/2-1/N times are trained for a sample by the BBT SVMs and The OBT SVMs, respectively, and the average training times for the UBT SVMs is between of them. We select the BBT SVMs to classify the fault classes of analog circuits, so that real-time fault diagnosis is promising.

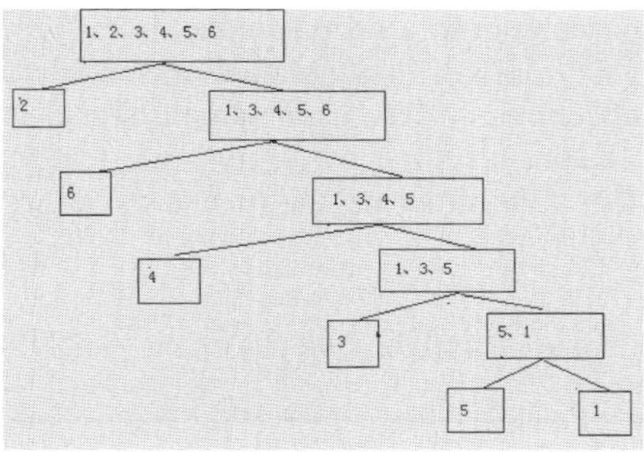

Fig. 5. The structure of the OBT about the example

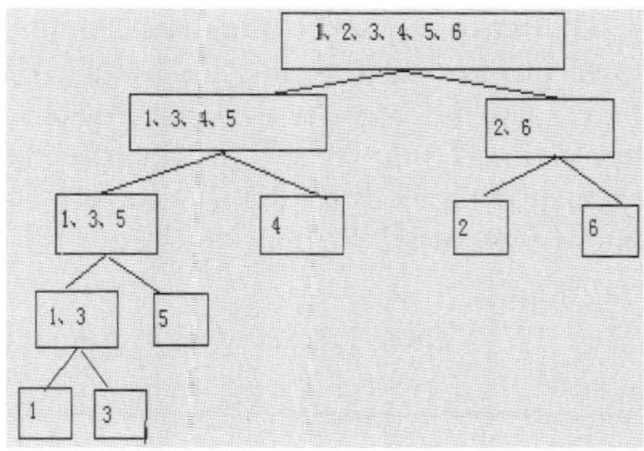

Fig. 6. The structure of the UBT about the example

4 The Simulations and Results Analysis

4.1 The Circuit Under Test (CUT)

Fig.7 shows a two-class video frequency enlarges electric circuit which is more complex circuit used in our study. The faults associated with this circuit are assumed to be R2O, R3S, R4O, R5S, R6↑, R7 ↓, C1↑, L3 ↓, C2 ↓, C3↑, L3↑, C4 ↓, R7 ↓, R8↑and C9↑. In this notation, ↑and↓ imply significantly higher or lower than nominal values and the letters S and O represent resistors shorted out and opened, respectively. In order to generate training data for the fault classes, we set the value for faulty components and vary other resistors and capacitors within their standard tolerances of 5% and 10%, respectively. Fault diagnosis of this circuit requires recognizing 16 fault classes, 15 fault classes indicated above and the fault-free class (NFT). 500 sample signals are obtained from 6500 Monte Carlo PSpice simulations.

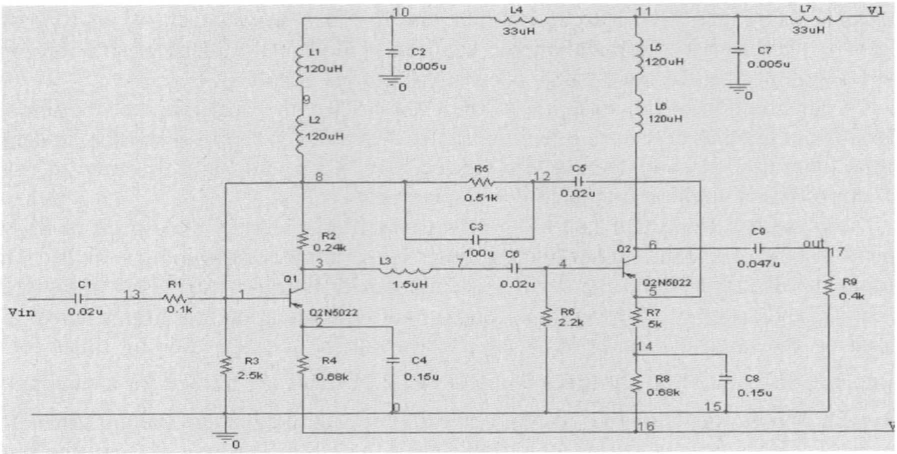

Fig. 7. Two-class video frequency enlarges electric circuit

The wavelet packet transform is applied to decompose all the samples by a tree-level wavelet packet using the MATLAB Wavelet toolbox 3.0.4 (MathWorks, Natick, MA), and the wavelet packet belongs to Daubechines series. Each sample signal is transformed into a set of five-dimension feature vectors based on the feature extraction methods the IWPT. Table 1 shows that a set of normalization energy values out of 500 for all the fault classes.

Table 1. A set of normalization energy values for all the fault classes

Energy values / fault classes	E_{10}	E_{34}	E_{35}	E_{36}	E_{37}
R2O	0.6995	0.0410	0.0333	0.0381	0.0431
R3S	0.6668	0.0520	0.0633	0.0461	0.0469
R4O	0.6747	0.0521	0.0535	0.0545	0.0396
R5S	0.6790	0.0444	0.0515	0.0441	0.0506
R6↑	0.9766	0.0039	0.0025	0.0029	0.0038
R7 ↓	0.9762	0.0047	0.0026	0.0031	0.0039
C1↑	0.9768	0.0040	0.0031	0.0024	0.0031
L3 ↓	0.5936	0.0679	0.0595	0.0566	0.0540
C2 ↓	0.6780	0.0404	0.0575	0.0491	0.0516
C3↑	0.5618	0.0533	0.0533	0.0550	0.0615
L3↑	0.6257	0.0622	0.0579	0.0738	0.0476
C4 ↓	0.5693	0.0809	0.0536	0.0781	0.0502
R7 ↓	0.9799	0.0029	0.0028	0.0029	0.0027
R8↑	0.9788	0.0026	0.0036	0.0032	0.0030
C9↑	0.9775	0.0036	0.0044	0.0031	0.0022
NFT	0.6257	0.0622	0.0579	0.0738	0.0476

The wavelet packet transform is applied to decompose all the samples by a tree-level wavelet packet using the MATLAB Wavelet toolbox 3.0.4 (MathWorks, Natick, MA), and the wavelet packet belongs to Daubechines series. Each sample signal is transformed into a set of five-dimension feature vectors based on the feature extraction methods the IWPT. Table 1 shows that a set of normalization energy values out of 500 for all the fault classes.

4.2 Construction of the BBT SVMs for the CUT

We implemented the algorithm for constructing the imprcved balance binary trees on a Pentium 4 machine with MATLAB 6.5. Calculating relative distances between two classes arbitrarily, which are based on separability measure in feature space, a separability matrix is generated. By the algorithm of the improved balanced binary tree, the structure of the BBT for the CUT is formed (Fig 8.)

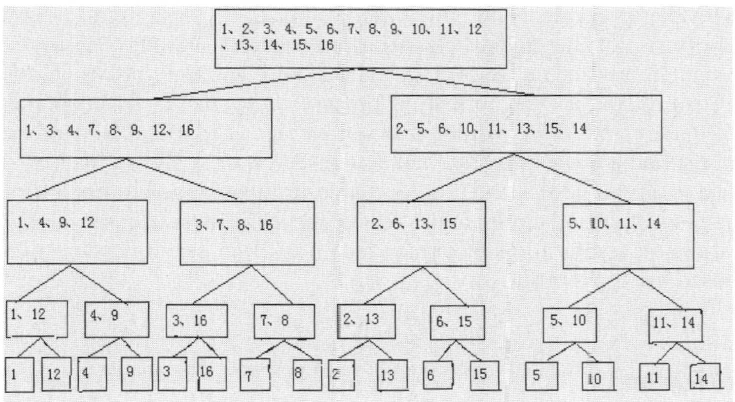

Fig. 8. The structure of the balanced binary tree for the CUT

Table 2. Comparison of several multi-class classification methods

Methods	No. of training samples	No. of testing samples	time (s)	accuracy (%)
o-a-a	300	200	5.18	92.67
o-a-o	300	200	4.73	93.33
DDAG	300	200	4.61	95.78
UBT	300	200	2.91	95.90
OBT	300	200	3.01	98.02
BBT	300	200	1.71	97.91

4.3 Simulation Results and Analysis

All the training feature vectors from the incomplete wavelet packet transform are inputted into the BBT SVMs, and then all the testing feature vectors are tested. In order to compare performance of six multi-classifiers: one-against-all SVMs, one-against-one SVMs, DDAGSVMs, the BBT SVMS, the OBT SVMs and the UBT SVMs, all the trainings are under the same condition. The average correct classification rates and times for the testing data of the fault classes are given in Tab.2.

From Table 2 we can know that the average classification time of the multi-classifiers based on binary trees and SVM is 2.91s, 3.01s and 1.71s, respectively, which is shorter obviously than other methods, for there is no unclassifiable region in the feature space in this way and the formation process of trees are based on the separability measure. Because the smallest average training and testing times for all the samples, the proposed multi-classifier obtains shortest testing time, 1.71s. Due to all the SVMs at every decision node based on maximal separability between classes, The UBT SVMs gets the highest classification accuracy, 98.02%. However, aiming to save computation time, a few classes classified are not based on the maximal separability principle in the formation process of the BBT, so it obtains relatively higher accuracy, 97.91%. Although the average classification accuracy of the proposed method is lower slightly than that based on the UBT SVMs by 0.11%, 1.20s is saved. It is advisable that the proposed algorithm is applied to implement real-time fault diagnosis of analog circuit at the cost of the little lower accuracy.

Classification results for the data set also prove the performance improvement of the proposed multi-classifier.

5 Conclusions

In this paper, we focus on several multi-classifiers based on binary-tree and SVMs and analyze their classification performance. Aiming at to the characteristic of fault diagnosis of analog circuit with tolerances, we propose a novel algorithm combining balanced binary tree multi-classifier based on the separability measure and SVMs classification. Separability between pattern classes is considered in constructing binary tree, which reduced the number of support vectors when SVM are adopted to classify these classes, and the smallest training and testing time are required in the structure of the balanced binary-tree SVMs, so that the proposed algorithm obtains higher classification accuracy and speed.

References

1. Selected Papers on Analog Fault diagnosis: IEEE press, New York (2000)
2. Liu, R.-W.: Testing and Diagnosis of Analog Circuit and Systems. Van Nostrand, New York (2003)
3. Aminian, F., Aminian, M., Collin, H.W.: Analog fault diagnosis of actual circuit using neural network. IEEE Trans. on Circuits. Syst.-II 5, 544–550 (2004)

4. Vapnik, V.: The Nature of Statistical Learning. Spring Verlng, New York (1995)
5. Hsu, C.-W., Lin, C-J.: A comparison of methods for multi-class support vector machines [J]. IEEE Transaction on Neural Network, 415–425 (2002)
6. Coifman, R.R., Wickerhauser, M.V.: Entropy-based algorithms for best basis selection. IEEE Transactions on Information Theory 38(2), 713–718 (1992)
7. Plattjc, Cristianinin, Shawe-Yaylorj.: Larg Margin DAGs for multi-class classification Processing classification. Advances in Neural information Systems, pp. 547–553 (2006)
8. Wang, X.: Using improved SVM decision tree to classify HRRP. In: Proceeding of the Fourth International Conference on Machine Learning and Cybernetics, Guangzhou, pp. 18–21 (2005)
9. Hui, Q.: Fault diagnosis of the rotating machinery based on improved wavelet packet transform and SVMs ensemble. Mechanical Systems and Signal Processing 21, 688–705 (2007)
10. Liu, B.: Selection of wavelet packet basis for rotating machinery fault diagnosis. Journal of Sound and Vibration 284, 567–582 (2005).

A VSC Algorithm for Nonlinear System Based on SVM

Yibo Zhang and Jia Ren

Department of Automation,
Zhejiang Sci-Tec University, Hangzhou, 310018, P. R. China
Zhangy41@yeah.net

Abstract. A variable structure control (VSC) algorithm for nonlinear system based on feed-forward and feedback technology is developed. A method based on feed-forward SVM (FFSVM) is brought forward to eliminate external disturbance. In order to reduce difference and to track desired trajectory, a feedback SVM (FBSVM) is introduced into the feed-forward system by adopting VSC algorithm. And so recognition and control designing are combined, and recognition of system is avoided. High track speed and robustness are granted. Simulation shows the effectiveness of the scheme.

1 Introduction

Over the past two decades, increasing attention has been drawn to sliding mode control schemes, which warrant the system robustness in the presence of norm-bounded system uncertainties or disturbances. The sliding mode control theory has been widely and successfully applied in many real systems such as robotics, servomechanisms, power systems, circuits systems, etc. In order to perform perfect tracking, from the viewpoint of internal model principle, the acquisition of the equivalent control signals is imperative in the sliding mode control.

SVM is a kind of Generalized Portrait algorithm, which has attracted increasing interest in recent years [1]-[5]. Originally put forward for classification, it has been generalized to regression, and applied to various problems. Generally the solution of SVM can be expressed by a small number of support vectors. The most important feature of SVM is that SVM is equivalent to minimize regularized risk. Vapnik's theory shows that generalization error is bounded by sum of the training error and a term depending on the Vapnik-Chervonenkis (VC) dimension of learning machine. By minimizing this upper bound, high generalization can be achieved, and the main question is addressed with a mathematical background. Unlike other learning machine such as neural networks, SVM's generalization error is not dependent on dimensionality of the input data, but the margin it separates the data instead. SVM can convert linear inseparable input data into a high dimensional linear separable feature space via a nonlinear mapping technique. So the regression of SVM is widely used in recognition of black-box systems. SVM can track arbitrary curves with arbitrary precisions, which means it can be easily applied in the recognition of linear and nonlinear systems. SVM is used in control system to combine recognition of system and design of controller, so input of system can be obtained directly from output of system.

K. Li et al. (Eds.): LSMS 2007, LNCS 4688, pp. 494 – 501, 2007.

Feed-forward control technology can eliminate effect of disturbance in advance, while feedback control technology ensure the stabilization of system. When they are combined, better control performance is achieved. SVM is an excellent tool to combine recognition and control technology, but it has not been applied in feed-forward and feedback system yet. So it is necessary to introduce SVM and feed-forward-feedback technology into system to ensure performance of system.

A VSC algorithm of feed-forward and feedback for black-box nonlinear system based on SVM is developed. Input and output of system and desired trajectory are trained by FFSVM to eliminate external disturbance. Then inverse model of the feed-forward system is trained by FBSVM to reduce difference and track desired trajectory. Simulation shows the scheme has higher tracking speed, and robustness is granted.

2 Support Vector Machine

The basic idea of SVM is to map $x \in R^n$ into a high dimensional space and construct an optimal hyper plane in that space. Different mapping $x \to \Phi(x) \in H$ construct different SVM.

The decision function is:

$$f(x) = \omega \cdot \Phi(x) + b = \sum \alpha_i^0 y_i K\langle x_i, x\rangle - b \tag{1}$$

where $y_i \in \{-1,+1\}$, α_i^0 are non-negative parameters learned from the training data and $K\langle \cdot,\cdot \rangle$ is a kernel function which defines an inner product in H.

The optimal hyper plane is the maximal-margin one [1], which maximizes the margin between the two classes of points (as measured in H space). This reduces to maximizing the following quadratic optimization problem:

$$\begin{gathered} W(\alpha) = \sum_{i=1}^{l} \alpha_i - \frac{1}{2}\sum_{i,j=1}^{l} y_i y_j \alpha_i \alpha_j K\langle x_i, x_j\rangle \\ \text{subject to } \sum_{i=1}^{l} y_i \alpha_i = 0 \\ \alpha_i \geq 0 \quad i = 1,2,\ldots,l \end{gathered} \tag{2}$$

In the feature selection problem, for a fixed $\sigma \in \{0,1\}$, the $W(\alpha)$ of equation (4) becomes:

$$W(\alpha,\sigma) = \sum_{i=1}^{l} \alpha_i - \frac{1}{2}\sum_{i,j=1}^{l} y_i y_j \alpha_i \alpha_j K_\sigma\langle x_i, x_j\rangle \tag{3}$$

where

$$K_\sigma\langle x, y\rangle = K\langle (x*\sigma), \langle y*\sigma\rangle\rangle$$

3 Control Scheme Based on FFSVM

Consider the following discrete black-box nonlinear system:

$$y(k+1)=\begin{bmatrix} y_1(k+1) \\ y_2(k+1) \\ \dots \\ y_n(k+1) \end{bmatrix}=\begin{bmatrix} f_1\big(y(k),x(k),\dots,x(k-m)\big) \\ f_2\big(y(k),x(k),\dots,x(k-m)\big) \\ \dots \\ f_n\big(y(k),x(k),\dots,x(k-m)\big) \end{bmatrix}+\begin{bmatrix} \delta_1(k+1) \\ \delta_2(k+1) \\ \dots \\ \delta_n(k+1) \end{bmatrix}$$
$$=f\big(y(k),x(k),\dots,x(k-m)\big)+\delta(k+1) \tag{4}$$

Where $x\in R$ denote input and $y\in R^n$ is output, $m,n\in R$, $f_i \quad (i=1,2,\dots,n)$ denotes nonlinear functions, and $\delta\in R^n$ denotes external disturbance. The feed-forward control system is shown in Fig.1.

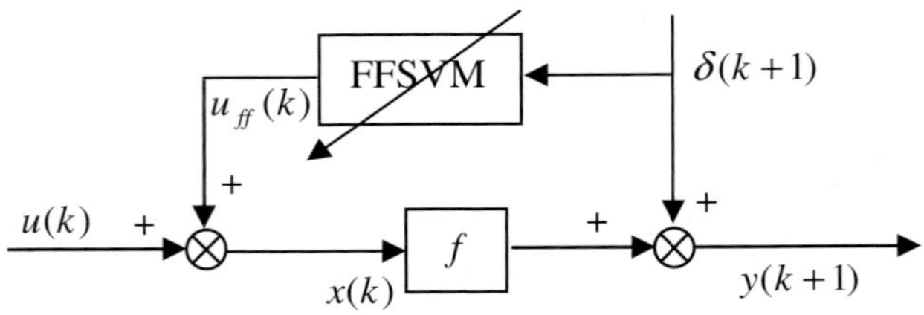

Fig. 1. Feed-forward control system

Function of $u_{ff}(k)$ is to eliminate the effect of $\delta(k+1)$, that is:

$$-\delta(k+1)=f\big[\delta(k),\dots,\delta(k-n),u_{ff}(k),\dots,u_{ff}(k-m)\big] \tag{5}$$

If inverse model of (4) exists [6], there must exist a mapping g that can satisfy:

$$u_{ff}(k)=g\big[\delta(k+1),\dots,\delta(k-n),\dots,u_{ff}(k-1),\dots,u_{ff}(k-m)\big] \tag{6}$$

Training (6) by SVM, $u_{ff}(k)$ can be obtained direct from the trained SVM. By submitting (6) into (4), the feed-forward control system can be expressed as:

$$\begin{aligned} y(k+1)&=f\big[y(k),\dots,y(k-n),u(k)+u_{ff}(k),\dots \\ &\quad ,u(k-m)+u_{ff}(k-m)\big]+\delta(k+1) \\ &=f_1\big[y(k),\dots,y(k-n),u(k),\dots,u(k-m), \\ &\quad \delta(k+1),\dots,\delta(k-n-m)\big]+\delta(k+1) \end{aligned} \tag{7}$$

where f_1 denotes a nonlinear function different with f.

4 Scheme Based on FBSVM

In order to eliminate the difference and track desired trajectory, a feedback based on FBSVM is introduced to the feed-forward control system. The control system based on FBSVM is shown in Fig2.

Let desired trajectory is $d \in R$, and

$$e(k) = d(k) - y(k) \tag{8}$$

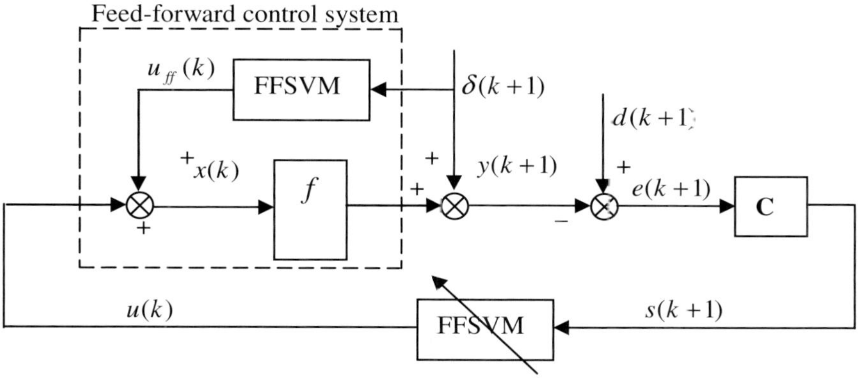

Fig. 2. Control system based on FFSVM and FBSVM

Define

$$s(k) = ce(k) \tag{9}$$

where $c = [c_1 \quad c_2 \quad \dots \quad c_n] \in R^n$ is parameter of switching surface. Then:

$$\begin{aligned} s(k+1) &= ce(k+1) \\ &= cd(k+1) - cf_1[y(k),\dots, y(k-n), u(k),\dots, u(k-m), \\ &\quad \delta(k+1),\dots\delta(k-n-m)] - c\delta(k) \\ &= f_2[e(k),\dots, e(k-n), u(k),\dots, u(k-m), \\ &\quad \delta(k+1),\dots, \delta(k-n-m), d(k+1),\dots, d(k-n)] \end{aligned} \tag{10}$$

where f_2 is a nonlinear function.

Consider $u(k)$ as solution of (10), it can be expressed as:

$$\begin{aligned} u(k) = h[&s(k+1),\dots, e(k-n), u(k-1),\dots, u(k-m), \\ &\delta(k+1),\dots, \delta(k-n-m), d(k+1),\dots, d(k-n)] \end{aligned} \tag{11}$$

where h denotes an unknown nonlinear mapping.

5 VSC Algorithm for Nonlinear System Based on FFSVM and FBSVM

After both of SVMs are trained, they can be introduced to the control system. Because disturbance $\delta(k+1)$ and $s(k+1)$ are not known in the k th sample time, they should be estimated. Let $\hat{\delta}(k+1)$ and $\hat{s}(k+1)$ denotes estimate of $\delta(k+1)$ and $s(k+1)$ separately, then (6) and (11) become:

$$u_{ff}(k) = g\left[\hat{\delta}(k+1), \delta(k), \ldots, \delta(k-n), \ldots, u_{ff}(k-1), \ldots, u_{ff}(k-m)\right] \tag{12}$$

and

$$\begin{aligned} u(k) = h[&\hat{s}(k+1), e(k), \ldots, e(k-n), u(k), \ldots, u(k-m), \\ &\hat{\delta}(k+1), \delta(k), \ldots, \delta(k-n-m), d(k+1), \ldots, d(k-n)] \end{aligned} \tag{13}$$

Suppose dynamic character of disturbance is much slower than sampling frequency, then $\delta(k+1)$ can be considered equal to $\delta(k)$, that is, $\hat{\delta}(k+1) = \delta(k)$. Then (12) becomes:

$$\begin{cases} u_{ff}(k) = g\left[\hat{\delta}(k+1), \delta(k) \ldots, \delta(k-n), \ldots, u_{ff}(k-1), \ldots, u_{ff}(k-m)\right] \\ \hat{\delta}(k+1) = \delta(k) \end{cases} \tag{14}$$

Suppose $\hat{s}(k+1)$ can satisfy the reaching law: $\hat{s}(k+1) = ps(k) - q\,\mathrm{sgn}\, s(k)$, where $|p| < 1$ and $q > 0$. According to (13), $u(k)$ can be expressed as:

$$\begin{cases} u(k) = h[\hat{s}(k+1), e(k), \ldots, e(k-n), u(k), \ldots, u(k-m), \\ \qquad\qquad \hat{\delta}(k+1), \delta(k), \ldots, \delta(k-n-m), d(k+1), \ldots, d(k-n)] \\ \hat{s}(k+1) = ps(k) - q\,\mathrm{sgn}\, s(k) \\ \hat{\delta}(k+1) = \delta(k) \end{cases} \tag{15}$$

6 Simulation

Consider a Van der Pol circuits system with an additive disturbance $\eta = 10\sin(5\pi t) + 5\sin(2\sqrt{t+1})$ [9]. By states transformation, the system dynamics becomes:

$$\begin{cases} \dot{x}_1 = x_2 \\ \dot{x}_2 = -2x_1 + \mu(1 - x_1^2)x_2 + u + \eta \end{cases}$$

where $x = [x_1 \quad x_2]^T$ is output, $\mu = 3$, $x_1(0) = 1.5$, $x_2(0) = 0$.

Its discrete-time form is:

$$\begin{cases} x_1(k+1) = x_1(k) + Tx_2(k) \\ x_2(k+1) = -2Tx_1(k) + \left[1 + T\mu - T\mu x_1^2(k)\right]x_2(k) + Tu(k) + T\eta(k+1) \end{cases}$$

The system is to track a model of oscillator---another Van der Pol equation (discrete form):

$$\begin{cases} x_{1d}(k+1) = x_{1d}(k) + Tx_{2d}(k) \\ x_{2d}(k+1) = -2Tx_{1d}(k) + \left[1 + T\mu_d - T\mu_d x_1^2(k)\right]x_2(k) \end{cases}$$

where $\mu_d = 1$, $x_{1d}(0) = 2$, $x_{2d}(0) = 1$. The parameter of switching surface is chosen as: $c = [1 \quad 1]^T$. Choose Gaussian Kernel and its parameter as 1. $p = 0.3$.

It can be seen from Fig 3 that when the scheme of feed-forward and feedback SVM is adopted, tracking time is shorter than that of [9]. And states are shown in Fig4, which show the states track desired trajectory.

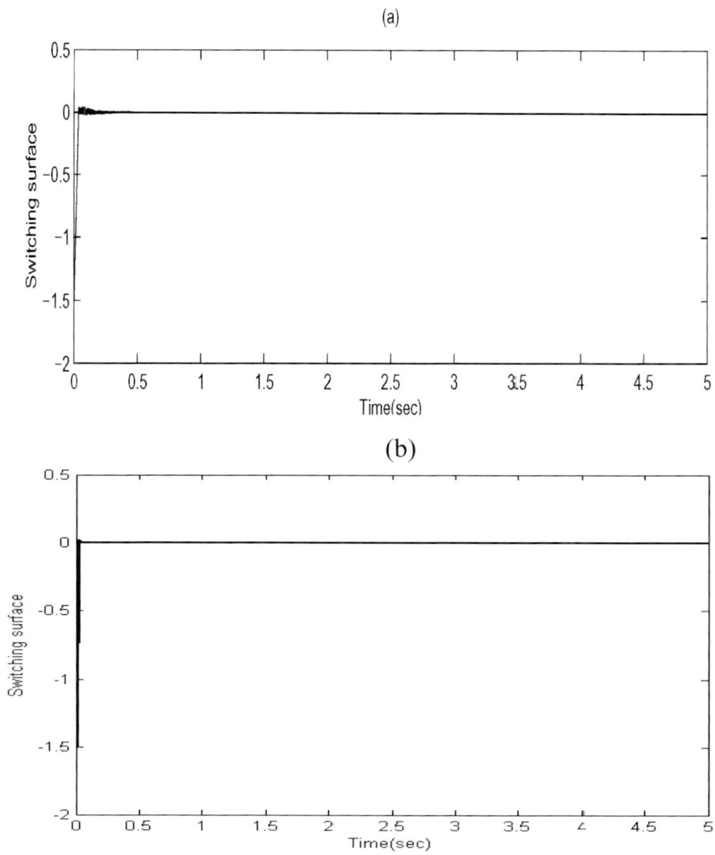

Fig. 3. (a) is the curve for scheme of [9]. (b) is the curve by adopting scheme of this paper.

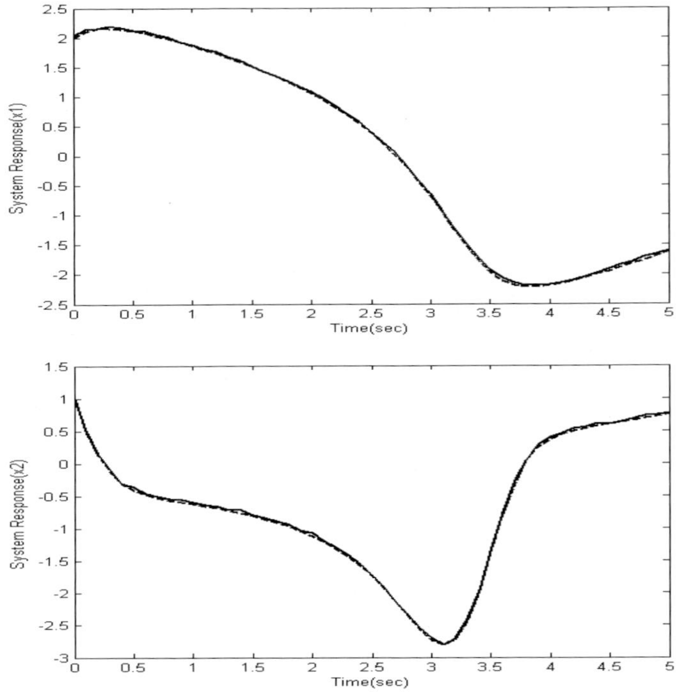

Fig. 4. System responses by adopting scheme. It can be seen that the scheme track desired trajectory very well. Tracking differences are less than 0.05.

7 Conclusion

A scheme for black-box nonlinear system based on FFSVM and FBSVM is developed. After adopting a FFSVM to compensate difference between desired trajectory and output of system, an inverse model is trained by FBSVM so that input of feedforward system can be obtained from the desired output. The effectiveness is proven in simulation.

References

1. Vapnik, V.N.: The nature of statistical learning theory. Springer, New York, USA (1995)
2. Zhong, W., Pi, D., Sun, Y.: An Approach of Nonlinear Model Multi-step-ahead Predictive Control Based on SVM [J]. In: Sunderam, V.S., van Albada, G.D., Sloot, P.M.A., Dongarra, J.J. (eds.) ICCS 2005. LNCS, vol. 3516, pp. 1036–1039. Springer, Heidelberg (2005)
3. Drezet, P.M.L., Harrison, R.F.: Support vector machines for system identification. In: [J] in UKACC International Conf. on control, UK, pp. 668–692 (1998)
4. Sanchez-Fernandez, M., Arenas-Garcia, J.: SVM Multiregression for Nonlinear Channel Estimation in Multiple-Input Multiple-Output Systems [J]. IEEE transactions on signal processing 52(8), 2298–2307 (2004)

5. Ribeiro, B.: Kernelized based functions with Minkovsky's Norm for SVM Regression [J]. IEEE Trans. on Fuzzy Systems, 2198–2203 (2002)
6. Yongji, W., Jian, T.: Neural network control [M], vol. 2. China Machine Press, Beijing (1998)
7. Shu, D.Q.: Predictive control system and its application (in Chinese). China Machine Press, Beijing (1996)
8. Xi, Y.G.: Predictive Control (in Chinese). National Defence Industry Press, Beijing (1993)
9. Xu, J.-X., Pan, Y.-J., Lee, T.-H.: Sliding Mode Control With Closed-Loop FilteringArchitecture for a Class of Nonlinear Systems. IEEE Trans. on Circus. and Systems 51(4) (2004))

IAOM: An Integrated Automatic Ontology Mapping Approach Towards Knowledge Integration

Jiangning Wu and Yonggui Wang

Institute of Systems Engineering, Dalian University of TechnologyLinggong Road 2, Dalian 116024, Liaoning, P.R. China
jnwu@dlut.edu.cn, ygwang@student.dlut.edu.cn

Abstract. Ontology mapping, with the purpose of finding the semantic correspondences between two existed overlapped ontologies, has been the key technique for solving the integration of ontology-based knowledge systems. Currently, ontology mapping is largely determined manually by domain experts, thus a time-consuming and labor-intensive process. In this paper, we propose an integrated automatic ontology mapping algorithm based on three dimensions of linguistics, structure and instance combining commonsense knowledge with domain knowledge from the systematic point of view. Experimental results on two course ontologies are presented, and show that the algorithm discovers semantic mappings with a high degree of accuracy.

1 Introduction

In the past decade, in order to better represent and share the semantics of the knowledge, ontology has been introduced and extensively applied in the domain of biology, medicine, artificial intelligence and so on. However, with the diverse purposes of different organizations and in terms of different tools, languages, etc., there exist many overlapped ontologies in the same or similar domains above-mentioned. Ontology mapping, with the purpose of finding the semantic correspondences between two overlapped ontologies, has been the key technique for solving the integration of ontology-based knowledge systems, such as bioinformatics integration aiming at intelligent bioinformatics processing. Currently, ontology mapping is largely determined manually by domain experts, thus a time-consuming and labor-intensive process. Therefore, (semi-)automated ontology mapping approaches are needed.

The researches on ontology mapping mainly focus on the domain of Semantic Web. Many foreign agents or institutes have done a lot of work on it [1-6]. However, almost all the methods are developed based on some special requirements and don't take the common knowledge into consideration. In our last paper [7], we proposed a framework of knowledge integration based on ontology mapping and presented a mapping algorithm for the lightweight ontology. In this paper, from the systematic point of view, we present an integrated automatic ontology mapping algorithm (IAOM) towards complex ontology based on three dimensions of linguistics, structure and instance which combines the commonsense knowledge with domain knowledge.

K. Li et al. (Eds.): LSMS 2007, LNCS 4688, pp. 502 – 509, 2007.

The reminder of the paper starts with a discussion of some related work in Section 2. Section 3 describes details of IAOM algorithm. Section 4 carries out an experiment for the proposed ontology mapping algorithm. At last, Section 5 arrives at conclusions and suggests some possible future works.

2 Related Works

Lots of prototypes related to ontology mapping have been developed. The state of the art of ontology mapping has been reviewed and critiqued in [8-9]. They scrutinized the literature and critically reviewed the works originating from a variety of fields. Specially, five related to ontology mapping prototypes including GLUE [2], MAFRA [4], IF-Map [3], SF [5] and OMEN [6] are reviewed here.

The GLUE system [2] developed by university of Washington introduces machine learning technique into the process of ontology mapping. Given any two concepts, the similarity between them is calculated by their joint probability distribution, taking place of the other approaches that use single or several fixed methods. Besides, multiple strategies are introduced and each of them exploits a different type of information either in the data instances or in the taxonomic structures of ontologies. In the end, they incorporate commonsense knowledge and domain constraints to further improve matching accuracy. But this approach only focuses on taking instances information into account and would be invalid without enough instances. MAFRA [4] is an interactive, incremental and dynamic ontology mapping framework for distributed ontologies in the Semantic Web. The framework is composed of five horizontal modules and four vertical modules. It exploits a multi-strategy process that the similarities between ontology entities are calculated using lexical similarity, property similarity, bottom-up similarity and top-down similarity. Specially, it introduces the concept of “Semantic Bridge” to semantically relate entities from the source and target ontologies. IF-Map [3] is based on the theory of information flow. Several local ontologies, global ontologies and reference ontologies are introduced into its architecture. The global ontology is logically existed. Its focus is on the use of information-flow theory as the underpinning mathematical foundation for establishing mappings between ontologies. SF [5] is based on the idea that two concepts are similar if the neighbors of them are similar. Firstly, it computes the linguistic similarity of the name of the instances as the initial similarity value. Then it recalculates the new similarity according to their neighbors' initial similarity values. The iteration will be terminated until the difference of two similarity values in two sequential iterations is less than a threshold. The tool OMEN [6], an ontology mapping enhancer, uses a set of meta-rules to capture the influence of the ontology structure and the semantics of ontology relations and matches nodes that are neighbors of already matched nodes in the two ontologies. Furthermore, it uses existing (semi-) tools to come up with the initial probability distributions for mappings, and then based on them to infer probability distributions for other mappings.

3 IAOM: An Integrated Automatic Ontology Mapping Algorithm

With the purpose of providing solutions for ontology mapping in our knowledge integration framework [7], we propose an IAOM (Integrated Automatic Ontology

Mapping) algorithm based on three dimensions of linguistics, structure and instance which combine commonsense knowledge with domain knowledge from the systematic point of view. The dimension of linguistics shows the semantics of commonsense knowledge, and the other two dimensions embody the semantics of the domain knowledge.

3.1 Linguistic Similarity Based on WordNet

Traditionally, the linguistic similarity is based on the edit distance [10-11] and basic distance [12]. The two ideas are derived from the angle of the syntax of the concept and statistic means respectively. Here we present a linguistic similarity approach based on WordNet [13]. For given two concepts, it takes both the shortest path between them and the depth of the common hypernym concepts of them into consideration. The detailed algorithm is shown as follows.

Algorithm 1: Linguistic similarity between entities of source ontology *A* and target ontology *B*

Input: *Entity A*—E_A in source ontology *A* and *entity B*—E_B in target ontology *B*

Output: $Sim_L(E_A, E_B)$—the linguistic similarity between E_A and E_B based on WordNet

Begin

Step 1: Initialize the linguistic similarity between E_A and E_B—$Sim_L(E_A, E_B)=0$.

//Assume *term A*—T_A in *A* and *term B*—T_B in *B*

Get the *synset* of T_A and T_B respectively.

Step 2: Similarity between T_A and T_B based on the semantic distance in WordNet—$Sim_1(T_A, T_B)$.

i) If (*synset A*==*synset B*) $Sim_1(T_A, T_B)=1$;

Else get all paths between synset *A* and synset *B*.

ii) Compute the lengths of all paths and get the shortest path $l(T_A, T_B)$.

iii) Calculate $Sim_1(T_A, T_B)$:

$$Sim_1(T_A, T_B) = e^{-\alpha l(T_A, T_B)} \tag{1}$$

Step 3: Similarity between T_A and T_B based on the depth of term in WordNet— $Sim_2(T_A, T_B)$.

i) If (*synset A*==*synset B*) $Sim_2(T_A, T_B)=1$;

Else get the hypernym term sets of synset *A* and synset *B* respectively—$h(A)$, $h(B)$.

ii) If ($h(A) \cap h(B)$==NULL) $Sim_2(T_A, T_B)=0$;

Else calculate the depths of all terms in set of $h(A) \cap h(B)$.

iii) Get the biggest depth *h* and compute $Sim_2(T_A, T_B)$:

$$Sim_2(T_A, T_B) = \frac{e^{\beta h} - e^{-\beta h}}{e^{\beta h} + e^{-\beta h}} \tag{2}$$

Step 4: Term similarity combination (where $0 < \omega < 1$).

$$Sim(T_A, T_B) = \omega \cdot Sim_1(T_A, T_B) + (1-\omega) \cdot Sim_2(T_A, T_B) \tag{3}$$

Step 5: The linguistic similarity between E_A and E_B— $Sim_L(E_A, E_B)$.
//Assume that E_A and E_B include m and n terms respectively
For (int i=1; i<=m; i++)
For (int j=1; j<=n; j++)

Calculate Sim_j (T_i, T_j) by Equation (3) and

$$Sim_i(T_i,T_j) = \max_{1\le j\le n} Sim_j(T_i,T_j) \tag{4}$$

Then the linguistic similarity between E_A and E_B can be computed as follows:

$$Sim_L(E_A,E_B) = \frac{1}{m}\sum_{i=1}^{m} Sim_i(T_A,T_B) \tag{5}$$

Return $Sim_L(E_A, E_B)$;
End;

3.2 Structural Similarity Based on Linguistic Similarity

The semantics of the concept entity in ontology is more embodied by its structure, that is, by the context of the concept entity. Many existing such principles have been proposed. For example, two concepts entities are similar if their super classes are uniform. Our structural similarity approach is just based on such principles. We contend that the structure of the concept entity should be classified as two categories. One is its internal structure, which is characterized by its properties. Another is its external structure, which is expressed by all concepts entities directly relevant to it. Below is the corresponding algorithm, where both the similarities among the concept entities' properties and the similarities among the concept entities' directly relevant concept entities are calculated by the linguistic similarity algorithm mentioned above.

Algorithm 2: Structural similarity between entities of source ontology *A* and target ontology *B*
Input: *Entity A*—E_A in source ontology *A* and *entity B*—E_B in target ontology *B*
Output: $Sim_S(E_A, E_B)$—the structural similarity between E_A and E_B
Begin
Step 1: Initialize the structural similarity between E_A and E_B— $Sim_S(E_A, E_B)$ =0;
Set the internal similarity and external similarity as 0—$Sim_{in}(E_A, E_B)$ =0, $Sim_{out}(E_A, E_B)$ =0.
Step 2: Get all properties and directly relevant entities of E_A and E_B.
Step 3: //Internal similarity calculations, assuming that E_A and E_B include s, t
//properties respectively.
For (int i=1; i<=s; i++)
For (int j=1; j<=t; j++)

$$Sim_{in}(E_A,E_B) = \frac{1}{st}\sum_{i=1}^{s}\sum_{j=1}^{t} Sim_L(P_i,F_j) \tag{6}$$

Step 4: //External similarity calculations, assuming that E_A and E_B include u, v
//directly relevant entities
For (int i=1; i<=u; i++)
For (int j=1; j<=v; j++)

$$Sim_{out}(E_A, E_B) = \frac{1}{uv}\sum_{i=1}^{u}\sum_{j=1}^{v} Sim_L(E_{Ai}, E_{Bj}) \tag{7}$$

Step 5: Then the structural similarity between E_A and E_B can be computed as follows

$$Sim_S(E_A, E_B) = \omega' Sim_{in}(E_A, E_B) + (1-\omega')Sim_{out}(E_A, E_B) \tag{8}$$

Where $0 < \omega' < 1$.

End;

3.3 Instance Similarity Derived from GLUE System

Besides the structure of the concept entity, instance is another important component for embodying its semantics. The corresponding principle is that two concept entities are similar if their instances are uniform. Here we adopt the famous instance-based similarity approach of GLUE system [2]. It uses Jaccard coefficient which corresponds well to our intuitive understanding of similarity as the similarity measure approach.

$$Sim_I(E_A, E_B) = \frac{P(E_A \cap E_B)}{P(E_A \cup E_B)} = \frac{P(E_A, E_B)}{P(E_A, E_B) + P(E_A, \bar{E}_B) + P(\bar{E}_A, E_B)} \tag{9}$$

Where P (E_A, E_B) is the joint probability distribution of the instances of E_A and E_B, and the other probability distributions can be acquired by P (E_A, E_B).

Let U_i be the instances set of O_i, $N(U_i)$ be the size of U_i and $N(U_i^{EA, EB})$ be the number of instances in U_i belonging to both E_A and E_B (i=1, 2). Then $P(E_A, E_B)$ can be computed:

$$P(E_A, E_B) = \frac{N(U_1^{E_A, E_B}) + N(U_2^{E_A, E_B})}{N(U_1) + N(U_2)} \tag{10}$$

Where $N(U_i^{EA, EB})$ (i=1, 2) is acquired by machine learning approach and the detailed information is described in [2].

3.4 IAOM: An Integrated Automatic Ontology Mapping Algorithm

From Sections 3.1 to 3.3, we give the detailed algorithms that are linguistics-based, structure-based and instance-based respectively. Now we combine these three dimensions by means of weight approach and Sigmoid function approach. Below are the explicit expressions.

$$Sim\,(E_A, E_B) = w_1 * Sim_L(E_A, E_B) + w_2 * Sim_S(E_A, E_B) + w_3 * Sim_I(E_A, E_B) \tag{11}$$

$$Sim\,(E_A, E_B) = Sig_1(Sim_L(E_A, E_B) - 0.5) + Sig_2(Sim_S(E_A, E_B) - 0.5) + Sig_3(Sim_I(E_A, E_B) - 0.5) \tag{12}$$

Where $Sim_L(E_A, E_B)$, $Sim_S(E_A, E_B)$, $Sim_I(E_A, E_B)$ can be gotten from Sections 3.1 to 3.3. According to the actual circumstances, the two approaches are selected properly.

4 Experimental Evaluations

We carry out our algorithm on two course ontologies [14]. The two course ontologies have approximately 340 concept entities and 12,000 instance entities, and their max level is 4. Their basic architectures are shown by Figure 1. Besides, in order to evaluate our approach and compare with the others, we choose three indexes: precision, recall and F-measure.

$$precision = \frac{N(accurate\_found\_mapping)}{N(found\_mapping)} \times 100\% \tag{13}$$

$$recall = \frac{N(accurate\_found\_mapping)}{N(actual\_mapping)} \times 100\% \tag{14}$$

$$F-measure = \frac{2precision * recall}{precisioin + recall} \tag{15}$$

The experimental results are shown in Figure 2, and IAOM (W), IAOM (S) represent the similarities' combining approach of weight and sigmoid function respectively. From experiment, we know that the OWSW [7] just uses linguistic similarity and only hypernym-hyponym structure similarity and gets an F-measure value about 59%. The GLUE system which is just based on instance-based similarity gains an F-measure value about 63%. Whereas our integrated automatic algorithm IAOM, which combines three kinds of similarities: linguistics-based similarity, structure-based similarity, and instance-based similarity, acquires a high F-measure value increased to about 74%. This proves that the IAOM algorithm can highly improve the mapping accuracy. Besides, considering that the linguistic similarity based on WordNet is involved, IAOM is also valid without enough instance information.

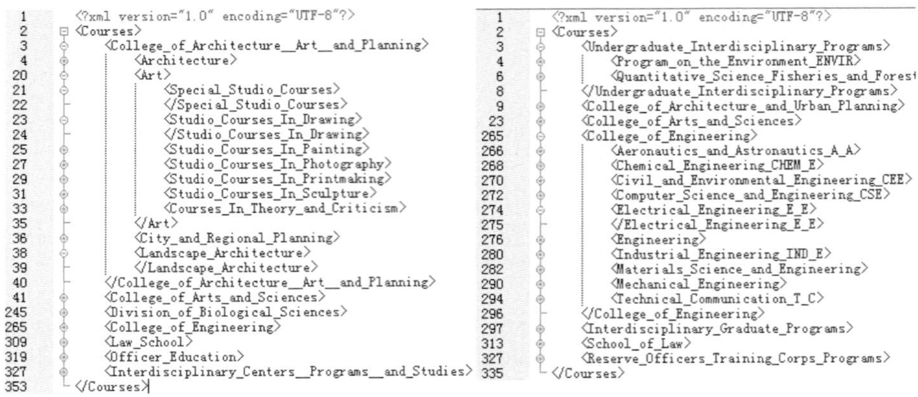

Fig. 1. The basic architecture of the two course ontologies

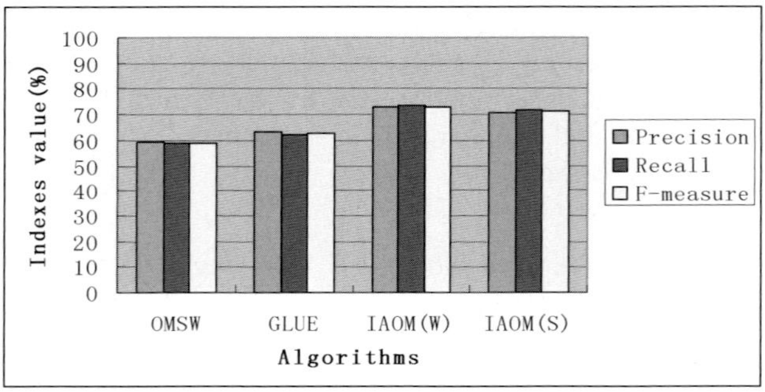

Fig. 2. Comparisons of several algorithms5 Conclusion and future work

5 Conclusions and Future Work

In this paper, we propose an integrated automatic ontology mapping algorithm for the problem of ontology mapping, especially for domain knowledge integration. Our algorithm is based on three dimensions of linguistics, structure and instance combining commonsense knowledge and domain knowledge from the systematic point of view. The results show that our approach works well with a high accuracy. Besides, it's also effective without enough instances involved.

In the future, our work mainly focuses on twofold. On one hand, although the accuracy of our algorithm is high, its efficiency isn't satisfying. So we should find a balance point between the efficiency and the accuracy by some optimization means. On the other hand, we plan to embed our algorithm into the knowledge integration system and provide the online ontology mapping service by setting some constraints.

Acknowledgements. This work is sponsored by the National Natural Science Foundation of China (NSFC) under Grants 70431001 and 70620140115.

References

1. Kent, R.: The Information Flow Foundation for Conceptual Knowledge Organization. In: Dynamism and Stability in Knowledge Organization. Proceedings of the Sixth International ISKO Conference. Advances in Knowledge Organization, pp. 111–117 (2000)
2. Doan, A., Madhavan, J., Dongmingos, P., Halevy, A. Learning to Map between Ontologies on the Semantic Web. In: Proceedings of the Eleventh International World Wide Web Conference, Honolulu, Hawaii, USA, pp. 662–673 (2002)
3. Kalfoglou, Y., Schorlemmer, M.: IF-Map: An Ontology Mapping Method Based on Information Flow Theory. Int. J. on Data Semantics 1, 98–127 (2003)
4. Macedche, A., Motik, B.: MAFRA-A Mapping Framework for Distributed Ontologies. Web Intelligence and Agent System 1, 235–248 (2003)
5. Melnik, S., Garcia-Molina, H., Rahm, E.: Similarity Flooding: A Versatile Graph Matching Algorithm. In: Proc. of the 18th International Conference on Data Engineering (ICDE), San Jose,CA, pp. 117–128 (2002)
6. Mitra, P., Noy, N.F., Jaiswal, A.R.: OMEN: A Probabilistic Ontology Mapping Tool. In: McIlraith, S.A., Plexousakis, D., van Harmelen, F. (eds.) ISWC 2004. LNCS, vol. 3298, Springer, Heidelberg (2004)
7. Wang, Y.G., Wu, J.N.: Knowledge Integration through Ontology Mapping Based on the Thesaurus and Multi-similarity Measure. In: Proceedings of the 3rd Asia-Pacific International Conference on Knowledge Management, Hong Kong, China (2006)
8. Kalfoglou, Y., Schorlemmer, M.: Ontology mapping: the State of the Art. The Knowledge Engineering Review 18, 1–31 (2003)
9. Euzenat, J., et al.: State of the Art on Ontology Alignment. Knowledge Web Deliverable 2.2.3, University of Karlsruhe (2004)
10. Diogene Ontology Mapping Prototype: http://diogene.cis.strath.ac.uk/prototype.html
11. Maedche, A., Staab, S.: Measuring Similarity between Ontologies. In: Gómez-Pérez, A., Benjamins, V.R. (eds.) EKAW 2002. LNCS (LNAI), vol. 2473, pp. 251–263. Springer, Heidelberg (2002)
12. Sekine, S., Sudo, K., Ogino, T.: Statistical Matching of Two Ontologies. In: The Proceedings of the SIGLEX99: Standerdizing Lexical Resources, Maryland, USA, pp. 69–73 (1999)
13. Wordnet: http://www.cogsci.princeton.edu/ wn
14. Illinois Semantic Integration Archive: http://anhai.cs.uiuc.edu/archive

Symbolic Model Checking Temporal Logics of Knowledge in Multi-Agent System Via Extended Mu-Calculus*

Lijun Wu and Jinshu Su

School of Computer Science, National University of Defense and Technology, 410073 Changsha , China
gzzsudocwlj@yahoo.com

Abstract. Clarke and McMillan presented symbolic approaches to model check temporal logics via mu-calculus and OBDDs. These approaches are very efficient and can be applied to verify many practical systems with extremely large state spaces in excess of 1020 states. However, these approaches cannot model check knowledge logics. But temporal logics of knowledge can describe more accurately the desirable specification of systems and protocols in distributed systems. In this paper, the symbolic approaches to model check the temporal logic of knowledge via extended mu-calculus and OBDDs are discussed mainly. First the mu-calculus is extended. Then the symbolic approaches to model check temporal logics of knowledge via extended mu-calculus and OBDDs are presented.

1 Introduction

Model checking has been used mainly to check if a system satisfies the specifications expressed in temporal logic. People pay little attention to the problem of model checking logics of knowledge. However, in the distributed systems, the desirable specifications of systems and protocols have been expressed widely in the temporal logics of knowledge. Temporal logics of knowledge can express more accurately the security properties of systems and protocols. Therefore, model checking temporal logics of knowledge is a new important research domain.

Ron van der Meyden and N.V.Shilov discussed the problem of model checking knowledge and time in systems with perfect recall [1,2]. Wiebe van der Hoek and Woolridge presented the algorithm based on the logic of local propositions, which can model check the temporal logics of knowledge. But it was limited in the logic combining LTL and knowledge components and the theoretical foundations of their

* Supported by the National Natural Science Foundation of China under Grant No.90604006 and No.60496327.

K. Li et al. (Eds.): LSMS 2007, LNCS 4688, pp. 510 – 522, 2007.

work still needed to be further investigated [3]. F.Raimondi, A.Lomuscio and Kaile Su et al have done much work in the temporal logic of knowledge [4,5,13,14,15,16]. Lijun Wu, et al discussed and researched approach to model check the temporal logics of knowledge based on SMV, but did not solve commendably the problem of system with large state space[6].

Clarke and McMillan, et al presented symbolic approaches to model check temporal logics via mu-calculus and OBDDs (Ordered Binary Decision Diagrams)[9,10]. These approaches are very efficient and can be applied to verify many practical systems with extremely large state spaces in excess of 10^{20} states. However, they cannot model check epistemic logics.

In the paper, according to the semantics of knowledge, we extend the mu-calculus of Kozen[7] and OBDDs, and exploit them to verification of temporal logics of knowledge. These approaches can handle security verification of systems with extremely large state spaces. As a case study, we verify the security properties of SSL 3.0 protocol.

2 The Temporal Logics of Knowledge and Extended Mu-Calculus

2.1 The Temporal Logics of Knowledge

1. Syntax of the temporal logics of knowledge

In the paper, we research the model checking of the logic CKKn that combines CTL and epistemic logic[6] via extended mu-calculus. The formulae of *CKKn* are composed of proposition operators and path quantifiers and temporal operators and knowledge operators. There are five proposition operators $\neg$（not）、$\vee$(or)、$\wedge$(and)、$\rightarrow$(implies)、$\leftrightarrow$(if and only if). The path quantifiers are used to describe the branching structure in the CTL. There are two such quantifiers A (for all computation paths) and E (for some computation path). The temporal operators describe properties of a path through the tree. There are five basic operators: X (next time), F (eventually or in the future), G (always or globally), U (until) and R (release). There are two epistemic operators K (know) and C_G (common knowledge of G, G is the set of agents). There are ten basic CTL operators:

AX and EX,

AF and EF,

AG and EG,

AU and EU,

AR and ER.

Each of the ten operators can be expressed in terms of three operators EX, EG and EU[9]. So the set *CKKnf* of formulae of *CKKn* can be defined as follows:

<*CKKnf*>::=true

| *p* /**p* is atomic proposition */

| ¬<*CKKnf*>

| <*CKKnf*>∨<*CKKnf*>

| EX<*CKKnf*>

| EG<*CKKnf*>

| E[<*CKKnf*>U<*CKKnf*>]

| K_i<*CKKnf*>

| C_G<*CKKnf*>

2. Semantics of the temporal logics of knowledge

We assume that Kripke structure $M=(S,T,L)$, here, S is a set of states, T is a set of transition relations such that for each transition a, $a \subseteq S \times S$, and L is a map $L{:}S\rightarrow 2^{AP}$(AP is an atomic proposition set), each state corresponds to the set of the atomic propositions that are true at the state. Suppose that there are n agents. Any state s is expressed as $s=(s_1, s_2, \ldots, s_n)$ which s is a global state of system and s_i is a local state of agent i. A trace is a finite sequence of states $u_1u_2\ldots u_m$ such that $(u_i,u_{i+1})\in a_i$, $a_i\in T$, $u_i\in S$ for all $1\leqslant i\leqslant m-1$. A run is an infinite sequence of states $r{:}N\rightarrow S$ such that every finite prefix of r is a trace [3]. Let $s=(s_1,s_2,\ldots,s_n)$ and $t=(t_1,t_2,\ldots,t_n)$ be two global states in S, then we say that s and t are indistinguishable to agent i, and write $s\sim_i t$, if $s_i=t_i$. Let $R_i=\{(s,t)|s\in S,t\in S,s\sim_i t\}(i=1,2,\ldots,n)$, then R_i is an equivalence relation of M concerned with agent i. So Kripke structure M can be expanded to $(S,T,L,R_1,\ldots,R_n)$. For convenience, we still write the expanded structure as M. According to semantics of knowledge and common knowledge in [11], we can define semantics of knowledge and common knowledge in M as follows:

$(M,s)|=K_i\varphi$ iff $(M,t)|=\varphi$ for all t such that $(s,t)\in R_i$.

$(M,s)|= E_G\varphi$ iff $(M,s)|= K_i\varphi$, for k=1,2,….

$(M,s)|= C_G\varphi$ iff $(M,s)|=E_G^{\,k}\varphi$, for k=1,2,….

Where $E_G^{\,0}\varphi$ is an abbreviation for φ, $E_G^{\,1}\varphi$ is an abbreviation for $E_G\varphi$, and $E_G^{\,k+1}\varphi$ is an abbreviation for $E_GE_G^{\,k}\varphi$ (that is, every agent in G knows $E_G^{\,k}\varphi$). The semantics of other operators are similar to that of [9].

Note that T is a set of transition relations concerned with system and R_i is an equivalence relations concerned with agent i , hence T is not different from R_i .

2.2 The Extended Mu-Calculus

In this paper we extend slightly the mu-calculus of Kozen[7] and exploit it to verification of temporal logics of knowledge. Let the Kripke structure of system be M= $(S,T,L,R_1,\ldots,R_n)$ and VAR=$\{Q_1,Q_2,\ldots,\}$ be a set of relation variables. Each relation variable $Q\in$ VAR can be assigned a subset of S. The extended mu-calculus formulas are constructed as follows:

If $p \in$ AP, then p is a formula.
If f is a relation variable, then f is a formula.
If f and g are formulae, then $\neg f$, $f\wedge g$, $f\vee g$ are formulae.
If f is a formula, and $a\in T$, then $[a]f$ and $\langle a\rangle f$ are formulae.
If f is a formula, then $[R_i]f$ is a formula (i=1,2,…,n).
If $Q\in$ VAR and f is a formula, then $\mu Q.f$ and $\nu Q.f$ are formulae, provided that f is syntactically monotone in Q, that is, all occurrences of Q within f fall under an even number of negations in f.

The intuitive meaning of the formula $\langle a\rangle f$ is " it is possible to make the transition a to a state where f holds." Similarly, $[a]f$ means that "f holds in all states reachable (in one step) by making the transition a " and $[R_i]f$ means that "f holds in all states by relation R_i ". μ and ν operators are used to express the least and great fixpoints, respectively. The empty set of states is denoted by False, and the set of all sates is denoted by True.

Formally, a formula f is interpreted as a set of states in which f is true. We write such set of states as $[f]_M e$, where M is a transition system and e: VAR$\rightarrow 2^s$ is an environment. We denote by $e[Q\leftarrow W]$ a new environment that is the same as except that $e[Q\leftarrow W](Q)=W$. Hence, the set $[f]_M e$ can be defined recursively as follows:

$[\neg f]_M e=S\backslash[f]_M e$.
$[f\wedge g]_M e=[f]_M e\cap[g]_M e$.
$[f\vee g]_M e=[f]_M e\cup[g]_M e$.
$[\langle a\rangle f]_M e=\{s|\ \exists t[(s,t)\in a \text{ and } t\in[f]_M e]\}$.
$[[a]f]_M e=\{s|\ \forall t[(s,t)\in a \text{ implies } t\in[f]_M e]\}$.
$[[R_i]f]_M e=\{s|\ \forall t[(s,t)\in R_i \text{ implies } t\in[f]_M e]\}$.
$[\mu Q.f]_M e$ is the least fixpoint of the predicate transformer $\tau:2^S\rightarrow 2^S$ defined by $\tau(w)=[f]_M e[Q\leftarrow W]$.
$[\nu Q.f]_M e$ is the greatest fixpoint of the predicate transformer $\tau:2^S\rightarrow 2^S$ defined by $\tau(w)=[f]_M e[Q\leftarrow W]$.

$[\nu Q.f]_M e$ is the greatest fixpoint of the predicate transformer $\tau:2^S \to 2^S$ defined by $\tau(w)=[f]_M e[Q \leftarrow W]$.

3 Symbolic Model Checking Temporal Logics of Knowledge Via Extended Mu-Calculus

We will complete the model checking in two steps. First, we show how to translate CKKn into extended mu-calculus. Then we show how to model check temporal logics of knowledge via extended mu-calculus and OBDDs.

3.1 Translating CKKn into Extended Mu-Calculus

In this section, we mainly discuss how to translate CKKn into mu-calculus.

Each operator of ten basic CTL operators can be expressed in terms of three operators EX, EG, and EU. And EG, EU can be characterized as a least or greatest fixpoint of an appropriate predicate transformer[9]. Now we state them by Proposition 1 and Proposition 2.

Proposition 1. EGf is the greatest fixpoint of the predicate transformer $\tau(Q)=f \wedge \mathrm{EX}Q$ [9].

Proposition 2. E$(f \cup g)$ is the least fixpoint of the predicate transformer $\tau(Q)=g \vee (f \wedge \mathrm{EX}Q)$[9].

So we only need to discuss how to translate $K_i f$ and $C_G f$ into extended mu-calculus formulas.

Let $\tau(Q)=\{s | s \in S, (M\ s) |= f \wedge K_i Q\ \ i=1,2,\ldots,n\} = \{s | s \in S, (M\ s) |= f \wedge E_G Q\}$.

Proposition 3. The extended mu-calculus formula corresponding with $K_i f$ is $[R_i]f$. The proposition can derive directly from the semantics of $K_i f$ and $[R_i]f$.

Proposition 4. If $[Q_1]_M e \subseteq [Q_2]_M e$, then $[E_G Q_1]_M e \subseteq [E_G Q_2]_M e$.

Proof. Let $s \in [K_i Q_1]_M e$ ($i=1,2,\ldots,n$), then $(M, s) |= K_i Q_1$. By the semantic of $K_i Q_1$, it follows that $(M, t) |= Q_1$ for all t such that $(s, t) \in R_i$. Since $[Q_1]_M e \subseteq [Q_2]_M e$, $(M, t) |= Q_2$ for all t such that $(s, t) \in R_i$. By the semantic of knowledge, it follows that $(M, s) |= K_i Q_2$. So $s \in [K_i Q_2]_M e$. Hence, $[K_i Q_1]_M e \subseteq [K_i Q_2]_M e$, $i=1,2,\ldots,n$. Therefore, $[E_G Q_1]_M e \subseteq [E_G Q_2]_M e$. ∎

Proposition 5. The predicate transformer τ is monotonic.

Proof. Let $[Q_1]_M e \subseteq [Q_2]_M e$. By proposition 4, it follows that $[E_G Q_1]_M e \subseteq [E_G Q_2]_M e$. By $\tau(Q_1)=\{s \mid s \in S, (M, s) \models f \wedge K_i Q_1, i=1,2,\ldots,n\}=\{s \mid s \in S, (M, s) \models f \wedge E_G Q_1\}$ and $\tau(Q_2)=\{s \mid s \in S, (M, s) \models f \wedge K_i Q_2, i=1,2,\ldots,n\}=\{s \mid s \in S, (M, s) \models f \wedge E_G Q_2\}$, it follows that $\tau(Q_1) \subseteq \tau(Q_2)$. ▮

Proposition 6. If R_i (i=1,…,n, n≥1) is equivalence relation, then $[C_G f]_M e \subseteq [f]_M e$.

Proof. Let $s \in [C_G f]_M e$. Then it follows that $s \in [K_i f]_M e$ $(i=1,2,\ldots,n)$. And by the semantic of knowledge, it follows that $(M,t) \models f$ for all t such that $(s,t) \in R_i$. Since R_i (i=1,…,n) is equivalence relation, $(s,s) \in R_i$. Hence, $(M,s) = f$. therefore, $s \in [f]_M e$ and $[C f] e \subseteq [f] e$. ▮

Proposition 7. $C_G f$ is a fixpoint of the predicate transformer $\tau(Q)=\{s \mid s \in S, (M,s) \models f \wedge E_G Q\}$.

Proof. By semantic of $C_G f$, it follows that $E_G(C_G f)= C_G f$. Hence, $f \wedge E_G(C_G f)= f \wedge C_G f$ and $[f \wedge E_G(C_G f)]_M e= [f \wedge C_G f]_M e= [f]_M e \cap [C_G f]_M e=[C_G f]_M e$. That is, $[f \wedge E_G(C_G f)]_M e=[C_G f]_M e$. Therefore, $C_G f$ is a fixpoint of the predicate transformer $\tau(Q)=\{s \mid s \in S, (M,s) \models f \wedge E_G Q\}$. ▮

Because the predicate transformer τ is monotonic and S is finite, from [9], it follows that the predicate transformer τ always has a greatest fixpoint. We prove below that $C_G f$ is the greatest fixpoint of the predicate transformer τ.

Proposition 8. $C_G f$ is the greatest fixpoint of the predicate transformer $\tau(Q)=\{s \mid s \in S, (M,s) \models f \wedge E_G Q\}$.

Proof. Let Q_1 be a fix point of the predicate transformer τ, then $[Q_1]_M e=[f \wedge E_G Q_1]_M e \subseteq [f]_M e$ and $[Q_1]_M e=[f \wedge E_G Q_1]_M e \subseteq [E_G Q_1]_M e$. By Proposition 4, it follows that $[E_G Q_1]_M e \subseteq [E_G f]_M e$. The same reason, by $[E_G Q_1]_M e \subseteq [E_G f]_M e$ and Proposition 4, it follows that $[E^2_G Q_1]_M e \subseteq [E^2_G f]_M e, \ldots$. It is obvious that $[E^k_G Q_1]_M e \subseteq [E^k_G f]_M e, \ldots$. According to the semantics of $C_G Q_1$ and $C_G f$, it follows that $[C_G Q_1]_M e \subseteq [C_G f]_M e$. By proposition 4 and $[Q_1]_M e=[f \wedge E_G Q_1]_M e \subseteq [E_G Q_1]_M e$, it follows that $[Q_1]_M e \subseteq [E Q_1]_M e \subseteq [E^2_G Q_1]_M e \ldots$. According to the semantic of $C_G Q_1$, it follows that $[Q_1]_M e \subseteq [C_G Q_1]_M e$. Hence, $[Q_1]_M e \subseteq [C_G f]_M e$. Therefore, $C_G f$ is the greatest fixpoint of the predicate transformer $\tau(Q)=\{s \mid s \in S, (M,s) \models f \wedge E_G Q\}$. ▮

Now, on basis of eight propositions above, we present algorithm for translation of CKLn into extended mu-calculus as follows.

```
Function CKLtomu(f: CKLn formula):extended mu-formula;
 Case
  f is an atomic proposition p
      Return p ;
    f is of the form ¬f1
      Return ¬CKLtomu(f1) ;
    f is of the form f1∧f2
      Return CKLtomu(f1)∧CKLtomu(f2) ;
    f is of the form f1∨f2
      Return CKLtomu(f1)∨CKLtomu(f2) ;
    f is of the form EXf
      Return ⟨a⟩CKLtomu(f);
    f is of the form EGf
      Return νQ.( CKLtomu(f) ∧⟨a⟩Q);
    f is of the form E(f∪g)
      Return μQ.( CKLtomu(g)∨(CKLtomu(f)∧⟨a⟩Q));
    f is of the form Ki f (i=1,…,n)
      Return [Ri]CKLtomu(f);
    f is of the form CGf
      Return νQ.( CKLtomu(f) ∧(∧^n_{i=1}[Ri]Q));
 End case
 End function
```

Fig. 1. The function for translation of CKLn into extended mu-calculus

3.2 Model Checking Temporal Logics of Knowledge Via Extended Mu-Calculus

In this section, we will present the algorithm for model checking extended mu-calculus. First we discuss how to encode a Kripke structure of multi-agent system $M=(S,T,L,R_1,\ldots,R_n)$ into OBDDs. This encoding is similar to the encoding of the Kripke structure presented in Section 5.2 of [9]. The domain S is encoded by the set of values of the n Boolean variables $x_1,\ldots,x_n$, that is, S is now the space of Boolean vectors of length n. Each variable x_i has corresponding primed variable x_i'. Instead of writing $x_1,\ldots,x_n$, we sometimes use vector notation x. Given an interpretation we build the OBDDs corresponding to closed extended mu-calculus formulas in following manner.

1 Each atomic proposition p has an OBDD associated with it. We will denote this $OBDD_p(x)$. $OBDD_p(x)$ has the property that $y\in\{0,1\}^n$ satisfies $OBDD_p$ if and only if $y\in L(p)$.

2 The transition T has an $OBBD_T(x,x')$ associated with it. A Boolean vector $(y,z) \in \{0,1\}^{2n}$ satisfies $OBDD_T$ if and only if $(y,z) \in T$.

3 Each relation R_i (i=1,…,n) has an $OBDD_{Ri}$ (x,x') associated with it. A Boolean vector $(y,z) \in \{0,1\}^{2n}$ satisfies $OBDD_{Ri}$ if and only if $(y,z) \in R_i$.

Assume that we are given an extended mu-calculus formula f with free relation variables $Q_1,\ldots,Q_k$. The assoc[Q_i] is the OBDD corresponding to the set of states associated with the relation variable Q_i. assoc$\langle Q \leftarrow B_Q \rangle$ creates a new association by adding a relation variable Q and associating an OBDD B_Q with Q. Now we describe the algorithm for model checking extended mu-calculus formula f. The algorithm consists mainly of a function OBDDF( ), which has two arguments f (extended mu-calculus formula) and assoc (association list) and returns an OBDD corresponding to the semantics of f (the set of states that satisfy f in M). The first and second and sixth and seventh cases in the function can derive directly from the respective semantic definitions for OBDDs and mu-calculus formulas. The eighth case can derive directly from the semantic definitions of knowledge and extended mu-formula $[R_i]f$ (i=1,…,n). Sub-functions OBDD-NEGATE( ) and OBDD-AND( ) and OBDD-OR( ) were described in [8] by Bryant. The implementation of OBDD-EXIST( ) and OBDD-ANY ( ) in term of disjunction and restriction were discussed in [8,10]. Sub function fixedpoint( ) were similar to [9]. The OBDDs for the Boolean functions True and False are denoted by TrueOBDD and FlaseOBDD respectively.

The algorithm is similar to [9, 10] and is described as follows:

```
Function OBDDF(f: extended mu-formula, assoc: association list):OBDD;
 Case
  f is an atomic proposition
    Return OBDD_f ;
  f is an relation variable Q
    Return assoc[Q] ;
  f is of the form ¬f1
    Return OBDD-NEGATE(OBDDF(f1)) ;
  f is of the form f1∧f2
    Return OBDD-AND(OBDDF(f1),OBDDF(f1)) ;
  f is of the form f1∨f2
    Return OBDD-OR(OBDDF(f1),OBDDF(f2)) ;
  f is of the form ⟨a⟩f
    Return OBDD-EXIST(x', OBDD-AND (OBBD_T(x,x'),OBDDF(f,assoc)( x'));
```

Fig. 2. Function for model checking via extended mu-calculus

```
    f is of the form [a]f
      Return OBDDF(¬⟨a⟩¬f,assoc);
    f is of the form [R_i]f (i=1,…,n)
      Return OBDD-ANY(x', OBDD-AND(OBDD_Ri (x,x'),OBDDF(f,assoc)( x'));
    f is of the form μQ.f
      Return fixedpoint(f,assoc,FalseOBDD);
    f is of the form νQ. f
      Return fixedpoint(f,assoc,TrueOBDD);
  End case
  End function
  Function fixedpoint(f: extended mu-formula, assoc: association list, Obdd:OBDD)
    Resultobdd:=Obdd;
    Oldobdd:= OBDDF(f, assoc(Q← Resultobdd));
    While (Resultobdd≠ Oldobdd)
      Resultobdd:= Oldobdd;
      Oldobdd:=OBDDF(f, assoc(Q← Resultobdd));
    End While;
    Return (resultobdd);
  End function
```

Fig. 2. (*Continued*)

Notice that if f is of the form $\mu Q.f$ or $\nu Q.\ f$, then we assume that f is syntactically monotone in Q.

Now we consider the complexity of our algorithm. Assume Kripke structure of multi-agent system M= $(S,T,L,R_1,\ldots,R_n)$.

Proposition 9. The time complexity of symbolic model checking temporal logics of knowledge via extended mu-calculus is $O(n^k)$ where $n=|S|$ and k is the maximum nesting depth of fixpoint operators in the checked formula, that is, polynomial in the number of states of system when the maximum nesting depth of fixpoint operators in formula is fixed.

The proposition can derive directly from [9,10].

From the section 3 and [9,10], it is not difficult to see the algorithms can applied to verify models (or systems) with extremely large state space in excess of 10^{20} states. So the algorithms are very efficient in practice.

4 A Case Study

We use SSL 3.0(Secure Sockets Layer) handshake protocol as a case study.

4.1 SSL 3.0 Handshake Protocol

SSL 3.0 protocol includes SSL 3.0 record protocol and SSL 3.0 handshake protocol. In the section, we model check mainly SSL 3.0 handshake protocol using our approach. The protocol is as follows[12]:

$C \rightarrow S$: $C, SuiteC, N_C$
$S \rightarrow C$: $SuiteS, N_S, \mathrm{Sign}_{CA}\{S, K_S^+\}$
$C \rightarrow S$: $\{SecretC\}_{KS+}$
$C \rightarrow S$: $\{N_S\}_{\mathrm{Master}(SecretC)}$
$S \rightarrow C$: $\{N_C\}_{\mathrm{Master}(SecretC)}$
$S \rightarrow C$: $\{SessionId\}_{\mathrm{Master}(SecretC)}$

Here, C is the client, S is the sever, $SuiteC$ and $SuiteS$ are cryptographic preferences of C and S respectively, N_C and N_S are nonces generated by C and S respectively, K_S^+ is the public encryption key of S, $SecretC$ is the random secret generated by C. $SessionId$ is the identifier of a particular session. Master($SecretC$) is the master secret (secret key) generated from $SecretC$. $\mathrm{Sign}_{CA}\{S, K_S^+\}$ is the signature of S. The purpose of the protocol is to establish a secret key between client C and sever S, which is used later when they communicate.

4.2 The Model of the System

We assume there is an intruder I. The possible messages in the protocol include:

Names and public encryption keys of three parties C, S, I --- C, S, I , K_C^+, K_S^+, K_I^+;
Cryptographic preferences of three parties C, S, I--- $SuiteC$, $SuiteS$, $SuiteI$;
Nonces of three parties C, S, I----- N_C, N_S, N_I;
Secrets of three parties C, S, I----- $SecretC$, $SecretS$, $SecretI$;
Master secrets of three parties C, S, I----- Master($SecretC$), Master($SecretS$), Master($SecretI$);
Private encryption keys of three parties C, S, I----- K_C^-, K_S^-, K_I^-;
Signatures of three parties C, S, I----- $\mathrm{Sign}_{CA}\{C, K_C^+\}$, $\mathrm{Sign}_{CA}\{S, K_S^+\}$, $\mathrm{Sign}_{CA}\{I, K_I^+\}$;
The identifier of the session---- $SessionId$;

We first consider the messages that client C can observe. The initial message is $\{C, S, I, K_C^+, K_S^+, K_I^+, N_C, SecretC$, Master($SecretC$), K_C^-, Sign$_{CA}\{C, K_C^+\}\}$. So the number of possible messages that client C can observe is less than ten. Hence, there are much less than 2^{10} observable local states for client C.

Based on the same reason, sever S and intruder I has less than 2^{10} observable local states respectively.

Therefore, the set $(S_C \times S_S \times S_I)$ of global states of system includes at most $2^{10} \times 2^{10} \times 2^{10}$ possible states, that is, 2^{30} states. Every global state can be expressed as $s=(s_c, s_s, s_I)$.

Now we describe partly the state transition R. Before client C sends message $\{C, SuiteC, N_C\}$, the local state of C is $s_{C1}=\{K_C^-\}$, the local state of S is $s_{S1}=\{K_S^-\}$, the local state of I is $s_{I1}=\{K_I^-\}$, and so the global state of system is (s_{C1}, s_{C1}, s_{I1}), after client C sends message $\{C, SuiteC, N_C\}$, the local state of C is $s_{C2}=\{K_C^-, C, SuiteC, N_C\}$, the local state of S is $s_{S2}=\{K_S^-, C, SuiteC, N_C\}$, the local state of I is $s_{I2}=\{K_I^-, C, SuiteC, N_C\}$, and so the global state of system is (s_{C2}, s_{S2}, s_{I2}). Hence the global state of system transfers from (s_{C1}, s_{S1}, s_{I1}) to (s_{A2}, s_{B1}, s_{I2}).

4.3 The Specification of the System

On the basis of [12], we can describe more accurately the security properties that SSL 3.0 protocol should satisfy the following temporal logics of knowledge specifications.

1 There is a run path on which "the share secret *SecretC* between C and S is *SecretC* " is eventually common knowledge between client C and sever S.

2 To all run paths, the intruder I always does not know "the share secret *SecretC* between C and S is *SecretC* ".

3 The cryptographic algorithms selected by the parties for encryption and authentication of handshake messages are the strongest ones that are supported by both C and S. We model this by requiring that the SuiteC stored by S as C's cryptographic preferences is identical to the one actually sent by C, and vice versa. Hence, there is a run path on which "Suite$C$ is cryptographic preferences of C " and "Suite$S$ is cryptographic preferences of S " are eventually common knowledge between client C and sever S.

Assume we express "the share secret between C and S is *SecretC* " as f, express "f is the common knowledge between client C and sever S" as $C_G f$ (here G={client C, sever S}), express "Suite$C$ is cryptographic preferences of C " as g, express "Suite$S$ is cryptographic preferences of S " as h, express "g is the common knowledge between client C and sever S " as $C_G\, g$, express "h is the common knowledge between client C and sever S " as $C_G\, h$. Then the first and third properties can be expressed as EF($C_G(f \wedge g \wedge h)$) (here G={client C, sever S}). The second property can be expressed as AG$\neg K_I f$. Therefore, the specification of the system can be expressed as EF($C_G(f \wedge g \wedge h)) \wedge$AG$\neg K_I f$.

4.4 Model Checking Specifications of the System

EF($C_G(f \wedge g \wedge h)$) can be translated into extended mu-calculus formula $\mu Z.(\nu Q.(f \wedge g \wedge h \wedge [R_C]Q \wedge [R_S]Q) \vee \langle a \rangle Z)$. And AG$\neg K_I f$ can be translated into extended mu-calculus formula $\neg\mu Z.([R_I] f) \vee \langle a \rangle Z)$. Hence, the specification EF($C_G(f \wedge g \wedge h)) \wedge$AG$\neg K_I f$ can be translated into extended mu-calculus formula $\mu Z.(\nu Q.(f \wedge g \wedge h \wedge [R_C]Q \wedge [R_S]Q) \vee \langle a \rangle Z) \wedge (\neg\mu Z.([R_I] f) \vee \langle a \rangle Z))$. Though there are 2^{30} possible states in the system as described above, the most states are not reachable. To be similar to [10], the most interesting result is the number of nodes in the transition relation OBDD is polynomial in the number of steps of the protocol. As a result, the verification time is polynomial in the number of steps of the protocol. Using our approach to model check temporal logics of knowledge by extended mu-calculus, we can verify that the system (SSL 3.0 protocol) does not satisfy the specification EF($C_G(f \wedge g \wedge h)) \wedge$AG$\neg K_I f$.

5 Conclusions

In the paper, we first extend mu-calculus and OBDDs, and then using them, on the basis of [9,10] and according to the semantics of knowledge and common knowledge, present an efficient approach to model check temporal logics of knowledge via extended mu-calculus. These approaches can handle security verification of systems with extremely large state spaces. As a case study, we verify the security properties of SSL 3.0 protocol.

References

1. van der Meyden, R., Shilov, N.V.: Model Checking Knowledge and Time in Systems with Perfect Recall(Extended Abstract). In: Pandu Rangan, C., Raman, V., Ramanujam, R. (eds.) Foundations of Software Technology and Theoretical Computer Science. LNCS, vol. 1738, pp. 432–445. Springer, Heidelberg (1999)
2. van der Meyden, R., Shilov, N.V.: Model Checking Knowledge and Time in Systems with Perfect Recall. In: Pandu Rangan, C., Raman, V., Ramanujam, R. (eds.) Foundations of Software Technology and Theoretical Computer Science. LNCS, vol. 1738, Springer, Heidelberg (1999)

3. van der Hoek, W., Wooldridge, M.: Model Checking Knowledge and Time. In: Stefan Leue, C.C (ed.) 9th international SPIN workshop on model checking of software, pp. 1–16 (2002)
4. van der Meyden, R., Su, K.: Symbolic model checking the knowledge of the dining cryptographers. In: The 17th IEEE Security Foundation Worhshop, pp. 280–291. IEEE Computer Society Press, Los Alamitos (2004)
5. Su, K.: Model Checking Temporal Logics of Knowledge in Distributed Systems. In: The Nineteenth National Conference on Artificial Intelligence (AAAI-04) (2004)
6. Li-Jun, W., Kai-Le, S.: A Model Checking Algorithm for Temporal Logics of Knowledge in Multi-Agent System. Journal of Software 15(7), 1012–1020 (2004)
7. Kozen, D.: Results on the prepositional mu-calculus. Theoretical Computer Science 27, 333–354 (1983)
8. Bryant, R.E.: Graph-based algorithms for boolean function manipulation. IEEE Transaction on computers 35(8), 687–691 (1986)
9. Clarke, E.M., Grumberg, O., Peled, D.A.: Model checking, 1st edn. MIT Press, Cambridge (1999)
10. Burch, J.R., Clarke, E.M., McMillan, K.L.: Symbolic model checking: 10^{20} states and beyond. Information and Computation 2, 142–170 (1998)
11. Fagin, R., Halpern, J.Y., Moses, Y., Vardi, M.Y.: Reasoning about Knowledge, 1st edn., pp. 111–120. MIT Press, Cambridge (1995)
12. Mitchell, J.C., Shmatikov, V., Stern, U.: Finite-state analysis of SSL 3.0. In: Proc. 7th USENIX Security Symposium, pp. 201–215 (1998)
13. Wozna, B., Lomuscio, A., Penczek, W.: Bounded Model Checking for Knowledge and Real Time. In: Proceedings of the 4th International Joint Conference on Autonomous Agents and Multi Agent Systems (AAMAS 2005), ACM Press, New York (2005)
14. Raimondi, F., Lomuscio, A.: MCMAS: a tool for verifying multi-agent systems. In: Hermanns, H., Palsberg, J. (eds.) TACAS 2006 and ETAPS 2006. LNCS, vol. 3920, Springer, Heidelberg (2006)
15. Lomuscio, A., Raimondi, F.: Model checking knowledge, strategies, and games in multi-agent systems. In: Proceedings of the 5th International Conference on Autonomous Agents and Multi-Agent systems (AAMAS 2006), ACM Press, New York (2006)
16. Su, K., Sattar, A., Wang, K., Luo, X., Governatori, G., Padmanabhan, V.: Observation-based model for BDI-agents. In: AAAI-05 (2005)

A Computer Security Model of Imitated Nature Immune and Its FSM

Zhenpeng Liu[1,3], Ailan Li[2], Dongfang Wang[1], and Wansheng Tang[3]

[1] Network Center, Hebei University, 071002 Baoding, China
{lzp,wdf}@hbu.edu.cn
[2] Department of Foundation Teaching, Hebei Jiaotong Vocational & TechnicalCollege, 050091 Shijiazhuang, China
[3] Institute of System Engineering, Tianjin University, 300072 Tianjin, China
tang@tju.edu.cn

Abstract. According to the nature immunology, This paper describe a computer security model imitated the principle of nature immunity, GECISM (GEneral Computer Immune System Model). The model is structured by agents. The agent imitates the immune cells. Through rules and co-operation the agent discriminate "self" and "non-self". Further it eliminates "non-self". The FSM model of GECISM accurately describes this system and meanwhile is the basement of performance analysis and system test about GECISM.

1 Introduction

The nature immune system (NIS) is a complex adaptive system and it cognizes *self* and *non-self* in the body and then eliminates *non-self* by immune responses in order to get the homeostasis of body. NIS includes the immune organism, the immunocyte and the immune molecule. The immunocyte is created in the immune organism and matures in it. After maturation the immunocyte act when they circulate in the blood and body fluid. The immune organism doesn't control the mature immunocyte. One immunocyte excites or inhibits another immunocyte through different immune molecule. NIS get homeostasis through the interaction of immunocytes.

There are some models about the computer immune system but they all pay more attention to the detection. In the nature immune system there also are the immune stabilization, the immune defense, the immune response, besides the immune detection. This paper put forward a GEneral Computer Immune System Model (GECISM) imitated the nature immune system, which includes some kinds of intelligent agents. The immune function is based on the co-operation of the multi-agents. With the start point of the detect agent, and point to the different results, GECISM activates the different immune response to eliminate the "non-self".

The nature immune system consists of the immune organism, the immunocyte, the immune molecule. The immunocytes are created, differentiated, matured in the immune organism and then move in the body fluid and function. In the immunocytes the MC cell, the B cell, the T cell is the key cell. The MC cell has the function of

K. Li et al. (Eds.): LSMS 2007, LNCS 4688, pp. 523 – 532, 2007.

presenting antigen. When meet the “non-self” cell it swallows, dissolves that cell and extracts the epitopes-----which indicates the “non-self” cells character and makes up of the protein segment. And then put the proteide on the cell membrane in order to be receptive with the B cell or the T cell. After binding with the epitope, the B cell creates antibody and activates the immune response of body fluid; and when combining with the epitope, the T cell activates un-specific immnuocyte or the killer cell to eliminate the antigen. Here the T cell includes three sub group-----helper T lymphocytes(TH), delayed type hypersensitivity T lymphocytes(TD), cytotoxic T lymphocytes(TC). TH accelerate the immune response; and TD arouse the un-specific immune response; TC kill the antigen cell through secreting the perforin after combine with the epitope. In the natural immune system, one type of detected cells detects more than one type of antigens, and there is only one type cell combines with one type of antigen, but the types of cell which combine with the antigens are more, and every type of immunocytes has their own lifecycle.

After matured the immunocytes circulate in the body, and the immune organism do not control the matured cells. The immune response and its strength is defended on the adjustment of the different types of immunocytes. Every type of immunocytes could accelerates or suppresses the other type cells, and the natural immune system get the homeostasis through the adjustment itself.

2 The General Computer Immune System Model (GECISM) Imitated NIS

The computer immune system is a computer secure system imitated the principles of NIS what include the immune detection and the immune response. The functions of GECISM include:

1. to cognize self and non-self.
2. to eliminate non-self.
3. to adjust the level of response.

Here self is defined as the coding string which is aroused by the normal user activities; and non-self is defined as the coding string which is aroused by the anomalous activities. The architecture of GECISM is given in fig. 1.

The aim of model is to define exactly the system attributes and the system features what need be built or be analyzed, and the model also offers the basement of testing those attributes and features. The finite automation is used to setup the model of GECISM.

The finite status automation includes a set of input, a set of status, a begin point and a set of function, and this set of function are in charge of the change of status when it get a certain input and then to get some certain outputs. This model can explain the database processing in distribute computer environment. The processing of database is related to the input and the change of status in NFS and the database are related to the status of NFS and the codes are related to the change of status.

Define 1. FSM of GECISM is five-tuple as follows:

$M=(Q, \sum, \delta, q_0, F)$,

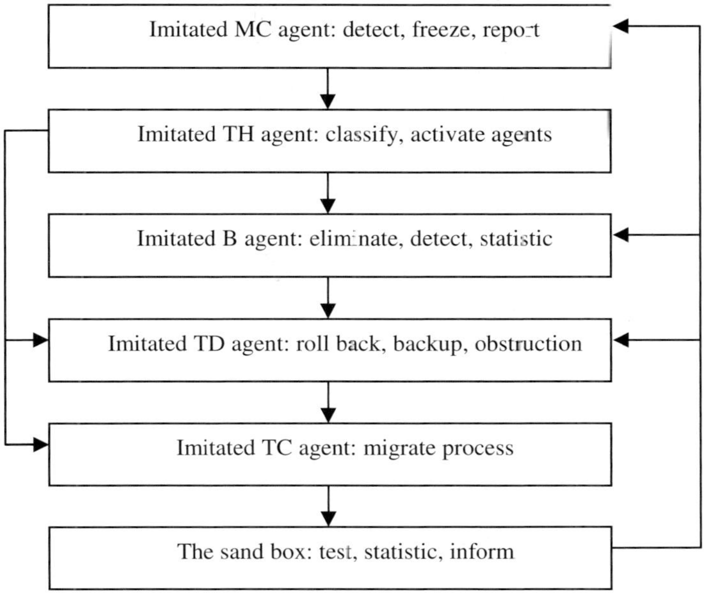

Fig. 1. The Architecture of GECISM

Whose components have the following meanings:

1. Q is a finite set of states. It is described by the Cartesian product of the status of the six agents, i.e. $Q = \{q_i | q_i \in a_1 \times a_2 \times a_3 \times a_4 \times a_5 \times a_6\}$. a_1,a_2,a_3,a_4,a_5,a_6 describe the status of MC,TH,B,TD,TC,SB.

In simples suppose GECISM always acts, i.e. MC has only busy status and TH,B,TD,TC,SB has two status: busy or free. In NFA M, the status of busy is defined as 0 and the status of free is defined as 1, i.e. a1∈{1}, a2∈{0,1}, a3∈{0,1}, a4∈{0,1}, a5∈{0,1}, a6∈{0,1}.

2. $\sum$ is a finite set of input symbols. The status of system is changed as soon as the status change in agents, and so the events that change the status of agents are also change the system. The set of events that can be accepted are defined as $\sum$={o,a,b,c,d,e,f,g1,g2, g3,g4, g5}. The meaning of the events are given below:

o: MC has detected self;
a: MC has detected non-self;
b: TH get the result;
c: B get the result ;
d: TD get the result;
e: TC get the result;
f: SB get the result;
g_1: TH is free;
g_2: B is free;
g_3: TD is free;
g_4: TC is free;
g_5: SB is free.

3. q_0, a member of Q, is the start state. q_0=(1,0,0,0,0,0) describes that the system does not detect non-self, i.e. MC is busy and other agents is free.

4. F, a subset of Q, is the set of accepting states. F={q_0} only has one status and describes non-self is eliminated by the system.

5. δ, the transition relation is a function that takes a state in Q and an input symbol in Σ as arguments and returns a subset of Q. The function set of the status change is given with the function description of the agents.

3 The Function of Agents in GECISM

3.1 Imitated MC Agent

In the natural immune system, the MC cells present the antigens. After phagocytized the antigen cell, the antigen is dissolved into epitopes and these epitopes indicate the type of antigen. The MC cells carry the epotipes to the cell membrane in order to combine with the B cells or T cells. In the further the combination activates the immunity of body fluid or cells.

In current systems [3,4], some of imitation about presenting antigen defends on the string match and some defend on rules. In GECISM, the presenting antigen defends on imitated MC agent (abbr. MC) and the GECISM judges “self” and ”non-self” by using self-rules and non-self-rules. MC architecture is given in fig. 2.

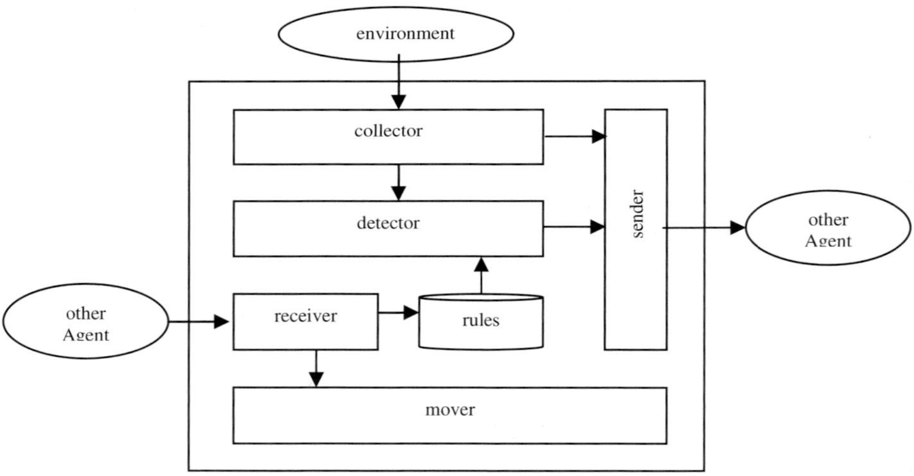

Fig. 2. Imitated MC Agent

When MC detects self, it act continue. And when detect non-self, MC stops this process and activates TH to class the kind of process. MC always is in busy. The rule of self and non-self is created and updated in sand box, and when that task is finished the sand box sends the rule set to MC.

The function of the status change is given below:

(1) δ $(q_0, o)=\{q_0\}$; When MC detects self it return.

(2) δ $(q_0, a)=\{(1,1,0,0,0,0)\}$; When MC detects non-self in q_0 it activates TH.

(3) δ (q_i, a)={q_j}; When MC is not in q_0 and detects non-self it activates TH if TH is free. And the n the other agents don't change. where:

$q_i \in \{(1,X_2,X_3,X_4,X_5,X_6)\}$, $q_j \in \{(1,1,X_3,X_4,X_5,X_6)\}$; X_k=0|1, $2 \le k \le 6$.

3.2 Imitated TH Agent

The helper T cell is a subgroup in T cells, accelerates the differentiation and breeding of itself and other cell through secreting the cytokine. In the further, it fast the speech of the response. The architecture of TH is given in fig. 3.

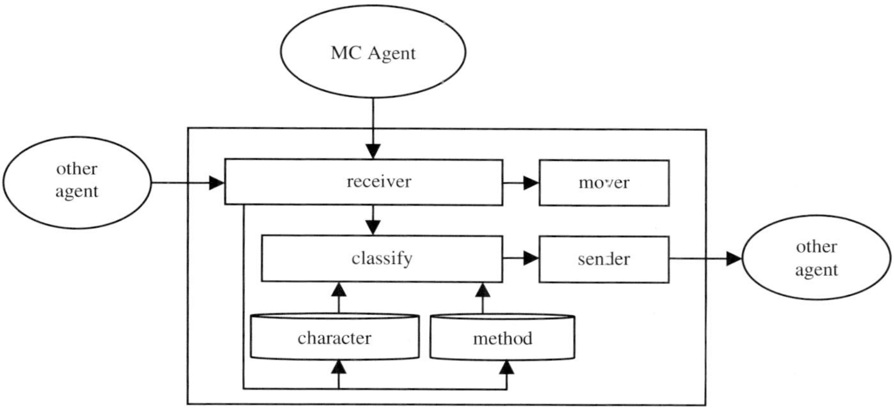

Fig. 3. Imitated TH Agent

In GECISM, imitated TH agent (abbr. TH) classifies the "non-self" samples have been detected. TH is free when not receive message from MC. But as soon as MC detects non-self TH becomes busy. And when finish the task and does not receiver the new message TH becomes free again. TH class the process through their feature into three classes. The one class includes the process with known features and can be eliminated. As to these process TH activates B to eliminate. The second class includes the process with known features and can't be eliminated. As to these process TH activates TD to obstruction. The third class includes the process with unknown features. As to these process TH activates TC to migration these process to SB in order to test and to find a method of elimination.

The function of the status change is given below:

(4) δ (q_i, b)={q_j}; When TH gets the result it choices one of B, TD, TC according to the result. If that agent is free it becomes busy and other agents don't change. where:

$q_i \in \{(1,1,X_3,X_4,X_5,X_6)\}$,$q_j \in \{(1,1,1,X_4,X_5,X_6), (1,1,X_3,1,X_5,X_6), (1,1,X_3,X_4,1,X_6)\}$; X_k=0|1, k=3,4,5,6.

(5) δ (q_i, g_1)={q_j}; When TH does not receiver the message it becomes free again. where:

$q_i \in \{(1,1,X_3,X_4,X_5,X_6)\}$, $q_j \in \{(1,0,X_3,X_4,X_5,X_6)\}$; X_k=0|1, k=2,3,4,5.

3.3 Imitated B Agent

In the natural immune system, the B cell change to a specific one which secrets antibody and eliminates the antigen after the B cell's receiver combines with the epitope. This process is called the immunity of body fluid. Every kind of B cell generates multi-antibody, and one type of antibody kills one type of antigen. At the same time every type of B cell presents the antigen that can be eliminated high efficiently.

In GECISM, imitated B agent (abbr. B) does the same things like the B cell in the natural system, and B eliminates non-self. B is divided into several subgroup based on the character and the eliminated method, the B subgroup is defined as B_i (i =1,2,3….); In Bi the eliminated method is defined as R_i ($R_i \subseteq R$), and R indicates the set of eliminated methods. The architecture of B is given in fig. 4.

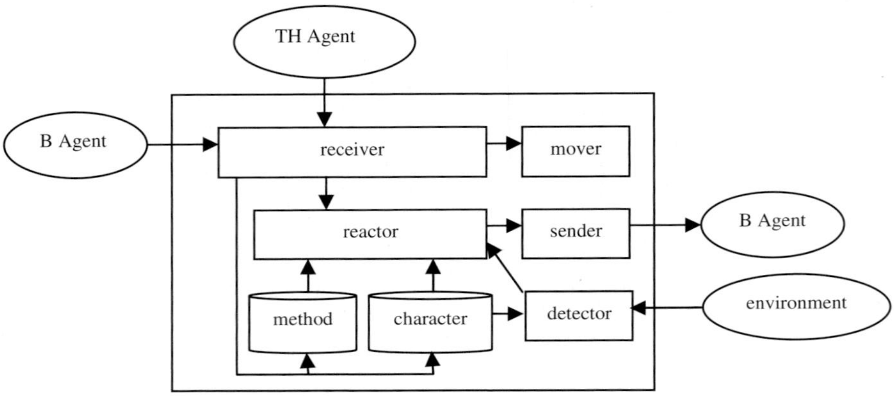

Fig. 4. Imitated Sub-B Agent

B is free when it does not receive the message from TH. And it is activated by TH when TH finishes its work and as soon as busy B eliminates the process include non-self. If the elimination failures B activates TD to obstruction. When B has not new work it becomes free again.

The function of the status change is given below:

(6) δ (q_i, c)={q_j}; If B eliminates non-self successfully it eliminates the next one, and if not it activates TD. Other agents don't change. where:

$q_i \in \{(1,X_2,1,X_4,X_5,X_6)\}$, $q_j \in \{(1,X_2,1,X_4,X_5,X_6),(1,X_2,X_3,1,X_5,X_6), (1,X_2,X_3,X_4,1, X_6)\}$; X_k=0|1, $2 \leqslant k \leqslant 6$.

(7) δ (q_i, g_2)={q_j}; When B has not new work it becomes free. where:

$q_i \in \{(1,X_2,1,X_4,X_5,X_6)\}$, $q_j \in \{(1,X_2,0,X_4,X_5,X_6)\}$; X_k=0|1, k=2,4,5,6.

3.4 Imitated TD Agent

The TD lymphocyte arouses the un-special immune response of monocyte phagocytosis after combination in the natural system. I.e. activate by special combination and response in un-special path.

In GECISM the function of the imitated TD agent (abbr. TD) likes to the TD cell. TD is activated by the "non-self" samples that have the known character but unknown eliminated method, and it also becomes busy when receives the "non-self" samples that cannot be eliminated in the B subgroup. After be activated TD rolls back the processes with "non-self", and find the nearest clean point and then back up the states of this point and at the last put a obstruction in here in order to stop the same process coming here next time (this course is called freeze). And after has done that all TD informs the TC agent the information of the process being in freezing. When TD has not new work it becomes free.

The function of the status change is given below:

(8) δ (q_i, d)={q_j}; If TD finishes its task successfully it continue to obstruction the next process. And if TD does not success it activates TC. Other agents don't change. where:

$q_i \in \{(1,X_2,X_3,1,X_5,X_6)\}, q_j \in \{(1,X_2,X_3,1,1,X_6), (1,X_2,X_3,1,0,X_6)\}; X_k=0|1, k=2,3,5,6$

(9) δ (q_i, g_3)={q_j}; When TD has not new work it becomes free. where:

$q_i \in \{(1,X_2,X_3,1,X_5,X_6)\}, q_j \in \{(1,X_2,X_3,0,X_5,X_6)\}; X_k=0|1, k=2,3,5,6.$

3.5 Imitated TC Agent

The TC lymphocytes kill the target cell that brings antigen directly. After abuts against the target cell, the TC lymphocyte secrets perforin and make the stances inside the target cell outflow. And then the target cell die.

In GECISM, imitated TC agent (abbr. TC) receives the process that is unknown or cannot be eliminated, and then migrates this process to the sand box. Although the function of TC is not same to the TC lymphocyte, we use this name in the natural system in order to consistent. TC is activated by TD and if it has not new work it becomes free.

The function of the status change is given below:

(10) δ (q_i, e)={q_j}; When TC finishes the process movement it activates SB. Other agents don't change. where:

$q_i \in \{(1,X_2,X_3,X_4,1,X_6)\}, q_j \in \{(1,X_2,X_3,X_4,1,1)\}; X_k=0|1, k=2,3,4,6.$

(11) δ (q_i, g_4)={q_j}; When TC has not new work it becomes free. where:

$q_i \in \{(1,X_2,X_3,X_4,1,X_6)\}, q_j \in \{(1,X_2,X_3,X_4,0,X_6)\}; X_k=0|1, k=2,3,4,5.$

3.6 Sand Box HC (Host Computer)

The natural immune system arouse the primary immune response when meets the new antigen. In the primary immune response the speech of response is slow, and the time to eliminate non-self is long. But when meet the same antigen next time, i.e. second immune response, the time to eliminate non-self will be very short.

The sand box can be a physical computer or a virtual one.

In GECISM, the sand box HC (abbr. sand box) tests the process of the unknown character and eliminated method "non-self" samples and decides weather this process is malicious or can be eliminated. And these methods imitate the primary immune response in NIS for short the time to eliminate non-self.

After testing if the process is not malicious SB informs MC to stop freezing. And if the process is malicious and it can be eliminated by the system SB informs B to eliminate. And if the process is malicious and it can not be eliminated by the system SB informs TD to obstruct.

The function of the status change is given below:

(12) δ $(q_i, f)=\{q_j\}$; When SB finishes the work successfully, it selects one of MC, B and TD to activate. And other agents don't change. where:

$q_i \in \{(1,X_2,X_3,X_4,X_5,1)\}$, $q_j \in \{(1,X_2,X_3,X_4,X_5,1), (1,X_2,1,X_4,X_5,1), (1,X_2,X_3,1,X_5,1)\}$; X_k=0|1, k=2,3,4,5.

(13) δ $(q_i, g_5)=\{q_j\}$; When SB has not new work it becomes free. where:

$q_i \in \{(1,X_2,X_3,X_4,X_5,1)\}$, $q_j \in \{(1,X_2,X_3,X_4,X_5,0)\}$; X_k=0|1, k=2,3,4,5.

4 The Analyze of GECISM

Based on the definition q_0 and F in the FSM of GECISM, the system is not in F if any one of TH, B, TD, TC and SB is busy. In the model of GECISM, the string that FSM of GECISM can accept means the sequences of events that change the agent status. And all of these events happen between the time that find the first non-self and the time that process all non-self successfully.

Definition 2. In the FSM of GECISM, if a input string $\omega \in \Sigma^*$, $\delta(q_0, \omega) \cap F \neq \phi$, then ω is an **incident** that can be accepted by FSM of GECISM.

Definition 3. The **feature of including backword** Suppose S is a string, $S=s_1s_1s_2s_1s_2s_3$......, and suppose exists a plus integer n , if the number of s_1 is greter than equal to the number of s_2 (i.e. $\Sigma s_1 \geqslant \Sigma s_2$) in the n length that is from the first character of S then name s_1 includes s_2 (i.e. $s_1 \succ s_2$). If $\Sigma s_1 \geqslant \Sigma s_2 + \Sigma s_3$, then s_1 includes s_2 and s_3 (i.e. $s_1 \succ (s_2 + s_3)$).

As to the string S, known from definition 3, the ability of including backword has the ability of self-return, i.e. $s_1 \succ s_1$, and has the ability of transaction, i.e. if $s_1 \succ s_2$ 、$s_2 \succ s_3$, then $s_1 \succ s_3$, and has the ability of asymmetric, i.e. if $s_1 \succ s_2$ is right, then $s_2 \succ s_1$ is wrong. And so the ability of including backword is a relationship of partial ordering on the set of the words number in a string S.

Definition 4. The **tail point**: In a string $S=s_1s_1s_2s_1s_2g_1$......, if there are not s_1 after g_1, then name g_1 the tail point of the s_1 string.

Definition 5. The **incident set** that FSM of GECISM can accept is denoted L(M),and is definite as the inside events sequence that happen when the system eliminates the non-self successfully. i.e.

$$L\quad M = \{\omega \mid \delta(q_0, \omega) \cap F \neq \phi,\ \omega \in \Sigma^*\}$$

From the definition of FSM of GECISM, the length of events sequence is varied, and the events sequence owns the features given below:

1. The number of event a is equal to the number of event b, and a includes b backword, i.e.$\sum a=\sum b \wedge a \succ b$.
2. The number of event e is equal to the number of event f, and e includes f backword, i.e. $\sum e=\sum f \wedge e \succ f$.
3. If non-self is eliminated successfully in one time, i.e. B does not activate TD and TD does not activate TC, then the number of event b is equal to the sum of event c and events d and event e. i.e. b includes c, d and e backword. Written as $\sum b=\sum c+\sum d+\sum e \wedge b \succ (c+d+e)$. if non-self is not eliminated successfully in one time then $\sum b \le \sum c+\sum d+\sum e$.
4. g_1,g_2,g_3,g_4,g_5 is the tail point of b,c,d,e,f.

5 Correlation and Conclusion

GECISM comes of the natural immune system, but not limit to it. It divides the "non-self" samples by the character and the eliminated method, and then deals with them in different policy. Meanwhile puts forward the sand box in order to test the unknown "non-self" samples-----that imitates the primary immune response. The correlation between GECISM and the natural immune system is given as in table 1.

Table 1. The correlation between GECISM and the natural immune system

	The Natural Immune System	GECISM
Distinguish "self" and "non-self"	Chemistry combination	Rule match
Presenting antigen	Macrophages, B cell	MC, B
Immune response	B cell, T cell	B, TD
Lymphocyte lifecycle	Natural life	Update
Movement	Lymphocyte circulation	Movement of agents
Cooperation	Lymphocyte secret cytokine	Communication of agents
Unknown antigen	Primary response	The sand box
Known antigen	Second response	B

Based on the definition q_0 and F in the FSM of GECISM, the system is not in F if any one of TH, B, TD, TC and SB is busy. In the model of GECISM, the string that FSM of GECISM can accept means the sequences of events that change the agent status. And all of these events happen between the time that find the first non-self and the time that process all non-self successfully.

The main features of GECISM are given as follow:

1. exact identify: The identification is based on the multi-agent cooperation. The false probability of the whole system is lower than the false probability of one agent.

2. real time: Because of the less time for detection the system is real time.

3. high efficiency: Because that the character for detection is less and the eliminated method database exists, the B subgroup can eliminates the "non-self" in high efficiency.

4. extension: The prober in the sand box can generates new eliminated method, and that upgrades the power of the system.

5. openness: The generation of the B subgroup artificially makes the achievement of others into GECISM.

6. flexible setting: The sensitivity of the system, the number of the B subgroups and the leak samples all can be set in dynamic change.

7. the performance depend on the initialization: If the initial rule database covers the less samples the system would be busy in the rules update. But after some time the coverage of rules is large the performance will up.

8. the new eliminated method depend on the manual: GECISM generates the "new" eliminated method, but the "new" one is only the pile of the method in database. The extension of eliminated method database depends on man.

The FSM of GECISM definite exactly its attributes and its features, and it is the basement of system analyze and the basement of testing the attributes and features of GECISM.

Acknowledgments. Supported by the National Science Foundation of China (70571056) and the High Technology Research and Development Program of Hebei Province (04213534, 04213529).

References

1. 1. Tong, J.-Y.: The Medicine Immunity and Microbiology, 3rd edn. The Press of People Health, Beijing (1996)
2. Hofmeyr, A., Forrest, S.: Architecture for an Artificial Immune System. Evolutionary Computation 1, 45–68 (2000)
3. Somayaji, A.B.:Operating System Stability and Security through Process Homeostasis. Albuquerque, The University of New Mexico, New Mexico (2002)
4. Warrander, C.: Effective Feedback in the Immune System. In: Genetic and Evolutionary Computation Conference Workship Program, pp. 329–332. Morgan Kaufman, San Francisco (2001)
5. Chao, D.L., Forrest, S.: Information Immune Systems. In: Proceedings of the First International Conference on Artificial Immune Systems (ICARIS) (2002)
6. Balthrop, J., Esponda, F., Forrest, S. (eds.): Coverage and Generalization in an Artificial Immune System. Proceedings of the Genetic and Evolutionary Computation Conference (GECCO 2002). Morgan Kaufmann, New York (2002)
7. Forrest, S., Hofmeyr, S., Somayaji, A.: Computer Immunology. Communications of the ACM 10, 88–96 (1997)
8. Forrest, S., Perelson, A., Allen, L. (eds.): Self-Nonself Discrimination in a Computer. Proceedings of the 1994 IEEE Symposium on Research in Security and Privacy, pp. 202–212. IEEE Computer Society Press, Los Alamitos, CA (1994)
9. Liu, M., Ling, T.W.: A Data Model for Semistructured Data with Partial and Inconsistent Information. In: Zaniolo, C., Grust, T., Scholl, M.H., Lockemann, P.C. (eds.) EDBT 2000. LNCS, vol. 1777, pp. 27–31. Springer, Heidelberg (2000)
10. Steven, A., Hofmeyr, S., Forrest, S.: Immunity by Design: An Artificial for Immune System. In: Proceedings of the 1999 Genetic and Evolutionary Computation Conference (GECCO), pp. 1289–1296 (1999)

Study on Signal Interpretation of GPR Based on Support Vector Machines

Huasheng Zou [1,2] and Feng Yang [1]

[1] China University of Mining and Technology (Beijing), Beijing, 100081, PRC
brihs@163.com
[2] College of Computer Science and Engineering , Wenzhou University, Wenzhou, 325027, PRC
zoufan99@163.com

Abstract. In the process of the roadbed disease recognition, as a result of the complexity and immaturity of the roadbed medium, the ground penetrating radar signal explanation has multi-solutions. It affects the application of ground penetrating radar and the roadbed disease detection. The support vector machines is one algorithm based on the structural risk minimization principle, it can obtain the overall optimal solution in sample less situations. It has solved the inevitable partial minimum problem and overcome the disadvantage, which the traditional neural network cannot avoid. In this paper the ground penetrating radar signal explanation model is established and applied in recognition to the roadbed disease using the support vector machine theory and methods. The result has proved that this method may enhance ground penetrating radar signal explanation precision and efficiency.

1 Introduction

Ground Penetrating Radar signal explanation is the last link of target examination analysis processes, also is an extremely important link, because it is the final goal that using Ground Penetrating Radar to carry on the goal interpretation. Usually, according to the radar image section plane get from the processed data and the reflection wave group profile and intensity characteristic, through alignment the same track, the Ground Penetrating Radar signal interpretation is to determine the concrete meaning of the reflection wave group, to construct the corresponding physical interpretation section plane (either model) and according to it obtain the final result chart of the measured region, for the engineering design, the construction and the quality appraisal provides the scientific basis [1,2]. Due to the complexity and the uncertainty of underground medium, and the physical prospecting technology itself to have the multi-solutions, creating inevitably Ground Penetrating Radar signal explanation system research duty to be extremely arduous. But, the Ground Penetrating Radar's interpretation effect does affect directly to the Ground Penetrating Radar examination result reliability and the accuracy, then relates to the examine result the further application[10]. Presently, domestic and foreign Ground Penetrating Radar explanation system research progress

K. Li et al. (Eds.): LSMS 2007, LNCS 4688, pp. 533 – 539, 2007.

to be slow, the research results still were not mature and perfect, this kind of condition restricted inevitably Ground Penetrating Radar hardware performance display and limit Ground Penetrating Radar technology application. Therefore, develops to Ground Penetrating Radar explanation system research, explore effectively Ground Penetrating Radar explanation method already extremely to be urgent[1,3,4,5].

Support vector machines (Support vector machines is abbreviated as SVM) is one kind new study machine based on the statistical study Vapnik-Chervonenkis Dimension theory and the structure risk minimum principle. The statistical study theory is a kind of new machine learning method which proposed by Vapnik et al., it mainly aims at the small sample the study question, finds optimal solution in existing information situation. Therefore, by using the support vector machiness method, to process and analysis and explain the Ground Penetrating Radar examination data, may obtain the satisfying explanation effect[6, 7].

2 Support Vector Machines

The basic idea of Support vector machines pattern recognition method is to reflect the sample space to the characteristic space, extracts the most superior classified plane of original sample collection in the characteristic space, discovers support vector machines to carry on the pattern classification[13, 14]. Using the SVM classifier, may display it fully to the small sample data processing superiority, obtain the most superior classified plane by using a few samples as far as possible, make the forecast cost minimum through this superior classified plane, thus obtain the finest promotion. The following is the brief introduction of pattern classification by using SVM.

Suppose the sample collection is:

$$D=\{(xi, yi), i=1,2,..,l\}, xi \in Rn, \ yi \in \{+1,-1\}, \quad i=1,2,\dots,l$$

The superior classified plane may be represented as below

$$(w \cdot x)+b=0 . \tag{1}$$

Satisfied by:

$$y_i((w \cdot x)+b) \geq 0, i=1,2,...,l . \tag{2}$$

Then Solute the superior classified plane question may transform to the following optimization problem:

$$\begin{aligned} &\min \tfrac{1}{2} \ w^{2} \\ &\text{s.t. } y_i((w \cdot x)+b) \geq 1, \ i=1,2,...,l . \end{aligned} \tag{3}$$

It s antithesis question is

$$\max \mathrm{Q}(\alpha)=\sum_{i=1}^{l}\alpha_i-\frac{1}{2}\sum_{\mathrm{i,j=1}}^{\mathrm{l}}\alpha_\mathrm{i}\alpha_\mathrm{j}\mathrm{y}_\mathrm{i}\mathrm{y}_\mathrm{j}(x_i\cdot x_j)$$

$$\text{s.t.}\ \sum_{\mathrm{i=1}}^{\mathrm{l}}\alpha_\mathrm{i}\mathrm{y}_\mathrm{i}=0,\ \alpha_\mathrm{i}\geq 0,\ \mathrm{i}=1,2,\ldots,\mathrm{l} \tag{4}$$

Through solution the expression (4), we may get the most superior classified function as follow:

$$f(x)=\mathrm{sgn}(\sum_{i,j=1}^{l}\alpha_i\alpha_j y_i y_j(x\cdot x_i)+b) \tag{5}$$

The sample corresponding $\alpha_i \neq 0$ is called the support vector, to solute b in the formula, we may chooses any a support vector, then solute b by the constraint condition (this time takes the equal sign) in formula (2).

In non-linear separable situation, we can introduce relaxation variable ξ_i and the penalty parameter c, then formula (3) becomes

$$\begin{aligned}&\phi(w,\xi)=\tfrac{1}{2}\|w\|^2+c\sum_{i=1}^{l}\xi_i\\&\text{s.t. } \mathrm{y}_\mathrm{i}((w\cdot x_i)+b)\geq 1-\xi\\&\xi_i\geq 0, i=1,2,\ldots,l\end{aligned} \tag{6}$$

Its antithesis question is

$$\begin{aligned}&\max Q(\alpha)=\sum_{i=1}^{l}\alpha_i-\tfrac{1}{2}\sum_{i,j=1}^{l}\alpha_i\alpha_j y_i y_j(x_i\cdot x_j)\\&\text{s.t. } \sum_{i=1}^{l}\alpha_i y_i=0\\&0\leq\alpha_i\leq c, i=1,2,\ldots,l\end{aligned} \tag{7}$$

Regarding the nonlinear separable question, by introducing the kernel function to map lower dimension input space to higher dimension characteristic space through the nonlinear transformation, then we can solute classification question in characteristic space[11, 12]. Vapnik calls this kind of nonlinear mapping function as the kernel function. So long as the function $k(x, y)$ satisfies the Mercer condition, all may take as kernel function, here $k(x, y) = (\phi(x) \cdot \phi(y))$ is a Mercer kernel, while $\phi(x)$ is a mapping function. After introducing the kernel function, the inner product in above formula all may use the kernel function to replace. This time the SVM policy-making function is:

$$f(x) = sgn(\sum_{i=1}^{l} \alpha_i y_i k(x_i, x) + b) \tag{8}$$

If the kernel functions select suitably, we may transform the nonlinear separable question in input space to the linear separable question in the characteristic space.

3 The Method of Ground Penetrating Radar Signal Explanation with SVM

3.1 The Principle of Ground Penetrating Radar

Ground Penetrating Radar is the powerful tool and method for detecting the shallow layer underground target. We may obtain the examination report of the detecting target by using the SVM method to manipulate and explain the data getting from the Ground Penetrating Radar.

Ground Penetrating Radar launches wide band short pulse high frequency electromagnetic wave to underground, then obtains amplitude, phase and frequency spectrum kinematics and dynamics characteristics from underground medium reflection waves. These detection data will be used as sample data, through processing and explanation the sample data, we can analyze and inference underground medium structure and the formation lithology characteristics, then realize the goal of recognition to the underground target[8, 9].

3.2 The Experiment and Result of Ground Penetrating Radar

We used Ground Penetrating Radar to detect the disease in the roadbed. In chart 1 we use the Ground Penetrating Radar of 100MHz antenna to detect the concealed disease in the roadbed, the testing point is the asphalt pavement constructed newly in Tsinghua University. Among them the chart (a) is the actual detection time section plane, the chart (b) is the logarithm power spectrum section plane. If we analysis roadbed diseases only from the time section plane, it will be difficult greatly, because the loose region is not obvious in time section plane corresponding to the changing. We may see clearly from the spectrum section plane that the concealed loose region also has the characteristic below: low-frequency signal is strong, high frequency signal is weak. Its antenna basic frequency signal receives the serious destruction. We may obtain that the frequency section plane has the very good response characteristic to the loose region through the contrast of time section plane and logarithm power spectrum.

Instrument: the Ground Penetrating Radar made by China University of Mining Technology(Beijing)

Gathering time: 120ns

Sampling spot: 512 spots/channel

Sampling frequency: 100MHz

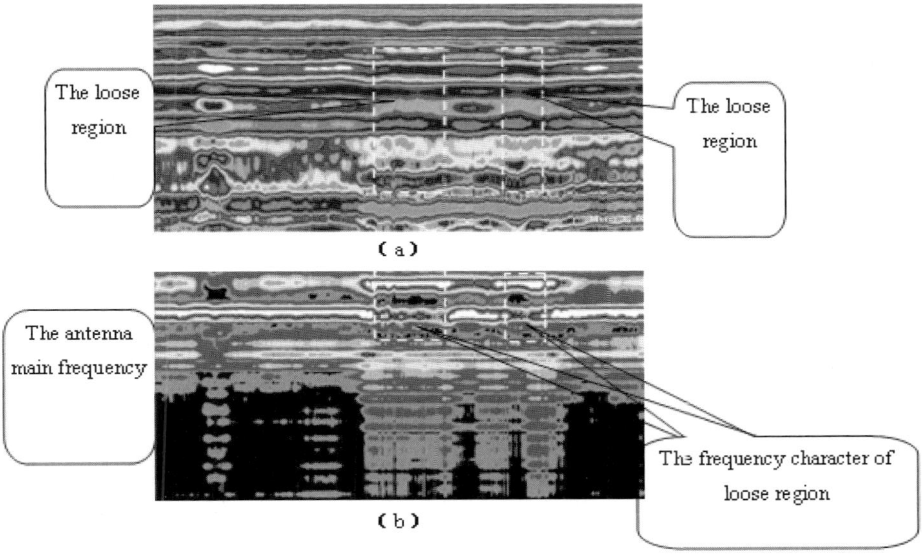

Fig. 1. The survey result to the roadbed. (a) The time section plane. (b) The logarithm power section plane.

3.3 Setting Up Model and Process the Detection Data Using Support Vector Machines

Data resources: making use of the detection power spectrum data to set up model and process data.

The experiment design:

(1) Selecting part of data from the detection data to use it in model training, testing and examination, another part to be used as forecasting data.

(2) Kernel function choice. This article uses the Gauss radial function as the kernel function.

(3) Parameters choice: To parameters g and c, choose three group parameters separately, the parameters choice and circulation establishment show in table 1.

(4) Model establishment and the forecasting. We have set up three models separately and used the models to forecast the detection data.

Table 1. The parameters choice and circulation setting

Groups	Starting Value of c	Step Value of c	Loop Time of c	Starting Value of g	Step Value of g	Loop Time of g
First Group	100	10	10	0.001	0.001	10
Second Group	10	1	10	0.01	0.001	10
Third Group	1	1	10	0.1	0.001	10

3.4 Experimental Result and Result Analysis

We use three group parameters to set up model separately, get three models, then carry on the forecast data, obtain three group forecast results, showing in figure 2. Through comparing and analyzing to the results, the correct rate of three models is as follow: 97.9%, 95.9%, 93.9%, the average of forecast correct rate is 95.9%. Comparing with the Neural Network methods[10,15], the average of forecast correct rate at most is 87%, so the Support vector machines methods is has higher forecast correct rate than the Neural Network methods. At the same time, the parameter choice has the certain influence to the model but not extremely big, the model we have set up is stable generally, the forecast results getting from the models is basically consistent with the actual examination results, therefore this method has a better effect and practical application value [10].

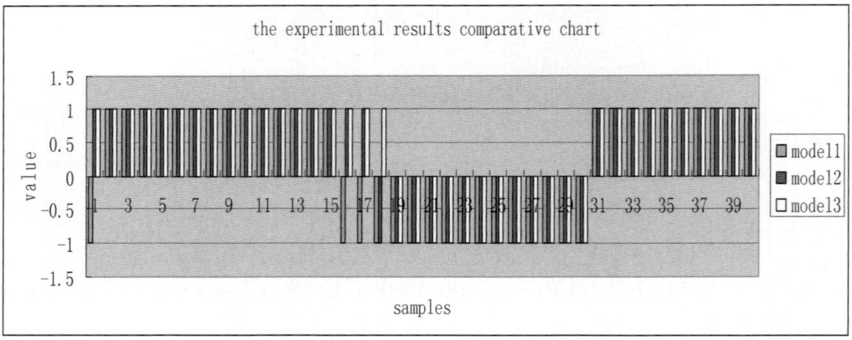

Fig. 2. The experimental results comparative chart

4 Conclusions

By using support vector machines method to interpret quantitatively the detection data of roadbed, we may obtain the quantitative interpretation result of the roadbed, it provides the new methods to the roadbed detection[16]. It is worth to note that in this method the selection of the kernel functions and the parameters has the direct influence to the model establishment, how to choose the kernel function and the parameters according to the actual problem to establish the quite reliable model still needs further studies.

Acknowledgements. This research has been undertaken within the Key Laboratory of Coal Resources and Safety Mining, China University of Mining and Technology (Beijing), and has been supported by the National Natural Science Foundation of China (Grant No. 50490271) .

References

1. Dunwen, L.: Study on Detecting and Controlling of Underground crag body project disaster. Central South University doctoral dissertation (2001)
2. Dunwen, L., Guoyuan, X., Rendong, H.: Application of Geophysical Prospecting Radar in Prospecting Integrality of Ancient Tombs. Progress in Gophysics 17(1), 96–101 (2002)
3. Libing, G., Bin, W., Mingjun, X.: The Applications of Ground-Penetrating Radar in Archaeological Prospection. The Geophysics Progresses 17(1), 61–69 (2000)
4. Jiaguang, Y., Saholin, L.: The Technology of Digital Signal Processing and Application in GPR. Progress in Gophysics 17(1), 171–175 (2002)
5. Qingyun, D., Miaoyue, W.: 2D Finite Element Modeling for Radr Wave. Chinese Journal of Geophysics 42(6), 818–825 (1999)
6. Vapnik, V.N.: Statistical Learning Theory. John Wiley & Sons, Inc , New York (1998)
7. Vapnik, V.N.: The Nature of Statistical Learning Theory. Springer, New York (2000)
8. Daxin, L.: The Ground Penetrating Radar Method and Application. Beijing Geology Publishing Press, Beijing (1994)
9. Chuanxiao, L.: Research on the Mechanism of Probing Cavity With Ground Penetrating Radar and Analysis on Application Examples. China Journal of Rock Mechanics and Engineering 19(2), 231–241 (2000)
10. Dunwen, L., Guoyuan, X., Rendong, H., Desheng, G.: Study on a Signal Interpretation of GPR Based on BP Neural Network. Progress in Geogphysics 19(1), 179–182 (2004)
11. Amari, S., Wu, S.: Improving support vector machines classiers by modifying kernel functions. Neural Networks 12(6), 783–789 (1999)
12. Cristianini, N., Shawe-Taylor, J.: An Introduction to Support vector machiness and Other Kernel-Based Learning Methods. Cambridge University Press, Cambridge (2000)
13. Campbell, C.: An introduction to kernel methods. In: Howlett, R.J., Jain, L.C. (eds.) Radial Basis Function Networks: Design and Applications, pp. 155–192. Physica Verlag, Wurzburg (2000)
14. Chang, C.-C., Lin, C.-J.: Training-support vector classiers: Theory and algorithms. Neural Computation 13(9), 2119–2147 (2001)
15. Qun, W., Yun-long, H., Chun-he, W., Xu-min, F., Hong-wei, N.: Study of Mine Detection with GPR Based on Neural Networks. Chinese Journal of Science 16(3), 400–403 (2001)
16. Burbidge, R., Trotter, M., Buxton, B., Holden, S.: Drug design by machine learning: Support vector machiness for pharmaceutical data analysis. Comput. Chem. 26(1), 5–14 (2001)

A Two-Phase Chaotic Neural Network Algorithm for Channel Assignment in Cellular Systems

Tinggao Qin, Xiaojin Zhu, Yanchun Chen, and Jian Wang

School of Mechatronics Engineering and Automation, Shanghai University
Shanghai, 200072, P. R. China
mgzhuxj@staff.shu.edu.cn

Abstract. In this paper, a new chaotic neural network algorithm for channel assignment in cellular networks is proposed as an optimization algorithm, and used in CAP. There are two phases in the new model. According to the difficulty measure of each cell, the first phase is executed to assign channels cell by using the frequency exhaustive strategy. If the optimum assignment solution is not obtained in the first stage, another phase which generated by adding chaotic noise to each neuron of the Hopfield neural network (HNN) called DCN-HNN is taken, and the DCN-HNN stage is then applied to continue the channel assignment until the optimum assignment is made or a maximum number of iteration is reached. The Kunz benchmark test problem is also used to compare the performance of the new algorithm against other algorithms. Simulation results show that the new method has a higher searching ability and lower computing time in searching the global minimum.

1 Introduction

In recent years, the demand for mobile telephone services has been rapidly increased, while the allocation of spectrum for this purpose is limited. So the channel assignment problem has become an important issue to be solved. From a mathematical point of view, the channel assignment problem(CAP) can be described as a combinatorial optimization problem with constrains, since this problem is NP-complete, an exact search for the best solution is practically impossible due to an exponentially growing time for large-scale channel assignment problems. Various novel algorithms have been proposed to solve this problem.

Neural network is an intrinsically parallel algorithm, and one of the typical models of artificial neural networks is the Hopfield neural network. The Hopfield neural network was first applied to solve the CAP by Kunz in 1991[1], and then used by many researchers to solve the CAP [1-4] in the following years. Although these neural networks guarantee convergence to a stable equilibrium point due to their gradient descent dynamics, the main drawback is that they suffer from the local minimum problems whenever applied to the CAP.

To alleviate the shortcomings of the conventional networks described above, nonlinear or chaotic dynamics has been introduced by various researchers. One approach to solve the combinatorial optimization problems is injecting chaotic noise

K. Li et al. (Eds.): LSMS 2007, LNCS 4688, pp. 540 – 550, 2007.

into the network. Hayakawa et al. [6] used the logistic map with different values of the bifurcation parameter as noise, and some values of the bifurcation parameter were found to offer better optimization ability. Asai et al [7] injected various types of chaotic noise into the Hopfield network. When compared to random noise, an improved wandering among local minima and an ability to trace the optimal solution was observed when correlated noises generated by chaotic maps are used.

In this paper, a new two-phase channel assignment algorithm is proposed to solve the CAP. According to the difficulty measure of each cell, the first stage is executed to assign channels cell by cell using the frequency exhaustive strategy. If the optimum assignment solution is not obtained in the first stage, the DCN-HNN stage is then applied to continue the channel assignment until the optimum assignment is made or a maximum number of iterations is reached. The channel assignment problem is described in Section 2. In Section 3, the DCN-HNN model is discussed, and then the Improved Frequency Exhaustive Strategy with DCN-HNN (IFES-DCN) is proposed. Energy function and the weight ω are also obtained In Section 3. In Section 4, these models are applied to the CAP. Simulation results based on three main sets of test data are discussed to compare the performance of the IFES-DCN with existing methods. Section 5 presents concluding remarks.

2 Channel Assignment Problem

The channel assignment problem (CAP) in a cellular mobile communication system usually follows the common model. It must satisfy the three constraints to avoid the interference between channels [8].

(1) Co site Constraint (CSC): any pair of frequencies assigned to a cell should have a minimal distance between frequencies.

(2) Co channel Constraint (CCC): for a certain pair of radio cells, the same frequency cannot be used simultaneously.

(3) Adjacent Channel Constraint (ACC): the adjacent frequencies in the frequency domain cannot be assigned to adjacent radio cells simultaneously.

The goal of the channel-assignment problem in a cellular radio network is the assignment of the required channel number (RCN) to each radio cell such that the above constraints are satisfied. Gamst [8] defined the compatibility matrix $C = (c_{ij})$, which is an $n \times n$ symmetric matrix, where n is the number of cells in the mobile network and c_{ij} is the minimum frequency separation between a frequency in cell i and one in cell j. For example, $c_{ij} = 0$ indicates that the same frequency can be used in cell i and cell j. Therefore, CCC and ACC can be represented in matrix C by $c_{ij} \geq 1$ and $c_{ij} \geq 2$, respectively. CSC can be represented by c_{ii} and CCC, and ACC can be represented by c_{ij}. The number of channels needed for each cell is presented by the demand vector $D = (d_i)$, where $1 \le i \le n$ and n is the number of cells in the network. Let f_{ik} indicate the assigned frequency for the k th call in the cell, where $1 \le i \le n$ and $1 \le k \le d$ The condition imposed by the compatibility constraints between f_{ik} and f_{jl} is given by $| f_{ik} - f_{jl} | \ge c_{ij}$ where $1 \le i, j \le n$ and $1 \le k, l \le d$ The channel-assignment problem is to find f_{ik}'s that satisfy the constraint conditions when the number of cells n in the mobile network are given along with the compatibility matrix C and the RCN vector D.

3 The Neural Network Model

3.1 Chaotic Simulated Annealing with Decaying Chaotic Noise (DCN-HNN)

The neural network with chaotic noise is defined as follows:

$$y_i(t+1) = \phi y_i(t) + \alpha(\sum_j w_{ij} v_j + I_i) + \gamma(z_i(t) - h) \tag{1}$$

$$v_i(t) = f(u_i(t)) = \frac{1}{1 + e^{-u_i(t)/\varepsilon}} \tag{2}$$

$$z_i(t) = a(t) z_i(t)(1 - z_i(t)) \tag{3}$$

$$a(t+1) = (1 - \beta) a(t) + \beta a_0 \tag{4}$$

$$h = 1 - 1/a_0 \tag{5}$$

Where
$v_i(t)$ Output of neuron;
y_i Internal state of neuron;
w_{ij} Connection weight from the j th chaotic neuron to the i th chaotic neuron;
ϕ Memory constant keeping chaotic behavior;
α Positive scaling parameter;
ε Steepness parameter of the output function;
I_i Threshold value of the i th chaotic neuron;
$z_i(t)$ Chaotic noise;
γ Positive scaling parameter for the chaotic noise
$1 \le a_0 \le 2.9$, $0 < \beta < 1$. With $z_i(t) - h = 0$, the network described by (1)–(5) reduces to the HNN. The following will show that the transient chaos induced by the decaying chaotic noise is very useful in preventing the network from being trapped in local minima.

3.2 Improved Frequency Exhaustive Strategy with DCN-HNN (IFES-DCN)

In this section, based on frequency exhaustive strategy, the implementation procedure of progressive initialization for CAP is presented. The procedure of progressive initial assignment for CAP is as follows.

(1) At the start of the assignment, all cells have the equal initial usage of all channels. So, at the beginning, the initial usage of each channel in each cell is set to zero.

(2) Based on RCN vector $D=(d_i)$ and compatibility matrix $C=(c_{ij})$, the assignment difficulty measure for each cell in an N-cell network is described by an N-element vector, which is called the assignment difficulty measure vector DM. DM_i in the vector DM for cell i is calculated by the following equation:

$$DM_i = d_i((d_i - 1)d_{CSC} + \sum_{\substack{j=1 \\ j \neq i}}^{N} d_j C_{ij}) \tag{6}$$

(3) Repeat the following step cell by cell until all cells are assigned according to vector DM.

(a) Select a cell that is not assigned according to the descending order of vector DM, that is, the largest one in DM.

(b) Assign the selected frequency to the cell, consistent with previous assignments, without violating the separation constraints.

If the optimal assignment solution is not obtained in the stage above, the DCN-TCNN stage is applied to continue the assignment of channels until the optimum assignment is made or a pre-specified maximum number of iteration is reached.

Take 4 cells system for example, the comparability matrix is shown in fig.2, and the demand vector D= [1, 1, 1, 3].the distribute sequence of the first stage (a)-(d) in IFES-DCN is shown in fig.1.

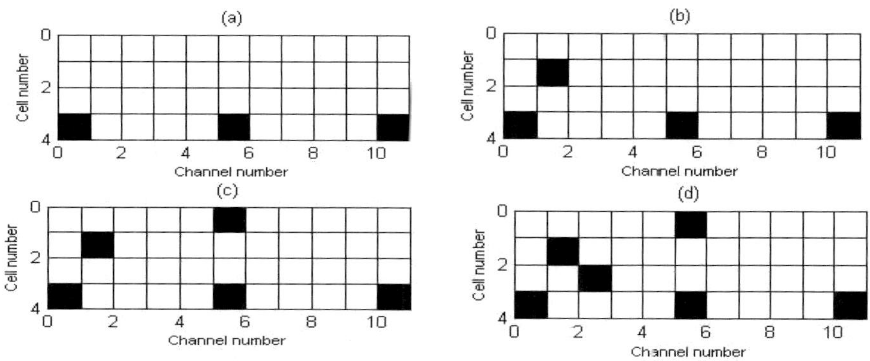

Fig. 1. The distribute sequence of the first stage in IFES-DCN for 4 cells CAP

3.3 Energy Function and the Weight ω

Suppose that there are N cells and M channels available in the network. We define: V_{ij} =1, if channel j is assigned to cell i, otherwise, V_{ij} =0. When i=1… N and j=1… M.

A cost function employed in the CAP encompasses the problem requirements and three channel constraints. Such a function can be defined as:

$$E = \frac{A}{2}\sum_{i=1}^{n}\left(\sum_{j=1}^{m} v_{ij} - d_i\right)^2 + \frac{B_1}{2}\sum_{i=1}^{n}\sum_{j=1}^{m}\sum_{\substack{q=j-(c_{ii}-1)\\ q\neq j}}^{j+(c_{ii}-1)} v_{iq}v_{ij} + \frac{B_2}{2}\sum_{i=1}^{n}\sum_{j=1}^{m}\sum_{\substack{p=1\\ p\neq i\\ c_{ip}>0}}^{n}\sum_{q=j-(c_{ip}-1)}^{j+(c_{ip}-1)} v_{pq}v_{ij} \tag{7}$$

The first term in equation (7) penalizes for the difference between the actual assigned channels and the required channels. Thus, for example, any number of assigned channels other than the number of required channels at cell i will cause the first term to be a positive integer rather than zero. The second and the third term penalize for interference violations where summations take into account all interference constraints. Then the CAP problem can be formulated to find an individual binary variables matrix V that minimizes E.

From equation (7) the interconnection weight is defined as follows:

$$w_{ijpq} = -A\delta_{ip} - B_1\delta_{ip}(1-\delta_{jq})\alpha_{jq}(c_{ii}-1) - B_2(1-\delta_{ip})\alpha_{jq}(c_{ip}-1) \tag{8}$$

And I_{ij} can be defined as:

$$I_{ij} = Ad_i \tag{9}$$

Where

① $\delta_{ij} = \begin{cases} 1, i=j \\ 0, i\neq j \end{cases}$ ② $\alpha_{ij}(x) = \begin{cases} 1, |i-j| \leq x \\ 0, |i-j| > x \end{cases}$

4 Simulation Results

4.1 Results for 4-Cell Cellular Network

The comparability matrix of 4-cell cellular network is defined as below:

$$\begin{array}{c|cccc} & C_1 & C_2 & C_3 & C_4 \\ \hline C_1 & 5 & 4 & 0 & 0 \\ C_2 & 4 & 5 & 0 & 1 \\ C_3 & 0 & 0 & 5 & 2 \\ C_4 & 0 & 1 & 2 & 5 \end{array}$$

Fig. 2. The 4-cell system (cell number is inside the cell)

Table 1 and table 2 show the simulation results according to different parameters used the DCN-HNN model. Let us varying β and γ to investigate the dynamics of the proposed chaotic neural-network, the results imply that the parameter affects not only the iteration number but also the convergence rate of the network. (GM=Rate of reaching the global minimum, NI=Number of iterations required for the network to converge). All of the NI are shown in round figures.

Table 1. Results of DCN-HNN model when γ=0.2

β	0.006	0.009	0.02
GM	100%	98%	96%
NI	184	146	82

In table 1, the decaying rate β varies from a smaller value of 0.006 to a larger value of 0.02 with γ=0.2.The number of iterations required for the network to converge decreases as the β increases.

Table 2. Results of DCN-HNN model when β=0.006

γ	0.1	0.2	0.4	0.8
GM	92%	100%	98%	93%
NI	189	184	217	269

Table 3 shows the IFES-DCN has a higher searching ability and lower computing time in searching the global minimum.

Table 3. Results of IFES-DCN model when β=0.006

γ	0.1	0.2	0.4	0.8
GM	100%	100%	100%	100%
NI	45	47	89	116

4.2 Results for 7-cell Cellular Network

The comparability matrix of 7-cell cellular network is defined as below:

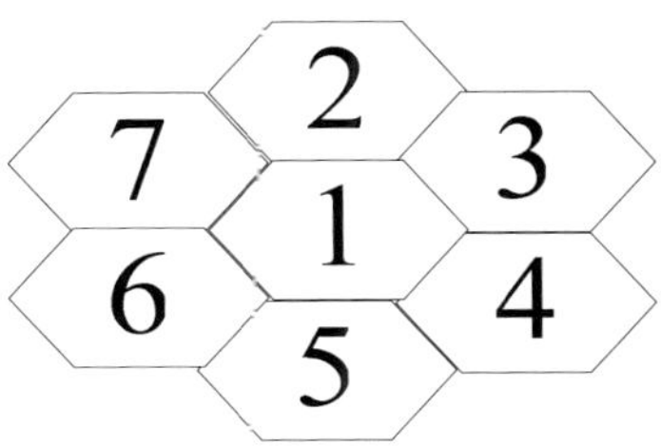

Fig. 3. The 7-cell system (cell number is inside the cell)

$$\begin{array}{c|ccccccc} & C_1 & C_2 & C_3 & C_4 & C_5 & C_6 & C_7 \\ \hline C_1 & 5 & 3 & 3 & 3 & 3 & 3 & 3 \\ C_2 & 3 & 5 & 3 & 0 & 0 & 0 & 3 \\ C_3 & 3 & 3 & 5 & 3 & 0 & 0 & 0 \\ C_4 & 3 & 0 & 3 & 5 & 3 & 0 & 0 \\ C_5 & 3 & 0 & 0 & 3 & 5 & 3 & 0 \\ C_6 & 3 & 0 & 0 & 0 & 3 & 5 & 3 \\ C_7 & 3 & 3 & 0 & 0 & 0 & 3 & 5 \end{array}$$

Fig. 4. Comparability matrix of 7-cell system

In table 4, β=0.006 is fixed and γ is changed from 0.1 to 0.3. As γ gets larger, the chaotic noise become stronger and the network needs more iteration to converge to the global minimum. The focus of the simulations is on the optimization performance (the rate of global optimization) and efficiency (the average number of iterations) with various values of parameter β and γ.The simulation results of IFES-DCN model is also shown in table 5.

Table 4. Results of DCN-HNN model when β=0.006

	γ		
	0.1	0.2	0.3
GM	80%	97.5%	70%
NI	223	233	216

Table 5. Results of IFES-DCN model when β=0.006

	γ		
	0.1	0.2	0.3
GM	90%	100%	85%
NI	127	154	138

Fig.5 and Fig.6 show the time evolutions of the continuous energy function and the time evolutions of a single neuron in IFES-DCN model.

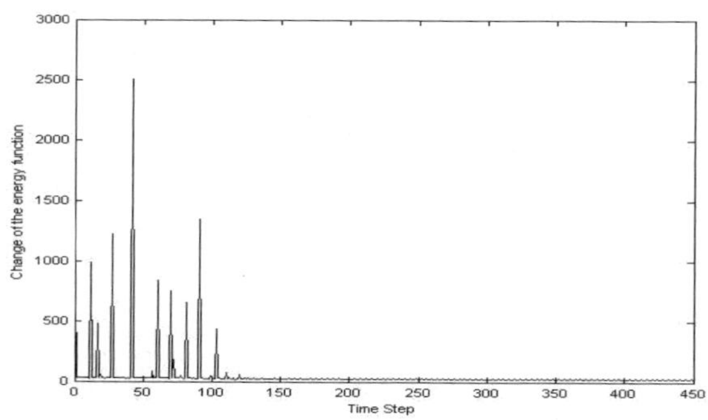

Fig. 5. Time evolutions of the continuous energy function in IFES-DCN model

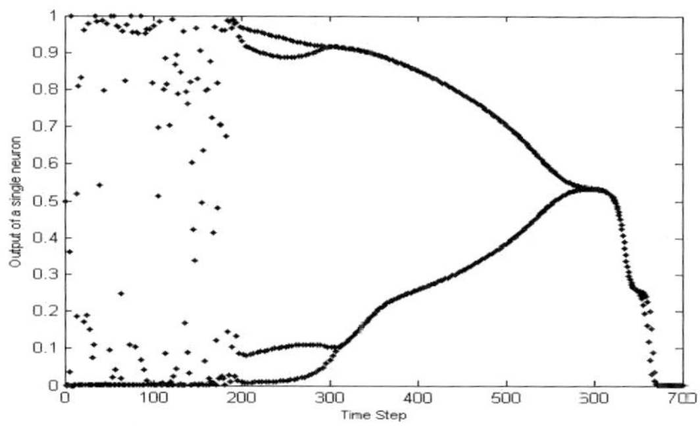

Fig. 6. Time evolutions of a single neuron in IFES-DCN model

4.3 Results for Kunz Benchmark Problems

Benchmark problems of mobile systems consisting of 25 cells in reference [1] used to evaluate the DCN-HNN and IFES—DCN in this paper Kunz's test problems is a practical CAP derived from traffic density data of an actual 24×21 km area around Helsinki, Finland. The shape of the 25-cell cellular network is shown in Fig.7 and the compatibility matrix is shown in Fig.8. The number of channels is taken to be a fraction of the 73 channels available for the entire 25 regions. The results are shown in table 6.

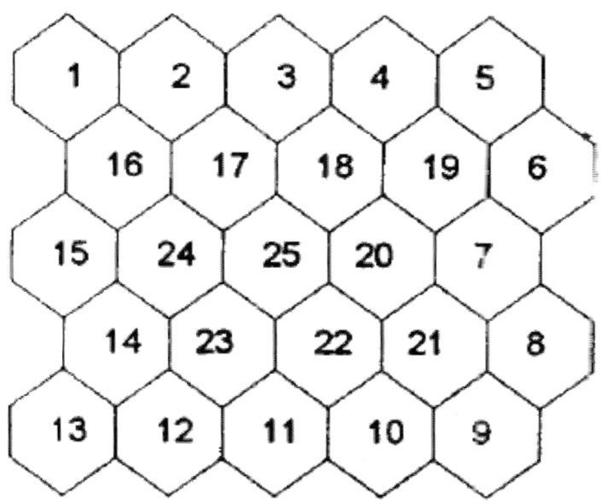

Fig. 7. The 25-cell system (cell number is inside the cell)

In reference [1], a typical demand vector D is proposed, that is:
D1=[10,11,9,5,9,4,5,7,4,8,8,9,10,7,7,6,4,5,5,7,6,4,5,7,5]. Fig.9 shows the time evolutions of a single neuron in IFES-DCN model in the 25-cell system. Fig.10 shows the time evolutions of the continuous energy function in IFES-DCN model according to the demand vector D1.

$$\begin{bmatrix}
2&1&1&0&1&0&1&1&1&1&0&1&1&1&1&0&0&0&0&0&0&0&0&0&0\\
1&2&1&0&1&0&1&1&0&1&0&1&1&1&1&0&0&0&0&0&0&0&0&0&0\\
1&1&2&1&1&1&1&1&1&1&1&1&1&1&0&0&0&0&0&0&0&0&0&0&0\\
0&0&1&2&0&0&1&1&1&1&1&1&1&0&0&0&0&0&0&0&0&0&1&1&1\\
1&1&1&0&2&0&0&0&0&1&1&1&1&1&1&1&0&0&0&0&0&0&0&0&0\\
0&0&1&0&0&2&1&1&1&1&0&0&0&0&0&0&0&0&0&0&0&0&0&0&0\\
1&1&1&1&0&1&2&1&1&1&1&1&1&0&0&0&0&0&0&0&0&0&0&0&0\\
1&1&1&1&0&1&1&2&1&1&1&1&1&0&0&0&0&0&0&0&0&0&0&1&0\\
1&0&1&1&0&1&1&1&2&1&1&1&0&0&0&0&0&0&0&0&0&0&0&1&1\\
1&1&1&1&1&1&1&1&1&2&1&1&1&1&1&1&0&0&0&0&0&1&0&1&0\\
0&0&1&1&1&0&1&1&1&1&2&0&1&1&1&1&0&1&1&1&1&1&1&1&1\\
1&1&1&1&1&0&1&1&1&1&0&2&1&1&0&0&0&0&0&0&0&0&0&0&0\\
1&1&1&1&1&0&1&1&0&1&1&1&2&1&1&1&1&1&1&1&0&0&0&0&0\\
1&1&1&0&1&0&0&0&0&1&1&1&1&2&1&1&1&1&1&1&0&0&0&0&0\\
1&1&0&0&1&0&0&0&0&1&1&0&1&1&2&1&1&1&1&1&1&1&0&0&0\\
0&0&0&0&1&0&0&0&0&1&1&0&1&1&1&2&1&1&1&1&0&0&0&0&0\\
0&0&0&0&0&0&0&0&0&0&0&0&1&1&1&1&2&1&1&0&0&0&0&0&0\\
0&0&0&0&0&0&0&0&0&0&1&0&1&1&1&1&1&2&1&1&0&0&0&0&0\\
0&0&0&0&0&0&0&0&0&0&1&0&1&1&1&1&1&1&2&1&1&1&1&0&0\\
0&0&0&0&0&0&0&0&0&0&1&0&1&1&1&1&0&1&1&2&1&1&1&0&0\\
0&0&0&0&0&0&0&0&0&0&1&0&0&0&1&0&0&0&1&1&2&1&1&0&0\\
0&0&0&0&0&0&0&0&0&1&1&0&0&0&1&0&0&0&1&1&1&2&1&1&1\\
0&0&0&1&0&0&0&0&0&0&1&0&0&0&0&0&0&0&1&1&1&1&2&1&1\\
0&0&0&1&0&0&0&1&1&1&1&0&0&0&0&0&0&0&0&0&0&1&1&2&1\\
0&0&0&1&0&0&0&0&1&0&1&0&0&0&0&0&0&0&0&0&0&1&1&1&2
\end{bmatrix}$$

Fig. 8. Comparability matrix of 25-cell system

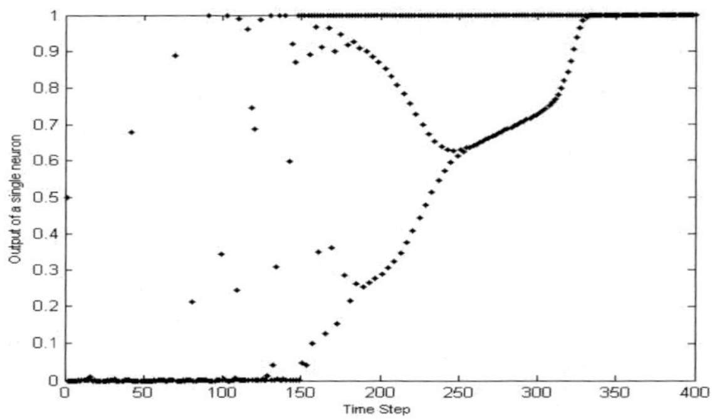

Fig. 9. Time evolutions of a single neuron in IFES-DCN model

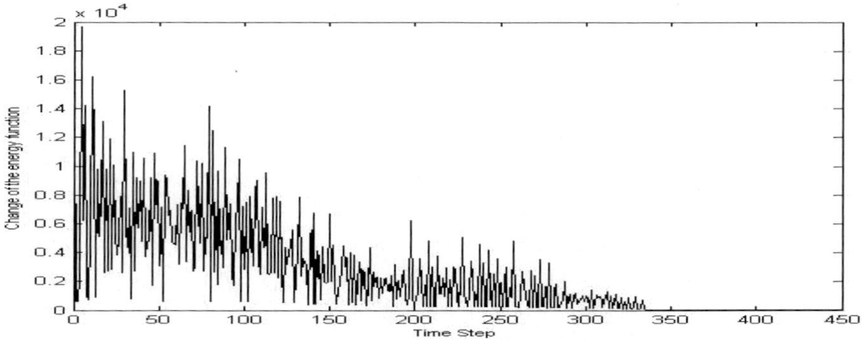

Fig. 10. Time evolutions of continuous energy function in IFES-DCN model according to the demand vector D1

Table 6. Table 6 Results of cap for various methods

	IFES-DCN	DCN-HNN	Results in[3]	Results in [2]
GM	80%	72%	62%	9%
NI	364	375	279.9	294

Table 6 summarizes the assignment results and shows the convergence rate compared with recently reported results, and also shows that IFES-DCN performs better than others. The IFES-DCN substantially improves performance through solving well-known benchmark problems, while the rate of reaching the global minimum is comparable with most existing algorithms.

5 Conclusions

The channel assignment problem in a cellular mobile communications network is very important in the telecommunications industry, finding application not just in cellular networks but also in satellite and other systems where the available frequency spectrum is a limited resource.

In this paper, we have developed and evaluated an efficient algorithm which has two stages called IFES-DCN for solving the CAP. According to the difficulty measure of each cell, the first stage is executed to assign channels cell by cell using the frequency exhaustive strategy. If the optimum assignment solution is not obtained in the first stage, the DCN-HNN stage is then applied to continue the channel assignment until the optimum assignment is made or a maximum number of iterations is reached. The assignment results on several benchmark problems indicate that the IFES-DCN performs extremely well when compared to other methods known so far to obtain optimum solutions for CAP.

Acknowledgments. This work was supported by the Scientific Research Foundation for the Returned Overseas Chinese Scholars, Shanghai Leading Academic Disciplines (under Grant No.T0103), and Shanghai Key Laboratory of Power Station Automation Technology.

References

1. Kunz, D.: Channel assignment for cellular radio using neural networks. IEEE Trans. veh. Technol. 40, 188–193 (1991)
2. Funabiki, N., Takefuji, Y.: A neural network parallel algorithm for channel assignment problems in cellular radio networks. IEEE Trans. veh. Technol. 41, 430–436 (1992)
3. Kim, J.S., et al.: Cellular Radio channel assignment using a modified Hopfield network. IEEE Trans. veh. Technol. 46, 957–967 (1997)
4. Smith, K., Palaniswami, M.: Static and dynamic channel assignment using neural networks. IEEE J. Select. Areas Commun. 15, 238–249 (1997)
5. Chen, L., Aihara, K.: Transient chaotic neural networks and chaotic simulated annealing. In: Yamaguti, M. (ed.) Toward the Harnessing of Chaos, pp. 347–352. Elsevier, Amsterdam, The Netherlands (1994)

6. Hayakawa, Y., Marumoto, A., Sawada, Y.: Effects of the chaotic noise on the performance of neural-network model for optimization problems. Phys. Rev. E 51, 2693–2696 (1995)
7. Asai, H., Onodera, K., Kamio, T., Ninomiya, H.: A study of Hopfield neural networks with external noises. In: Proc. IEEE Int. Conf. Neural. Net., vol. 4, pp. 1584–1589. IEEE Computer Society Press, Los Alamitos (1995)
8. Gamst, A., Rave, W.: On frequency assignment in mobile automatic telephone systems. In: Proc GLOBECOM'82, pp. 309–315 (1982)
9. He, Y.: Chaotic Simulated Annealing With Decaying Chaotic Noise. IEEE Transactions on Neural Networks 13 (2002)
10. Lu, H., van Leeuwen, C.: Synchronization of chaotic neural networks via output or state coupling. Chaos, Solitons and Fractals 30, 166–176 (2006)

A Novel Object Tracking Algorithm Based on Discrete Wavelet Transform and Extended Kalman Filter

Yinghua Lu[1], Ying Zheng[1,2], Xianliang Tong[1,2], Yanfen Zhang[1,2], and Jun Kong[1,2,*]

[1] Computer School, Northeast Normal University, Changchun, Jilin Province, China
[2] Key Laboratory for Applied Statistics of MOE, China
{luyh,zhengy043,Kongjun}@nenu.edu.cn

Abstract. A new method for detecting and tracking multiple moving objects based on discrete wavelet transform and Extended Kalman filter is proposed in this paper. Although Kalman filter tracks moving objects accurately, it requires a heavy computational burden. Discrete wavelet transform has a nice property that it can divide a frame into four different frequency bands without loss of the spatial information, we use Kalman filter on low frequency sub-band so it can reduce the computational burden and remove most of the fake motions in the high frequency sub-band. In tracking multiple moving objects, many applications have problems when objects pass across each other. We exploit pattern matching in a simple feature space for solving this problem. The experimental results prove the feasibility and usefulness of the proposed method.

1 Introduction

Moving object detection and tracking is one of the key tasks in computer vision. The detecting, recognizing, and tracking of moving-object are much more important in many situations, such as real-time surveillance, unpiloted vehicle control and so on [1] [2]. Approaches about moving-object detecting and tracking can broadly be classified into two classes: Camera moves with the moving-object and always keeps the object in the central of the image. Camera is fixed and tracks the moving-objects in the visual area of its own. This paper mainly concerns with multi-target detecting and tracking in stationary scene.

There are several ways to tracking vehicles and resolving occlusion from traffic video streams. In [3] tracking is performed by template matching, with a probabilistic model based on a robust error norm. Matching is performed by finding the image region that yields the maximum likelihood with respect to the calculated probability distribution. Templates are matched by translation, rotation and scaling parameters; during occlusions, a pixel is regarded as an outlier if the measurement error exceeds a predefined threshold. Templates cannot be updated during occlusion, so it is necessary to detect when overlapping ends. This algorithm can manage complete occlusions, but it cannot withstand severe occlusions lasting more than 25 frames, due to wrong Kalman filter estimation.

---

* Corresponding author. This work is supported by science foundation for young teachers of Northeast Normal University, No. 20061002, China.

K. Li et al. (Eds.): LSMS 2007, LNCS 4688, pp. 551 – 560, 2007.

Authors in [4, 5] propose an occlusion robust method based on special-Temporal Markov Random Field Model. Each frame is divided in blocks of 8*8 pixels and the feature correlation between blocks of consecutive images is performed by exploiting motion vectors for each block. In order to determine which object a block belongs to, a complex minimization problem must be solved. To this purpose a relaxation algorithm (Metropolis) is employed, thus preventing the feasibility for the system to work in real time.

In [6] authors propose to track vehicles as a set of parts to resolve partial occlusions. To fulfill the tracking purpose a specific segmentation method is proposed, which make use of active contours ('snakes'). Moving objects are initially segmented into a square grid, and then contours are moved to minimize an energy function. The goal is to obtain parts that are the most suitable for a gradient-based tracking algorithm. However, presented results show only occlusions for cars that move out of the field of view and only for one car at a time. Segmentation should also be repeated as vehicles change their appearance, resulting in a consistent computational overhead.

In this paper, we integrated discrete wavelet transform and Extended Kalman filter for reducing the computational burden and removing most of the fake motions. Although Kalman filter tracks moving objects accurately, it requires a heavy computational burden. Discrete wavelet transform has a nice property that it can divide a frame into four different frequency bands without loss of the spatial information, we use Kalman filter on low frequency sub-band, so it can reduce the computational burden and remove most of the fake motions in the high frequency sub-band. In tracking multiple moving objects, many applications have problems when objects pass across each other, we exploit pattern matching in a simple feature space for solving this problem. The experimental results prove the feasibility and usefulness of the proposed method.

In Section 2, the vehicle detection method based on background subtraction of one-scale DWT is presented. In Section 3, the extended Kalman filter tracking method which includes occlusion removal based on features matching is described. Experimental results and conclusion are given in Section 4.

2 Vehicle Detection

In the following sub-sections, we will present the vehicle detection approach. Firstly, the background is extracted, then the background subtraction method based on wavelet transform is proposed, and then propose a background updating method which can resolve the background updating deadlock problem, at last the centroids of the moving vehicles are located.

2.1 Background Extraction

To accommodate the small variations of background pixel values due to these disturbances, gray-level quantification is introduced to aggregate several gray values which in a small range into one range, so 256 gray levels can be regrouped into Θ ranges, $\Theta = 256/s$, where s is the size of the range, the choice of the parameter s is done empirically (in our experiment $s = 4$), so it can make the proposed algorithm more robust. The range of each gray-level is:

$$[l * s, (l+1) * s - 1] \quad l = 0,1,2,...,\Theta - 1,$$

where s is the size of the range, l is gray-level after quantification. We adopt the method used in [7] to introduce two parameters, each region has two parameters: one is 'regional mean' μ_i, which is the weighted average of all pixel intensities in the gray level range i; the other is 'regional counter' C_i, which is the numbers of video sequence in the gray-level range i. They both update when the video sequence changes. Considering the heavy weight of current frame, two attenuation weights are introduced to reduce the impact of the environment lighting condition. We update μ_i and C_i by formulas (1) and (2). Regard μ_i belonging to the range which has the biggest C_i as background pixel value of pixel (x, y). For all pixels of frames in video sequence, repeat the work of above, we can get a background image. The extracted background image is show in Fig.1.

$$\mu_{i,n} = \begin{cases} \dfrac{\mu_{i,n-1} * \sum_{i=1}^{C_{i,n-1}} \omega_1 + F_n}{\sum_{i=0}^{C_{i,n-1}} \omega_1} & F_n \in [i * s, (i+1) * s - 1], \\ \mu_{i,n-1} & F_n \notin [i * s, (i+1) * s - 1], \end{cases} \tag{1}$$

$$C_{i,n} = \begin{cases} \omega_2 * C_{i,n-1} + 1 & F_n \in [i * s, (i+1) * s - 1], \\ \omega_2 * C_{i,n-1} & F_n \notin [i * s, (i+1) * s - 1], \end{cases} \tag{2}$$

where $\omega_1, \omega_2 \in [0, 1]$ are used to reduce the impact of the history information. The slower the illumination changes between the frames, the bigger values of the speed coefficient are selected.

(a) (b)

Fig. 1. (a) one frame of video sequence (b) extracted background

2.2 Background Subtraction Based on Discrete Wavelet Transform

When digital images are to be viewed or processed at multiple resolutions, the discrete wavelet transform (DWT) is the mathematical tool of choice [8]. In addition to being an efficient, highly intuitive framework for the representation and storage of multi-resolution images, the DWT provides powerful insight into an image's spatial and frequency characteristics. The Fourier transform, on the other hand, reveals only an image's frequency attributions. One dimensional discrete wavelet transform (DWT) can be used to decompose an image into four sub-images (LL, HL, LH and HH). An example of one-scale DWT is shown in Fig .2. We can see that the high frequency sub-images, such as HH and HL, contain the detailed information like the edge of leaves. Because the moving vehicles belong to low frequency part and we don't take more attention to the detailed information of the frame, so our background subtraction is performed in the low frequency sub-image LL of the one-scale DWT (the left corner in Fig. 2 (b)) due to the consideration of low computing cost and noise reduction issue.

(a) (b)

Fig. 2. DWT: (a) Original image (b) the result of one-scale DWT

After the background extracted, the foreground pixels can be extracted by the following equation:

$$D_{x,y} = \begin{cases} 1 & if \left| F^{LL}_{x,y} - B^{LL}_{x,y} \right| \geq T1 \\ 0 & otherwise \end{cases} \tag{3}$$

where $F^{LL}_{x,y}$ and $B^{LL}_{x,y}$ are the low frequency sub-image of the one-scale DWT corresponding to the current input image and background image. $T1$ is a threshold which is acquired by ostu method.

2.3 Background Updating

Traditional background updating algorithms only update background pixels expect the pixels of moving objects, they can make the system model more stable, but they only consider the pixel-level updating, so they may cause the problem of background updating deadlock (when the objects, which belong to the background, start moving, the method of background subtraction using the traditional background updating algorithm will get a false object. The selectivity background updating methods are part of pixel-level updating, so it is helpless for regional updating, and then the false object will be detected as a moving object. This is called 'background updating deadlock problem'). A novel method is presented for solving this problem. In what follows, details of the presented updating method are described. Firstly, in order to understand the method easier, two discriminant functions T and S in literature [9] are introduced to distinguish the false moving objects and the moving objects.

$$T = \max\{| D_n - D_{n-j} |\}, \ j \in [1, 5] \tag{4}$$

T is the maximum of absolute value which is from the gray difference between current difference image foreground regions and previous 5 difference images foreground regions.

$$S = \frac{k\sum_{j=0}^{k} D_{(n-j)}^2 - (\sum_{j=0}^{k} D_{n-j})^2}{k(k-1)}, \quad k \in [5, 10] \tag{5}$$

where S denotes the degree of difference image foreground region gray variety in a period of time, the choice of the parameter k is done empirically. We use background subtraction to get foreground objects, calculate T and S of each foreground object, a small T and S indicate that it is a false moving object; otherwise it is a moving object. In what follows, details of our Background updating method are described.

2.3.1 Background Updating Algorithm

Step.1. Extract the perfect initialization background using the proposed method.

Step.2. Use DWT and background subtraction to gain a difference image on the low frequency sub-image LL of the one-scale DWT, after that, a series of simple morphological operations and mean filter are applied for noise removing.

Step.3. Distinguish the moving objects and false objects using the two discriminant functions (4) and (5).

Step 4. Use formula (6) to update the moving objects and the false objects separately.

$$B^{LL}{}_n(x, y) = \begin{cases} \omega * B_{n-1}^{LL}(x,y) + (1-\omega) * F_n^{LL}(x, y) & \{F_n^{LL}(x, y) \in \textit{moving objects}\}, \\ (1-\omega) * B_{n-1}^{LL}(x,y) + \omega * F_n^{LL}(x, y) & \{F_n^{LL}(x, y) \in \textit{false objects}\}, \\ B_{n-1}^{LL}(x, y) & \{\textit{others}\}, \end{cases} \tag{6}$$

where $B_n^{LL}(x, y)$, $B_{n-1}^{LL}(x, y)$ are the intensities of background of frame n and the frame

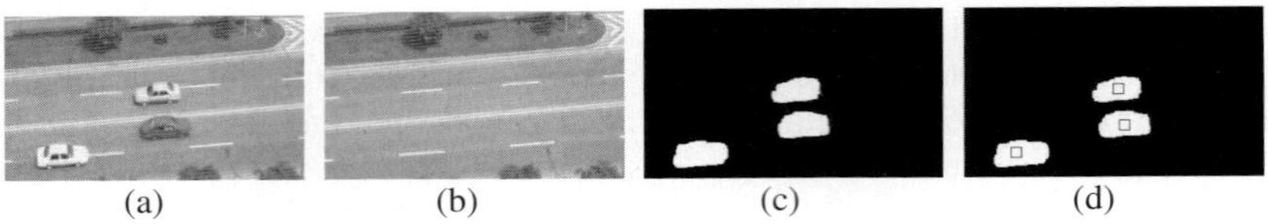

Fig. 3. (a) Image of current frame, (b) Background image after background updating, (c) The difference image of background subtraction, (d) Centroids of the moving vehicles

Fig. 4. Vehicle detection which exists vehicles from stillness to moving. Fig.4. (a) (e) Image of current frame, Fig.4. (b) Background image after traditional updating method, Fig.4. (c) (d) False detected vehicles using background (b), Fig.4. (f) Background image after proposed updating method, Fig.4. (g) (h) True detected vehicles using background (f).

$n-1$ after discrete wavelet transform. $F_n^{LL}(x, y)$ is the intensity of the frame n after discrete wavelet transform , where $\omega \in [0,1]$ is a constant that controls the rate of the adaptation of the background to the current frame and determines the sensibility of the update to the variations. The higher the illumination changes (or the presence of moving objects) between the frames, the lower value (close to zero) of the speed coefficient is selected, so that the background model is mostly influenced by the current image.

3 Object Tracking

3.1 Extended Kalman Filter

For tracking objects in a sequence of frames, the relation between objects in two consecutive frames must be found and recorded. Doing this we developed three modules. The first one is a complete image search, in which the whole image is searched. In the second module, only the area on the road (i.e. excluding the background) is searched. And in the last module, a small area in which the vehicle might be seen in next frames was determined and the trajectory of each vehicle was tested. These three modules were tested. The first two modules were very time consuming and could not be used in real time, because their processing time were 3 and 6 frames per second, respectively. But the last module could be run in 25 frames per second. Therefore, it was used in our tracking procedure. In order to reduce the cost of search operation, we used the extended Kalman filter (yongkang Chen 2005) on the low frequency sub-image LL of the one-scale DWT to predict the location of trajectory of a vehicle in frame sequences. First, we should calculate vehicle's centroid coordinate. Use extended Kalman filter to predict the position of vehicle in next frame. And then adjust its parameters for reducing its search area to achieve fast tracking. This filter composes six parameters in x axis direction.

$$X_k = (x_k, \hat{x}_k, \Delta x_k, v_{x,k}, \hat{v}_{x,k}, \Delta v_{x,k})$$

where x_k is the centroid position of moving object at time k, $\hat{x}_k$ is the predict centroid position of moving object at time k, Δx_k is the search area, $v_{x,k}$ is the centroid measure velocity of moving object. $\hat{v}_{x,k}$ is the centroid estimate velocity of moving object. $\Delta v_{x,k}$ is the error of velocity. The process of tracking are described as follows:

Step.1. Initialize position and velocity of moving objects.
Step.2. Predict moving state of next time.

$$\hat{x}_{k+1} = \hat{x}_k + \hat{v}_k * \Delta t, \quad \Delta x_{k+1} = \Delta x_k + \Delta v_{x,k} * \Delta t$$

Step.3. Matching

After predict the centroid of the vehicle by Kalman filter, then calculate distance between predicted centroid and detected centroids in searched areas, if the distance is equal to zero or less than a threshold, we consider it as '1' which means that detected centroid is match for predicted centroid, otherwise, regard as '0' which means that their cannot match. After that, we can construct an index table like table.1. Ideally, one detected centroid corresponds to one predicted centroid, If one detected centroid have no predicted centroid to match (example centroid 5'), it means that there is a new vehicle coming, and then establish a new tracking object, if there have no detected centroid to match (example centroid 5), it means that there is a vehicle leaving, tracking will be stop. If one detected centroid corresponds to multi-predicted centroids, we consider as occlusion occurring, if one predicted centroid corresponds to multi-detected centroids, then consider as occlusion releasing.

Use index table to judge occlusion occurring or occlusion releasing, when occlusion occur, extract their features respectively, in this paper, we use mean and variance. We track occlusion vehicles as one during occlusion, when occlusion releasing, extract their features again, then match in the feature space, and record their matching centroids, after that, we can remove the occlusion.

After all vehicle matched, record their true positions, calculate their measure velocity $v_{x,k+1}$

$$v_{x,k+1} = (x_{k+1} - x_x) / \Delta t$$

Table 1. Matching matrix index table

Detected centriod / Predicted centroid	1’	2’	3’	4’	5’
1	0	1	1	0	0
2	0	0	0	1	0
3	0	1	0	0	0
4	0	0	0	1	0
5	0	0	0	0	0

Step.4. Correct filter parameters using the following formulas.

$$\hat{v}_{x,k+1} = \partial 1 * v_{k+1} + (1 - \partial 1) * \hat{v}_{x,k}$$

$$\Delta v_{x,k+1} = \partial 1 * \left| \hat{v}_{k+1} - v_{k+1} \right| + (1 - \partial 1) * \Delta v_{x,k}$$

$$\Delta x_{k+1} = \beta 1 * \left| \hat{x}_{k+1} - x_{k+1} \right| + (1 - \beta 1) * \Delta x_k$$

where $\partial 1$, $\beta 1$ are constants between [0 1] choiced by experience. We can design another filter in y axis direction, therefore, to track in two-dimension.

4 Experimental Results and Conclusions

4.1 Experimental

In order to test our algorithm, we used image sequences of about 400 frames of a video and another video sequence of about 500 frames tape captured by a common CCD camera. This camera was installed on the side-view of the road. The result of the tracking without occlusion removal is shown in Fig. 5. , the result of the tracking with occlusion removal is shown in Fig.6.

4.2 Conclusions

A new method for detecting and tracking multiple moving objects based on discrete wavelet transform and Extended Kalman filter is proposed in this paper. Although Kalman filter tracks moving objects accurately, it requires a heavy computational burden. Discrete wavelet transform has a nice property that it can divide a frame into four different frequency bands without loss of the spatial information, we use Kalman filter on low frequency sub-band so it can reduce the computational burden and remove most of the fake motions in the high frequency sub-band. In tracking multiple moving objects, many applications have problems when objects pass across each other. We exploit pattern matching in a simple feature space for solving this problem. The experimental results prove the feasibility and usefulness of the proposed method.

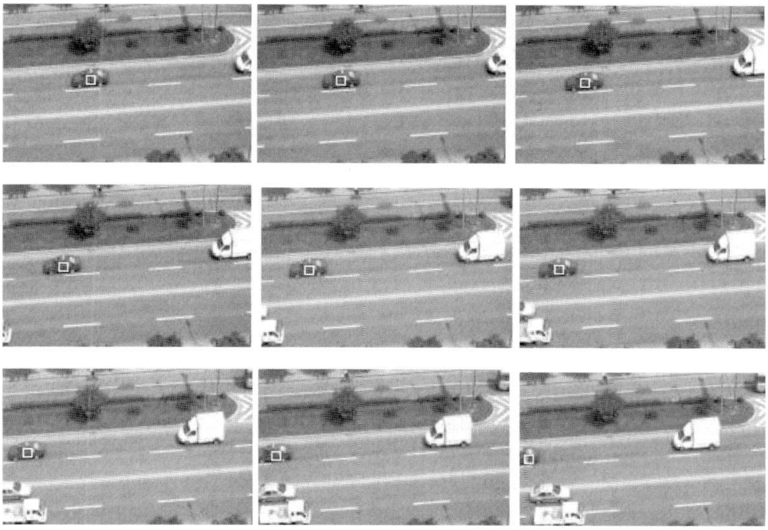

Fig. 5. The object tracking results sequence of the frames 10, 14, 18, 22, 26, 30, 34, 38 and 42 are shown

Fig. 6. The object tracking result sequence with occlusion of the frames 1, 5, 9, 13, 17, 21, 25, 29 and 32 are shown

References

1. Boghossian, B.A: Motion-based image processing algorithm. [Ph.D.Thesis], King's College, London (2000)
2. Collins, R., Lipton, A., Kanadeand, T., Fujiyoshl, H., Duggins, D., Tsin, Y.: A System for Video Surveilance and Monitoring. VSAM Final Report, Carnegie Mellon University, CMU-RI-TR-00-12 (2000)
3. Nguyen, H.T., Smeuklers, A.W.M.: Fast Occluded Object Tracking by a Robust Appearance Filter. In: Pattern Analysis and Machine Intelligence, IEEE Transactions, pp. 1099–1104. IEEE Computer Society Press, Los Alamitos (August 2004)
4. Kamijo, S., Matsushita, Y., Ikcuchi, K., Sakauchi, M.: Occlusion Robust Tracking utilizing Spatio-Temporal Markov Random Field Model. In: International Conference on Pattern Recognition, ICPR, Barcelona, vol. 1, pp. 142–147 (2000)
5. Kamijo, S., Nishida, T., Sakauchi, M.: Occlusion robust and Illumination Invariant Vehicle Tracking for Acquiring Detailed Statistics from Traffic Images. IEICE TRANSACTIONS on Information and Systems E85-D(11), 1753–1764 (2002)
6. Gentile, C., Camps, O., Sznaier, M.: Segmentation for Robust Tracking in the Presence of Severe Occlusion. IEEE Transaction On Image Processing 13(2) (February 2004)
7. Bo, L., Ming-Xu, E., He-Qin, Z.: A Zone-Distribution Based Adaptive Background Abstraction Algorithm. Pattern Recognition and artificial intelligence 18(3), 316–321 (2005)
8. Gonzalez, R.C., Woods, R.E., Eddins, S.L.: Digital Image Processing Using MATLAB (2004)
9. Xiao-peng, D., Hua-wei, W., Huang, H.: Novel Method for Background Modeling and Update. Application Research of Computers, 141–239 (May 2005)

Shallow-Water Bottom Target Detection Based on Time, Frequency Dispersive Channel and Adaptive Beamforming Algorithm

Qiang Wang[1] and Xianyi Gong[2]

[1] College of Quality and Safety Engineering, China Jiliang University, 310018 Hangzhou, P.R.China
wangtulip@yahoo.com.cn
[2] Department of Information and Electronic Engineering, Zhejiang University, 310027 Hangzhou, P.R.China
qiangwang@cjlu.edu.cn

Abstract. In active sonar system, the detection of bottom object in shallow water is usually a difficult task since the reverberation, which consisting of a large number of spurious echoes, and generates a lot of false alarm. The propagation channel exhibits time-spreading distortion(TSD) and fast fading distortion(FFD) in ocean. The distortion mechanisms degrade optimum detector performance and reduce the performance of high resolution detector in shallow water. The recursive adaptive beamforming based on the recursive least square(RLS) and the generalized sidelobe canceller(GSC) structure is constructed for uniform circular receiving array application. Then we combined detectors based TSD/FFD channel model with RLS-GSC, to enhance bottomed object detection performance in littoral area. The approach is evaluated experimentally using real-data set and in simulation test. It improves detection signal-to-reverberation ratio(SRR) and also reduce false alarm probability in real conditions.

1 Introduction

Active sonar involves transmitting an acoustical signal from a source and receiving reflected echoes from the object of interest to detect, classify, and localize underwater objects. In active sonar, underwater target detection is often limited by the presence of reverberation. A common problem in active sonar is that of multipath propagation in the ocean environment. As a result, target reflections may appear. Reverberation is a special interference, which adjont transmitted signal, and have relationship with transmitted waveform [1]. Reverberation is caused by the reflections, the diffractions, and/or the diffusions of the transmitted signal on the surface or the bottom of ocean interfaces [2]. It resultes in several problems for target detection. First, the signal-to-reverberation-noise ratio (SRR) is very weak because reverberation is composed of many strong echoes. Another problem is that the reverberation duration is much more important than the transmitted signal duration. But the most disturbing problem

K. Li et al. (Eds.): LSMS 2007, LNCS 4688, pp. 561 – 570, 2007.

comes from the strong correlation between reverberation and the transmitted signal, which makes classical detection methods, such as matched filtering, inefficient [2-3].

However, a real ocean has many dispersive transmission characteristics in both time and frequency. The utility of using broadband waveforms has been investigated as a means to improve active sonar detection performance against slow moving or stationary targets in shallow water [4]. The motivation for the deployment of broadband active system is their potential to enhance underwater objects detection performance in reverberation-limited area through increased signal bandwidth [5].

Broadband waveforms achieve their advantage by spreading their energy over a wide frequency band. This lowers the reverberation spectral level while allowing the target echo energy to be recovered via coherent integration (i.e. replica correlation) at the receiver. Ideally, fully coherent broadband processing provides significantly improved sonar performance in reverberation-limited environments [6]. That is to say, a higher SRR, finer range resolution, and improved bearing resolution and Doppler sensitivity as a consequence of operating at higher frequencies [7]. In reference [8], authors investigate the effects of coherence loss on detection in reverberation-limited environments, and provide insight into the optimum sonar receiver design.

Receiving arrays are linear assembles of hydrophones designed to increment signal-to-noise ratio(SNR) and directionality [9]. We designed circular array(CA) to receiving bottomed targets echoes. The advantages of CA: it is circular symmetrical, owns equal main beam width in 360° and exhibits vertical spatial discrimination capability [11-12]. The two main types of signals utilized are continuous wave(CW) and frequency modulated(FM) pulses. The FM pulses are characterized by poor Doppler resolution but high range resolution and reverberation suppression [3], [10]. In this paper only linearly frequency modulated(LFM) pulses are considered.

In this paper, to reduce transmitted waveform bandwidth and enhance sonar penetration ability in sediment, a uniform circular array is designed for bottom object detection. The sonar processing scheme combined RLS-GSC beamforming with segmented replica correlation (SRC) and replica correlation integration (RCI) detectors is proposed to improve underwater object detection performance in reverberation-limited condition.

2 Time and Frequency Spread Distortion Channel

Real ocean has many dispersive transmission characteristics in both time and frequency, such as turbulence, currents, and internal waves that cause time varying fluctuations in the index of refraction [7]. This results in time spreading and coherence loss of the signal. The primary effect is distortion to the received target echo as a result of time varying multipaths, micro-path spreading, and Doppler shifts. After propagating through a time-varying random ocean, the target echo has become "mismatched" to the transmitted replica, which results in SRR degradation and non-idealized signal ambiguity function properties [2].

2.1 TSD and FFD

In Fig. 1, the depiction of time and frequency spreading presents time and frequency spread effect of transmitted waveform and echo. Transmitted LFM signal parameters are: bandwidth B=1250Hz–2000Hz, pulse duration T=512ms. The echo and its spectrum is presented in Fig. 1. It is clear that the echo waveform time spread is 552ms, frequency bandwidth is 1100Hz–2000Hz. It turns out the echo pulse duration spread, and exists frequency band shift owing to propagation channel effects.

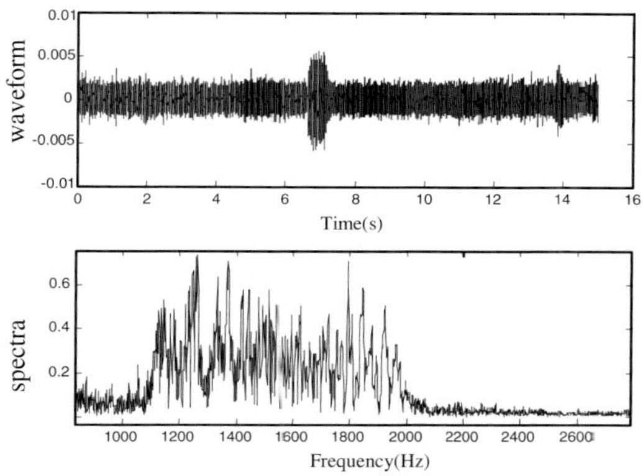

Fig. 1. Received echo waveform and its spectra

During propagation and reflection from the target, the transmitted signal is subject to two types of distortions: time spreading distortion(TSD) and fast fading distortion(FFD) [12]. The TSD usually occurs in multipath channels and is caused by multiple propagation paths. In this case, the received echo from the target can be represented as the convolution of the transmitted signal with a time spreading function $h(t)$ of effective length T_S [13].

$$x(i) = \frac{a}{\sqrt{f_s T_s}} \sum_{l=0}^{f_s T_s - 1} h(l) s(i - n_0 - l) \tag{1}$$

where a denotes the echo attenuation, n_0/f_s is the two-way propagation delay, and $h(l)$ is assumed as a realization of a random zero mean complex white Gaussian process of unity variance [14].

The FFD is caused by the dynamics of the ocean environment. The received echo from the target is modelled by modulation of the transmitted waveform with a time-varying function $w(t)$, which is usually characterized by its bandwidth $B_c = 1/T_c$, where T_c is the coherence time. The received echo from the target is given by

$$x(i)=aw(i)s(i-n_0) \tag{2}$$

$$w(i) = \sum_{k=0}^{M_c-1} \tilde{w}_k \{U[i - kN/M_c] - U[i-(k+1)N/M_c]\}, \quad kN/M_c \le i < (k+1)N/M_c \tag{3}$$

where $\tilde{w}_k$ is a zero mean complex white Gaussian random process of unit variance, and $U(k)$ is the unit step function, and M_c=T/ T_c is the number of segments.

The TDS and FFD distortions have an equivalent effect on the waveform, there is a Fourier pair relationship between the time spreading function and the corresponding frequency spreading modulation function [14].

2.2 Likelihood Rate Test of TSD and FFD

In the case of an FFD channel, the likelihood rate test(LRT) statistic is segmented replica correlation (SRC) detector, which is given by

$$y_{src}(n) = \sum_{k=0}^{M_{src}-1} \left| \sqrt{\frac{2M_{src}}{N}} \sum_{i=0}^{\frac{N}{M_{src}}-1} s^*\left(i + k\frac{N}{M_{src}}\right) r\left(n + i + k\frac{N}{M_{src}}\right) \right|^2 \tag{4}$$

Using the results in [14], the LRT detector for a TSD channel is replica correlation integration(RCI). The replica length is $N{=}Tf_s$, where $M_{rci}{=}T_sf_s$, T_s is time-spreading width. The discrete time RCI is given by

$$y_{rci}(n) = \sum_{k=0}^{M_{rci}-1} \left| \sqrt{\frac{2}{N}} \sum_{k=0}^{N-1} s^*(i-k) r(n+i) \right|^2 \tag{5}$$

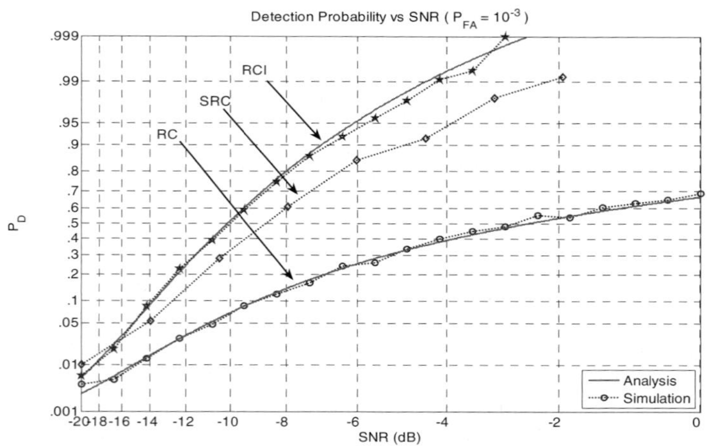

Fig. 2. Detection probability curves of three detectors in FFD channels

The simulation and theoretical analysis of LFM signal in FFD channel with replica correlation(RC), SRC and RCI detectors are presented in Fig. 2. We performed Monte Carlo simulations in order to obtain some experimental receiver operating characteristics curve(ROC curve). These curves represent the detection probability versus the false-alarm probability under constant false alarm probability P_{FA}=0.001. We evaluate these curves experimentally on the three detectors. The parameters of the

LFM signal are B=1kHz, T=128ms, sample frequency f_s=1kHz. The FFD coherence time is T_c=16ms. The parameters of SRC and RCI: M_{src}=8, M_{rci}=8, respectively. The same to FFD channel parameter M_c=8. The SRC is FFD channel LRT statistics, owns the best performance, but RC gains worse performance.

The ROC curves of three detectors detection probability in TSD channel are shown in Fig. 3. The parameters of the LFM signal, SRC and RCI, are all the same as in the previous example. The TSD channel effective length T_s=8ms. So The RCI detector exhibits the best performance in TSD channels, while RC detector has high false alarm probability.

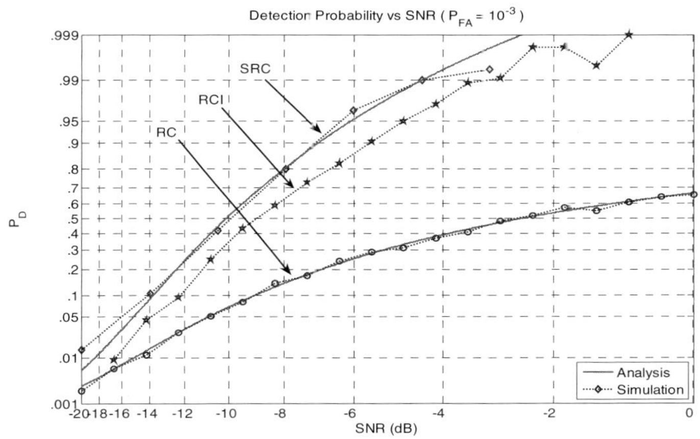

Fig. 3. Detection probability curves of three detectors in TSD channels

3 Adaptive Beamforming of Circular Array

3.1 Conventional Beamforming of Circular Array

For a uniform circular array(CA), considering a CA of radius r with M sensors, symmetrically placed on the circumferencce, see Fig.4. Let a plane wavefront be incident on the receiving array at azimuth angle θ_s and elevation angle φ_s. The wavenumber vector **k** is written as [15]:

$$\mathbf{k} = \frac{2\pi}{\lambda}[\sin\theta_s \cos\varphi_s \ \sin\theta_s \sin\varphi_s \ \cos\theta_s]^T \tag{6}$$

The array position parameter **P** is :

$$\mathbf{P} = R[\cos\varphi_n \ \sin\varphi_n \ 0]^T \tag{7}$$

The receiving signal of the transmitted signal $s(t)$ is presented as:

$$x(\varphi,t)=s(t)\exp\left[j\frac{2\pi}{\lambda}\sin\theta_s\cos\left(\varphi_s-\varphi_n\right)\right] \tag{8}$$

The conventional beamforming(CBF) is presented as:

$$B_{cir}(\theta,\varphi)=\sum_{n=0}^{N-1}w(\varphi_n)s(t)\exp\left[j\frac{2\pi}{\lambda}\sin\theta_s\cos\left(\varphi_s-\varphi_n\right)\right] \tag{9}$$

In Eq. (9), $w(\varphi_n)=\exp\left(-j\frac{2\pi}{\lambda}\sin\theta\cos(\varphi-\varphi_n)\right)$.

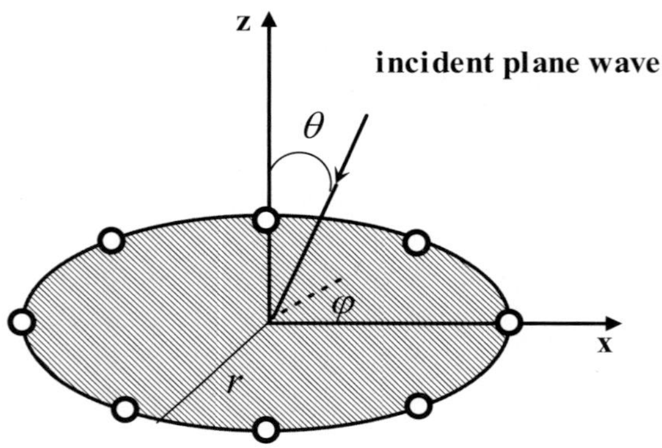

Fig. 4. Geometric configuration and coordinate system for circular array

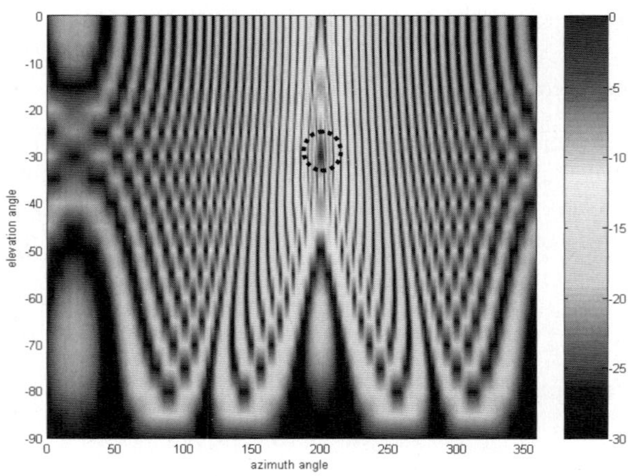

Fig. 5. The beampattern of uniform CA

In simulation test, we assume sound source location is (ϑ_s=200°, φ_s =−30°), the uniform circular array radius r is 0.55m, which contain 90 hydrophones, simulation frequency is 17kHz. The sound speed in water c=1500m/s. In Fig. 5, the the main beam is high than the first side-lobe about 8 dB. The dashed circle indicates true target location.

The Fig. 6 is the simulation beampattern of circular array, beam vs azimuth angle, when elevation angle is -30^0.

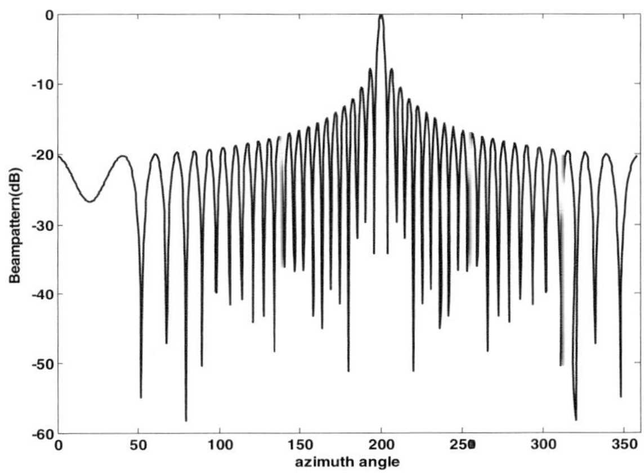

Fig. 6. Azimuth power pattern from beamforming(elevation angle=-30^0)

3.2 Adaptive Beamforming Based on RLS-GSC

A key component of modern sonar processing schemes is adaptive beamforming which enables the simultaneous formation of a beam maximum toward the signal of interest and ideal nulls toward directions of interfering signals [16]. Traditional adaptive broadband beamformer usually employs tapped-delay lines or linear transversal filters with adaptive coefficients to generate appropriate beam patterns for suppressing undesirable interference. This usually requires considerable number of adaptive coefficients resulting in a rather long convergence time and high implementation complexity.

Assume that the desired signal impinges the array at an azimuth angle, to recover the desired signal from the array out, the minimum variance distortion-less response (MVDR) beamformer is often considered. The basic idea of MVDR is to choose the weight vector **w** such that the output energy of the array is minimized, while requiring the response of the array in the looking direction to be 1. Thus, optimum **w** can be obtained by sample matrix inversion (SMI) method. The SMI method, however, is very computational expensive because it requires the inversion of the autocorrelation matrix. Alternatively, the weight vector **w** can be solved recursively using recursive least squares (RLS) algorithm.

The generalized sidelobe canceller (GSC) is an alternative approach to the MVDR [18]. It reduces the adaptive problem to an unconstrained minimization process. The

GSC formulation produces a much less computationally intensive implementation. The problem is reduced to the following [12].

Minimize: $$\sigma_u^2 = \{[\overline{w} - \mathbf{C}\overline{u}]^* \mathbf{R}[\overline{w} - \mathbf{C}\overline{u}]\} \tag{10}$$

which is satisfied by

$$\overline{u}_{opt} = (\mathbf{C}^* \mathbf{R} \mathbf{C})^{-1} \mathbf{C} \mathbf{R} \overline{w} \tag{11}$$

with $\mathbf{u}_{opt}$ being the value of the weight at convergence. The matrix **C** may be computed from **CI**=0 where **I** is a vector of ones. This matrix will satisfy **v=Cu**. The filter vector is defined as $\overline{W} = \overline{w} - C\overline{u}$ [15].

In reference [17], beamformer based GSC-RLS algorithm is applied in smart antenna, which requires much fewer variable taps than CBF of CA for the same steady-state performance, while offering much faster convergence speed. In the paper, we use RLS-GSC algorithm to implement uniform circular array beamformer in underwater target detection.

4 Discussion of Real Data Analysis

The test was conducted in a lake. Reverberation is dominated zero Doppler interference at short range but as range increases and reverberation dies out to the ambient level, and the signal becomes noise limited. The bottomed mine-like object detection test is conducted in reverberation-limited area. Receiver is a circular array with 91 hydrophones.

The CBF plus matched filter(CBF+MF), and RLS-GSC with RCI(RLS-GSC+RCI) were compared against each other to determine if there was any improvement in the bottom object detection performance in shallow-water. The RC detector is equivalent to matched filter. The RLS-GSC algorithm is presented in [17]. The transmitted LFM signal bandwidth is 20-25kHz, and pulse duration is 10ms. From the outputs of CBF+MF, the target is not obvious because of strong bottom reverberation interference. The CBF+MF is not able to discriminate bottom target at low SNR.

The RLS-GSC+RCI and RLS-GSC+SRC are applied to deal with the same gathered data. The results of RLS-GSC+RCI show the RCI processor outperforms the matched filter in reverberation limited area. The hgih resolution adaptive beamforming algorithm (RLS-GSC) gains better reverberation suppression effect than CBF. The output of RLS-GSC+SRC shows it enhances detection performance and improves SRR about 3dB.

On the whole, the high-resolution adaptive beamforming(RLS-GSC) owns capability to suppress reverberation, compared with CBF. Moreover, SRC and RCI detectors reduces the false alarm probability, smooth the pulse of false, and outperform the matched filter. So RLS-GSC+SRC and RLS-GSC+RCI gain good detection performance and are suitable for circular array implementation.

5 Conclusions

The paper discussed the time and frequency dispersive channels model in shallow water. Based on TSD and FFD, a modern experimental sonar system suitable for active detection in the littoral area is presented. The system consists of a circular receiving array with vertical discrimination capability. The adaptive beamforming principle of circular array is discussed. The standard processing chain associated with a system is improved to include RLS-GSC beamforming, SRC and RCI detection and bottomed mine forming to accommodate detection and localization. The real data analysis proves RLS-GSC+SRC and RLS-GSC+RCI are better than CBF+MF detector, even in the low SNR. The high-resolution adaptive beamforming with capability to suppress reverberation, compared with CBF, also SRC and RCI detectors reduce the false alarm. Evaluated theoretically detection performance of three detectors in simulation. Finally the detection outputs of bottomed mine-like in reverberation-limited conditions have confirmed the efficiency of the scheme. It not only indicate that TSD and FFD channel models are quite effective in reverberation suppression but also demonstrate that the adaptive beamformer proposed improves SRR in bottom target detection.

Acknowledgments. The project supported by National Natural Science Foundation of China (No. 50676088). And partially supported by Project of the National Ocean Acoustics Key Lab (No. 514460704).

References

1. Kinsler, L.E., Frey, A.R., Coppens, A.B., Sanders, J.V.: Fundamentals of Acoustics, 4th edn. John Wiley & Sons, New York (2000)
2. Etter, P.C.: Underwater Acoustic Modeling and Simulation, 3rd edn. Spon Press, London (2003)
3. Waite, A.D.: Sonar for Practising Engineering, 3rd edn. John Wiley & Sons, London (2002)
4. Ginolhac, G.: Morphological and Statistical Approaches to Improve Detection in the Presence of Reverberation. IEEE Trans. on Oceanic Engineering 30, 881–899 (2005)
5. Baggenstoss, P.M.: Adaptive pulselength correction(APLECORR): A strategy for waveform optimization in ultrawideband active sonar. IEEE Trans. On Oceanic Engineering 23, 1–11 (1998)
6. Van Trees, H.L.: Detection, Estimation, and Modulation Theory: Part I. John Wiley & Sons, New York (2001)
7. Lynch, R., Kirsteins, I., Kelly, J.: The Detection and Classification of Broadband Signals in Reverberant Dispersive Environments. In: IEEE Aerospace 2005 Conference (March 2005)
8. Lynch, R., Kirsteins, I., Kelly, J., Willett, P.: Analytical Results in Detecting Broadband Signals in Reverberation Dispersive Environments. In: IEEE Aerospace 2006 Conference (March 2006)
9. Owsley, N.L.: Sonar Array Processing. In: Haykin, S. (ed.) Array Signal Processing, Prentice-Hall, Inc., Englewood Cliffs (1985)
10. Urick, R.J.: Principles of Underwater Sound, 3rd edn. McGraw-Hill, New York (1983)

11. Van Trees, H.L.: Optimum Array Processing. John Wiley & Sons, New York (2002)
12. Stergiopoulos, S.: Advanced Beamformers. In: Stergiopoulos, S. (ed.) Advanced Signal Processing Handbook: Theory and Implementation for Radar, Sonar, and Medical imaging, CRC Press, Boca Raton (2001)
13. Baggenstoss, P.M.: On Detecting Linear Frequency-Modulated Waveforms in Frequency- and Time-dispersive Channels: Alternatives to Segmented Replica Correlation. IEEE Trans. On Oceanic Engineering 19, 591–598 (1994)
14. Friedlander, B., Zeira, A.: Detection of Broadband Signals in Frequency and Time Dispersive Channels. IEEE Trans. On Signal Processing 7, 1613–1622 (1996)
15. Naidu, P.S.: Sensor Array Signal Processing. CRC Press, Boca Raton (2001)
16. Ioannides, P., Balanis, C.A.: Uniform Circular and Rectangular Arrays for Adaptive Beamforming Applications. IEEE Antennas and Wireless Propagation Letters 4, 351–354 (2005)
17. Chen, H.H., Chan, S.C., Ho, K.L.: New Recursive Adaptive Beamforming Algorithms for Uniform Concentric Circular arrays with Frequency Invariant Characteristics. In: 14th European Signal Processing Conference, Florence, Italy (September 2006)
18. Van Veen, B.D., Buckley, K.M.: Beamforming: A Versatile Approach to Spatial Filtering. IEEE Signal Processing Mag. 5, 4–24 (1988)

Classification Rule Acquisition Based on Extended Concept Lattice

Yan Wang and Ming Li

College of Computer and Communication, Lanzhou University of Technology, Lanzhou, Gansu, 730050, China
wangyan@lut.cnj,Lim3076@163.com

Abstract. Classification is an important task in data mining. The number of rules which are obtained through traditional classification rule acquisition algorithm is much enormous. Concept lattice is powerful tool for data mining and rule acquisition. Through analyzing characteristic of concept in concept lattice, extended concept lattice and classification rule acquisition based on extended concept lattice are proposed. Experiment results show that this algorithm can obtain simple and understandable rule set.

1 Introduction

Concept Lattice (CL) is the heart of FCA (Formal Concept Analysis) [1].CL structure is an effective tool for data analysis, knowledge discovery, machine learning, information retrieval and document browsing, and more especially for classification and association rules. Theoretical foundation of CL rests on the mathematical lattice theory [2~5].

Classification is an essential task of machine learning and data mining. Classification aims to find the rules that predict the class label of an object. Once the data set is fixed, classification process consists of two phases: learning and classification. Many classification algorithms have been developed and widely applied in practice. For example, decision tree, neural network and rough set, etc. Each method has its characteristics which can be employed to certain classification task. Experiment results show concept lattice based on classification method is as same as traditional method in classification effect [3,4,6-7].

Extracting knowledge from Concept lattice has been studied by many machine learning researchers. Knowledge acquisition based on concept lattice consists of mining association rules and classification rules [8]. Some existing concept lattice based on machine learning algorithms build complete lattice to obtain classification rule. These algorithms can not extract probability rules. Some other concept lattice based on machine learning algorithms can process uncertainty of data. However their learning parameters have to be set by users in advance. It is very difficult to set these parameters unless the whole domain knowledge is known. GALOIS [9] and RULEARNER [10] belong to the former kind. They employed an incremental method to build complete lattice and then to extract rules. LEGAL [11], LACS [12] and CBALattice [13] belong to the latter one. They use the learning parameters provided

K. Li et al. (Eds.): LSMS 2007, LNCS 4688, pp. 571–578, 2007.

by users. If one user does not understand the related domain knowledge quite well, it will be much difficult to set appropriate learning parameters. Otherwise, whether classification rule is obtained by which method, each rule could be extracted from the intent of the concept. According to the characteristic of the formal concept, the concept intent is maximum attribute set corresponding to the concept extent. So, rule extracted from concept intent will not be an optimal rule.

In this paper, through analyzing characteristic of concept in concept lattice, extended concept lattice and classification rule acquisition based on extended concept lattice are proposed. And in the algorithm we employ rule novelty to prune rule sets. Consequently simple and understandable rule can be obtained. Experiment results show that this algorithm is valid for acquiring classification rule.

The rest of this paper is organized as follows. In section 2 we describe basic definition. In section 3, classification rule acquisition algorithm based on extended concept lattice is proposed. Experiment results are reported in Section 4, and show that this algorithm is valid for acquiring classification rule. A conclusion is given in Section 5.

2 Basic Definition

Definition 1[14]. A formal context $K=(U,M,R)$ consists of two sets U and M and a relation R between U and M. The elements of U are called the objects and the elements of M are called the attributes of the context. In order to express that an object x is in a relation R with an attribute m, write xRm and read it as "the object x has the attribute m". f and g are two mapping relations defined in $P(U)$ and $P(A)$, denoted as:

$$\forall X \subseteq U : f(X)=\{m\in M \mid xRm, \forall x\in X\}$$

$$\forall B \subseteq M : g(B)=\{x\in U \mid xRm, \forall m\in B\}$$

According to above mapping relations, formal concept and concept lattice can be defined.

Definition 2[14]. Given a pair (X,S) on the formal context K, where $X\subseteq U, S\subseteq M$ .If (X,S) satisfies with $f(X)=S$, $X=g(S)$, then (X,S) is called a formal concept, X and S are called the extent and the intent of the concept respectively. The set of all concepts of formal context and order relation between formal concept are denoted as $L(U,M,R)$ which is called the concept lattice of the context (U,M,R).

Definition 3. Given a pair (X,S) on the formal context K, where $X\subseteq U, S\subseteq M$ .If (X,S) satisfies with $X=g(S)$ and $S\subseteq f(X)$, then (X,S) is called extended concept formal context K.

Theorem 1. Given a formal context $K = (U, A, R)$, if (X, S) is an extended concept on the formal context, then there must be a concept (X, T) on the formal context such that $S \subseteq T$.

Proof. According to the definition of the formal concept, the concept intent is maximum attribute set corresponding to the concept extent. Because (X, S) is an extended concept, S is not maximum attribute set corresponding to the object set X. Then there must be an attribute set T with $X = g(T)$ and $T = f(X)$, where (X, T) is a concept on the formal context. In terms of definition 3, $X = g(S)$ and $S \subseteq F(X)$, then $S \subseteq T$.

As a result, on the formal context, an extended concept has same object set with certain concept, but its attribute set of objects is a part of attribute set of objects of this concept. So, the rules obtained by this extended concept can become simpler, and it can be proved that the confidence of these rules is as same as that of the rules obtained by original concept.

3 Classification Rule Acquisition Based on Extended Concept Lattice

In the classification rule algorithm LACS, which is based on concept lattice and deals with uncertainty of data using concept purity and concept intensity to control building concept lattice and pruning rules for multi-classification problem, the parameter used needs to be set initially. And rules are obtained from formal concept, so rule is not optimal rule. Classification rule acquisition based on extended concept lattice (CAECL) improves LACS, which is shown in the following. The first, the local minimum certainty of dataset α_c proposed in [15] replaces concept purity in LACS to deal with uncertainty of data. Rule novelty replaces concept intensity in LACS pruning rule. Because α_c is from data, it is not necessary to set it initially. The second, when concept lattice is built, we employ top-bottom to research concept or extended concept of data. Through this method, obtained rule will be simple and understandable.

Generally, classification rule is obtained from decision table. Obviously, a decision table can be a multi-value system. In other words, one attribute can have several attribute values. However, the formal context of concept lattice is a binary system generally. So, before algorithm is executed, a decision table has to be converted from multi-value to binary value when building the concept lattice based on multi-value decision table. Concretely, attribute values in the decision table are viewed as a new attributes in the formal context. And, the condition and decision part of decision table are converted to condition formal lattice and decision formal lattice, respectively, denoted as $(U, ConCorrespond(C), R_C)$ and $(U, DesCorrespond(D), R_D)$, where U is object set, C and D are condition attribute set and decision attribute set, respectively. $ConConversion(C)$ and $DesConversion(D)$ are new attribute sets in the formal context obtained through attribute conversation. Rule novelty is defined in the following.

Definition 4. Given condition formal context $(U, ConCorrespond(C), R_C)$ and decision formal context $(U, DesCorrespond(D), R_D)$. $CoverSet$ is the object set covered by the rules from rule set. For a new rule $S \rightarrow T$, $S \subset ConCorrespond(C)$, $T \subset DesCorrespond(D)$, if $g(S) \not\subset CoverSet$, then the rule $S \rightarrow T$ is called novelty, otherwise is called redundancy.

According to the definition of novelty, if a new rule is implicated by other rule in the rule set, this new rule is viewed as redundancy, so it should not be added to the rule set. This method not only can simplify the rule set, but can improve efficiency of the algorithm.

In terms of above definition, we search concepts or extended concepts satisfying certain condition with top-down method in the algorithm. At each level, the quadruple (X, S, Y, T) is used to denote the concepts or extended concepts, where X, Y are the sets of objects, $S \subset ConConversion(C)$, $T \subset ConConversion(D)$, $X = g(S)$, $S \subseteq f(X)$, $Y = g(T)$, $T = f(Y)$. If concept purity of a concept or an extended concept is larger than α_c, it will be used to obtain a rule, otherwise it will be ignored. After the processing finished at each level, we test whether the objects covered by the present rule set contain all objects in the original decision table already and then forecast if some novelty rules are obtained by the concepts or extended concepts in the next. If yes, the subsequent step begins, otherwise the algorithm ends. In the algorithm, α_c is calculated by training data. It is not necessary to set it initially, so data-driven knowledge acquisition can be realized. At the same time, by choosing novelty rules, the search space of the concepts will be reduced and the efficiency of the algorithm will be improved. In the following description of the algorithm, both of concept and extended concept are called concept.

Algorithm: Classification Rule Acquisition Based on Extended Concept Lattice(CAECL).

Input: Decision table $S = \langle U, R, V, f \rangle$, condition formal context $(U, ConCorrespond(C), R_C)$ and decision formal context $(U, DesCorrespond(D), R_D)$. Let $RuleSet = \phi$ which will be used to denote the classification rule set.

Output: The rule set $RuleSet$.

Step1: according to Reference [15], calculate threshold α_c and use α_c as threshold to control the generating of the rules.

Step2: initialize the top level concept: $L_1 = \{(g(S), S, g(T), T)\}$,where $g(S), g(T) \subset U$, $S \subset ConConversion(C)$, $T \subset DesConversion(D)$,

$(g(S),S)$ and $(g(T),T)$ are concepts of the condition formal context and decision formal context, respectively, satisfying $g(S)=\max_{M\subset ConConversion(D)}(\frac{|g(S)\cap g(M)|}{|g(S)|})$ .For the concepts in the L_1 , if $\frac{|g(S)\cap g(T)|}{|g(S)|}>\alpha_c$, then the rule $S\rightarrow T$ is obtained, and $RuleSet=RuleSet\cup\{S\rightarrow T\}$.

Step 3: build next level concept lattice, and to obtain rules, initialize $i=1$.

① If the number of the objects covered by the present rule set is equal to that of all objects in the original decision table and then forecast if some novelty rules are obtained by the concepts in the next level. If yes, the subsequent step begins, otherwise the algorithm ends.

② When the number of concepts is larger than 1 in the L_i level, suppose $L_{i+1}=\phi$. Denote the object set of union of two concepts $(g(S),S,g(T),T)$ and $(g(S'),S',g(T'),T')$ in L_i as $g(Z)=g(S)\cap g(S')$.

③ For $g(Z)$, if $g(Z)\neq\phi$, then the condition part of new concept is denoted as $(g(Z),S\cup S')$; and decision part is denoted as $(g(T),T)$, where $g(T)=\max_{M\subset ConConversion(D)}(g(Z)\cap g(M))$.

Then $L_{i+1}=L_i\cup(g(Z),S\cup S',g(T),T)$.

④ If $|g(Z)|\neq|g(S)|$ and $|g(Z)|\neq|g(S')|$, $\frac{|(g(Z)\cap g(T)|}{|g(Z)|}>\alpha_c$,

then the rule $S\cup S'\rightarrow T$ is obtained,

and $RuleSet=RuleSet\cup\{S\cup S'\rightarrow T\}$.

⑤ If $|g(Z)|=|g(S)|$ or $|g(Z)|=|g(S')|$, then no rules will be obtained, and next combination continues.

⑥ If the pairwise combination of concepts in L_i is finished, then $i=i+1$, and go to ②.

4 Experiment

In order to verify this algorithm, the UCI data set [16] is employed to test the algorithms LACS and CAECL in the following. In order to increase uncertainty of data, some attributes are deleted from the data set randomly. The value of default attribute is complemented with the attribute value occurring most frequently. The continuous attributes is discretized by using the method, "discretization in rough set theory" [17]. We use 50 percent data for training set and 50 percent for testing. The rule selection complying with high confidence priority is exploited to avoid confliction. The basic information of the data sets is listed in Table 1:

Table 1. Data sets

Name of datasets	Total number of sample	Number of the training sample	Number of test sample
Breast-Cancer	699	349	350
Balance-Scale	625	312	313
Letter	20000	2000	18000
Mushroom	8124	4012	4012
Car	1728	691	1037
Iris	150	75	75
Solar-flare	1066	533	533

We test algorithm LACS in the above data sets. The value of the concept purity starts from 0.1and the shift step is 0.1. The value of concept intensity starts from 0.01, the shift step is 0.01. The results with larger accuracy and fewer rules from test results of LACS are shown in Table 2.

Then we employ α_c as threshold of concept purity to test algorithm CAECL in the above data sets. The results are shown in Table 3.

Table 2. The results of algorithm LACS

Data set	LACS algorithm				
	concept purity	concept intensity	accuracy	number of rules	average length of rule
Breast-Cancer	0.8	0.02	92.8	66	2.08
Balance-Scale	0.5	0.03	78.3	63	2.01
Letter	0.1	0.01	40.1	63	1
Mushroom	0.8	0.01	60	33	3.57
Car	0.3	0.02	64	9	1
Iris	0.6	0.02	90.7	10	1
Solar-flare	0.7	0.04	97.6	40	2.59

From Table 2 and Table 3, we can see that for LACS, the value of concept purity and concept intensity are different for different data sets for guaranteeing higher accuracy and fewer rules. So, if one user does not understand the related knowledge quite well, it is much difficult to set appropriate learning parameters. And average length of rules is longer. Although the highest accuracy could not be reached by using CAECL, the results are close to the maximal accuracy, and the number of rules is relatively smaller, average length of rules is shorter as well. As the experiment result shows, it is valid to acquire rules from the extended concept lattice by employing the local minimum certainty and novelty of rules.

Table 3. The results of algorithm CAECL

Data set	CAECL algorithm			
	α_c	accuracy	number of rules	average length of rule
Breast-Cancer	0.5	92.3	39	1
Balance-Scale	0.33	73.1	21	1
Letter	0.2	40.1	40	1
Mushroom	0.65	74.5	10	1
Car	0.6	64	6	1
Iris	0.5	89.3	12	1
Solar-flare	0.5	97.4	21	1

5 Conclusions

In this paper, an algorithm of classification rule acquisition from extended concept lattice is proposed. The parameters used in the algorithm can be calculated by training data, so it is not necessary for users to initiate by hand. At the same time, by estimating the novelty of rules the redundant rules can be deleted, and hence the efficiency of the algorithm is improved. Rule is obtained from extended concept, average length of rules is shorten and rules is more simple. Experiment results show that this algorithm is valid for acquiring classification rules.

References

1. Wille, R.: Recon structing lattice theory: an approach based on hierarchies of concepts. In: Rival, I. (ed.) Ordered Sets, pp. 445–470. Reidel, Dordrecht-Boston (1982)
2. Godin, R.: Incremental concept formation algorithm based on Galois(concept) lattices. Computational Intelligence 11(2), 246–267 (1995)
3. Njiwoua, P., Nguifo, E.M.: Forwarding the choice of bias LEGAL-F:Using feature selection Toreduce the complexity of LEGAL. In: Walter, D. (ed.) Proceedings of the BENELEARN-97, ILKandINFOLAB, pp. 89–98. Tiburg Univer-sityPress, Netherlands (1997)
4. Carpineto, C., Romano, G.: Galois: anorder-heoretic approach to conceptual clustering. In: Utgoff, P. (ed.) Proceedings of the ICML 93, pp. 33–40. ElsevierSciencePublishers, Amhers (1993)
5. Oosthuizen, G.D.: The application of concept lattice to machine learning. Technical Report, University of Pretoria, South Africa (1996)
6. Sahami, M.: Learning classification rules using lattices. In: Lavran, N., Wrobel, S. (eds.) Proceedings Of the ECML-95, pp. 343–346. ElsevierSciencePublishers, Grete (1995)

7. Godin, R., Missaoui, R.: An incremental concept formation approach for learning from databases. Theoretical Computer Science 133, 387–419 (1994)
8. Fu, H.Y., Fu, H.G., Njiwoua, P., Nguifo, E.M.: A Comparative Study of FCA-Based Supervised Classification Algorithms. In: Proceedings of the Second International Conference on Formal Concept Analysis, Sydney,Australia, February 23-26, 2004, pp. 313–320 (2004)
9. Carpineto, C., Romano, G.: Galois: An order-theoretic approach to conceptual clustering. In: Proceedings of the Tenth International Conference on Machine Learning, University of Massachusetts, Amherst, MA, USA, June 27-29, 1993, pp. 33–40 (1993)
10. Sahami, M.: Learning Classification Rules Using Lattices. In: Proceedings of the Eighth European Conference on Machine Learning, Heraclion, Crete, Greece, pp. 343–346 (April 1995)
11. Mephu-Nguifo, E.: Galois Lattice:A Framework for Concept Learning.Design, Evaluation and Refinement. In: Proceedings of the Sixth International Conference on Tools with Artificial Intelligence, New Orleans, Louisiana, USA, November 6-9, 1994, pp. 461–467 (1994)
12. Xie, Zh.P., Liu, Z.T.: Research on Classifier Based on Lattice Structure. In: 16# World Computer congress 2000. Proceeding of conference on Intelligent Information Processing, Beijing, china, August 21- 25, 2000, pp. 333–338 (2000)
13. Anamika, G., Naveen, K., Vasudha, B.: Incremental Classification Rules Based on Association Rules Using Formal Concept Analysis. In: Proceedings of the 4th International Conference on Machine Learning and Data Mining in Pattern Recognition, Leipzig, Germany, July 9-11, 2005, pp. 11–20 (2005)
14. Ganter, B., Wille, R.: Formal concept analysis. Springer, Heidelberg (1999)
15. Wang, G.Y., He, X.: A self-Learning Model under Uncertain Conditions. Journal of Software (in Chinese) 14, 1096–1102 (2003)
16. http://www.ics.uci.edy/~mlearn/MLRepository
17. Hou, L.J., Wang, G.Y., Nie, N., Wu, Y.: Discretization in rough set theory. Computer Science 27, 89–94 (2000).

Modeling and Real-Time Scheduling of Semiconductor Manufacturing Line Based on Simulation

Zhongjie Wang[1], Xinhua Jiang[2], and Qidi Wu[1]

[1] Department of Control Science and Engineering, Tongji University, Shanghai, China
[2] Huawei Technology Co. Ltd
wang_zhongjie@mail.tongji.edu.cn

Abstract. Due to the difficulties of establishing a model of semiconductor manufacturing line, a real-time scheduling system of semiconductor manufacturing line is established with discrete event simulation technique. By employing the object-oriented modeling method, this system is implemented by simulation. An actual system with 40 machines and 60 processing steps is employed to demonstrate the validity of this system. Experiments indicate that the modeling and real-time scheduling system is accurate and reliable.

1 Introduction of Semiconductor Manufacturing Line

Semiconductor products are widely applied to civil field, communication, industry, space flight and military field etc. The procedure of produce semiconductor products are divided into three parts, i.e. IC design, wafer manufacturing and packaging & test. The semiconductor manufacturing line is used to fabricate wafers, which is divided into 6 independent sections [1], including diffusion, photolithography, etching, masking, iron planting and polishing.

Semiconductor manufacturing line is the most complicated process, which is beyond of the traditional system classification. In 1993, Kumar named it after reentrant system [2]. Semiconductor manufacturing line has some inherent characteristics.

(1) High reentrance, i.e. jobs may revisit the same machine in different steps during being processed.
(2) The manufacturing process is quite complex, the flowing path of wafers is very long. The total manufacturing process usually consists of several hundreds to thousand steps, the number of machines is at least more than 100, even to several thousands.
(3) More than one kind of products are flowing in the production line at the same time, each product may rework or be scrapped.
(4) Stochastic machine failure and maintenance, urgent orders and many other unpredictable factors increase the randomness and dynamic property.

2 Scheduling of Semiconductor Manufacturing Line

It is worthy of investigating the scheduling of semiconductor manufacturing line not only in theory but also in practice. A good scheduling system can greatly enhance the

K. Li et al. (Eds.): LSMS 2007, LNCS 4688, pp. 579–589, 2007.

profit of enterprises, decrease the production cost, satisfy the needs of customers quickly and occupy the market. It is not very long for the researchers to investigate the scheduling of semiconductor manufacturing line, but with the fast development of semiconductor industry, scheduling of semiconductor manufacturing line is becoming the hotspot of research for both industry and academy [3-7]. There are several methods to investigate the scheduling of semiconductor manufacturing line.

(1) Traditional methods. The traditional methods are usually pure mathematical methods, e.g., linear programming [8], stochastic optimization. In this method, the function describing the relationship between resources and jobs is established firstly, and then this problem is solved with mature theories to get the optimal solution to allocate jobs and resources.

(2) Intelligent methods. Intelligent methods include artificial neural network, genetic algorithms, fuzzy algorithms and multi-agents etc. The common characteristic of intelligent methods is to simulate the intelligence in nature (including human brain) to solve complex problems [4-5].

(3) Simulation methods. With the development of computer technology, computer simulation is playing more and more important role as the third research means besides theoretical deduction and physical experiment [9]. Thompson [10] pointed that, scheduling software verified properly is able to predict performance with an error less than 2% compared with the practical system, scheduling based on simulation can meet the challenge of semiconductor production. Miller [11] employs the simulation result to optimize scheduling of semiconductor manufacturing line without increasing any equipment and operators, which shorten the making span by 16% and decrease the work in process by 20% with the throughput being guaranteed.

Scheduling of semiconductor manufacturing line is very complex. Because the mathematical methods has limitations in model description and much simplification must be made when modeling the real circumstance, pure mathematical method is only applied to small scale system. Furthermore, it is NP-hard to get the optimal solution, the solving period increases exponentially with the problem scale. For a practical problem (e.g., more than ten machines, less than one hundred jobs), it is impossible to get the optimal solution in finite period no matter how fast the computer is.

Computer simulation overcomes the difficulty of getting the optimal solution of complex problem by seeking feasible solution. Computer simulation is able to compare different schemes and checks different decision rules. Computer simulation has strong ability of model description and is able to modeling semiconductor manufacturing line in detail, which can describe the semiconductor manufacturing line veritably. Combined with the intelligent methods and other efficient methods in other discipline, computer simulation is bringing bright foreground.

3 Real-Time Scheduling System of Semiconductor Manufacturing Line Based on Simulation

The developed real-time scheduling system of semiconductor manufacturing line is composed of four independent and cooperative modules and three correlative databases, as shown in figure 1.

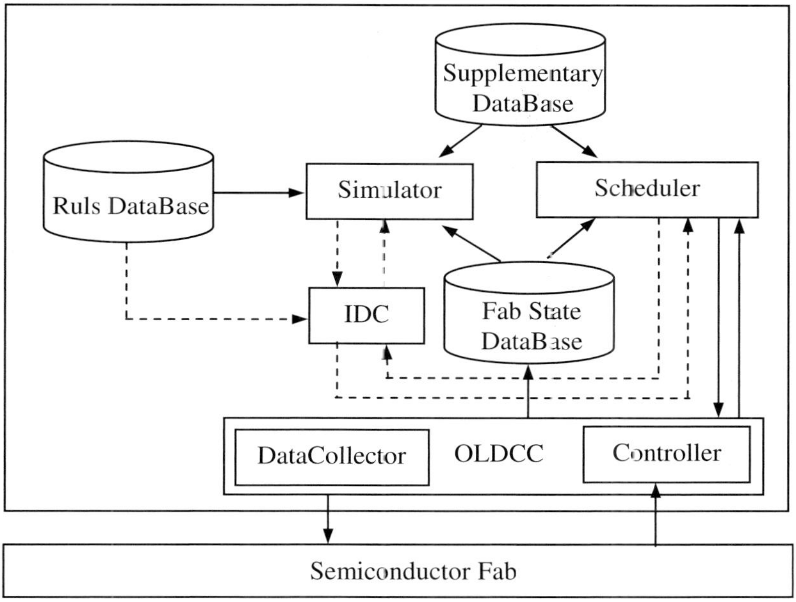

Fig. 1. Diagram of the real-time scheduling of semiconductor manufacturing line

The three databases are respectively state database, rule database and supplement database. In state database, there is mainly the data concerning wafer and machine states. The rule database includes three types of rules, i.e., wafer release rules, single scheduling rules and batch scheduling rules. The data that is not in MES system but indispensable are mainly in the supplement database, e.g. the preventative maintenance plan and machine standards [12].

The four modules are respectively on-line data collector and controller (OLDCC), Simulator , Scheduler and intelligent decision-making center (IDC). Usually, MRP II/ERP, MES or both of them serve as the on-line data collector, and MES acts as the on-line controller. The function of scheduler is to make new scheduling policy in time. The simulator has two functions. One is to establish simulation model according to the datum in state database, rule database and supplement database, the other is to run a series of simulation programs as soon as it receives request from the scheduler or intelligent decision-making center. Only one rule in the rule database is estimated for each simulation, the simulation will not stop until all the rules in the rule database or the rule set selected by the intelligent decision-making center are estimated. Finally, the simulator returns the simulation results to the scheduler. Actually, the real-time scheduling system is implemented in two phase, the intelligent decision-making center is only employed in the second phase to increase the flexibility, performance and intelligence of the simulation system. At this time, the simulator is triggered accordingly by the intelligent decision-making center instead of the scheduler. Although there are many intelligent methods, such as genetic algorithm, evolutional algorithm, simulation annealing, ant colony algorithm and so on, here we advise to employ neural network for the intelligent decision-making center. This is because of the purpose of shortening the period needed to produce new scheduling policy so that real-time scheduling is satisfied and increasing the intelligence and adaptability of the simulation system.

4 Object-Oriented Simulation Model of Semiconductor Manufacturing Line

Since 1980s, object-oriented technology is becoming dominant and maturely developed. Object-oriented technology is employed to investigate the above real-time scheduling simulation system.

4.1 Requirement for the Simulation System of Semiconductor Manufacturing Line

The requirement for the simulation system of semiconductor manufacturing line is shown in figure 2.

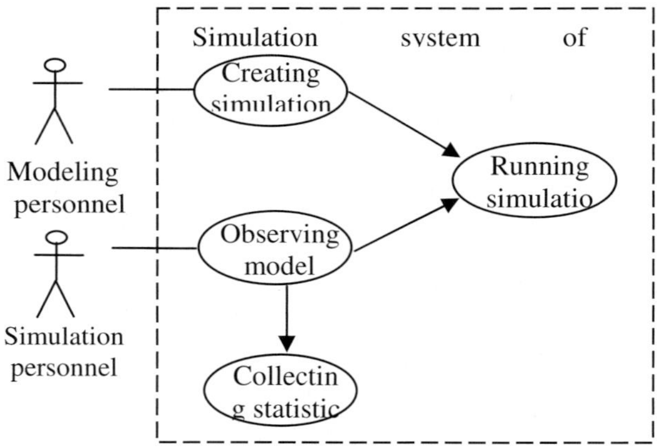

Fig. 2. Requirement for the simulation

In figure 2, there are two participants, modeling personnel and simulation personnel. The modeling personnel will establish the simulation model of semiconductor manufacturing line and run it to guarantee its correctness. The simulation personnel will run the simulation model and collect usable statistic datum for further analysis. Four components are needed to construct the simulation model of semiconductor manufacturing line, i.e. creation of simulation model, running of simulation model, observation of model behavior and collection of statistic datum. Creation of simulation model and observation of model behavior are both dependent on running of simulation model, so the relationship between them is the same as that between employer and employee. Collection of statistic datum is the special case of observation of model behavior, so the former extends the latter.

4.2 Structure of the Simulation Model of Semiconductor Manufacturing Line

According to logic function, the simulation system is composed of two Namespaces, i.e. Tools.Graphs and Fab. The Tools.Graphs Namespace provides specially the function of drawing. Tools.Graphs Namespace is shown in figure 3. The Fab Namespace is the main part of this simulation system and provides most functions, which is further divided into three Namespaces, i.e. Fab.Simulation, Fab.Distribution and Fab.UI, as shown in figure 4.

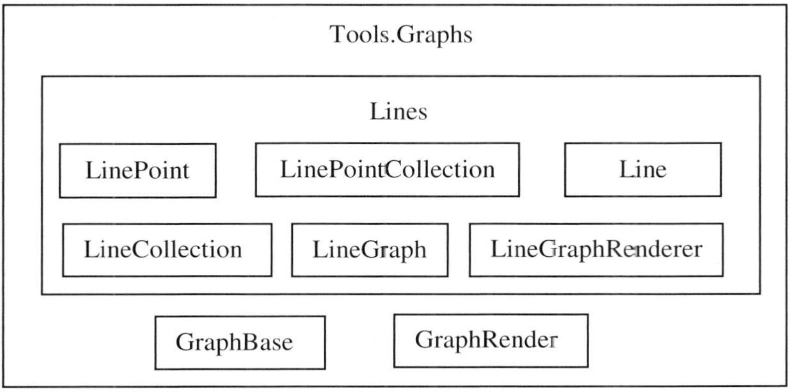

Fig. 3. Diagram of Tools. Graphs Namespace of the simulation system.

In this simulation system, because the input and output datum are stored with XML files and TXT files instead of in databases, the system is not divided into database layer, data visiting layer, apply logic layer and representation layer. The function of the data visiting layer is implemented by Fab.SemiFab-Class. The representation layer is in Fab.UI Namespace. The function of the apply logic layer is further divided into simulation supporting layer and physical layer of semiconductor manufacturing line. The simulation supporting layer is in Fab.Simulation Namespace, the physical layer of semiconductor manufacturing line is in Fab Namespace. Furthermore, the random data generators are separated from the simulation supporting layer and form Fab.Distribution Namespace.

4.3 Simulation Supporting Layer of the Simulation System

The notion of Process is mainly employed in the simulation supporting layer. Process is implemented by Common-Class, the main problem to be solved is the definition of Process-List, Process-queuing, Simulation-controller-Class, Resource-Class, Random-data-generator- Class and Statistics-Class.

Usually, Process-queuing simulates the buffers of a real system. For example, there is a buffer in the front of each working center in semiconductor manufacturing line, in which wafers are waiting to be manufactured. More than one Process is in the simulation program, and there are many restriction conditions among them. If the

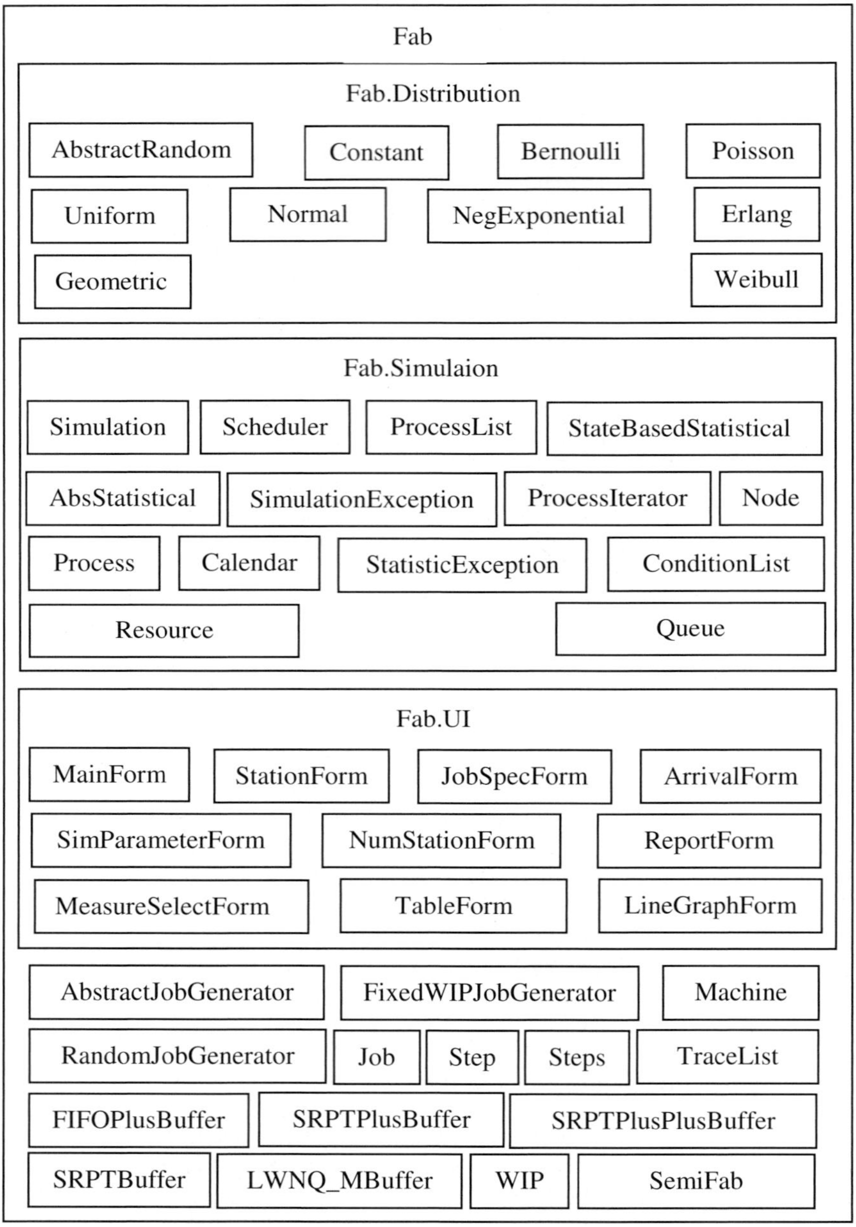

Fig. 4. Diagram of Fab Namespace of the simulation system

Machine-Process is not idle, the Wafer-Process has no choice but to wait in the Queuing-Process. A wafer is allowed to be processed for the next step only when it has finished the last step. The Simulator-Class provides necessary mechanism to harmonize the collaboration among different Processes. The Resource-Class represents the reusable resources in practice, e.g., machines. The Random-data-generator-Class is employed to generate random data series with certain distribution. The common discrete random data series include Bernoulli, Poisson and Geometric distribution, the common continuous random series include Uniform,Weibull, Normal, NegExponential and Erlang distribution. The ultimate purpose of simulation is to get the important performance index of the actual system. Otherwise, the simulation system has no value no matte how perfect it is. The Statistics-Class is used to collect the real-time datum and calculate performance index.

4.4 Physical Layer of the Simulation System

The physical layer of the simulation system is defined by different Classes. Wafer-Class, Wafer-release-Class, Machine-Class and Buffer-Class are respectively designed in the simulation system.

Wafer-Class inherits from Process-Class, which has its own characteristics, including wafer identification number, wafer type, working procedure table and phase variable. Wafer- release-Class is employed to perform different release policies in semiconductor manufacturing line. The simulation system includes three Wafer-release-Classes, i.e. AbstractJobGenerator-Class, RandomJobGenerator-Class and FixedWIPJobGenerator-Class. AbstractJobGenerator-Class inherits from Process-Class, it is an abstract Class from the point view of role it played, which provides the construction function for sub-Classes to indicate the wafer type and its working procedure table to be released. RandomJobGenerator-Class and Fixed WIP Job Generator-Class both inherit from AbstractJobGenerator-Class. By employing different random data generator, RandomJobGenerator-Class performs the deterministic release policy and eight types of random release policies, such as possion release policy, normal release policy and so on. FixedWIPJobGenerator-Class performs the closed-loop release strategy. Given a value of the expected working-in-process (WIP), if the actual WIP is less than the expected value, Fixed WIP Job Generator-Class employs the deterministic release policy. Once the actual WIP equals to the given value, FixedWIPJobGenerator-Class releases a wafer into the manufacturing line if and only if a wafer leaves. Machine-Class inherits from Process-Class, which has its own properties, including the working station ID, machine ID, resources possessed by the machine and jobs to be processed on this machine. There are 6 Buffer-Classes in this simulation system, i.e. Queueing-Class, FIFOPlusBuffer-Class, SRPTBuffer-Class, SRPTPlusBuffer-Class, SRPTPlusPlusBuffer and LWNQ_Mbuffer-Class. These six Classes correspond to FIFO (first-in-first-out), FIFO+ (first-in-first-out+), SRPT (shortest-remaining-process-time), SRPT (shortest-remaining-process-time+), SRPT (shortest-remaining-process-time++) and LWNQ (least-work-next-queue) dispatching rules respectively. Buffer-Classes are employed to perform different dispatching rules in the simulation system.

5 Simulations

The demonstration example comes from the literature [13], the semiconductor manufacturing line to be simulated is composed of 12 working stations and 40 machines. The number of machines at each working station and the processing time is listed in table 1. Only one kind of wafer will be manufactured for this system, 60 steps will be needed to finish manufacturing a wafer, the technological path of wafer being manufactured is shown in figure 5.

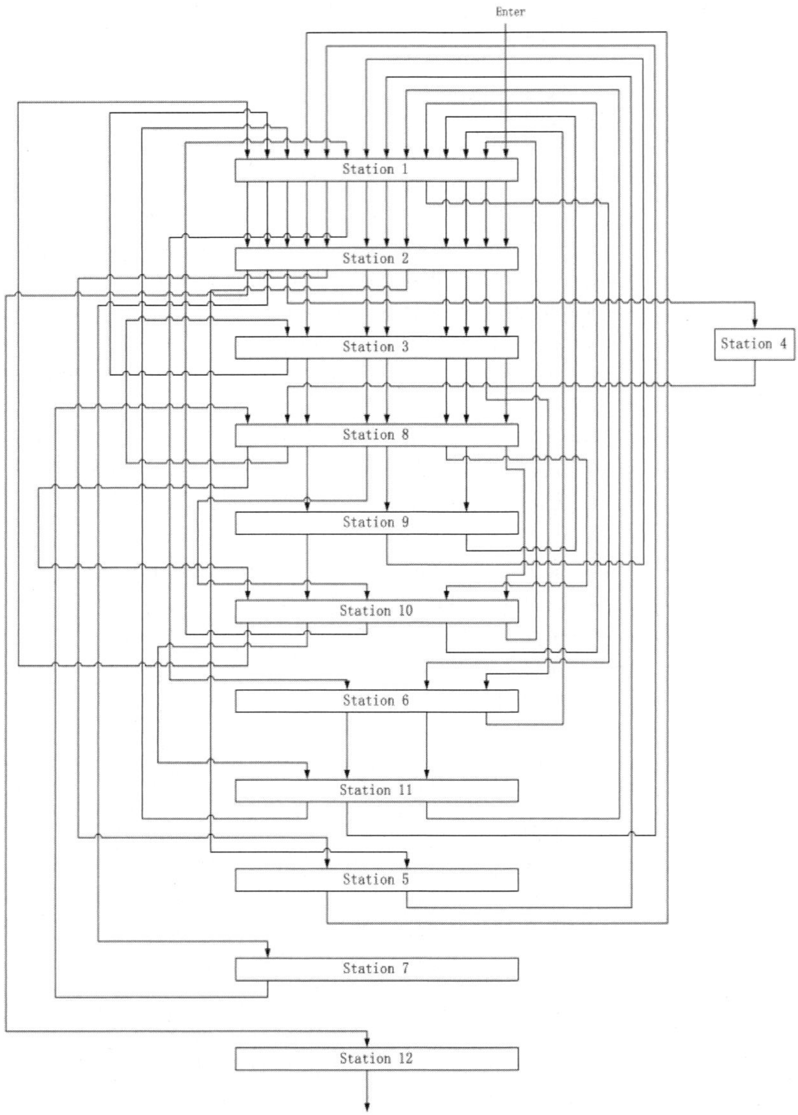

Fig. 5. Technological path of wafer being manufactured

Table 1. Configuration of the semiconductor manufacturing line

Working station	Machine number	Visiting number	Processing time (min)
1	4	14	30
2	3	12	22.5
3	10	7	150
4	1	1	108
5	1	2	54
6	2	3	72
7	1	1	108
8	4	8	48
9	1	3	36
10	9	5	180
11	2	3	72
12	2	1	150

We employ C# to make the program of the simulation model and real-time scheduling system. In this simulation experiment, the module of IDC (intelligent decision-making center) is not implemented, but we have reserve the interface through which it communicates with the rule database, the simulator and the scheduler. The simulation lasts 60000 minutes, i.e. 42.7 days, FIFO dispatching rule is employed for each buffer. The wafer release policy is deterministic with an interval of 115.4 minutes. The simulation result is listed in Table 2.

Table 2. Simulation result

Working station	Average utility of machines	Queuing length
1	0.896	0.64
2	0.746	0.10
3	0.913	0.08
4	0.932	0.14
5	0.926	0.24
6	0.927	0.33
7	0.932	0.05
8	0.823	0.15
9	0.919	0.33
10	0.862	0.04
11	0.929	0.27
12	0.649	0

The simulation result further demonstrates the correctness of the simulation system proposed in this paper. Of course, the correctness of a simulation system can not be verified by only one example. During the whole developing process, the simulation system is verified from easy example to complex ones. First of all, the system is verified by examples with only one machine-one kind of product-one step, then with one machine-one kind of product-multiple steps, next with more than one machine-one kind of product-one step, next with more than one machine-two kinds of product-more than one step, and lastly with the large scale. The verification proceeds with a circulate procedure of finding bugs. It is through this procedure that the correctness of the simulation system is verified.

6 Conclusions

The semiconductor industry is very important for the national economy, and it is developing with full of vitality. The scheduling of semiconductor manufacturing line is becoming more and more important. In order to find better scheduling solutions, researchers have made great efforts, but there are lack of eficient methods due to the inherent characteristics of semiconductor manufacturing line, such as large scale, complex technology, serious randomness and high reentrance and so on. With the fast development of computer technology, it is not only feasible to investigate the scheduling of semiconductor manufacturing line by simulation technology, but also save a lot of manpower and material resources, which will increase the profit of enterprises. A simulation system of semiconductor manufacturing line is developed with great flexibility and easy expansion by object-oriented technology in this paper, and some achievement is obtained. Of course, there are still many works to be done for perfection.

(1) To further improve the lower data structure to satisfy the rapidity that large scale simulation requires.
(2) To further improve the physical layer of the simulation system. More factors, e.g. machine failure and operators, should be taken into account in order that the simulation model describes the actual system more precisely.
(3) To further expand the statistics function, the simulation system is already able to collect sampled data and get the max-value, min-value, average value, covariance and the standard variance. Whether the system is stable should be further identified automatically.
(4) To further expand the rule base of the simulation system. At present, the system support only three release policies and six dispatching rules. The rule base should be enriched, and the interface to the MES system should be added.

Acknowledgments. Supported by Project of Fok Ying Tong Education Foundation (104030), Important Project and Project of New Century Excellent Person of Education Ministry of China (306023, NECT-06-0832).

References

1. Quirk, M., Serda, J.: Semiconductor manufacturing technology. Electronic industry publisher (2004)
2. Kumar, P.R.: Re-entrant lines. Queuing Systems 13(1-2) (1993)

3. Wang, Z., Wu, Q.: Multi-object optimal scheduling for semiconductor manufacturing line based on satisfactory control. In: Proceedings of the American Control Conference, vol. 2, pp. 1603–1608 (2002)
4. Sikora, R., Shaw, M.J.: Coordination mechanisms for multi-agent manufacturing systems: applications to integrated manufacturing scheduling. IEEE Transactions on Engineering Management 44(2), 175–187
5. Wang, X.-r., Wu, T.-J.: Ant colony optimization for intelligent scheduling. Proceedings of the 4th World Congress on Intelligent Control and Automation 1, 66–70 (2002)
6. Zhongjie, W., Qidi, W.: Investigation of control and scheduling of semiconductor manufacturing line. Computer Integrated Manufacturing System-CIMS, pp. 607–611 (2003)
7. Zhongjie, W.: Research of optimal scheduling of semiconductor manufacturing line based on hierarchical control. Post Doctorate Report, Shanghai, Tongji University (2002)
8. Qi, L., Dali, L., Li, Z.: Optimal decision in production plan of semiconductor industry: linear programming. Semiconductor technology 7, 35–39 (2004)
9. Chong, C.S.: Simulation-based scheduling for dynamic discrete manufacturing. In: Proceedings of the 2003 Winter Simulation Conference, vol. 2, pp. 1465–1473 (2003)
10. Thompson, Michael: Simulation-based scheduling: meeting the semiconductor wafer fabrication challenge Industrial Engineer (May 1996)
11. Miller, D.J.: Simulation of a SML Communications of the ACM 33, 98–108 (1990)
12. Watt, D.G.: Integrating simulation based scheduling with MES in a semi-conductor fabrication Winter Simulation Conference Proceedings. vol. 2, pp. 1713–1715 (1998)
13. Lu Steve, C.H., Ramaswamy, D., Kumar, P.R.: Efficient scheduling policies to reduce mean and variance of cycle-time in semiconductor manufacturing plants. IEEE Trans. Semiconductor Manufacturing, 373–388 (1994)

Application of Bayesian Network to Tendency Prediction of Blast Furnace Silicon Content in Hot Metal

Wenhui Wang

Basic Department, Zhejiang Water Conservancy and Hydropower College, 310018 Hangzhou, China
wangwh@mail.zjwchc.com

Abstract. This paper proposes a new method for predicting the change tendency of silicon content in hot metal based on Bayesian networks. Firstly, some important factors that affect silicon content are selected out using grey relationship analysis (GRA). Secondly, a Bayesian network (BN) model is constructed to predict silicon content in hot metal based on the causal relationship of the factors. The model shows good performance due to the high percentages of prediction hitting the target, and can help blast furnace (BF) foreman acquaint himself with the status of BF.

1 Introduction

In blast furnace ironmaking process, the control of the thermal state of a BF is one of the important factors ensuring the BF stable operation. The silicon content of hot metal is directly related to the BF thermal regime [1, 2, 3], thus the prediction of the silicon content of BF hot metal is used to estimate the thermal state. However, in the past years, accurate prediction of hot metal silicon content has not been solved in world by now. The traditional models for Si prediction were established based on time series method such as AR model and neural network models such as BP network model [4].

On the other hand, Bayesian networks also called causal probabilistic networks, belief network*s*, can be used to discover and learn causal relationships, and hence can be used to gain understanding about a problem domain and to predict the consequences of intervention [5]. In point of probability theory, Bayesian networks are computational architectures that guarantee the consistency and coherence of a causal probabilistic model of a given problem domain and maintain the probabilistic equilibrium of the model upon arrival of new information. In this work, we construct a BN model and predict BF silicon content in hot metal with it.

2 Bayesian Networks

Bayesian network is one of methods for representing and reasoning with uncertain knowledge, which is a significant area of research in Artificial Intelligence [4]. A Bayesian network is defined over a finite domain V of variables, each of which may

K. Li et al. (Eds.): LSMS 2007, LNCS 4688, pp. 590 – 597, 2007.

be discrete or continuous-for ease of exposition we will restrict attention to discrete variables that take a finite number of values. A Bayesian network consists of a DAG whose nodes are the variables in V; a probability specification *S* which contains the probability distribution $p(V_i \mid pa_i)$ of each variable $V_i \in V$ conditional on its parents pa_i in *G*; and an assumption, called the Markov condition, which states that each variable $V_i \in V$ is probabilistically independent of its non-descendants, ND_i, conditional on its parents.

A causally interpreted Bayesian network, or causal network, is a Bayesian network in which the graph G represents the causal relations amongst the variables in *V*, with an arrow from V_i to V_j if V_i is a direct cause of V_j. There are arguments to the effect that, if a Bayesian network is causally interpreted then the Markov condition – now called the causal Markov condition – is a valid assumption: i.e. that any variable is probabilistically independent of its non-effects, conditional on its direct causes[5]. Thus in cases where causal relations are known, the graphical structure G in a Bayesian network can just be taken to be the causal graph. This reduces the problem of constructing a Bayesian network to that of determining the probability specification *S*, and this is normally done by taking the corresponding sample frequencies from a database of past case data, or by eliciting degrees of belief from experts.

3 Variable Analysis Using Grey Relationship Analysis

To construct a Bayesian network model, we should select the appropriate variables first. Incorporating variables that have little relevance to the particular output variable would cause excessive noise in the model. Not all variables of the blast furnace are useful to predict the hot metal silicon content, due to structural and chemical complexities. In this section, some important factors that affect silicon content are selected out using grey relationship analysis.

Deng [6] pioneered grey system theory in 1982. By system relational analysis, model constructing, forecasting, or decision analysis, grey system theory can effectively resolve various problems with uncertainty and multivariable or discrete data. In this study, we used grey relationship analysis to select the key variables.

The relationship coefficient is defined by following equation:

$$\gamma(x_0(k), x_i(k)) = \frac{\min_i \min_k |x_0(k) - x_i(k)| + \varsigma \max_i \max_k |x_0(k) - x_i(k)|}{|x_0(k) - x_i(k)| + \varsigma \max_i \max_k |x_0(k) - x_i(k)|}. \tag{1}$$

Where, $\varsigma \in [0,1]$, distinguishing coefficient, usually, $\varsigma = 0.5$. And the grey relationship grade (GRG) between x_0 and x_k is defined as the following equation:

$$\gamma_i = \frac{1}{n} \sum_{k=1}^{n} \gamma(x_0(k), x_i(k)). \tag{2}$$

Table 1. Some Variables affecting the silicon content

Sign	*VC*(t/h)	*PI*(m^3/min.kPa)	*ME*(t)	$[Si]_{n-1}$(%)
Input variable	Charging mixture velocity	Permeability index	molten iron error	Last [Si]
Sign	*PC*(t/h)	*BT*(°C)	*BQ*(m^3/min)	*CO*
Input variable	Pulverized coal injection	Blast temperature	Blast quantity	CO content of top gas

Table 2. Corresponding grey relationship grades of each variable

Variables	γ_{VC}	γ_{PI}	γ_{ME}	$\gamma_{[Si]_{n-1}}$	γ_{PC}	γ_{BT}	γ_{BQ}	γ_{CO}
GRG	0.530	0.612	0.526	0.931	0.644	0.758	0.773	0.508

Table 3. Nodes of Bayesian network and their states

Node	Values
BT	LL, LN, LH, NL, NN, NH, HL, HN, HH
BQ	LL, LN, LH, NL, NN, NH, HL, HN, HH
PC, PC_{n-1}	LL, LN, LH, NL, NN, NH, HL, HN, HH
VC, VC_{n-1}	LL, LN, LH, NL, NN, NH, HL, HN, HH
PI, PI_{n-1}	LL, LN, LH, NL, NN, NH, HL, HN, HH
ME	Low, Normal, High
[*Si*]n-1	Low, Normal, High
$\Delta[Si]$	DA, DS, EA, RS, RA

There are many factors, which affect the thermal state of a BF, including the physical and chemistry characteristics and the charging method, and also including the bottom parameters such as charging mixture velocity, permeability index, and so on. Table 1 gives some variables affecting hot metal silicon content. In table 1, $ME = TPI - API$, where TPI is the theoretic production of molten iron, API is the actual production of molten iron.. Table 2 gives the corresponding grey relationship grade between the variable and silicon content. In table 2, we can obtain the grey relationship sequence as follows:

$$\gamma_{[Si]_{n-1}} > \gamma_{BQ} > \gamma_{BT} > \gamma_{PC} > \gamma_{PI} > \gamma_{VC} > \gamma_{ME} > \gamma_{CO} \tag{3}$$

Hence, we select the first seven variables, $[Si]_{n-1}$, *BQ*, *BT*, *PC*, *PI*, *VC*, *ME* as the nodes of Bayesian network.

On the other hand, the seven variables are all continuous, so we should convert them into discrete ones before establishing the Bayesian network model with them. Table 3 shows all the Bayesian network nodes and their values. In table 3, PC_{n-1} denotes the last PC, and node 'BT' = 'LN' means that variable 'BT' = 'low' and '$\Delta[BT]$' = 'normal', and so on.

4 Constructing Bayesian Networks

We will construct the Bayesian network for predicting blast furnace silicon content in hot metal in this section. Taking the graph of a Bayesian network, also called causal probability network, to be the causal graph can simplify the problem of network construction. Evidently, there are causal relations, which might be strongly or weakly, amongst the nodes in table 3. For instance, $\Delta[Si]$ will fall when *VC* rises, and will rise when *BT* rises, and the range of *PC* always can result in the change of *VC*, and so on. Then we can connect the corresponding nodes by an arc that shows causal relationship. Some nodes, on the other hand, have weakly causal dependencies of each other, just as *ME* and *PI*, $[Si]_{n-1}$ and *ME*, etc. Then we can assume that they are independent of each other. We can learn these relations from expert knowledge about ironmaking processing.

Having finished the above analysis of the nodes, we then construct a causal network model from causal knowledge [3], which is shown in figure 1.

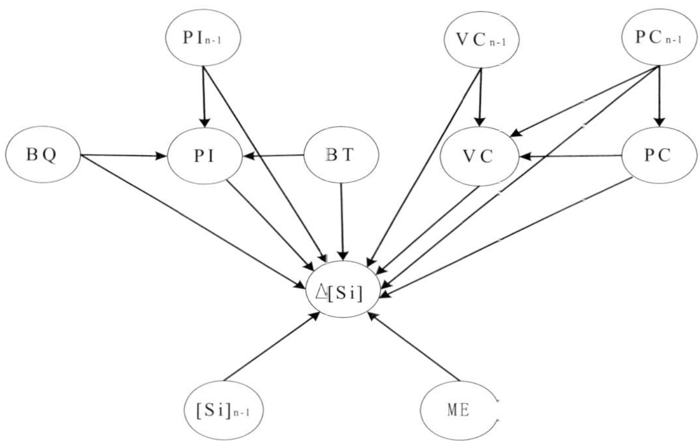

Fig. 1. The Bayesian network constructed for predicting BF silicon content in hot metal

The arcs between the nodes show causal dependencies and are quantified by conditional probability matrices that reflect the rules the experts use when making similar decisions. In order to specify the constructed Bayesian network above, we need to assign prior probabilities to all root nodes and conditional probabilities for all states of all non-root nodes given all possible combinations of states of their parents.

5 Learning Probabilities in Bayesian Network

We assume that the physical joint probability distribution for $X = (x_1, x_2, \cdots, x_n)$ can be encoded in some network structure *S*. We write

$$p(X \mid \theta_s, S^h) = \prod_{i=1}^{n} p(x_i \mid pa_i, \theta_i, S^h) \tag{4}$$

where θ_i is the vector of parameters for the distribution $p(x_i \mid pa_i, \theta_i, S^h)$; θ_s is the vector of parameters $(\theta_1, \theta_2, \cdots, \theta_n)$; pa_i denotes all the parent nodes of x_i; and S^h denotes the event that the physical joint probability distribution can be factored according to S. In addition, we assume that we have a random sample $D = \{X_1, \cdots, X_n\}$ from the physical joint probability distribution of X. We refer to an element X_i of D as a case. Thus the problem of learning probabilities in a BN can now be stated simply: Given a random sample D, compute the posterior distribution $p(\theta_s \mid D, S^h)$. We assume that each variable is discrete, having r_i possible values $x_i^1, \cdots, x_i^{r_i}$, and each local distribution function is collection of multinomial distributions, one distribution for each configuration of pa_i. Namely, we assume

$$p(x_i^k \mid pa_i^j, \theta_i, S^h) = \theta_{ijk} > 0 \tag{5}$$

where $pa_i^1, \cdots, pa_i^{q_i}$ $(q_i = \prod_{X_i \in pa_i} r_i)$ denote the configurations of pa_i, and $\theta_i = ((\theta_{ijk})_{k=2}^{r_i})_{j=1}^{q_i}$ are the parameters. For convenience, we define the vector of parameters $\theta_{ij} = (\theta_{ij1,} \cdots, \theta_{ijr_i})$ for all i and j.

Given a random sample D, we can computer the posterior distribution efficiently and in closed form under two assumptions [3, 4, 6]. The first assumption is that the random sample D is complete. The second assumption is that the parameter vectors are mutually independent, and have the prior Dirichlet distribution. Thus the posterior distribution can be given with the following formula:

$$P(\theta_{ij} \mid D, S^h) = Dir(\alpha_{ij1} + N_{ij1}, ..., \alpha_{ijr_i} + N_{ijr_i}) \tag{6}$$

where N_{ijk} is the number of cases in D in which $X_i = x_i^k$ and $pa_i = pa_i^j$.

If we let X_{N+1} denote the next case to be seen after D, and suppose that, in case X_{N+1}, $X_i = x_i^k$ and $pa_i = pa_i^j$, where k and j depend on i, we can give the prediction formula as following [3]:

$$P(X_{N+1} \mid D, S^h) = \prod_{i=1}^{n} \prod_{j=1}^{q_i} \frac{\alpha_{ijk} + N_{ijk}}{\alpha_{ij} + N_{ij}} \tag{7}$$

where $\alpha_{ij} = \sum_{k=1}^{r_i} \alpha_{ijk}$ and $N_{ij} = \sum_{k=1}^{r_i} N_{ijk}$.

6 Prediction with the Constructed Bayesian Network

Time series data of the nodes, namely $[Si]_{n-1}$, *BQ, PI, BT, VC, PC, ME*, observed in No.1 blast furnace at Laiwu Iron and Steel Group Co. are taken as the random sample *D* with the capacity of 2000 tap numbers, including heat No.28872~30871. Suppose that, in equation 6, $\alpha_{ij1} = \cdots = \alpha_{ijr_i} = 1$ for all *i* and *j*. Thus Equation 7 becomes

$$P(X_{N+1} \mid D, S^h) = \prod_{i=1}^{n} \prod_{j=1}^{q_i} \frac{1 + N_{ijk}}{1 + N_{ij}} \tag{8}$$

And parameter θ_{ijk} can be gotten by the following formula:

$$\theta_{ijk} = \frac{1 + N_{ijk}}{r_i + \sum_k N_{ijk}} \tag{9}$$

We assume that x_{i_0} denotes the nodes $\Delta[Si]$, and $P_k = \theta_{i_0 jk} (k = 1,2,3,4,5)$ denotes the conditional probability of $\Delta[Si]$ equal to 'DA', 'DS', 'EA', 'RS', or 'RA' given the states of all parent nodes pa_{i_0}, respectively. According to $P_k (k = 1,2,3,4,5)$, we could establish the rules for the tendency prediction of silicon content, which are shown in the following list:

1. $\Delta[Si] = \text{'DA'}$, if $\max_i \{P_i\} = P_1 > 0.52$;
2. $\Delta[Si] = \text{'DS'}$, if $\max_i \{P_i\} = P_2 > 0.52$ and $P_3 < 0.24$;
3. $\Delta[Si] = \text{'EA'}$, if $\max_i \{P_i\} = P_3 > 0.52$ and $P_2 (or\ P_4) < 0.24$;
4. $\Delta[Si] = \text{'RS'}$, if $\max_i \{P_i\} = P_4 > 0.52$ and $P_3 < 0.24$;
5. $\Delta[Si] = \text{'RA'}$, if $\max_i \{P_i\} = P_5 > 0.52$;
6. $\Delta[Si] = \text{'ED'}$, if $\max_i \{P_i\} = P_2 (or\ P_3) \le 0.52$ and $P_2 + P_3 \ge 0.65$;
7. $\Delta[Si] = \text{'ER'}$, if $\max_i \{P_i\} = P_3 (or\ P_4) \le 0.52$ and $P_3 + P_4 \ge 0.65$;
8. $\Delta[Si] = \text{'ED'}$, if $\max_i \{P_i\} = P_2 (or\ P_3) > 0.52$ and $P_3 (or\ P_2) \ge 0.24$;
9. $\Delta[Si] = \text{'ER'}$, if $\max_i \{P_i\} = P_3 (or\ P_4) > 0.52$ and $P_4 (or\ P_3) \ge 0.24$.

In the above equations, 'ED' denotes 'EA' or 'DS', and 'ER' denotes 'EA' or 'RS'.

7 Results and Discussion

With the above Bayesian network model, we predict the testing data set, including 50 tap numbers (from No.30872 to 30921), collected online from No.1 BF at Laiwu Iron and Steel Group Co. Table 4 gives a part of the predictive results and table 5 gives the statistic results.

Table 4. Comparison between the observed states and predicted states

Heat No.	$\Delta[Si]$	predicted states	P_1	P_2	P_3	P_4	P_5
30872	DA	DS	0.351	0.533	0.096	0.012	0.008
30873	DA	DA	0.413	0.309	0.180	0.089	0.009
30874	RS	ER	0.001	0.028	0.596	0.327	0.048
30875	EA	EA	0.059	0.153	0.527	0.206	0.053
30876	DS	DS	0.173	0.608	0.124	0.077	0.017
30877	EA	EA	0.003	0.082	0.607	0.286	0.023
30878	EA	EA	0.006	0.136	0.547	0.200	0.111
30879	EA	EA	0.003	0.085	0.601	0.234	0.078
30880	RS	ER	0.010	0.055	0.421	0.502	0.012
30881	EA	ER	0.005	0.067	0.363	0.556	0.009
30882	DS	ED	0.017	0.392	0.441	0.136	0.013
30883	EA	ER	0.000	0.069	0.502	0.416	0.013
30884	EA	EA	0.001	0.031	0.667	0.279	0.022
30885	EA	EA	0.000	0.020	0.551	0.158	0.271
30886	EA	EA	0.007	0.166	0.597	0.172	0.059
30887	EA	EA	0.000	0.025	0.723	0.222	0.030
30888	RS	ER	0.000	0.054	0.572	0.371	0.004
30889	EA	EA	0.001	0.143	0.644	0.169	0.043
30890	EA	EA	0.004	0.182	0.743	0.042	0.030
30891	RS	ER	0.005	0.118	0.312	0.488	0.077

Table 4 shows that we can forecast $\Delta[Si]$ with Bayesian network not numerically but qualitatively. Table 5 and the comparison between the observed and predicted states of $\Delta[Si]$ both tell us that the prediction of $\Delta[Si]$ is very successful. That is, the Bayesian network model we constructed can effectively learn the causal relations between the nodes, and quantify them by conditional probability matrices. The inference in Bayesian network is visible. Table 4 also shows that the results of prediction including not only the states of $\Delta[Si]$ but also the values of five parameters, namely, P_1, P_2, P_3, P_4 and P_5, which are also helpful for BF foreman. On the other hand, the inference of BP network [7] is not visible at all.

The obtained results above show that the approach is very potential. Nevertheless, there are some defects, for example, the consequence of the time lags of variable, which can affect the veracity of the constructed Bayesian model, and the structure of the Bayesian network should be improved given the random sample D. Improvement should be made toward the following aspects: a) to estimate the time lag of each variable; b) to choose the better Bayesian network structure using the given data set D; c) to get a new numerical prediction model by defuzzification the prediction results of $\Delta[Si]$, and so on. Certainly, acquiring more deep knowledge about the blast furnace process is also necessary.

Table 5. Statistic results of forecasting

Heat No.	successful forecast of $\Delta[Si]$	percent	successful forecast of varying trend	percent
29072~29131	43	86.0%	46	92.0%

8 Conclusions

(1) Bayesian networks can effectively learn the causal relations amongst the variables given the random sample *D*. And the constructed BN model shows good performance due to the high percentages of prediction hitting the target.

(2) In comparison with that in BP network, the inference in Bayesian network is visible, which is useful for BF operations, and makes the results of Bayesian network more convictive.

(3) When we predict hot metal silicon content with Bayesian network, it is not necessary for us to establish exact mathematical model of BF ironmaking, which is very difficult due to the complexity of BF process.

References

1. H. J. Bachhofen.: The application of modern process control technology in hot metal production. Ironmaking Conference Proceedings. (1991) 703-708
2. B. Yao and T. J. Yang.: Optiming generation of an expert system based on neural network by genetic algorithm to predict the silicon content in hot metal. Iron and Steel. 4 (2000) 13-18
3. Chen J.: A predictive system for blast furnaces by integrating a neural network with qualitative analysis. Engineering Application of Artificial Intelligence. 14 (2001) 77-85
4. Liu Xiangguan, Liu Fang.: Optimization and Intelligent Control System of Blast Furnace Ironmaking Process. Beijing Metallurgical Industry Press. (2003)
5. David Heckman.: Bayesian networks for data mining. Data Mining and Knowledge Discovery. 1 (1997) 79-119
6. Deng J.L.: Introduction to grey system theory. Journal of Grey System. 1 (1989) 1-24
7. Stephenson T A.: An introduction to Bayesian network theory and usage. IDIAP-RR (2000) 00-33
8. J. Peal.: Causality: models, reasoning, and inference. Cambridge University Press. (2000)
9. Jensen, F.V., Lauritzen, S.L., and Olesen, K.G.: Bayesian updating in causal probabilistic networks by local computations. Computational Statistics Quarterly. 4 (1997) 269-282
10. Singh H, Sridhar N V, Deo B.: Artificial neural nets for prediction of silicon content of blast furnace hot metal. Steel Research. 67 (1996) 521-527

Global Synchronization of Ghostburster Neurons Via Active Control

Jiang Wang, Lisong Chen, Bin Deng, and Feng Dong

School of Electrical engineering and automation, Tianjin University, 300072 Tianjin, China
jiangwang@tju.edu.cn

Abstract. In this paper, active control law is derived and applied to control and synchronize two unidirectional coupled Ghostbursters neurons under external electrical stimulation. Firstly, the dynamical behavior of the nonlinear Ghostburster model responding to various external electrical stimulations is studied. Then, using the results of the analysis, the active control strategy is designed for global synchronization of the two unidirectional coupled neurons and stabilizing the chaotic bursting trajectory of the slave system to desired tonic firing of the master system. Numerical simulations demonstrate the validity and feasibility of the proposed method.

1 Introduction

Motivated by potential applications in physics, biology, electrical engineering, communication theory and many other fields, the synchronization of chaotic systems has received an increasing interest [1-6]. Experimental studies [7-9] have pointed out that the synchronization is significant in the information processing of large ensembles of neurons. In experiments, the synchronization of two coupled living neurons can be achieved when depolarized by an external DC current [10-11]. Synchronization control [12-14] which has been intensively studied during last decade is found to be useful or has great potential in many domains such as in collapse prevention of power systems, biomedical engineering application to the human brain and heart and so on.

Ghostburster model [15-16] which is a two-compartment of pyramidal cells in the electrosensory lateral line lobe (ELL) from weakly electric fish. In this paper, we will investigate the relationship between the external electrical stimulations and the various dynamical behavior of the Ghostburster model.With the development of control theory, various modern control methods, such as feedback linearization control [17], backstepping design [18-19], fuzzy adaptive control [20] and active control [21], have been successfully applied to neuronal synchronization theoretically in recent years. These control methods have been investigated with the objective of stabilizing equilibrium points or periodic orbits embedded in chaotic attractors [22]. In this paper, based on Lyapunov stability theory and Routh–Hurwitz criteria, some criteria for globally asymptotical chaos synchronization are established. The active controller can be easily designed on the basis of these conditions to synchronize two unidirectional coupled neurons and convert the chaotic motion of the slave neuron into the tonic firing as the master neuron.

K. Li et al. (Eds.): LSMS 2007, LNCS 4688, pp. 598 – 607, 2007.

The rest of the paper has been organized as follows. In Sec. 2, dynamics of the two neurons in external electrical stimulation is studied. In Sec. 3, the global synchronization of two unidirectional coupled Ghostbursters neurons under external electrical stimulation is derived and numerical simulations are done to validate the proposed synchronization approach. Finally, conclusions are drawn in Sec. 4.

2 Dynamics of Nonlinear Ghostburster Model for Individual Neuron

Pyramidal cells in the electrosensory lateral line lobe (ELL) of weakly electric fish have been observed to produce high-frequency burst discharge with constant depolarizing current (turner et al., 1994). We investigate a two-compartment model of an ELL pyramidal cell in Fig.1, where one compartment represents the somatic region, and the second the entire proximal apical dendrite.

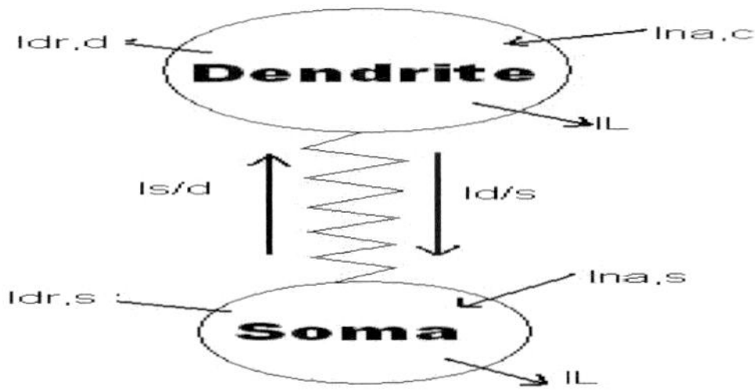

Fig. 1. Schematic of two-compartment model representation of an ELL pyramidal cell

In Fig.1, Both the soma and dendrite contain fast inward Na^+ current, I_{Na} and $I_{Na,d}$, and outward delayed rectifying K^+ current, respectively I_{Dr} and $I_{Dr,d}$. In addition, I_{leak} is somatic and dendritic passive leak currents, the Vs is somatic membrane potentials, the V_d is dendritic membrane potentials. The coupling between the compartments is assumed to be through simple electrotonic diffusion giving currents from soma to dendrite $I_{s/d}$, or vice versa $I_{d/s}$. In total, the dynamical system comprises six nonlinear differential equations, Eqs.(1) through (6), we refer to Eqs.(1) Through (6) as the ghostburster model.

Soma:

$$\frac{dV_s}{dt} = I_s + g_{Na,s} \cdot m_{\infty,s}^2(V_s) \cdot (1-n_s) \cdot (V_{Na} - V_s) + g_{Dr,s} \cdot n_s^2 \cdot (V_K - V_s) + \frac{g_c}{k} \cdot (V_d - V_s) + g_{leak} \cdot (V_l - V_s) \quad (1)$$

$$\frac{dn_s}{dt} = \frac{n_{\infty,s}(V_s) - n_s}{\tau_{n,s}} \quad (2)$$

Dendrite:

$$\frac{dV_d}{dt} = g_{Na,d} \cdot m_{\infty,d}^2(V_d) \cdot h_d \cdot (V_{Na} - V_d) + g_{Dr,d} \cdot n_d^2 \cdot p_d \cdot (V_K - V_d) + \frac{g_c}{1-k} \cdot (V_s - V_d) + g_{leak} \cdot (V_l - V_d) \quad (3)$$

$$\frac{dh_d}{dt} = \frac{h_{\infty,d}(V_d) - h_d}{\tau_{h.d}} \quad (4)$$

$$\frac{dn_d}{dt} = \frac{n_{\infty,d}(V_d) - n_d}{\tau_{n.d}} \quad (5)$$

$$\frac{dp_d}{dt} = \frac{p_{\infty,d}(V_d) - p_d}{\tau_{p.d}} \quad (6)$$

In Table.1, each ionic current ($I_{Na,s}$; $I_{Dr,s}$; $I_{Na,d}$; $I_{Dr,d}$) is modeled by a maximal conductance g_{max} (in units of mS/cm^2), infinite conductance curves involving both $V_{1/2}$ and k parameters $m_{\infty,s}(V_s) = \frac{1}{1+e^{-(V_s - V_{1/2})/k}}$, and a channels time constant τ (in units of ms). *x*/*y* correspond to channels with both activation (*x*) and inactivation (*y*), *N/A* means the channel activation tracks the membrane potential instantaneously.

Table 1. Parameter values have been introduced in Eqs. (1) through (6)

Current	g_{max}	$V_{1/2}$	K	τ
$I_{Na,s}(m_{\infty,s}(V_s))$	55	-40	3	N/A
$I_{Dr,s}(n_s(V_s))$	20	-40	3	0.39
$I_{Na,d}(m_{\infty,d}(V_d)/h_d(V_d))$	5	-40/-52	5/-5	N/A
$I_{Dr,d}(n_d(V_d)/p_d(V_d))$	15	-40/-65	5/-6	0.9/5

Other parameters values are k=0.4, V_{Na}=40mV, V_K=-88.5mV, V_{leak}=-70mV, g_c=1, g_{leak}=0.18, $C_m = 1\mu F/cm^2$. The values of all channel parameters used in the simulations.

3 Global Synchronization of Two Ghostbursters Systems Using Active Control

3.1 Synchronization Principle

Consider a general master–slave unidirectional coupled chaotic system as following:

$$\begin{cases} \dot{x}_m = Ax_m + f(x_m) \\ \dot{x}_s = Ax_s + f(x_s) + u(t) \end{cases} \tag{7}$$

where $x_m(t), x_s(t) \in R$ is n-dimensional state vectors of the system, $A \in R^n$ is a constant matrix for the system parameter, $f: R^n \to R^n$ is a nonlinear part of the system, and $u(t) \in R^n$ is the control input. The lower scripts m and s stand for the master (or drive) systems, the slave (or response) one, respectively. The synchronization problem is how to design the controller u(t), which would synchronize the states of both the driving and the responding systems. If we define the error vector as e=y-x, the dynamic Eq. of synchronization error can be expressed as:

$$\dot{e} = Ae + f(x_s) - f(x_m) + u(t). \tag{8}$$

Hence, the objective of synchronization is to make $\lim_{x\to\infty} \|e(t)\| = 0$.

The problem of synchronization between the driving and the responding systems can be transformed into a problem of how to realize the asymptotical stabilization of the error system (8). So our aim is to design a controller u(t) to make the dynamical system (8) asymptotically stable at the origin.

Following the active control approach of [23], to eliminate the nonlinear part of the error dynamics, we can choose the active control function u(t) as:

$$u(t) = -Be - f(x_s) + f(x_m) \tag{9}$$

where $B \in R^n$ is a constant feedback gain matrix.

Then the error dynamical system (8) can be rewritten as:

$$\dot{e} = Me \tag{10}$$

where $M = A - B$ and $M \in R^n$.

3.2 Synchronization of the Ghostburster Systems Via Active Control

In order to state the synchronization problem of two Ghostbursters neurons, let us redefine the Eqs. of unidirectional coupled system based on the Ghostburster model which has been stated in Sec. 2.

Master system:

$$\frac{dV_{s,m}}{dt} = I_{s,m} + g_{Na,s} \cdot m_{\infty,s}^2(V_{s,m}) \cdot (1-n_{s,m}) \cdot (V_{Na} - V_{s,m}) + g_{Dr,s} \cdot n_{s,m}^2 \cdot (V_K - V_{s,m}) + \frac{g_c}{k} \cdot (V_K - V_{s,m}) + g_{leak} \cdot (V_l - V_{s,m}) \tag{11}$$

$$\frac{dn_{s,m}}{dt} = \frac{n_{\infty,s}(V_{s,m}) - n_{s,m}}{\tau_{n,s}} \tag{12}$$

$$\frac{dV_{d,m}}{dt} = g_{Na,d} \cdot m_{\infty,d}^2(V_{d,m}) \cdot h_{d,m} \cdot (V_{Na} - V_{d,m}) + g_{Dr,d} \cdot n_{d,m}^2 \cdot p_{d,m} \cdot (V_K - V_{d,m}) + \frac{g_c}{(1-k)} \cdot (V_{s,m} - V_{d,m}) + g_{leak} \cdot (V_l - V_{d,m}) \tag{13}$$

$$\frac{dh_{d,m}}{dt} = \frac{h_{\infty,d}(V_{d,m}) - h_{d,m}}{\tau_{h,d}} \tag{14}$$

$$\frac{dn_{d,m}}{dt} = \frac{n_{\infty,d}(V_{d,m}) - n_{d,m}}{\tau_{n,d}} \tag{15}$$

$$\frac{dp_{d,m}}{dt} = \frac{p_{\infty,d}(V_{d,m}) - p_{d,m}}{\tau_{p,d}} \tag{16}$$

Slaver system:

$$\frac{dV_{s,s}}{dt} = I_{s,s} + g_{Na,s} \cdot m_{\infty,s}^2(V_{s,s}) \cdot (1-n_{s,s}) \cdot (V_{Na} - V_{s,s}) + g_{Dr,s} \cdot n_{s,s}^2 \cdot (V_K - V_{s,s}) + \frac{g_c}{k} \cdot (V_K - V_{s,s}) + g_{leak} \cdot (V_l - V_{s,s}) + u_1 \tag{17}$$

$$\frac{dn_{s,s}}{dt} = \frac{n_{\infty,s}(V_{s,s}) - n_{s,s}}{\tau_{n,s}} \tag{18}$$

$$\frac{dV_{d,s}}{dt} = g_{Na,d} \cdot m_{\infty,d}^2(V_{d,s}) \cdot h_{d,m} \cdot (V_{Na} - V_{d,s}) + g_{Dr,d} \cdot n_{d,s}^2 \cdot p_{d,m}(V_K - V_{d,s}) + \frac{g_c}{(1-k)} \cdot (V_{s,s} - V_{d,s}) + g_{leak} \cdot (V_l - V_{d,s}) + u_2 \tag{19}$$

$$\frac{dh_{d,s}}{dt} = \frac{h_{\infty,d}(V_{d,s}) - h_{d,s}}{\tau_{h,d}} \tag{20}$$

$$\frac{dn_{d,s}}{dt} = \frac{n_{\infty,d}(V_{d,s}) - n_{d,s}}{\tau_{n,d}} \tag{21}$$

$$\frac{dp_{d,s}}{dt} = \frac{p_{\infty,d}(V_{d,s}) - p_{d,s}}{\tau_{p,d}} \tag{22}$$

The added term u in Eq.(8) is the control force (synchronization command). Let $e_1 = V_{s,s} - V_{s,m}$, $e_2 = V_{d,s} - V_{d,m}$, We define the nonlinear functions f_1, f_2, f_3 and f_4 in Eq.(11), (13), (17) and (19), respectively as follows:

$$f_1 = g_{Na,s} \cdot m_{\infty,s}^2(V_{s,m}) \cdot (1 - n_{s,m}) \cdot (V_{Na} - V_{s,m}) + g_{Dr,s} \cdot n_{s,m}^2 \cdot (V_K - V_{s,m}) \tag{23}$$

$$f_2 = g_{Na,d} \cdot m_{\infty,s}^2(V_{d,m}) \cdot h_d \cdot (V_{Na} - V_{d,m}) + g_{Dr,d} \cdot n_{d,m}^2 \cdot p_d \cdot (V_K - V_{d,m}) \tag{24}$$

$$f_3 = g_{Na,s} \cdot m_{\infty,s}^2(V_{s,s}) \cdot (1 - n_{s,s}) \cdot (V_{Na} - V_{s,s}) + g_{Dr,s} \cdot n_{s,s}^2 \cdot (V_K - V_{s,s}) \tag{25}$$

$$f_4 = g_{Na,d} \cdot m_{\infty,s}^2(V_{d,s}) \cdot h_d \cdot (V_{Na} - V_{d,s}) + g_{Dr,d} \cdot n_{d,s}^2 \cdot p_d \cdot (V_K - V_{d,s}) \tag{26}$$

The error dynamical system of the coupled neurons can be expressed by:

$$\begin{cases} \dot{e}_1 = I_{s,s} - I_{s,m} + f_3 - f_1 + \dfrac{g_c}{k} \cdot (e_1 - e_2) - g_{leak} \cdot e_2 + u_1 \\ \dot{e}_2 = f_4 - f_2 + \dfrac{g_c}{1-k} \cdot (e_2 - e_1) - g_{leak} \cdot e_1 + u_2 \end{cases} \tag{27}$$

We define the active control functions $u_1(t)$ and $u_2(t)$ as follows:

$$\begin{cases} u_1 = I_{s,m} - I_{s,s} - f_3 + f_1 + V_1 \\ u_2 = -f_4 + f_2 + V_2 \end{cases} \tag{28}$$

$$\begin{cases} \dot{e}_1 = \dfrac{g_c}{k} \cdot e_1 - (\dfrac{g_c}{k} + g_{leak}) \cdot e_2 + V_1 \\ \dot{e}_2 = -(\dfrac{g_c}{1-k} + g_{leak}) \cdot e_1 + \dfrac{g_c}{1-k} \cdot e_2 + V_2 \end{cases} \tag{29}$$

The system (29) describes the error dynamics and can be interpreted as a control problem, where the system to be controlled is a linear system with a control input $V_1(t)$ and $V_2(t)$ as functions of e_1 and e_2. As long as these feedbacks stabilize the system, e_1 and e_2 converge to zero as time t goes to infinity. This implies that two systems are synchronized with active control. There are many possible choices for the control $V_1(t)$ and $V_2(t)$. We choose:

$$\begin{pmatrix} V_1(t) \\ V_2(t) \end{pmatrix} = B \begin{pmatrix} e_1 \\ e_2 \end{pmatrix} \quad (30)$$

Where $B = \begin{pmatrix} k_1 & k_2 \\ k_3 & k_4 \end{pmatrix}$ is a 2×2 constant matrix. Hence the error system (29) can be rewritten as:

$$\begin{pmatrix} \dot{e}_1 \\ \dot{e}_2 \end{pmatrix} = M(t) \begin{pmatrix} e_1 \\ e_2 \end{pmatrix} \quad (31)$$

where $M(t) = \begin{pmatrix} k_1 + \frac{g_c}{k} & k_2 - (\frac{g_c}{k} + g_{leak}) \\ k_3 - (\frac{g_c}{1-k} + g_{leak}) & k_4 + \frac{g_c}{1-k} \end{pmatrix}$ is the coefficient matrix.

According to Lyapunov stability theory and Routh–Hurwitz criteria, if

$$\begin{cases} (k_1 + \frac{g_c}{k}) + (k_4 + \frac{g_c}{1-k}) < 0 \\ (k_2 - (\frac{g_c}{k} + g_{leak})) \cdot (k_3 - (\frac{g_c}{1-k} + g_{leak})) < 0 \end{cases} \quad (32)$$

Then the eigenvalues of the error system (31) must be negative real or complex with negative real parts. From Theorem 1, the error system will be stable and the two Ghostbursters systems are globally asymptotic synchronized.

3.3 Numerical Simulations

In this subsection, Numerical simulations were carried out for the above Ghostburster neuronal synchronization system. We choose the system of tonic firing behavior as the master system (at I_s=6.5) and the chaotic bursting behavior as the slave system (at I_s=9). All the parameters and the initial conditions are the same as them in Sec. 2. The control action was implemented at time t_0=500ms. And a particular form of the matrix B is given by $B=\begin{pmatrix}-3 & 0\\ 2 & -2\end{pmatrix}$. For this particular choice, the conditions (32) are satisfied, thus leading to the synchronization of two Ghostburster systems.

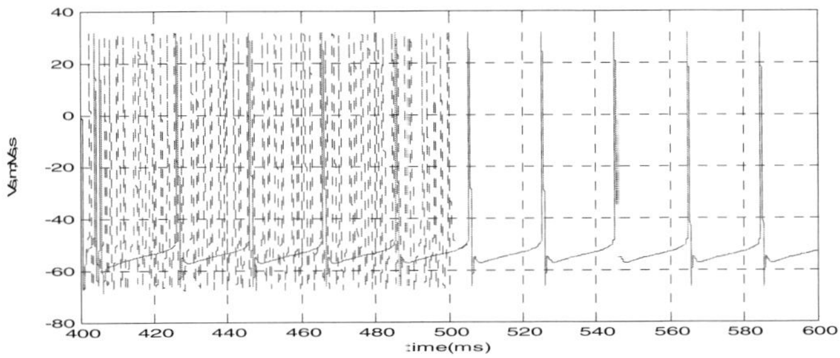

Fig. 2. Regular synchronized state of the action potentials $V_{s,m}$ and $V_{s,s}$ under controller (28)

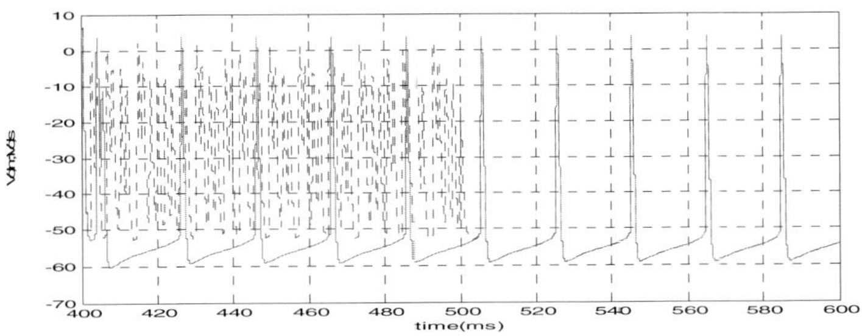

Fig. 3. Regular synchronized state of the action potentials $V_{d,m}$ and $V_{d,s}$ under controller (28)

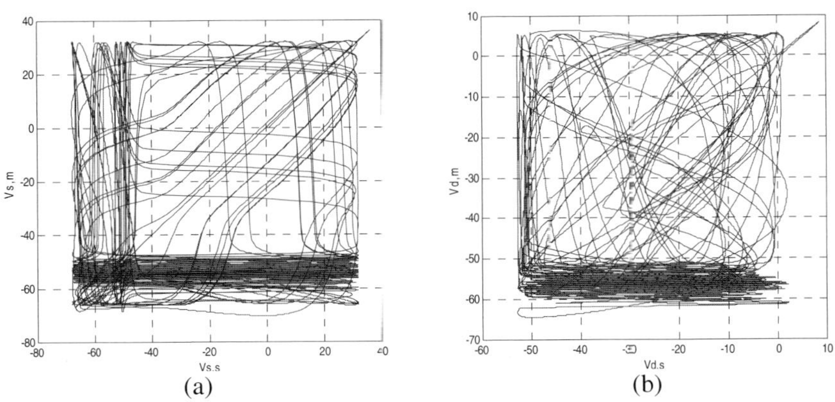

Fig. 4. $V_{s,m}$— $V_{s,s}$ and $V_{d,m}$ — $V_{d,s}$ phase plane, before the controller (28) is applied. (a) $V_{s,m}$— $V_{s,s}$ phase plane, (b) $V_{d,m}$ — $V_{d,s}$ phase plane.

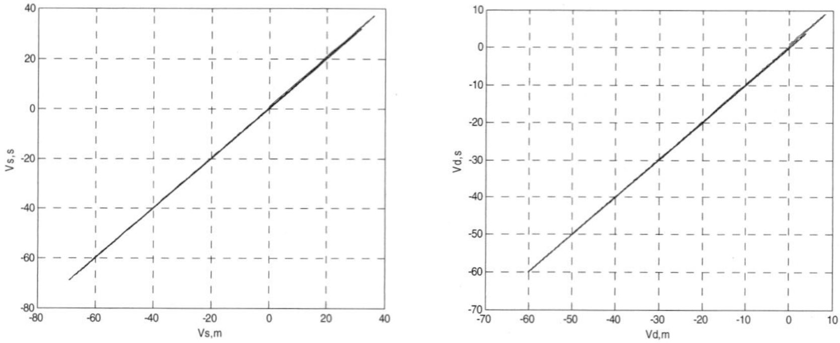

(a) (b)

Fig. 5. $V_{s,m}$ — $V_{s,s}$ and $V_{d,m}$ — $V_{d,s}$ phase plane, after the controller (28) is applied. (a) $V_{s,m}$ — $V_{s,s}$ phase plane, (b) $V_{d,m}$ — $V_{d,s}$ phase plane.

In Fig.2 and Fig.3, the initial state of the master (solid line) and the slave (dashed line) systems is tonic firing and chaotic bursting, respectively. After the controller (28) is applied, the slave system transform from chaotic bursting state into tonic firing synchronizing with the master one. in Fig.4 and Fig.5, respectively ,$V_{s,m}$——$V_{s,s}$ and $V_{d,m}$——$V_{d,s}$ phase plane diagram of master and slave system, before and after the controller (28) is applied, which shows the globally asymptotic synchronization.

4 Conclusions

In this paper, synchronization of two Ghostbursters neurons under external electrical stimulation via the active control method is investigated. Firstly, the different dynamical behavior of the ELL pyramidal cells based on the Ghostburst model responding to various external electrical stimulations is studied. The tonic firing or chaotic bursting state of the trans-membrane voltage is obtained, as shown in Fig.2, (a) and (b), respectively. Next, based on Lyapunov stability theory and Routh–Hurwitz criteria, this paper offers some sufficient conditions for global asymptotic synchronization between two Ghostbursters systems by the active control method. The controller can be easily designed on the basis of these conditions to ensure the global chaos synchronization of the action potentials as shown in Fig.3 — Fig.5. Moreover, numerical simulations show that the proposed control methods can effectively stabilize the chaotic bursting trajectory of the slave system to tonic firing.

Acknowledgments. The authors gratefully acknowledge the support of the NSFC (No. 50537030).

References

1. Caroll, T.L., Pecora, L.M.: Synchronizing chaotic circuits. IEEE Trans. Circ. Syst. I 38, 453–456 (1991)
2. Liao, T.L.: Adaptive synchronization of two Lorenz systems. Chaos, Solitons & Fractals 9, 1555–1561 (1998)

3. Restrepo, J.G., Ott, E., Hunt, B.R.: Emergence of synchronization in complex networks of interacting dynamical systems. Physics D: Nonlinear Phenomena 224, 114–122 (2006)
4. Yang, J.Z., Hu, G.: Three types of generalized synchronization. Physics Letters A 361, 332–335 (2007)
5. David, I., Almeida, R., Alvarez, J., Barajas, J.G.: Robust synchronization of Sprott circuits using sliding mode control. Chaos, Solitons & Fractals 30 (2006)
6. Cuomo, K.M., Oppenheim, A.V., Strogatz, S.H.: Synchronization of Lorenz-based chaotic circuits with applications to communications. IEEE Trans. Circ. Syst. I 40, 626–633 (1993)
7. Meister, M., Wong, R.O., Baylor, D.A., Shatz, C.J.: Synchronous bursts of action potentials in ganglion cells of the developing mammalian retina. Science 252, 939–943 (1991)
8. Kreiter, A.K., Singer, W.: Stimulus-dependent synchronization of neuronal responses in the visual cortex of the awake macaque monkey. J. Neurons 16, 2381–2396 (1996)
9. Garbo, A.D., Barbi, M., Chillemi, S.: The synchronization properties of a network of inhibitory interneurons depend on the biophysical model. BioSystems 88, 216–227 (2007)
10. Elson, R.C., Selverston, A.I., Huerta, R., Rulkov, N.F., Rabinovich, M.I., Abarbanel, H.D.I.: Synchronization behavior of two coupled biological Neurons. Physical Review Letters 81, 5692–5695 (1999)
11. Gray, C.M., Konig, P., Engel, A.K., Singer, W.: Oscillatory responses in cat visual cortex exhibit inter-columnar synchronization which reflects global stimulus properties. Nature 338, 334–337 (1989)
12. Yu, H., Peng, J.: Chaotic synchronization and control in nonlinear-coupled Hindmarsh–Rose neural systems. Chaos, Solitons & Fractals 29, 342–348 (2006)
13. David, I., Almeida, R., Alvarez, J., Barajas, J.G.: Robust synchronization of Sprott circuits using sliding mode control. Chaos, Solitons & Fractals 30, 11–18 (2006)
14. Longtin, A.: Effect of noise on the tuning properties of excitable systems. Chaos, Solitons & Fractals 11, 1835–1848 (2000)
15. Doiron, B., Laing, C., Longtin, A.: Ghostbursting: A Novel Neuronal Burst Mechanism. Journal of Computational Neuroscience 12, 5–25 (2002)
16. Laing, C.R., Doiron, B., Longtin, A., Maler, L.: Ghostburster: the effects of dendrites on spike patterns. Neurocomputing 44, 127–132 (2002)
17. Liao, X.X., Chen, G.R.: Chaos synchronization of general Lur'e systems via time-delay feedback control. Int. J. Bifurcat Chaos 13, 207–213 (2003)
18. Bowong, S., Kakmeni, F.M.M.: Synchronization of uncertain chaotic systems via backstepping approach. Chaos, Solitons & Fractals 21, 999–1011 (2004)
19. Wang, C., Ge, S.S.: Adaptive synchronization of uncertain chaotic systems via backstepping design. Chaos, Solitons & Fractals 12, 1199–1206 (2001)
20. Wang, L.X.: Adaptive Fuzzy System and Control: Design and Stability Analysis. Prentice-Hall, Englewood Cliffs, NJ (1994)
21. Bai, E.W., Lonngran, E.E.: Synchronization of two Lorenz systems using active control. Chaos, Solitons & Fractals 8, 51–58 (1997)
22. Wang, C.C., Su, J.P.: A new adaptive variable structure control for chaotic synchronization and secure communication. Chaos, Solitons & Fractals 20(5), 967–977 (2004)
23. Bai, E., Lonngrn, K.E.: Sequential synchronization of two Lorenz systems using active control. Chaos, Solitons & Fractals 11, 1041–1044 (2000)

Research of Sludge Compost Maturity Degree Modeling Method Based on Wavelet Neural Network for Sewage Treatment

Meijuan Gao[1,2], Jingwen Tian[1,2], Wei Jiang[1], and Kai Li[2]

[1] Beijing Union University, Department of Automatic Control, Postfach 100101, Beijing, China
ttgg500@163.com

[2] Beijing University of Chemical Technology, School of Information Science, Postfach 100029, Beijing, China
ttjjww999@163.com

Abstract. Because of the complicated interaction of the sludge compost components, it makes the compost maturity degree judging system appear the nonlinearity and uncertainty. According to the physical circumstances of sludge compost, a compost maturity degree modeling method based on wavelet neural network is presented. We select the index of compost maturity degree and take high temperature duration, moisture content, volatile solids, the value of fecal bacteria, germination index as the judgment parameters. We adopt a method of reduce the number of the wavelet basic function by analysis the sparse property of sample data, and use the learning algorithm based on gradient descent to train network. With the ability of strong function approach and fast convergence of wavelet network, the modeling method can truly judge the sludge compost maturity degree by learning the index information of compost maturity degree. The experimental results show that this method is feasible and effective.

1 Introduction

Along with the development of economy and increasing of population, the sewage quantity of city is increasing. The centralized biological treatment of urban sewage is the trend at home and abroad. At the present, the sludge of the sewage treatment brings on the second pollution to the human environment by sewage treatment process. Moreover, the sludge quantity is increasing with the sewage treatment depth magnification. The technologies of sanitary landfill, incineration and compost are suit for sludge disposal in urban sewage treatment plant. Among of those the compost technology is one kind of method of which energy consumption is comparatively low and resources can be reutilization.

The compost technology is a new biologic treatment method. It can decompose the multiphase organic matter by means of the mixed microbial community in special condition, and change the organic solid wastes into the stabile humus which used for fertilizing the field or improving soil. Because the compost technology can realize the effects of harmless, resourceful realization and reduce quantification in the practice

K. Li et al. (Eds.): LSMS 2007, LNCS 4688, pp. 608 – 618, 2007.

application, hurtles decrement and callback resource effect. So it obtains wide attention and has become the research hotspot in the field of environmental protection.

How to correctly judge the maturity degree of sludge compost is always the key problem of the sludge compost research and the engineering practice. Whether the compost could be safely used in soil is mainly decided by the maturity degree and the stability of sludge compost, and this acquires a stable amount of organism, and with no plant toxicity compound, pathogens and parasitic ovum. The maturity degree judgment parameters of compost include three kinds that are physical, chemical and biological indexes. The typical physical indexes include temperature, smells, moisture content etc. The typical chemical indexes are w(C)/w(N), organism, N content, humus, organic acid etc. And the biologic indexes include respiration, germination index, growth amount of plants, microbial activity, and pathogenic bacteria. Moreover, the relationship between the maturity degree of sludge compost and judgment parameters is a complicated and nonlinear, the traditional comprehensive analysis methods such as the exponential method [1] and the fuzzy mathematics method [2] are limited [3].

Artificial neural network is a new method in recent years, its ability to impend nonlinear function has been proved in theory also have been validated in actual applications [4]. But the BP network has some problems such as converge to local minimum and the speed of network learning slowly. Wavelet neural network (WNN) is new kinds of network based on the wavelet transform theory and the artificial neural network [5]. It utilizes the good localize character of the wavelet transformation and combines the self-learning function of the neural network. The confirmation of the wavelet base and the whole network has dependable theory warranty, so it can avoid the sightlessness in structural design of BP network. The learning object function which concerned to weights is convex, therefore the global minimum solution is exclusive. It has the ability of strong adaptive learning and function approach [6]. Meanwhile the WNN has the simple implementation process and fast convergence rate. The wavelet neural network is widely used in the field of pattern recognition. Actually the judgment of sludge compost maturity degree is a kind of pattern recognition problem. So in order to truly judge the maturity degree of sludge compost, a sludge compost maturity degree modeling method based on WNN for sewage treatment is presented in this paper.

2 Method Study

2.1 Wavelet Neural Network

The wavelet neural network is a neural network model which based on wavelet analysis. It replaces the common non-linear sigmoid function with non-linear wavelet basic function, the corresponding weights from the input layer to the hidden layer and the threshold value of the hidden layer are replaced respectively by the scale parameter and the translation parameter of the wavelet. The output of network is the linear superposition of the chosen wavelet base, namely the output of output layer is a linear neuron output.

To $\forall \psi \in L^2(R)$, if $\hat{\psi}(\omega)$ which is Fourier transform of $\psi(t)$ satisfy condition:

$$c_{\psi} = \int_{-\infty}^{+\infty} \frac{|\hat{\psi}(\omega)|^2}{|\omega|} d\omega < +\infty \tag{1}$$

We name $\psi(t)$ is a basic wavelet or wavelet generating function. One function family can be build by telescoping and translation the $\psi(t)$ as follow:

$$\psi_{a,b}(t) = |a|^{-\frac{1}{2}} \psi\left(\frac{t-b}{a}\right) \quad a, b \in R, \quad a \neq 0 \tag{2}$$

Here, $\psi_{a,b}(t)$ is named as wavelet, a is the scaling parameter, b is the translation parameter.

Wavelet transform of signal $f(t) \in L^2(R)$ is defined as follow:

$$W_f(a,b) = < f, \psi_{a,b} > = |a|^{-\frac{1}{2}} \int_{-\infty}^{+\infty} f(t) \psi\left(\frac{t-b}{a}\right) dt \tag{3}$$

According to the wavelet transform principle: In the Hilbert space, select a wavelet generating function $\psi(x_{t-1}, x_{t-2}, \cdots, x_{t-p})$, make it satisfy the admissibility condition:

$$c_{\psi} = \int_{L^2} \frac{|\hat{\psi}(\omega_{x_{t-1}}, \omega_{x_{t-2}}, \dots \omega_{x_{t-p}})|}{|\omega_{x_{t-1}}|^2 + |\omega_{x_{t-2}}|^2 + \cdots + |\omega_{x_{t-p}}|^2} d\omega_{x_{t-1}} \dots d\omega_{x_{t-p}} < +\infty \tag{4}$$

Here, $\hat{\psi}(\omega_{x_{t-1}}, \omega_{x_{t-2}}, \dots \omega_{x_{t-p}})$ is the Fourier transform of $\psi(x_{t-1}, x_{t-2}, \dots x_{t-p})$.

Doing the telescoping, translation and rotational transform to the $\psi(x_{t-1}, x_{t-2}, \dots x_{t-p})$, we can obtain the wavelet basic function as follow:

$$\psi_{a,\theta,\bar{b}}(x_{t-1}, \dots, x_{t-p}) = a^{-1} \psi(a^{-1} r_{-\theta}(x_{t-1} - b_{x_{t-1}}, \dots, x_{t-p} - b_{x_{t-p}})) \tag{5}$$

Be abbreviated to $\psi_{a,\theta,\bar{b}}(\cdot)$. Here $a \in R$, and $a \neq 0$ is the scaling parameter, $\bar{b} = (b_{x_{t-1}}, \dots, b_{x_{t-p}}) \in R^p$ is the translation parameter.

The rotating vector is defined as follow:

$$r_{\updownarrow\div}(x_{t\updownarrow 1}, \cdots x_{t\updownarrow i}, \cdots x_{t\updownarrow j}, \cdots, x_{t\updownarrow p}) = x_{t\updownarrow i} \cos\div \updownarrow x_{t\updownarrow j} \sin\div \quad 1 \le i \le j \le p \tag{6}$$

The $\psi_{a,\theta,\bar{b}}(\cdot)$ can satisfy the framework property of the function space by proper selecting the scaling parameter, translation parameter $a > 0, \bar{b}, \theta$.

$$A\| f \|^2 \le \sum_{a,b} |\langle \psi_{a,\theta,\bar{b}}, f \rangle|^2 \le B \| f \|^2 \quad 0 \le A \le B < \infty \tag{7}$$

Rearrange the set of wavelet functions $\{ \psi_{a,\theta,\bar{b}}(\cdot) \}$ into $\psi_1(\cdot), \psi_2(\cdot), \ldots, \psi_n(\cdot)$, and replace the common non-linear sigmoid function of the single hidden layer with it, we can obtain the corresponding wavelet neural network. The topology structure diagram of the wavelet neural network is shown as Fig.1.

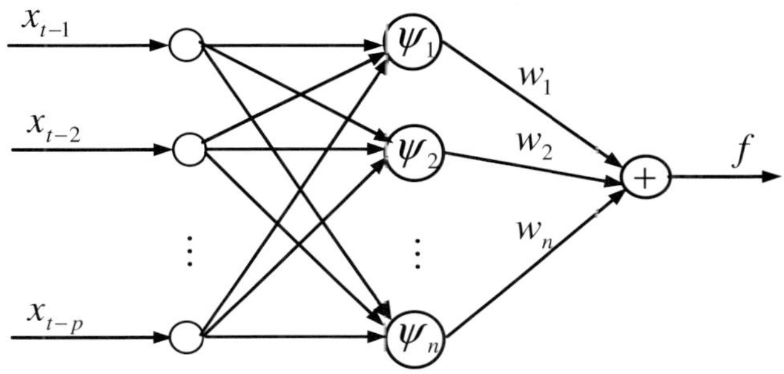

Fig. 1. The topology structure diagram of the wavelet neural network

2.2 The Ascertain of the Number of Wavelet Basic Function

The number of the wavelet basic function is a key parameter, it can influence the capability and the operational speed of the network. We adopt a method of reduce the number of the wavelet basic function by analysis the sparse property of sample data in this paper. The sample data that we can obtain is bounded and sparse in most high-dimension actual problem, maybe that some wavelet functions do not cover any sample data in the wavelet network. These wavelet basic functions have no use to reconstructed function, so they could be deleted. Therefore, the first step of the algorithm is to delete the wavelet basic function which does not cover any sample data.

Provided that S_i is supported set of wavelet function $\psi_i(x)$. Using (X, Y) express the sample data set which include M pair of sample data. For every $x_k \in X$, the order number assemblage I_k of the wavelet basic function which include the x_k in the compactly supported can express as follow:

$$I_k = \{ i : x_k \in S_i \} \qquad (k = 1, 2, \cdots, M) \tag{8}$$

The number of the wavelet basic function is as follow:

$$T = \{ \psi_i : i \in I_1 \bigcup I_2 \bigcup \cdots \bigcup I_M \} \tag{9}$$

Suppose that L is the number of the wavelet basic function of T, then

$$T = \{\psi_1, \psi_2, \cdots, \psi_L\} \tag{10}$$

Actually, some items of T have no use to reconstruct the original function. Because we only take the input sample data X into account, but do not consider the output sample data Y. So the second step of algorithm is to delete the useless items in T.

Deleting the useless items is equal to choose a subset in T, just make its span space as close as possible to the output vector Y. The matter could be divided into two subproblem: one is how to ascertain the size of the subset, the other is how to choose the item in the subset.

For the problem of choosing the item in the subset, we suppose the subnet size is s, finally we choose s items from the T. In order to attain this object, firstly, we choose an item of most eligible sample data from the T, then choose the fittest item from the residual terms repeatedly. For calculation convenience, the item that is chosen later should be perpendicular to the former ones.

For the problem to ascertain the size of the subset s, as the size of the T is L, so the value of s should be a number between 1 and L. We test every feasible value, namely choose s wavelet basic function to construct the wavelet neural network with the algorithm mentioned above for every feasible s value. Then evaluate the performance of network using the least mean square deviation method. The s value of the network which has the least value of mean square error between the network output and the sample actual output corresponding is the s value that we need gain.

The mean square error of the wavelet neural network is defined as follow:

$$MSE = \frac{1}{2M} \sum_{i=1}^{M} \sum_{k=1}^{K} (f_k(x_i) - y_{ik})^2 \tag{11}$$

Here, M is the amount of training samples; K is the number of network output layer neuron; $f_k(x_i)$ is the network output of output layer k neuron corresponding to the i sample; y_{ik} is the actual output of output layer k neuron corresponding to the i sample.

The number of wavelet basic function that we expected is as follow:

$$s = \arg\min_{s=1,2,\cdots,L} MSE \tag{12}$$

2.3 Learning Algorithm

When the number of wavelet basic function was ascertained, the output of wavelet neural network was expressed as follow:

$$\hat{f}(\cdot) = \sum_{i=1}^{n} w_i \psi_{a,\theta,\bar{b}}(x_{t-1}, x_{t-2}, \ldots, x_{t-p}) \tag{13}$$

$$\text{or } \hat{f}(\cdot)=\sum_{i=1}^{n} w_i \psi_i(\cdot) \tag{14}$$

Here, w_i was the weight between the hidden layer node and the output layer node; $\psi_{a,\theta,\bar{b}}(\cdot)$ was the output value of the hidden layer node.

Gather together all the parameters of formula (13), and go by the general name of ϕ, replaced the $\hat{f}(\cdot)$ of formula (13) use $f_\phi(\cdot)$. We adopted the learning algorithm based on the gradient descent to train the wavelet neural network. The object function was constructed as follow:

$$E=c(\phi)=\frac{1}{2}\sum_{p}[f_\phi(\cdot)-y]^2 \tag{15}$$

Here, p was the number of the training samples.

Recurrence decrease the formula (15) with the in-out data pair. Through the gradient algorithm of every measured value, the parameter ϕ was attenuated along the gradient direction of function.

$$c_K(\phi)=\frac{1}{2}[f_\phi(\cdot)-y_k]^2 \tag{16}$$

Our goal was to ascertain $w_i, a_i, b_i, \gamma_{-\theta}$, and make the fitting of that between the predicted value $f_\phi(\cdot)$ sequence of the wavelet network and the actual value y_k sequence was optimum. Here, $w_i, a_i, b_i, \gamma_{-\theta}$ can be optimized by the least square error energy function formula (16).

We adopted the Morlet wavelet as the excitation function of the hidden layer nodes of wavelet neural network, $\psi(t)=(e^{-i\omega_0 t}-e^{-\omega_0^2/2})e^{-t^2/2}$. When $\omega_0 \geq 5$, $e^{-i\omega_0/2}\approx 0$, the second item can be ignored, generally the approximate representation was $\psi(t)=e^{-i\omega_0 t}e^{-t^2/2}$.

Let $\psi(x)=d\psi(x)/dx$, $e_K=f_\phi(\cdot)-y_K$, $Z_i=a(x-t_i)$, then the partial differential of the function formula (16) corresponding to every component of every parameter vector ϕ was as follow:

$$\frac{\partial c}{\partial f}=e_K \tag{17}$$

$$\frac{\partial c}{\partial w_i}=e_K.\psi(Z_i) \tag{18}$$

$$\frac{\partial c}{\partial b_i} = -e_K.w_i.\gamma_{-\theta}.\psi'(\gamma_{-\theta}(Z_i)) \tag{19}$$

$$\frac{\partial c}{\partial a_i} = -e_K.w_i.\gamma_{-\theta}^2.\psi'(\gamma_{-\theta}(Z_i)) \tag{20}$$

$$\frac{\partial c}{\partial \gamma_{-\theta}} = -e_K.w_i.(x - t_i).\psi'(\gamma_{-\theta}(Z_i)) \tag{21}$$

Here, the modified value of weight was: $\Delta w_i = \frac{\partial c}{\partial w_i}$

The computation procedure was described as follow:

1) Initialization w_i, b_i, a_i and $\gamma_{-\theta}$. $i = 0$;

2) Compute the parameter in the step 1) according to the formula that from (18) to (21);

3) Substitute these parameters that calculated in the step 2) into formula (13), and compute $\widehat{f}(\cdot)$;

4) Compute the error $err = \sqrt{\frac{\sum_{i=1}^{n}(f_i(\cdot) - y_i)^2}{\sum_{i=1}^{n} y_i}}$;

5) If the error could satisfy the accuracy requirement and stop; else $i = i + 1$, turn to the step 2).

3 The Compost Maturity Degree Affect Factors

The maturity degree judgment indexes of compost include three kinds that are physical, chemical and biological indexes. The typical physical indexes include temperature, smells, moisture content etc. The temperature index is always used as compost processing control index to judge the producing processing course of compost, for instance, the compost temperature higher than 55 ^{O}C should be keep on no less than 3 day. Smell index can only present the maturity degree of compost from the apparent. The variety of moisture content shows a well regularity in the sludge compost and the value of moisture content impacts the sales and use of sludge compost products. So the moisture content can be used as judgment index of sludge compost, it can characterize the maturity degree of sludge compost in some extent.

The typical chemical indexes are w(C)/w(N), organism, N content, humus, organic acid etc. The w(C)/w(N) is a mostly used index for compost maturity degree judgment, but w(C)/w(N) is related to the compost material, and is impacted seriously by the amendment and the sort of compost. The extract and test processes of humus and organic acid are relatively complex, and the N component varies relatively fast in the sludge compost production. These indexes are not adapted for judgment the maturity degree of sludge compost. The volatility solid (VS) in sludge is the reduced part of dry sludge after been iglossed in high temperature and its testing is relatively simple, which can basically reflect the content of organism in sludge. The range of variation of organism in different compost material is big, and the change of VS can reflect the maturity degree of compost.

The biologic indexes include respiration, germination index, growth amount of plants, microbial activity, and pathogenic bacteria. The index of respiration has high requirement for instrument. The growth amount of plants can characterize the plant toxicity of compost well, but it needs a too long foster period. The testing for the index of microbial activity is complex, and with an inadequate accuracy. In order to prevent the pollution of environment, there should be some certain requirements for the containing of pathogenic bacteria and other microorganism in the compost. The germination index is a typical index for judgment the maturity degree of compost, it can indirectly characterize the biology toxicity action of compost. The main intention of sludge compost is resource utility, after compare, we select the value of fecal bacteria and the germination index as the biology indexes for the judging the maturity degree of compost.

On the basis of deeply analysis of the above indexed, we select the high temperature duration, moisture content, volatile solids, the value of fecal bacteria and germination index five kinds parameters as the judgment indexes of compost maturity degree.

There is no unanimous standard to judge sludge compost maturity degree till now. In this paper, the reference standards of compost maturity degree index referred to the provision and several correlative research findings about agricultural utilization of sewage sludge, dejection and organic matter at home and abroad. The maturity degree of sludge compost can be classified into 4 categories: full maturity, preferable maturity, general maturity and immaturity. The judgment criterion is shown as Table 1.

Table 1. The judgment criterion of compost maturity degree

judgment parameter / maturity degree level	High temperature duration (Day)	Moisture content (%)	Volatile solids (%)	The value of fecal bacteria (/g)	Germination index (%)
full maturity	>7	<40	<55	<300	>80
preferable maturity	5-7	40-50	55-60	300-700	60-80
general maturity	4-5	50-65	60-65	700-1000	50-60
immaturity	<3	>65	>65	>1000	<50

The full maturity means that the sludge compost production can be vended directly. The preferable maturity production mainly are used to the areas where people may contacted the sludge compost production such as parks, gardens, green belt and golf field. The sludge compost production of general maturity are used to the areas where people could not contacted to mature the compost such as forest, repairing mine site and sludge private disposal site. The immaturity compost production could not apply to any field, because its property is not steady enough or the pathogenic bacteria content is too high.

4 The Sludge Compost Maturity Degree Modeling Based on Wavelet Neural Network

4.1 The Characteristic Parameters

The first task of judgment the sludge compost maturity degree using wavelet neural network is to collect sample data. In order to collect the training sample, we must extract the characteristic parameters that correlate with the sludge compost maturity degree. In a general way, we can extract multi characteristic parameters which correlate with the sludge compost maturity degree of sewage treatment. Then these characteristic parameters are analyzed and optimized. Using the forward model, we can obtain the knowledge of all kinds of characteristic parameters in reflecting the sludge compost maturity degree, and combining the expert experience, we confirm the key parameters that have bigger degree of correlation with the sludge compost maturity degree, the key parameters act as the input neurons of wavelet neural network.

4.2 Application Study

Through the correlative analysis between the characteristic parameters and the sludge compost maturity degree, we find the characteristic parameters which have the maximum correlativity with the sludge compost maturity degree. Combining expert experience, finally, we select 5 kind of characteristic parameters to as input neurons of wavelet neural network, these parameters are the high temperature duration, moisture content, volatile solids, the value of fecal bacteria and germination index. The output layer of wavelet neural network contains four nodes, which stand for 4 kind of output mode, namely 4 kind of maturity degree level of sludge compost. the expectant output of "full maturity" maturity degree sample is [1,0,0,0], the expectant output of "preferable maturity" maturity degree sample is [0,1,0,0], the expectant output of "general maturity" maturity degree sample is [0,0,1,0], the expectant output of "immaturity" maturity degree sample is [0,0,0,1].

We can get many group training samples and testing samples, and every training sample is composed of 5-input and 4-output. We ascertain the number of the basic function of the wavelet neural network according to the sparseness of the sample data and ascertain the initial value of the others every parameter of network. When the structure and every parameter initial value of the wavelet neural network is ascertained, using the learning algorithm to train the network in order to obtain every parameter value.

We write the corresponding algorithm program using C++. In order to show the advantage and feasibility of wavelet neural network, we adopt wavelet neural network and BP network to judge the sludge compost maturity degree. The structure of BP is the same as wavelet neural network, namely BP network also adopt the three layer structure, and has one hidden layer. The input layer of the wavelet neural network and the BP network has the same number of input neurons, that is 5, and the output layer has the same four output neurons. According to the actual sample data and the program computing, the number of the basic function of wavelet neural network is 8, the hidden layer of BP network has 8 neurons also. So the topology structure both of wavelet neural network and the BP network is 5-8-4. The learning parameters of network are selected as follow: the learning efficiency is 0.9, the inertia coefficient is 0.7, the system maximum error is 0.01, the maximum error of single sample is 0.001, the iteration times of network is 5000.

We could random separate the 500 samples into 5 groups, and take out 80 sample data as training samples each time, while the other 20 samples as testing sample. After the normalized process of the sample set, the wavelet neural network and BP network was trained 5 times respectively using 5 groups different training sample, thenceforth we used the corresponding testing samples to test. The training error and testing error were respectively the average value of the 5 times training error and 5 times testing error. The result of train and test by wavelet neural network and BP network is shown as Table 2.

Table 2. Contrast of training and testing results by wavelet neural network and BP network

Method	Training error	Testing error	Average iteration times
Wavelet neural network	0.066	0.125	351
BP network	0.085	0.221	1082

From the table 2, we can see the mean squared error of training samples of wavelet neural network is smaller than that of BP network. Moreover the mean squared error of testing samples of wavelet neural network is also smaller than that of BP network. Furthermore the iteration times of the wavelet neural network is obvious smaller than that of BP network with the same system error. It shows that the wavelet neural network is all superior to the BP network at the estimation accuracy, detection accuracy and convergence rate aspects. So the compost maturity degree modeling method based on wavelet neural network has higher stability, and can obtain higher judging and recognition accuracy.

5 Conclusions

Because the wavelet neural network has the advantages of both wavelet analysis and neural network, so it has the faster convergence rate and the stronger approaching ability in relation to the BP network. So the maturity degree modeling method of sludge compost based on wavelet neural network is presented and applied to maturity degree judgment of sludge compost in this paper. In the maturity degree judgment process, considering the constraint between the model and the judgment index, we take the maturity degree level and the judgment index as a complex mapping relation, and use wavelet neural network model that possessed of global approach and fast convergence to establish the mapping, then the judgment accuracy and speed of network can be enhanced. With the ability of strong self-learning and function approach and fast convergence rate of wavelet neural network, the maturity degree modeling method can truly judge the maturity degree level of compost by learning the maturity degree judgment information of sludge compost. The experimental results show that this method is feasible and effective and this modeling method has extensive application foreground in the sewage treatment.

References

1. California Compost Quality Council Compost Maturity Index [EB/OL] (2001)
2. Qian, X.L., Sun, Y., Li, D.T.: Study on Degree of Maturity of Compost Using Fuzzy Comprehensive Assessment [J]. Shanghai Environmental Sciences 20, 85–87 (2001)
3. Zeng, W.H., Chen, R.C., Shi, Y., et al.: A model for evaluating ability of professional innocent production based on artificial neural network [J]. Environmental Engineering 21, 70–72 (2003)
4. Licheng, J.: Neural Network System Theory [M]. Xian: Xi an Electronic Science and Technology University Press (1995)
5. Pati, Y.C., Krishnaprasad, P.S.: Analysis and sythesis of feedforward neural networks using discrete affine wavelet transformations. IEEE Trans. Neural Networks 4, 73–85 (1993)
6. Zhang, Q., Benveniste, A.: Wavelet networks. IEEE Trans. Neural Networks 3, 889–898 (1992)
7. Zhenning, Z., Yongmao, X.: Introduction to fuzzy theory and neural networks and their application [M]. Tsinghua University Press, Beijing (1996)
8. Huashuai, C., Xuya, P., Ming, L.: Applacation of Fuzzy Mathematics in the Evaluation of MSW Compost quality. Journal of Chongqing Jianzhu University 11, 87–89 (2006)
9. Dominic, H., Josef, B., Enzo, F., et al.: Review of compost standards in New Zealand. The Waste and Resources Action Programme (2002)

Neural Network Inverse Control for Turning Complex Profile of Piston

Tingzhang Liu, Xiao Yang, and Jian Wang

Shanghai University, Department of Automation,
200072 Shanghai, China
liutzh@staff.shu.edu.cn

Abstract. Advanced internal combustion engine piston is designed with middle-convex contour and variable ellipticity section for better working performance. Machining system shows significantly multi-input multi-output and nonlinear behaviors when turning this piston profile on NC lathe. These behaviors lead to an obvious error between machined parts to the desired profile. Traditional error compensation method for linear system cannot assure the required accuracy. Therefore, an artificial neural network (NN) inverse control method is proposed for high profile accuracy. A multilayered feedforward neural network is designed for this multi-input multi-output system and trained off-line to identify the inverse model of the machining system. Then the inverse model is employed to tune the system online. It is shown, by experiments of piston turning, that this control scheme can improve the machining accuracy effectively.

1 Introduction

Advanced internal combustion engine piston is designed with middle-convex contour and variable ellipticity section for better working performance. Machining these parts needs special turning with a high-frequency reciprocating motion in radial direction and contour motion in longitudinal direction for cutting tool [1]. Different machining methods were developed to turn this kind of complex profile, such as sound coil linear machine [2], linear servo motor [3], electrohydraulic and piezoelectric actuator [4]. In this machining process, cutting tool works in a variant velocity and inertia at different point, and the machining system shows significantly multi-input multi-output and nonlinear characteristics when turning this piston profile on NC lathe. This characteristics lead to the difficulties in achieving high performance with conventional control methods. Obviously, it needs advanced control methods to improve turning accuracy of complex profile. A robust repetitive control method through a sequence of two single-input-single-output (SISO) designs is proposed for noncircular machining application. This method is used to control a dual-stage actuator which consists of an electrohydraulic actuator for 25-mm-gross motion and piezoelectric actuator for 40-μm fine motion [4]. Considering the machining system as a linear, periodic, time-varying systems, a periodic sampling interval repetitive

K. Li et al. (Eds.): LSMS 2007, LNCS 4688, pp. 619 – 627, 2007.

control method based on gain scheduling is proposed to the control of a piezoelectric fast-tool stage for variable spindle speed noncircular turning process [5].

Neural network (NN) has a multi-input multi-output structure and process information in distributed mode, so it is suitable for multi-input multi-output system. NN possesses an inherent nonlinear ability, so it is suitable to express nonlinear systems. Furthermore, NN has capability of continuous learning. When the NN model cannot meet the demands, it can learn continuously to adjust the old model to follow the varied system by enhancing the learning accuracy or supplying new samples. To reduce the large transient errors introduced by sudden jumps and discontinuities in tracking noncircular piston profiles, a fast converging repetitive learning control scheme by feedforward neural network based on stable plant inversion is proposed. Simulations show that the learning scheme results in monotonic error convergence [6]. Another work uses neural network to capture the complex relationship between the errors on the component and the input process conditions such as tool/work deflections, machine tool accuracy, machining conditions etc. Significant improvement in part accuracy has been obtained using the neural network to predict these errors and correct CNC code during machining [7].

In this paper, a neural network inverse control method is proposed for high profile accuracy. A multilayered BP neural network is designed for this multi-input multi-output system and trained off-line to identify the inverse model of the machining system. Then the inverse model is employed to tune the system online.

2 The Architecture of the Experimental System

The complex profile of advanced internal combustion engine piston is shown in Fig. 1. In Fig. 1, Dx is the long axis of elliptical section at longitudinal position x, Rx is the radius at angle α of elliptical section at longitudinal position x, Gx is the ellipticity of elliptical section at longitudinal position x.

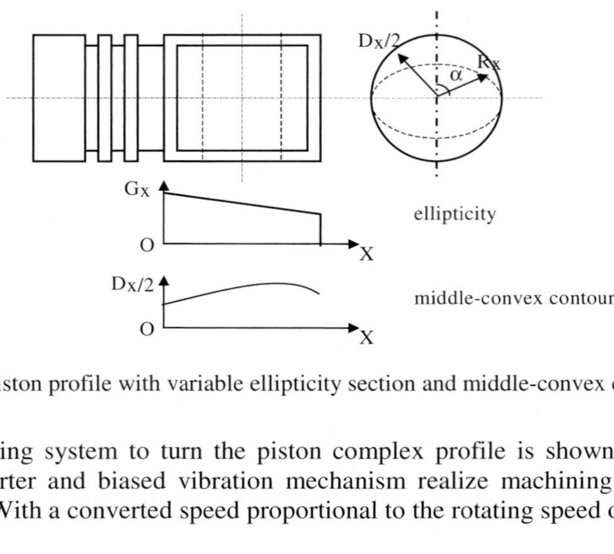

Fig. 1. Piston profile with variable ellipticity section and middle-convex contour

The machining system to turn the piston complex profile is shown in Fig. 2. A velocity converter and biased vibration mechanism realize machining for elliptical cross section. With a converted speed proportional to the rotating speed of the spindle, the velocity converter drives the biased vibration mechanism to produce a reciprocating motion, whose frequency is 2 times to the speed of the spindle. This motion drives the cutting tool reciprocating 2 times in every piston rotation. In this way, an elliptical cross section is machined. The longitudinal contour machining is realized by a stepmotor and a cam mechanism. Firstly, the piston contour is fit with the cam contour by interpolation to form a soft template. Secondly, the stepmotor adjusts its speed and steps according to the soft template and then drives the cam to rotate in a variant speed. Finally, the cam pushes a swing bar to form a micro displacement in radial direction which is delivered to the cutting tool. When the cutting tool moves at a given constant speed in longitudinal direction, the longitudinal contour is machined. The reciprocating motion for the elliptical cross section and the longitudinal contour motion are integrated to machine the desired piston profile. The synchronization of the cross section and longitudinal contour machining is implemented by a photoactor.

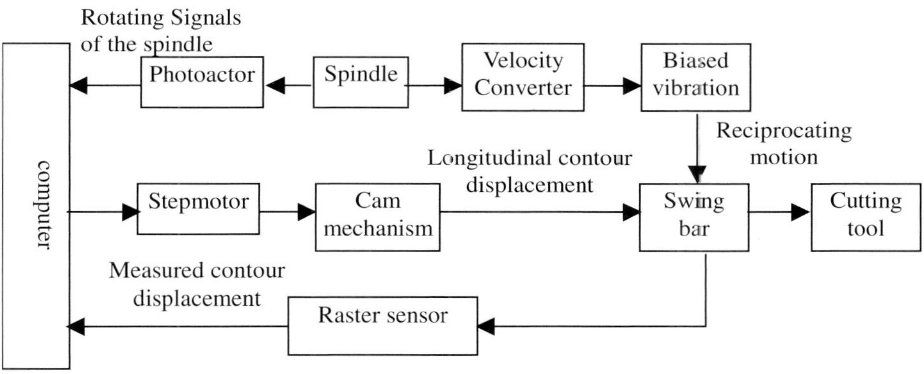

Fig. 2. The machining system for turning piston complex profile

In above system, the elliptical section machining is realized by the special designed mechanical mechanism, and the longitudinal contour machining is realized by a controllable stepmotor. In this paper, a NN controller is used to adjust the displacement of stepmotor to improve the machining accuracy for the longitudinal contour. The displacement of the swing bar is measured by a raster sensor and fedback to the controller as an indirect value of piston contour to realize the continuous online identification for the machining system.

3 Neural Network Inverse Control Method for Machining the Complex Profile

It is proved that a multilayered feedforward neural network with single hidden layer using the back propagation algorithm can approximate a wide range of nonlinear functions to any desired accuracy. Therefore, a multilayered neural network with the feedforward structure is used to fulfill the nonlinear modeling in this paper. This neural network is a complex nonlinear mapping from n-dimensional space to m-dimensional space, and the total network is equivalent to an implicit Function F:

$$Z=F(V)\ . \tag{1}$$

Where $V\in R^n$ is a n-dimension input vector.
$Z\in R^m$ is a m-dimension output vector.

The idea to control the machining errors of complex profile by NN is that NN can learn from the inverse–dynamic characteristics of the machining system with the historical input and output datum which are generally measurable. The NN model Z=F(V) can map the relation of input and output of the machining system after enough learning. If the NN model mapping the inverse relation of machining system is applied to tune the machining system, then the nonlinear characteristics of system can be compensated and controlled effectively.

3.1 Inverse Model Identification of the Machining System

When the complex profile is turned, the MIMO nonlinear characteristics of the machining system can be defined as:

$$Y=G(U)\ . \tag{2}$$

Where $U\in R^m$ is a m-dimension input vector.
$Y\in R^n$ is a n-dimension output vector.
$G: R^m\rightarrow R^n$ is a nonlinear function which maps the inputs to the system outputs.

The method identifying the inverse model of controlled system by means of NN is shown in Fig. 3. For NN is defined as a function which maps the inputs to outputs of the system, the input of NN can be acquired from the system output Y of practical machining system, the tutorial signal of NN can be acquired from the practical input U, and the input of the machining system can be simulated by the NN output Z. Assume G is invertible, then NN output Z is:

$$Z=F(Y)=F(G(U))=(FG)U\ . \tag{3}$$

With the backward Propagating algorithm, the connected weight of NN can be adjusted continuously up to that Z equals to U. From (3), we have FG=1 or $F=G^{-1}$. It is to say, the NN function F is equivalent to G^{-1} which is the transfer function of inverse model of controlled system.

Assume the ideal middle-convex contour of piston profile can be represented by n discrete points $\{(x_1,r_1),(x_2,r_2),\ldots(x_n,r_n)\}$, the outputs of the system can be written as $\{(x_1,y_1),(x_2,y_2),\ldots(x_n,y_n)\}$ which are the practical measured values of the turned workpiece at corresponding points, and the outputs are determined completely by n controllable inputs $\{(x_1,u_1),(x_2,u_2),\ldots(x_n,u_n)\}$ at these corresponding points, where x_i is the longitudinal position for point i, r_i, y_i, u_i is the radial coordinates for point i. Upon condition that the rotating speed of the spindle and the cutting tool speed in longitudinal direction are constant, the x_i can be simplified by subscript i. So the input vector in equation (2) is $U=(u_1,u_2,\ldots,u_n)$, the output vector $Y=(y_1,y_2,\ldots,y_n)$, where input vector and the output vector have the same dimensions (m=n). When adjusting the system input U, system output Y can be measured. The sample data set $(U^{(p)},Y^{(p)})(p=1,2,\ldots L)$ is formed by L times experimental results.

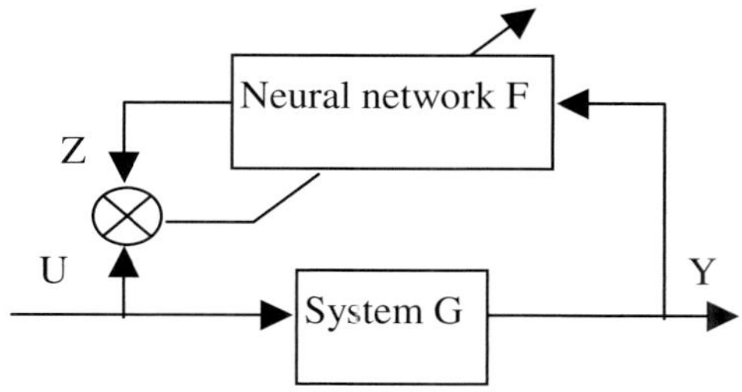

Fig. 3. Inverse model identification of system by NN

A NN structure is designed with n input layer nodes, q hidden layer nodes and n output layer nodes. Using above sample data set as the training sample for NN model identification, we can get the NN model P_{Lmod}, in which system output Y is used as NN input and system input U is used as the tutorial signal of the NN output. After identification, the inverse model of the system can be approximated by the NN model P_{Lmod}. Since BP network has a longer training time that is unlikely to meet the demand of real-time control, the identification of the inverse model is generally obtained by off-time learning.

3.2 Control of the Machining System

When the inverse model is available, it can be applied to tune the machining system. The basic concept of error control based upon NN is shown in Fig. 4. If the trained NN is equivalent to the inverse model of controlled system, whose transfer function is G^{-1}, then the output Y of controlled system is:

$$Y=G(U)=G(G^{-1}R)=(G\,G^{-1})R=R\,. \quad (4)$$

Where U is the controllable vector of the machining system, $U=(u_1,u_2,\ldots,u_n)$
R is the ideal output vector, $R=(r_1,r_2,\ldots,r_n)$

In practical machining process, if we give the ideal value R of discrete contour points to nodes of input layer of the trained model P_{Lmod} as the reference input for the NN controller, then, after the adjusting of the NN inverse model, the NN will output the optimal value Zr which can minimize the machining system output error if Zr is used as system input U. in this way, the machining error can be better controlled. The NN model P_{Lmod} is an identified inverse model, so it has a fast forward propagation speed and can control the machining error on-line.

In order to keep Y as close as possible to the desired value in spite of the parameter variations and external disturbances, the NN should has the ability to continuously identify the inverse model of the machining system. For this consideration, we can

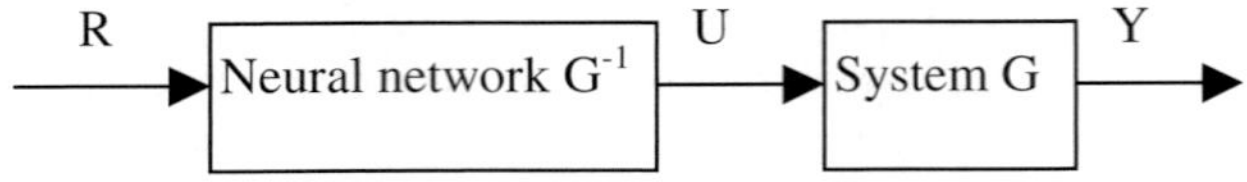

Fig. 4. NN inverse model control

take the present on-line measured datum as the new samples to finish the off-time system identification in non-machining time interval (such as the time to remove, load a part and adjust the lathe, etc), then the effect of the parameter variations and disturbances can be compensated in next turning. In this way, the dynamic characteristics of the machining system can be continuously identified and controlled.

4 Experimental Results

The desired relative longitudinal middle-convex contour R of a typical aluminum piston is shown in Table 1. The length of piston skirt is 68mm, and it has 16 discrete eigenpoints. The complex profile of the piston is machined on the machining system shown in Fig. 2. The rotating speed of the spindle is 1120r/min, the cutting tool speed in longitudinal direction is 0.1mm/r, and the demanded contour accuracy is ±0.01mm. With the NN inverse control method stated above, the high accuracy machining for the longitudinal contour of piston with middle-convex contour and variable ellipticity section is carried out.

Table 1. Contour machining results for 31 pistons without NN control and 1 piston with NN control

X mm	R μm	Y0 μm	U1 μm	Y1 μm	U2 μm	Y2 μm	U3 μm	Y3 μm	---	U30 μm	Y30 μm	Ur μm	Y31 μm
0	0	0	0	0	0	0	0	0	0	0	0	0	0
2.5	5	8	8	6	6	4.9	6	5.5		3.5	8.3	3.2	8
5	10	16.5	13	9	11	8	12	11.3		9.5	12.9	8.2	11
10	8.5	13.5	11	7.5	12.5	8.5	9.5	9.6		8.4	10.1	7.1	7.5
15	6.5	14	8	4	9	8	6.5	5.3		6.0	9.1	4.3	7
20	4	10.5	4	0.5	5	7.3	3.5	3.3		3.8	4.6	2.9	3
25	1	11	3	-0.5	4	4.5	1.5	2.4		1.8	3.4	0.1	0
30	-3	8	-1	-4.3	-7.5	-2	-1	-1		-1.2	-1	-3.8	-5
35	-8	0	-6.5	-12.5	-9	-7.5	-7.5	-7.5		-7	-5.3	-7.7	-8.8
40	-14.5	-9.5	-12.5	-19	-12	-17	-11.5	-12.5		-13	-13	-13.3	-17
45	-23	-17	-18	-26.5	-21.5	-27	-19	-22		-20.8	-19.5	-22.4	-25.5
50	-33.5	-29.5	-28.5	-38	-27.5	-38.5	-25	-30.8		-30	-30.3	-31.7	-38
55	-47.5	-39	-42.5	-51.8	-45	-55.5	-42	-47.4		-42.4	-42	-44.7	-49
60.5	-67.5	-57.2	-62.5	-72.8	-66.5	-75.3	-60.5	-62.4		-62.2	-62.1	-64.8	-71.3
65	-88.0	-76.7	-82	-93	-84	-100.5	-80	-82.6		-82	-79.6	-85.1	-90.8
68	-105	-93.2	-100	-111	-103	-108	-102	-101		-104	-102.9	-104.4	-108.5

Contour machining results for 31 pistons without NN control are shown in Table 1. In Table 1, X is the longitudinal coordinate (longitudinal height from the piston bottom) at 16 discrete eigenpoints of piston skirt, R is the desired relative radial dimensions at 16 discrete eigenpoints which shows the desired contour given by the blueprint, Y0 is the measured relative contour dimensions of machined piston corresponding to R as the system input directly. The max contour error is 11.8μm. The result shows that the direct machining cannot meet the accuracy demand. Changing the system inputs, we got other 30 groups of machining results without NN control. Measured relative contour dimensions of 30 machined piston Y1~Y30 corresponding to different system input U1~U30 are shown in Table 1.

Above 30 groups of machining results are used to identify the system inverse model by NN. We designed a BP NN structure with three layers, in which the number of input nodes, hidden layer nodes and output nodes is 16,15,16 respectively. 30 groups of practical datum are chosen as the learning samples to identity the system, where the practical measured value Y is used as NN input and the system input U is used as the tutorial signal of the NN. After 5000 times learning, the normalized error E of total system is decreased to 0.00025, and the normalized error of single point ε is 0.001 responding to the practical learning error 0.12μm. Thus the model P_{30mod} is obtained.

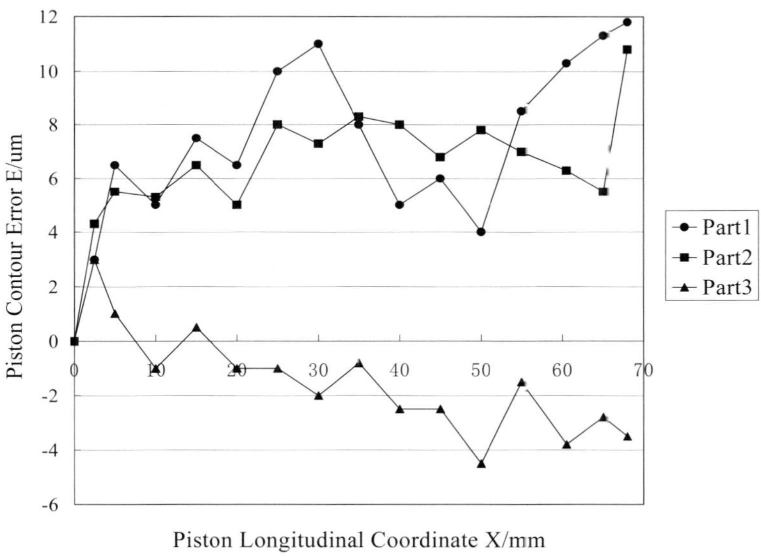

Fig. 5. Machining error using different methods.

Furthermore, with the NN inverse control method, we put the desired relative contour R into P_{30mod} to obtain the optimal input Ur of the machining system. Using Ur to control the stepmotor to turn the piston contour, the measured relative contour dimensions Y31 and optimal input Ur are shown in Table 1. To compare the effectiveness of different methods, the measured contour errors of 3 machined pistons is shown in Fig. 5, where the abscissa is the longitudinal coordinate X, the ordinate is the error. Part1 is from direct machining result using R as system input directly, part2 is from machining result using SISO linear compensation method, and part3 is from

NN inverse control method using Ur as system input. From Fig. 5, the contour machining errors using NN inverse control method is ≤±4.5μm which meets excellently the blueprint demand. Compared to NN control, direct machining method and SISO linear compensation method got the max contour machining errors at 11.8μm and 10.7μm respectively. It is shown that the NN inverse control method can improve the machining accuracy significantly and control the machining error effectively.

5 Conclusion

During the machining process of advanced internal combustion engine piston with middle-convex contour and variable ellipticity section, there exist difficulties such as multi-input multi-output nonlinear characteristics and slow time-varying dynamics. These factors lead to an obvious error between machined parts to the desired profile. Traditional error compensation method for linear system cannot assure the required accuracy.

This paper introduced a neural network inverse control method for turning the complex profile of piston by the idea of NN nonlinear modeling. A multilayered feedforward neural network is designed to identify the inverse model of machining system offline, then this identified NN is employed as the controller to adjust the nonlinear characteristics of system online. To follow the system parameter variations and disturbances, NN can identify the inverse model of practical system continuously by using present on-line measured datum as the new samples in non-machining time interval. The experiments of piston turning show that the NN inverse control method can improve the machining accuracy significantly and control the machining error effectively.

Acknowledgments. Supported by Shanghai Leading Academic Discipline Project, Project Number: T0103.

References

1. Liu, T.Z., Li, B.M., Ye, B., et al.: Complex Profile of Engine Piston Skirt Produced by a Technology with Vibration. Journal of Xi'an Jiaotong University 32(10), 47–50 (1998)
2. Li, L.C., Lu, B.H.: The Elimination of DC-component of the Sound Coil Current in a Piston Lathe Tool Holder. Journal of Xi'an Jiaotong University 29(7), 49–53 (1995)
3. Jiang, S.M., Wang, X.K., Shi, Z.D., et al.: Study on the High Performance Linear Servo System for Middle-convex and Varying Ellipse Piston Machining. Chinese Journal of Mechanical Engineering 37(9), 58–61 (2001)
4. Kim, B.S., Li, J.W., Tsao, T.C.: Two-parameter Robust Repetitive Control with Application to a Novel Dual-stage Actuator for Noncircular Machining. ASME Transactions on Mechatronics 9(4), 644–652 (2004)
5. Hanson, R.D., Tsao, T.C.: Periodic Sampling Interval Repetitive Control and Its Application to Variable Spindle Speed Noncircular Turning Process. Journal of Dynamic Systems, Measurement and Control Transactions of the ASME 122(3), 560–566 (2000)

6. Krishnamoorthy, K., Tsao, T.C.: Repetitive Learning Control for Precision Machining of Complex Profiles. ASME Dynamic Systems and Control Division DSC 73(1) (Part A), 63–69 (2004)
7. Suneel, T.S., Pande, S.S.: Intelligent Tool Path Correction for Improving Profile Accuracy in CNC Turning. International Journal of Production Research 38(14), 3181–3202 (2000)

The Modeling and Parameters Identification for IGBT Based on Optimization and Simulation*

Yanxia Gao, Nan Li, Shuibao Guo, and Haijian Liu

Shanghai Key Laboratory of Power Station Automation Technology, Shanghai University
200072, Shanghai, China
gaoyanxia@shu.edu.cn

Abstract. IGBT (Insulated Gate Bipolar Transistor) is becoming more and more popular in many power applications, since it offers a good compromise between on-state loss, switching loss and easy of use. To develop circuits and systems with the devices, model and model parameters are necessary in circuit simulations. This paper presents a procedure for identifying the most important model parameters of IGBT. As an example, the results of identification for BUP302 are given.

1 Introduction

With the development of the power electronic system, the simulation becomes more and more important. In recent years, the prototype-less design of power electronic system has appeared, so more accurate simulation results are required. The accuracy of simulation results depends on the power electronic devices model and its parameters. Therefore the modeling of device and its parameter identification is always one of the most important research directions in power electronic technology.

Having the virtues of high current density and low drive power, the insulated gate bipolar transistors (IGBT) are widely used in the area of power electronic converters and electronic drives. Generally the models are categorized into two different classes [1]. The first one is "behavioral" or "empirical" model [2], this kind of model is based on equivalent circuit, simple but is not enough for accuracy. The second is mathematical-physical model. It refers to mathematics equation based on IGBT physical structures. The model of this kind can precisely describe the physical characteristics of IGBT, but it is complex. According to the different model, there are different methods for parameters identification. Literature [3] proposes the parameter extraction based on behavioral model. The identification of model parameters is the main challenge with math-physics based models. In [4], the parameter extraction is based on mathematical-physical model, but the procedure is too complex to use for common electrical engineers. In this paper, we proposed a method of parameters identification based on experiment, simulation and optimization algorithm. The identification procedure is quite easy for general users. The simulation software used

* The work was supported by Shanghai Leading Academic Discipline Project(Project Number ：T0103)

K. Li et al. (Eds.): LSMS 2007, LNCS 4688, pp. 628 – 638, 2007.

is PACTE developed by CEGELY, specified for the simulation of power electronics. As an example, the paper presents the identification results of BUP302.

2 The Mathematical-Physical Model of NPT IGBT

As mentioned in section one, the mathematical-physical model can describe the characteristic of IGBT well. So the proposed extraction method is based on this model. The basic structure of the non punch-through (NPT) IGBT is shown in Fig.1. Being a conductivity-modulated power device, the behavior of an IGBT depends heavily on the carrier distribution in its wide drift region. Usually we use one-dimensional (1-D) ambipolar diffusion equation to describe it [3-5].

According to semiconductor physical theory, Hefner built the static and dynamic model of IGBT, the equivalent circuit as Fig. 2. This model has been replanted to the circuit simulation software such as Saber, IG-Spice and PACTE.

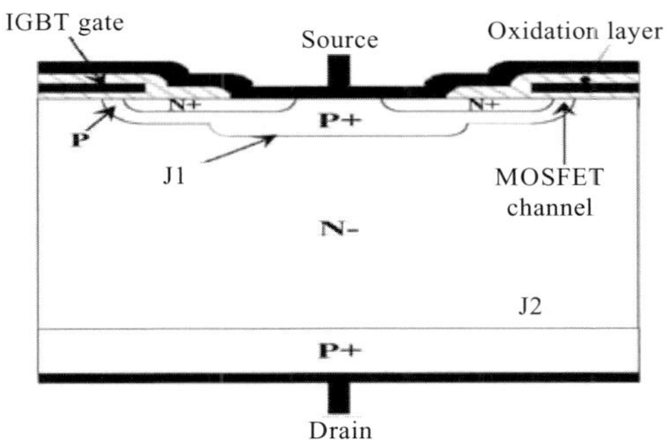

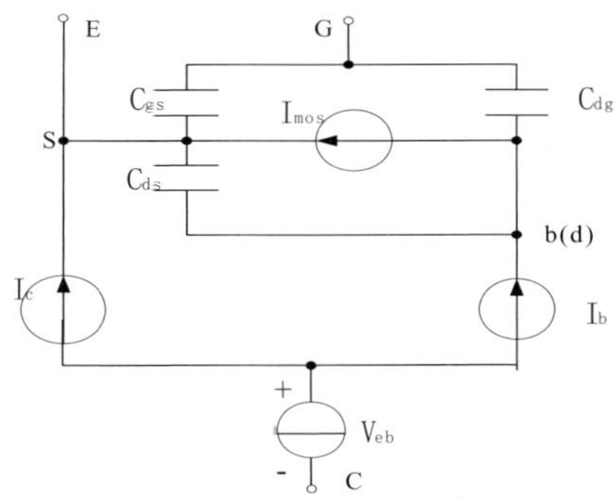

Fig. 1. The basic structure of NPT IGBT model **Fig. 2.** The equivalent circuit of NPT IGBT

The important equations to describe IGBT's model are as equations (1)-(8). Symbols used here by are defined in the appendix.

1) I_T is the current which flows from MOSFET drain region to PNP-base as shown in equation (1).

$$I_T = I_n(W) + I_p(W) \tag{1}$$

2) PNP-base current I_n, show in equation (2)

$$I_n(W) = I_{mos} + C_{ds}\frac{dV_{ds}}{dt} + C_{dg}\frac{dV_{dg}}{dt} \tag{2}$$

3) I_p is composed of electron and couple current of holes carrier, re-distribution current, etc. shown in (3)

$$I_p(W)=\left(\frac{1}{1+b}\right)I_T+\left(\frac{b}{1+b}\right)\frac{4}{W^2}Q+\frac{C_{bcj}}{3}\frac{Q}{Q_B}\frac{dV_{bc}}{dt} \tag{3}$$

4) IGBT gate current I_g, show as equation (4)

$$I_g=(C_{gs}+C_{gd})\frac{dV_{gs}}{dt}+C_{gd}\frac{dV_{bc}}{dt} \tag{4}$$

5) I_{MOS} MOSFET channel current as equation (5)

$$I_{mos}=\begin{cases}0\cdots\cdots V_{gs}<V_T\\ \dfrac{K_{Plin}\left[(V_{gs}-V_T)V_{ds}-\dfrac{K_{Plin}V_{ds}^2}{2K_{Psat}}\right]}{1+\theta(V_{gs}-V_T)}\cdots V_{ds}\le(V_{gs}-V_T)\dfrac{K_{Plin}}{K_{Psat}}\\ \dfrac{K_{Psat}(V_{gs}-V_T)^2}{2[1+\theta(V_{gs}-V_T)]}\cdots\cdots V_{ds}>(V_{gs}-V_T)\dfrac{K_{Plin}}{K_{Psat}}\end{cases} \tag{5}$$

6) The excess majority carrier charge in the bipolar transistor base (equal to the excess minority carrier charge to maintain quasi neutrality) is supplied by the electron current that enters the collector end of the base (base current), and decays by recombination in *HDB*, recombination in the *LDB*, and injection of electrons into the emitter.

$$\frac{dQ}{dt}=I_n(W)-\frac{Q}{\tau_{HL}}-\frac{Q^2}{Q_B^2}\frac{4N_{scl}^2}{n_i^2}I_{sne} \tag{6}$$

7) The V_{eb} can be written as equation (7)

$$V_{eb}=V_{ebq}+I_TR_b \tag{7}$$

8) For forward conduction, V_{ebq} is equal to V_{ebd} , and for reverse blocking, V_{ebq} is equal to V_{ebj} , as shown in

$$V_{ebd}=\begin{cases}V_{eb}\cdots\cdots Q<0\\ \min(V_{ebj},V_{ebd})\cdots\cdots Q_0>Q\ge 0\\ V_{ebd}\qquad Q\ge Q_0\end{cases} \tag{8}$$

Other equations of IGBT model refer to reference [3-5]. Table 1 lists IGBT model parameters. Some parameters (signed with *) determine static characteristic of the device and others determine the dynamic process, therefore the identification procedure is composed of two sub procedures, one for static and the other for dynamic.

Table 1. The model parameters of NPT IGBT

Symbol	Description	Unit
A^*	Device active area	cm^2
A_{gd}	Gate-drain overlap area	mm^2
W_B	Metallurgical base width	μm
V_{TH}^*	MOS threshold voltage	V
V_{Td}	Gate-drain overlap depletion threshold	V
K_{plin0}^*	MOS transconductance coefficient	A/V^2
K_{psat0}^*	Saturated region MOS transconductance parameter	A/V^2
C_{gs}	Gate-source capacitance	nF
C_{oxd}	Gate-drain overlap oxide capacitance	nF
I_{sne}^*	Emitter electron saturation current	A
tau_HL	Minority carrier lifetime in buffer layer	μs
N_B	Base doping concentration	cm^{-3}
$tetal^*$	Proofed coefficient allowed by calculate the range of MOSFET cross section	V

3 The Parameter Identification and Optimization Algorithm

The optimization procedure requires suitable experimental waveforms to compare with the simulated waveforms. These waveforms should be captured using a digital oscilloscope and transferred to the host computer, allowing a direct comparison with the simulated waveforms to produce an error figure which is then used to guide the optimization [10]. The test beach for the waveform acquisition is shown as Fig.3.

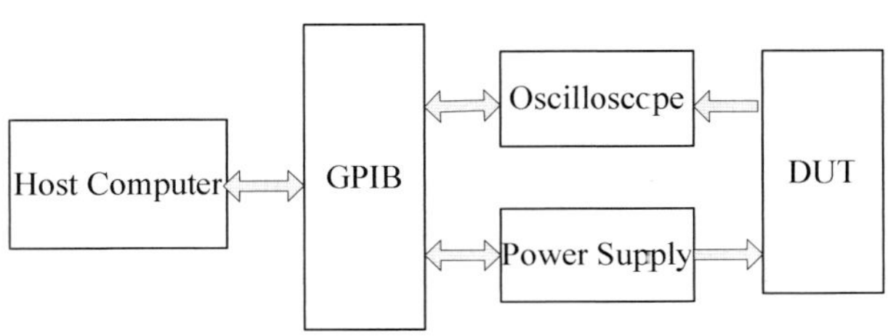

Fig. 3. The beach for waveform acquisition

The proposed parameter identification procedure consists of two separate steps[10].

1) Initial Parameter Extraction—This is based on device datasheets to get the initial values for the optimization.

2) Optimization Procedure—Refinement of parameters through a formal optimization procedure.

Some of the IGBT parameters listed in Tables 1 can be obtained from the datasheet or calculated based on textbook equations, so there is no need to use measurement for their extraction. Extra measurement can be used to verify the mathematical results. But some parameters, like IGBT carrier lifetime, have to be determined by experiment. Accordingly, the clamped inductive load test experiment is used. The second step in the procedure is a formal optimization which can be completely automated. An advantage of the procedure is that inaccuracies in the initial parameter estimation may be corrected by the optimization procedure.

3.1 Optimization Procedure

Fig.4 is the skeleton of the parameters identification proposed. The process can be summarized in the following:

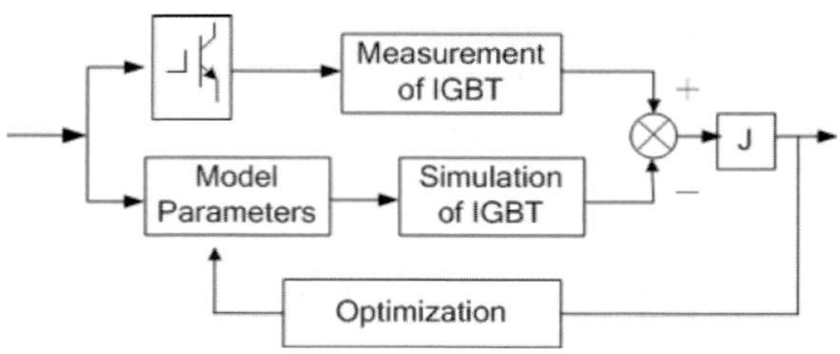

Fig. 4. The procedure of identification

a) Initial estimates of parameters made from device datasheets and basic measurements;

b) Measurement of device switching behavior, i.e., acquisition of switching waveforms;

c) Simulation of circuit behavior using parameters estimated;

d) Comparison of simulated and measured waveforms to produce an error value;

e) Variation of the parameter values to minimize the error value.

The process of identification is the process that the simulated waveforms approaching to the measured waveforms via optimizing parameter. The IGBT model parameters are divided into the static parameter and the dynamic parameter, so the process of identification also is divided into the static model parameter identification and the dynamic model parameter identification.

3.2 Static Parameters Identification

Typical static parameter of IGBT consist of transfer characteristic and output characteristic, and we take output characteristic for example here for parameter extraction. The IGBT output characteristic is defined as the relationship between I_C and V_{CE} while V_{GE} as a control variable.

The object function of static parameters extraction is made up of the difference between static output characteristics of measurement and simulation, shows as equation(2).

$$J = \sqrt{\sum_{m=1}^{m=M} \int_{V_{CE}=0}^{V_{CE}=V_{CE\max}} \delta_{V_{CE}}^{2} \, dV_{CE}} \tag{9}$$

Where

$$\delta_{V_{CE}} = \left| \frac{i_C^s(v_{CE}^s) - i_C^e(v_{CE}^e)}{i_C^e(v_{CE}^e)} \right|$$

i_c^s and i_c^e are the drain-source currents of simulation and experiment. v_{ce}^s and v_{ce}^e are drain-source voltages . *m* is the number of static output characteristics in different gate currents.

As an example, the static parameters identified of BUP 302 are in Tab.2, and Fig.4 shows the comparison of simulation and experimental results. Fig.5 shows the object function in the procedure of optimization.

Table 2. The results of idetification of static parameters (for BUP302)

Parameters	A	Vth	Kplin0	Kpsat0	Isne	tau_HL	teta1
unit	Mm^2	V	A/V^2	A/V^2	e^{-14}A	μs	V
Min.	8	5	0,5	0,5	100	10	50
Max.	16	7	5	5	1000	15	150
Optimized	**15.0**	**6.0**	**1.8**	**0.8**	**552.9**	**12.48**	**110.9**
Object function	$J = 0.7123$ (V_{CEmax}=20V, M=4)						

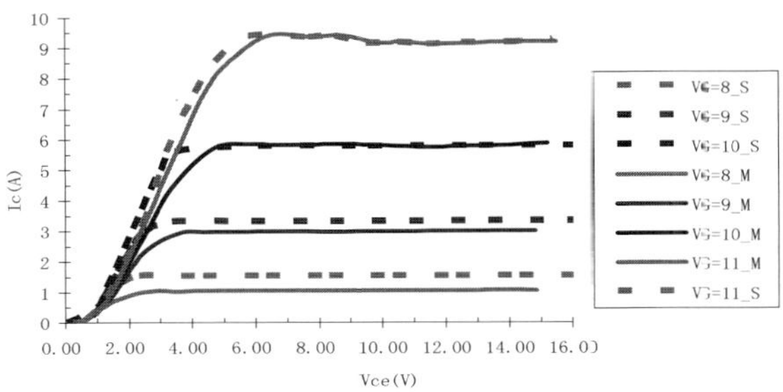

Fig. 5. Waveforms Comparison of measurement (solid) and simulation (dashed)

3.3 Dynamic Parameters Identification

For the dynamic parameters extraction, we use the transient waveform fitting method. The object function for dynamic parameters extraction is shown as

$$J = \sqrt{\sum_{i=1}^{n} k_i \Delta_i^2} \tag{10}$$

Where

$$\Delta_i = \left| \frac{x_i^m - x_i^s}{x_i^m} \right|$$

x_i^m and x_i^s denote some measured and simulated critical values of transient waveforms separately, such as , in the transient process. Total eighteen critical values for turn-on and turn-off waveforms are used to build the object function. Table 3 shows the identified results of dynamic parameters for BUP302. The objection function during the optimization procedure is shown in Fig.7. Fig. 8 to Fig.15 give a comparison of experimental and simulated results of V_{GE} , I_g , V_{DS} and I_D , for turn-on and turn-off commutations.

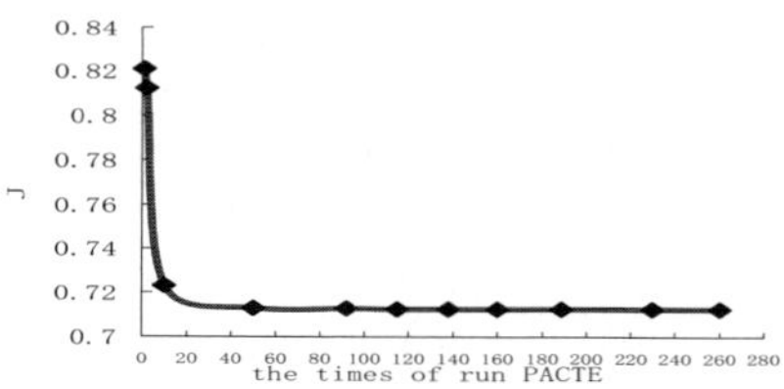

Fig. 6. J in static identification

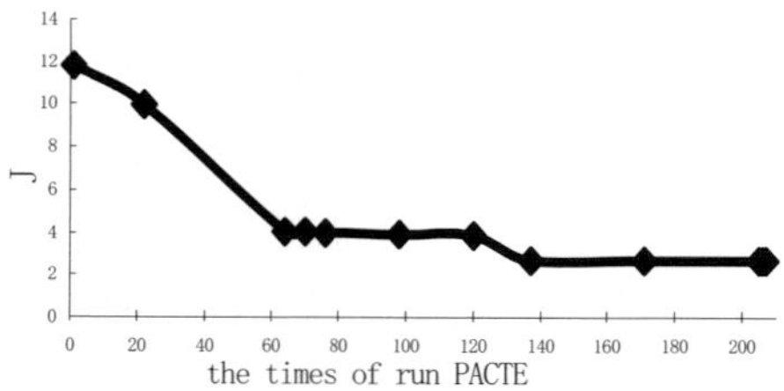

Fig. 7. J in dynamic identification

Table 3. The dynamic parameters of BUP302 and the identified results

Parameter unit	Agd mm^2	Wb um	Vtd V	Cgs nF	Coxd nF	T_{au_HL} μs	Nb $10^{14}cm^{-3}$
Min.	4	50	−0,5	0,t5	0,5	12	10
Max.	13	200	1	2	3	15	25
Optimazied	11.94	90	-0.247	0.533	0.3203	12.48	15.0
Object function	J= 2.48						

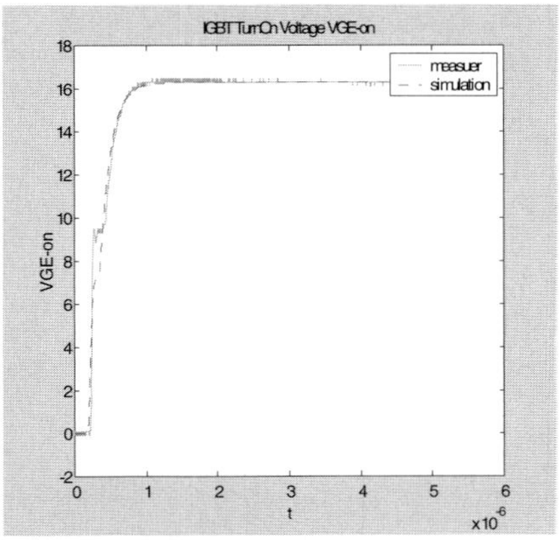

Fig. 8. V_{GS} when turn-on

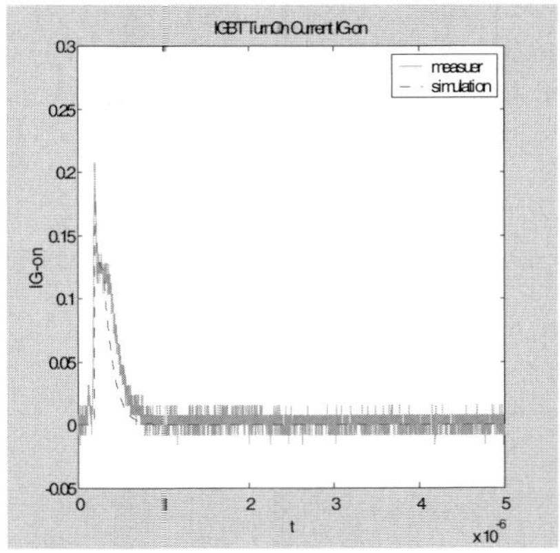

Fig. 9. I_g when turn-on

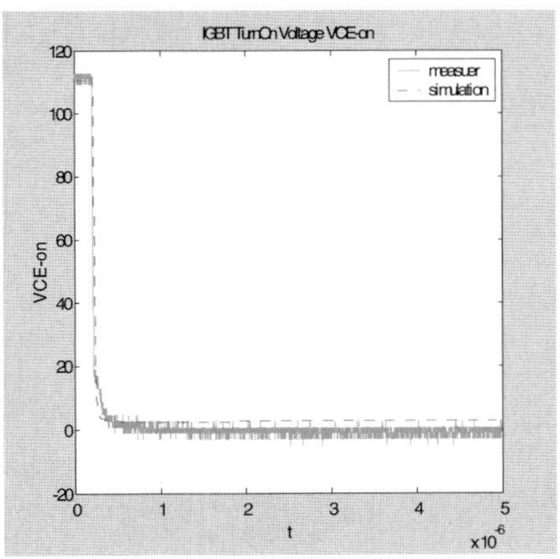

Fig. 10. V_{CE} when turn-on

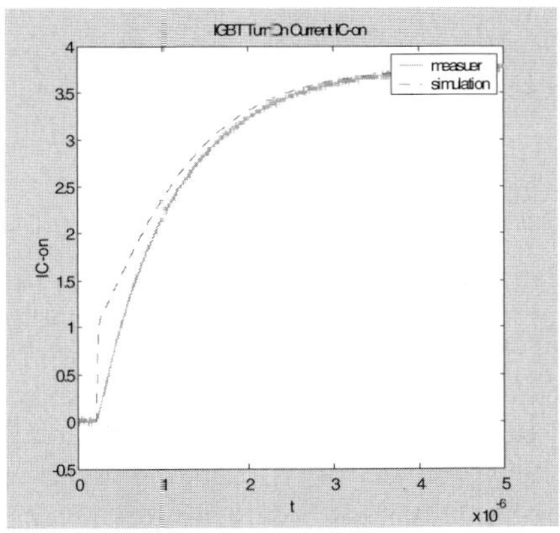

Fig. 11. I_C when turn-on

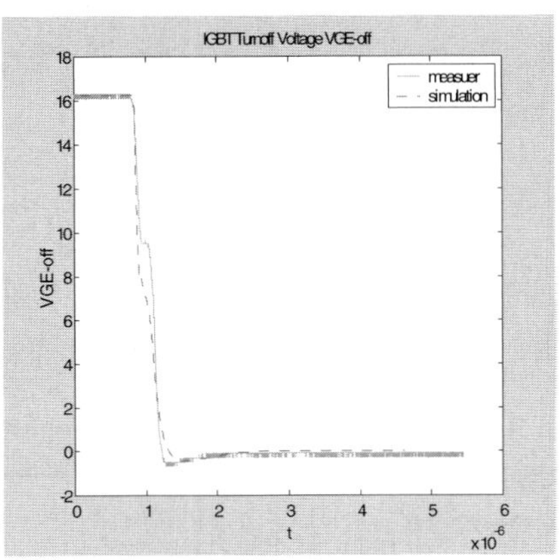

Fig. 12. V_{GE} when turn-off

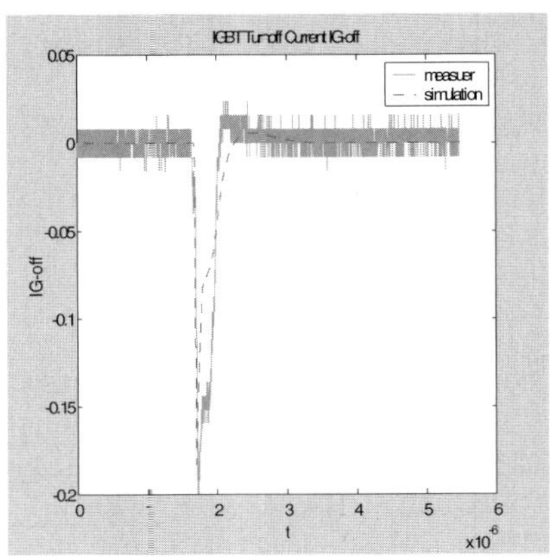

Fig. 13. I_g when turn-off

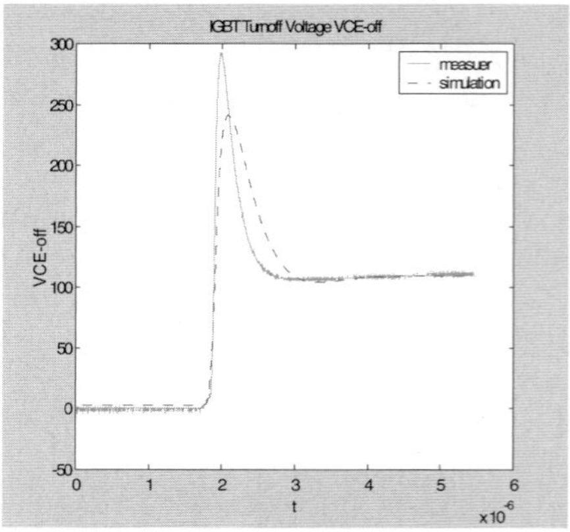

Fig. 14. V_{CE} when turn-off

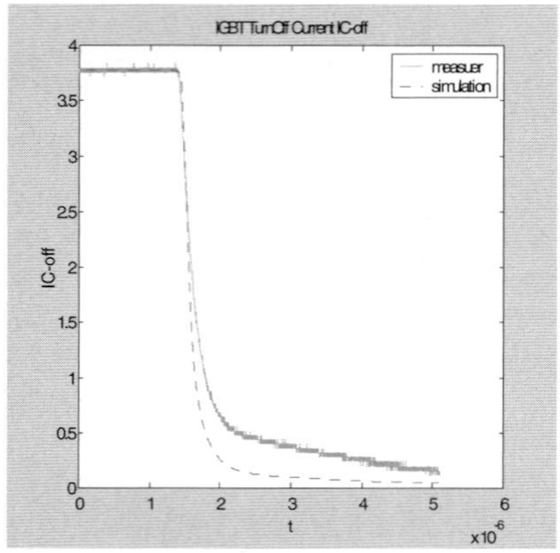

Fig. 15. I_C when turn –off

The waveforms in Figs. 8-15 show that the optimization procedure improves the matching of the device waveforms. The waveforms are shown after synchronization.

4 Conclusions

A practical parameter identification method based on optimization and simulation is proposed and the results of identification for BUP302 are presented. As the results of the parameter identification is appropriate indeed. It can reflect the device's physical characteristics factually due to the mathematical-physical model. The process of identification is a process of optimization. After the simulation software and the model of device are chosen, the procedure is independent of the device's structure and is very convenient for no specified users.

Acknowledgements. The authors would like to acknowledge Prof. H. Morel, for his support of PACTE, the simulation software.

References

1. Sheng, K., Williams, B.W., Finney, S.J.: A Review of IGBT Models[J]. IEEE Trans. on Power Electronics 15(6), 1250–1266 (2000)
2. Oh, H.S., El-Nokali, M.A.: A new IGBT Behavioral Model[J]. Solid-State Electron. 45(11), 2069–2075 (2001)
3. Hefner Jr., A.R.: Modeling Buffer Layer IGBTs for Circuit Simulation[J]. IEEE Trans. Power Electron. 10(3), 111–123 (1995)
4. Hefner, A.R., Blackburn, D.L.: An Analytical Model for the Steady-State and Transient Characteristics of the Power Insulated-Gate Bipolar Transistor[J]. Solid-State Electron. 31(10), 1513–1532 (1988)
5. Hefner, A.R.: Analytical modeling of device-circuit interactions for the power insulated gate bipolar transistor (IGBT)[J]. IEEE Trans. on Industrial Application 35(6), 606–614 (1988)

6. Kang, W.Y., Ahn, H.: A Parameter Extraction Algorithm for IGBT Behavioral Model[J]. IEEE Trans. on Power Electronics 19(6), 1365–1371 (2004)
7. Musumeci, S., Raciti, A., Sardo, M., et al.: PT-IGBT PSpice Model with New Parameter Extraction for Life-time and Epy Dependent Behavior Simulation[C]. IEEE PESC Annual Mtg. Rec., 1682–1688 (1996)
8. Purkes, S.J., Kraus, R.: Parameter Extraction Methodology and Validation for an Electro-Thermal Physics-Based NPT IGBT Model. IEEE IAS Rec., 1166–1173 (October 1997)
9. Hefner, A.R., Bouche, S.: Automated Parameter Extraction Software for Advanced IGBT Modeling. In: Computer in power electronics 2000. COMPEL 2000, The 7th workshop, 16-18 July, 2000, pp. 10–18 (2000)
10. Bryant, A.T., Kang, X.S., Santi, E., Palmer, P.R., Hudgins, J.L.: Two-Step Parameter Extraction Procedure With Formal Optimization for Physics-Based Circuit Simulator IGBT and p-i-n Diode Models. IEEE Transactions on Power Electronics 21(2), 295–308 (2006)
11. Penny, J.E.T., Lindfield, G.R.: Numerical Methods Using MATLAB, 2nd edn. Prentice-Hall, Englewood Cliffs, NJ (2000)
12. Allard, B., Garrab, H., Ammous, W.K., Morel, H.: Switching parameter maps-A new approach to the validity domain of power device models. In: The proceedings of EPE Conference, Toulouse, France, September 2003, pp. 1220–1225 (2003)

Appendix: Nomenclature

A	Device area (cm^2)
A_{gd}	Gate-drain overlap area (cm^2)
BV_f	Breakdown voltage nonplanar unction factor
BV_n	Avalanche multiplication exponent
C_{ds}	Drain-source depletion capacitance (F)
C_{gd}	Gate-drain capacitance (F)
C_{gs}	Gate-source capacitance (F)
C_{oxd}	Gate-drain overlap oxide capacitance(nF)
D	Ambipolar diffusivity($cm^2 \cdot s^{-1}$)
HDB	High-doped base region (buffer layer)
I_C	collector current (A)
I_E	emitter current (A)
I_g	IGBT gate current (A)
I_{MOS}	IGBT MOSFET channel current (A)
I_{sne}	Emitter electron saturation current (A)
K_{plin0}	Linear region MOSFET transconductance parameter (A/V^2)
K_{psat0}	Saturated region MOSFET transconductance parameter (A/V^2)
LDB	Low-doped base region

N_B Drift region (base) doping concentration (cm^{-3})

p ambipolar carrier density (cm^{-3})

V_{CE} collector–emitter voltage (V)

V_{DS} IGBT MOSFET drain–source voltage (V)

V_{eb} IGBT emitter-base voltage (V)

V_{ebd} V_{ebj} Emitter-base diffusion, depletion voltage

V_{ebq} Emitter-base capacitor voltage (V)

V_T MOSFET channel voltage (V)

V_{TH} IGBT MOSFET threshold voltage (V)

W_B Drift region (base) width (μm) (cm^{-3})

Automated Chinese Domain Ontology Construction from Text Documents

Yu Zheng[1], Wenxiang Dou[1], Gengfeng Wu[1], and Xin Li[2]

[1] School of Computer Engineering and Science, Shanghai University, Shanghai 200072, P. R. China
[2] School of Electromechanical Engineering and Automation, Shanghai University, Shanghai 200072, P. R. China

Abstract. Ontology as an important knowledge representation tool is widely used in many fields. Constructing domain ontology is a lengthy, costly task. Rapid, accurate construction of ontology has thus become an important topic. In this paper, a method that automates construction of the ontology is proposed. The method integrates text analysis, TF/IDF calculation, association rules extraction, pattern rules matching and RDF technologies. The ontology construction method does not require expenditure of time to select keywords and to define the relations by human edit or expert assistance. The method facilitates user understanding of the content of data and its relevancy, and is able to suggest content that is highly relevant. Experimental results show that the proposed approach can effectively construct Chinese domain ontology from text documents.

1 Introduction

Ontology as an important knowledge representation tool is widely used in many fields such as enterprise integration, translation of natural language, mechanical engineering, E-Commerce, geographical information systems, legal information systems and biological information systems[1] and etc. Research on ontology construction is becoming increasingly widespread in computer science community. At present, ontology may be constructed either by full manual construction and or semi-automatic construction, each with its own limitations and operating methods [2-4]. In full manual construction method, human experts can be important in creating ontology, but have difficulty in coming up with widely recognized impersonal perspectives. It is difficult to modify the concepts of the extant domain ontology for a semantic web [5].

On the other side, there are some semi-automatic methods used to construct ontology such as the dictionary method, the knowledge base method etc. If the ontology is generated via the dictionary method, unrelated specific domains will result. Therefore, this method must be combined with other methods or experts to construct a meaningful domain ontology [6]. Constructing an ontology based on a knowledge base is limited by the size of rules. Further, the knowledge base must be constructed in advance. A small set of knowledge base rules permits the formation of only a small ontology [7].

In this paper, an approach to automatically construct domain ontology from Chinese textual documents is proposed. We integrate natural language processing (NLP) tools,

K. Li et al. (Eds.): LSMS 2007, LNCS 4688, pp. 639 – 648, 2007.

statistical principles and data mining technologies to acquire concepts and relations from text. The objectives of this study include: (1) Pre-processing; (2) Extracting terminologies and concepts; and (3) Learning the relationships of concepts.

The remainder of the paper is organized as follows: Section 2 introduces the domain ontology construction model. In Section 3, we describe pre-processing on Chinese text documents. Section 4 specifies the method of concepts extraction. Then concept relation learning algorithm is presented in Section 5. Section 6 presents the results of the experiments. Some conclusions will be made in Section 7.

2 Domain Ontology Construction Model

For ontology constructing, discovering conceptual tags and their relationships is the fundamental step in the early stage of constructing. This study integrates statistical techniques, NLP tools and data mining techniques to build domain ontology. As illustrated in Fig. 1, the inputs of the model can be Chinese text documents of specific domain and the final result is a conceptual system with a hierarchical structure, which is represented by RDFs (Resource Description Frames).

The first stage is pre-processing. It includes sentence segmentation, Chinese word segmentation , Chinese Part-of-Speech (POS) Tagging and stop words filter etc. The second stage is concepts extraction. Concepts extraction consists of terminologies extraction and synonymous concept incorporating. Much of the previous research on extracting conceptual structures has studied the refinement processes from text to concepts [8-9]. Generally, these studies find that statistical and linguistic techniques are useful for terminologies recognized processes. Since the textual sources are mass data, efficient modules and tools are essential. Additionally, different aspects of analysis approaches may derive distinct results from the same corpus. This study considers two parallel modules, statistical and data mining for extracting terminologies. These modules are complementary each other in the theoretical practical sphere. Synonymous concept incorporating will be executed after extracting terminologies. The next stage asks whether a hierarchically arranged scheme can be identified by calculating relations among concepts. The final stage represents the ontology as RDFs by Jena package.

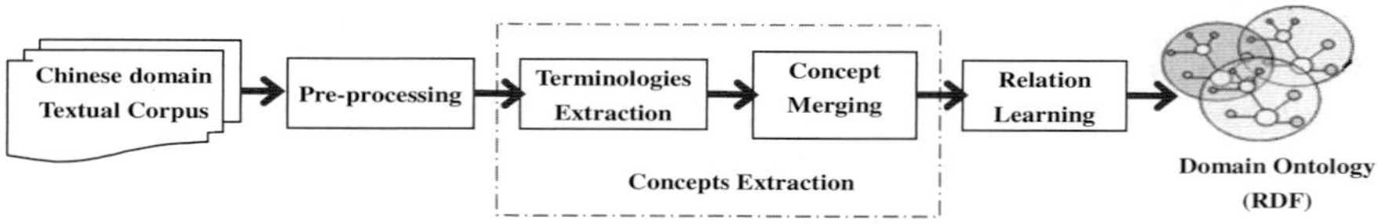

Fig. 1. Architecture of Automated Domain Ontology Construction Model

3 Documents Pre-processing

Documents pre-processing includes sentence segmentation, word segmentation, Part-of-Speech tagging, stop word filter and unknown words recognition. It illustrates in Fig.2.

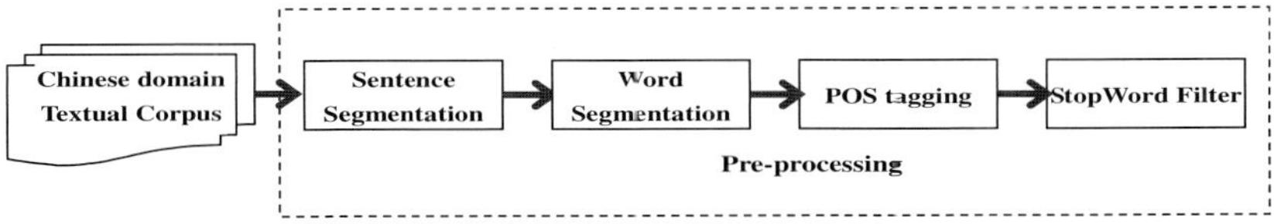

Fig. 2. The flow chart of Pre-processing

(1) Sentence Segmentation: We'll analyze syntax in succeeding stages, in this study we simply segment document into sentences by Chinese punctuations.

(2) Word Segmentation: Not like English, there are no separators between words in Chinese text. In this study we use ICTCLAS (Chinese Lexical Analysis System research by Institute of Computing Technology, Chinese Academy of Sciences) as a word segmentation parser.

(3) Part-of-Speech (POS) Tagging: Generally, most terminologies and concepts of ontology are nouns or partial nouns. In pre-processing stage, we also use ICTCLAS as a POS tagger.

(4) Stop word Filter: There are many words such as "的", "地", "得", and so on in Chinese text. These words usually are high frequency but have no contribution to the subject of text. Stop word list is a list that contains many conjunctions, particles, prepositions, articles and auxiliary words . Stop word filter is used to reduce the noise words in stop word list.

4 Concepts Extraction

In this study, we will extract concepts from textual documents. The flow chart is illustrated in Fig.3.

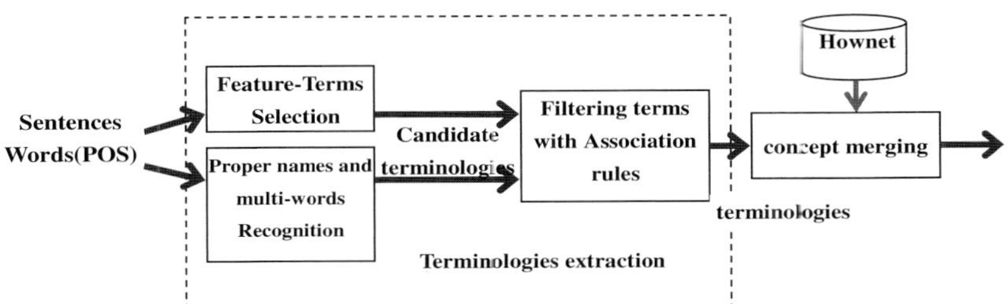

Fig. 3. The flow chart of extracting concepts

4.1 Feature-Terms Selection

This process aims to select important terms from documents. We do it based on the information retrieval measure tf/idf (Term Frequency/Inverse Document Frequency):

$$tfidf(T,D_i) = tf(T,D_i) * log(\frac{N}{df(T)}) \tag{1}$$

Where *tf(T,Di)* is the number of term *T* appeared in document *Di*, *N* is the number of documents in corpus. *df(T)* is the number of documents that term *T* appears in corpus. $\frac{N}{df(T)}$ is the *idf* of term *T*.

Tf/idf weighs the frequency of a term in a document with a factor that discounts its importance when it appears in almost all documents. Therefore terms that appear too rarely or too frequently are ranked low.

In this study, Consider that the documents in corpus belong to identical domain, we collect about 8,000 documents of various subjects from Internet. We calculate the idf value of every word occurred in the corpus in advance and write it to an idf file. When we calculate tf/idf value of a term, we get it's idf from the idf file, if it not exist in file, that means the term is not common and may be important to the domain, so we give it a large idf value. The nouns with the high tf/idf values in every document are preserved and put into candidate terminologies set.

4.2 Proper Names and Multi-words Recognition

Proper names and multi-words are pervasive in texts, especially in specific domain texts. Most of them are the instances of domain concepts, but word segmentation parser can not segment them correctly. For example in communication domain,"蓝 牙" (Bluetooth) is a wireless communication protocol, but word segmentation parser segments the word into "蓝"(adjective) and "牙"(noun). In this study, we use co-occurrence frequency of the connected terms to recognize these Proper names and multi-words.

Definition 1. Term Pair's Co-occurrence Weight (TPCW)

If T_{ij}, T_{ik} denote two connected terms in document *i,* The TPCW of term pair (T_{ij},T_{ik}) is a modified Dice coefficient :

$$TPCW(T_{ij},T_{ik}) = \frac{2tf(T_{ij} \cap T_{ik})}{tf(T_{ij}) + tf(T_{ik})} \tag{2}$$

Where $tf(T_{ij})$ denotes the number of T_j occurring in document *i* and $tf(T_{ij} \cap T_{ik})$ denotes the number of $T_{j,}T_k$ consecutive co-occurring in document *i*. The TPCW larger than a threshold are selected and put into the candidate terminologies set.

4.3 Filtering Terms with Association Rules

Terms are concepts whose meaning is agreed upon large user communities in a given domain. A more selective analysis should take into account not only its appearance in single documents but also the overall occurrence in the target corpus [10].. In terms selection and multi-words recognition phase, we mainly focus on extracting terms from single documents. There are some redundant terms in candidate terminologies set. In this phase we analyze the relevance of terms by identifying association rules. A typical association rule takes the form of an "if–then" statement. In the rule format, the antecedent (the "if" part) and consequent (the "then" part) are disjoint sets of items [11-12]..For example, Let $T=\{t_i|i=1,2,......,n\}$ denote a set of transactions, where each transaction t_i consists of a set items. $I = \{i_1, i_2, . . . , i_n\}$ denotes a set of items. An association rule states X associates with Y, denoted by $X\Rightarrow Y$, where $X,Y\subseteq I$, and $X\cap Y =\emptyset$. An association rule generally is accompanied by two numbers expressing the efficiency and accuracy of the rule. The first number is known as the support for the rule. The support simply indicates how often items occur together, as a percentage of the total transactions [13]. A typical support expression is given by

$$support(X \Rightarrow Y) = \frac{(t_i \mid X \cup Y \subseteq t_i \}}{n} \tag{3}$$

The second number is termed the confidence of the rule. Confidence measures how much a particular item is dependent on another [13]. The confidence expression is given by

$$confidence(X \Rightarrow Y) = \frac{support(X \Rightarrow Y)}{support(X)} \tag{4}$$

In this study, a document is regarded as a transaction, a term is regarded as an item. The number of association rules can be controlled by setting minimum support and (or) confidence values. We gather both the antecedent and consequent from the rules and organized them to produce a list of final terminologies.

4.4 Synonymous Concept Merging

[14] noted a COCO describing four dimensions in text analysis, including consensus (where same exist in both terminology and concepts), conflict (where the same terminology is applied for different concepts), correspondence (where different terminology is applied for the same concepts), and contrast (where differences exist in both terminology and concepts). In this study, concept merging stage is used to incorporate synonymous concepts.

HowNet is a bilingual general knowledge-base describing relations between concepts and relations between the attributes of concepts. The latest version covers over 65,000 concepts in Chinese and close to 75,000 English equivalents. The relations include hyponymy, synonymy, antonymy, meronymy, attribute-host, material-product, converse, dynamic role and concept cooccurrence. In this study, HowNet is utilized to merge terms of concepts and collect synonyms and Antonyms.

5 Concepts Relation Learning Algorithm

In this phase, we extract every sentence in the documents pre-processed by step3 and parse the structure of these sentences. Then we define some rules according with the structure for learning relation.

5.1 Parsing Structure of the Sentence

Syntax of the Chinese Language has some basic rules. These rules are mainly composed of noun, verb, and adjective and reflect the relation among nouns. The concepts belong to nouns of the document. If the pattern of Chinese sentence is discovered, the rules to find the relation can be defined. For example,

Input text:

> 索爱K810还具备出色音乐播放功能，兼容MP3等格式音乐文件播放，支持立体声蓝牙耳机并增加了曲目识别功能。

Tagging the Structure of this Sentence:

[索爱K810 noun]还[具备 verb][出色的 adjective][音乐播放功能 noun]，[兼容 verb][MP3 noun]等格式[音乐文件 noun]播放，[支持 verb][立体声 adjective][蓝牙耳机 noun]并[增加 verb]了[曲目识别功能 noun]。

This hand-tagged instance can be used as seed for learning pattern. And in our system, pattern is showed as follows:

(Adjective Phrase)(Noun Phrase)*(Verb: reflect the relation)*(Adjective Phrase)*(Noun Phrase)

5.2 Rules Definition

According to the pattern, the rules to match sentences are defined. If the sentence satisfies one of these rules, it should be kept. Rules defined as follows:

Parataxis = (,) ∨ (和) ∨ (、);

Hold = (的) ∨ (有);

NounGroup = [Noun(Parataxis)] ∨ [Noun];

AdjectiveGroup = [Adjective (Parataxis)] ∨ [Adjective];

VerbNoun = (Verb) ∧ (Noun ∨ (AdjectiveGroup ∧ Noun));

VerbNounGroup = [VerbNoun(Parataxis)] ∨ [VerbNoun];

Rule1 = (AdjectiveGroup) ∧ (NounGroup) ∧ (Hold) ∧ (NounGroup);

Rule2 = (AdjectiveGroup) ∧ (NounGroup) ∧ (VerbNounGroup);

We will add these rules into Rules DB for the relation learning.

5.3 Relation Learning

Now the rules and concepts to match the structure of sentences can be used. When concepts and rules match with the sentence of the documents, the relation between concepts from this sentence is extracted. The relation will be stored by the ternary group such as [concept, concept, verb]. The verb is between two concepts and expresses the relation of them.

Relation Learning Algorithm:

```
For (every document has been tagged) {
    For (every tagged sentence in document) {
        Let the tagged sentence to match rules
        If (this sentence contains one of these rules) {
            Let nouns in this sentence to match the concept-set
            If (two or more nouns are in this concept-set and at least one verb
                among these nouns) {
                Parse this sentence and produce the ternary groups of relation
            }
        }
    }
}
```

Then the relation that has the higher frequency (that means the relation must appear in term of some proportion in documents) will be kept.

Finally, the ontology will be built by using the concepts and their relations will be stored as RDFs.

6 Experiments and Evaluation

To design the experiments and to evaluate the process for automatic ontology construction is not a thing easy to do. Since the process which consists of several sub-process with each has separate goals is difficult to be evaluated. From this aspect, the ontologies are built for specific purposes [15].

6.1 Standards of Evaluation

In order to evaluate the effects of concepts extraction and domain ontology, precision and recall of information retrieval as the evaluation of concepts extraction are adopted.

Definition 1. Precision

Let ST be the golden standard of terms, T be the terms set we learned. Term Precision is defined as:

$$Precision = \frac{ST \cap T}{T} \tag{5}$$

Definition 2. Recall

Let ST be the golden standard of terms, T be the terms set we learned. Term Recall is defined as:

$$Recall = \frac{ST \cap T}{ST} \tag{6}$$

6.2 Experiments

The system was implemented on a PC Pentium 4 2.4 G with 1G RAM, Java programming language and Jena package as the RDF ontology language.

In this experiment, we select an expert mobile phone ontology that an expert has constructed. The web pages in experiments are collected by Google (http:// www. google.com) and Baidu (http://www.baidu.com). Then extract text from these web pages using page parser program developed by ourselves. After new ontology has been created, we compare the new ontology with the expert ontology to demonstrate the precision.

The result of concepts extraction is list in table1.

Table 1. The result of concepts extraction

Support threshold	Extraction rate (%)	Precision/Recall (%)
0.3	10.39	82.35/71.42
0.35	5.81	84.2/57.14
0.4	3.98	92.3/52.38

Here , extraction rate is the percent of concepts extracted to candidate terms.

Then the learning algorithm of ontology relations can be executed. A part of the mobile ontology learning is illustrated in Fig.4.

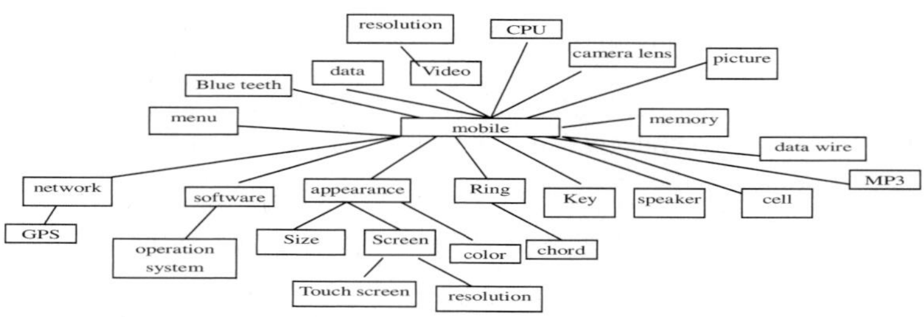

Fig. 4. Part of mobile ontology

An example of RDF is described as follow

```
<rdf:RDF
    xmlns:rdf="http://www.w3.org/1999/02/22-rdf-syntax-ns#"
    xmlns:property="http://information/手机/property/" >
  <rdf:Description rdf:about="http://information/手机/property/外观">
    <property:体积>体积</property:体积>
    <property:屏幕>屏幕</property:屏幕>
  </rdf:Description>
……
</rdf:RDF>
```

Experimental results indicate the precision and recall is good for an automated ontology approach. Some intermediate results of this approach can help domain experts validate the domain ontology and discover further domain knowledge.

7 Conclusions

The ontology can help a user to learn and search for relevant information. Rapid, accurate construction of ontology has thus become an important topic. In this paper, a method that automates construction of the ontology is proposed. The method integrates text analysis, TF/IDF calculation, association rules extraction, pattern rules matching and RDF technologies. The ontology construction method does not require expenditure of time to select keywords and to define the relations by human edit or expert assistance. The method facilitates user understanding of the content of data and its relevancy, and is able to suggest content that is highly relevant.

Acknowledgements. This paper is supported by the project of Shanghai Municipal Education Commission Youth Fund (Grant No. 04AC84)

References

1. Guarino, N.: Formal ontology and information system. In: Proceedings of FOIS'98 (FOIS'98), pp. 3–15 (1998)
2. Gillam, L., Tariq, M., Ahmad, K.: Terminology and the construction of ontology. Terminology 11(1), 55–81 (2005)
3. Maedche, A., Staab, S.: Ontology learning for the semantic web. IEEE Intelligent Systems 16, 72–79 (2001)
4. Perez, A., Corcho, O.: Ontology languages for the semantic web. IEEE Intelligent Systems 17, 54–60 (2002)
5. Noy, N.F., McGuinness, D.L.: Ontology development 101: A guide to creating your first ontology. Stanford knowledge System Laboratory Technical Report KSL-01-05 (2001)
6. Chen, R.-C., et al.: Using recursive ART network to construction domain ontology based on term frequency and inverse document frequency, Expert Systems with Applications, doi:10.1016/j.eswa.2006.09.019

7. Alani, H., Kim, S., Millard, D., Weal, M., Hall, W., Lewis, P., et al.: Automatic ontology-based knowledge extraction from Web documents. IEEE Intelligent Systems 18, 14–21 (2003)
8. De Bruijn, B., Martin, J.: Getting to the (c)ore of knowledge:mining biomedical literature. International Journal of Medical Informatics 67(1-3), 7–18 (2002)
9. Scharl, A., Bauer, C.: Mining large samples of web-based corpora. Knowledge-Based Systems 17(5-6), 229–233 (2004)
10. Velardi, P., Fabriani, P., Missikoff, M.: Using text processing techniques to automatically enrich a domain ontology. In: Proceedings of the International Conference on Formal Ontology in Information Systems, Ogunquit, pp. 270–284. ACM Press, New York (2001)
11. Han, J., Kamber, M.: Data mining: Concepts and techniques. Morgan-Kaufman, San Mateo, CA (2001)
12. Srikant, R., Agrawal, R.: Mining generalized association rules. Future Generation Computer Systems 13(2-3), 161–180 (1997)
13. Chi, Y.-L.: Elicitation synergy of extracting conceptual tags and hierarchies in textual document. Expert SYtems with Applications 32, 349–357 (2007)
14. Shaw, M.L.G., Gaines, B.R.: Comparing conceptual structures: consensus, conflict, correspondence and contrast. Knowledge Acquisition 1(4), 341–363 (1989)
15. Noy, N.F., Hafner, C.: The state of the art in ontology design: A survey and comparative review. AI Maganize 18, 53–74 (1997)

Robust Stabilization of Discrete Switched Delay Systems

Yang Song [1,2], Jian Fan [1,2], and Minrui Fei [1,2]

[1] Shanghai Key Laboratory of Power Station Automation Technology, Shanghai, PRC
[2] Department of Automation, Shanghai University, PRC, 200072
y_song@shu.edu.cn

Abstract. This paper considers the robust stabilization of discrete-time switched systems with time delay. The methodology of robust control is proposed by using state feedback and constructing admissible switching sequences of which the average dwell time (ADT) is limited within a certain bound. The bound of ADT can be calculated by the formulations presented in this paper. Finally, a numerical example is given to illustrate the effectiveness of the proposed method.

1 Introduction

Switched systems are composed of a set of subsystems and a switching rule designating which subsystem to be actuated. Switched systems are commonly found in engineering practice, such as in automotive engine control systems, chemical process control, robotic manufacture, air traffic control. Recent years have witnessed much progress in the study of switched systems [1-8]. An important issue in the research of switched systems is related to the stability and stabilization[1-6]. Common Lyapunov function technique[1] and multiple Lyapunov functions technique[2-3] were firstly introduced for checking the stability of switched systems. Ye et al[4] extended the results in [2] by using so-called weak Lyapunov-like functions that means the energy function does not need to decrease while no switching occurs. Narendra[7] and Ooba[8] studied the stability under arbitrary switching for several special classes of switched systems. For those switched systems which can not be stable under arbitrary switching sequence, in most cases there still exist lots of practical useful switching sequences could make them stable. Hence the concepts of dwell time and average dwell time (ADT) were proposed[9][10] and were applied to stabilize switched systems. The stabilization by average dwell time, in short, is to make switched systems stable by constructing admissible switching sequences of which the average dwell time is within a certain bound. Zhai et al[11] investigated the disturbance attenuation properties of time-controlled switched systems by using average dwell time approach. Peng & Ozaby[12] proposed an method for design switching H_∞ controllers for a class of LPV systems in terms of average dwell time. Vu et al [13] investigated the input-to-state stability properties of a switched nonlinear

K. Li et al. (Eds.): LSMS 2007, LNCS 4688, pp. 649 – 657, 2007.

system based on average dwell time approach and then applied the results to stabilize uncertain nonlinear systems via switching supervisory control.

On the other hand, it is well known that time delay exists extensively in various engineering systems, such as long transmission lines, pneumatic or hydraulic systems, and network-based systems. In practical engineering, time delay often results in poor performance, oscillations and even instability to a control system. The study on delay systems has been a hotspot for a long time in control community. In recent years, some researchers began to pay attention to switched delay systems[14-16]. Yuan et al[14] studied the asymptotic stability of switched delay systems. Wang et al[15] proposed a sufficient and necessary condition of the controllability of switched systems with input delay under constrained switching. Hetel et al [16] investigated the stabilization of the switched linear systems with unknown time varying delays by transforming a continuous unknown time delay into a polytopic robust design problem. In this paper, we focus on the robust stabilization of discrete-time switched system with state delay by average dwell time scheme and develop a method for evaluating the bound of average dwell time.

2 Problem Preliminaries

Consider linear switched system in the domain of discrete time:

$$x_{k+1} = (A_i + \Delta A_{ik})x_k + (B_i + \Delta B_{ik})x_{k-\tau} + (C_i + \Delta C_{ik})U_{ik} + D_i w_k \quad (1\text{-}1)$$

$$z(k) = L_i x_k + G_i U_{ik} \quad (1\text{-}2)$$

$$x_s = 0, \quad s = -\tau, \ -\tau+1, \ \cdots, -1 \quad (1\text{-}3)$$

where $x_k \in R^n$ is the state vector at time k, τ is the fixed time delay of state, x_s denotes the state vector prior to the initial instant. $U_{ik} \in R^m$ is the control input vector at time k of subsystem i. $i(k): Z^+ \to M = \{1, 2, \cdots, N\}$ specifies the switching sequence (Z^+ is the set of positive integers). $N > 1$ denote the number of subsystems. A_i, B_i, C_i , D_i, L_i, G_i are a set of known real constant matrices with appropriate dimensions, moreover, (A_i, B_i) is controllable pairs. $w_k \in R^p$ is the exogenous disturbance input which belongs to $L_2[0,\infty]$, $z_k \in R^q$ is the controlled output. It is assumed that time varying parameter uncertainties $\Delta A_{ik}, \Delta B_{ik}, \Delta C_{ik}$ satisfy:

$$[\Delta A_{ik} \quad \Delta B_{ik} \quad \Delta C_{ik}] = M_i E_{ik} [N_{Ai} \quad N_{Bi} \quad N_{Ci}]$$

where $E_{ik}{}^T E_{ik} \le I$.

The switching sequence orchestrating the trajectory is expressed by

$$\sigma = (i_0, t_0), (i_1, t_1), (i_2, t_2), \cdots, (i_n, t_n), \cdots \tag{2}$$

where $t_n \in Z^+$ denotes the switching instant, the pair (i_n, t_n) means that subsystem i_n is active during the interval $t_n \le k \le t_{n+1}$.

The minimal dwell time of switching sequence σ is defined as:

$$\tau := \min\left\{t_{j+1} - t_j, j = 0, 1, \cdots, n, \cdots\right\} \tag{3}$$

It is obvious that the interval between any two consecutive switching instants in switching sequence σ is no smaller than τ.

Hespanha proposed average dwell time (ADT)[10] in continuous time domain by extending the concept of minimal dwell time. Correspondingly, the definition of ADT for discrete-time switched systems is given in what follows.

Definition 1. Given two integers $k_2 > k_1 > 0$, let $N_\sigma(k_1, k_2)$ denote the total number of the switches during the interval $[k_1, k_2]$ for switching sequence σ, then the average dwell time τ_σ of discrete-time switched systems is defined as:

$$\tau_\sigma = \begin{cases} \dfrac{(k_2 - k_1)}{N_\sigma(k_1, k_2)}, & \text{if } \dfrac{(k_2 - k_1)}{N_\sigma(k_1, k_2)} \text{ is an integer} \\ \left[\dfrac{(k_2 - k_1)}{N_\sigma(k_1, k_2)}\right], & \text{else} \end{cases}$$

where $[\cdot]$ denotes the integral part of a real number in the square bracket.

Assumption 1. The time used to achieve the switching action is infinite small.

From this assumption, we know that if the time k is a switching instant, x_k is both the *switching out* state of a certain subsystem and also the *switching in* state of the subsequent subsystem.

Applying the state feedback control laws $U_i(k) = F_i x(k)$ to (1), we get

$$x_{k+1} = [\overline{A}_i + (\Delta A_{ik} + \Delta C_{ik} F_i)] x_k + (B_i + \Delta B_{ik}) x_{k-\tau} + D_i w_k \tag{4-1}$$

$$z_k = \overline{L}_i x_k \tag{4-2}$$

where $\overline{A}_i := A_i + C_i F_i$, $\overline{L}_i = L_i + G_i F_i$

3 Main Results

In this section, we deal with the stabilization of the discrete-time switched delay systems by using the ADT scheme.

Theorem 1. The switched system (4) with $\Delta A_i = \Delta B_i = \Delta C_i = 0$ is asymptotically stable for the admissible switching sequence σ if there exists matrices $P_q > 0$, $Q_q^{\,i} > 0$, F_q and scalars $\mu_{qr} > 0$, $-1 < \alpha < 0$, such that:

$$Q_q^{\,j+1} \le (\alpha+1) Q_q^{\,j}, \quad j = 1, 2, \cdots, \tau - 1 \tag{5}$$

$$\begin{bmatrix} -(\alpha+1)P_q + Q_q^{\,1} & 0 & (A_q + C_q F_q)^T P_q \\ 0 & -(\alpha+1)Q_q^{\,\tau} & B_q^{\,T} P_q \\ P_q(A_q + C_q F_q) & P_q B_q & -P_q \end{bmatrix} < 0 \tag{6}$$

$$P_q \le \mu_{rq} P_r, \quad q \ne r \tag{7}$$

$$Q_q^{\,i} \le \mu_{rq} Q_r^{\,i}, \, q \ne r \tag{8}$$

where $q \in M$, $r \in M$, $i = 1, 2, \cdots, \tau$, the average dwell time τ_σ satisfies that,

a. For the finite switching sequence

$$\tau_\sigma \ge \varepsilon \tag{9-1}$$

b. For the infinite switching sequence

$$\tau_\sigma \ge \begin{cases} \varepsilon & \text{if } \bar{\mu} \le 1 \\ [-\ln \bar{\mu} / \ln(\alpha+1)] + 1 & \text{if } \bar{\mu} > 1 \end{cases} \tag{9-2}$$

where $\bar{\mu} = \prod_{\substack{r,q=1 \\ q \ne r}}^{N} \mu_{rq}^{\;p_{rq}}$, ε is an arbitrary small constant.

Proof. Without loss of generality, we assume that subsystem r and subsystem q operate at the interval $[k - n_r, k-1]$, $[k, k + n_q]$, respectively.

Choose Lyapunov functional candidate for subsystem q as:

$$V_q(x_k) = x_k^{\,T} P_q x_k + \sum_{i=1}^{\tau} x_{k-i}^{\,T} Q_q^{\,i} x_{k-i}$$

The proof is addressed by distinguishing between the infinite and the finite switching sequence.

a. For the infinite switching sequence, we have

$$V_q(x_{k+1}) - V_q(x_k)$$

$$= x_{k+1}^{\,T} P_q x_{k+1} + \sum_{i=1}^{\tau} x_{k+1-i}^{\,T} Q_q^{\,i} x_{k+1-i} - x_k^{\,T} P_q x_k - \sum_{i=1}^{\tau} x_{k-i}^{\,T} Q_q^{\,i} x_{k-i}$$

$$= x_{k+1}{}^T P_q x_{k+1} + x_k{}^T (Q_q{}^1 - P_q) x_k + \sum_{i=1}^{\tau-1} x_{k-i}{}^T (Q_q{}^{i+1} - Q_q{}^i) x_{k-i} - x_{k-\tau}{}^T Q_q{}^\tau x_{k-\tau}$$

$$= x_k{}^T (\bar{A}_q{}^T P_q \bar{A} - P_q + Q_q{}^1) x_k + x_k{}^T \bar{A}_q{}^T P_q B_q x_{k-\tau} + x_{k-\tau}{}^T B_q{}^T P_q \bar{A}_q x_k + \sum_{i=1}^{\tau-1} x_{k-i}{}^T (Q_q{}^{i+1} - Q_q{}^i) x_{k-i} - x_{k-\tau}{}^T (Q_q{}^\tau - B_q{}^T P_q B_q) x_{k-\tau}$$

From (5), it follows that

$$\begin{aligned} & V_q(x_{k+1}) - V_q(x_k) \\ & \le x_k{}^T (\bar{A}_q{}^T P_q \bar{A} - P_q + Q_q{}^1) x_k + x_k{}^T \bar{A}_q{}^T P_q B_q x_{k-\tau} + x_{k-\tau}{}^T B_q{}^T P_q \bar{A}_q x_k \\ & \quad + \alpha \sum_{i=1}^{\tau-1} x_{k-i}{}^T Q_q{}^i x_{k-i} - x_{k-\tau}{}^T (Q_q{}^\tau - B_q{}^T P_q B_q) x_{k-\tau} \end{aligned} \tag{10}$$

On the other hand, applying Schur Complement to (6), we obtain

$$\begin{bmatrix} -(\alpha+1)P_q + Q_q{}^1 + \bar{A}_q{}^T P_q \bar{A}_q & \bar{A}_q{}^T P_q B_q \\ B_q{}^T P_q \bar{A}_q & -(\alpha+1)Q_q{}^\tau + B_q{}^T P_q B_q \end{bmatrix} < 0$$

which implies

$$\begin{aligned} & x_k{}^T (\bar{A}_q{}^T P_q \bar{A} - P_q + Q_q{}^1) x_k + x_k{}^T \bar{A}_q{}^T P_q B_q x_{k-\tau} + x_{k-\tau}{}^T B_q{}^T P_q \bar{A}_q x_k \\ & \quad - x_{k-\tau}{}^T (Q_q{}^\tau - B_q{}^T P_q B_q) x_{k-\tau} < \alpha x_k{}^T P_q x_k + \alpha x_{k-\tau}{}^T Q_q{}^\tau x_{k-\tau} \end{aligned} \tag{11}$$

Substituting (11) into (10) yields

$$V_q(x_{k+1}) - V_q(x_k) < \alpha V_q(x_k) \tag{12}$$

Therefore,

$$\begin{aligned} V_q(x_{k+1}) &< (\alpha+1) V_q(x_k), \\ V_q(x_{k+2}) &< (\alpha+1) V_q(x_{k+1}) \le (\alpha+1)^2 V_q(x_k) \\ V_q(x_{k+n_q}) &< (\alpha+1)^{n_q} V_q(x_k) \end{aligned} \tag{13}$$

From (7) and (8), we know that

$$V_q(x_k) < \mu_{rq} V_r(x_k)$$

Thus

$$V_q(x_{k+n_q}) < (\alpha+1)^{n_q} \mu_{rq} V_r(x_k) < \mu_{rq} (\alpha+1)^{n_q+n_r} V_r(x_{k-n_q}) < \cdots$$

Then, for the following admissible switching sequence (K represents an instant large enough that statistical characteristic of the switching sequence is shown fully),

$$\sigma = (i_0, 0),\ (i_1, t_1),\ (i_2, t_2), \cdots,\ (i_n, K)$$

It follows that

$$\begin{aligned} V(x_K) &< (\alpha+1)^K (\mu_{i_{n-1}i_n} \mu_{i_{n-2}i_{n-1}} \cdots \mu_{i_1 i_2}) V(x_0) \\ &= (\alpha+1)^K \left(\prod_{\substack{r,q\in M \\ r\neq q}}^{\sigma} \mu_{rq}{}^{N_\sigma(0,K)p_{rq}} \right) V(x_0) \\ &= (\alpha+1)^K \bar{\mu}^{N_\sigma(0,K)} V(x_0) = (\alpha+1)^{K+N_\sigma(0,K)\ln^{\bar{\mu}}/\ln^{(\alpha+1)}} V(x_0) \end{aligned} \tag{14}$$

Where

$$\bar{\mu} = \prod_{\substack{r,q=1 \\ q\neq r}}^{N} \mu_{rq}{}^{p_{rq}} \tag{15}$$

Noticing that $\ln(\alpha+1)<0$, we know that if $\bar{\mu}\le 1$, τ^* can be arbitrary small; else if $\bar{\mu}>1$, note that

$$K \ge \tau_\sigma N_\sigma(0,K) \tag{16}$$

Therefore

$$\begin{aligned} V(x_K) &< (\alpha+1)^{K+N_\sigma(0,K)\ln^{\bar{\mu}}/\ln^{(\alpha+1)}} V(x_0) \\ &\le (\alpha+1)^{N_\sigma(0,K)\left(\tau_\sigma+\ln^{\bar{\mu}}/\ln^{(\alpha+1)}\right)} V(x_0) \end{aligned} \tag{17}$$

From (9-2), it follows that

$$\tau_\sigma + \ln\bar{\mu}/\ln(\alpha+1) > 0 \tag{18}$$

Noticing that when $K \to +\infty$, there must be

$$N_\sigma(0,K) \to \infty$$

we obtain

$$V(x_K) \to 0$$

Thus

$$\|x_K\| \to 0.$$

b. For the finite switching sequence, from (12), it is easy to see that the switched system is asymptotically stable since $\alpha<0$.

This completes the proof.

Remark 1. Eq.(15) holds with the potential assumption that the arbitrary switch is possible in the switching sequence σ. For the constrained switching, (15) should be revised by eliminating those impossible switches.

Remark 2. From (17), we can see that for an instant *K*, if $\bar{\mu}<1$, Lyapunov functional $V(x_K)$ has a faster converging rate with the decrease of average dwell time τ_σ; on the contrary, if $\bar{\mu}>1$, the smaller average dwell time τ_σ implies the slower converging rate of $V(x_K)$.

Remark 3. Since that the switching sequence may be finite and the subsystem actuated ultimately can become of any subsystem, it is clear that each subsystem should be asymptotically stable, which is fulfilled by $\alpha<0$. If we only consider the infinite switching sequence, the constraint $\alpha<0$ is not necessary, as corollary 1 show.

4 Numerical Example

Consider the following switched systems composed of two subsystems:

Subsystem 1: $A_1=\begin{bmatrix}-0.35 & -0.1\\ 0.21 & 0.2\end{bmatrix}$, $B_1=\begin{bmatrix}0.02 & -0.03\\ 0 & -0.05\end{bmatrix}$, $C_1=\begin{bmatrix}0\\1\end{bmatrix}$,

Subsystem 2: $A_2=\begin{bmatrix}-0.2 & 0.1\\ 0.1 & 0.1\end{bmatrix}$, $B_2=\begin{bmatrix}0.04 & -0.02\\ 0 & 0.05\end{bmatrix}$, $C_2=\begin{bmatrix}0\\1\end{bmatrix}$,

Time delay $\tau=1$. $p_{12}=p_{21}$, $x_0=\begin{bmatrix}-2 & 1\end{bmatrix}^T$.

This switched system can not be stabilized by the proposition 3.2 in [12], which is based on common Lyapunov function technique. By theorem 1, we can obtain

$$F_1=\begin{bmatrix}-0.099 & -0.5\end{bmatrix},\ F_2=\begin{bmatrix}-0.0498 & -0.3123\end{bmatrix},\ \alpha=-0.5,$$

$$\tau\geq\left[-\ln\bar{\mu}/\ln(\alpha+1)\right]+1=3$$

$$P_1=\begin{bmatrix}42.5397 & 0.4681\\ 0.4681 & 41.2709\end{bmatrix},\ Q_1^1=\begin{bmatrix}10.7642 & 0.0421\\ 0.0421 & 11.3753\end{bmatrix}$$

$$Q_1^2=\begin{bmatrix}3.5580 & 0.0053\\ 0.0053 & 3.7690\end{bmatrix},\ P_2=\begin{bmatrix}51.8773 & -1.0137\\ -1.0137 & 53.1403\end{bmatrix},$$

$$Q_2^1=\begin{bmatrix}16.3510 & 0.7942\\ 0.7942 & 16.2763\end{bmatrix},\ Q_2^2=\begin{bmatrix}5.3711 & 0.2188\\ 0.2188 & 5.3672\end{bmatrix},$$

$$u_{12}=7.8785,\ u_{21}=5.8481,\ \bar{\mu}=6.7878$$

Construct the infinite switching sequences

$$\sigma=(2,0),(1,4),(2,6),(1,10),\cdots,(1,4+6n),(2,6+6n),\cdots$$

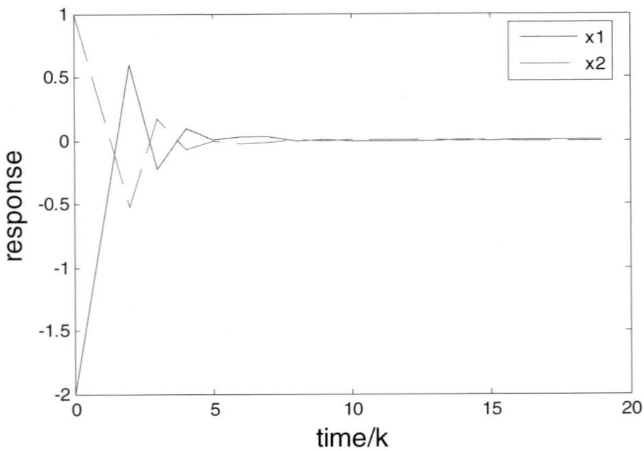

Fig. 1. Response under switching sequence σ

5 Conclusions

In this paper, we discussed the robust stabilization of discrete-time switched systems with time delay by average dwell time (ADT) approach incorporated with multiple Lyapunov functionals. The key point of the ADT approach is to ascertain the bound of average dwell time of the admissible switching sequence, which can be obtained by the formulation proposed in this paper. Finally, a numerical example is given to illustrate the applications of the proposed results.

Acknowledgments. This work is supported by Key Project of Science & Technology Commission of Shanghai Municipality under Grant 061107031 and 06DZ22011, and Shanghai Leading Academic Disciplines under Grant T0103.

References

1. Dogruel, M., Ozgunzer, U.: Stability of hybrid systems. In: Proceedings of IEEE International Symposium on Intelligent Control, Monterey, California, USA, pp. 129–134. IEEE Computer Society Press, Los Alamitos (1995)
2. Branicky, M.S.: Multiple Lyapunov functions and other analysis tools for switched and hybrid systems. IEEE Transaction on Automatic Control 43, 475–482 (1998)
3. Peleties, P., Decarlo, R.A.: Asymptotic stability of m-switched systems using Lyapunov-like functions. In: Proceedings of conference on American Control Conference, Boston, MA, USA, pp. 1679–1684 (1991)
4. Ye, H., Michel, A.N., Hou, L.: Stability theory for hybrid systems. IEEE Transaction on Automatic Control 43, 461–474 (1998)
5. Lin, H., Antsaklis, P.J.: Switching stabilizability for continuous time uncertain switched linear systems. IEEE Transactions on Automatic Control 52, 633–646 (2007)

6. Sun, Z., Ge, S.S.: Analysis and synthesis of switched linear control systems. Automatica 41, 181–195 (2005)
7. Narendra, K.S., Balakrishnan, J.: A common Lyapunov function for Stable LTI systems with commuting A-Matrices. IEEE Transaction on Automatic Control 39, 2469–2471 (1994)
8. Ooba, T., Funahashi, Y.: Two conditions concerning common quadratic Lyapunov functions for linear Systems. IEEE. Transaction on Automatic Control 42, 719–721 (1997)
9. Morse, A.S.: Supervisory control of families of linear set-point controllers-Part 1: exact matching. IEEE Transaction on Automatic Control 41, 1413–1431 (1996)
10. Hespanha, J.P., Morse, A.S.: Stability of switched systems by average dwell time. In: Proceedings of the 38th IEEE Conference on Decision and Control, Arizona, USA, pp. 2655–2660. IEEE Computer Society Press, Los Alamitos (1999)
11. Zhai, G., Hu, B., Yasuda, K., Michel, A.N.: Disturbance attenuation properties of time-controlled switched systems. Journal of Franklin Institute 338, 765–779 (2001)
12. Yan, P., Ozbay, H.: On switching H controllers for a class of LPV systems. In: Proceedings of the 42nd IEEE Conference in Decision and Control, Howail, USA, pp. 5644–5650. IEEE Computer Society Press, Los Alamitos (2003)
13. Vu, L., Chatterjee, D., Liberzon, D.: Input-to-state stability of switched systems and switching adaptive control. Automatica 43, 639–646 (2007)
14. Yuan, R., Jing, Z., Chen, L.: Uniform asymptotic stability of hybrid dynamical systems with delay. IEEE Transaction on Automatic Control 48, 344–368 (2003)
15. Wang, Y., Xie, G., Wang, L.: Controllability of switched time-delay systems under constrained switching. Journal of Mathematical Analysis and Applications 286, 397–421 (2003)
16. Hetel, L., Daafouz, J., Iung, C.: Stabilization of Arbitrary Switched Linear Systems With Unknown Time-Varying Delays. IEEE Transaction on Automatic Control 51, 1668–1674 (2006)

A Model of Time Performance of a Hybrid Wired/Wireless System

Weiyan Hou [1,2], Haikuan Wang[2], and Zhongyong Wang [1]

[1] Univ.Zhengzhou, School of Information Engineering, Zhengzhou 450052
[2] Univ. Shanghai, School of Mechatronical Engineering and Automation, Shanghai 200072, China

Abstract. The PHY layer in RFieldbus system, which has hybrid wired and wireless segment, integrates two different traditional Asynchrony + DSSS in wired and wireless segment respectively, but with the unique FDL protocol of token-Passing. In order to evaluate the transmission time-performance in such a hybrid system, the protocol stack has been modeled as a multi- server- single-queue, which its target token-rotation time meets a dynamic boundary. The polling-time on the slaves was transformed into an internal token walking time, and the rest two lower priority queues were transformed into two $M/G/1/\infty$ systems with two different Poisson arrival rates, so that the complexities of the time properties when transmission happens in different segment can be depicted easily by 7 different parameters. Then, a simulator RF-P based on OMNet++ platform is constructed with a 4-layers infrastructure. The top layer is the system-module inclosing 4 compound modules, which was composed by different simple modules on bottom layer. The comparison with the real measuring in 13 typical network infrastructures of RFieldbus show: 1, the error deviation of RF-P model is not more then 25%, and is about 10% when there is no mobile segment or stations in the measuring scenarios. 2, Error bias is smaller when the system load is lower. 3, The accuracy of the simulator is better when there is no TCP/IP traffic in the system.

1 Introduction

With the developing of the wireless communication technology it is possible and necessary to integrate wireless technology in the field of the industrial control application.

RFieldbus which was launched in 2000 under the support of European Committee 5th IST-plan [1] is the 1st real hybrid Fieldbus system which integrates the PROFIBUS and wireless LAN (WLAN) technology. Token-Passing which is similar to the IEEE 802.4 Token-Bus has been adopted on FDL in RFieldbus and its PHY layer integrates the traditional Asynchrony + DSSS in wired and wireless segment respectively. Furthermore, the transmission of TCP/IP and control data is merged in RFieldbus. The

K. Li et al. (Eds.): LSMS 2007, LNCS 4688, pp. 658 – 665, 2007.

integration of wired and wireless segment in RFieldbus may bring much new delay on the message transaction circle, especially in a harsh industrial field control environment.

As for this mixed transmission protocol on PHY under a unique FDL protocol, the normal ways to evaluate the PROFIBUS real time-performance cannot be adapted directly in RFieldbus.

In order to study the real-time property of RFieldbus a simulator based on a model of Token-Passing is built on OMNet++ discrete event simulation platform.

In the following section the RFieldbus infrastructure and the Token-Passing protocol is briefly introduced. And in section 3 a Model about the Token-Passing in RFieldbus is introduced also. Section 4 presents a simulator which is developed from the OMNet++. Section 5 gives the comparison of the simulation with the real measuring values in 13 measuring scenarios.

2 Property of Token-Passing in RFieldbus

All the nodes in RFieldbus can be classified into two kinds: Master, Slave nodes. They have been connected to the wired and wireless Bus medium. Its infrastructure was showed in Fig.1.All the Master station make a logic domain, called Logic-Ring. Every Master on the Logic-Ring will catch and resent a unique special frame, called Token which runs around all the system, from the previous to the next master. When it catches the Token, it has the bus-accessing right to control the system bus, and send its Data_requery to the Slave to exchange data.

In RFieldbus, the transmission queues on the Master have 3 different priorities. 1, High priority tasks have the strict response time limits, e.g. Alarm, denoted as P1; 2, IP traffic with QoS demands and the Low priority cyclic transmission tasks, denoted as P2; 3, IP traffic without QoS demands and the ring management tasks, denoted as P3.

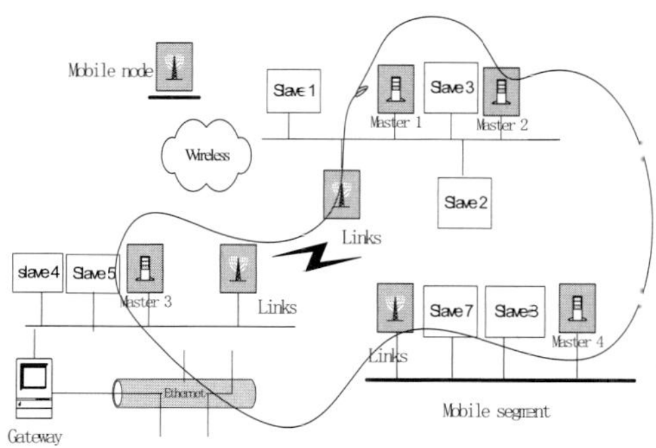

Fig. 1. Infrastructure of RFieldbus[2]

In Token-Passing protocol the parameter Token Target Rotation Time (T_{TR}) regulates the estimated token rotation time on the logic ring. T_{RR} is the real token rotation time. Therefore, Token Hold Time is $T_{TH} = T_{TR} - T_{RR}$ When Token comes latterly, TS(this station) will accomplish at least one task in P1 waiting-queue, even if $T_{TH} <= 0$. If $T_{TH} > 0$, TS will continue the transactions for P1 queue and then P2 queue It will resend the Token to the NS (next station) if TTH is decreased to zero, otherwise the tasks in P3 queue will proceed continuously. So, every Master has a variable Token Hold Time which depends on the burdens, transmission speed, re-transmit of the system [3]. The above protocol can be depicted in Fig.2, denoted as P-Token-Passing.

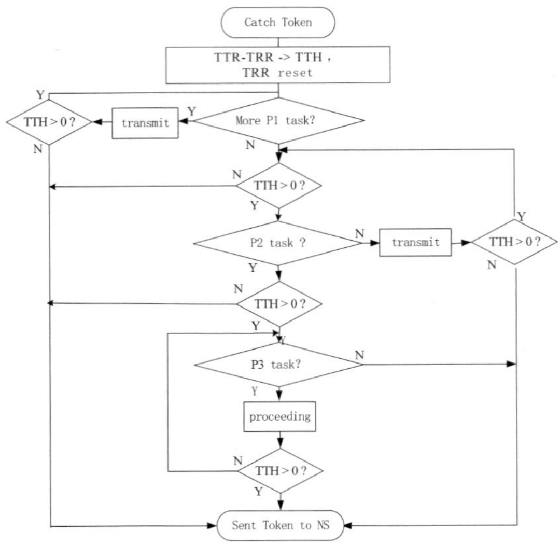

Fig. 2. Token-Passing in RFieldbus

Message transaction circle is noted as *Tcycle*, (also as *Ttrac*) includes 4 portions[4], as follows:1, *Tg*: Tasks building and be added to the waiting queue;2, T_Q: waiting time in the queue;3,T_C: from transmission beginning to get the response;4, Td: data be resent to the application layer.

Generally to say, *Tg*, *Td* are associated with the performance of the host resourc、the front sensors ,etc. T_Q, T_C reflect the communication performance of the system. T_Q is the main factor in *Tcycle* and can influence greatly the real-time performance of a distributed control system, e.g.RFieldbus.

3 Modeling of Token-Passing in RFieldbus

Depend upon the queue-theory Token-passing can be looked as a queue system with single servant on multi server + multi queues, as Fig.3, denoted as a M1+M2+ …+MN / G / 1 / ∞system.

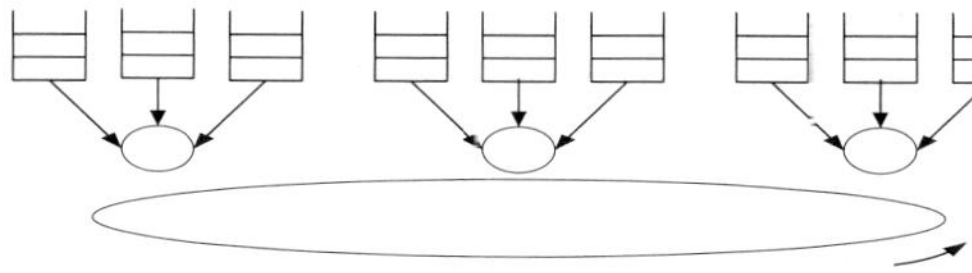

Fig. 3. Waiting-queue model of Token-Passing

M means the interval of "customer" arrival time is negative exponential distributed (Poisson process). G indicates that the service time is generally distributed. "1" means there is only one service platform in one station. "∞" means the length of waiting queue has no limit.

$$\begin{cases} \rho < 1 \\ \rho_a T_{RR} + T_{RR} \le T_{TR},\ a \in 1,\ldots,N \end{cases} \quad (1)$$

When TTR meets the above dynamic boundary for the stability of the system, the multi-server-multi-queue model can be transformed into a multi-server-single-queue[5]. It means M1 + M2 + …+MN/G/1/∞ will be disassembled to M1/G/1/∞, M2/G/1/∞,…, MN/G/1/∞ queues[5], as in followed Fig.4.

Every Master Station works as a server with 3 queues, and the Token is as the single Servant on the Token-Ring over the hybrid wired and wireless segments. The above model can be denoted RF-P.

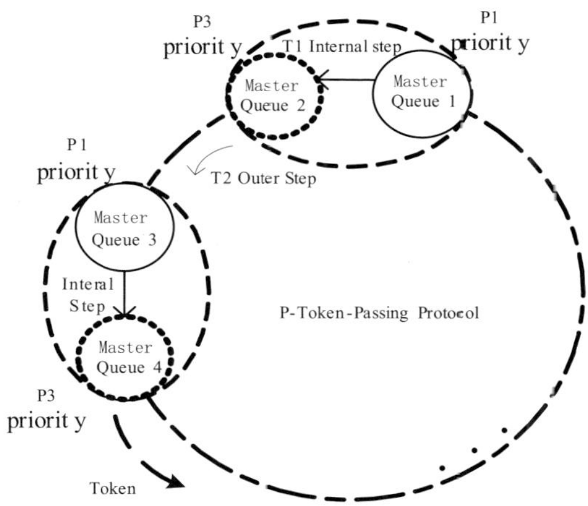

Fig. 4. Multi-Server-Single-Queue Model

In this model, the polling-time on the slaves was transformed into an internal token walking time, and the rest two queues were transformed into two $M/G/1/\infty$ systems with two different Poisson arrival rates, so that the complexities of the time properties in RFieldbus can be depicted easily by 7 different parameters. It must be mentioned that the extra Idle-time, transmission delay brought by the integration of heterogeneous segments in RFieldbus, and even the BER in wireless transmission can be described by respective parameters.

4 A Simulator Upon the OMNet++

Upon the RF-P model a simulator prototype is constructed. The simulator prototype which is based on OMNet++, an object-oriented discrete-event platform designed for the simulation of communication protocols, communication networks [6], has the modular infrastructure as in Fig.5.

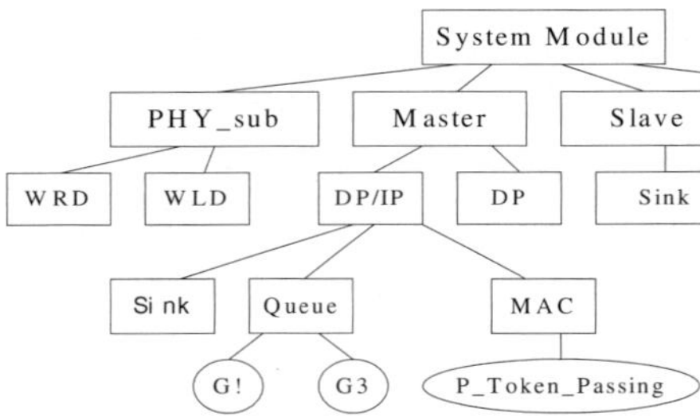

Fig. 5. Modular Infrastructure of simulator

The simulator can be classed into 4 layers which was build up by separated modules. Between the modules there is messages transmission cannel. The top layer is the system-module inclosing 4 compound modules: PHY_sub, Master, Slave, IS, which was composed by different simple modules on bottom layer. All the activities of the Token-Passing protocol algorithm will be implemented in the subroutine written in C++.

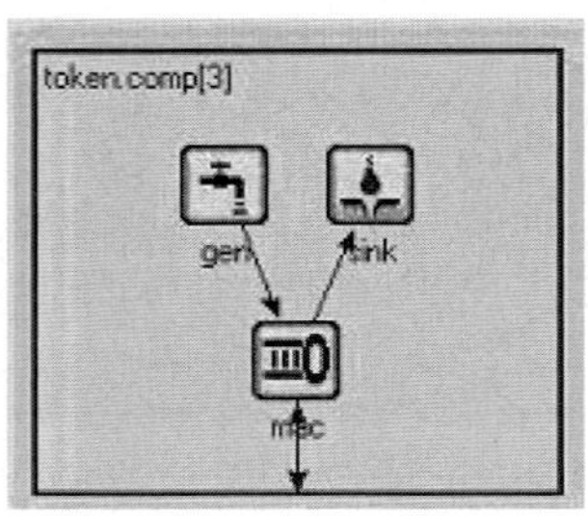

Fig. 6. Structure of master

The PHY_sub includes two simple modules: WRD, WLD that reflect the wired and wireless segments in RFieldbus respectively. It is provided that the TRX power of antenna is big enough to cover the whole cell, in order to simplify the wireless signal shield-radium problem. And the Handoff time will be considered as a fixed δ, so that it has no influence on the average *Tcycle*. The length, and transmission rate is set as a random integer using the proprietary random generate function in OMNet++, in order to meet the demands which is regulated in IEC 61158.

The compound Master-Station modules have been divided into: DP and DP/IP module which compose 3 simple modules. Based on the RF-P model in Fig.4, the task with P1,P3 priority on every master station are transformed to two task-queues with different Poisson arrival rate. DP Master structure can be referred by Fig.6.

IS: Intermediate Station module have 3 different kinds: LS,LBS,BS, which were implemented with 3 different activity C++ classes.

5 Comparison

To test the effectiveness of this simulator, 13 typical network infrastructures of RFieldbus was adopted as the testing scenarios accord to 4 following criteria. The detailed description of the measuring scenarios can be referred in [7].

- With or without wireless connection
- Intracellular or intercellular mobility
- Single Master or Multimaster (with or without Beacon master)
- One point, or a segment mobility

Every scenario has its corresponding simulation running configuration file: omnetpp.ini, in which different sets of parameters, environment were designed.

The results of the simulation have two formats: text and vector files, e.g. as Fig.7. It can be exported directly into the Matlab, Octave software.

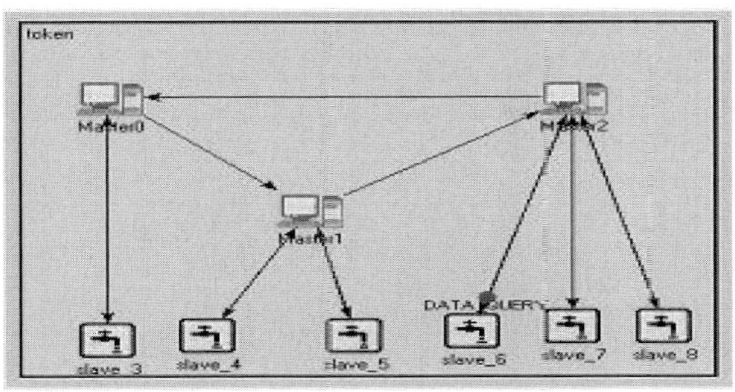

Fig. 7. Simulation running

All the simulations will run for 15 minutes respectively. The running results will be verified through the following form.

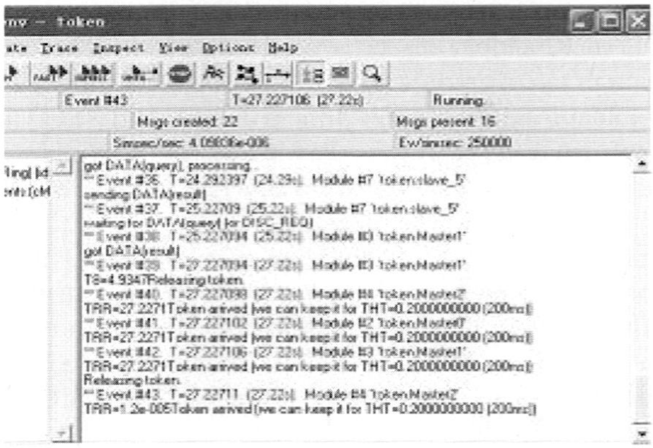

Fig. 8. Plain text output window

$$\text{Bias} = \frac{T_{\text{cycle - simulation}} - T_{\text{cycle - real}}}{T_{\text{cycle - real}}} \cdot 100\% \tag{2}$$

In the test the default settings of DP-Bus parameters defined in configuration tools were used, as following setting:

min.Tsdr = 1000 bittimes	Tslot = 5000 bittimes
TTR = 50000 bittimes	Min.Slave.Int. =1 ms

* Datarate =1.5 Mbits/s one bittime =0.667 ns.

Table 1. Comparison with the real T_{cycle}

	Measured		Simulation	
	No IP	With IP	No IP	With IP
Scenario 1	11.2	11.5	13.2	15.3
Scenario 2	11.6	12.4	12.1	15.5
Scenario 3	11.1	11.9	14.0	10.7
Scenario 4	11.1	11.7	11..3	8.5
Scenario 5	12.4	12.2	16.9	11.9
Scenario 6	11.9	12.2	9.99	13.7
Scenario 7	12.6	13.4	14.9	11..3
Scenario 8	18.5	20.8	22.2	18.1
Scenario 9	18.5	14.2	20.3	15.1
Scenario 10	12.8	13.9	13.9	12.7
Scenario 11	12.4	12.6	15.5	12.5
Scenario 12	15.0	73.3	14.4	79.0
Scenario 13	14.7	76.3	16.4	58.9

From the examination and analysis it can be concluded that: 1, The error deviation is not more then 25%, and is about 10% when there is no mobile segment or stations in the measuring scenarios. 2, Error bias is smaller when the system load is lower. 3, The accuracy of the simulator is better when there is no TCP/IP traffic in the system.

6 Conclusions

With the above simulator which is developed from the token-passing model RF-P, the complex time-properties in RFieldbus can be simulated and studied. Upon the simulator platform it is easy to change the parameters in the protocol configuration, in order to optimize the real-time performance in the design of system structure, and to assign load in every stations before a real system-building.

The comparison with the real measuring in 13 scenarios shows the accuracy of this simulator is good enough, especially when the system load is low and there is no TCP/IP traffic. The reason is the TCP/IP traffic cannot be expressed thoroughly in RF-P model. The next step of work is to improve the model and develop more standard omnetpp.ini file as templates for different system structure.

Acknowledgments. Supported by: 1,Chinese Educational Ministry Overseas-Returnee Academy Foundation Program(2006-331); 2, Shanghai Leading Academic Disciplines Project(T0103); 3, European Committee 5th IST-Plan No. 199911316.

References

[1] Rauchhaupt, L.: High Performance wireless fieldbus in Industrial Related Multimedia environment. In: Proceedings of 4th IFAC Conference, Sweden, pp. 184–192 (2002)

[2] Hou, W.Y., Fei, M.R., Chen, B.S.: Industrial environment oriented wireless-fieldbus with high performence. DIANCE YU YIBIAO (in Chinese) 39(12), 5–9 (2002)

[3] General structure of the RFieldbus: White Paper, D1.3 of the IST Report (2000)

[4] Tovar, E., Vasques, F.: Cycle time properties of the PROFIBUS timed-token protocol. Computer Communications 22, 1206–1216 (1999)

[5] Klehmet, U.: Mathematisch-heuristische Leistung- sbewertung des Medienzugangs verfahren beim PROFIBUS. at-Automatisierungstechnik 44, 108–120 (1996)

[6] www.omnetpp.org

[7] Hou, W.Y., Adamzierk, H., Rauchhaupt, L.: Performance Evaluation of RFieldbus. In: Proc. of the 5th IFAC Conference on Fieldbus Systems Technology (FET'03), Aveiro, Portugal, pp. 123–129 (2003)

Robust Auto Tune Smith Predictor Controller Design for Plant with Large Delay

Huang Ziyuan[1], Zhen Lanlan[1], and Fei Minrui[2]

[1] Information and Control Department, Shanghai Electric Power College, Shanghai, 200090
[2] Automation Department, Shanghai University, Shanghai, 200072

Abstract. A robust Smith controller for plant with large delay is proposed. A nonlinear PD controller is design to auto tune on-line the parameters of robust controller such that the closed system is of robust stability as well as good dynamic performance. The simulation results show that the method can control the plant with large time-variable delay properly.

1 Introduction

With the development of Sci. & Tech., there appear a few of difficult controlled plants, which are nonlinear, time-variable and with large time delay. And the parameters of the plant are uncertain because working environment has been changed, or nonlinear object is linearized and so on. It is a difficult problem in the control theory and control engineering how to control the plant with time delay, because there being time delay, the system may be unstable.

Smith O J designed a predictor to control the plant with time delay [1]. But, when there is difference between the plant and its identified model, Smith predictor is of poor robust performance. So, the method is difficult to be used in practical engineering. Then, a few of improved Smith predictor controllers have been proposed.

In 1997, R.F.Giles and T.M.Bartley [2] designed a gain adaptive compensation controller. The division operator and multiplication operator are introduced. When the gain of the plant varies with time, the closed system can track the reference input properly. But, the system response may be poor when the other dynamic parameters of the plant vary with time.

Hang C.C [3] proposed another Smith predictor control strategy, in which the second controller $D_2(s)$ was introduced. And $D_2(s)$ usually was designed in proportional and integral controller (Abbr. PI). When the dynamic parameters (including gain) of the plant vary with time, the system response is relatively good. But, it is difficult to tune the PI parameters of $D_2(s)$.

Cunyan Liang and Jianying Xie [4] proposed a adaptive compensator for the plant with time delay, in which an error compensation operator was introduced. However, the compensation effect may be poor if the appropriate gain of compensator is not given.

Based on the Smith predictor, Brosilow proposed an Internal Model Control (abbr. IMC) control structure in 1978. And in 1982, a general IMC structure was given by Garcia and Morari, which was shown in SISO system diagram [5]. This method is suitable for plant with time delay, because it is easy to design, of good robust

K. Li et al. (Eds.): LSMS 2007, LNCS 4688, pp. 666 – 678, 2007.

performance and of strong ability to eliminate noise. According to the principle of IMC, Weidong Zhang etc.[6] and Zhang Jin etc.[7] designed a novel robust control scheme.

But, the controller is all certain in Weidong Zhang etc.[6] and Zhang Jin etc.[7] such that it is difficult to get good dynamic performance in the closed system. In this paper, a robust auto-tune Smith controller for large delay plant is proposed. A nonlinear PD controller is designed to auto tune on-line the parameters of robust controller such that the closed system is of robust stability and good dynamic performance.

2 Robust Auto-Tune Smith Predictor Controller Design

2.1 Structure of Controller

In general, PD controller is of fast tracking property. However, there exists static error. Robust Auto-tune Smith Predictor Controller consists of paralleled nonlinear PD controller and PID robust controller. It can realize error-free control because it has integrating operator. The K of nonlinear PD in this composite controller is usually relatively small, and therefore the robust stability is mainly decided by the robust controller (Ref. Diagram 1).

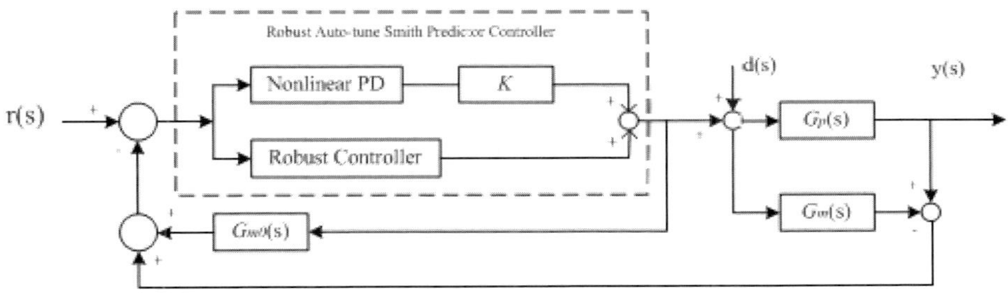

Fig. 1. Robust Auto-tune Smith Predictor Controller

2.2 Robust Controller Design

IMC Structure is shown in figure 2, and $Q(s)$ stands for IMC controller. Obviously,

$$G_c(s) = \frac{Q(s)}{1 - G_m(s)Q(s)} \text{。}$$

In the nominal status (when model is precise), the system sensitivity function is:

$$S(s) = 1 - G_P(s)Q(s) \tag{1}$$

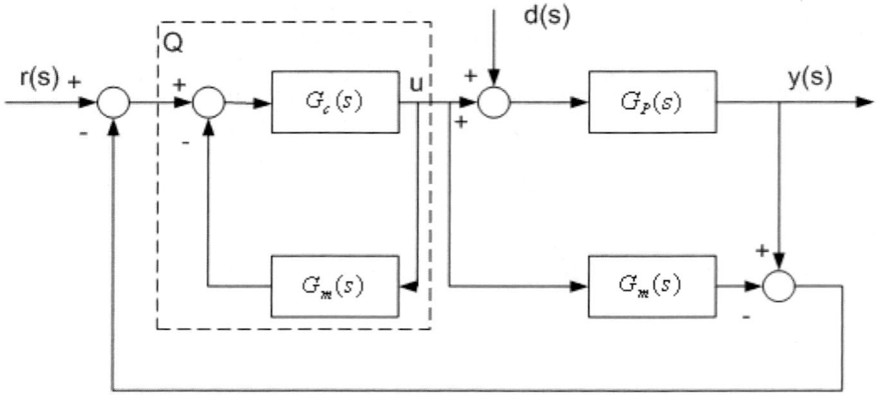

Fig. 2. IMC Structure

Supplementary sensitivity function is:

$$T(s) = G_P(s)Q(s) \tag{2}$$

The transfer-function matrix from [r, d]T to [y, u]T is:

$$\begin{bmatrix} G_P(s)Q(s) & G_P(s)(1-G_P(s)Q(s)) \\ Q(s) & -G_P(s)Q(s) \end{bmatrix}$$

Definition 1. The closed loop system is stable if all transfer functions in above matrix are stable.

Take second-order plus delay object $G_P(s) = \dfrac{K_P}{(\tau_1 s+1)(\tau_2 s+1)} e^{-\theta_P s}$

(τ_1、 $\tau_2 > 0$) as example to deduce the robust control rule.

Lemma 1. [8] Presumed that there exists the feedback control loop in figure 2 and $G_P(s) \in RH_\infty$ (which is stable and canonical rational function), then all the limited dimension linear steady controllers, which guarantee that the closed loop is stable, can be expressed as:

$$G_c(s) = Q(s)[I - G_P(s)Q(s)]^{-1}, \text{ and}$$

$$Q(s) \in RH_\infty, \det[I - G_P(\infty)Q(\infty)] \neq 0 。$$

According to Lemma 1, for the object that the open loop is stable, if there is a stable and canonical proper $Q(s)$, then the satisfying robust controller $G_c(s)$ can be designed so as to guarantee that the closed loop system is stable.

Define the design index as the best sensitivity, that is:

$$\min\|W(s)S(s)\|_\infty \tag{3}$$

Here, W(s) is the performance weighting function, and controller is designed for step-function signal, that is $W(s) = 1/s$.

Introducing one-order Taylor series approximation for time delay, then $G_P(s) = \dfrac{K_P(1-\theta_P s)}{(\tau_1 s+1)(\tau_2 s+1)}$. According to the definition of H_∞ norm, then

$$\|W(s)S(s)\|_\infty = \|W(s)(1-G_P(s)Q(s))\|_\infty \geq |W(1/\theta_P)|$$

To minimize formula (3), we can conclude that

$$W(s)(1-G_P(s)Q(s)) = \theta_P \tag{4}$$

So that the best Q(s) is:

$$Q_{\min}(s) = \frac{(\tau_1 s+1)(\tau_2 s+1)}{K_P} \tag{5}$$

Obviously, $Q_{\min}(s)$ is impossible to be realized in physics. Therefore, a low band filter $J(s)$ is introduced so that $J(s)Q_{\min}(s)$ is realizable. Assuming $J(s) = \dfrac{1}{(\lambda s+1)^2}$ ($\lambda > 0$), then

$$Q(s) = Q_{\min}(s)J(s) = \frac{(\tau_1 s+1)(\tau_2 s+1)}{K_P(\lambda s+1)^2} \tag{6}$$

Obviously $Q(s)$ is canonical and stable. Correspondingly,

$$G_c(s) = \frac{1}{K_P} \cdot \frac{(\tau_1 s+1)(\tau_2 s+1)}{s[\lambda^2 s + (2\lambda + \theta_P)]} \tag{7}$$

This is a PID-type controller with a filter.

For the controlled plant $G_P(s) = \frac{k_P}{(\tau_1 s+1)(\tau_2 s+1)} e^{-\theta_P s}$, whose nominal model is $G_m(s) = \frac{k_m}{(\tau_{m1} s+1)(\tau_{m2} s+1)} e^{-\theta_m s}$, a controller was designed on the basis of equivalent generalized model in literature [4]. The equivalent generalized model is $G_w(s) = G_{m0}(s) + G_{P0}(s)e^{-\theta_P s} - G_{m0}(s)e^{-\theta_m s} \approx G_{m0}(s).G_n(s)$, whose equivalent gain k_n and equivalent delay time θ_n can be approximated via the Michael Lauren Series of $G_n(s) = k_n e^{-\theta_n s}$, and they are

$$\left.\begin{aligned} k_n &= \frac{k_p}{k_m} \\ \theta_n &= (\tau_{P1} - \tau_{m1}) + (\tau_{P2} - \tau_{m2}) + \theta_P - \frac{k_m}{k_P}\theta_m \end{aligned}\right\} \tag{8}$$

Presuming the uncertainty of the parameters in the controlled process is defined within below range

$$\left.\begin{aligned} k_m - \Delta_k \le k_P \le k_m + \Delta_k \\ \tau_{m1} - \Delta_\tau \le \tau_{P1} \le \tau_{m1} + \Delta_\tau \\ \tau_{m2} - \Delta_\tau \le \tau_{P2} \le \tau_{m2} + \Delta_\tau \\ \theta_m - \Delta_\theta \le \theta_P \le \theta_m + \Delta_\theta \end{aligned}\right\} \tag{9}$$

Here, Δ_k、 Δ_τ *and* Δ_θ separately stands for uncertain tolerance of the gain, time constant and delay time.

If the controller is designed for the worst situation, then the system will be robust stable under other circumstances. At that time, $G_n(s)$ should be of maximum gain and time delay. From formula (8) and (9), we can conclude that

$$\left.\begin{aligned} k_n &= \frac{k_m + \Delta_k}{k_m} \\ \theta_n &= \frac{\Delta_k}{k_m + \Delta_k}\theta_m + \Delta_\theta + 2\Delta_\tau \end{aligned}\right\}$$

Therefore the equivalent generalized model is

$$G_w(s) = \frac{k_n k_n}{(\tau_{m1}s+1)(\tau_{m2}s+1)} e^{-\theta_n s} \tag{10}$$

From formula (7), the controller can be expressed as

$$G_c(s) = \frac{1}{K} \cdot \frac{(\tau_{m1}s+1)(\tau_{m2}s+1)}{s[\lambda^2 s + (2\lambda + \theta_n)]} \tag{11}$$

Here, $K = k_m + \Delta k$.

This controller is designed under the worst circumstances. Although it can guarantee that the closed system is robust stable, but it is difficult to get good dynamic performance. Therefore, based on the thought of analytic fuzzy control, a non-linear PD controller is designed in this paper. And PD controller parameters can be auto tune on-line according to the control performance of the system. Then the composite controller, which consists of paralleled nonlinear PD controller and PID robust controller, can both guarantee the system's robust stability as well as good dynamic performance, such as rapid response and small overshoot, etc.

2.3 Nonlinear PD Controller Design

Fuzzy control is a branch of intelligent control, whose application and research have been developed quickly within the recent 20 years. And it has been successfully applied in refrigerator, washing machine etc. While it is one of the hot research focuses how to adjust the fuzzy control rule so as to improve system stability, robust and adaptive capacity. Long Shengzhao and Wang Peizhuang[9] has proposed an analytic description method as below

$$u = < \alpha E + (1-\alpha)EC >$$

Here, E stands for error, while EC is error variation and α is weight parameter. The fuzzy rule produced by this method reflects the continuity, veracity and regularity of human brains' reasoning process. However, α is fixed. In order to improve system performance, the weights for error and error variation should vary with different working status. Thus, Li Shiyong etc.[10] have given a fuzzy control rule which adjusts factors within the whole scope, that can be expressed as

$$\begin{cases} u = < \alpha E + (1-\alpha)EC > \\ \alpha = \dfrac{1}{N}(1-\alpha_0)\,|E| + \alpha_0 \end{cases}$$

Here, α_0 is a constant in (0,1). The above formula reflects the thought of auto-adjusting the weights for error and error variation according to error. Based on the literature [7], a nonlinear PD controller was designed by Peng Xiaohua, etc.[11].

Set R as the system's set value, yt is the output of the system at time t, ut is the output of controller, $\overline{e}_t = \dfrac{R-y_t}{R} = \dfrac{e_t}{R}$ is the relative error, $\overline{C}_t = \dfrac{e_t - e_{t-1}}{M} = \dfrac{y_t - y_{t-1}}{M}$ is the relative error variation, while M is the maximum absolute value of the system output increment in unit interval during the on-line control process. The control algorithm can be described as

$$\begin{cases} u_t = \alpha \overline{e}_t + (1-\alpha)\overline{C}_t \\ \alpha = (1-\alpha_0)\,|\overline{e}_t| + \alpha_0 \end{cases} \tag{12}$$

This controller is of fuzzy PD controller's good character such as rapidity and robust, etc, while it doesn't need to fuzzify E and EC.

In this paper, an improved analytic nonlinear weighting PD controller has been put forward. According to the system control requirements, set the system error upper bound as M（>0）. Nonlinear PD controller output u_{PD} is:

$$\begin{cases} u_{PD} = K.(\alpha e_t + (1-\alpha)\dot{e}_t) \\ \alpha = (1-\alpha_0)\beta + \alpha_0 \end{cases} \tag{13}$$

And $0 < \alpha_0 < 1$, $K > 0$, calculate β as below:

$$\beta = \begin{cases} 1, & if\ |e_t| \geq M \\ \left|\frac{|e_t| - M}{M}\right|, & else \end{cases} \tag{14}$$

The α_0 in nonlinear PD controller can be valued according to the system time delay parameter and inertia character, when the delay time is larger and the inertia is bigger, α_0 is less, and vice versa. In this way, the overshoot can be reduced in the initial control period.

2.4 Robust Stability Analysis

Take second-order process $G_P(s) = \frac{K_P}{(\tau_1 s+1)(\tau_2 s+1)} e^{-\theta_P s}$ as example, and based on the equivalent generalized object design and control strategy, the robust stability of closed loop system is illustrated. $G_P(s)$、$G_C(s)$、$G_w(s)$ stands separately for the transfer function of control object, equivalent generalized object and controller, then the transfer function of the closed loop system is

$$\frac{y(s)}{r(s)} = \frac{G_c(s).G_{P0}(s)}{1+G_c(s).G_{m0}(s)k_n e^{-\theta_n s}} e^{-\theta_P s}.$$

Here, $G_{P0}(s)$、$G_{m0}(s)$ separately stands for the controlled object and the its none time delay part of nominal model, $G_w(s) = G_{m0}(s)k_n e^{-\theta_n s}$. Because θ_n is far less than θ_P, the impact of delay time upon closed loop system can be reduced and the system performance can also be improved on the basis of generalized object design controller.

Under the circumstances that the model matches precisely, the controller is designed according to (11), and the transfer function of closed loop system is

$$\frac{y(s)}{r(s)} = \frac{1}{(1+\lambda s)^2} e^{-\theta_P s} \tag{15}$$

Thus, the parameter λ can be adjusted to assign the poles of the system. So, it can be guaranteed that the system is stable. Besides, except for a time delay θ_P, the object output will trace the reference input.

The robust controller is designed under the worst circumstances. So, the system will be robust stable although there is difference between object and its model.

3 Simulation Test

Take a second-order object as example, it can be expressed as

$$G(s) = \frac{3.67}{(0.0714s+1)(0.1505s+1)} e^{-0.9375s} \tag{16}$$

To consider the uncertainty caused by the modeling error and working environment etc., we presumed that the uncertainty of the K is $\pm 20\%$, the time constant uncertainty of time constant is $\pm 80\%$. In the simulation test, the object will be controlled by three strategies, which are Smith predictor control, Smith on-line identification predictor control(see Literature[12]) and robust auto-tune Smith predictor, under the circumstances that the delay time $\theta(t)$ with modeling error and without modeling error.

(A) Delay time without modeling error（ $\theta(t) = 0.9375$ ）

The object's nominal model is given according to formula（16）, while the object modeling is under the worst circumstances, that is

$$G(s) = \frac{4.404}{(0.1285s+1)(0.2709s+1)} e^{-0.9375s} .$$

In Smith predictor control and Smith on-line identification predictor control, the controller adopts PI form, whose proportion parameter $Kp = 0.2$ and integral parameter $Ki = 0.03$.

The equivalent generalized controlled object from formula (11) is

$$G_w(s) = \frac{4.404}{(0.0714s+1)(0.1505s+1)} e^{-0.3338s} .$$

Robust auto-tune Smith predictor controller is designed based on the equivalent generalized controlled object. Adjustable parameter $\hat{\lambda}$ is set to 1.5, and robust controller in composite controller is

$$G_c(s) = 0.2271 . \frac{(0.0714s+1)(0.1505s+1)}{2.25s^2 + 3.3338s}$$

In the nonlinear PD controller, $\alpha_0 = 0.8$, $K = 0.02$.

At $t = 25ms$, the constant input interference of 0.1 is adopted. The simulation result is shown in figure 3, where the Curve 1, 2 and 3 for the unit step response curve of Smith predictor control, Smith on-line identification predictor control and robust auto-tune Smith predictor control.

From the simulation results, we can draw conclusions as follows.

(1) From Curve 1 & 2, when there is not modeling error for the delay time, the on-line identification algorithm RLS has relative quickly reduced the error between the object and its nominal model (model gain, time constant etc). So, the system is of a good tracing and strong ability to eliminate noise by Smith on-line identification predictor control method.

(2) From Curve 2 and Curve 3, when there is not modeling error for the delay time, although the robust auto-tune Smith predictor control method is poor than the Smith on-line identification control at the transition period, it's still basically competent for the control task. Especially, the robust auto-tune Smith predictor control is of good response such as better tracing ability and stronger ability to eliminate noise and the smaller overshoot etc.

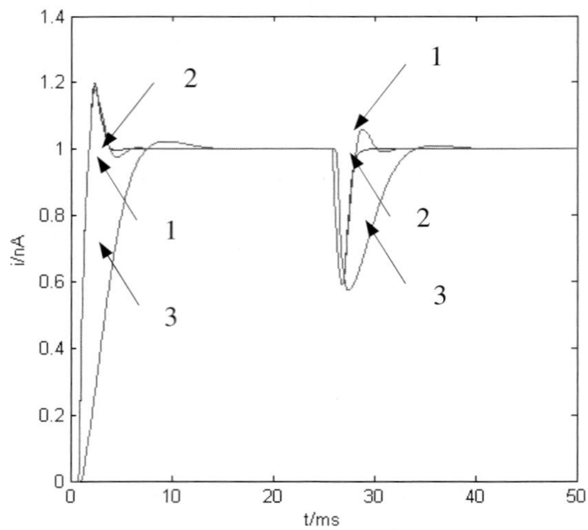

Curve 1. Smith Predictor Control

Curve 2. Smith on-line predictor control

Curve 3. Robust auto-tune predictor Controller

Fig. 3. Unit step response curve（ $\theta(t) = 0.9375$ ）

(B) Delay time with modeling error（ $\theta(t) = 1.4063$ ）

Set the uncertainty of delay time coefficient θ as $\pm 50\%$， and the simulation is performed under the worst circumstance. The object nominal model is expressed a formula (16), and the object model is

$$G(s) = \frac{4.404}{(0.1285s+1)(0.2709s+1)} e^{-1.4063s} .$$

Set the adjustable parameter λ as 1.5, then the robust controller in the composite controller is

$$G_c(s) = 0.2271.\frac{(0.0714s+1)(0.1505s+1)}{2.25s^2+3.8025s}$$

The other parameters of the composite controller are the same as the (A). At $t = 25ms$, the constant input interference of 0.1 is adopted. The simulation result is shown in figure 4, where the Curve 1, 2 and 3 for the unit step response curve of Smith predictor control, Smith on-line identification predictor control and robust auto-tune Smith predictor control.

From the simulation results, when there is modeling error to some extent to the delay time, the Smith predictor control and Smith on-line identification predictor control are not all competent for the control task, while the robust auto-tune Smith predictor control method can make the system with ideal response characters such as good tracing ability and strong ability to eliminate noise and small overshoot etc.

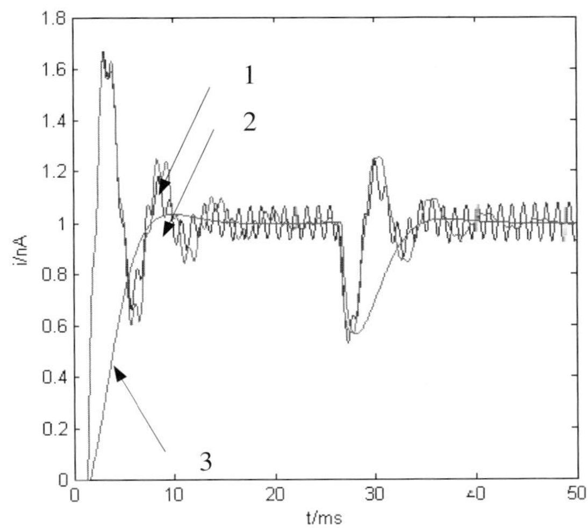

Curve1. Smith Predictor Control

Curve 2. Smith on-line discrimination predictor Control

Curve 3. Robust auto-tune Smith Preditcor Control

Fig. 4. Unit step response（$\theta(t) = 1.4063$）

4 Conclusions

In this paper, a robust auto-tune Smith controller for large delay plant is proposed, based on the equivalent generalized object. By this method, the system is not only robust stable, but also of the good dynamic performance, because that the controller parameters are auto-tunned on-line according to the control error. From simulation results we can see that the system is of relative quick response speed, small overshoot, as well as relative strong ability to eliminate noise and tracing capacity.

Acknowlegements. The paper is supported by Shanghai Colleges Excellent Youth

Teacher Promotion Science and Research Foundation No. Z-2006-10/Z-2006-82 and Shanghai Natural Science Foundation No. 06ZR14133.

References

1. Smith, O.J.: A Controller to Overcome Dead Time. ISA J. 6, 28–33 (1959)
2. Giles, R.F., Gaines, L.D.: Contrlling Inerts in Ammonia Synthesis. Instrumentation Technology 24, 41–45 (1977)
3. Hang, C.C.: A performance study of control system with dead time. IEEE IECI 27, 234–241 (1980)
4. Liang, C., Xie, J.: Adaptive compensation control for the plant with large time delay. Control theory and applications 18, 176–180 (2001)
5. Carcia, C.E., Morari, M.: Internal Model Control: A Unifuing Review and Some New Results. IEC Process Des. Dev. (1982)
6. Zhang, W., Xu, X., Sun, Y.: Quantitative performance design for integrating processes with time delay. Automatica 35, 719–723 (1999)
7. Zhangjin, Liping, Yangzhi: Great time delay process without self-balancing Smith predictor robust control method. In: Zhejiang University Proceedings, vol. 36, pp. 490–493 (2002)
8. Sanchez, P.S.: Robust systems theory and applications [M]. John W Wiley and Sons Inc., New York (1998)
9. Shengzhao, L., Peizhuang, W.: The self-adjusting problem of fuzzy control rule. Fuzzy Mathematics 4, 105–182 (1982)
10. Shiyong, L., et al.: Fuzzy control and intelligent control theory and application. Harbin institute of Technology Publishing House (1989)
11. Xiaohuang, P., Sizong, G., Fangchun, Y.: A nonlinear PD controller and silmulation based on the fuzzy control thought. Liaoning Engineering Technology University Proceedings (Science Edition) 17, 89–91 (1998)
12. Ziyuan, H., Minrui, F., Zhiqiang, H.: Smith Online Identification Prediction Control and Its Application in STM. In: WCICA, Shanghai, China, pp. 2254–2257 (2002)

Research on Applications of a New-Type Fuzzy-Neural Network Controller

Xiucheng Dong[1,2], Haibin Wang[1], Qiang Xu[1], and Xiaoxiao Zhao[1]

[1] School of Electrical and Information Engineering, Xihua University, Chengdu, Sichuan, 610039, China
dxc136@163.com

[2] State Key Laboratory of Intelligent Technology and Systems, Tsinghua University, Beijing, 100084, China

Abstract. A new fuzzy neural network is introduced in this paper which employs self-organization competition neural network to optimize the structure of the fuzzy neural network, and applies a genetic algorithm to adjust the connection weights of the fuzzy neural network so as to get the best structure and weights of the fuzzy neural network. Simulations are made when the pole becomes 2 meters and the random white noise is added in the cart-pendulum system, and control effects of the Adaptive Neural Fuzzy Illation System (ANFIS) and Genetic Algorithm Fuzzy Neural Network (GAFNN) are analyzed. Simulation results indicate that GAFNN controller has greater control performance, high convergence speed, strong robustness and better dynamic characteristics. The effectiveness of the method introduced in this paper is demonstrated by its encouraging study results.

1 Introduction

In resent years, researches on combinations of neural network and fuzzy logic control attract people's attentions. The architecture of neural network-based fuzzy logic controller has a function of fuzzy logic reasoning, and the weights of w the network have a definite fuzzy logic meaning at the same time. By this way, e can get the best design result by merging their capabilities in designing. However, no matter what training means you adopt, the weights you gain is only the optimization under the condition that the structure of the fuzzy neural network has been fixed in advance. The number of the weights which can represent the fuzzy labels (namely: Positive small, negative big etc.) has been given by the designer according to the error accuracy or his experiences before the training begins. It is very clear that we can hardly get optimal value or so-called minimum value: If the number of this fuzzy labels is too small, we can hardly achieve the desired target; whereas it is too big, fuzzy control rule number of the output layer will increase in exponent form, which will inevitably bring about huge network, and inconvenience in the training of the weights and the realization of the network.

As a typical example of multivariable, nonlinear and unsteady non-minimum phrase system, pendulum has been the hot topic in the research area of control.

K. Li et al. (Eds.): LSMS 2007, LNCS 4688, pp. 679–687, 2007.

It is a typical experiment equipment that can verifies modern control theory, and its control methods have been used extensively in dealing with the general industry process.

This paper describes a new design method to solve all the problems above. In order to acquire FNN (Fuzzy-Neural Network) with the best structure and parameters, we employ SCNN (Self-organization Competition Neural Network) to optimize the structure of the network. Its design procedures can be divided into three steps:

First, we design a FFN with more weights, and determine the number of fuzzy labels (take 7 for example), then, optimize its weights by training; Secondly, we choose the weights (include the deviations) which present the fuzzy membership function in the FNN as the inputs of the SCNN. Through the network competition and training procedure, we group the input vectors according to their categories automatically, and the combination number attained after competition will become the optimal number of fuzzy labels of FNN, and we can reduce the number of fuzzy labels to be minimum in this way; At the end, to retrain the FNN weights of the minimum fuzzy labels got in the second step of the competition.

2 Adaptive Neural Fuzzy Illation System(ANFIS)

2.1 Structure of ANFIS

The structure of a simple Sugeno fuzzy system ANFIS is shown in Fig1.

First layer: Fuzzification of the input variable. Outputs are the membership function of the fuzzy variable, and the transfer function of one node can be represented as follows:

$$o_i^1 = \mu_{A_i}(x). \tag{1}$$

Second layer: Carry out the operation of the conditional part Outputs is the apply degrees of every rule. Usually, we adopt multiplication:

$$o_i^2 = \omega_i = \mu_{A_i}(x) \times \mu_{B_i}(y). \tag{2}$$

Third layer: Normalize the apply degree of every rule.

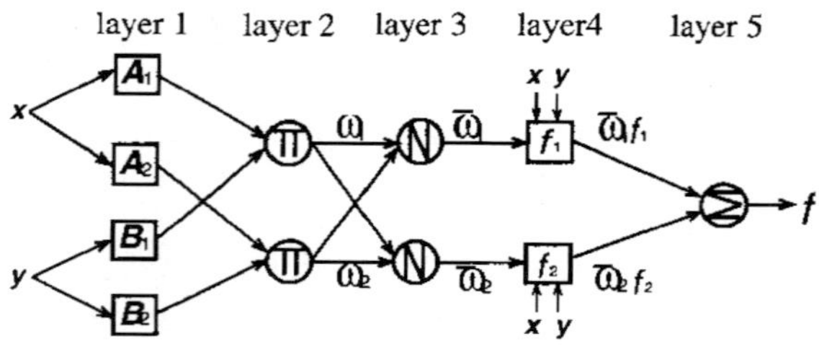

Fig. 1. Structure of ANFIS

$$o_i^3 = \overline{\omega_i} = \frac{\omega_i}{\omega_1 + \omega_2}. \tag{3}$$

Fourth layer: Transfer functions of every node are linear functions, and this expresses that they are local linear model. Calculate outputs of every rule:

$$o_i^4 = \overline{\omega_i} f_i = \overline{\omega_i}(a_i x + b_i y + c_i). \tag{4}$$

Parameter gather composed of all $\{a_i, b_i, c_i\}$ is called conclusion parameter. Fifth layer: Calculation of the outputs of all rules.

$$o_i^5 = f = \sum \overline{\omega_i} f_i = \frac{\sum_i \omega_i f_i}{\sum_i \omega_i}. \tag{5}$$

We can see the relations of inputs and outputs of above network that the network shown in Fig1 and the fuzzy reasoning system expressed in equation are equal. Parameter learning of the fuzzy reasoning system can sum up to be the adjust of the condition parameter (nonlinear parameter) and the conclusion parameter (linear parameter).

2.2 Mixed Learning Algorithm

Back propagation algorithm of the grads decline can be adopted to adjust parameter for all parameters. Here, a mixed algorithm is adopted for the training and confirming of the ANFIS parameters. The aim is to increase learning speed. Condition parameters in the mixed algorithm adopt Back propagation algorithm, and conclusion parameter adopts linear least squares estimate algorithm to adjust parameter. According to ANFIS system shown in Fig1, outputs of the system can be expressed as follows:

$$\begin{aligned} f &= \sum_{i=1}^{2} \overline{\omega_i} f_i = \sum_{i=1}^{2} \overline{\omega_i}(a_i x + b_i y + c_i) \\ &= (\overline{\omega_1} x) a_1 + (\overline{\omega_1} y) b_1 + (\overline{\omega_1}) c_1 + (\overline{\omega_2} x) a_2 + (\overline{\omega_2} y) b_1 + (\overline{\omega_2}) c_2. \end{aligned} \tag{6}$$

Adopt mixed learning algorithm, so that we can get optimization point of the conclusion parameters for given condition parameters. It can reduce the dimension of the search space of the grads algorithm, and can increase the convergence speed of parameters.

3 FNN Controller

First, design a FNN controller with 7 fuzzy labels, namely with 49 control rules. It is a three-layer neural network including input layer, hidden layer and output layer. In function, three-layer nodes corresponds to three steps of fuzzification, rule reasoning and defuzzification of the fuzzy logic control, hence it has definite fuzzy logic meanings. First, the inputs to the network are the error and the change in error relatively; Secondly, what the transfer function of the input layer's node represents is just the membership function of the fuzzy variable. The differences of this layer's weights $\omega_{ij}^{(1)}$ and θ_{ij} mean the differences of the shape and position of the membership function. What the output of this layer $Y^{(1)}$ represents is the degree of membership. Furthermore, the function of the hidden layer is to multiply the degrees of membership attained in fuzzy action. The output of the hidden layer represents the strength of the fuzzy rule, and we can carry out defuzzification if we deliver these strengths to the next layer. Finally, each weight $W^{(2)}$ of the output layer represents the fuzzy rule, according to the centroid defuzzification method, and consider them to be weights and inputs, we sums all rule strengths to obtain the final output results.

In summary, the input/ output of the FNN are represented as:

$$x = [e \quad ec]^T. \tag{7}$$

$$y_{ij}^{(1)} = \exp[-(\omega_{ij}^{(1)} \cdot x_j + \theta_{ij})^2]. \tag{8}$$

$$y_k^{(2)} = y_{i1}^{(1)} \cdot y_{i2}^{(2)} = \exp\{-[(\omega_{i1}^{(1)} \cdot x_1 + \theta_{i1})^2 + (\omega_{i2}^{(1)} \cdot x_2 + \theta_{i2})^2]\}. \tag{9}$$

$$u = W^{(2)} \cdot Y^{(2)} = \sum_{k=1}^{49} y_k^{(2)} \cdot \omega_k^{(2)}. \tag{10}$$

Where $j = 1, 2; I = i1, i2$, $i1, i2 = 1, 2, \ldots, r1$, $K = 1, 2, \ldots, r1 \times r1$, $r1$ is the number of the fuzzy labels.

In order to acquire the optimal network weights and desired control accuracy of the closed loop system, indirect approach is adopted to train the weights, namely series the controller and controlled process, and establish a negative feedback control loop. In order to guarantee the total optimal quality, we consider the square of the subtraction of the parameter inputs and the outputs of the closed loop to be target function. Moreover, considering that it is time-consuming to use Back-propagation algorithm to train weights and that there exists the shortage of being trapped into partial minimum, we can adopt improved genetic algorithm to optimize weights.

Assume that the controlled object is a one-level pendulum system. The equation of the single level pendulum can be represented as:

$$\ddot{\theta} = \frac{m(m+M)gl}{(M+m)I + Mml^2}\theta - \frac{ml}{(M+m)I + Mml^2}u. \tag{11}$$

$$\ddot{x} = \frac{m^2gl^2}{(M+m)I + Mml^2}\theta - \frac{I + ml^2}{(M+m)I + Mml^2}u. \tag{12}$$

Here, $I = \frac{1}{12}ml^2$, $l = \frac{1}{2}L$. Assume $x_1 = \theta$, $x_2 = \overline{\theta}$, $x_3 = x$, $x_4 = \overline{x}$,then the equation(5) and(6) can be expressed to be equation (7).

$$\bar{x} = Ax + Bu. \tag{13}$$

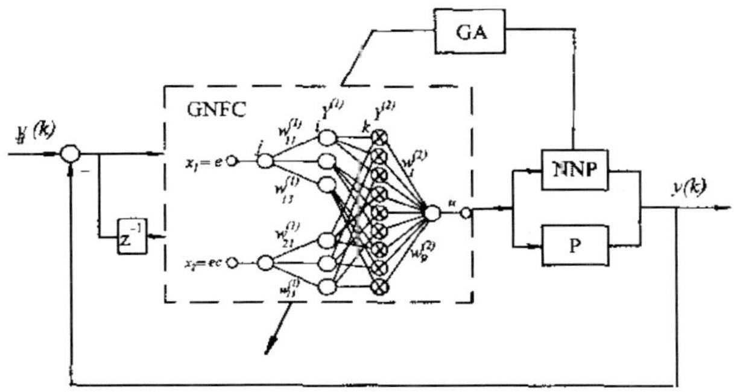

Fig. 2. The optimization control system structure of the pendulum angle

Because the controlled object has a nonlinear characteristic, artificial neural network is utilized to carry out model training of the input and output characteristics of the nonlinear model. In order to get an identification of the dynamic characteristics, we adopt a forward network of one nodal lead forecast model. Sigmoidal activation function and the linear activation function are adopted in internal structure of network, and 10 nodes are included in hidden layer.

Before the identification, the I/O data test should be operated on the controlled object in order to get the input vector and target vector approached by the neutral network function$[Y(k), u(k), Y(k+1)]$.

Here, $Y(k) = [x_1(k), x_2(k), x_3(k), x_4(k)]$.

3.1 The Genetic Algorithm Optimization of the FNN Controller

Because the FNN is not the standard full-link network any more, the speed of the BP training method that has been adopted is very slow. In order to accelerate the training speed and avoid sinking into the partial minimum, we adopt the genetic algorithm to adjust and optimize the parameters.

First, identified process model is series with unknown neutral network to get negative feedback control loop, then, dynamic simulation is made by improved genetic algorithm to seek for the optimized membership function of fuzzy control and combination control effect of control rules.

The definition includes the variables. In order to meet the needs of the operation conveniences and the accuracy, we adopt coding of real number to arrange the unknown weights one by one directly into digital strings to form the individual of the solution. N individuals form species. Adaptive degree F_i can be defined as follows:

$$F_i = \sum_{k=1}^{M} E_i^2(k) = \sum_{k=1}^{M} [Y_d - Y_i(k)]^2. \tag{14}$$

Where $i = 1, 2, \ldots, N$ is the number of the individuals, k is the number of the unknown variables in individuals. The target of the optimization is to make the target function F_i achieve a satisfying index.

In the copy process of the genetic operation, we eliminate $0.25N$ inferior solution under the average adaptive degree, and supplement N solutions in a random way to maintain the diversification of the solution in the exchange processes of the colonies. In order to keep the overall optimization in the search process, we should improve the optimal solution of the sample of this generation into that of next generation directly before the exchange. In addition, each individual should do one-bit or multi-bit exchanges mutually at some proportion to form new individuals. The random individuals added in the replication, particularly after several generations' evolution, exchange with the inferior individuals to gain new individuals, and these individuals can delay the emergency of prematurity effectively. When the evolution nears to the superior solution, the adaptive value of the next generation generated through exchange operation may no longer be better than their elder generation. At this time, changing a natural number of one position into another nature number will increase the random and diversification. Moreover, with the increment of the evolution, the proportion of mutation increases subsequently, so that new individuals will occur in colonies approaching optimal solution. The mutation proportion has been changed from 0.05 to 0.12 in the whole search process.

Optimized weight can be got after training is over. This is a huge artificial neural network with 77 parameters. The weight optimization is not the optimization value of the simplest network structure, but the one under the condition that the number of fuzzy labels is set to be 7. It means that we may use a simpler network to accomplish the design of the controller under the same index.

3.2 Adopting the SCNN to Optimize the Number of Fuzzy Labels

As far as the FNN with 7 fuzzy labels above is concerned, there are 28 weights and deviations represent membership functions of different inputs among 77 parameters. They can be divided into two groups: one group is composed of 7 weights and 7 deviations of the error variable e, and the other includes 7 weights and 7 deviations of the change in error ec. Now the number of the competition layer's nodes of the input vector of SCNN in the two groups is set to be 5. The largest circle number $N = 500$, the deviation of similarity $b = -0.9$. After carry out normalization processes to the input vectors, we begin to do trainings of competition and weight. The competition learning rules are described in nodes, and the training goal is to integrate the similar input vectors into the same categories in order to reduce the array of the input vector. Using MATLAB and its Neural Network Toolbox completes all design and training procedure.

The minimum number of fuzzy target is acquired after SCNN competition and trainings, then, 4 fuzzy labels make up into a new FNN, and a new network weight can be gained by the same progress above.

4 Simulation Results Analyse and Compare of ANFIS Control and GAFNN Control Equations

In order to compare the control effects of ANFIS control and GAFNN control, simulations are made when the pole becomes 2 meters and the random white noise is added, as shown in Fig 3.

Real line is the curve of the ANFIS control, broken line is the curve of the GAFNN, and the guideline of the control performances are shown in table 1.

We can see from Fig 3 and Table 1 that model prognostication and identification technique based on neural network are adopted in ANFIS and GAFNN control system in the condition of no disturb, added random white noise and the length of the pole becomes 2 meters. Parameters of FNN controllers are modified depending on model parameters of identification process. Steady state error of the pole angle can be controlled in a small range when added random white noise, the pole angle can be stable quickly, and two of them have strong

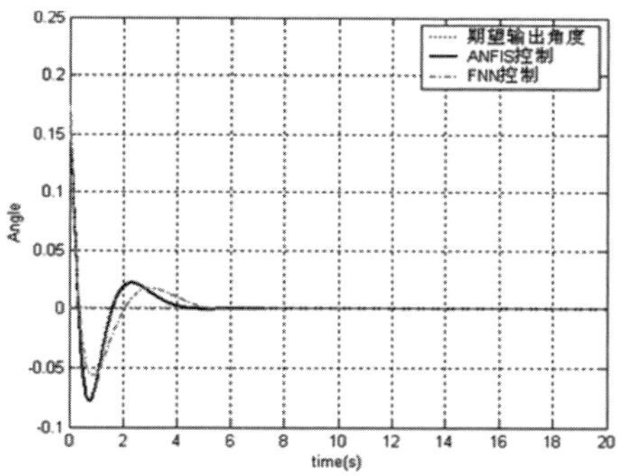

(a) No disturbing(L=1 meter)

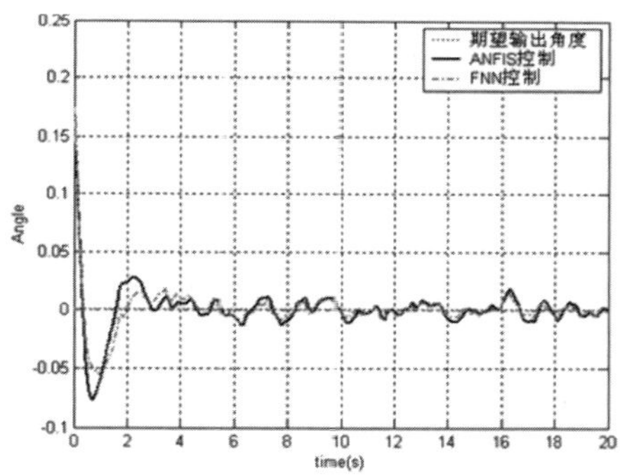

(b) Added random white

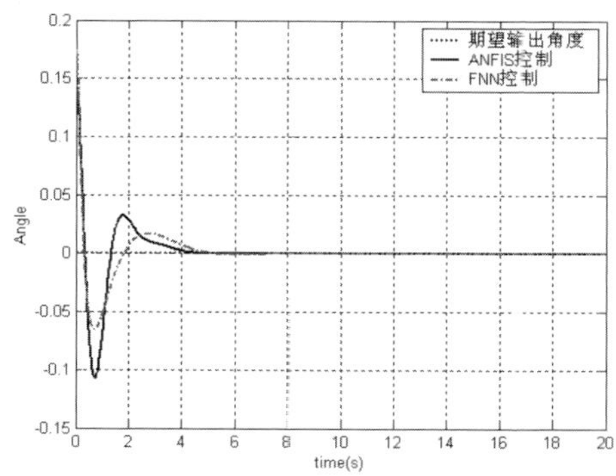

(c)The length of the pole becomes 2 meters

Fig. 3. Curves of the angle of the pole of two FNN control

Table 1. The simulation result and characteristic index

	Adjust time(s)	$\delta\%$ (rad)	error	Adjust time(s)	$\delta\%$ (rad)	error	Adjust time(s)	$\delta\%$ (rad)	error
ANFIS	4	0.075	0	4	0.075	< 0.025	4.5	0.11	0
AFFNN	5	0.06	0	4	0.06	< 0.02	6	0.06	0
	no disturbing			added random white noise			the length of the pole becomes 2 meters		

adaptive abilities and strong robustness. But GAFNN controller has smaller error and overshoot when added disturb and the length of the pole becomes 2 meters.

5 Conclusion

With the increase of the input values and the fuzzy division, the number of the state of the fuzzy space will increase, and the calculation number will be very large when the self-adaptive FNN carry out training. A new fuzzy neural network is introduced in this paper which employs self-organization competition neural network to optimize the structure of the fuzzy neural network, and applies a genetic algorithm to adjust the connection weights of the fuzzy neural network so as to get the best structure and weights of the fuzzy neural network. The effectiveness of the method introduced in this paper is demonstrated by contrasting of the control effects after competition and training. Simulations are made when the pole becomes 2 meters and the random white noise is added in the cart-pendulum system, and control effects of the ANFIS and GAFNN are analyzed. Simulation results indicate that GAFNN controller has greater control performance, high convergence speed, strong robustness and better dynamic characteristics. The effectiveness of the method introduced in this paper is demonstrated by its encouraging study results.

Acknowledgments. This project is supported by the key project of science and technology bureau of Sichuan Province (Grant No. 06209084), and partially supported by the Foundation of State Key Laboratory of Intelligent Technology and Systems in Tsinghua University.

References

1. FengJuang, C.: A TSK-Type Recurrent Fuzzy Network for Dynamic Systems Processing by Neural Network and Genetic Algorithms. IEEE Transactions on Fuzzy Systems 10(2) (2002)
2. Renhou, L.: The Intelligent Control Theories and methods. Electronics science and technology university of Xian (1999)

3. Hsiung Hung, T., Ming-Feng, Y., Hung-Ching, L.: A pi-link fuzzy controller implementation for the inverted pendulum system. In: Processing of IEEE Conference on Intelligent Processing System, pp. 218–222. IEEE Computer Society Press, Los Alamitos (1997)
4. Mario, E.M., Holzapfel, F.: Fuzzy-logic control of an inverted pendulum with vision feedback. IEEE Trans. on Education 14(2), 165–170 (1998)
5. Miyamoto, M., Kawato, M., Setoyama, T., Suzuki, R.: Feedback Error Learning. IEEE Trans. on Neural Networks 1, 251–265 (1998)
6. Sigeru, Omatu, T.F., Michifumi, Y.: Neuro-pid Control for Inverted Single and Double Pendulums. In: IEEE Conference, pp. 2685–2690. IEEE Computer Society Press, Los Alamitos (2000)

A Robust Approach to Find Effective Items in Distributed Data Streams

Xiaoxia Rong[1] and Jindong Wang[2]

[1] School of Mathematics and System Sciences, Shandong University, 250100, Jinan, China
rongxiaoxia@sdu.edu.cn
[2] Shandong Computer Science Center, 250014, Jinan, China

Abstract. A data stream is a massive unbounded sequence of data elements continuously generated at a rapid rate. Consequently, the knowledge embedded in a data stream is more likely to be changed as time goes by. Data items that appear frequently in data streams are called frequent data items, which often play a more important role than others in data streams management system. So how to identifying frequent items is one of key technologies. As distributed data streams management system is concerned, there are many input data streams having different effect on result, the pure character of frequency is unfit for finding the important data. To solve this problem, effective data of distributed data streams is defined in this paper, which combines the frequency of items and the weight of streams. Based on an optimization technique that is devised to minimize main memory usage, a robust mining approach is proposed. According to this algorithm, the effective data can be output with limited space cost. At the same time, the sensitivity of algorithm is analyzed which shows the output result is within the error given by the user. Finally a series of experiments show the efficiency of the mining algorithm.

1 Introduction

In a data stream, data items that appear frequently are called frequent data items, which are important to the result and should be processed emphatically. Therefore, the study on mining frequent data is very important about data processing. Facing the infinite, high speed input data stream, the system often is incapable of processing so many data in required time, besides this, the huge number of input data often surpasses the handling ability of the system. In order to decrease processing time and improve the stability of system, we must mine the important data to process emphatically and discard part data items that are less important. So the study of mining frequent items is one of the key technologies in data streams (see [1] and [2]), which will take great effect on the field such as data stream mining and network monitor et al and the relevant applications are found in [3] and [4].

In reference [5], G. Manku statistics the data items that appear frequently using some counters, the number of which varied with demand. In [6] a certain

K. Li et al. (Eds.): LSMS 2007, LNCS 4688, pp. 688–696, 2007.

number of counters were set for the threshold $0 < \theta < 1$ and total number N of data items, the given algorithm used $1/\theta$ counters to find the data items of which appearing number is larger than $\theta \times N$. The authors of [7] and [8] use Hash functions to statistic all data items, on the basis of which they mined data items with frequency exceed certain degree.

However methods above take the same object that single data stream is studied only. In practical applications the distributional data streams management system often include many input data streams, the traditional methods for single data stream are unsuitable for finding the important data. The main difficulty is in two aspects:

(1) The frequent data of one input data stream is not necessarily frequent for the system, therefore the mining methods for single input data stream is not used here directly. We should integrate the results of all input data streams to obtain frequent data items of the system.

(2) In distributional data streams management system, for each input data stream has different influence on the result, the pure character of frequency is incapable of finding important data items.

To solve the problems above, we define the effective data of distributional data streams management system and propose a mining algorithm in this paper.

The remaining of the paper is organized as follows: In the next section the problem is described and definition of effective data is introduced. In Section 3 the algorithm is proposed based on sketch method. Subsequently we discuss how to choose b and t to optimize the space cost, and do robust analysis to maintain ϵ-approximate of the algorithm in section 4. Section 5 is the simulation, which is helpful to get the character of algorithm such as efficiency and relationship of parameters. The paper ends with conclusions.

2 Problem Setting and Definition of Effective Data

Assuming there are m input data streams in distributional data streams management system: $S = \{S_1, S_2, ..., S_m\}$ $(m \geq 2)$, S_i is a sequence made of data item q, that is $S_i = \{q_{i1}, q_{i2}, q_{i3}, ...\}$. Where $q_{ij} \in D$, $D = \{d_1, d_2, ...d_l\}$ and d_i is the value of data item that is known already. The importance of S_i to the system is expressed by w_i and S_i comes into the system through the node M_i that is called input-node.

Supposing l_{ij} is the appearance number of data item d_i in data stream S_j, $d_i^{'} = \sum_{j=1}^{m} l_{ij} w_j$ and $N_j = \sum_{d_i \in S_j} l_{ij} w_j$ is defined as the frequency weight (fre-weight in short) of d_i and of S_j separately. $N = \sum_{i=1}^{m} N_i$ is the total fre-weight of the input data streams.

Definition 1 (Effective Data). *for threshold $s \in (0, 1)$ given in advance, if the inequality $d_i^{'} \geq sN$ holds , then d_i is defined as effective data item (effective data in short).*

Three parameters are given in advance: threshold $s \in (0,1)$, error $\varepsilon \in (0,1)$, probability $\delta \in (0,1)$ and $\varepsilon \ll s$ is also required. The objective of mining effective data is as follows:

After mining computing, all data items that satisfy $d_i^{'} \geq sN$ are output while data items satisfying $d_j^{'} \leq (s-\varepsilon)N$ are not. Further more, the error of estimated fre-weight with real fre-weight is no more than ε with probability δ near to 1.

3 Mining Algorithm

On basis of problem describing in the second part, the following requests are concerned in algorithm designing:

(1) The algorithm can count efficiently the fre-weight of data item in single data stream.

(2) The algorithm can integrate the result of (1) and get the fre-weight of data item.

(3) The output by algorithm should be ensured with the error given by the user.

According to demands above and the relevant study in [7] and [8], sketch method is adopted to design algorithm because it can integrate statistical information of single data stream without loss. The mining algorithm is composed of three steps as follows.

3.1 Statistic Items of Single Data Stream

The first step of the algorithm is to statistic fre-weight of data items in single data stream. Known from the discussion above, S_i comes into the system through the node M_i , so the statistic work of S_k is done by M_k. In this paper, we design statistic table according to s, ε, δ given by the user and a set D known already. Supposing there are t Hash functions adopted in the table and each of them has b kinds of value, the framework is in table 1:

Here each unit of statistic table is called a counter.

TS_k is denoted as the statistic table of data stream S_k and $TS_k(i,j)$ is the statistic value of line i column j in TS_k with primal value 0. In this section, we will propose mining algorithm basing on statistic table above, in which the computing of t and b is described in detail in section 4.

The statistic pseudocode for single data stream is follows:

Table 1. Statistics Table

\	1	2	...	b
$Hash_1()$	$Counter_{11}$	$Counter_{12}$	...	$Counter_{1b}$
$Hash_2()$	$Counter_{21}$	$Counter_{22}$	...	$Counter_{2b}$
...	...	...	...	...
$Hash_t()$	$Counter_{t1}$	$Counter_{t2}$	...	$Counter_{tb}$

```
STATISTIC ( DataType NewData )
(1) { Count = Count + w_k;
(2)     for(i = 1; i ≤ t; i++)
(3)         { j = Hash_ki(DataTypeNewData);
(4)           TS_k(i, j) = TS_k(i, j) + w_k }}
```

3.2 Integrating Information over All Data Streams

The second step of the algorithm is to integrate the result of single data stream. In general, there are two relevant kinds of methods. One is setting a central node and another method is based on a tree composed of all nodes with leaf of input-node. In the former, the central node takes charge of result integration from all data streams, which need a large of memory and CPU time. At the same time it should communicate with many nodes. Too much work often overloads the nodes, so we take the second method in this paper.

In our algorithm, a k-ary tree is set to integrate result from all data streams. Within each period of time leaves send statistic result to parent nodes, which send the integrated information to higher lever and at last the information is got by root node. For all statistic tables of single data stream are designed according to the same parameter of s, ε, δ and D, we can get the integrated result by adding all statistic tables.

The integration pseudocode for data streams is follows:

```
INTEGRATION()
(1) { for (i = 1; i ≤ t; i++)
(2)       for (j = 1; j ≤ b; j++)
(3)           TS(i, j) = 0;
(4)   Count=0;
(5)   for (m = 1, m ≤ k, m++)
(6)       if (TS^m ≠ NULL )
(7)           {for (i = 1; i ≤ t; i++)
(8)               for (j = 1; j ≤ b; j++)
(9)                   TS(i, j) = TS(i, j) + TS^m(i, j);
(10)              Count = Count + Count^m; }}
```

Where, TS is statistic table and TS^m is got from the mth child node. $Count$ records the number of nods and $Count^m$ is got from the mth child node. The root node not only integrates information but also finds the effective data items from TS.

3.3 Finding Effective Items

The last step of the algorithm is to find the effective data items basing on the work above.

The finding pseudocode is follows:

```
FINDING()
{for (i = 1; i ≤ l; i++)
    {W = ∞ ;
     for (j = 1; j ≤ t; j++)
         {m = Hash_j(d_i)
          if (TS(j, m) < W) W = TS(j, m) ;}
     if (W > s × Count) output (d_i); }}
```

4 Algorithm Analysis

4.1 Space Costs Analysis

Given threshold $s \in (0,1)$, error $\varepsilon \in (0,1)$ ($\varepsilon \ll s$ is demanded), probability $\delta \in (0,1)$ near to 1 and data set D, let random variable Y denotes the error of fre-weight in data stream S_i, there is formula as follows using a Hash computing.

$$E[Y] \leq (N_i/b) \tag{1}$$

According to of *Markov's inequality*, we have:

$$Pr[|Y| - \lambda E[|Y|] > 0] \leq 1/\lambda$$

Where λ is positive. Known from the function STATISTIC() in section 3, Y is none negative. In addition, based on (1), the formula above can be rewritten as

$$Pr[Y - \lambda(N_i/b) > 0] \leq 1/\lambda \tag{2}$$

This inequality means the event that $Y \geq \lambda(N_i/b)$ happens with probability is no more than $1/\lambda$ after a Hash computing.

Let Y_{min} is the minimal error of t times Hash computing, the opposite event is thought and the following can be obtained

$$Pr[Y_{min} < \lambda(N_i/b)] \geq 1 - 1/\lambda^t$$

Then the probability that all errors in l kinds of data items satisfy the formula above is:

$$\delta = (1 - 1/\lambda^t) \approx e^{(-l/\lambda^t)} \tag{3}$$

Let $\varepsilon = \lambda/b$, then we get the memory of each Hash table

$$V = t \times b = \frac{1}{\varepsilon}\ln(-\frac{l}{\ln\delta})\frac{\lambda}{\ln\lambda}$$

Here λ is positive. Because $\min_{\lambda>0}(\frac{\lambda}{\ln\lambda}) = e$, $V = \frac{e}{\varepsilon}\ln(-\frac{l}{\ln\delta})$ holds. So there is $t = \ln(-\frac{l}{\ln\delta})$, $b = \frac{e}{\varepsilon}$ and $\lambda = e$ when Hash table has the minimal memory.

Basing on analysis above, we can get theorem 1.

Theorem 1. *Using our algorithm to statistic single data stream, such as* S_i, $O(\frac{e}{\varepsilon}\ln(-\frac{l}{\ln\delta}))$ *space is cost at most to estimate effective data items.*

Thereinafter we analyze the error about fre-weight in integration table.

The structure of integration table is the same as the statistic table facing single data stream, in other words, they have the same $t = \ln(-l/\ln\delta)$ and $b = e/\varepsilon$. Assuming the total fre-weight is $N = \sum_{i=1}^{m} N_i$, the average value of b counters about each Hash function is N/b. When one Hash function is used to estimate, there holds $E(Y) \le N/b = N\varepsilon/e$, where Y is defined before.

Known from analysis above, $\lambda = 1$ holds when statistic table has the minimal memory. So according to *Markov's inequality*, we try once, then

$$Pr[Y - \varepsilon N > 0] \le 1/e$$

If we try t times Hash computing, there is

$$Pr[Y_{min} < \varepsilon N] \ge 1 - 1/e^t \tag{4}$$

Where Y_{min} is the minimal error of t times Hash computing. Then the probability that the error about l kinds of data satisfies formula (4) is $(1 - 1/e^t)^l$, which is equal to δ according to formula (3), so the event happens with the probability:

$$Pr[Y_{min} < \varepsilon N] \ge \delta \tag{5}$$

Theorem 2. *By our algorithm to statistic all data streams, $S = \{S_1, S_2, ...S_m\}$, space of $O(q\frac{e}{\varepsilon}\ln(-\frac{l}{\ln\delta}))$ is consumed at most to find effective items. For k-ary tree, q is no more than $\frac{k(m-1)}{k-1}$.*

4.2 Stability Analysis

However, in an environment where multiple data streams arrive at a set of distributed observation points, sketches must be computed remotely and then must be integrated through a hierarchy before queries may be conducted. As a result, many sketch-based methods for the single stream case do not apply directly, as either the error introduced becomes large, or because the methods assume that the streams are non-overlapping. These limitations hinder the application of these techniques to practical problems in network traffic monitoring and integration in sensor networks. To address this, we introduce a framework for robust analysis over data produced by distributed sources. From reference [10,11], we introduce the concepts of sensitivity as follows.

Definition 2 (Order Insensitivity). *Consider a data stream S_i of distinct items $S_i = \{q_{i1}, q_{i2}, q_{i3}, ...\}$, and let π denotes an arbitrary permutation on m items, and $\pi(S_i)$ denotes the stream of items after applying permutation π. A sketching algorithm is said to be order-insensitive if $\forall\pi, T(S_i) = T(\pi(S_i))$. Similarly, a sequence of union operations is order-insensitive if the result is not affected by reordering.*

Definition 3 (Duplicate Insensitivity). *Consider a data stream S_i of distinct items $S_i = \{q_{i1}, q_{i2}, q_{i3}, ...\}$, Now consider another stream S_i^+ which consists of S_i as a subsequence but with arbitrary repetition of items. A sketching algorithm is said to be duplicate-insensitive if $\forall S_i, S_i^+, T(S_i) = T(S_i^+)$.*

So the sketch adopt in this paper has the property of order and duplicate insensitivity. Ordering of insertions does not affect the structure. From formula (3) and (5), then we have

Theorem 3. *For a data stream S_i with the fre-weight $N_i\omega_i$, error ε, threshold s, and probability δ that is near 1, the frequent items computed by this algorithm can be found with probability δ, and the error is no more than $\varepsilon N_i\omega_i$.*

Then we sketch error proof as follows. In the integration table, assuming n_i is the fre-weight of data item d_i in stream S_i, the statistical fre-weight is $\tilde{d}_i^{'} = \sum_{i=1}^{m} n_i$. If the real fre-weight of d_i is $d_i^{'}$, according to statistic step in section 3 and theorem 2, we know that $\tilde{d}_i^{'} > d_i^{'}$ and $\tilde{d}_i^{'} \leq d_i^{'} + \varepsilon N$ hold at the same time. So it is reasonable that $\tilde{d}_i^{'} = d_i^{'} + \sigma$, here $0 \leq \sigma \leq \varepsilon N$. Then according to finding step in section 3, all data items that satisfying $\tilde{d}_i^{'} > sN$ should be output. Because the condition is equal to $d_i^{'} + \sigma > sN$, that is $d_i^{'} > (s - \varepsilon)N$, hence the data items that satisfying $d_i^{'} \leq (s - \varepsilon)N$ can't be found. From discussion above, the main result of robust mining approach is given as follows.

Theorem 4. *For distributed data streams $S = \{S_1, S_2, ...S_m\}$ with error ε, the threshold s, the total fre-weight N, and probability δ that is near 1, let the real value of effective item is $d_i^{'}$ and $\tilde{d}_i^{'}$ is the estimation computed using this algorithm, then $d_i^t \leq \tilde{d}_i^{'} \leq d_i^{'} + \varepsilon N$ and $d_i^{'} \geq sN$ hold with the probability δ.*

5 Simulation

In this section, we take simulation to test the precision of our algorithm.

Supposing d_{out} is the output data set, d_{real} is the data set satisfying demand, e_i is the estimated fre-weight of $d_i \in d_{out}$ and r_j is the fact fre-weight of $d_j \in d_{real}$, we define the precision of simulation as:

$$\tau = \sum_{d_j \in d_{real}} r_j / \sum_{d_i \in d_{out}} r_i .$$

The simulation is taken in LAN made of 30 computers with P4 1.7G CPU and 512M memory. $D = \{1, 2, ..., 100000\}$ and 10,000,000 data items are produced that obey Zipf distribution in each test. 3 ary-tree is adopted. The simulation shows the relation of precision with integration point, counter number, threshold value and parameter α of Zipf distribution.

Figure 1 shows the relation of precision with integration point when $Zipf$ $\alpha = 1$, $s = 0.01$, and 5 Hash functions with 200 counters for each. When the number of integration point increases, the precision change slightly. Figure 2 shows the relation of precision with number of Hash functions, when $Zipf\alpha = 1$, $s = 0.01$, and 200 counters for each Hash function. Figure 3 shows the relation of precision with number of counters, when $Zipf\alpha = 1$, $s = 0.01$, and the number of Hash function is 5. We can see from Fig.2 and Fig.3 that the increase of the number of Hash functions and the counters of each Hash function will increase

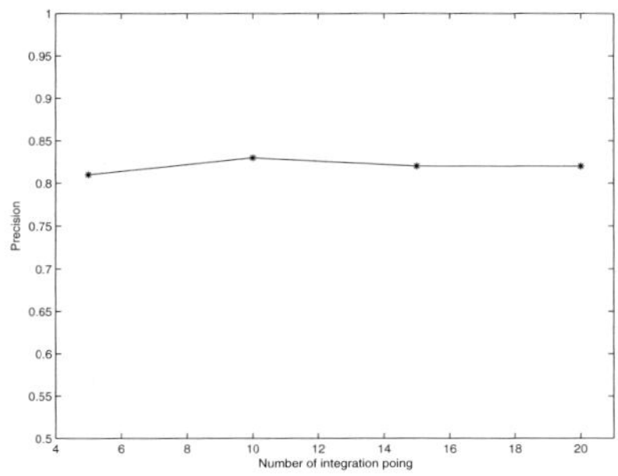

Fig. 1. Relationship between precision and number of integration points

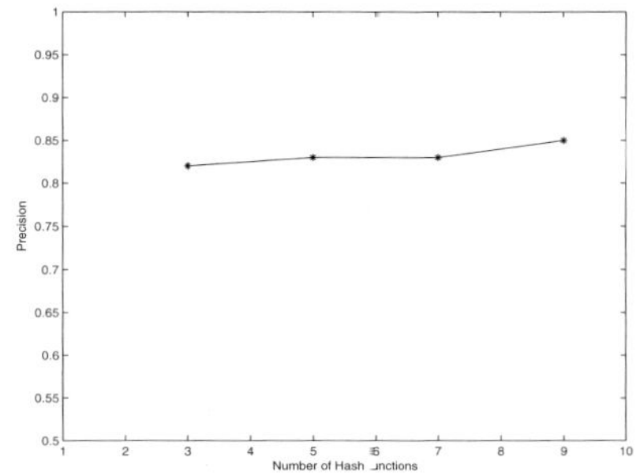

Fig. 2. Relationship between precision and number of Hash functions

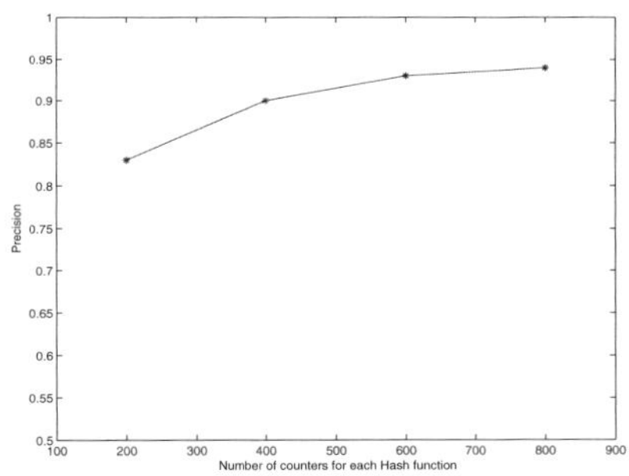

Fig. 3. Relationship between precision and number of counters for each Hash function

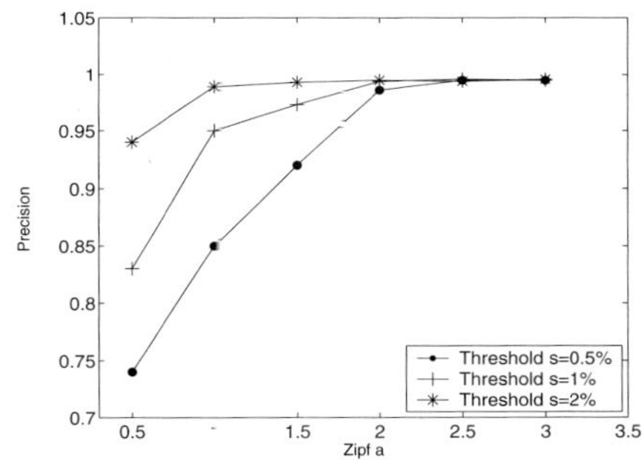

Fig. 4. The relationship between precision and α

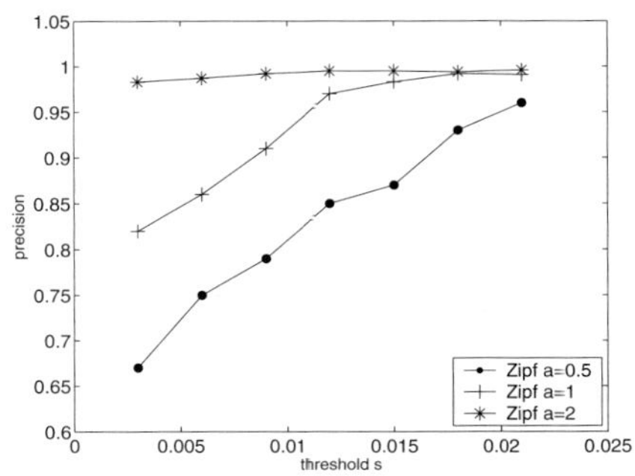

Fig. 5. The relationship between precision and threshold

the precision of result, and the later do much better. Figure 4 shows the relation of precision with α when the threshold has different values. If α becomes large, the set made of data items with high fre-weight will have lesser elements. So the effect aroused by data items with low fre-weight is weakened, which arouses the improvement of precision.

Figure 5 shows the relation of precision with threshold when α has different values. When the threshold increases, the fre-weight of data items that are output will becomes large. So there will be great difference between data items that with high or low fre-weight. For the data items with low fre-weight take little influence on the result, the precision is improved.

6 Conclusions

To satisfy the demands of mining effective data in distributed data stream system, we propose a robust approach based on sketch method. The algorithm includes three key steps: statistic in single data stream, integrating statistic result of all streams and finding effective data at last. The memory consumption and robust analysis and a series of simulation shows the character of algorithm.

References

1. Muthukrishnan, S.: Data Streams: Algorithms and Applications. In: Proceedings of the fourteenth annual ACM-SIAM symposium on Discrete algorithms table of contents, pp. 413–413. ACM Press, New York (2003)
2. Phillip, B., Gibbons, Tirthapura, S.: Distributed Streams Algorithms for Sliding Windows. In: Proceedings of the fourteenth annual ACM symposium on Parallel algorithms and architectures, pp. 63–72. ACM Press, New York (2002)
3. Lin, X., Lu, H., Xu, J., Yu, J.X.: Continuously Maintaining Quantile Summaries of the Most Recent N Elements over a Data Stream. In: Proceedings of the 20th International Conference on Data Engineering, pp. 362–373 (2004)
4. Brian, Babcock, Surajit, Chaudhuri, Gautam, Das: Dynamic Sample Selection for Approximate Query Processing. In: SIGMOD Conference, pp. 539–550 (2003)
5. Manku, G., Motwani, R.: Approximate Frequency Counts over Data Streams. In: Proceedings of the 28th International Conference on Very Large Data Bases, pp. 346–357 (2002)
6. Karp, R., Papadimitriou, C., Shenker, S.: A simple algorithm for finding frequent elements in sets and bags. ACM Transactions on Database Systems (TODS) 28, 51–55 (2003)
7. Graham, Cormode, Muthukrishnan, S.: An Improved Data Stream Summary: The Count-Min Sketch and Its Applications. In: Farach-Colton, M. (ed.): LATIN 2004. LNCS, vol. 2976, pp. 29–38. Springer, Heidelberg (2004)
8. Charikar, M., Chen, K., Farach-Colton, M.: Finding frequent items in data streams. Theoretical Computer Science. 312, 3–15 (2004)
9. Flajolet, P., Martin, G.N.: Probabilistic counting algorithms for data base applications. Journal of Computer and System Sciences 31, 182–209 (1985)
10. Marios, Hadjieleftheriou, John, W., Byers, George, Kollios: Robust Sketching and Aggregation of Distributed Data Streams. Boston University Computer Science Technical Report (2005)
11. Nath, S., Gibbons, P.B., Seshan, S., Anderson, Z.R.: Synopsis diffusion for robust aggregation in sensor networks. In: SenSys '04: Proceedings of the 2nd international conference on Embedded networked sensor systems, pp. 250–262 (2004)

Two-Layer Networked Learning Control of a Nonlinear HVAC System

Minrui Fei[1], Dajun Du[1], and Kang Li[2]

[1] School of Mechatronical Engineering and Automation,
Shanghai University, Shanghai, 200072, China
mrfei@staff.shu.edu.cn, ddj559@163.com

[2] School of Electronics, Electrical Engineering and Computer Science,
Queen's University Belfast, United Kingdom
k.li@ee.qub.ac.uk

Abstract. Networked control system (NCS) has become the focus of many recent researches in the control field, with topics covering from scheduling methods, modeling and compensation schemes for the network-induced delay, to stability analysis and controller design strategies, etc. While the majority of researches are carried out on linear NCS, little has been done so far to investigate nonlinear NCS with more complex NCS architecture for nonlinear plants. The main objective of this paper is to propose two-layer networked learning control system architecture that is suitable for complex plants. In proposed architecture, the local controller communicates with the sensors and actuators that are attached to the plant through the first layer communication network, and the network is typically some kind of field bus dedicated to the real-time control. Through the second layer communication network, local controller also communicates with computer systems that typically functions as learning agent. For such a learning agent, firstly, a packet-discard strategy is developed to counteract network-induced delay, data packets that are out of order, and data packet loss. Then cubic spline interpolator is employed to compensate lost data. Finally, the output of learning agent using Q-learning strategy dynamically is used to tune the control signal of local controller. Simulation results of a nonlinear heating, ventilation and air-conditioning (HVAC) system demonstrate the effectiveness of the proposed architecture.

1 Introduction

Networked control system has been widely used in car suspension system [1], large pressurized heavy water reactor [2] and mobile robot [3] due to various advantages including low cost of installation, ease of maintenance and greater flexibility. In the last decade, researches have been mainly carried out linear NCS, i.e. control system communicates with the sensors and actuators over one-layer communication network and the controlled plant is linear. Recently, researches on nonlinear NCS have attracted significant amount of interests from both the industry and academics. In [4], a nonlinear system with time-varying delay was

K. Li et al. (Eds.): LSMS 2007, LNCS 4688, pp. 697–709, 2007.

investigated, which was approximated by Takagi-Sugeno (T-S) fuzzy models with time-varying delays and a delay dependent sufficient condition for ensuring the stability was derived. In [5], T-S fuzzy model with parametrical uncertainties was used to approximate nonlinear NCS, and feedback gain of a memoryless controller could be obtained by solving a set of LMIs.

To improve NCS performance, some learning and intelligent control algorithms have been proposed. In [6], previous cycle based learning (PCL) method was incorporated into networked control for a nonlinear system, and the convergence in the iteration domain could be guaranteed. A model predictive path tracking control methodology over an IP network was proposed in [7], where the parameter was adjusted externally with respect to the current network traffic conditions. A novel feedback scheduler based on neural networks was suggested in [8], then the control period was dynamically adjusted in response to estimated available network utilization. In [9], with the delays estimated by a delay window (DW), a novel fuzzy logic method was proposed to calculate the LQR parameter online, which could not only save the power of control node but also preserve NCS performance.

For nonlinear NCS, conventional modeling and control methods may be difficult to apply as the adverse effects of network-induced delay and data packet loss make the control of nonlinear plant more complicated. Although learning control is able to improve control performance by on-line mining of valuable knowledge and potential laws, accumulating experience and adapting to environment, it is difficult to design and implement such control strategy using embedded controller or program logic controllers (PLC) due to high computational complexity in typical NCS. However, two-layer networked control system architecture was introduced in [10]-[11], which provided a path for the implementation of learning strategies due to strong computational ability of the second layer controller. Motivated from the above observations, two-layer nonlinear networked learning control systems is investigated in this paper. Comparing with typical NCS, this system is characterized by two-layer network, local controller, learning agent, and nonlinear plant.

This paper is organized as follows. Section 2 describes the two-layer networked learning control systems architecture, and a discard-packet strategy is proposed and a cubic spline interpolation is used to compensate data packet loss. Reinforce learning and Q-learning algorithm in learning agent are introduced in section 3, and simulations are given in section 4. Section 5 is the conclusion.

2 Two-Layer Networked Learning Control Systems

2.1 The Introduction of Two-Layer Networked Learning Control Systems Architecture

In this paper, two-layer networked learning control system architecture is introduced as shown in Fig.1, with the objectives to achieve better control performance, better interference rejection and to increase the adaptability to varying environment. In this architecture, local controller communicates with

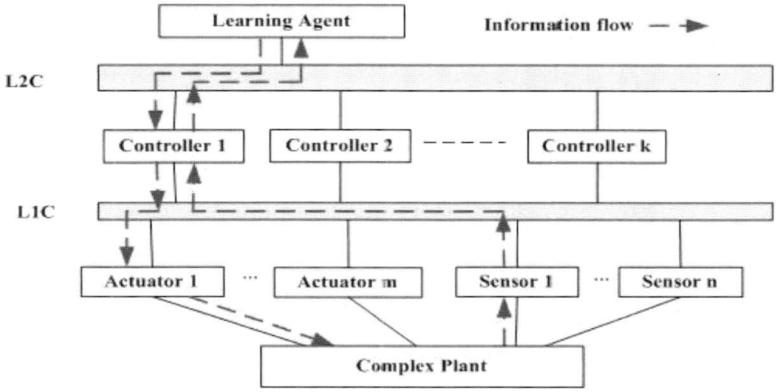

Fig. 1. Two-layer networked learning control system

the sensors and actuators, that are attached to the plant through the first layer communication network called L1C, typically some kind of field bus dedicated to real-time control. Local controller also communicates with computer system which typically functions as an learning agent through the second layer communication network called L2C. This network is typically some kind of existing local area network, wide area network (WAN), or possibly the Internet. Control signal traffic at L2C shares the available network bandwidth with other data communications. The above general architecture can be used in many industrial applications for distributed plants. For example, an industrial process control systems such as SIMATIC PCS 7 process control system where a number of embedded controllers or PLC through L1C control a number of sub-Processes or units, and supervisory computers through L2C co-ordinate various low-level control actions carried out by local controllers.

2.2 Characteristics of Two-Layer Networked Learning Control Architecture

From Fig.1, L1C uses typically some kind of field bus, and network-induced delay and data packet loss can be minimized. Local area network is used as L2C, and network-induced delay and data packet loss are unavoidably introduced into L2C. Network-induced delay is a critical issue in NCS. In [10], strategies to deal with short network-induced delay were summarized. While methods to cope with large network-induced delay were surveyed in [12]. In this paper, the sensor node was sampled periodically, and the sampling period was h, and $h = 15$ ms. Measurements character of network-induced delay is illustrated in Fig.2.

On the other hand, it is well known that the network transmission may not be reliable and data packets can suffer from network-induced delays, out-of-order, in the worst situation, they can get lost during the transmission pathway. Those cases are illustrated in Fig.3. The time instants at which the data packets reach the learning agent are random due to the random nature of network-induced

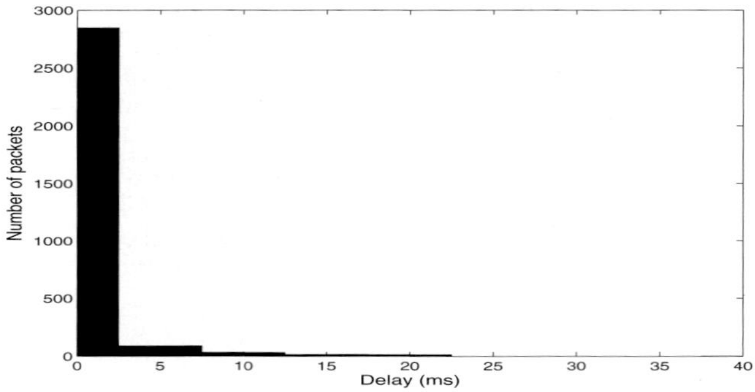

Fig. 2. Network-induced delay from local controller to the learning agent

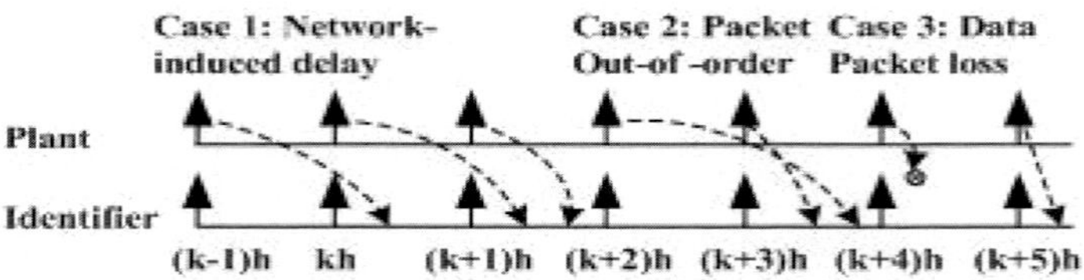

Fig. 3. Timing diagram

delay. Let T_i^l denote those instants relative to initial time, where the subscript 'i' is the serial number of data packets received by the learning agent and the superscript 'l' means the learning agent. Note that we have $T_{i+1}^l > T_i^l$. Therefore, corresponding to the arrival data packets by the learning agent, there exists a serial pairs $\{y(i),\, T_i^l\}$, where, '$y(i)$' is a subset of $\{\mathbf{y}(k)\}$. Several methods have been proposed to deal with packet loss. In [13], a zero order holding (ZOH) at the receiving side of a communication medium was used, i.e., when an actuator or sensor fails to access the medium, the value stored in a ZOH is fed into the plant or controller $(y(i) = y(i-1))$. In [14], a value of zero was fed into the plant or controller when an actuator or sensor fails to access the medium $(y(i) = 0)$, which however was inappropriate due to the real system output unequal zero.

Unlike the above mentioned methods in dealing with networked-induced delay and data packet loss separately, we take into account the networked-induced delay, data packet out-of-order and packet loss within one framework in this paper. It is well known that in many control strategies, only the 'newest' data points are used to achieve the best control performance and guarantee the system stabilization. Therefore, it is fairly reasonable to remove a long delayed data packet and out-of-order data packet. Then, a data packet processing strategy, called packet-discard strategy, is used in this paper. Here, if a data packet is unable to arrive at the learning agent within the corresponding sampling period, it will be discarded. This strategy treat long delayed data packets and out-of-order data packets as packet loss such that the 'newest' data samples are used. Furthermore, to compensate the data packet $(y(i))$ loss and improve the control performance, the cubic spline interpolator is designed ($y(i) = \hat{y}(i)$) and is particularly described in the following section.

2.3 Cubic Spline Interpolator

A cubic spline [15] is a spline consisting of piecewise cubic polynomials $g(x)$ that interpolates the real-valued function $f(x)$ at the points $x_0 < x_1 < \cdots < x_N$ where the values of $f(x)$ are known. So, we construct $g(x)$ at each interval $[x_i, x_{i+1}]$, $i = 0, \ldots, N-1$, as a cubic polynomial

$$V_i(x) = c_{0,i} + c_{1,i}(x - x_i) + c_{2,i}(x - x_i)^2 + c_{3,i}(x - x_i)^3 \tag{1}$$

Once we find the coefficients $c_{j,i}$, $j = 0, \ldots, 3; i = 0, \ldots, N-1$, we can evaluate $g(x)$ for any point x in $[x_0, x_N]$.

To guarantee $g(x)$ to be continuous on $[x_0, x_N]$, it is assumed that

$$V_{i-1}(x_i) = V_i(x_i) = f_i(x_i) \tag{2}$$

Each $V_i(x)$ is interpolated at only two points. Since a cubic polynomial can interpolate a function at four points, we have freedoms in choosing the $V_i(x)$. To make use of the freedom, further constraints are imposed such that $V_i(x)$ must agree with $V_{i-1}(x_i)$ in both slope and curvature; that is,

$$V_i'(x_i) = V_{i-1}'(x_i) \tag{3}$$

and

$$V_i''(x_i) = V_{i-1}''(x_i), i = 1, \cdots, N-1 \tag{4}$$

We assume that the curvatures k_i at x_i and k_{i+1} at x_{i+1} are dependent on $V_i(x)$. Since $V_i(x)$ is a cubic polynomial, $V_i''(x)$ is a linear polynomial constrained such that $V_i''(x_i) = k_i$ and $V_i''(x_{i+1}) = k_{i+1}$, $i = 0, \ldots, N-1$. That is,

$$V_i''(x) = k_i \frac{x_{i+1} - x}{x_{i+1} - x_i} + k_{i+1} \frac{x - x_i}{x_{i+1} - x_i} \tag{5}$$

Integrating (5) gives

$$V_i'(x) = -\frac{k_i}{2} \frac{(x_{i+1} - x)}{x_{i+1} - x_i} + \frac{k_{i+1}}{2} \frac{(x - x_i)^2}{x_{i+1} - x_i} + a_i \tag{6}$$

Integrating (6) and using (2), it holds that

$$\begin{aligned} V_i(x) = &-\tfrac{k_i}{6} \tfrac{(x_{i+1}-x)^3}{x_{i+1}-x_i} + \tfrac{k_{i+1}}{6} \tfrac{(x-x_i)^3}{x_{i+1}-x_i} + A_i(x_{i+1} - x) + B_i(x - x_i) \\ &i = 0, \ldots, N-1 \end{aligned} \tag{7}$$

where $A_i = \frac{f(x_i)}{x_{i+1}-x_i} - \frac{k_i}{6}(x_{i+1} - x_i)$; $B_i = \frac{f(x_{i+1})}{x_{i+1}-x_i} - \frac{k_{i+1}}{6}(x_{i+1} - x_i)$.

Applying the constraint given by (3), yielding

$$\begin{aligned} &k_{i-1}(x_i - x_{i-1}) + 2k_i(x_{i+1} - x_{i-1}) + k_{i+1}(x_{i+1} - x_i) \\ &= 6[\tfrac{f(x_{i+1})-f(x_i)}{x_{i+1}-x_i} - \tfrac{f(x_i)-f(x_{i-1})}{x_i-x_{i-1}}],\ i = 1, \cdots, N-1 \end{aligned} \tag{8}$$

Equation (8) represents a system of $N-1$ linear equations with $N+1$ unknowns k_i, $i=0,\ldots,N$. Obviously, if the boundary conditions at the endpoints x_0 and x_N are given, (8) can be solved, thus the coefficients $c_{j,i}$ of the $V_i(x)$ in (1) can be found, which in turn construct our interpolator $g(x)$. If cubic spline interpolation has no boundary conditions, then the not-a-knot end conditions are commonly used. This is equivalent to the condition that the first two cubic polynomials equals in the cubic derivative, in the mean time the last two cubic polynomials also equals in the cubic derivative.

3 Reinforcement Learning and Q-Learning Algorithm

3.1 Overview of Reinforcement Learning

Reinforcement learning is to learn a mapping from situations to actions so as to maximize a scalar reward or reinforcement signal. In the standard reinforcement learning model [16], an agent is connected to its environment, as depicted in Fig.4. Reinforcement learning promise beguiles the programming agents by reward and punishment without specifying how the task is to be achieved. An agent must learn through trial-and-error interactions with a dynamic environment. At each step of interaction, the agent receives as input, i, some indication of the current state, s, of the environment, the agent then choose an action, a, to generate the output. The action changes the state of the environment, and the value of this state transition is communicated to the agent through a scalar reinforcement signal, r. The agent's behavior should choose actions that tend to increase the long-run sum of values of the reinforcement signal. It can learn to do this over time by systematic trial and error.

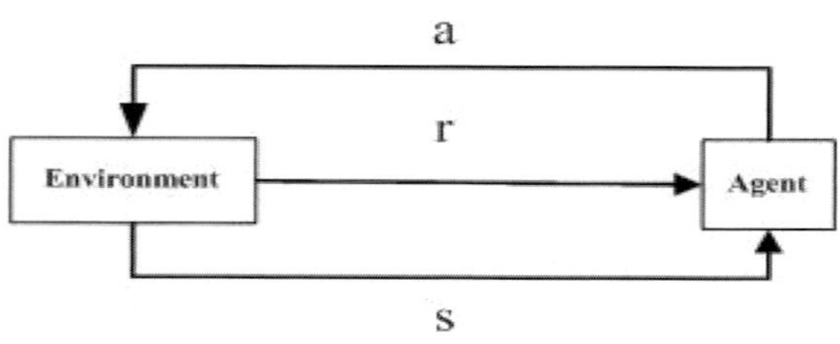

Fig. 4. The standard reinforcement learning model

3.2 Q-Learning Algorithm

A significant advance in the field of reinforcement learning is the Q-learning algorithm [16]-[17]. To define the Q-learning algorithm, a system is to be controlled as consisting of a discrete state space, S, and a finite set of actions, A, that can be taken in all states in Q-learning algorithm. A function $Q_\pi(s_t, a_t)$ of the observed state, s_t, and action, a_t, at time t for a policy π is learned whose value is the expected value of the sum of future reinforcement. Let the reinforcement resulting from applying action a_t while the system is in state s_t be $R(s_t, a_t)$. Then, the desired value of $Q_\pi(s_t, a_t)$ is

$$Q_\pi(s_t, a_t) = E_\pi \left\{ \sum_{k=0}^{T} \gamma^k R(s_{t+k}, a_{t+k}) \right\} \tag{9}$$

where γ is a discount factor between 0 and 1 that weights reinforcement received sooner more heavily than reinforcement received later.

To improve the action-selection policy and to achieve the optimal control, the dynamic programming method called value iteration can be applied. This method combines the steps of policy evaluation with policy improvement. Therefore, the agent's experience consists of a sequence of distinct stages of episodes in Q-learning. In the nth episode, the agent:

1. observes its current state x_n,
2. selects and performs an action a_n,
3. observes the subsequent state y_n,
4. receives an immediate payoff r_n,
5. adjusts its Q_{n-1} values using a learning factor a_n, according to:

$$Q_n(x, a) = \begin{cases} (1 - \alpha_n)Q_{n-1}(x, a) + \alpha_n[r_n + \gamma V_{n-1}(y_n)] \ if \ x = x_n, a = a_n \\ Q_{n-1}(x, a) \end{cases} \tag{10}$$

where

$$V_{n-1}(y) = \max_b \{Q_{n-1}(y, b)\} \tag{11}$$

is the best the agent thinks it can do from state y. The Q function implicitly defines the policy, π, defined as

$$\pi(s_t) = \arg\max_{a \in A} Q(s_t, a_t) \tag{12}$$

However, on the learning process of Q, π will not be optimal. Therefore we introduce a method to force the variety of actions for every state in order to learn sufficiently accurate Q values for the state-action pairs that are encountered. In the following experiment, a random action was taken with probability p_t for step t, where $p_{t+1} = \lambda p_t$, $p_0 = 1$, and $0 < \lambda < 1$. Thus, as the value of p_t approaches 0 with time and the policy slowly shifts from a random policy to one determined by the learned Q function.

4 Simulation Example

A schematic diagram of HVAC system [18] is shown in Fig.5. The states of the modeled system are the input and output temperatures of air and water, and the air and water flow rate, T_{ai}, T_{ao}, T_{wi}, T_{wo}, f_a, f_w. The control signal, C,

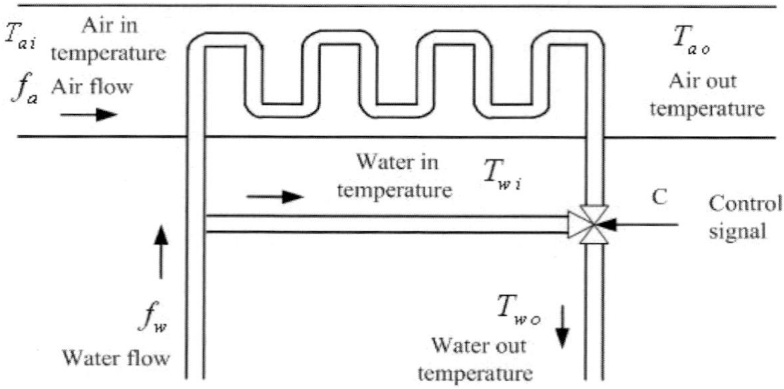

Fig. 5. Schematic diagram of heating,ventilation and air-conditioning systems

affects the water flow rate. The model is given by the following equations with constants determined by a least-square fit to data from an HVAC system :

$$\begin{aligned} f_w(t) &= 0.008 + 0.00703(-41.29 + 0.30932c(t-1) \\ &\quad - 3.2681 \times 10^{-4}c(t-1)^2 + 9.56 \times 10^{-8}c(t-1)^3) \end{aligned} \tag{13}$$

$$\begin{aligned} T_{wo}(t) &= T_{wo}(t-1) + 0.64908f_w(t-1)(T_{wi}(t-1) - T_{wo}(t-1)) \\ &\quad + (0.02319 + 0.10357f_w(t-1) + 0.02806f_a(t-1)) \\ &\quad \times (T_{ai}(t-1) - \tfrac{(T_{wi}(t-1)+T_{wo}(t-1))}{2}) \end{aligned} \tag{14}$$

$$\begin{aligned} T_{ao}(t) &= T_{ao}(t-1) + 0.19739f_a(t-1)(T_{ai}(t-1) - T_{ao}(t-1)) \\ &\quad + (0.03184 + 0.15440f_w(t-1) + 0.04468f_a(t-1)) \\ &\quad \times (\tfrac{(T_{wi}(t-1)+T_{wo}(t-1))}{2}) + 0.20569(T_{ai}(t) - T_{ai}(t-1)) \end{aligned} \tag{15}$$

The variables T_{ai}, T_{wi}, and f_a were modified by random walks to model the disturbances and changing conditions that would occur in actual heating and air conditioning systems. The bounds on the random walks were $5 \leq T_{ai} \leq 6°C$, $77 \leq T_{wi} \leq 78°C$, and $0.7 \leq f_a \leq 0.9kg/s$.

4.1 Local Control (PI Control) of HVAC System

In the first experiment, a PI controller was design as a local controller, which was used to control the HVAC system. The PI control law was

$$C'(t) = k_p e(t) + k_i \int e(t)dt$$

where $e(t)$ was the difference between the set point and the actual output air temperature at time t, k_p was the proportional gain given as 0.1, and k_i was the integral gain given as 0.03; $C'(t)$ was a normalized control value that ranged from 0 to 1 to specify flow rates from the minimum to the maximum allowed values. This normalized control signal was converted to the control signal for the model by

$$C(t) = 1400 - 730C'(t)$$

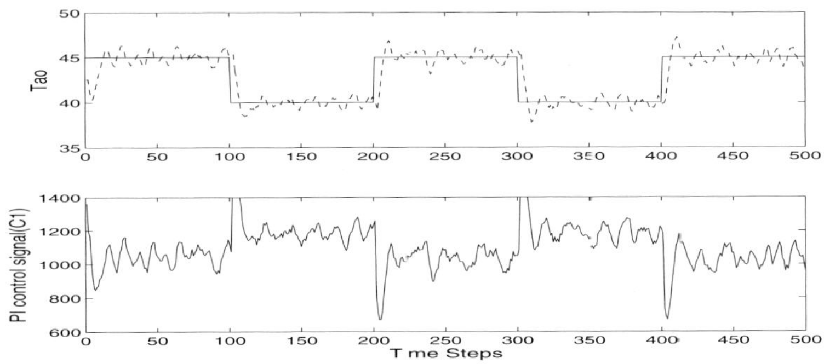

Fig. 6. Performance of PI controller

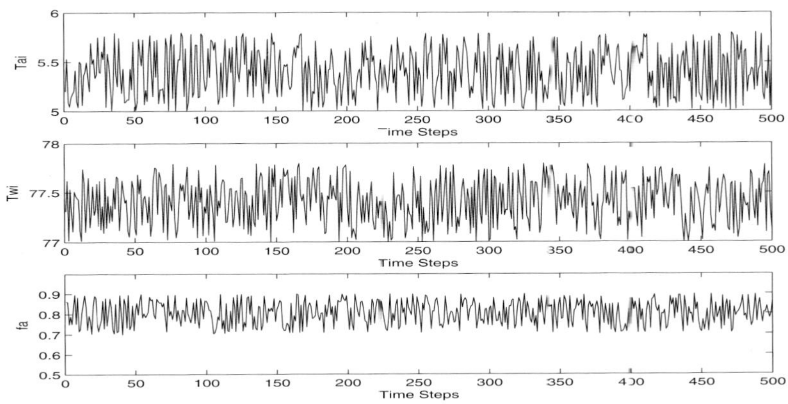

Fig. 7. Disturbances

The control signal $C(t)$ ranged from 670 for the maximal opening position to 1400 for the maximal closing position. Fig.6. showed the out temperatures of HVAC just using PI controller and the PI output control signal. The disturbance trajectories for T_{ai}, T_{wi} and f_a were illustrated in Fig.7.

4.2 Two-Layer Networked Learning Control of HVAC System

In the second experiment, except for a local controller (PI controller), a reinforcement learning agent was combined with the local controller through the second layer communication network as shown in Fig.8.

The experiment had some noticed detail as follow:

1. Shanghai University campus-wide network was used as the second layer network in the experiment. Communication between nodes used TCP/IP sockets. The loss rate of data packet using a discard-packet strategy was 8.5% in the experiment. Data packet dropouts as PI controller communicates with

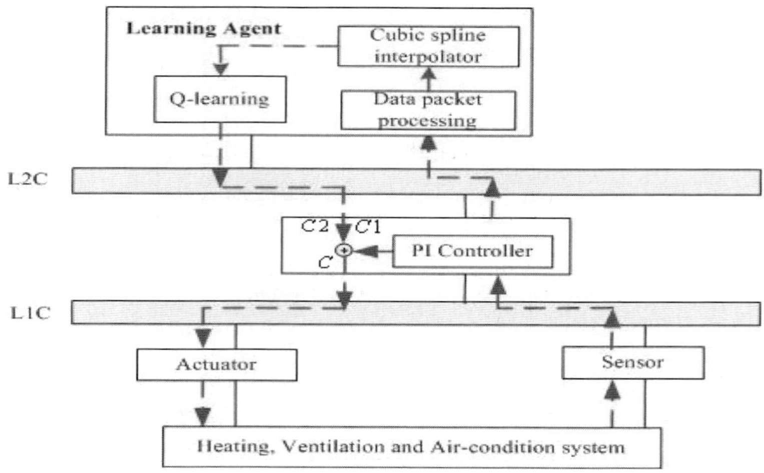

Fig. 8. The HVAC by sum of a PI Controller and Learning agent in two-layer network architecture

learning agent were compensated by cubic spline interpolator, whereas lost control signals ($C1$) were compensated by ZOH due to limited computational ability of local controller. The proportional gain (k_p), the integral gain given (k_i) and (T_{ai}, T_{wi}, f_a) were the same as experiment 1.

2. Inputs to learning agent were T_{ai}, T_{ao}, T_{wi}, T_{wo}, f_a, f_w and the set point ($T_{ao}(t)^*$). Total control signal (C) consisted of control signal of learning agent ($C1$) plus control signal of PI controller ($C2$), where control signal of learning agent ($C1$) dynamically tuned total control signal (C). The allowed output actions of learning agent was the set of discrete actions $A = \{-100, -50, -20, -10, 0, 10, 20, 50, 100\}$. The Q functions were implemented using a table look-up method. Each of the seven input variables were divided into five intervals, which quantized the 7-dimensional input space into 5^7 hypercubes. Unique Q value for the nine possible actions were stored in each hypercube.
3. Reinforcement R at time step t was determined by the squared error between the controlled variable and its set point plus a term proportional to the square of the action change form one time step to the next:

$$R(t) = (T_{ao}(t)^* - T_{ao}(t))^2 + \beta(a_t - a_{t-1})^2 \quad (16)$$

The action update term was introduced to reduce the fluctuation in the control signal to minimize the stress on the control mechanism. The values of a_t were indices of the set A, rather than actual output values.
4. The parameters a_t, γ, β ,λ were given as 0.1, 0.95, 0.15, 0.95, respectively. Learning agent was trained for 1200 episodes, each of which involved 500 steps interaction with the HVAC.

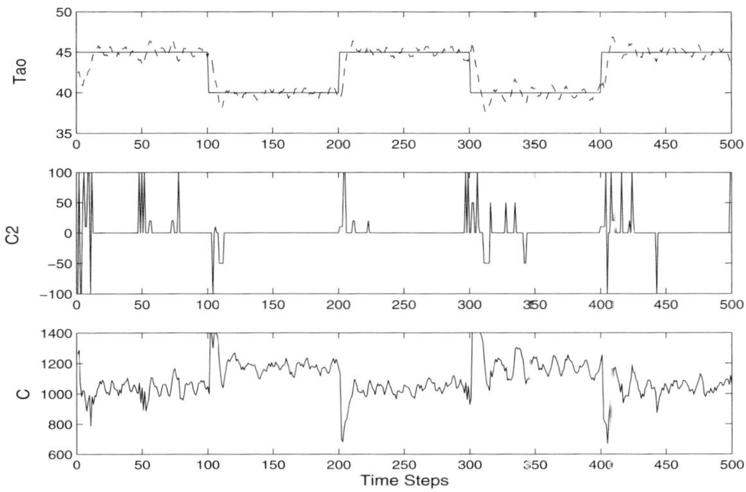

Fig. 9. Performance of combined PI Controller and Learning agent in two-layer network architecture

The response of the controlled air temperature was shown in the first graph of Fig.9. The second graph showed the output of learning agent. The output of learning agent plus PI controller were shown in the third graph. The RMS error between the set point and actual output air temperature over these 500 steps for the PI controller in the experiment 1 was 1.1896. However, two-layer networked control system with learning agent achieved an RMS error of 1.1103 in experiment 2, which indicated that learning agent was able to dynamically tune the total control signal and reduce the RMS error of the controlled temperature by about 6.66%. Comparing with one-layer networked control system with PI controller, two-layer networked control system with learning agent can improve the control performance and interference rejection ability.

4.3 Two-Layer Networked Learning Control of HVAC System Under Different Network Load

It is well known that network load is an main indicator for network-induced delay and data packet loss. In order to specifically evaluate the performance of the developed two-layer networked learning control system, several corresponding loss rates of data packets using a discard-packet strategy, 2.88%, 5.12%, 8.50%, 15.5%, were performed under L2C with different categories of network loads. The RMS error over different loss rates of data packets was shown in Table 1. According to Table 1, it was found that although data packet loss degraded the condition, learning agent was able to reduce the RMS error of the controlled temperature. It further indicated that learning agent could effectively tune the total control signal by data compensation using the cubic spline interpolator.

Table 1. RMS only using local controller and two-layer networked learning control under different loss rate of data packet

RMS only using local controller	two-layer networked learning control loss rate of data packet	RMS	RMS reduction Rate
1.2177	2.88%	1.1104	8.81%
1.2154	5.12%	1.1306	7.01%
1.1896	8.50%	1.1103	6.66%
1.2176	15.50%	1.1402	6.36%

5 Conclusions

In this paper, two-layer networked learning control system has been introduced, which uses both local controller (PI controller) and learning agent to improve the networked control performance and interference rejection capacity. To compensate the data packet loss in reinforcement learning agent, the cubic spline interpolator has been used. The experimental results under different categories of network loads have shown the effectiveness of the proposed scheme. The future work will include various learning agent design, and other compensation approaches for data pocket loss.

Acknowledgments. This work is supported by Key Project of Science & Technology Commission of Shanghai Municipality under Grant 061111008 and 06DZ22011, Program for New Century Excellent Talents in University under Grant NCET-04-0433, the Sunlight Plan Following Project of Shanghai Municipal Education Commission and Shanghai Educational Development Foundation, and Shanghai Leading Academic Disciplines under GrantT0103.

References

1. LoonTang, P., Silva, C.W.: Compensation for Transmission Delays in an Ethernet-Based Control Network Using Variable-Horizon Predictive Control. IEEE Transactions on Control Systems Technology 14(4), 707–718 (2006)
2. Monotosh, D., Ghosh, R., Goswami, B., Gupta, A., Tiwari, A.P., Balasubrmanian, R., Chandra, A.K.: Network Control System Applied to a Large Pressurized Heavy Water Reactor. IEEE Transactions on Nuclear Science 53(5), 2948–2956 (2006)
3. Wei, W.B., Pan, Y.D., Furuta, K.: Internet-based Tele-control System for Wheeled Mobile Robot. In: Proceedings of the IEEE International Conference on Mechatronics & Automation, Niagara Falls, pp. 1151–1156. IEEE Computer Society Press, Los Alamitos (2005)
4. Tanaka, K., Ohtake, H., Wang, H.O.: Shared Nonlinear Control in Internet-Based Remote Stabilization. In: The 2005 IEEE International Conference on Fuzzy Systems, pp. 714–719. IEEE Computer Society Press, Los Alamitos (2005)
5. Wang, Y., Hu, W.L., Fan, W.H.: Fuzzy Robust H_∞ Control for Networked Control Systems. In: Proceedings of the 6th World Congress on Intelligent Control and Automation, pp. 4448–4452 (2006)

6. Pan, Y.J., Marquez, H.J., Chen, T.: Sampled-data Iterative Learning Control for a Class of Nonlinear Networked Control Systems. In: Proceeding of the 2006 American Control Conference Minneapolis, pp. 3494–3499 (2006)
7. Tipsuwan, Y., Chow, M.Y.: Neural Network Middleware for Model Predictive Path Tracking of Networked Mobile Robot over IP Network. In: IECON Proceedings, pp. 1419–1424 (2003)
8. Feng X., L, S.b., S, Y.x.: Neural Network Based Feedback Scheduler for Networked Control System with Flexible Workload. In: Proc. of ICNCOS (2005)
9. Li, Z., Fang, H.J.: A novel controller design and evaluation for networked control systems with time-variant delays. Journal of the Franklin Institute, 161–167 (2006)
10. Yang, T.C.: Networked control system: a brief survey. IEE Proc. Control Theory Appl. 153(4), 403–412 (2006)
11. Yang, T.C., Yu, H., Fei, M.R., Li, L.X.: Networked Control Systems: A Historical Review and Current Research Topics. Measurement + Control 38(1) (2005)
12. Li, Y.Q., Fang, H.J.: Control Methodologies of Large Delays in Networked Control Systems. In: 2005 International Conference on Control and Automation (ICCA 2005), pp. 1225–1230 (2005)
13. Nilsson, J.: Real-time control systems with delays. Ph.D. thesis Department of Automatic Control, Lund Institute of Technology, Lund, Sweden
14. Lei, Z., Dimitrios, H.V.: Communication and control co-design for networked control systems. Automatica 42, 953–958 (2006)
15. Dyer, S.A., He, X.: Cubic-spline interpolation: Part2. IEEE Instrumentation and Measurement Magazine 4(2), 34–36 (2001)
16. Leslie, K.P., Michael, L.L., Andrew, W.M.: Reinforcement learning: A survey. Journal of Artificial Intelligence Research 4, 237–285 (1996)
17. Walkins, C.J.C.H.: Technical Note, Q-learning. Machine learnig 8, 279–292 (1992)
18. Tashtoush, B., Molhim, M., Al-Rousan, M.: Dynamic model of an HVAC system for control analysis. Energy 30, 1729–1745 (2005)

In Silico Drug Action Estimation from Cardiac Action Potentials by Model Fitting in a Sampled Parameter Space

Jianyin Lu[1], Keichi Asakura[2], Akira Amano[3], and Tetsuya Matsuda[3]

[1] Cell/Biodynamics Simulation Project, Kyoto University, Japan
[2] Research Laboratories, Nippon Shinyaku Co. Ltd., Japan
[3] Graduate School of Informatics, Kyoto University, Kyoto

Abstract. Model-based predictive approaches have been receiving increasing attention as a valuable tool to reduce high development cost in drug discovery. Recently developed cardiac cell models are integrated with major ion channels and capable of reproducing action potentials (AP) precisely. In this work, a model-fitting-based approach for estimating drug action from cardiac AP recordings is investigated. Giving a test AP, the activities of involved ion channels can be determined by fitting the cell model to reproduce the test AP. Using experimental APs recordings both before and after drug dose, drug actions can be estimated by changes in activity of corresponding ion channels. Local-gradient-based optimization methods are too time-consuming due to the high computational cost of cardiac cell models. A fast approach using only pre-calculated samples to improve computational efficiency is addressed. The searching strategy in the sampled space is divided into two steps: in the first step, the sample of best similarity comparing with the test AP is selected; then response surface approximation using the neighboring samples is followed and the estimation value is obtained by the approximated surface. This approach showed quite good estimation accuracy for a large number of simulation tests. Also results for animal AP recordings from drug dose trials were exemplified in which case the ICaL inhibition effect of Nifedipine and IKr inhibition effect of E4031 were correctly discovered.

1 Introduction

Supported by recent progress in system biology research, model-based predictive approaches have been receiving increasing attention as a valuable tool to reduce high development cost in drug discovery [1,2]. Early stage drug action/risk factor evaluation is among many potential applications that model-based predictive approach can contribute. While the cardiac electrograms are widely used in late phase of drug discovery for verification, Action potentials (APs) are used in relatively early stage for inspection of risk factors such as QT prolongation. Dynamics of cardiac AP in different phases of membrane excitation is mainly contributed by different ion currents. Therefor drugs acting on different ion channels

K. Li et al. (Eds.): LSMS 2007, LNCS 4688, pp. 710–718, 2007.

tend to cause different changes on the shape of AP. An experienced inspector is able to make qualitative judgments of the drug effect/risk based on such shape changes. On the other hand, AP reconstruction has been a basic research target for cardiac cell modeling from the beginning [3]. Recently developed cardiac cell models [4,5,6] are integrated with most main ion channels and capable of reproducing AP waves very precisely. Giving a test AP, it is possible for a model-based approach to inversely estimate the activities of involving ion channels using some optimization techniques. Using experimental APs recordings both before and after drug dose, drug actions can be estimated by changes in activity of involving ion channels. This approach is very useful for early stage drug decision making not only because it provides a more quantitative answer for drug action and risk evaluation but also in the great potential hiding in the obtained cell models. Using the estimated models, cellular condition after drug dose can be simulated completely in silico. It is also possible to evaluate the heart function by applying the cell model to a whole heart contraction model as shown in [8].

General optimization approaches rely greatly on local gradient to gradually approach to the best answer. As a result, a large number of model calculations are always necessary. However the computational cost of a comprehensive cell models is generally too high for a typical gradient-based searching approach to perform efficiently. In this work, a fast method which utilizes only a pre-calculated sample set of the parameter space is investigated. The searching strategy in the sampled space is divided into two steps: in the first step, the sample of highest similarity among the training set with test one is selected; then a curved surface is fitted in the neighbors of the matched sample and the optimized answer is calculated according to the expression of surface. The efficiency of this sample based approach depends on the selected sample set. An iterative boosting algorithm for adaptive sampling which appends failed tests to the sample set is presented.

2 Method

In this section, the proposed fast optimization approach utilizes only pre-calculated training samples is addressed. A brief description for cardiac cell model and the properties of cardiac APs is given first. Then the optimization strategy and boosting algorithm are discussed.

2.1 Cardiac Cell Model and Its APs

Cardiac action potentials are known to be effected by many interactions between involved ion currents during the membrane excitation process. When a cardiac cell is induced by a stimulus current over a certain threshold level, the opening of voltage-gated ion channels causes the positively charged ions to move out of the cell, i.e. the cell shifts from the resting state to depolarization stage. The opening of voltage-gated calcium channels in the depolarization stage induces release of Ca^{2+} from the t-tubules. The influx of Ca^{2+} further causes calcium-induces calcium release from the sarcoplasmic reticulum, and such free Ca^{2+} causes muscle to contract. After a certain period of delay, potassium channel

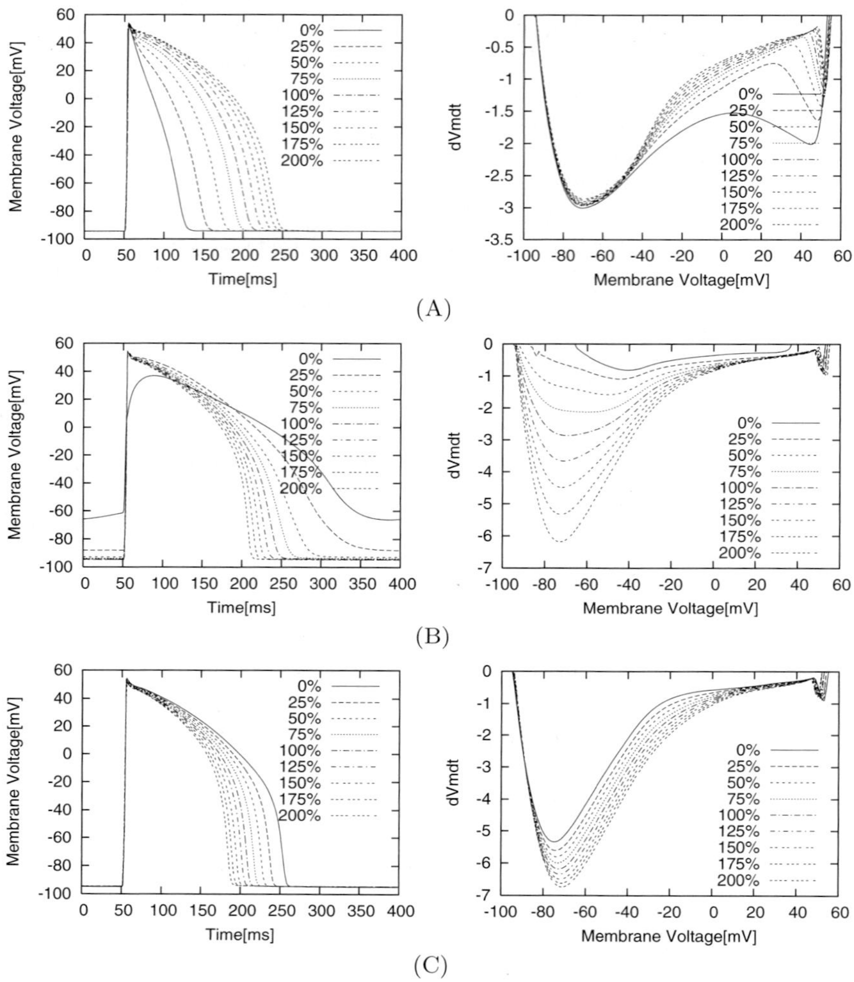

Fig. 1. Deformation of AP (left) and dVmdt (right) with respect to different channel activities of KM: (A) ICaL, (B) IK1 , and (C) IKr

reopens and the resulting flow of K^+ out of the cell (repolarization) finally causes the cell to return to its resting state.

The contribution of major ion currents such as the calcium channel and potassium channel in cardiac membrane excitation has been intensively studied. Cardiac cell models developed recently are capable of integrating all major channels and reproducing APs very precisely. In this work, Kyoto Model (KM) [6,7] becomes our choice because of its accuracy as well as its ability to simulate mechanical contraction. KM is a comprehensive ventricular cell model for guinea pig and its major currents affecting cellular repolarization stage are the L-type calcium channel (ICaL), the inward rectifier current (IK1), and the rapidly activating potassium channel (IKr). In Fig. 1 the deformation of AP (Vm) and its differential tranjectory (dVmdt) with respect to different channel activities are illustrated. The ranges of channel activity parameter are from 0 to 200 percent of the initial steady state value of KM. Note that changes in shape of dVmdt are more distinguishable than that of AP. For the complete mathematical expressions and other model details of KM, refer to the original paper.

The computational cost of a comprehensive cardiac cell model is general too high for an ordinary gradient-based optimization approach to accomplish efficiently. In the case of KM, simulation of the cellular state after drug dose over 5 minutes takes roughly 3 and half minutes on a PentiumIV 3.0GHz intel machine with 2.0G bytes of memory. Optimization approaches that need thousands calculation of model will take weeks. We discuss a quite simple but fast optimization strategy in a sampled parameter space.

2.2 Optimization Strategy in a Sampled Parameter Space

The problem of model fitting for AP can be defined as: giving an unknown test AP u_i and a cardiac model M, find the ion channel parameters $(ICaL, IK1, IKr)$ of M that bestly reproduces u_i. The optimization strategy (Fig. 2) using a sample set $s_i(i = 1, \ldots, N)$ and a similarity evaluation function $Sim(p_1, p_2)$ is as below:

1. Find sample s_i in the sample set that of highest similarity with the test AP u_i.
2. For samples in the neighbor of s_i, take the three channel parameters as input and the similarity $Sim(s_i, u_i)$ as output and approximate this relation using a surface of second order. The optimized answer can be obtained using the appoximated surface.

Step 2 is actually the response surface method (RSM) [9] for finding the local extreme which can be used here totally without new model calculations if only the neighbouring samples are used.

The similarity evaluation function is a weighted sum of the normalized correlation of AP and dVmdt waves

$$Sim(p_1, p_2) = w_{vm} Corr(p_1, p_2) + w_{dvm} Corr(d_1, d_2),$$

where d_i is the corresponding dVmdt wave of p_i, $Corr(p_1, p_2)$ is the normalized correlation of p_1, p_2, and the value of weight coefficients are $w_{vm} = 0.25, w_{dvm} = 0.75$.

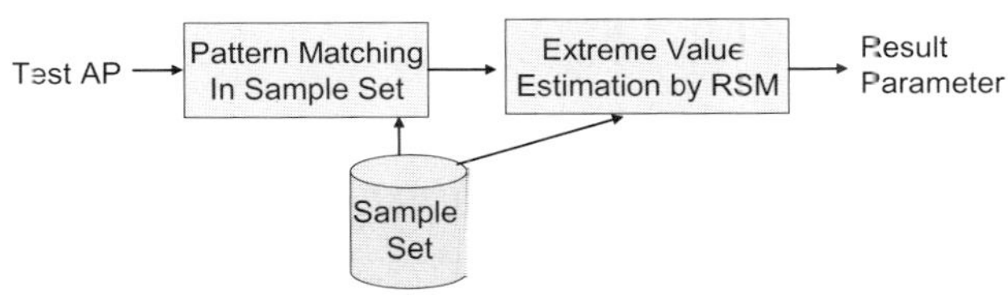

Fig. 2. Optimization strategy in a sampled parameter space

2.3 Adaptive Sampling by Boosting

Since the proposed optimization approach uses only pre-calculated sample set, its efficiency depends heavily on the sample set used. Generally the local property of the considering parameter space has to be thoroughly studied to perform an adaptive sampling, which is of formidable computational load. A noval idea from statistical learning theory is to use an iterative boosting technique for adaptive sampling. The following iterative boosting process is used:

1. Collect the initial training sample set equally spaced for each parameter.
2. Using the training sample set to solve a random test AP. If the error of estimated parameters is over a threshold then the test random sample is appended to the training sample set.
3. Test resultant sample set using an independent random test set. Terminate the boosting process if the result accuracy is good enough or if a number of iterative steps are reached. Otherwise go back to step 2.

3 Results

3.1 Sample Set and Similarity Distribution of the Parameter Space

Ranges for activity value of channel ICaL,IK1 and IKr are from 0 to 200 percent of its initial steady state value of KM. The initial sample set equally divides the ranger of each parameter into 32 regions. As a result a sample set of number $33 \times 33 \times 33 = 35937$ is created. AP cycle are 400ms for gunie pig and the range of dVmdt signal is from -100mV to 60mV. Dimension of AP and dVmdt signals are 400 and 160 respectively. It takes nearly 60 hours to create the sample set on a IBM P690 machine with 30 CPUs. In the boosting process follows, 100000

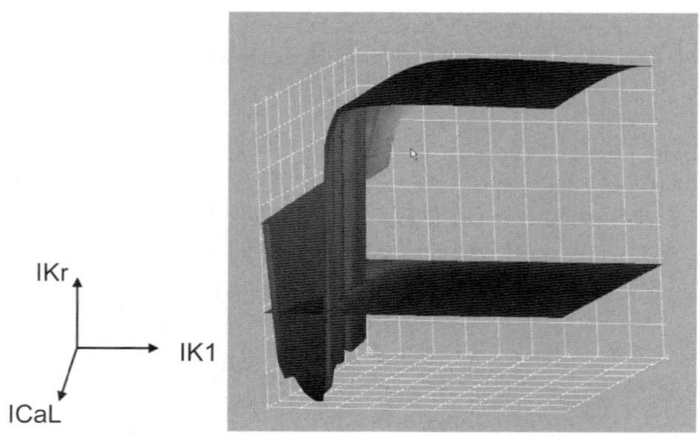

Fig. 3. Cross-section view of distribution of similarity of each sample in train set with the steady state AP of KM: The lower flat plane indicates the position of the cross section, and the height/color of the upper surface stands for similarity values. (Graph axes: ICaL is from back to front, IK1 from left to right, and IKr from bottom to top).

number of random samples are used, and tests result in total estimation error of three channel parameters over 0.05 are appended. This boosting process ends up with nearly 5000 samples being added to the initial sample set.

The efficiency of a sub-sampling approach depends largely on basic properties of the parameter space. Using the steady state parameter of KM (the middle point in the figure) as a reference, the distribution of similarity between the reference and sample set is illustrated in Fig. 3. The lower flat plane indicates the position of the cross section. The height/color of the upper surface stands for similarity of the cross section. Though only one cross-section view with fixed IKr is shown, the trend of the distribution is similar across the whole range. The cardiac APs are observed to deform slowly in the most region(region of red) which is desirable for a sub-sampling optimization approach. Problem exists in small region with IK1<15% (region of blue) where the shape of AP changes rapidly. Actually these are the region of KM where abnormal automaticity occurs. Since such parameters are obvious not acceptable in practical drug development, they are simply discard.

3.2 Results for Random Simulation Tests

The efficiency of the proposed optimization approach is tested firstly using simulated APs. 36000 APs with random channel values are prepared and tested. Besides the time spend in preparing the sample set, the processing time for one test is about 1 seconds, and it takes almost 10 hours to finished the total test.

Shown in Tab. 1 is the result estimation error, which is the difference between real channel activity parameters and estimated ones, and the fraction of error distribution. The error values in the table are normalized with respect to the whole 0-200% range being 1. The result is general good (below 0.02) in terms of mean error. As for individual channels, IK1 is much better behaved and the best accuracy was obtained. For ICaL and IKr, though the fraction is very low, i.e. nearly 0.5% for error greater than 0.05, some difficult tests remain with estimation error up to 0.08. It is also observed that such sample of failure distributes mainly in the areas with both very high ICal and IKr activity near 200%, where the AP is much less sensitive to the change of activity of channels. Comparing the result without boosting process, the mean errors almost remain unchanged but the worst error for ICaL and IKr both increased up to 0.11.

3.3 Results for Drug Dose Experimental AP Recordings

In this experiment, AP recordings of isolated guinea-pig papillaly muscles are tested. The AP assays are originally used for inspection of drug-induced QT

Table 1. Result for Random Simulation Tests

	Mean Error	Worst Error	Error$\geq$ 0.05
ICaL	0.013	0.07	0.55%
IK1	0.009	0.03	0%
IKr	0.017	0.08	0.61%

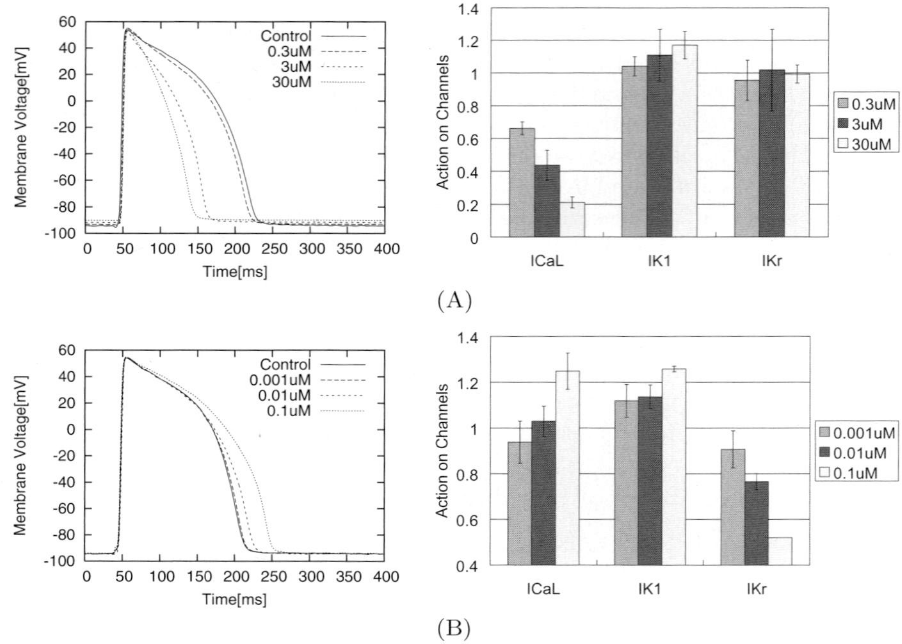

Fig. 4. Drug dose experiment: action potentials and estimated drug actions: (A) Nifedipine (ICaL inhibitor) (B) E4031 (IKr inhibitor)

interval prolongation [10]. Generally the papillary muscles are firstly prepared in a tyrode solution, then during the perfusion of a test drug compond the stimulation frequency is kept at the base level of 1.0Hz for 30 minutes, finally the stimulation frequency is inceased to 2.5Hz and APs are recorded after the stabilizing period. Two compounds affects the ion channels under consideration are examined: Nifedipine, a pure ICaL inhibitor; and E4031, an IKr inhibitor. Dependening on the recording devices used, the original AP recordings can be quite noisy and of different sampling rates comparing with simulation. Preprocesses like denoising and resampling are neccesary for such signals.

The AP recordings of control condition and several drug doses as well as estimated channel activities for the two drugs are shown in Fig. 4. Whenever multiple APs recordings are available for each compound and dose concentration, both the average value and the standard deviation are depicted. As shown in Fig. 4(A), the ICaL inhibition effect of Nifedipine was discovered as 66%,43%,21% for three levels of dose ($0.3\mu M, 3\mu M, 30\mu M$) respectively, while parameter of unaffected channels remain almost unchanged at 1.0. These values are near the experemental data from guinea pig ventricular myocytes in [11] where the concentration for 50% inhibition (IC_{50}) is $0.3\mu M$. In Fig. 4(B), the IKr inhibition effect of E4031 was discovered as 90%,76%, and 52% for $0.001\mu M$, $0.01\mu M$, and $0.1\mu M$ dose respectively. These values are also near the experimental results from cultured cardiomyocytes in [12] where the IC_{50} values are $0.045\mu M$.

3.4 Discussion

The proposed model-based approach for estimating drug action from cardiac AP recordings was shown to be effective. Using only pre-calculated sample set, this approach was very fast and achieved quite good estimation accuracy generally. The limitation was at the area with very high channel activity near 200% of ICaL and IKr where estimation error increased. This is relatively not a critical issue because drugs increase both ICaL and IKr activity by times are very unusual. The iterative boosting method for adaptive sampling is very time consuming, but does improve estimation accuracy. More sophisticated algorithms are possible with a complete investigation of the parameter space of KM.

For verification using actual drug dose experiment APs, two cases (Nifedipine and E4031) from the QT-PRODACT data set were used. Estimation result above showed a 10%-15% difference with that from reference literature. The cultured tissues are generally considered more sensitive to drug dose than isolated in vivo ones. Also the difference in detailed experimental protocol and measurement devices may contribute in the result of the estimation. For early stage drug action estimation, such result should still be quite useful in practice.

4 Conclusion

In this work, a model-based fast optimization approach using only a pre-calculated sample set is proposed to estimate drug actions from cardiac AP recordings. The approach is extremely fast and achieved quite good accuracy using a set of 40000 samples. The inhibitation effects of experimental AP after Nifedipine and E4031 dosing were also verified. Using the cardiac model estimated, kinds of the effect of the drug can be examined by simulation. Some attempts have been done on using the models to evaluate proarrhythmic effect of drugs [13]. Potential future efforts include a thorough verification using a large number of drug tests and then some new ideas for pharmacological applications will be investigated.

Acknowledgments. This work was supported by the Leading Project for Biosimulation Project and Grant-in-Aid for Scientific Research (C) No.16500186, MEXT. The authors also want to thank the pharmaceutical companies participating in this project for kindly providing the experimental action potential data.

References

1. Michelson, S.: Assessing the impact of predictive biosimulation on drug discovery and development. J. Bioinformatics and Computational Biology 1, 169–177 (2003)
2. Zhang, L., Sinha, V., Forgue, S.T., Callies, S., Ni, L., Peck, R., Allerheiligen, S.R.B.: Model-Based Drug Development: The Road to Quantitative Pharmacology. J. Pharmacokinetics and Pharmacodynamics 33, 1567–1573 (2006)

3. Beeler, G.W., Reuter, H.: Reconstruction of the action potential of ventricular myocardial fibres. J. Physiol. 268, 177–210 (1977)
4. Luo, C.H., Rudy, Y.: A dynamic model of the cardiac ventricular action potential I. Simulation of ionic currents and concentration changes. Circ Res. 74, 1071–1096 (1994)
5. Faber, G.M., Rudy, Y.: Rudy: Action Potential and contractility Changes in [Na+]i; Overloaded Cardiac Myocytes: A Simulation Study. Biophys. J. 78, 2392–2404 (2000)
6. Matsuoka, S., Sarai, N., Kuratomi, S., Ono, K., Noma, A.: Role of individual ionic current systems in ventricular cells hypothesized by a model study. Jpn. J. Physiol. 53, 105–123 (2003)
7. Sarai, N., Matsuoka, S., Noma, A.: simBio: a Java package for the development of detailed cell models. Prog. in Bioph. and Mol. Biol. 90, 360–377 (2006)
8. Smith, N., Mulquiney, P.J., Nash, M.P., Bradley, C.P., Nickerson, D., Hunter, P.J.: Mathematical modelling of the heart: Cell to organ. Chaos Solitons Fractals 13, 1613–1621 (2002)
9. Myers, R., Montgomery, D.: Response Surface Methodology: Process and Product Optimization Using Designed Experiments. John Wiley Sons. Inc., Chichester (1995)
10. Yamazaki, R., Yamada, M., Kobayashi, K., Kitani, S., Shimada, C., Shimosato, T., Mori, T., Suganami, H., Yamamoto, K.: QT PRODACT: Inter- and Intra-facility Variability of the Action Potential Assay Using Isolated Guinea-Pig Papillary Muscles. J. of Phar. Sci. 99, 439–447 (2005)
11. Shen, J., Jiang, B., Pappano, A.J.: Comparison of L-Tyle Calcium Channel Blockage by Nifedipine and/or Cadmium in Guinea Pig Ventricular Myocytes. J. Pharmacol Exp Ther. 294, 562–570 (2000)
12. hERG channel currents measured with the IonWorks HT system: `http://www.nihonmdc.com/pages/instruments/electro/ionworks_herg.html`
13. Matsuda, H., Matsuoka, S., Sarai, N., Noma, A.: Proarrhythmic mechanisms of HERG channel block: A Simulation Study. In: The 84th Annu. Meeting of the Physi. Soc. of Jpn. (March 2007)

Intelligent Knowledge Capsule Design for the Multi Functional Aspects

JeongYon Shim

Division of General Studies, Computer Science, Kangnam University
San 6-2, Kugal-ri, Kihung-up,YongIn Si, KyeongKi Do, Korea
Tel.: +82 31 2803 736
mariashim@kangnam.ac.kr

Abstract. Every Intelligent technique has its strong and weak point and its appropriate application area is different. For making more intelligent system, sometimes it needs to assemble the different technology. On this purpose, Intelligent Knowledge Capsule assembling connective learning of neural network and logical conceptual learning was developed for multiple aspects in this paper. Focusing on the structural memory and knowledge retrieval We apply this mechanism to virtual memory and test the retrieving process.

1 Introduction

It is known that the brain is an assembling nervous system structured by multi-area taking part in the different functions and it works cooperatively. Because human being faces on the various situation and should solve the problem from the external, he has an assembling brain system performing the multiple functions. This system is very appropriate for surviving in the dynamic complex environment.

As the paradigm of modern society is changing to computer based system more and more, the complexity of information society is getting higher. In the dynamic huge amount of data base environment, the design of more intelligent system assembling the multiple functions is essential. For these several decades, many studies adopting the intelligent mechanisms of a human brain have been made and various systems have been implemented.

As one of these studies, we design Intelligent Knowledge capsule assembling connective learning of Neural network and logical conceptual learning. The connective learning methodology has a strong point that it is successfully applied to the area of perception or pattern recognition in the dynamic circumstance because it has a good generalization ability. But this technology has a week point of bad explanation. On the other side, logical conceptual learning is good for symbol based logical processing and explanation ability, while it has a bad generation. In this study, combining two methodologies we designed the structured intelligent system for efficient learning and memory. It can be a basic frame of intelligent system for the multiple purposes and aspects. In this paper, the structure of intelligent system for learning and memory is designed mainly and focusing on knowledge retrieval process it was tested with sample data of virtual memory.

K. Li et al. Eds.): LSMS 2007, LNCS 4688, pp. 719–725, 2007.

2 Intelligent Knowledge Capsule Design

2.1 The Structure of Intelligent Knowledge Capsule

As shown in figure 1, there is a data passing path in the centering part vertically. The incoming and out-coming processes of data are occurred through this central vertical part including I/O interface, sensing, controlling. temporal memorizing space. Memory is divided into connective learning part, logical conceptual learning part, index for connection and meta knowledge. Each part of memory communicates with each other and cooperates for performing multiple processes. In this design, Back Propagation mechanism takes part in the connective learning part and Knowledge network consisting of conceptual knowledge cells and associative relations take part in logical conceptual learning part[1]. The data processing in this model is as follows: When the data come into I/O interface, it is filtered by controlling mode and filtering mechanism in sensory module. Then according to the control signal in control point this data is propagated to the particular memory part. In the case of connective learning process, the data is passed to BP part and in the case of symbolic processing the data is passed to logical conceptual learning part. During the process of Inference and decision making, these learning memory parts are connected and communicate through index line. In this point Index takes charge of a bus line of data passing for searching,connection or associative correlation. The combination of connective learning part and logical conceptual learning part is performed in this location.

2.2 Control Modes in Control Point

Control point offers the selection plate with several control button similar to remote control pad of TV as depicted in figure. The incoming data or queries are controlled by the control mode signal.

The detailed description about learning process by BP algorithm, memory structuring are referred in the precious papers[1][2]. In this study, we focus on the the configuration of knowledge network in memory and knowledge retrieving process based on this structure.

2.3 Knowledge Network in Memory

Knowledge network in memory consists of nodes and arcs. one node have knowledge and connected to other nodes with an arc of associative relation.

Association are formed in the classical paradigm. System have a bias for associating stimuli that are likely to be related in the environment. The system is also sensitive to the likelihood of an association between the two stimuli(associative bias). It is might be characterized as forming a statistical inference. Bayesian statistical models for inference about probabilistic relationships consider not only the data but also prior beliefs about what relationships are likely. The fundamental relationship in Bayesian statistics is

$$PosteriorBelief = PriorBelief * Evidence$$

Table 1. Associative list

node1	Relation	strength	node2
C_1	IS-A	1.0	C_2
C_1	IS-A	1.0	C_3
C_2	MADE-OF	0.9	C_3
C_3	AFFECT	0.7	C_4
C_4	SELF	0.0	NULL
C_3	AFFECT	0.3	C_5
C_5	SELF	0.0	*NULL*

That is,

$$P(h|e) = \frac{P(e|h)P(h)}{p(e)}$$

Associative strength,R_{ij}, between C_i and C_j in the knowledge graph is represented by Bayesian Posterior belief and calculated by equation (2).

$$R_{ij} = P(a_i|a_j)D \quad (1)$$

where D is the direction arrow, $D = 1 or - 1$, $i = 1, \ldots, n$, $j = 1, \ldots, n$.

The relation is characterized by the associative strength of relation which is not fixed but changed by incoming evidence.

Knowledge Network is transformed to Associative Knowledge list of Table 1 for the efficient memory management.

3 Multiple Aspect Knowledge Retrieval from the Combined Structure of Memory

3.1 Knowledge Extraction by Combining Multiple Sections

This system provides two main passing ways for extracting the related knowledge cells in memory : first way is the path from connective learning memory part to logical concept learning part and second way is the path from logical concept learning part to connective learning memory part. Two learning parts can be used for knowledge retrieving process after the learning process is finished because the learned connected network contains useful knowledge.

We represents two knowledge retrieving algorithms of two directions as follows.

Algorithm 1 : Knowledge Retrieval Algorithm in the First Direction

```
STEP 1: Input input values for perception

perception in connective learning memory:
STEP 2: intermediate output values in hidden layer
STEP 3 : get the output value in output layer

STEP 4: interpret the result and add the perception result to extracting list.
STEP 5 : extract the keyword

Index search and move:
STEP 6: Search the corresponding keyword in Index
STEP 7: If (not found)
    Print ("There is no corresponding keyword!")
    Else (found)
    associative-knowledge-extraction in logical conceptual learning memory

associative-knowledge-extraction:
    next=associative-list;
    i = found-i;
    while( (next.node2[i]!= null) and (i !=N))
    { Put next.node1[i] to F-list;
    Put next.R[i] to F-list;
    Put next.node2[i] to F-list ;
    i=i+1; }


STEP 8: Stop.
```

Algorithm 2 : Knowledge Retrieval Algorithm in the Second Direction

```
STEP 1: Input the Keyword.
STEP 2: Search the corresponding keyword in Index
STEP 3: If (not found)
    Print ("There is no corresponding keyword!")
    Else (found)
    associative-knowledge-extraction in logical conceptual learning memory

associative-knowledge-extraction:
    next=associative-list;
    i = found-i;
    while( (next.node2[i]!= null) and (i !=N))
    { Put next.node1[i] to F-list;
```

```
    Put next.R[i] to F-list;
    Put next.node2[i] to F-list ;
    i=i+1; }

Index search and move:
STEP 4: If (not found)
    Print ("There is no corresponding keyword!")
    Else (found)
    Then goto connection learning memory.

perception in connective learning memory:
STEP 5: back propagate the data from output layer to input layer
STEP 6 : extract the nodes in Input layer.
STEP 7: add the extracted data to extracting list.
STEP 8: Stop.
```

4 Experiments

In this experiment, because we are focusing on the knowledge retrieval process, the knowledge structure of memory is supposed to be fixed by finishing the learning process. Figure 1 shows the structure of connective learning memory and logical conceptual learning memory. The knowledge nodes are not specified but represented as symbols.

Figure 2 represents the retrieved knowledge list of algorithm1. This process starts from the perception in connective learning memory as propagating the input data. Then the keyword is selected and searching process is performed in INDEX. If the keyword is found, another extracting process in logical conceptual learning memory. Figure 3 also shows the extracted results by algorithm2. Two ways of knowledge retrieving process produced the output successfully.

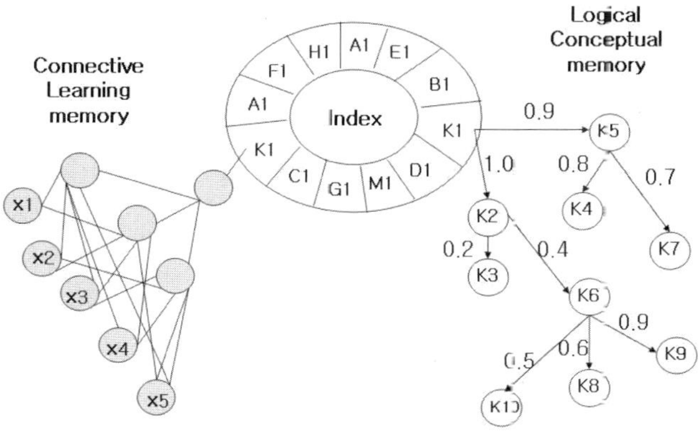

Fig. 1. The structure of knowledge network

```
Knowledge retrieval test
1: starting from connective learning memory
2: starting point logical conceptual memory
3: quit
... select a key(1/2/3)?1

knowledge retrieving...
x1 x2 x3 x4 x5 k1
searching the keyword in Index...
found...
knowledge retrieving...
k1 1.0 k2 0.2 k3 0.0 null
k1 1.0 k2 0.4 k6 0.5 k9 0.0 null
k1 1.0 k2 0.4 k6 0.6 k8 0.0 null
k1 1.0 k2 0.4 k6 0.5 k10 0.0 null
k1 0.9 k5 0.8 k4 0.0 null
k1 0.9 k5 0.7 k7 0.0 null
```

Fig. 2. The retrieved knowledge algorithm1

```
Knowledge retrieval test
1: starting from connective learning memory
2: starting point logical conceptual memory
3: quit
... select a key(1/2/3)?2

knowledge retrieving...
k1 1.0 k2 0.2 k3 0.0 null
k1 1.0 k2 0.4 k6 0.5 k9 0.0 null
k1 1.0 k2 0.4 k6 0.6 k8 0.0 null
k1 1.0 k2 0.4 k6 0.5 k10 0.0 null
k1 0.9 k5 0.8 k4 0.0 null
k1 0.9 k5 0.7 k7 0.0 null
searching the keyword in Index...
found...
knowledge retrieving...
k1 x1 x2 x3 x4 x5
```

Fig. 3. The retrieved knowledge extracted by algorithm2

5 Conclusions

We proposed knowledge capsule which has a hierarchical structure and many functions of selection, learning,storing and data extraction. We tested associative knowledge retrieval function with virtual memory. As a result of experiments, we could find this system has a flexible memory and can produce the extracted data according to two way retrieving process. This function is applicable to build the flexible memory in intelligent brain system.

References

1. Shim, J.-Y.: Knowledge Retrieval Using Bayesian Associative Relation in the Three Dimensional Modular System. In: Yu, J.X., Lin, X., Lu, H., Zhang, Y. (eds.) APWeb 2004. LNCS, vol. 3007, pp. 630–635. Springer, Heidelberg (2004)
2. Anderson, J.R.: Learning and Memory. Prentice Hall, Englewood Cliffs
3. Fausett, L.: Fundamentals of Neural Networks. Prentice Hall. Englewood Cliffs
4. Haykin, S.: Neural Networks. Prentice Hall. Englewood Cliffs
5. Shim, J.-Y., Hwang, C.-S.: Data Extraction from Associative Matrix based on Selective learning system. In: IJCNN '99, Washongton, DC (1999)
6. Anderson, J.R.: Learning and Memory. Prentice Hall. Englewood Cliffs

Robust Orientation Diffusion Via PCA Method and Application to Image Super-Resolution Reconstruction

Liang Xiao, Zhihui Wei, and Huizhong Wu

School of Computer Science and Technology, Nanjing University of Science and Technology, Nanjing 210094, China
xtxiaoliang@163.com

Abstract. An efficient method with systems of nonlinear partial difference equations (PDE) for image super-resolution was presented, in which geometric edge orientation estimated by the principle component analysis (PCA) method. Using the local and contextual discontinuity of image pixel's neighborhood, the robust orientation diffusion processes were exploited to enhance the sharpness of edges without incurring any additional artifacts. We also discuss the connection between the new PDE framework and the classical variational PDE method. Furthermore, the implementation scheme of the proposed approach was given by a novel finite difference method. Simulation results show that the proposed approach exhibited better performance than the standard linear interpolation and classical PDE approaches, particularly in the edge regions.

1 Introduction

Super-Resolution (SR) is the process of reconstructing a higher resolution image from Low-Resolution(LR) input observations. These LR observations are acquired by either multiple sensors or by a single sensor imaging the scene over a period of time[1] . No matter the source of observation, the critical requirement for SR is that the observations contain different but related views of the scene. This problem of super-resolution reconstruction has been addressed using many different methods [1-3] including iterative back projection (IBP)method, projection of convex sets(POCS) method, Maximum like-hood (ML)/ maximum a posteriori estimation (MAP)method, and hybrid ML/ MAP/POCS methods. In mathematical points of view, this problem can be considered as an inverse problem. Hence a class of variational PDE (Partial Difference Equations) methods for this problem has been drawn many researchers' attention [4-6]. However, differing from the de-noising problem, the class of typical variational PDE methods cannot effectively reduce the disturbing artifacts on edge regions, e.g. blocking artifacts, ringing artifacts along SR image edges, thus can not reconstruct a good HR image with clear edges. In general, owing to over-locality of the class of typical variational PDE methods, they often use the pixel's spatial gradient as the discontinuity measure to provide a nonlinear diffusion process. However, when the nature images are degraded by motion warping and an incorrect lens adjustment, down sampling and noise addition, etc., then the gradient vectors of LR image are also degraded. Hence it is not suitable to choose the

K. Li et al. (Eds.): LSMS 2007, LNCS 4688, pp. 726 – 735, 2007.

pixel-wise gradient as the discontinuity measure. As we know, the edges in nature images are always scale-dependence, hence the nonlinear diffusion process must consider more contexture and robust discontinuity measure [7].

Inspired by such ideas, we propose a PCA (Principle Component Analysis) method to estimate the true edge orientation, which is robust to noise and down sampling effect, etc. Then we propose a robust orientation diffusion with systems of nonlinear partial difference equations, which is argued to provide a theoretical foundation for the class of typical variational PDE in image de-noising to be modified to give good solution for image super-resolution problems. Our experimental results show that in our framework, the robust orientation diffusion is very effective, where noise and any additional artifacts along LR image edges can be removed. Furthermore, compared with traditional SR method, our method can obtain good results.

2 Framework of Variational Approach to PDE

It is well known that during formulation, transmission and recording processes images degradation. Practically, this degradation is the result of two phenomena. The first one is deterministic and is related to the model of optical imaging system, including blur created by motion warping and an incorrect lens adjustment and down sampling, etc.. The second one is a random noise which usually characterized some probabilistic distribution for example Gaussian or Laplace distribution.

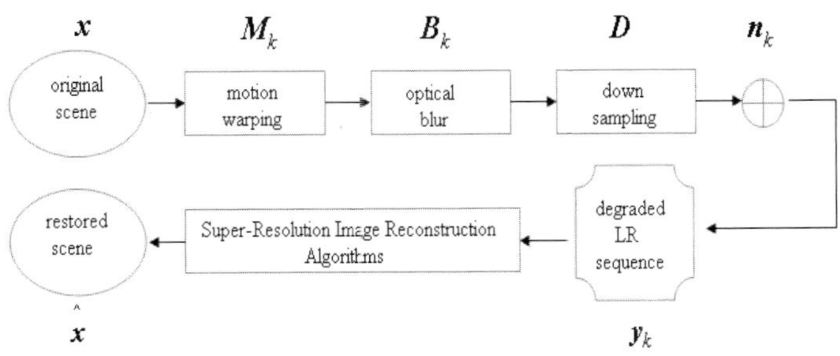

Fig. 1. Flow chart of super-resolution restoration

Shown as fig. 1, image degradation procedure can be formulated as the following degradation model [4]

$$u_k = D\, B_k M_k u + n_k, \quad k = 1, 2 \ldots, K. \tag{1}$$

where, u denote the original high resolution scene, yk denote the kth frame low resolution image and nk represent the random noise, while Mk denote the motion warping matrix with size of M×M between yk and x, Bk denote the optical blur matrix with size of L×L, and D denote the down sampling matrix with size of L×M.

When considering the signal model (1), our goal may be succinctly stated as one of estimating the underlying image u based on the observations uk and/or any potential knowledge of the noise statistics to further regularize the solution. Motivated by the MAP method [8-9], this yields the following fidelity-constrained optimization problem

$$\hat{u} = \arg\min_{u} \left\{ \int_{\Omega} \varphi(|\nabla u|) dx_1 dx_2 + \lambda \cdot \sum_{k=1}^{K} \| H_k u - u_k) \|_2^2 \right\}, \tag{2}$$

Where the first term is prior knowledge of the image,while the second is the data fidelity measure. The Euler-Lagrange equation of formulation (4) is the nonlinear parabolic PDE given by

$$\frac{\partial u}{\partial t} = \nabla \cdot \left(g\left(\|\nabla u\|\right) \cdot \nabla u \right) - \lambda \bullet \sum_{k=1}^{K} (H_k^T H_k u - H_k^T u_k) \tag{3}$$

where t is an artificial time parameter, $\nabla(\cdot)$ denotes the divergence operator, and $g(s) = \varphi'(s)/s$ is a nonnegative monotonically decreasing function. There are numerous functions of such a property. Two common forms are [9]

$$g(s,K) = \exp\left(-\frac{s^2}{2K^2}\right)$$

$$g(s,K) = \frac{1}{1 + s^2 / K^2}$$

Furthermore, the numerical iterative scheme can be designed to obtain the practical super-resolution algorithm.

3 Proposed Method

3.1 Motivation

In order to give our proposed method, let us first investigate the following PDE

$$\frac{\partial u}{\partial t} = \nabla \cdot \left(g\left(\|\nabla u\|\right) \cdot \nabla u \right) \tag{4}$$

We are going to show that it can be decomposed using the local image structure, that is, the tangent and normal directions to the grey-level lines. More precisely, for each point (x_1, x_2), we define the vectors $\mathrm{p} = \frac{\nabla u}{|\nabla u|}, \mathrm{q} = \frac{\nabla u^{\perp}}{|\nabla u|}$, and p is the normal vector and q is tangent vector. Then the above PDE can be rewritten as

$$\frac{\partial u}{\partial t} = \frac{\varphi'(\|\nabla u\|)}{\|\nabla u\|} \cdot u_{qq} + \varphi''(\|\nabla u\|) \cdot u_{pp} \tag{5}$$

Where we denote by u_{pp} and u_{qq} the second derivatives of u in the $\mathbf{T}$-direction and $\mathbf{N}$-direction, respectively:

$$u_{qq} = \mathrm{q}^T \mathbf{H}(u)\mathrm{q}, u_{NN} = \mathrm{p}^T \mathbf{H}(u)\mathrm{p}, \tag{6}$$

with $\mathbf{H}(u)$ is the Hessian Matrix of u in classical sense $[\mathbf{H}(u)]_{(i,j)} = \dfrac{\partial^2 u}{\partial x_i \partial x_j}$.

Although the above geometric PDE can obtain good effect for image de-noising, they may not be the good model for the super-resolution image reconstruction. Because from the LR image degradation model, the LR images are often created by motion warping and an incorrect lens adjustment, down sampling and noise addition, etc.. Thus the gradient vectors of LR image are also degraded, hence it is not suitable to choose p and q as the normal and tangent vector. The better definition of such orientation vectors must be modified to be robust to various degradation process.

3.2 PCA Based Robust Image Local Orientation Estimation

Let us assume that in the image of interest u, the orientation field is piecewise constant. Under this assumption, the gradient vectors in an image block should on average be orthogonal to the dominant orientation of the image pattern. In order to get the local orientation estimate, we first divide the region of interest into small blocks. Within each block, rearrange the gradient vectors $\mathrm{g}_i = (g_{i,x}, g_{i,y})^t$ (i = 1,2... n, within a local window). So orientation estimation can be formulated as the task of finding a unit vector $\tilde{q}$, to maximize the average of angles between q and gradient vectors g_i . Equivalently we can minimize [12]:

$$\sum_{i=1}^{n} (\tilde{\mathrm{q}}^T g_i)^2 = \tilde{\mathrm{q}}^T \sum_{i=1}^{n} (g_i g_i^t)^2 \tilde{\mathrm{q}} = \tilde{\mathrm{q}}^T C \tilde{\mathrm{q}} \tag{7}$$

Subject to $\|\tilde{q}\| = 1$, where

$$C = \begin{bmatrix} \sum_{i=1}^{n} g_{i,x}^2 & \sum_{i=1}^{n} g_{i,x} g_{i,y} \\ \sum_{i=1}^{n} g_{i,x} g_{i,y} & \sum_{i=1}^{n} g_{i,y}^2 \end{bmatrix}$$

Where $g_{i,x}$ and $g_{i,y}$ are derivatives in x and y direction at pixel location indexed by i, respectively. This problem is a constrained optimization problem and can be solved using Lagrange multipliers:

$$E(\mathrm{q},\varepsilon)=\tilde{\mathrm{q}}^T C\tilde{\mathrm{q}}-\varepsilon(\tilde{\mathrm{q}}^T\tilde{\mathrm{q}}-1)$$

Hence the solution can be obtained by the Euler equation

$$\frac{\partial E}{\partial \tilde{\mathrm{q}}}=0\Leftrightarrow (C-\varepsilon I)\tilde{\mathrm{q}}=0 \tag{8}$$

This is a standard eigenvalue problem. So the unit vector $\tilde{q}$ minimizing $\tilde{\mathrm{q}}^T C\tilde{\mathrm{q}}$ is the eigenvector of C corresponding to the smallest eigenvalue. Hence we can use SVD method to get the unit tangent vector vector $\tilde{q}$. Furthermore, the average normal vector $\tilde{\mathrm{p}}$ is easy to get by rotate $\tilde{q}$ by $\pi/2$.

3.3 Improved Model and Numerical Scheme

Substituting the robust estimate orientation $\tilde{\mathrm{p}}$ and $\tilde{q}$ into the (3), then we can obtain the improved PDE:

$$\frac{\partial u}{\partial t}=\frac{\varphi'(\|\nabla u\|)}{\|\nabla u\|}\cdot u_{\tilde{q}\tilde{q}}+\varphi''(\|\nabla u\|)\cdot u_{\tilde{p}\tilde{p}}-\lambda\bullet\sum_{k=1}^{K}(H_k^T H_k u-H_k^T u_k) \tag{9}$$

with

$$u_{\tilde{q}\tilde{q}}=\tilde{\mathrm{q}}^T\mathbf{H}(u)\tilde{\mathrm{q}}=\tilde{\mathrm{q}}^T\begin{bmatrix}u_{xx} & u_{xy}\\ u_{xy} & u_{yy}\end{bmatrix}\tilde{\mathrm{q}},\quad u_{\tilde{\mathrm{p}}\tilde{\mathrm{p}}}=\tilde{\mathrm{p}}^T\mathbf{H}(u)\tilde{\mathrm{p}}=\tilde{\mathrm{p}}^T\begin{bmatrix}u_{xx} & u_{xy}\\ u_{xy} & u_{yy}\end{bmatrix}\tilde{\mathrm{p}},$$

$$|\nabla u|=\sqrt{\left(\frac{\partial}{\partial\tilde{\mathrm{p}}}u\right)^2+\left(\frac{\partial}{\partial\tilde{\mathrm{q}}}u\right)^2}$$

Obviously, the orientation $\tilde{\mathrm{p}}$ and $\tilde{q}$ can be rewritten $\tilde{\mathrm{p}}=(\cos\theta,\sin\theta),\tilde{\mathrm{q}}=(-\sin\theta,\cos\theta)$,Then

$$u_{\tilde{q}\tilde{q}}=u_{xx}\cos^2\theta-2\cos\theta\sin\theta u_{xy}+u_{yy}\sin^2\theta$$

$$u_{\tilde{p}\tilde{p}}=u_{xx}\cos^2\theta+2\cos\theta\sin\theta u_{xy}+u_{yy}\sin^2\theta$$

In this paper, we adopt the following numerical scheme for the modified gradient decent flow. Let the radii of the local window of the current pixel (i, j) is R_{ij}, then the neighborhoods are defined by

$$N_{ij} = \{(v,w) | d_{vw;ij} = (v-i)^2 + (w-j)^2 \leq R_{ij}\}$$

The edge pixels in the neighborhoods must satisfy the constraints:

$$Q_{ij} = \{(v,w) | (v-i, w-j) \cdot \tilde{p} = 0, (v,w) \in N_{ij}\}$$

Similarity, the pixels in the normal vectors must satisfy the constraints:

$$P_{ij} = \{(v,w) | (v-i, w-j) \cdot \tilde{q} = 0, (v,w) \in N_{ij}\}$$

Therefore, $\frac{\partial}{\partial \tilde{p}} u$ and $\frac{\partial}{\partial \tilde{q}} u$ can use the following approximation

$$\left[\frac{\partial}{\partial \tilde{p}} u\right]_{ij} \approx \frac{\sum_{(i',j') \in P_{ij}} \left(u(i',j') - u(i,j)\right) d_{i'j';ij}}{\sum_{(i',j') \in P_{ij}} d_{i'j';ij}}, \quad \left[\frac{\partial}{\partial \tilde{q}} u\right]_{ij} \approx \frac{\sum_{(i',j') \in Q_{ij}} \left(u(i',j') - u(i,j)\right) d_{i'j';ij}}{\sum_{(i',j') \in Q_{ij}} d_{i'j';ij}}$$

Notice that the for digital image, in the area of Q_{ij} and P_{ij}, it may have not pixel's gray value, thus the interpolation are needed. Thus $\left[|\nabla u|\right]_{ij} = \sqrt{\left(\left[\frac{\partial}{\partial \tilde{p}} u\right]_{ij}\right)^2 + \left(\left[\frac{\partial}{\partial \tilde{q}} u\right]_{ij}\right)^2}$, and the definition of such above discontinuity has more contextual information. As for the u_{xx}, u_{xy}, u_{yy}, we use the following difference approximation, then

$$u_{xx} \approx u(i, j+1) + u(i, j-1) - 2u(i, j),$$

$$u_{yy} \approx u(i, j+1) + u(i, j-1) - 2u(i, j),$$

$$u_{xy} \approx \frac{1}{2}\left(u(i+1, j+1) - u(i-1, j+1) + u(i+1, j-1) - u(i-1, j-1)\right)$$

Thus, using the above difference approximation, the iterative update process is

$$u^{(n+1)}(i,j) = u^{(n)}(i,j) + \frac{\varphi'\left(\left|\nabla u^{(n)}\right|_{ij}\right)}{\left|\nabla u^{(n)}\right|_{ij}} \cdot \left[u^{(n)}_{\tilde{q}\tilde{q}}\right]_{ij} + \varphi''\left(\left|\nabla u^{(n)}\right|_{ij}\right) \cdot \left[u^{(n)}_{\tilde{p}\tilde{p}}\right]_{ij} \tag{10}$$

$$-\lambda \sum_{k=1}^{K} \left(H_k^T H_k u^{(n)}(i,j) - H_k^T u_k(i,j)\right)$$

4 Experimental Results and Performance Analysis

In this section, we first describe the methodology used in our simulation then report comparative results for two different tasks: image magnification and image super-resolution.

4.1 Methodology

Two image quality assessment measures are exploited to evaluate the performance our method. The first measure PSNR (Peak Signal Noise Ratio) is defined by

$$PSNR = 10\log_{10} \frac{u^{*\,2}_{\max}}{\frac{1}{(M\cdot N)^2}\sum_{i=1}^{M}\sum_{j=1}^{N}\left(u(i,j)-u^*(i,j)\right)^2},$$

The second measure is called PEPA (Peak-Edge Preserving Ability), which is defined by

$$PEPA = 10\log_{10} \frac{Num(edge(u^*))}{\sum_{(i,j)\in edge(u^*)} \left\|\nabla u(i,j)-\nabla u^*(i,j)\right\|^2}$$

Where u^* is the HR image, u is the evaluating image. And $edge(u^*)$ denotes the collection of the ideal edge pixels, which can be detected by Canny operators with optimal threshold, while $Num(edge(u^*))$ is the number of such edge pixels.

4.2 Comparative Results

In the following experiments, the edge stopping function is chosen as $g(s,K)=\frac{1}{1+s^2/K^2}$, the choice of K is discussed in [10], and the parameter $\lambda \approx 1.0$.

A. Image Magnification Experiments

In this experiment, an original HR Lena image is locally averaged and under-sampled by the factor of 4 in each direction (see Fig. 2(a)). Comparative visual results of three methods are given in Figs.2b, 2c and 2d. Fig.2b shows the result images obtained by the Bi-cubic interpolation, and Fig.2c is the result of original PDE (which corresponding to the equation (3)), while Fig.2d shows the result of our proposed method. From the image visual quality, it is easy to see that, our method can achieve better performance, which can effectively reduce the disturbing artifacts on edge regions, e.g. blocking artifacts, mosaic artifacts along SR image edges, thus the result image with clear edges. The values of PSNR and PEPA are listed in Table 1, our method also has better PSNR and PEPA.

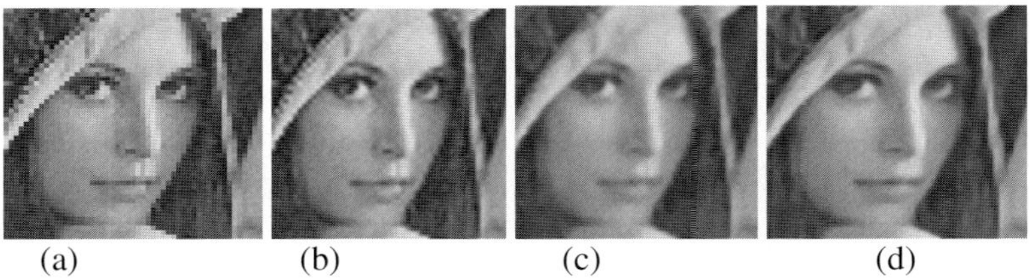

Fig. 2. Magnification results (q=4) (a) LR images (b) Bicubic, (c) Original PDE approach, (d) Proposed approach

Table 1. Comparison results of image magnification

Methods	PSNR	PEPA
Bicubic	25.24	23.55
Original PDE	27.89	25.43
Proposed approach	29.04	27.71

B. Image Super-Resolution Experiments

In the experiment, we create a sequence of LR frames by using one HR image [Fig. 3(a)] with size of 512×512. First, we shifted the HR image by a pixel in the vertical direction. Then, the shifted image was 3×3 Gaussian blur with variance of 0.5 averaged and under-sampled by the factor of 4 in each direction. The same operation with different motion vectors in vertical and horizontal directions was used to produce 16 LR frames from the original scene. Finally, the resulting LR frames were added with white Gaussian noise (The noise variance is 5) One of these LR frames is presented in Fig. 3(b). Fig.4 (a), (b) and (c) show the result of Bi-cubic interpolation + deconvolution, original PDE approach and our approach, respectively. Fig.4 (a1), (b1) and (c1) show the result of edge map detected by Canny operator with Optimal threshold corresponding to the Fig.4 (a), (b) and (c) respectively. Table 2 shows the calculated values of PSNR and PEPA. From the values listed in Table 2, we can see that our approach has better visual quality then other approach. Also, from the edge map of each result images, the Bi-cubic interpolation – deconvolution, original PDE approach can not effectively reduce mosaic artifacts along image edges, while our approach has better demosaicing effect.

Table 2. Comparison results of image SR.

Methods	PSNR	PEPA
Bicubic	23.75	22.23
Original PDE	25.64	24.56
Proposed approach	27.34	26.58

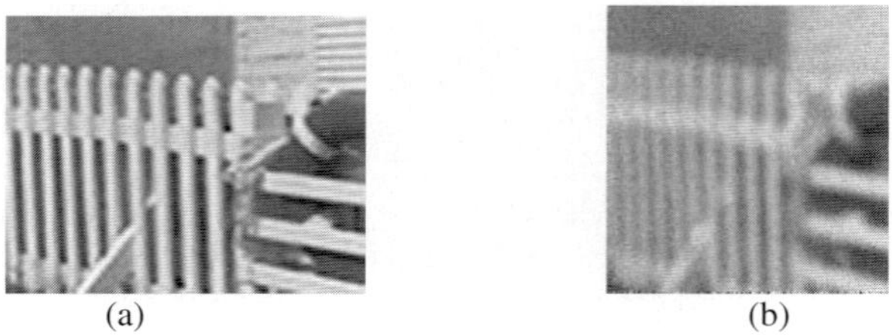

Fig. 3. (a) Original HR image (512×512). (b) LR image (64×64)

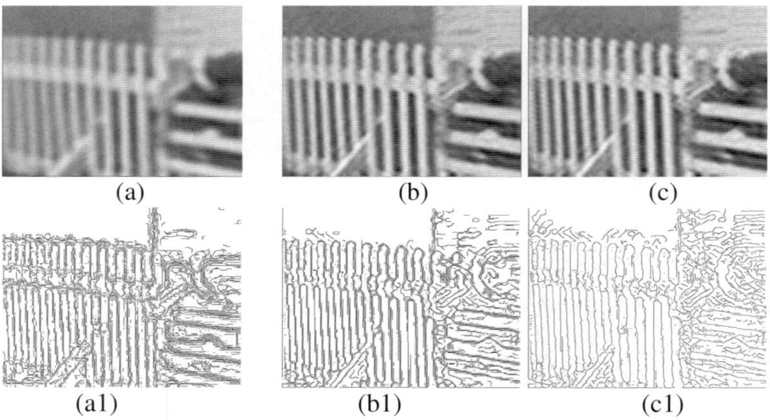

Fig. 4. (a) result of Bi-cubic interpolation + deconvolution, (b) original PDE approach and (c) our approach. Fig.4 (a1), (b1) and (c1) he result of edge map detected by Canny operator.

5 Conclusions

This paper focuses on the problem of image super-resolution reconstruction. Starting from the image degradation model, we have presented the framework of the variational approach to PDE method. The common drawbacks are that they have less local and contextual information and thus can not effective reduce the disturbing artifacts on edge regions, e.g. blocking artifacts, ringing artifacts along SR image edges, thus can not reconstruct a good HR image with clear edges. According to this analysis, an efficient method with systems of nonlinear partial difference equations for image super-resolution was presented, in which geometric edge orientation estimated by the PCA method. Experimental results have demonstrated the effectiveness of our approach.

Acknowledgments. This work is supported by Nature Science Foundation of China (No.60672074), Nature Science Foundation of Jiangsu province of China(No.BK2006569), Post-doctor Foundation of China(No.2006039028) and Jiangsu province(No.200601005B) respectively. Also we thanks the support of Youth Foundation of Nanjing university of science and technology (No.NJUST200401).

References

1. Borman, S.: Topics in multiframe superresolution restoration. [Ph.D. Thesis]. University of Notre Dame, Notre Dame, Indiana (2004)
2. Schultz, R.R., Stevenson, R.L.: Extraction of high-resolution frames from video sequences. IEEE Transactions on Image Processing 5(6), 996–1011 (1996)
3. Elad, M., Feuer, A.: Restoration of a single super-resolution image from several blurred, noisy and undersampled measured Images. IEEE Transactions on Image Processing 6(12), 1646–1658 (1997)
4. Borman, S., Stevenson, R.L.: Super-resolution for image sequences—a review. In: Proc. IEEE Int. Symp. Circuits and Systems, pp. 374–378. IEEE Computer Society Press, Los Alamitos (1998)
5. El-Khamy, S.E., Hadhoud, M.M., Dessouky, M.I., Salam, B.M., Abd, F.E.: Efficient implementation of image interpolation as an inverse problem. Digital Signal Processing 15, 137–152 (2005)
6. Farsiu, S., Dirk Robinson, M.: Fast and robust multiframe super resolution. IEEE Transactions on Image Processing 13(10), 1327–1344 (2004)
7. Chen, K.: Adaptive Smoothing Via Contextual and Local Discontinuities. IEEE Transactions on Pattern Analysis and Machine Analysis 27(10), 1552–1567 (2005)
8. Schultz, R.R., Stevenson, R.L.: A Bayesian approach to image expansion for improved definition. IEEE Transactions on Image Processing 3(3), 233–242 (1994)
9. Charbonnier, P., Blanc-Fe'raud, L., Aubert, G., Barlaud, M.: Deterministic edge-preserving regularization in computed imaging. IEEE Transactions on Image Processing 6(2), 298–311 (1997)
10. Perona, P., Malik, J.: Scale-Space and Edge Detection Using Anisotropic Diffusion. IEEE Transactions on Pattern Analysis and Machine Analysis 6(12), 629–639 (1990)
11. Feng, X.G.: Analysis and approaches to image local orientation estimation: [D]. University of California Santa Cruz (2003)

A Scalable Secret Image Sharing Method Based on Discrete Wavelet Transform

Jun Kong[1, 2], Yanfen Zhang[1, 2], Xiangbo Meng[1, 2], Ying Zheng[1,2], and Yinghua Lu[1,*]

[1] Computer School, Northeast Normal University, Changchun, Jilin Province, China
[2] Key Laboratory for Applied Statistics of MOE, China
{kongjun,zhangyf815,mengxb956,zhengy043,luyh}@nenu.edu.cn

Abstract. This paper presents a scalable secure scheme for sharing and hiding secret image. The given secret image is firstly divided into numerous non-overlapping blocks, and then transformed each block into wavelet frequency by one-level discrete wavelet transform. Second, quantization the wavelet coefficients into 256 gray level. Third, a bit-plane scanning method rearranges the gray value information of the quantized image. Four, share the rearranged image into n shadows by using multiple thresholds. Five, conceal each shadow image in R, G and B channels of the cover image, which doesn't attract Attracters' attention. Six, the more shadows the better quality of the retrieved secret image. The experimental results show that the retrieved secret image presents higher visual quality and the stego-images present better invisibility than traditional methods.

1 Introduction

In recent years, the transmission of digital media via Internet becomes more and more popular. Among the digital data transmitted, some of them are confidential and cannot be known by unauthorized users, secret information can be hidden in some original images. Image hiding schemes can be classified into two main categories: spatial domain [6-10] and frequency domain [11]. Generally, spatial-domain schemes provide large hiding capacities than those of the frequency-domain schemes. In addition, spatial-domain schemes are less complicated than frequency-domain schemes.

Among the spatial-domain image hiding schemes, the least-significant bits (LSB) substitution technique is the most simple and intuitive one. In this scheme, the secret data are directly embedded into the LSB of the cover pixels to form the stego-image. The main drawback of this scheme is that the image quality of the stego-image becomes poor when the number of LSB used is greater than or equal to 4. To solve this problem, the optimal LSB substitution [7], the approximately optimal LSB substitutions based on genetic algorithm [8] and the modulus LSB substitution [10] have been proposed.

[*] Corresponding author.This work is supported by science foundation for young teachers of Northeast Normal University, No. 20061002, China.

K. Li et al. (Eds.): LSMS 2007, LNCS 4688, pp. 736 – 745, 2007.

Blakley [1] and shamir [2] first proposed the idea of secret sharing, known as the (r, n) threshold scheme. in which, the secret image is shared into n shadow images, later the secret image can be revealed if r shadow images are received, and no information about the secret can be revealed if less than r shadow images are received [3]. However, the shadow images look like random noise rather than ordinary images. Therefore, some data-hiding methods [4] utilize stego-images instead of shadow images by hiding the shadow images in some ordinary images so as not to attract any attacker's attention. S.K. Chen, J.C. Lin [5] invents a progressive Fault-tolerant and transmission of images, which adapt multiple thresholds to control the quality of the revealed secret image in specifically circumstance.

In order to improve visual quality of the revealed secret image, a novel image sharing and hiding scheme is proposed in this paper. The secret image is firstly transformed into wavelet frequency, and then shared into several shadows. Finally, hiding the shadow images into R, G and B channels of every cover image, the numeral of bits are selected when the PSNR of the stego-image is the highest.

The rest of the paper is organized as follows: In Section 2, the secret image sharing method is briefly introduced. Section 3 describes the details of our scheme. The experimental results and comparisons are shown in Section 4. Section 5 gives the conclusions of our study.

2 Thien and Lin's Secret Image Sharing Method

Thien and Lin [4] proposed a method to share a secret image based on Shamir's [2] polynomial threshold scheme with a prime number p. The image is divided into several non-overlapping sectors, and each sector has r pixels. For each sector j, the r coefficients $a_0, a_1, ..., a_{r-1}$ of the corresponding polynomial are assigned as the r gray values of the r pixels in the sector:

$$q_j(x) = a_0 + a_1 x + ... + a_{r-1} x^{r-1} (\text{mod } p) \tag{1}$$

The wth shadow image is the collection $\{qj\ (w) \mid j = 1, 2, \ldots, \textit{original image size} / r\}$. Since each sector j, which has r pixels, contributes only one pixel $q_j(w)$ to the wth shadow image, the size of the wth shadow image is only 1/r of the secret image. This property holds for every $w \in \{1, 2, 3, \ldots, p - 1\}$. Only obtain r shadows can reveal the secret image, any shadows which num<r can reveal nothing.

3 The Proposed Method

As indicated in Fig. 1, the proposed method has four major parts: (a) Pre-Processing, (b) Sharing, (c) Hiding, (d) Retrieving .These four procedures are introduced in Section 3.1–3.4 respectively.

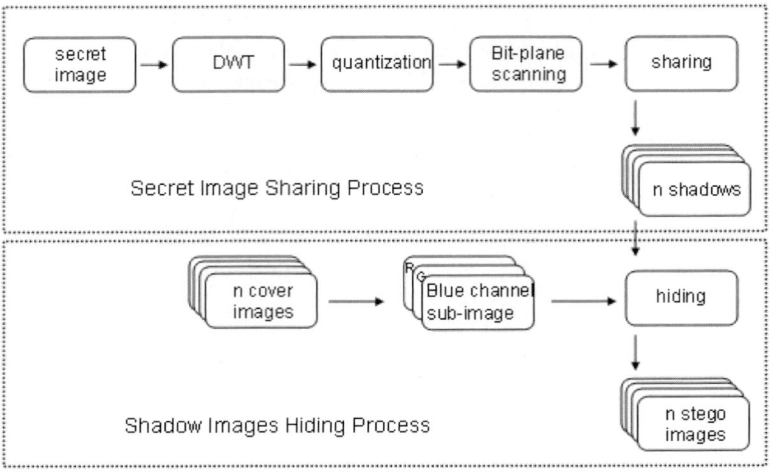

Fig. 1. The flowchart of the proposed method

3.1 Pre-processing

A. DWT (Discrete Wavelet Transform) and Quantization

In the DWT, an image is first decomposed into four subbands, LL1, LH1, HL1 and HH1 (L: i.e., low, H: i.e., high) by cascading horizontal and vertical two-channel critically subsampled filter banks. Each entry in subbands LH1, HL1 and HH1 represents the finest scale wavelet coefficients. And the LL1 contains the most energy. Fig.2 shows the image decomposed into four subbands for one scale. The low frequency subband is at the top left corner and the high frequency subband is at the bottom right corner.

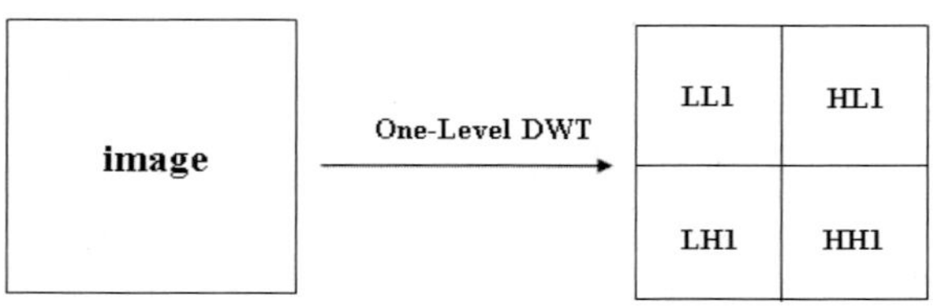

Fig. 2. The flowchart of Discrete Wavelet Transform

Let the k threshold values $r_1, r_2, \ldots, r_k$ be assigned and $r_1 \le r_2 \le \ldots \le r_k = r$ .The k thresholds denote the distinct number of shared results needed to recover the image with distinct quality levels. For example, a quite rough image can be recovered when r_1 shared results are received. Then, the image quality gradually improves when more shared results are received. Finally, the image can be recovered completely when r_k shared results are received.

First, divide the secret image into non-overlapping blocks of size 4×4, where $4\times4 = (r_1+r_2+...+r_k)$, and then transform each block into wavelet frequency adapting One-Level Discrete Wavelet Transform. Four sub images LL1, LH1, HL1 and HH1 are obtained. Adapt Equ. (2) to reset the coefficients of all sub images.

$$\begin{aligned} B_j(1,1:4) &= LL1 \qquad B_j(2,1:4) = LH1 \\ B_j(3,1:4) &= HL1 \qquad B_j(4,1:4) = HH1 \end{aligned} \tag{2}$$

Quantization the wavelet coefficients into 256 gray level (0-255). The formula is given as follows:

$$Max_j = \max(\sum_{x=1}^{4}\sum_{y=1}^{4} B_j(x,y)) \qquad Min_j = \min(\sum_{x=1}^{4}\sum_{y=1}^{4} B_j(x,y)) \tag{3}$$

Then adjust B_j (x, y) as:

$$\begin{aligned} B_j(x,y)' &= B_j(x,y) + |Min_j| \\ B_j(x,y)'' &= B_j(x,y)'/\max_j \times 255 \ (x,y=1,2,3,4) \end{aligned} \tag{4}$$

Finally, the quantized wavelet coefficients image is obtained.

B. Bit-Plane Scanning

The original image is divided into several non-overlapping sectors, and the sectors are processed one by one. Each sector has *Rsum* pixels ($Rsum = r_1 + r_2 + ... + r_k$). Since each pixel has 8 bits in the gray-level image, each sector has $8\times(r_1 + r_2 + ... + r_k)$ bits. We then rearrange these bits to get another *Rsum* values: each of them is still an 8-bit number ranging from 0 to 255. For example, let the 6 pixels of the sector have the gray values 166, 167, 164, 166, 168, 165, respectively. By the rearranging process, the 6 transformed values of the sector are 252, 16,192, 22, 221, 17. The details are shown in paper [5].

3.2 Secret Image Sharing

Divide the quantized image into non-overlapping blocks, each size is $n\times n$. And arrange the values into one stream. For *jth* block, use the Lagrange Interpolating Polynomial which is shown below to generate *k* pixel values. $f_j^i(1) \sim f_j^i(n)$ are the jth block pixel values for every shadow images $F_1 \sim F_n$.

$$\begin{aligned} F_j^1(x) &= (a_0 + a_1x + ... + a_{r1-1}x^{r1-1}) \bmod p \\ F_j^2(x) &= a_{r1} + a_{r1+1}x + ... + a_{r1+r2-1}x^{r1+r2-1} \bmod p \\ &\cdots\cdots \\ F_j^k(x) &= a_{rsum-rk+1} + a_{rsum-rk+2}x + ... + a_{rsum-1}x^{rsum-1} \bmod p \end{aligned} \tag{5}$$

Take (2,2,3,4,5,6) for example: We assumed that r_1=2, r_2=2, r_3=3, r_4=4, r_5=5, n=6. And therefore each sector has 16 pixels. They are: B_j= {104, 105, 103, 107, 108, 102, 101,109,100,110,111,112,113,141,144,142}. Now, the first r_1=2 transformed values {104, 105} form the first polynomial

$$F_j^1(x) = (104 + 105x) \bmod 257$$

for the current block. Then the rest r_2=2, r_3=3, r_4=4, r_5=5 transformed values {103,107}, {108,102,101}, {109,100,110,111}, {112,113,141,144,142}, form the second polynomial

$$F_j^2(x) = (103 + 107x) \bmod 257$$

$$F_j^3(x) = (108 + 102x + 101x^2) \bmod 257$$

$$F_j^4(x) = (109 + 100x + 110x^2 + 111x^3) \bmod 257$$

$$F_j^5(x) = (112 + 113x + 141x^2 + 144x^3 + 142x^4) \bmod 257$$

for the current block. Here x denotes the current shadow, j indicates jth block of the input image, and the value *1, 2, 3, 4, 5* denote the current position of each block for each shadow. Continue to the next block, until all the blocks have been processed. The size of every shadow image is $k/(r_1 + r_2 + ... + r_k)$ of the cover image.

3.3 Shadow Images Hiding

In this section the standard 256×256 color images are used as cover images. Adapt LSB substitution technique to conceal the information in the LSB of the cover images. According to HVS (Human Vision System), the capacity of hiding information in different color channel is varied with the different images. G channel is more sensitive to human eyes and B channel is not. So an adaptive hiding method is presented. The size of the shadow is equal to the cover image. Here N_R, N_G, N_B denotes the number of LSBs of every color channel. This is defined as follows:

$$F(N_R, N_G, N_B) = \{1 \le N_R, N_G, N_B \le 4 \wedge Sum(N_R, N_G, N_B) = 8\} \tag{6}$$

Decompose the color image into three sub-images indicated as R, G and B channels. Select one pair integers (N_R, N_G, N_B) randomly according to Esq. 6 as the bits of LSB. Transform each pixel of shadow into 8-bits binary streams. Adapt LSB substitution technique, imbedding every 8 bits into the LSB of the three channels. After all the pixels of shadows have been processed, compute the APSNR (which is described in Section 4) of every stego-image, and then save it in n arrays. Repeat the above operations, until all integer pairs have been processed. For every stego-image, saving the integer pairs as private key when the APSNR reaches the maximum value. n stego-images are generated based on the above technique. Notably, all of them still appear to be general images.

3.4 Secret Image Retrieving

The recovery of the original secret image includes three steps: First, extract the shared results from stego images. Second, recover the rearranged values from the shared results. Third, restore the pixel values from the rearranged values.

Step 1: *Extract the shared results from stego images*:

This step needs only simple operations such as division, addition, and multiplication. The procedure is similar to the one used in Section 3.4, and hence omitted.

Step 2: *Recover the value from the shadows*

To illustrate this step, let us inspect the example given in section 3.3 where k=5, n=6. r_1=2, r_2=2, r_3=3, r_4=4, r_5=5. Assume that we receive three shared results, $F1$, $F3$ and $F5$, for *jth* block $F1_j$ ={209,210,54,165,138}, $F2_j$ ={57.60,202,63,214} and $F5_j$ ={115,124,59,72,185}. According to Esq. (4) we can retrieve some values of original secret image. Using F1 and F3 we have:

$$F1_j^{\ 1} = (a_0 + a_1 \times 1) \bmod 257 = 209$$

$$F2_j^{\ 1} = (a_0 + a_1 \times 2) \bmod 257 = 57$$

$$F1_j^{\ 2} = (a_2 + a_3 \times 1) \bmod 257 = 210$$

$$F2_j^{\ 2} = (a_2 + a_3 \times 2) \bmod 257 = 60$$

According to the formula above we can retrieve the value a_0, a_1, a_2 and a_3. And then we use F1, F3 and F5 to retrieve the next three values: a_4, a_5 and a_6:

$$F1_j^{\ 3} = (a_4 + a_5 \times 1 + a_6 \times 1^2) \bmod 257 = 54$$

$$F2_j^{\ 3} = (a_4 + a_5 \times 2 + a_6 \times 2^2) \bmod 257 = 202$$

$$F3_j^{\ 3} = (a_4 + a_5 \times 3 + a_6 \times 3^2) \bmod 257 = 59$$

After that we have retrieved seven values of original jth block, we know that the front values in each block denote the low frequency information and contain most energy.

Step 3: *Restore the pixel values from the rearranged values*:

After obtaining the rearranged values in the previous step, the values can be transformed back to restore the pixels of the original secret image.

(1) Divide the input image into non-overlapping blocks, each size is $n \times n$ pixels. Rearrange the pixels into another $n \times n$ pixels as shown in Section 3.1

(2) Transmit the values into original wavelet coefficients according to Esq. (2).

(3) Perform the inverse of the wavelet transformation to reconstruct the original secret image.

4 Experimental Results

The *APSNR* (Average Peak Signal to Noise Ratio) is used to evaluate the visual quality of the stego image, which is defined as:

$$PSNR_{R,G,B} = 10 \times \log_{10}\left(\frac{M \times N \times \max(I_{R,G,B}{}^2(i,j))}{\sum_{i=1}^{M}\sum_{j=1}^{N}[I_{R,G,B}(i,j) - I_{R,G,B}'(i,j)]^2}\right)$$

$$APSNR = \frac{(PSNR_R + PSNR_G + PSNR_B)}{3} \tag{7}$$

M , N denote the size of the stego image, $I_{R,G,B}(i,j)$ indicates the cover image and $I_{R,G,B}'(i,j)$ denotes the stego image.

We adapt *LPSNR* (Local Peak Signal to Noise Ratio) to evaluate the similar degree between the original secret image and the retrieved secret image, which is defined as follows:

$$LPSNR = \frac{1}{S} \times 10 \times \log_{10}\left(\frac{4 \times 4 \times \max(B^2(i,j))}{\sum_{i=1}^{4}\sum_{j=1}^{4}(B'(i,j) - B(i,j))}\right) \tag{8}$$

S denotes the num of blocks, $S = M \times N / 4 \times 4$, $B(i,j)$ denotes each block of size 4×4.

Experiments are conducted to demonstrate the good invisibility of the stego-images and the perfect visual quality of the retrieved secret image using our scheme. Here we adapt $(2, 2, 3, 4, 5, 6)$ threshold values. In the experiments, the standard 256×256 gray-level image is adapted. The secret image can be reconstructed by collecting any five shadows, and the coarse information of the secret image can be retrieved under any two shadows, the more stego-images we obtained the better quality of the secret image, as shown in Fig. 3. And adapt RGB color image as cover image. The size of the cover image is $144 \times 144 \approx 256 \times 256 \times 5 \div (2+2+3+4+5)$, which is $5/16$ times of the secret image. Accordingly, six stego-images are generated as shown in Fig. 4. Table 1 indicates the visual quality of the stego images after hiding different bits in R, G and B sub-images. Here we adapt $(2,3,3)$, $(3,1,4)$ and $(2,2,4)$. Take $(3,1,4)$ for example, we select three bits each pixel of Red channel, select one bit each pixel of Green channel, and select four bits each pixel of Blue channel to hide secret information. All the *APSNR* are above 37, which are higher than S.K.Chen's.

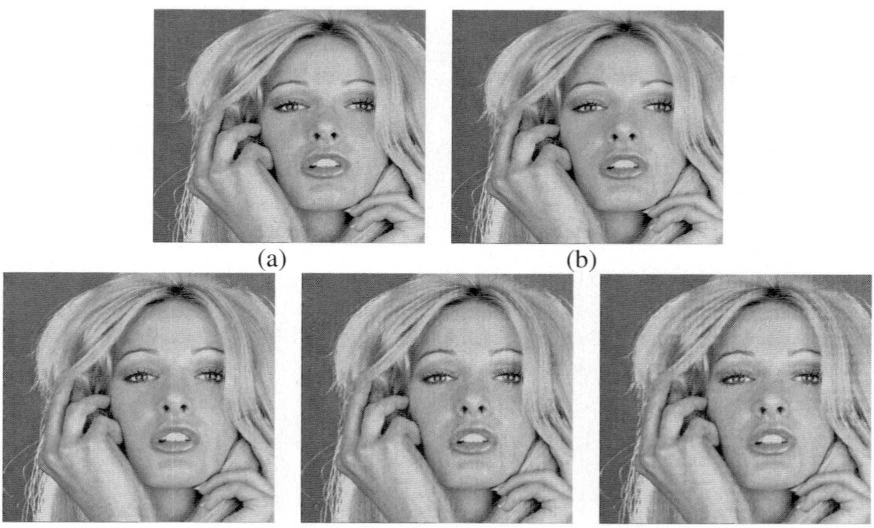

Fig. 3. The secret images revealed from different numbers of stego images: (a) the original secret image of size 256×256 pixels, (b) from any 5 stego-images, (c) from any 4 stego-images, (d) from any 3 stego-images, (e) from any 2 stego-image.

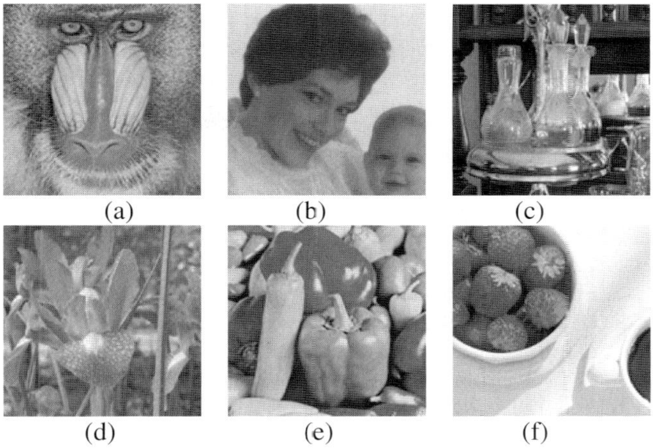

Fig. 4. The six stego-images of size 144×144 pixels

Fig. 5. The comparable results between the two methods: (a1)-(c1) are the retrieved secret images adapting our method, (a2)-(c2) are the retrieved secret images adapting S.K.Chen's, (a1),(a2) are the results of owning 4 stego-images, (b1),(b2) are the results of owning 3 stego images, (c1),(c2) are the results of owning 2 stego images.

In order to inspect the performance of our method, we simulate other experiments and compare our scheme with S.K.Chen's in [5]. Fig. 5 gives the comparison of the two experiments: (a1)-(c1) are the retrieved secret images adapting our methods, (a2)-(c2) are the retrieved secret images adapting S.K.Chen's. (a1),(a2) are the results of owning 4 stego images, (b1),(b2) are the results of owning 3 stego images, (c1),(c2) are the results of owning 2 stego images. Table 2 gives the *LPSNR* of the revealed secret image between our method and S.K.Chen's. the visual quality of the two methods are high when owning four or five shadow results. But when the shadow results become fewer, our method can still retrieve high quality secret image.

Table 1. The APSNR of the stego image in the case of selecting different bits in R, G and B channels for hiding

R G B	The APSNR of the stego-images					
	a	b	c	d	e	f
3 2 3	37.3	37.2	38.1	38.2	37.4	38.0
3 1 4	37.9	37.6	38.8	38.8	38.1	38.7
2 2 4	37.4	37.1	38.3	38.3	37.5	38.2

Table 2. The compared results of retrieved secret images between our method and S.K.Chen's method

The shadows	The LPSNR of the retrieved secret image			
	Lena		woman white	
	our method	*S.K.Chen's*	*our method*	*S.K.Chen's*
5	51.6	52.5	52.8	55.5
4	30.9	30.4	30.4	28.8
3	22.7	17.4	22.4	21.4
2	21.8	11.27	21.9	13.7

5 Conclusions

In this paper, a novel secret image sharing and hiding scheme based on DWT has been proposed. There are several characteristics in the proposed scheme, including: First, the hiding capacity is significantly improved by using color image as cover image. Second, the image quality of the stego-image obtained by the proposed scheme is perfect by hiding information in R, G and B channels separately. Third, the selection of multiple thresholds improves the probability of revealing the secret image. Four, the scheme is secure, that is to say, less than r_1 shared results cannot reveal any information about the image.

References

1. Blakley, G.R.: Safeguarding cryptographic keys. In: Proceedings AFIPS 1979. National Computer Conference, New York, USA, vol. 48, pp. 313–317 (June 1979)
2. Shamir, A.: How to share a secret. Commun. ACM 22(11), 612–613 (1979)
3. Thien, C.C., Lin, J.C.: Secret image sharing. Computer, Graphics 26, 765–770 (2002)
4. Wu, Y.S., Thien, C.C., Lin, J.C.: Sharing and hiding secret images with size constraint. Pattern Recognition 37(7), 1377–1385 (2004)
5. Chen, S.K., Lin, J.C.: Fault-tolerant and progressive transmission of images. Pattern Recognition 38, 2466–2471 (2005)

6. Wu, D.C., Tsai, W.H.: Spatial-domain image hiding using image differencing. IEE Proceedings: Vision, Image & Signal Processing 147(1), 29–37 (2000)
7. Chang, C.C., Lin, M.H., Hu, Y.C.: A fast and secure image hiding scheme based on LSB substitution. Int. J. Pattern Recognition Artif. Intel. 16(4), 399–416 (2002)
8. Wang, R.-Z., Lin, C.-F., Lin, J.-C.: Image hiding by optimal LSB substitution and genetic algorithm. Pattern Recognition 34(3), 671–683 (2001)
9. Wang, R.-Z., Lin, C.-F., Lin, J.-C.: Hiding data in images by optimal moderately significant-bit replacement. IEEElectron. Lett. 36(25), 2069–2070 (2000)
10. Thien, C.C., Lin, J.C.: A simple and high-hiding capacity method for hiding digit-by-digit data in images based on modulus function. Pattern Recognition 36, 2876–2881 (2003)
11. Chang, C.C., Chiang, C.L., Hsiao, J.Y.: A DCT-domain System for Hiding Fractal Compressed Images. In: Proceedings of the 19th International Conference on Advanced Information Networking and Applications (AINA'05), 1550-455X/05(C), IEEE, Los Alamitos (2005)

Two Phase Indexes Based Passage Retrieval in Biomedical Texts

Ran Chen, Hongfei Lin, and Zhihao Yang

Department of Computer Science and Engineering,
Dalian University of Technology, 116024 Dalian, China
ranhappy1983@163.com, {hflin,yangzh}@dlut.edu.cn

Abstract. The biomedical literature is growing at a double-exponential pace. Passage-level retrieval is more effective to provide the information section than document-level retrieval. This paper presents a method of two phase indexes based passage retrieval. First two phase indexes: paragraph index and sentence-level half-overlapped windows index are built. Then, BM25 model is used to retrieval on the two phase indexes. At last, the passage and paragraph retrieval results are combined as the result of the passage retrieval. The experiment result shows that the performance is improved 5% with two phase indexes than only with the paragraph index.

Keywords: Passage Retrieval, Sentence-level Half-overlapped Windows, Retrieval models.

1 Introduction

Nowadays, the biomedical literature is growing at a double-exponential pace, which becomes the huge knowledge treasure. It provides reliable and exact scientific basis for the clinical studies and the medical decision-making. For example, MEDLINE database(Medical Literature Analysis and Retrieval System Online), which is s the U.S. National Library of Medicine's(NLM) premier bibliographic database that contains over 15 million references with a concentration on biomedicine in life sciences.

To help biomedical researchers address the problem of information overload, information retrieval is to explore the relevant information in biomedical field from the huge literatures according to the needs of users. In information retrieval system, similarity is used to measure the relativity between the query terms and the documents. The higher the score is, the more valuable the document is to the query term. But when the full-text document is calculated as the unit of similarity, it could not locate exactly and search for the information that users need. Instead, it can search more suitable information in passage retrieval for the section of the document is calculated as the unit of similarity. Moreover, it has two advantages: on one hand, it provides users with the section but not a full-text of a document. Thus, users face to the focus segments instead of a superfluous document, in that the utilization of a document is raised. On the other hand, the relevant short text can be searched easier than the long text. Comparing with a document containing variety of topics, the section of document including only one topic is more exact in searching. So passage retrieval is more efficient than document retrieval.

K. Li et al. (Eds.): LSMS 2007, LNCS 4688, pp. 746–755, 2007.

In the Genomics Track of TREC (Text REtrieval Conference) 2006, a new task is presented that focuses on passage retrieval. Systems are required to return the passages of the document including the answer to the topic question from the biomedical literature [1]. The passage which is from the original document may be a sentence, some sentences or a paragraph which can answer the question. Different passage types include structural (such as sentences, paragraphs and sections), semantic [2, 3, 4, 5], window-based [5, 6] and arbitrary [7].

The method of passage retrieval is also various. For example, W. Bruce Croft presented the language modeling approach in 2002[6, 8]. A language model is built for each passage and passages are then ranked according to the probability that the query could have been generated by each of them. But the method needs some smoothing algorithms. Also, Jing Jiang and Chengxiang Zhai presented the Hidden Markov Model approach in 2004. It can improve the passage retrieval accuracy and recall, but at the price of precision sometimes [2, 9]. Otherwise, window-based retrieval approach is used to the passage retrieval. The window consists of a fixed number of words or bytes. The window's form can be fixed length, arbitrary length, overlapped and non-overlapped. All of these passage types are defined at indexing time, are indexed as independent units.

While structural properties of documents are not always explicit, semantic segmentation may be inaccurate. Previous window-based passages don't take the sentence integrality into account whether they are the fixed-length or the variable-length passages.

In this paper, the overlapped windows based method is used to segment the document. Since one sentence should convey a single idea and sentences should be about one issue [7], it may be more desirable that use the sentences to return as the result to the user. So in our experiment, complete sentences are the limited condition on the basic of the fixed-length passage, which is defined as sentence-level half-overlapped windows method: the first window starts at the first term in a document and sequent windows half-overlap in the middle of preceding ones. In addition, two phase indexes are built: paragraph index and sentence-level half-overlapped windows index. Then four different retrieval models (BM25, TFIDF, Indri and KL) are used to retrieve and their retrieval performances are compared. Next, the same retrieval model is used on the two phase indexes and the performances are compared. At last, we design a new retrieval strategy which combines the passage and the paragraph retrieval result to compute the passage retrieval score and rank the passage according the score as the passage final result. The idea is similar to the combine algorithm which York university experiment at TREC 2004[10].

The rest of the paper is organized as follows. A brief description of the retrieval models is given in section 2. Then the method and result of our experiment are presented in section 3 and 4. Finally, conclusions and future work are discussed in the section 5.

2 Retrieval Models

The paper uses the Lemur Toolkit. It is convenient for language modeling and information retrieval research. The application runs retrieval experiments to evaluate different retrieval models as well as different parameter settings for those models. Lemur supports major language modeling approaches such as Indri and KL-divergence, as well as vector space, TFIDF, Okapi and InQuery [3]. The four models: BM25, Indri, KL, TFIDF are used in our experiment.

2.1 BM25 Model

The BM25 formula is defined as

$$BM\,25(q,d)=\sum_{t\in Q} q_t \times \ln\left(\frac{D-df_t+0.5}{df_t+0.5}\right)\times \frac{freq_{t,d}\times(1+k_1)}{freq_{t,d}+k_1\left[(1-b)+b\times\frac{dl_d}{avdl}\right]} \tag{1}$$

Where $BM25(q,d)$ is the $BM25$ score of document *d* for the query q_t, q_t is the frequency of the term *t* in *q*, *D* is the total number of documents, df_t is the document frequency of *t*, $freq_{t,d}$ is the term frequency of *t* in *d*, k_1 is a parameter, *b* is a parameter, dl_d is the document length of *d* and *avdl* is the average document length of all the documents. Parameters of BM25 algorithms are adopted from Lemur's default setting [11].

2.2 KL Model

Kullback-Leibler (KL) divergence retrieval model is defined as

$$s(Q,D)=D(\theta_Q \,||\, \theta_D)=\sum_{w\in V} p(w\,|\,\theta_Q)\log\frac{p(w\,|\,\theta_Q)}{p(w\,|\,\theta_D)} \tag{2}$$

Where given a query *Q* and a document D, we estimate a query language model θ_Q and a document language model θ_D, and then simply score the document using the KL-divergence of θ_Q and θ_D. V is the vocabulary set and w is the query term [12].

2.3 Indri Model

The retrieval model implemented in the Indri search engine is an enhanced version of the model described in, which combines the language modeling and inference network approaches to information retrieval. As in the original inference network framework, documents are ranked according to $P(I|D, \alpha, \beta)$, the belief the information need *I* is met giving document *D* and hyper parameters α and β as evidence.[13]

2.4 TFIDF Model

TFIDF function is the classical algorithm and used to compute the feature weight (Salton, Buckley, 1988). Its definition is as follow:

$$TF * IDF = TF * 1/DF \tag{3}$$

Where TF is the term frequency, IDF is the inverse document frequency; DF is the document frequency.

3 Experiment Method

The aim of our experiment is to find the effective method of passage retrieval. A document is segmented into passages based on sentence-level half-overlapped windows method. To prove its superiority, the different two phase indexes are built: paragraph index and sentence-level half-overlapped windows index. The range of paragraph is wide, and the fixed-length windows limit the number of words strictly. By contrast, the method of sentence-level half-overlapped windows is presented more flexible. Then, we retrieve based on the four retrieval models on the same index and choose the model with the best performance as the tool in the next step. Then the chosen model is used to retrieve on the two phase indexes. Finally, combine the paragraph result and the passage result as the final passage result. Though comparing the passage MAP scores, it is proven that the sentence-level half-overlapped windows method is more desirable to the user's need. The idea of the retrieval strategy we used in the last step is: if a passage is hit by both searches, it should be assigned more weight than others that are hit by only one search. Thus, it considers not only the return passage but also the relevant sentences that are not in the passage. Then passages are ranked based on the score of the combine computation.

3.1 Experiment Data and Performance Measures

Experiment data is the test collection of the Genomics TREC (Text REtrieval Conference) in 2006, concluding 162,259 full-text documents from the biomedical literature. The words are extracted as the original query terms from the 28 topics which are given in the form of question (topic160-topic187). Then, some query terms are expanded using some biomedical databases such as Entrez Gene and UMLS Metathesaurus.

In our experiment, the retrieval results are evaluated against the gold standard using measures which are used in the Genomic track: mean average precision (MAP). The measure includes three levels of retrieval performances: passage retrieval, document retrieval, aspect retrieval. Correspondingly, there are three guide lines: passage MAP, document MAP, and aspect MAP.

Passage MAP is the precision which is computed as the fraction of characters overlapping with the gold standard passages divided by the total number of characters included in all nominated passages from this system for the topic up until that point.

Aspect level is used to normalize passages on the same answer, for most users prefer passages with different aspects. For each submitted run, the ranked passages will be transformed to two types of values, either the aspect of the gold standard or not-relevant.

Document level, any PMID that has a passage associated with a topic ID in the set of gold standard passages will be considered a positive document for that topic.

3.2 Building Index

To locate the position rapidly, index is built to the original data. In our experiment, two phase indexes are built: paragraph index and sentence-level half-overlapped windows index. Meanwhile, the query terms are preprocessed. Some terms have the metaplasm, such as "activate" has the form of "activates", "activating", "activated" and so on. Thus, these terms are reverted to their etyma. Moreover, some words appearance with high frequency, but they have nothing with the search. A list of stop words is used to filter out the words which will not be helpful for the topic searching. However, we don't find an effective automatic way to filter the "noise" terms induced by synonyms expansion.

3.3 Passage Retrieval

Retrieve on Paragraph Index. The retrieval result is ranked based on the BM25, TFIDF, KL, Indri models. This set of experiments focus on the following points: first, find the models with best performance; secondly, verify using the query expansion has more precision than the initial query.

Retrieve on Sentence-level Half-Overlapped Windows Index. In this paper, the definition of the window contains the following two conditions:

First, the number of words of all the sentences in one window is lager than the limitative range. As showed in Fig.1, the document is segmented into the sections of Part A, Part B, Part C.... Each part makes up of the integrated sentence. Every two parts constitute a window and every window starts at the middle of the previous one.

The formula is as follow:

Window one = Part A + Part B

Window two = Part B + Part C

Window three = Part D + Part E

Because the length of a sentence in the windows is not fixed, the windows' length may be larger than the given length.

Secondly, the windows can not cross the paragraphs. It is limited by the scoring system of the Genomic Track. Such as the Part D in Fig.1, it is a new start of a paragraph. So it is not half-overlapped with the Part C. Thus, it can assure the sentence's integrity and it is more desirable for the users' need.

Part A: PIES proceeds to obtain all abstracts from Medline satisfying the search specification "calyculin". Each abstract is analysed to identify sentences that mention interaction of proteins.[b] We consider only inhibit vs activate type of interactions.

Part B: Each such sentence is considered an "evidence" for an interaction. The "actor" and "patient" of each interaction are identified. These interaction evidence sentences are then grouped by actor and patient.

Part C: The result is then displayed in textual form. Figure 2 depicts a small fragment of this result. It shows an inhibition interaction, where "calyculin A" is the actor and "serine/threonine phosphatase" is the patient, and four of the many evidence sentences for this interaction.

Part D: PIES puts these interactions together into a pathway diagram where the nodes are the proteins and the directed edges denote interactions (green arrows for "activate" and red arrows for "inhibit") between the connected proteins.

Part E: PIES then performs a visually pleasant graphical layout of the pathway diagram and displays it as a GIF file, as depicted in Figure 3. In addition, PIES also produces a Postscript file of the diagram.

This Postscript file is "tiled" automatically by PIES so that a large pathway diagram can be printed on several pages that can be placed side-by-side to re-assemble the diagram.

Fig. 1. Definition of passage

According to the character of the corpus, the answers of the given topic question are very elaborate and have the flexible length. Also, using the segmentation method of fixed length is not adaptive to some special query terms and special document. The length of windows is improved, from the fixed length to variable length on sentence-level. Moreover, most existing approaches to passage retrieval rely on pre-segmentation of documents, but the method of the optimal boundaries of a relevant passage depends on both the query and the document [9]. Half-overlapped windows have been shown in standard IR setting to be at least as effective as and more efficient than other pre-defined passages [6].

Combine the result. The passage and paragraph score is combined by giving weights to them as the following function:

$$S = \lambda \times S_1 + (1-\lambda) S_2 \tag{4}$$

Where λ is a weighting parameter of a passage, $1-\lambda$ is a weighting parameter of a paragraph, S_1 is the score of a passage, and S_2 is the score of a paragraph. Analyzing the data collection, the correct answer maybe concentrate in one sentence or separate in a paragraph. It can not be a good answer that only revert one sentence or one paragraph. Considering the two aspects, the weight is introduced. For example, the sentences which are retrieved by using the sentence-level half-overlapped windows index are the part of the answer, but the other part is in the sentences which are retrieved by using paragraph index. Weights are given to the two parts and then the final score are computed as the passage score. Therefore the score of sentence-level half-overlapped windows result is higher than that of only using one phase index and we obtain the higher precision.

According to the formula (4) introduced in this section, the value of λ is adjusted in our experiment to find the best proportion coefficient. When the value of λ is small, the score can be seen as the single result using the sentence-level half-overlapped windows index; when $1-\lambda$ is small, the score can be seen as the single result using the paragraph index.

4 Experiment Result Analysis

4.1 Contrast of the Retrieval Results of Models

Retrieving all documents using the different models to the different indexes, then retrieving again using the expansion words and rank based on the scores.

Fig.2 shows the retrieval results on the two phase indexes. The passage MAP is mainly consult. From Fig.2 BM25 model has been shown to be a very good probabilistic weighting scheme in the two phase indexes in the retrieval. It reaches the better performance than other models. When the expansion words are used for searching, the score is 0.07388.

Next, from Fig.2, the retrieval results which using the query expansion method obtain the higher precision than that using the original terms. For example, when the TFIDF model is based on retrieval, the disparity between the highest and the lowest passage MAP is about 21%. In the searching process, if only use the original query terms, many synonyms or words whose meaning are similar of the query term may not be found. Thus, these documents containing these query terms may not be retrieved. Add the longer and more accurate new query words that are relevant with query terms, therefore, the results will have higher accuracy.

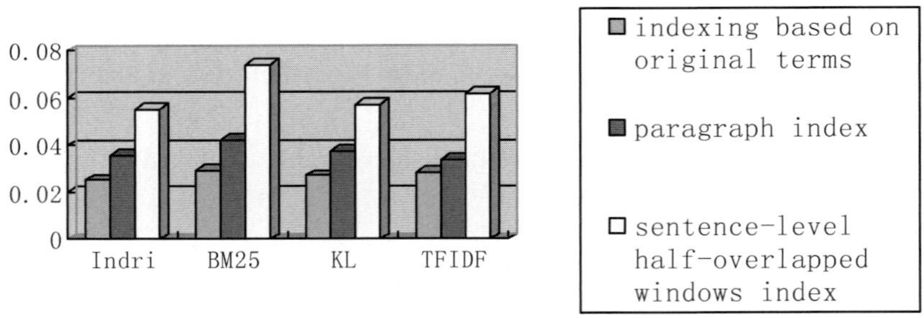

Fig. 2. Passage MAP based on the four models

4.2 Retrieve Based on Two Phase Indexes

In this set of experiments, BM25 model is used on the two phase indexes. Table 1 summarizes the results of these experiments and shows the three MAPs. The passage retrieval is based on half-overlapped windows of size 30. We observe that the performance of the two phase indexes is better than the paragraph index. The experiment shows that the performance enhances 5% in the passage retrieval than in the document retrieval.

Table 1. BM25 model, window size=30, λ=0.3, $1-\lambda$=0.7

	Document MAP	Passage MAP	Aspect MAP
Paragraph index	0.3608	0.0424	0.2397
Sentence-level index	0.3584	0.0739	0.1848

Table 2 summarizes the different results of the window size of 10 and 30 on the sentence-level half-overlapped windows index. If the paragraph is seen a bigger window and is compared with the size of window of 10 and 30, the setting of the length of window has obviously affect on the retrieval result, but it was not depend on the length of windows as increased as linear. So searching the suit window is benefit for the precision increase. In the experiment we observe that it reaches the best performance when the size of window is 30 based on the BM25 model.

Table 2. BM25 model, λ=0.3, $1-\lambda$=0.7

Size of windows	Document MAP	Passage MAP	Aspect MAP
Length of paragraph	0.29525	0.02509	0.1858
30	0.3584	0.07338	0.1848
10	0.2483	0.0659	0.1633

4.3 Combing Result Analysis

The effect is not very obviously when only using the half-overlapped windows index or the paragraph index. The different weights are given to the different retrieval results, and then through adding the two scores, a score is gotten as the passage retrieval result.

The set of experiments uses the sentence-level half-overlapped windows index which the size of the window is 30. Based on the BM25 model, the formula (4) above is used to adjust the weights of the paragraph and the passage. Table 3 summarizes that the three MAPs when the results have the different weights. In Table 3, the "Exp" in front of the model means the query expansion is used based on the model. The decimal fraction below the model name in the first line represents the value of $1-\lambda$ in the formula (4) and the decimal fraction below in the second line is the value of λ in the formula (4). Similarly, the "ExpBM25" expresses the query expansion is used based on the BM25 model. We can see that the passage MAP which reaches the highest is 0.07388 when λ=0.7.

In addition, when the weight of the paragraph score is well under the weight of the passage score, the final score approximately is seen as the score of using the single sentence-level half-overlapped windows index. Contrarily, the final score approximately is seen as the score of using the single paragraph index. Such as "bm25=0.9, Expbm25=0.1" in Table 3, the passage MAP 0.0655 is the score of using the single paragraph index. When we mainly focus on the passage MAP, the MAP with sentence-level half-overlapped windows index is observed higher than that with paragraph index. That is, the performance on the two phase indexes is superior to that on the paragraph index.

Table 3. BM25 model, window size=30

λ	Document MAP	Passage MAP	Aspect MAP
0.9	0.3536	0.0713	0.1839
0.8	0.3562	0.0726	0.1845
0.7	0.3584	0.0739	0.1848
0.6	0.3566	0.0738	0.1862
0.5	0.3645	0.0731	0.1855
0.4	0.3632	0.0714	0.1850
0.3	0.3610	0.0695	0.1857
0.2	0.3621	0.0687	0.1912
0.1	0.3564	0.0655	0.1922

5 Conclusions

In this paper, through different experiments, the effective method on the passage retrieval is studied. In our experiments, two phase indexes are built: paragraph index and sentence-level half-overlapped windows index. First, the different results are compared with the same index based on the different retrieval models (BM25, TFIDF, Indri, KL). BM25 model has been shown to be a very good probabilistic weighting scheme and has well performance. Then we retrieve on the two phase indexes based on the BM25 model. At last, a new retrieval strategy is designed which combines the passage and the paragraph retrieval result to compute the passage retrieval score and rank the passage according the score as the passage final result. The weights of the score of the two indexes are adjusted to make it work well. The return result with the tow phase indexes is observed closer to the gold standard than with paragraph index. So its precision is higher. With the various contrasts, we obtain that the precision of the passage retrieval improves about 5% than the precision of the paragraph retrieval. But there still exists some problems: the query terms are no matched to the content of a document; whether the weight of the query term gives suitably has some effect on the final retrieval result. In our future work, we will try to preprocess the paragraph before the searching. Such as getting rid of the sentence's beginning and ending which don't contain the query terms, and then the index is built to improve the retrieval precision. For our retrieval formula, different parameters will be introduced and adjusted to reach the perfect performance.

Acknowledgments. This work is supported by grant from the Natural Science Foundation of China (No.60373095 and 60673039) and the National High Tech Research and Development Plan of China (2006AA01Z151).

References

1. Hersh, W., Cohen, A.M., Roberts, P., Rekapalli, H.K.: TREC 2006 Genomics Track Overview. In: Proceedings of the 15th Text REtrieval Conference (2006)
2. Jiang, J., Zhai, C.: Extraction of coherent relevant passages using hidden Markov models. ACM Transactions on Information Systems 24(3), 295–319 (2006)

3. Ofoghi, B., Yearwood, J., Jan, R.G.: A semantic approach to boost passage retrieval effectiveness for question answering. In: Proceedings of the 29th Australasian Computer Science Conference, pp. 95–101 (2006)
4. Hearst, M.: TextTiling: Segmenting Text into Multi-Paragraph Subtopic Passages. Computational Linguistics 23(1), 33–64 (1997)
5. Callan, J.P.: Passage-level evidence in document retrieval. In: Proceedings of the 17th annual international ACM-SIGIR conference on research and developments in information retrieval(SIGIR), Dublin, Ireland, New York, pp. 302–310. ACM Press, New York (1994)
6. Liu, X., Croft, W.B.: Passage: Retrieval Based on Language Models. In: Proceedings of the 11th International Conference on Information and Knowledge Management, pp. 375–382 (2002)
7. Kaszkiel, M., Zobel, J.: Effective ranking with arbitrary passages. Journal of the American Society For Information Science and Technology 52(4), 344–364 (2001)
8. Ponte, J.M., Bruce Croft, W.: A language modeling approach to information retrieval. In: Proceedings of the 21st Annual International ACM SIGIR Conference on Research and Development in Information Retrieval (SIGIR), Melbourne, Australia, pp. 275–281. ACM Press, New York (1998)
9. Jiang, J., Zhai, C.: UIUC in HARD 2004–Passage retrieval using HMMs. In: Proceedings of the 13th Text REtrieval Conference2004 (2004)
10. Huang, X., Huang, Y., Wen, M., et al.: York University at TREC 2004: HARD and Genomics Tracks. In: Proceedings of the 13th Text REtrieval Conference (2004)
11. Lin, K.H.-Y., Hou, W.-J., Chen, H.-H.: Retrieval of Biomedical Documents by Prioritizing Key Phrases. In: Proceedings of the 14th Text REtrieval Conference, Gaithersburg, Maryland (2005)
12. Zhai, C., Lu, X., Ling, X., He, X., Velivelli, A., Wang, X., Fang, H., Shakery, A.: UIUC/MUSC at TREC 2005 Genomics Track. In: Proceedings of the 14th Text REtrieval Conference (2006)
13. Strohman, T., Metzler, D., Turtle, H., Croft, W.B.: Indri: A language model based search engine for complex queries. In: Proceedings of the International Conference on Intelligence Analysis (2004))

Global Translational Motion Estimation (GTME)*

Yang Tao [1], Zhiming Liu[2], and Yuxing Peng[1]

[1] College of Computer
National University of Defense Technology, Changsha, China
[2] College of Computer, Nanhua University, Hengyang, China
tao_yang76_8@hotmail.com

Abstract. Motion estimation and motion compensation are included as major technologies into the existing video coding standards, which generate the motion vectors that determine how each motion compensated prediction frame is created from the reference frame. As traditional motion vectors are 2-dimensional and translational, they could not represent the actual moving directions of objects or blocks in frames. This paper proposes the special idea of GTMV (global translational motion vector), and then presents us the nearly Full Search Motion Estimation and Motion Compensation algorithms based on GTMV. Experimental results based on the H.264/AVC reference software of JM10.1 show that the proposed algorithms could get a better video compression performance and could improve the coding efficiency despite increasing some coding complexity.

1 Introduction

Motion estimation in video coding is an important step in many video coding systems and video processing applications,such as biomedical video coding systems. Several methods for solving this problem have been proposed, but only a few manage to achieve a high level of accuracy without sacrificing processing speed. But motion estimation and motion compensation technologies are still the very important technologies in the model of temporal video coding, which are efficiently used to eliminate the large amount of temporal and spatial redundancy that exists in video sequences, especially for biomedical video sequences.

In conventional predictive coding the difference between the current frame and the predicted frame (based on the adjacent or nearby frames) is coded and transmitted. Among all the motion estimation algorithms, Block-Matching Algorithm (BMA) is the most widely used nowadays, which is adopted by many video coding standards, such as Mpeg1/2/4 [1], H.261 [2], H.263 [3] and the new JVT (or H.264, MPEG-4 AVC) [4] video-coding standard and AVS [5]. In addition, many researchers have focused only on the Block Matching Algorithms and fast motion estimation algorithms, such as Three Step Search (TSS) [6] and Four Step Search (FSS), HEXS [7] and so on. On the other hand, although each block within a given search window is compared to the current block and the best match is obtained by the Full Search or the Exhaustive

* The research work was supported by NSF project no. 60433040.

K. Li et al. (Eds.): LSMS 2007, LNCS 4688, pp. 756–766, 2007.

Search algorithm, the best exact match for block moticn is away from being obtained because of the 3-dimensional moving directions of objects in video sequences, which is described by a simple mismatching exmple as illustrated by Fig.1.

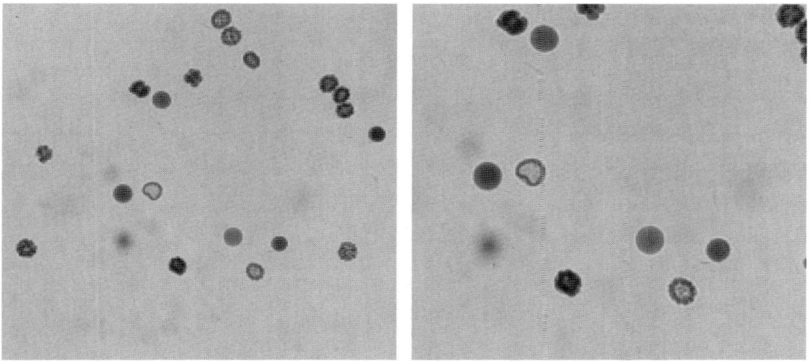

Fig. 1. A simple mismatching example

As illustrated by Fig.1, they are two frames in a biomedical sequence, and the right one could be gotten by zooming in some area inside the left one. Although each block within a given search window in the left frame is compared to the current block in the right frame, even if the best match is obtained by the Full Search or the Exhaustive Search algorithm, the best exact match for block motion is away from being obtained. Because the algorithm has not taken the zooming factor into account.

Thus, this paper presents a novel motion estimation algorithm, which gives excellent results on the block matching. The algorithm starts by computing global translational motion vectors (which is called GTMV) from the videc frames. And then it gets the better exact match. Experimental results show that the algorithm based on the GTMV improves the performance of video coding systems. Computationally, the algorithm idea is simple and can be implemented into the new video coding systems.

The rest of the paper is organized as follows: in the next sectien, we state the related work. Section 3 provides the proposed lcbal translational motion vector and section 4 presents us the Motion Vector Matching Criteria based on the GTMV. In addition, section 5 proposes the details of the motion estimation based on the special global translational motion vector. Section 6 presents experimental results to show the performance of the proposed motion estimation algorithms. Finally, we draw our conclusions in section 7.

2 Related Work

The Macro Block (MB) is the basic unit for motion compensation prediction technologies in many video coding standards. The motion estimation process based on MBs is to determine a referential macro block within the search window (also 16 by 16) in a referential frame. The reference frame comes from the encoded frames among the video sequences, which is before or after the current frame in the playing order. The searching procedure begins within the center of search window to find the smallest value according to the matching criterion. The block matching motion estimation is illustrated in Fig.2.

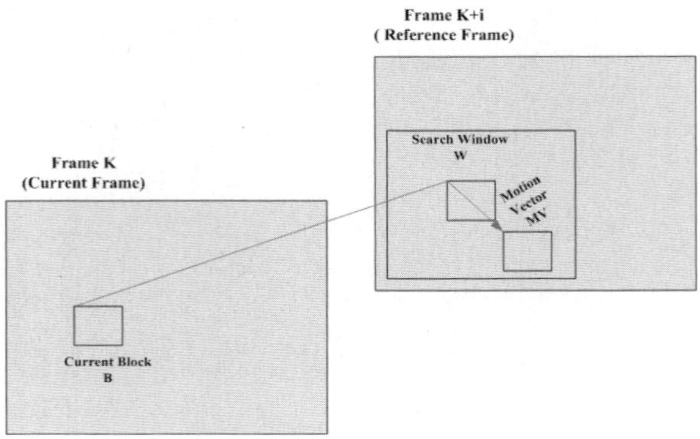

Fig. 2. Block Matching in Motion Estimation

Assume that the size of current block is MxN, and w is the size of search window, and $\vec{d}$ represents the motion vector for current block, then the criterion of Sum of Absolute Difference (SAD) can be presented in the following equation.

$$. \ \mathrm{SAD}_{i,j}(\vec{d}) = \sum_{p=0}^{M-1}\sum_{q=0}^{N-1} \left| I_k(i+p, j+q) - I_{k+z}(i+p+d_x, j+q+d_y) \right| \tag{1}$$

Where k is the subscript for current frame, and d_x and d_y are respective the horizontal and vertical elements of $\vec{d}$. In the search window, lower the $\mathrm{SAD}_{i,j}(\vec{d})$ is, better the matching block is. By the criterion, any block's motion vector is shown as Equation.2.

$$. \ \vec{d} = \arg \underset{\vec{d}_{m,n} \in W_{i,j}}{MinSAD_{i,j}}(\vec{d}_{m,n}) \tag{2}$$

According to Equation.2, $\vec{d}$ is the right motion vector in the search window $w_{i,j}$.

In order to get the motion vector quickly, many researchers have done a lot of work [6-11]. S. Zhu presented a new diamond search algorithm for fast block matching [7]. Based on an in-depth examination of the influence of the search pattern on speed performance, Ce Zhu proposed a novel algorithm using a hexagon-based search pattern [8] to achieve further improvement. The hexagon-based search pattern was investigated in comparison with diamond search pattern and demonstrated significant speedup gain over the diamond-based search. Cheung Chun-Ho proposed two cross- diamond - hexagonal search (CDHS) algorithms [9], which differed from each other by their sizes of hexagonal search patterns. These algorithms basically employed two cross-shaped search patterns consecutively in the very beginning steps and switch using diamond-shaped patterns. From above, many algorithms were used to reduce the computing complexity. But they didn't take more exact motion directions into account, thus the algorithms couldn't get more matching accuracy.

Although the authors in [10] presented a Kalman filter approach for motion estimation for accurately estimating the 3D position and orientation of a moving object from a sequence of stereo images, and Shackleton formed an interpolation image from the interpolation 3-D shape model and the image shaping values from the first and second 3-D shape models [11], the Kalman filter approach and the interpolation 3-D motion vector are not suitable to code blocks for classical video coding systems. By analysing the existing motion estimation technologies, we believe that they have never been taken the zoom factor into account in the video coding or image compression systems. Thus, in order to represent the moving directions (i.e. a vidicon moving up and down), this paper presents a kind of special global translational motion estimation algorithm based on the special three-dimensional motion vector, which will be described in the next section.

3 Global Translational Motion Vector

Compared with the traditional motion vector, the global translational motion vector has three elements, and one of them can represent the match block zoom direction.

The residual frame is also the difference between the current frame and the prediction frame. Now we present the description of the global translational motion vector as follows.

$$\vec{d} \in \{(d_x, d_y, d_z) | d_x, d_y \in \mathrm{w} \text{ and } 0 \le d_z \le \psi\}. \tag{3}$$

Where w is the size of search window, and ψ is the value of limitation for block zooms. One GTMV consists of one traditional MV and another dimensional vector. When the current block moves to a point within the search window in the reference frame, we will get a traditional MV with elements of (d_x, d_y). Then obtain another element of d_z by zooming in or out on the matching block. Because of the limitation of the size of current block and frame, we use the value of ψ to limit the zoom. Above all, the global translational MV is illustrated in Fig.3.

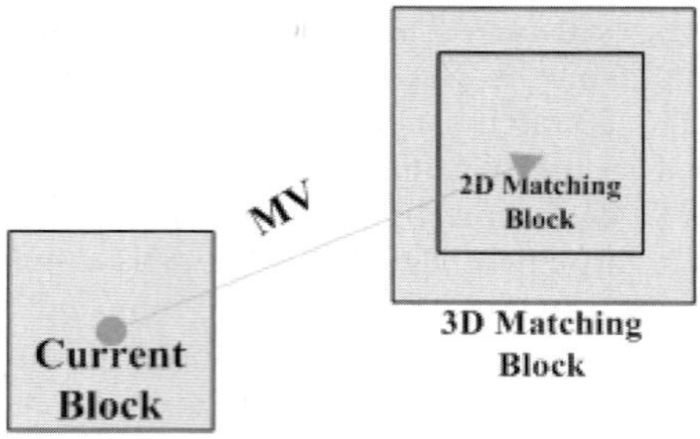

Fig. 3. Global Translational Motion Vector

As shown in Fig.3, the size of matching block obtained by traditional motion vector is the same as current one, but that of matching block got by GTMV is not always the same as current. Since it has three elements, one of which zooms in or out on the current, the size of matching block is more or less than that of current block.

4 Novel Motion Vector Matching Criteria

In order to improve the compression efficiency of video coding, there are a number of criteria to evaluate the "goodness" of a match, but we still adopt the Sum of Absolute Difference (SAD) as the matching criterion with the difference on computing the prediction blocks comparing with traditional matching.

According to Equation. 1, assume that the size of current block is MxN, and w is the size of searching window, and $\vec{d}$ represents the global translational motion vector for current block, then we can get the updated matching criterion as follows.

$$\mathrm{SAD}_{i,j}(\vec{d}) = \sum_{p=0}^{M-1}\sum_{q=0}^{N-1}\left|I_k(i+p, j+q) - I'_{k+z}(i+p+d_x, j+q+d_y, d_z)\right| \cdot \tag{4}$$

Where k is the subscript for current frame, and d_x, d_y and d_z are respective the horizontal, vertical and zoom elements of $\vec{d}$. I'_{k+z} is the mapping function of matching block to prediction one, which is got by the global translational motion vector transform. The mapping function is just treated as the process of upsampling or downsampling. As illustrated in Equation.4, if we compute the value of SAD, the current block and prediction block computed by the mapping function should have the same size.

Based on the main idea of video coding, matching blocks come from the coded frames, so given a global translational motion vector $\vec{d}\left(d_x, d_y, d_z\right)$, the mapping function I' should be satisfied with the following conditions.

1) I' may be a suitable surjection, and the elements of prediction blocks can be computed out through the matching blocks.

2) $I' : I \xrightarrow{d_z} I'$. d_x and d_y play the same role as traditional motion vector and only represent the position moving horizontally. d_z is the key element of global translational MV.

3) When d_z>1, I' is a downsampling mapping. When d_z equals 1, I' is a equal mapping and when 0<d_z<1, I' is a upsampling mapping, illustrated in Fig.4.

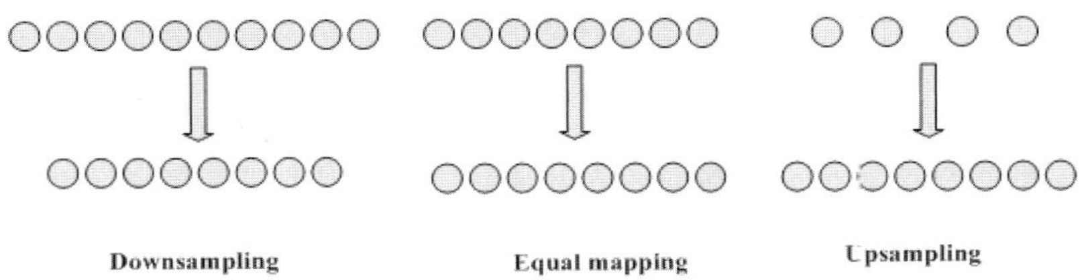

Fig. 4. Mapping Function I'

Based on the interpolation technologies of video coding, we adopt the following filters as the downsampling and upsampling filter, and then make the mapping function I' for global translational motion vectors.

$$H_{down} = \frac{[2,0,-4,-3,5,19,26,19,5,-3,-4,0,2]}{64}. \tag{5}$$

$$H_{up} = \frac{[1,-5,20,20,-5,1]}{32}. \tag{6}$$

5 Motion Estimation and Motion Compensation Algorithms

Although there are several approaches to reducing the computational complexity, we still adopt the full search algorithm in the proposed GTME firstly in order to show the excellence of GTMV.

We make the FSA updated from traditional one, and describe it as follows.

1) Initialization: initialize the environment. Set the size of current block to MxN, and set the prediction block P with motion vector of $\vec{d}(d_x, d_y, d_z)$ and search window W.

2) In the search window, collect all the global translational motion vectors into the set of *MV*, that is

$$MV = \left\{ \vec{d}(d_x, d_y, d_z) \middle| d_x, d_y \in W \text{ and } \frac{1}{M \times N} \le d_z \le \frac{W}{M \times N} \right\}. \tag{7}$$

3) Take one out of the set *MV*, and make the mapping I' from Equation.5 or 6.

4) According to the matching criterion of Equation.4, compute the SAD of the global translational motion vector, and then put it into the set S, i.e.

$$S = \left\{ SAD_{\vec{d}} \middle| \vec{d} \in MV \right\}. \tag{8}$$

5) Repeat the step 3 and 4 until all the vectors in the set *MV* are taken out.

6) Reorder all the elements in the set S and find out the motion vector $\vec{d}(d_x, d_y, d_z)$, which makes the lowest $SAD_{\vec{d}}$ value. Then return the vector, and finish the whole process.

We can get the best matching motion vector by the proposed algorithm and then go on with the process of motion compensation for prediction blocks. Thus, we have

$$R_{i,j}(p,q) = I_k(i+p, j+q) - I'_{k+z}(i+p+d_x, j+q+d_y, d_z) \,. \tag{9}$$

Where $R_{i,j}$ represents the residual block corresponding to the current block *k* and $R_{i,j}(p,q)$ is the value of pixel (*p*, *q*) in the residual block with the best matching motion vector $\vec{d}(d_x, d_y, d_z)$. The whole process of motion compensation will not finish until the residual block $R_{i,j}$ and motion vector $\vec{d}(d_x, d_y, d_z)$ are encoded. The proposed video coding syntax is illustrated in Fig.5.

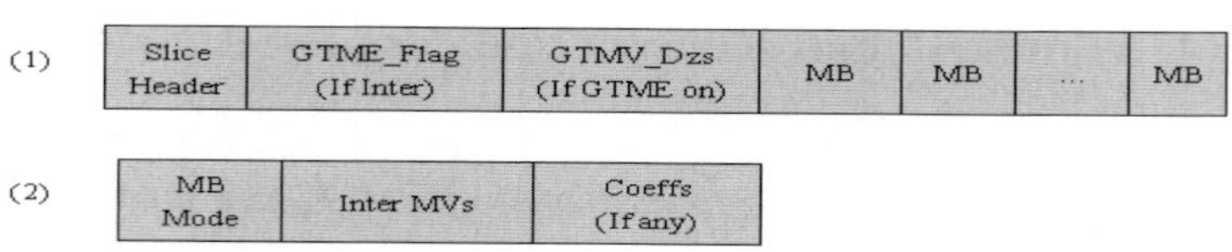

Fig. 5. Bit stream syntax of the proposed video coding method: (1) Slice-level syntax; and (2) MB-level syntax.

For each P frame, a GTME_flag symbol is added right after the slice header to signal if GTME is on or off for the current slice. The codeword number for GTME _flag is listed in Table 1. If GTME is off, then all following bit stream will be exactly same as that of current JM [12]; comparing to the original JM syntax, only 1 extra bit per inter frame is needed for the GTME _flag. If GTME is on, GTMVs are to be coded after the GTME _flag; then, for each MB, if the MB mode is determined to be one of the GTME modes, motion vector coding can be simply skipped; the MV values for all 4×4 blocks in that MB can be derived from GTMVs through the interpolation.

Table 1. Codeword number of GTME_flag

Coded	GTME_Flag	Comment
0	0	GTME off
1	1	GTME on

6 Experimental Results

In the experiment, the proposed motion estimation algorithm was integrated within the JVT reference software JM10.1 [12]. We updated motion estimation and motion compensation modules in the encoder and decoder of H 264 a little so as to make our proposal efficient. The test sequences were the classical sequences. The system platform was the Intel Pentium III Processor of speed 1.0GHz, 512MB SDRAM, and Microsoft Windows 2000 Server. To compare the bit rate reduction performance and the computational complexity of the proposed coding method with the traditional one of H.264, the PSNR and the bit rate are measured at different QP from 24 to 40. The parameter of FramesToBeEncoded for encoding is set to 100 and the parameter of IntraPeriod is set to 10. Experiments were carried out for comparing global translational motion vectors with motion estimation algorithms provided in the JM software and with one as shown in Section 5 only added the early-termination rule. Three test sequences were used, and simulations were performed using JM10.1 version with parameters and conditions as listed in Table 2.

Table 2. Parameters and Conditions

Contents	Descriptions
Reference Software	H.264 reference software model JM10.1
Video Sequences	highway.cif, mobile and calendar.cif,tempete.yuv
Frame Format	IPPPP
Frame Number	100
Environment	PC CPU Intel Celeron 1G, 512M SDRAM
PSNR Compare	Yes
Complexity Compare	Yes
Else	MatLab6.5

In the experiment, we got different PSNRs and bit rates by changing the Qp parameter.Fig.6, Fig.7 and Fig.8 show the RD performance for the three different sequences. The results for computing complexity are some times in time cost by the modified JM10.1 over the JVT JM10.1. If we adopt any fast algorithms, the encoding efficiency for video coding will be improved heavily. This paper is just to show the accuracy and validity for the idea of designing motion estimation systems based on the special Global Translational Motion Vector.

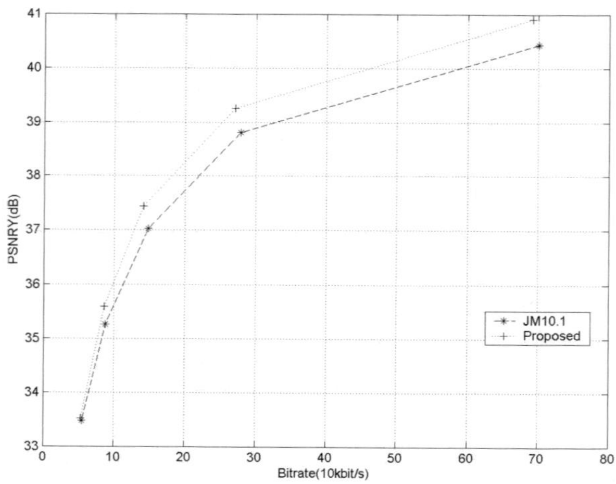

Fig. 6. RD-curve of highway.cif

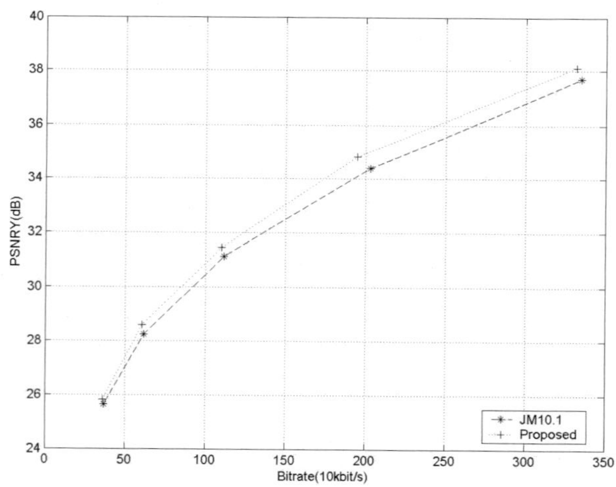

Fig. 7. RD-curve of mobile and calendar.cif

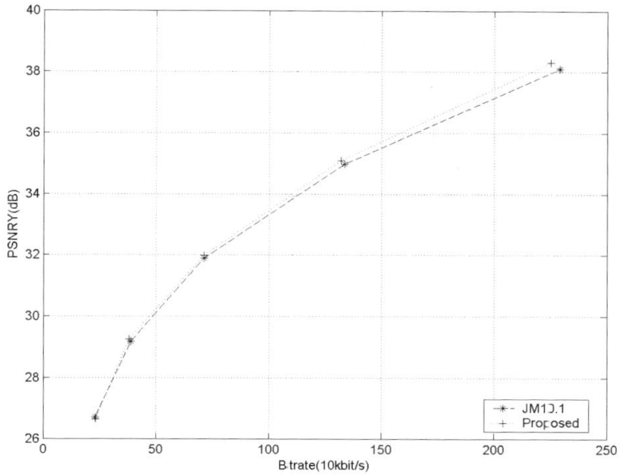

Fig. 8. RD-curve of tempete.yuv

The experimental results show that the proposed method could achieve a better coding performance and gain some computing complexity while encoder and decoder complexity is neglectable increased with the Full Search Algorithms.

7 Conclusions

In this paper, we have taken global translational motion vector into account to update the encoder and decoder for H.264. We first present the incomplete representation of block moving directions in the traditional video coding systems, which are only working with the traditional motion vector in the motion estimation technologies in the codec. Then propose the special global translational motion vector with corresponding motion estimation algorithms integrated into the video coding techniques and codec of H.264. According to experimental results, we analyze the performance of the proposed methods and prove that they are efficient and significant.

As part of future work, we would do much more on finding fast motion estimation algorithms based on GTMV and then propose them to JVT so as to applying to implement their video coding applications. More efficient motion estimation algorithms based the proposed GTMVs, more efficient and practice in video coding even for the newest video coding standard of JVT H.264.

References

1. ISO/IEC JTC1/SC29/WG11 N4668: MPEG-4 Overview (March 2002), http:// www. m4if. org/ resources/ Overview.pdf
2. Video Codec for Audiovisual Services at px64 kbit/s ITU-T Recommendation H.261: Version 1, ITU-T, ITU-T Recommendation H.261 Version 1 (1990)

3. Video Coding for Low Bitrate Communication: International Telecommunication Union, ITU-T Recommendation H. 263 (1995)
4. Draft ITU-T recommendation and fmal draft intemational standard of joint video specification (ITU-T Rec.H.264/ISO/IEC 14496-10 AVC): in Joint Video Team (JVT) of ISOiIEC MPEG and ITU-T VCEG, JVTGOSO (2003)
5. Audio Video Coding Standard Workgroup of China(AVS): Video Coding Standard FCDl.0 (November 2003)
6. Koga, T., Iinuma, K., Hirano, A., et al.: Motion Compensated InterFrame coding for Video Conferencing. In: Proceedings Nat. Telecommunications Conf. 81, pp. G5.3.5 (November 1981)
7. Zhu, S., Ma, K.: A new diamond search algorithm for fast block matching[J]. IEEE Transactions on Image Processing 9(2), 287–290 (2000)
8. Zhu, C., Lin, X., Chau, L.-P.: Hexagon-Based Search Patten for Fast Block Motion Estimation. IEEE Trans. On CSVT 12, 349–355 (2002)
9. Chun-Ho, C., Lai-Man, P.: Novel Cross Diamond Hexagonal Search Algorithms for Fast Block Motion Estimation. IEEE Transactions on Multimedia 7(1), 16–22 (2005)
10. Lee, S., Kay, Y.: A Kalman filter approach for accurate 3D motion estimation from a sequence of stereo images. Pattern Recognition (1990)
11. Shackleton, et al.: Method and device for frame interpolation of a moving image. WO93/02529 (February 04, 1993)
12. Joint Video Team (JVT): reference software JM10.1, http:// iphome. hhi.de/ suehring/ tml/ download/jm10.1.zi

Contented-Based Satellite Cloud Image Processing and Information Retrieval

Yanling Hao, Wei ShangGuan, Yi Zhu, and YanHong Tang

School of Automation, Harbin Engineering University
Harbin, Heilongjiang Province, China, 150001
sgw1998@163.com, shangguanwei@hrbeu.edu.cn

Abstract. Satellite cloud image is a kind of useful image which includes abundant information, for acquired this information, the image processing and character extraction method adapt to satellite cloud image has to be used. Content-based satellite cloud image processing and information retrieval (CBIPIR) is a very important problem in image processing and analysis field. The basic character, like color, texture, edge and shape was extracted from the cloud image, and then the satellite cloud image database was provided to store the basic character information. Since traditional image retrieval method has some limitation, for realized image retrieval accurately and quickly, the CBIR method is adaptive. On the basis of the key technology of CBIPIR was studied, we could obtain the better retrieval effect, and the image retrieval result was shown in detail. The experiment result proves that the research and application of content-based satellite cloud image processing is valuable, which could improve the professional image application efficiency more.

1 Introduction

With the development of remote sensing imaging technology, the satellite information could include more abundant data, and the quality of satellite cloud image had achieved great progress. While relative to the upper work like image analyzing, image reconstruction, the quality of image is not very well still. For avoiding the influence of disturb in process of image processing, the image pretreatment method have to be used to the image, so that we could acquire better image, then improve the precision and effect of image processing. The processing of satellite cloud image has become one of the most important tasks for using the satellite cloud image information for continuous research. The satellite cloud image character extraction and database system has become very important research contents; there are abundant image characters in the database, including the source image, contour, framework, acreage, center and texture of the cloud series, etc. How to apply plentiful information in the database is one important problem we must faced.

Content-based image retrieval (CBIR) is becoming increasingly important in visual information management [1]. Research is underway on CBIR systems that can retrieve images in an unconstrained environment. However, general solutions to the key problems relating to regions of interest, segmentation, and retrieval are still being

K. Li et al. (Eds.): LSMS 2007, LNCS 4688, pp. 767 – 776, 2007.

sought. Over the years various techniques have been proposed for retrieval in digital image databases [2]. Many of these techniques operate under specific constraints about the type of image data they operate upon. Frequently, such constrains have been put on the retrieval of images containing certain objects.

2 Satellite Cloud Image Pretreatment

Construct satellite cloud image database must do some processing to the image and obtain the image character data, image pretreatment mentioned in this paper means acquire necessary image character like gray/ color, texture, contour, etc, it could provide good preparation for image matching and image retrieval. Pretreatment of satellite cloud image is a key procedure before practical image processing and analysis, since the research aim of satellite cloud image processing is acquired more abundant information from this kind of useful data source. In this paper, several amendatory image pretreatment methods which adapt to satellite cloud image were applied and the experiment demonstration was shown in detail. The most valuable image pretreatment method is threshold processing, and some image filtering method have to be used for better experiment result before image processing and analysis [3-6].

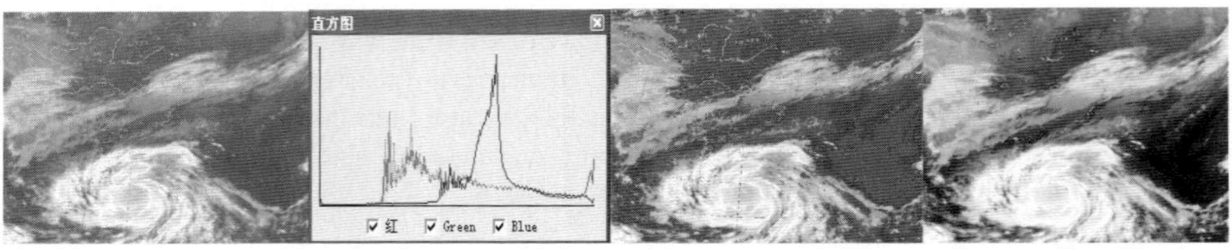

(a) The typhoon satellite cloud image (b) Image histogram (c) Big template median filtering (d) Adaptive weighted median filtering

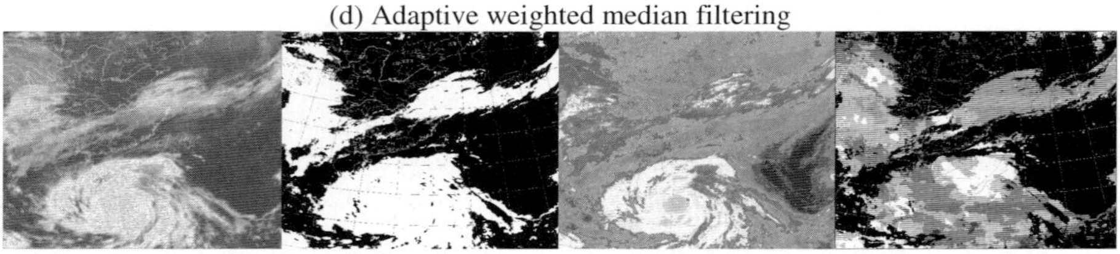

(e) Local statistic filtering (f) Threshold processing (g) Rainbow code (h) Remove the background of satellite cloud image

Fig. 1. Satellite cloud image pretreatment process

In practical image processing, for achieved better effect quickly and efficiently, some applied image pretreatment should be used. Applied satellite cloud image pretreatment method includes fake color processing and background elimination mainly. Fake color technology is one of the applied image enhancement technology, it was designed according to the theory that the people optical distinguish ability to the bight less than to the color [7]. This technology divided the pixel value of image into some layers, different layers of color was used to represent a layer of image region, then the original image become fake color image. In practical application, according to the weather demand, using the temperature for image degree partition. Below the zero degree, every 10 degree is a layer; above the zero degree, every 10 degree is a layer. According to this rule, we could carry through the density partition to infrared satellite cloud image. The common fake color processing to satellite cloud image is shown in Fig. 1(g).

Background elimination means remove the high temperature point which has small pixel value. According to weather demands and continuous processing and analysis demands, using the subsection threshold method to remove the image back ground. On the basis of image pretreatment, this image segmentation is a kind of multi-threshold adaptive segmentation. The segmentation procedure is like that:

(1) Find the maximum pixel value level a1 and center region pixel value level a2 in the pretreatment image, seen the a2 as threshold, decide the center region of different cloud series; (2) Partition these points according to the distance between the neighbor pixels, get the number of segmentation region and center position; (3) Carry through the erode processing to primary segmentation image according to different structure factor, then compute 'and' operation with the processing result and segmentation region center, remember the erode number and corresponding segmentation region, then acquired the initial segmentation template; (4) Using 'expand' operation to every segmentation region center using corresponding structure factor and computation number, integrate these different region. the adaptive segmentation template M of every region was obtained; (5) Using template M 'and' with preparatory segmentation image, then carry through the mathematical morphologic operation like 'open' or 'close' to the final result, then the main cloud series could recognition in the satellite cloud image. The experiment result of background elimination arithmetic to satellite cloud image is shown in Fig. 1 (h).

3 Satellite Cloud Image Character Extraction

In satellite cloud image database, cloud image character plays very important role, so how to acquire the necessary cloud image character accurately has become one important problem.

3.1 Color Character

As the first chosen character for contented-based image proceeing, color character is stable. The color character is a kind of vision character of object surface, every object has its own color character; at the same time, same kind of object has similar color character. So we could distinguish object according to different color character. Color character-based image retrieval solving three questions mainly: 1) spatial coordinate expression of color character; 2) extraction of color character; 3) similar degree measurement arithmetic base on color character [8].

The familiar color coordinate space includes: (1) The tricolor model ((RGB), (CMY), (I1, I2, I3)) faced hardware device;(2) The Munsell model, HSI (hue, saturation, intensity) model, HSV (hue, saturation, value), HSB (hue, saturation, brightness) model which faced vision perception;(3) The MTM (mathematical transform to munsell) model of equal color space, L*u*v* model and L*a*b* model of CIE.

There are many extraction method of color character, like histogram method, cumulative histogram method, local cumulative histogram method, color distribution method, etc. Because the data quantity of color image is abundant, the same aspect of every method is describing color information using efficient and compact method.

Color-based similar degree measurement arithmetic includes: histogram intersect method raised by Swain and Ballard, histogram weighted method, histogram distance comparison method, etc. For reduce operation complexity degree, we could quantity the color space, use the body part of histogram, or use low resolution histogram, etc.

3.2 Texture Character

Texture is the visual pattern of object surface structure, the research of image texture is very important to image understanding and analysis. Infrared satellite cloud image is a kind of texture image, so we could do some texture processing to the image. The general texture character extraction method includes co-occurrence matrix, K-L transform, texture table, etc [9].

The co-occurrence matrix method of texture description is based on the repeated occurrence of some gray-level configuration in the texture image; this configuration varies rapidly with distance in fine texture and slowly in coarse texture [10]. Suppose the part of a textured image to be analyzed in an M × N rectangular window. An occurrence of some gray-level configuration may be described by a matrix of relative frequency $P_{\phi,d}(a,b)$, describing how frequently two pixels with gray-level a, b appear in the window separated by a distance d in direction ϕ. These matrices are symmetric if defined as given below. However, an asymmetric definition may be used, where matrix values are also dependent on the direction of co-occurrence. A co-occurrence matrix computation scheme was given as follows, the following algorithm calculate the co-occurrence matrix C_r from the image $f(i, j)$.

1) Assign $C_r(z, y) = 0$ for all z, y ∈ [0, L], where L is the maximum brightness.

2) For all pixel (i1, j1) in the image, determine (i2, j2) which has the relation r with the pixel (i1, j1), and perform

$$C_r[f(i_1, j_1), f(i_2, j_2)] = C_r[f(i_1, j_1), f(i_2, j_2)] + 1 \tag{1}$$

Texture identification can be based on texture parameter criteria derived from the co-occurrence matrices. The parameter includes:1) Energy; 2) Entropy; 3) Contrast; 4) Correlation; 5) Local steady. GMS infrared satellite cloud image have 256 gray levels, if compute with original gray level, the computation quantity will be too enormous, so we need classify the gray level. We could classify the gray level to 32 levels, gray value 0~15 is 0, and gray value 16~31 is 1, as like this. Choose the 32 × 32 windows as the basic unit for computation; compute the texture feature of 120 cloud image samples with different cloud type, the texture character of four kind of object cloud is shown in table 1.

Table 1. The texture character of cloud image object. The pixel gray distance of cloudless sky and stratus is less, the texture is more symmetrical; the average gray value of cirrocumulus is higher, pixel gray distance is bigger, the texture is more compact; the pixel gray distance of cumulonimbus is biggest, the texture is coarse. These texture feature parameter computed by gray co-occurrence describe the change degree of different object and texture.

	Stratus	cirrocumulus	cumulonimbus	Cloudless sky
Energy	0.46	0.88	0.78	0.30
Entropy	2.12	3.17	3.62	0.90
Contrast	1.95	2.43	3.89	1.03
Correlation	2.06	1.92	1.90	2.31
Local steady	0.80	0.67	0.25	0.86

3.3 Shape Character

Image shape is an important character which could describe the object contour and physical structure. The describe method of image shape includes: 1) Shape character method; 2) shape changing method; 3) shape relation method. Traditional shape expression method often use a set of character, some character describe the global character of object shape, like acreage, stretch length, principal axis direction, etc; some character describe local character of object boundary, like corner point, character pint, etc [11]. The shape character could be shown with shape character spatial point, or a set of character vector.

Shape means the shape of target region commonly, the shape description of target region could study from two aspects, using pixel group in region and using pixel group on region boundary, the two pixel group is complementary each other. According to this definition, the character expression method could be divided into: 1) interior parameter method. Like center square or eccentricity degree, etc; 2) outer parameter method, like curve point, corner point, corner degree, etc.

Shape-based retrieval technology includes: 1) the character expression-based technology; 2) using the transform field technology; 3) in virtue of shape transform technology. But most of the method only adapt to lesser database, which made it impossible to search the bigger database because of the complexity degree.

3.4 Spatial Relation Character

Image is obtained from the information spatial scene projection, the spatial position and the correlation of all the object is one of the most important information of image too. The spatial correlation of the object reflects the image interior correlation. When there are many absolute parts or targets, the emphasis of retrieval is the position correlation of the target, especially when the number of the target is not only one, then we could ignore the shape and size of the object target relative to the target object, using spatial relation character retrieval method for image retrieval.

Spatial relation could be divided into connection or neighbor, superposition or overlap, contain, etc. In general, spatial position relation includes relative position information and absolute position information. The former information emphasize the relative condition of targets, such as up down or life right, the latter information emphasize the distance and direction between the targets [12].

Using spatial relation character could enhance the description and analysis ability to the image content, while the spatial relation is sensitive to the geometry transform of the targets, like rotation, reverse, zoom, etc, it could show the scene information accurately and efficiently. For better retrieval, except spatial relation character, we should use other character as the assistant character too.

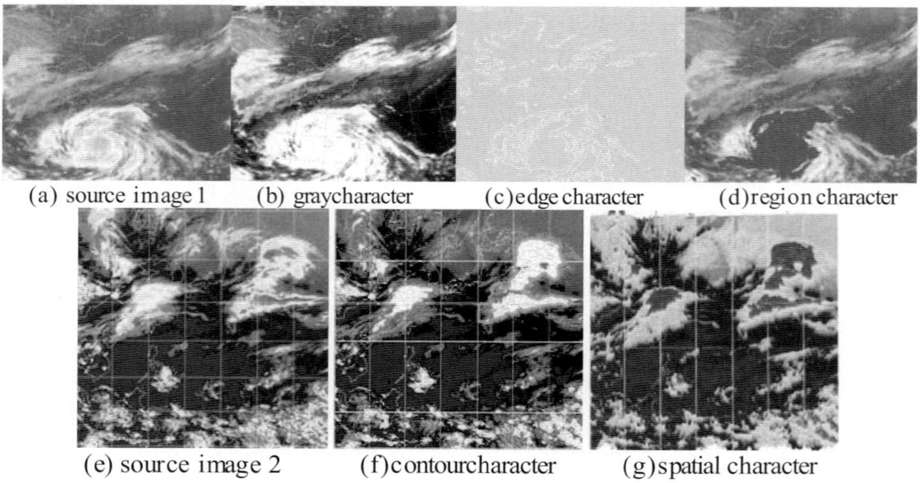

Fig. 2. The character extraction of satellite cloud image

4 Satellite Cloud Image Database System

The research of satellite cloud image database is a new subject, because the demands of constructing satellite cloud image database, the research have to be combined with weather field knowledge. In this paper, using digital image processing and pattern recognition technology, combine with weather field demands, do some segmentation processing and character extraction to source satellite cloud image, these characters includes configuration character, statistical character and texture character, etc, then we stored these characters in the database, the satellite cloud image database was constructed.

The satellite cloud image has three "uncertain" character, uncertain record item, uncertain record number, and uncertain record length. To the complex data format, it's difficult to realize operation work like data management or random search, for general relation database. We describe satellite cloud image using following data structure form:

- ✧ *{image}={date}{label}*
- ✧ *{region extension)={jumping-off {point}end-point}*
- ✧ *{region center}={point}*
- ✧ *{character}={statistic character}{texture character}{shape character}*

- ✧ *{statistic character}={Acreage}{Perimeter}{Length Width ratio}*
- ✧ *{texture character}={histogram}{Energy}{Entropy}{gray difference}*
- ✧ *{shape character}={contour}{framework}{isoline}*
- ✧ *{contour}={label}{point number}{point}*
- ✧ *{{framework)= {label}{point number}{point}*
- ✧ *{isoline}={gray level}{point number}{point}*
- ✧ *{point}={longitude}{latitude}*

According to above mentioned definition, the program could process every image data, provide efficient information for later research work.

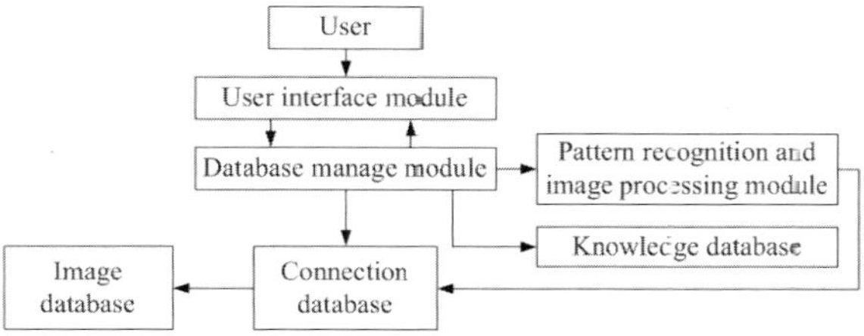

Fig. 3. The basic structure model of satellite cloud image database

The satellite cloud image database is an appropriative relation database practically. As the image database, it has some other special character. The database we have designed includes four modules: user interface module, image processing and pattern recognition module, database module and database management module. User interface module provide visual user interface mainly, finish the man-machine operation function; image processing and pattern recognition module finish the image processing, segmentation and acquire the character to the satellite cloud image; database module used for store the satellite cloud image, which includes image database, relation database and knowledge database (store the research field knowledge); database management module finished some maintenance functions like insert, delete, index, edit of database.

5 Content-Based Satellite Cloud Image Retrieval

The purpose of image retrieval is retrieve the corresponding image or image sequence from image library quickly and accurately. The earliest image retrieval research begins in 1979 on the Database Techniques for Pictorial Applications conference in Florence. Recently, the great application potential has attracted more attention of researcher.

5.1 Traditional Satellite Cloud Image Retrieval and Its Limitation

Traditional image management is file system, when user searches an image, they have to open the file and explore every image. With the increment of image quantity, the retrieval efficiency is very low. In many research fields, we have to deal with much visual information while the image processing ability is not enough; there are abundant image data without efficient use. For example, the satellite remote sensing is a classical field. Many data have been stored since it is received. The USA Geostationary Operational Environmental Satellite 9 (GOES9) transmit an hemispherical infrared cloud image every hour, if the image resolution is 1504x1301, the image size is 133KB to 155KB, 600 satellite cloud image every month, then the total image capacity is 80 to 100MB, there are nearly 1GB data in one year.

The traditional search method of satellite cloud image using image-based character remark. The file name of every satellite cloud image is date and name, and the search keywords is the file name too, then using traditional database management system for image retrieval. The image retrieval method will lead to inconvenience to satellite cloud image database. For example, the name of a GOES9 satellite cloud image is:

L1 GOES9 METEOSATS VISSR 03 MSIC 20070301 0051 RT EA NO V10.J PG

Seen form the image name, the file is the infrared cloud image in wee hours 0:51, March 1, 2007, while the image feature couldn't reflected by the image name. Not only the text description is incomplete, but the name couldn't express the changing weather phenomenon. To the satellite cloud image, only well-trained experts could understand the image information accurately, quick search a certain image need enough experience.

5.2 Content-Based Image Retrieval

With the quick increment of visual information, efficient data management pattern become more urgent, this made content-based image retrieval technology get more attention. In 1992, the American national science fund institute establishes an expert group for visual information management, aim at finding new research direction for image database manage system, the experts generally think more efficient and intuitionistic visual information retrieval method is based on image interior feature. There are many researchers in different fields like computer vision, database management, user-machine interface and information retrieval, etc, they attracted by this valuable task, then the research of content-based image retrieval get quick development. Several famous image retrieval systems include QBIC system of IBM, Visual SEEK system of Columbia University, and Virage system of Virage Company.

In contented-based image retrieval, using interior visual character like color, shape, texture and spatial character for image retrieval. In classical contented-based image retrieval system, the visual content of image was stored with multi-dimension eigenvector form in the image database, the image eigenvector make up of the character database. When retrieve the image, the user provide example image or sketch map to retrieval system, then calculate the correlation or distance between the image eigenvector, finished the retrieval work in virtue of the index pattern [13]. The index pattern has provided an efficient route for image database retrieval. The basic image retrieval flow of CBIR was shown in Fig.4.

Recognize the cloud layer information from satellite cloud image is valuable in many research fields, which includes remote sensing, resource detect and weather forecast, etc. If apply the new generation visual information retrieval in the satellite cloud image retrieval, we could acquire better retrieval result than traditional retrieval system. If combine with the character of satellite cloud image, acquire its interior character or model parameter, then choose appropriate correlation degree, we could find a preferable contented –based satellite cloud image retrieval methods, which made it become possible that assistant weather forecast by history data. The design and experiment using CBIR method for satellite cloud image is less, so it's meaningful to develop this research task.

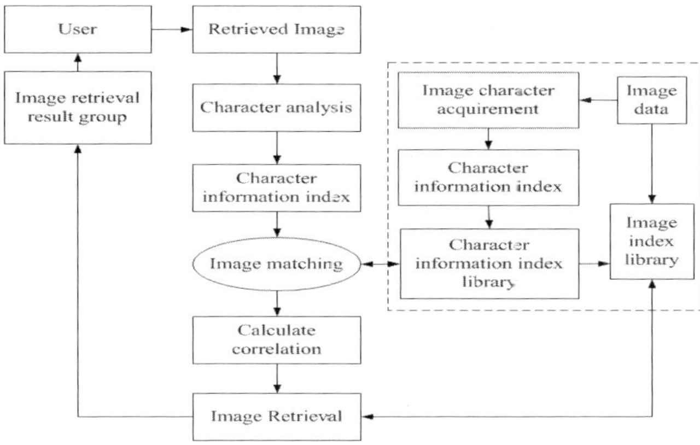

Fig. 4. The basic retrieval flow of CBIR

Fig. 5. Realization of satellite cloud image retrieval system. We could choose most appropriate image from the image database according to the correlation degree. The experiment proves that this method applied in satellite cloud image is effective, we could obtain the image we need.

On the basis of above mentioned idea, through analysis to the satellite cloud image, we choose the gray level, color, texture and spatial character as the main retrieval character, combine the image pretreatment result, the method for satellite cloud eigenvector group was studied. Then we could realize the image retrieval by using the synthetical character information we have obtained. Using above mentioned satellite cloud image database and corresponding image character, we could realize the quick search image and get accurate weather forecast by real-time and history data in same region. As well known, the main information in the cloud image is cloud layer information, earth and ocean is secondary information, in practical processing and analysis, we should consider all these information synthetically for better achievement.

6 Conclusions

The satellite cloud image is a valuable image data, how to use the information efficiently is a very important problem. Since the research aim of satellite cloud image processing is acquired more abundant information from this kind of useful data source, we do some processing work to image and extract the image character for future application, then the character was stored in the database. The basic theory and structure model of satellite cloud image database was studied, and the data model of satellite cloud image database was shown. The advantage and disadvantage of traditional and content-based image retrieval method were analyzed systematically, traditional image retrieval method has some limitation, and content-based satellite cloud image retrieval is a preferable method which could use the image information completely. For realized image retrieval accurately and quickly, the CBIR method is an adaptive method. While how to use the information retrieved from the database adequately still is a problem, so improve the second processing ability of image retrieval and database system is the key question in our future research work.

References

1. Picard, R.W.: Light years from Lena: Video and image libraries of the future. In: IEEE Int. Conference on Image Processing, vol. 1/3, pp. 310–313. IEEE Computer Society Press, Los Alamitos (1995)
2. Bimbo, A.D., Pala, P., Santini, S.: Visual image retrieval by elastic deformation of object sketches. In: IEEE Symposium on Visual Languages, St. Louis,MO, pp. 216–223. IEEE Computer Society Press, Los Alamitos (1994)
3. Sonka, M., Hlavac, V., Boyle, R.: Image Processing, Analysis and Machine Vision. People's Post Press (September 2003)
4. Pizer, S.: Adaptive histogram equalization and its variations. CVCIP 39(3), 355–368 (1987)
5. Lee, J.S.: Digital Image Enhancement and noise filtering by use of local statistics. IEEE Trans. PAMI-2, 165 (1980)
6. Zhang, Y.: Image processing and analysis, pp. 82–87. QingHua University Press (July 2002)
7. Welch, R.M., Sengupta, S.K., Goroch, A.K., et al.: Color cloud and surface classification using AVHRR imagery: An inter comparison of methods. J. Appl. Meteor., 405–420 (1992)
8. Iqbal, Q., Aggarwal, J.K.: Content-based Retrieval in Digital Image Databases Using Structure, Color and Texture. Pattern Recognition in Information Systems (PRIS), Portugal, pp. 1–2 (2004)
9. Haralick, R.M.: Statiscal and structured approaches to Texture. Proc.of the IEEE 67(5), 786–804 (1979)
10. Haralick, R.M., Shanmugam, K., Dinstein, I.: Textural features for image classification. IEEE Transactions on Systems, Man and Cybernetics, 610–621 (1973)
11. Zhou, X.S., Huang, T.S.: Edge/structural features for content based image retrieval. Pattern Recognition Letters 22, 457–468 (2001)
12. Li, Y., Shapiro, L.G.: Consistent line clusters for building recognition in cbir. IEEE International Conference on Pattern Recognition (ICPR) 3, 952–956 (2002)
13. Iqbal, Q., Aggarwal, J.K.: Retrieval by classification of images containing large manmade objects using perceptual grouping. Pattern Recognition 35, 1463–1479 (2002)

Biology Inspired Robot Behavior Selection Mechanism: Using Genetic Algorithm

Yiping Wang, Sheng Li, Qingwei Chen, and Weili Hu

Automation School, Nanjing University of Science & Technology,
210094 Nanjing, China
{wangyp1002,nonholonomic}@163.com
cqw1002@sina.com
hwl1002@mail.njust.edu

Abstract. Since behavior selection is a crucial issue not only in biology, but also in robotics, especially in behavior-based robotics, it is nature to consider the behavior selection problem both in biological view and robotic view. In recent years, accumulative evidences from neurobiology and anatomy have given rise to proposals that the basal ganglia-a group of subcortex nuclei in vertebrate brains- serve as a central selection mechanism. This paper introduces a robot behavior selection mechanism inspired by basal ganglia and makes explorations of applying genetic algorithm to the optimization of model parameters. The proposed method demonstrates its efficiency through a simulated robot foraging task and casts light on designing more intelligent and fluent behavior selection mechanism in the future.

1 Introduction

Behavior-based robotics (BBR) dates back to the early days of psychology, ethology, neuroscience and cybernetics. Ablib[1] is generally considered as a precursor in this area. In the middle 1980s, Brooks[2-3] introduced a complete new mode of thought for building intelligent mobile robots, which was later called behavior-based robotics.

In behavior-based robotics, the control of a robot is shared between a set of purposive perception-reaction modules called behaviors. Due to limited motor resources, behaviors with different and possibly incompatible objectives may produce conflicting control commands. Thus a major issue in the design of behavior-based control systems is the formulation of effective mechanisms for selecting behaviors.

Many behavior selection mechanisms have been proposed over the last two decades and they can be classified into two categories[4]: competitive mechanisms[2, 5 6] and cooperative mechanisms [7-9] respectively.

All the selection mechanisms mentioned above were considered in engineering view. But how do cockroaches, rats or even human beings select an appropriate behavior from a set of available behaviors? Why can they behave gracefully when encounter conflicts caused by two or more behaviors. What is the neural substrate of behavior selection? Recent progress in neurobiology and anatomy has indicated that

K. Li et al. (Eds.): LSMS 2007, LNCS 4688, pp. 777 – 786, 2007.

the basal ganglia play a crucial role in behavior selection for vertebrate and this is the primary focus of this paper.

This paper is organized as follows. In Sect. 2, two well-known computational models of basal ganglia are introduced. In Sect. 3, genetic algorithm optimized version of the two models mentioned above is proposed. Sect. 4, a robot behavior selection mechanism based on the optimized basal ganglia model is implemented and simulation experiments are carried out to demonstrate its efficiency. Conclusions are also provided in Sect. 5.

2 Models of the Basal Ganglia

2.1 Introduction to the Basal Ganglia

The basal ganglia are a collection of highly interconnected subcortex nuclei located in central brain, consisting of the striatum, the subthalamic nucleus (STN), the globus pallidus (GP, including the internal segment GPi and the external segment GPe), the substantia nigra (SN, including the pars reticulate SNr and the pars compacta SNc). Anatomical studies demonstrate that the basal ganglia take massive inputs originated from cerebral cortex, the brainstem, and the limbic system, process it in a complex way, and then pass it back to the cerebral cortex. Further more, studies from amphibiology suggest that basal ganglia are ancient structures and exist in the majority of vertebrate species. The above two facts imply that the basal ganglia can fulfill its role as an action selection mechanism (see Humphries[10] for details).

2.2 Qualitative Modeling

In order to support the hypothesis that the basal ganglia act as a central behavior selection mechanism in the brain, qualitative models of the basal ganglia should be constructed and analyzed. Although there are numerous models, only two of them will be discussed in this paper.

The first and most prevalent model of the basal ganglia termed direct-indirect pathway model (DIPM) was proposed by Albin et al.[11], and it had successfully explained some basal ganglia related movement disorders such as Parkinson's disease. In DIPM, the internal connections of basal ganglia were divided into two pathways: the inhibitory direct pathway projects from the D1-type cells in striatum to the GPi; the indirect pathway is inhibitory from the D2-type cells in striatum to the GPe, inhibitory from there to the STN, and excitatory to the GPi (Fig. 1 (a)).

As research goes on, some new connections in the basal ganglia have been found and confirmed. Gurney et al.[12], developed a new model called select-control pathway model (SCPM) of the basal ganglia to include those recently discovered connections such as the inhibitory connection from the GPe to the GPi and the excitatory connections from the STN to the GPe (Fig. 1 (b)). Humphries and Gurney[13] extended SCPM to include the thalamic complex, but since this extended model is rather complex and the thalamic complex are generally not regard as a part of the intrinsic basal ganglia, it will not be discussed here.

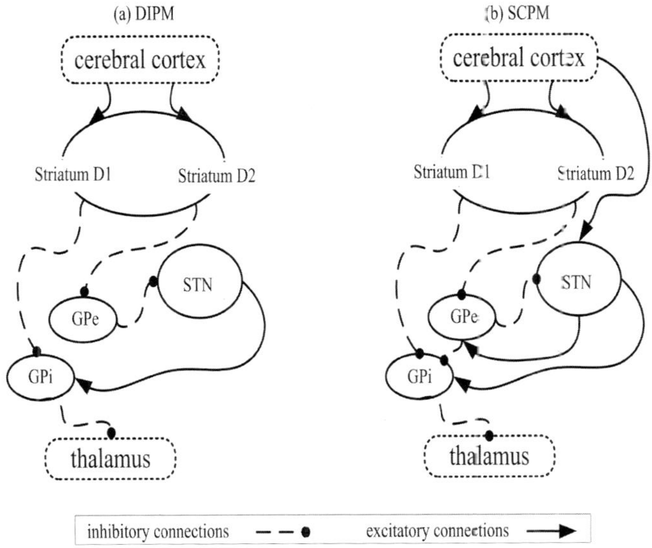

Fig. 1. Two qualitative models of the basal ganglia. (a) is DIPM, (b) is SCPM. Excitatory: solid line. Inhibitory: dashed line.

2.3 Quantitative Modeling

In quantitative models, leaky-integrator artificial neurons are used to simulate the real neurons. Let u be the total post-synaptic potential generated by the afferent input, and k be the constant determining the rate of activation, the activation a of a leaky integrator is then

$$\frac{da}{dt} = -k(a-u) \tag{1}$$

y is the output of the neuron and given by

$$y = m(a-\theta)H(a-\theta) \tag{2}$$

where θ is a threshold for activation, $H(\ )$ is the Heaviside step function and m is a slope parameter.

It is supposed that the basal ganglia are processing in a parallel manner, just like the other parts of the brain. When the numerous neural information streams or channels originated in the cerebral cortex, each channel may potentially request for a particular behavior to be expressed, reach the basal ganglia, only a small number of them will be released according to their importance or salience. In this way, the basal ganglia allow the most urgent behavior to be selected and executed.

	DIPM	SCPM
Cortex	$y_i^C = S_i$	$y_i^C = S_i$
StriatumD1	$\begin{cases} u_i^{d1} = w_{CS1} * y_i^C \\ y_i^{d1} = m\left(a_i^{d1} - \theta_{d1}\right) H\left(a_i^{d1} - \theta_{d1}\right) \end{cases}$	$\begin{cases} u_i^{d1} = w_{CS1} * y_i^C \\ y_i^{d1} = m\left(a_i^{d1} - \theta_{d1}\right) H\left(a_i^{d1} - \theta_{d1}\right) \end{cases}$
StriatumD2	$\begin{cases} u_i^{d2} = w_{CS2} * y_i^C \\ y_i^{d2} = m\left(a_i^{d2} - \theta_{d2}\right) H\left(a_i^{d2} - \theta_{d2}\right) \end{cases}$	$\begin{cases} u_i^{d2} = w_{CS2} * y_i^C \\ y_i^{d2} = m\left(a_i^{d2} - \theta_{d2}\right) H\left(a_i^{d2} - \theta_{d2}\right) \end{cases}$
GPe	$\begin{cases} u_i^{GPe} = -w_{Sd2-GPe} * y_i^{d2} \\ y_i^{GPe} = m\left(a_i^{GPe} - \theta_{GPe}\right) H\left(a_i^{GPe} - \theta_{GPe}\right) \end{cases}$	$\begin{cases} u_i^{GPe} = -w_{Sd2-GPe} * y_i^{d2} + w_{STN-GPe} * \sum_i y_i^{STN} \\ y_i^{GPe} = m\left(a_i^{GPe} - \theta_{GPe}\right) H\left(a_i^{GPe} - \theta_{GPe}\right) \end{cases}$
STN	$\begin{cases} u_i^{STN} = -w_{GPe-STN} * y_i^{GPe} \\ y_i^{STN} = m\left(a_i^{STN} - \theta_{STN}\right) H\left(a_i^{STN} - \theta_{STN}\right) \end{cases}$	$\begin{cases} u_i^{STN} = w_{C-STN} * y_i^C - w_{GPe-STN} * y_i^{GPe} \\ y_i^{STN} = m\left(a_i^{STN} - \theta_{STN}\right) H\left(a_i^{STN} - \theta_{STN}\right) \end{cases}$
GPi	$\begin{cases} u_i^{GPi} = -w_{Sd1-GPi} * y_i^{Sd1} + w_{STN-GPi} * \sum_i y_i^{STN} \\ y_i^{GPi} = m\left(a_i^{GPi} - \theta_{GPi}\right) H\left(a_i^{GPi} - \theta_{GPi}\right) \end{cases}$	$\begin{cases} u_i^{GPi} = -w_{Sd1-GPi} * y_i^{Sd1} + w_{STN-GPi} * \sum_i y_i^{STN} - w_{GPe-GPi} * y_i^{GPe} \\ y_i^{GPi} = m\left(a_i^{GPi} - \theta_{GPi}\right) H\left(a_i^{GPi} - \theta_{GPi}\right) \end{cases}$

Fig. 2. The net input and output for the ith channel of each component in DIPM and SCPM

Thus, according to Fig. 1, the net input u_i and output y_i for the ith channel in each component of DIPM and SCPM are computed in the equations shown in Fig. 2. w_{CS1}, w_{CS2}, $w_{Sd2\text{-}GPe}$, $w_{STN\text{-}GPe}$, $w_{C\text{-}STN}$, $w_{GPe\text{-}STN}$, $w_{Sd1\text{-}GPi}$, $w_{STN\text{-}GPi}$ and $w_{GPe\text{-}GPi}$ are the weights of the respective connections with the basal ganglia. θ_{d1}, θ_{d2}, θ_{GPe}, θ_{STN} and θ_{GPi} are the thresholds of the respective components in the basal ganglia.

2.4 Simulation Results

Two simulation experiments are conducted for each model. The first experiment uses three channels, and amplitude of the inputs on the three channels represented the salience of three behaviors which are competing for expressing themselves. At time t=1, the salience of behavior 1 increases from 0 to 0.4, when t=2, the salience of behavior 2 increases to 0.6, exceeds behavior 1, at t=3, the salience of behavior 1 increases to 0.6 and soon falls back to 0.4 at t=4, the salience of behavior 3 is always 0. If the channel's output is under the threshold, the corresponding behavior was selected (see Fig. 3).All the simulation parameters are the same as those used in Humphries in order to compare the simulation results with the GA optimized models proposed in Sect. 3.

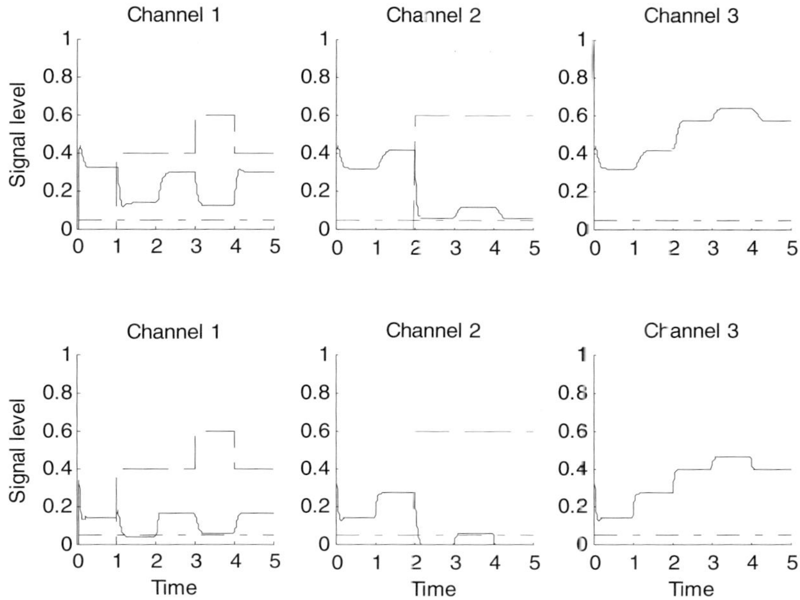

Fig. 3. Simulation results for DIPM (top) and SCPM (bottom) They show that DIPM doesn't have the selection (at t=1) and switching (at t=2) capabilities, while the SCPM possesses these capabilities. The simulation results of SCPM are the same as these in Humphries[10]. Output of GPi: solid line. Salience of each behavior: dashed line. Selection threshold: dash-dot line.

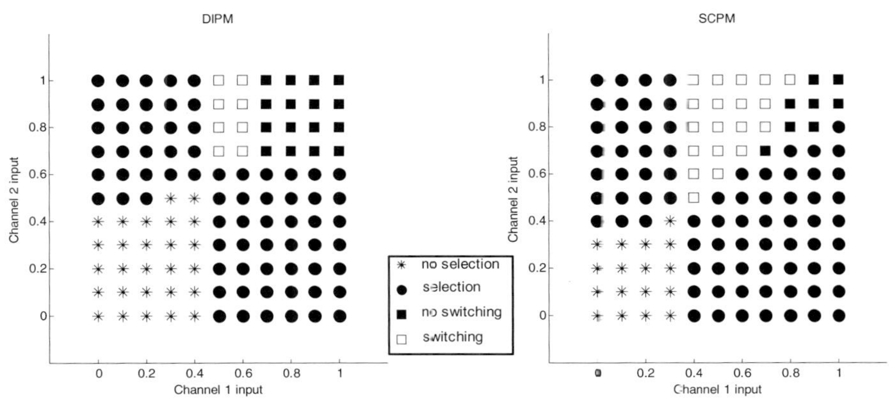

Fig. 4. Selection and switching abilities of DIPM (left) and SCPM (right)

It should be noted that DIPM was not originally proposed for studying basal ganglia's behavior selection properties, and has never being quantitatively simulated. But since the main purpose of this paper is to develop biology inspired robot selection mechanisms, it is important to remember that biological studies are not necessarily viewed as constraints for robotics, but nonetheless serve as inspirations.

In the second experiment, 121 simulations are run on each model, consisting of the salience input pairs (S1, S2) where S1 and S2 increase from 0 to 1 in the step of 0.1. The input to channel 1 started at time t=1; the input to channel 2 started at t=2. The simulation results are shown in Fig. 4. No selection means neither of the two channels becomes selected. Selection: only one channel becomes selected, for example, channel 1 becomes selected at t=1 or channel 2 becomes selected at t=2; No switching: channel 1 becomes selected at t=1 and remains selected from that time on and channel 2 also becomes selected at t=2; Switching: channel 1 fist becomes selected at t=1 but becomes deselected at t=2 while channel 2 becomes selected at t=2.

3 Genetic Algorithm Optimization of the Basal Ganglia Models

3.1 Details of the Genetic Algorithm

In order to enhance the performance of DIPM and SCPM, which considerably depends on the weights and thresholds, the genetic algorithm (GA) is applied as a means of optimization. The initial population is first generated at random. The genome of each individual in the population is simply a chain of all the weights and thresholds in the two models. Each weight or threshold is a single gene within the genome and is a real number in range [0, 1]. The fitness value of each individual is calculated in the below way: apply the parameters contained in the individual to DIPM and SCPM, then run the second simulation experiment discussed in section 2 to see how many selection results in the 121 simulations match an ideal selection results diagram, the more the larger fitness value will be. When all the individuals in the population have had their fitness evaluated, selection operator, crossover operator and mutation operator were performed in sequence. The whole process will be repeated until reach the maximum generation.

3.2 Optimization Results

Three genetic operators employed in this paper are roulette wheel selector, one-point crossover and simple mutation respectively. Elitism is also used to protect the best individual that had ever found. Parameters of GA are as follows: population size is 100, maximum generation is 100, crossover probability is 0.8 and mutation probability is 0.02. Simulation results of optimized DIPM and SCPM are shown Fig. 5 and Fig.6.

Compared with Humphries [10], the simulation results show that the GA optimized DIPM has the ability to select between the sample input signals. Thus DIPM is also a candidate for the selection model of the basal ganglia. Fig. 6 shows that both of the two GA optimized model have a considered improved selection and switching performance than the two original models shown in Fig. 4.

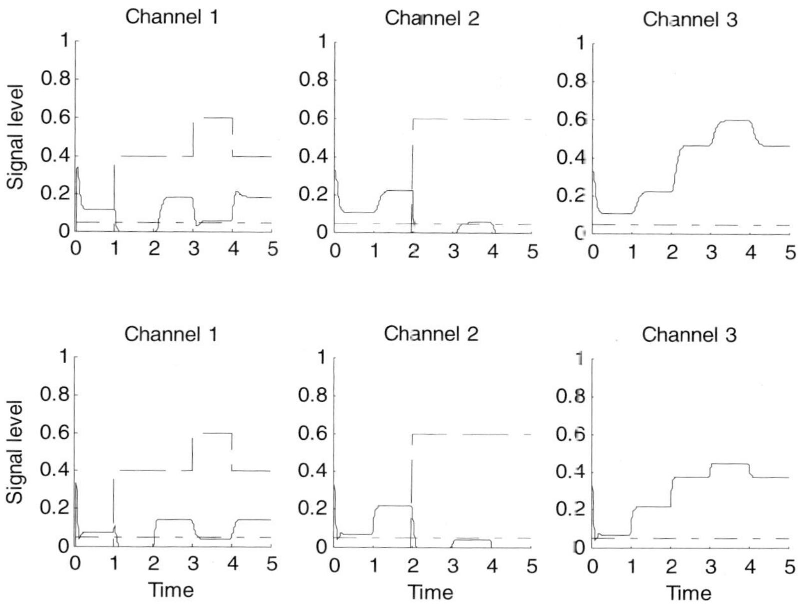

Fig. 5. Simulation results for GA-DIPM (top) and GA-SCPM (bottom). Output of GPi: solid line. Salience of each behavior: dashed line. Selection threshold: dash-dot line.

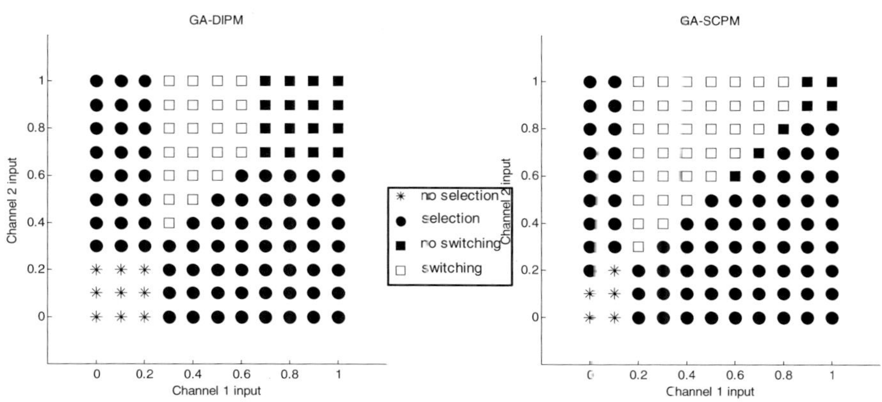

Fig. 6. Selection and switching ability of GA-DIPM (left) and GA-SCPM (right)

4 Robot Implementation and Simulation

4.1 Robot Implementation

Girard et al.[14] and Prescott et al.[15] had made early exploration of applying basal ganglia inspired behavior selection mechanism to robotics and got some valuable results. In this section, GA optimized basal ganglia models are embedded in a robot control system. The robot architecture is typical behavior based control architecture. It has four main parts: the robot sensory system, the behavior system, the basal ganglia and the robot motor system.

The robot motor system is equipped with two driven wheels and a gripper, which is used to pick up targets (food). The sensory system has five range scanners which allow the robot to know whether there are obstacles or targets around it. The gripper is also equipped with two sensors, one tells the robot if something locates in the gripper and the other indicates the gripper's state (closed or open). The behavior system is decomposed into six behaviors: *avoid obstacle*, *search*, *move to target*, *pick up*, *go home* and *put down*, respectively. Each behavior consists of a set of condition-action mappings. The salience signal for each behavior is computed in the below equations:

$$S_{avoid\_obstacle}=0.96*P_{obstacle} \tag{3}$$

$$S_{search}=0.6-0.5*P_{obstacle} \tag{4}$$

$$S_{move\_to\_target}=0.72*P_{target}-0.3*P_{obstacle} \tag{5}$$

$$S_{go\_home}=0.34*P_{target\_in\_grip}+0.56*P_{grip\_closed} \tag{6}$$

$$S_{put\_down}=0.17*P_{target\_in\_grip}+0.28*P_{grip\_closed}+0.5*P_{home} \tag{7}$$

Whether the robot is near to an obstacle, a target or the home area are indicated by $P_{obstacle}$ (0 or 1), P_{target} (0 or 1) and P_{home} (0 or 1) respectively. $P_{target\_in\_grip}$ (0 or 1) indicates whether there is a target in the gripper and P_{grip_closed} (0 or 1) indicates whether the gripper is closed. When the salience value has been computed, the basal ganglia will then select a suitable behavior based on these salience signals and command the robot motor system to execute it.

4.2 Simulation Results

The simulation environment is a 4.5m×6.0m rectangle shaped room, which is surrounded with wall and contains some obstacles. The home area is a circle area whose diameter is 0.9m. Some targets (small circles whose diameter is about 0.15m) are scattered in the room. The range of the scanners is about 0.8m.

Screenshots of the simulation are shown in Fig.7. The simulation results demonstrate that the robot can switch fluently between the six behaviors and validate that the basal ganglia can fulfill its role as a behavior selection mechanism.

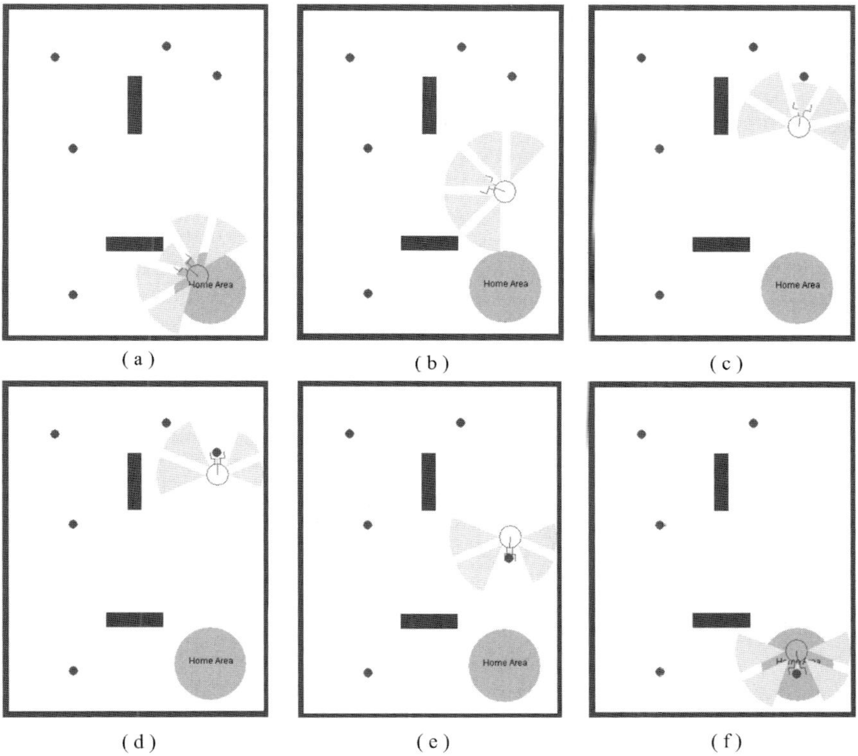

Fig. 7. Screenshots of the simulated foraging task. (a) avoid obstacle, (b) search, (c) move to target, (d) pick up, (e) go home, (f) put down .

5 Conclusions

As stated in the introduction, this paper first introduces a biology-inspired behavior selection mechanism based on the recent research achievements on the basal ganglia. Then, the genetic algorithm is utilized to optimize the weights and thresholds in two basal ganglia models, DIPM and SCPM respectively. A robot is also implemented for a simulated foraging task. Some conclusions drawn from this paper are the following:

-The DIPM can also be considered as a selection model of the basal ganglia although further studies and biological evidence are needed to support this conclusion.

-GA is an effective optimization method and can considerably improve the performance of the computational basal ganglia models.

-The biology inspired behavior selection mechanism can be successfully embedded in behavior-based robot control architecture, and can switch fluently between the robot behaviors.

References

1. Arbib, M.A., Liaw, J.-S.: Sensorimotor transformations in the worlds of frogs and robots. Artificial Intelligence 72, 53–79 (1995)
2. Brooks, R.: A robust layered control system for a mobile robot. Robotics and Automation, IEEE Journal of [legacy, pre - 1988] 2, 14–23 (1986)
3. Brooks, R.: Intelligence Without Representation. Artificial Intelligence, 139–160 (1991)
4. Arkin, R.C.: Behavior-Based Robotics. MIT Press, Cambridge (1998)
5. Thrun, S.: Probabilistic Algorithms in Robotics. AI Magazine, 93–109 (2000)
6. Maes, P.: How to do the right thing. Connection Science, 291–323 (1989)
7. Arkin, R.C.: Motor schema-based mobile robot navigation. The International journal of robotics research, 92–112 (1989)
8. Rosenblatt, J.K.: DAMN: a distributed architecture for mobile navigation. Journal of Experimental & Theoretical Artificial Intelligence 9, 339–360 (1997)
9. Aguirre, E., Gonzalez, A.: Fuzzy behaviors for mobile robot navigation: design, coordination and fusion. International Journal of Approximate Reasoning, 255–289 (2000)
10. Humphries, M.D.: The Basal Ganglia and Action: A Computational Study at Multiple Levels of Abstraction. PhD Thesis, University of Sheffield, Sheffield (2002)
11. Albin, R., Young, A., Penny, J.: Functional anatomy of basal ganglia disorders. Trends in Neurosciences 12, 366–375 (1989)
12. Gurney, K., Prescott, T.J., Redgrave, P.: A computational model of action selection in the basal ganglia. I & II. A new functional anatomy. Biological Cybernetics 84, 401–410 (2001)
13. Humphries, M.D., Gurney, K.: The role of intra-thalamic and thalamocortical circuits in action selection. Network: Computation in Neural Systems, 131–156 (2002)
14. Girard, B., Cuzin, V., Guillot, A., Gurney, K.N., Prescott, T.J.: A basal ganglia inspired model of action selection evaluated in a robotics survival task. Journal of Integrative Neuroscience, 179–200 (2003)
15. Prescott, T.J., Montes Gonzalez, F.M., Gurney, K., Humphries, M.D., Redgrave, P.: A robot model of the basal ganglia: Behavior and intrinsic processing. Neural Networks 19, 31–61 (2006)

Asymmetry Computing for Cholesteatoma Detection Based on 3-D CT Images

Anping Song, Guangtai Ding, and Wu Zhang

School of Computer Engineering and Science, Shanghai University, Shanghai, China
{apsong,gtding,wzhang}@shu.edu.cn

Abstract. Cholesteatoma is a destructive and expanding sac in the middle ear and/or mastoid process. If untreated, cholesteatoma can result in nerve deterioration, deafness, imbalance and vertigo. Traditional diagnose methods rely on doctors' experience and often misdiagnosed. In this paper, we propose a novel asymmetry-computing algorithm for cholesteatoma detection based on 3-D CT images. By applying this algorithm, we provide a complete numerical calculation framework and its simulation. The proposed algorithm is tested on real 3-D eardrum CT images. The diagnose accuracy rate of cholesteatoma is 72.73%, and the misdiagnose rate is only 27.27%. The result demonstrates the applicability of the proposed asymmetry-computing algorithm for cholesteatoma detection. Therefore the proposed algorithm is beneficial for clinic diagnose.

1 Introduction

Symmetry is widely exists in nature. When people come to recognize new thing, they often use the property of symmetry to recognize it and use symmetry to measure it. So the objects that people can direct feel and interested in are often consists of various forms of symmetry. Object has symmetry can generate harmony, and harmony can result the sense of beauty.

Symmetry one of the basic properties of an object, it is widely exists in all fields, for example in architecture, medicine and natural selection etc. Symmetry is an important approach to recognize object, identify object and depict object. A symmetric object can be depicted using only small amount data. It can also be easily detected in pattern reorganization [1,2].

In the mathematical description of symmetry, an object never has a certain degree of symmetry. Either an object is symmetric or asymmetric. With that definition, natural things are rarely symmetric. If an object is asymmetric, calculated as how far the object is from being symmetric is an unsolved problem[3].

The main research topic in medical image processing is study the images of human organs. There exist a lot of symmetries in human organs; for example, the five sense organs are symmetry. The asymmetry in human organ is often a sign of pathology. The symmetry of humans does not only occur on the outside, some of the bones and organs are close to being symmetric as well. As for the outside, the symmetry normally occurs as reflection symmetry. If an organ for a person appears to be more

K. Li et al. (Eds.): LSMS 2007, LNCS 4688, pp. 787–794, 2007.

asymmetric than for other people, this might be a sign of pathology. A method to measure asymmetry in medical imaging would make it possible to distinguish patients with normal asymmetry and unhealthy asymmetry. For some diseases, it will be a breakthrough if healthy and pathological organs are distinguished. For instance, the examination of hippocampi for a difference is used to distinguish patients with schizophrenia from healthy controls is done in various studies [4,5]. In other cases, it is more interesting to locate asymmetry regions and remediate them. Detection of the middle ears structure is used to compute the brain axial plane for further registration[6].

We can utilize the symmetry to visualize and intuitionisticly understand the human organs. If an organ for a person appears to be more asymmetric than for other people, this might be a sign of pathology. But how to effectively measure the discrepancies of the asymmetry? So a calculation method is needed to calculate the discrepancy of the asymmetry. Here the size of discrepancy of the asymmetry is called as the asymmetry degree.

Cholesteatoma is a destructive and expanding sac in the middle ear and/or mastoid process. Acquired cholesteatomas can be caused by a tear or retraction of the eardrum. The status of cholesteatomas as tumors is currently unsettled. If untreated, a cholesteatoma can eat into the three small bones located in the middle ear (the malleus, incus and stapes, collectively called ossicles), which can result in nerve deterioration, deafness, imbalance and vertigo. It can also affect and erode, through the enzymes it produces, the thin bone structure that isolates the top of the ear from the brain, as well as lay the covering of the brain open to infection with serious complications[7,8]. The diagnosis of Cholesteatoma mainly based on the CT images of the patient brain. Theoretically, the patient's current CT images should be compared to the CT images of the same area when the patient is in healthy condition. But, most patients don't have the CT images of the area. So in practice, doctors will use the CT images of current patient's contralateral temporal bone area to make diagnose. By judging whether the images is normal bilateral asymmetry or clinicopathological asymmetry to get the diagnosis result. Obviously, during the process of diagnosis, the doctor's personal experience is vital. No adequate experience will lead to misdiagnoses. Because the diagnosis mainly depended on the personal experience, the diagnose result is relatively subjective. Therefore it is of clinic value to introduce a way to assist diagnostic process, which provides academic foundation but external measurement as well. From the perspective of the statistics, for a no trauma imaging, the accuracy of diagnosis can be approximately 50% by traditional inquire， biochemistry measurement etc. Even with the aid of imaging means, the state of art diagnosis has been improved greatly, but misdiagnosis is still inevitable. More importantly, the diagnosis depends on the doctor's ability and experience; there exists difference among doctors. So utilize the medical digital images to aid disease diagnose is an emerging trend. Although asymmetry computing is widely used in disease diagnose. For example, measuring asymmetry on mandibles from children with cleft in the lip a palate and from children wnoneith plagiocephaly syndrome[2]. But to our best knowledge, no one has applied the asymmetry computing to diagnose cholesteatoma.

In this paper, we propose a calculation model of asymmetry degree of 3-D images, based this model, we present calculation method for measurement of asymmetry in

3D CT images and apply it to Cholesteatoma detection experiment. An algorithm for the calculation has been implemented.

The paper is organized as follows: In section 2 we detail the calculation model of asymmetry degree of 3-D CT images. In section 3, we present the calculation of asymmetry degree of 3-D CT images. In section 4 we apply our calculation algorithm to cholesteatoma detection experiment and analysis the results obtained. Last section concludes the paper the present some further works.

2 The Asymmetry Computation Model of 3-D Images

Suppose the 3D images to be expressed as $f(x, y, z)$ or $f(P)$, where $f(P)$ express the pixel values at the point P, $P = (x, y, z)$.

1) Compute the symmetric point S(P) of the point P by the plane $Ax + By + Cz + D = 0$

Two points are symmetric with regard to a plane, if and only if the plane is the central vertical plane of the link line of the two points; Two points are symmetric with regard to a curve surface, if and only if there exists a diffeomorphism mapping, which maps the curve surface into the plane, and the plane is the central vertical plane of the link line of the two points.

Let （x, y, z） be the coordinates of the perpendicular point. Then, from the following equation,

$$\begin{cases} Ax + By + Cz + D = 0 \\ (x, y, z) - P = t(A, B, C) \end{cases} \tag{1}$$

We get the parameter t, so we have

$$S(P) = P + 2t(A, B, C) \tag{2}$$

2) Compute the symmetric point S(P) of the point P by the central vertical plane of the link line of the two points: P_1, P_2

$$\begin{cases} \overrightarrow{P_0P_1} \bullet ((x, y, z) - (P_0 + \frac{\overrightarrow{P_0P_1}}{2})) = 0 \\ (x, y, z) = P + t\overrightarrow{P_0P_1} \end{cases} \tag{3}$$

From (3), we get the parameter t, then the coordinates of the S(P) is

$$S(P, P_0, P_1) = P + 2t\overrightarrow{P_0P_1} = P + 2\frac{\overrightarrow{P_0P_1} \bullet (P_0 + \frac{\overrightarrow{P_0P_1}}{2}) - P \bullet \overrightarrow{P_0P_1}}{|\overrightarrow{P_0P_1}|^2}\overrightarrow{P_0P_1} \tag{4}$$

3) The definition and calculation equation of 3-D images asymmetry degree is defined as follows:

A) Select a local parameter d>0;
B) Select an initial point P0 in the pathological area;
C) Select the region Ω0 which center is P0;
D) Select the symmetry point P1 of P0. So for the point P, its symmetry point S(P) is about P0 and P1, the function of P(S(P,P0,P1)), S(P) is the abbreviation form;
E) Define $|\Omega| = \int_\Omega d\mu$, μ is the distance measurement, then let

$$\rho_{P_1}(f) = \frac{\int_{\Omega_0} \frac{\int_{|P-(x,y,z)|\le d} |f(P) - f(S(P,P_0,P_1))| \, d\mu}{\int_{|P-(x,y,z)|\le d} (\max f(P) + \max f(S(P,P_0,P_1)))d\mu} d\mu}{|\Omega_0|} \tag{5}$$

The value of $\rho_{P_1}(f)$ must be in the range of $[0,1]$, The calculation range of the value max at equation is defined as: $|P-(x,y,z)| \le d$;

When P1 transverse in $S(\Omega_0)$, there should exist a best P_1^* , such that

$$\min_{P_1 \in Domain(f)} \rho_{P_1(f)} = \rho_{P_1^*}(f)$$

or

$$P_1^* = \arg \min_{P_1 \in Domain(f)} \rho_{P_1(f)}$$

F) The symmetric plane determined by P_0, P_1^* is called the best symmetric plane.
G) The value of $\rho_{P_1^*}(f) = \min_{P_1 \in Domain(f)} \rho_{P_1(f)}$ is the value of the asymmetric degree of the 3D space.

3 The Asymmetry Computation Algorithm of 3-D CT Images

Based on the 3-D CT images on the asymmetry computing theory, this section will discusses the detailed implementation algorithm.

3.1 The Read in of DICOM Medical Images

The CT image is represented by different gray scale, which reflects the tissues and organ's absorption to the X-ray. So as the same as the black and white images represented by X-ray images, dark area represent the low absorption area, that is low-density area, for example the parenchyma. The white area represents the high absorption area, which is high-density area, for example the bone. Because the CT has very high density resolving power, so it is widely used in medical clinic diagnoses.

By scanning we can get the CT layered images of the head area, its interval ranges from 0.75 mm to 2 mm. Because not each sectional layered image contains regional image (ROI) that we are interested in. So we do not calculate all the 3-D images, which is contained in the CT images, but select the image within the target area to calculate. By this way, not only it can reduce the other irrelevant image area to interference with the calculation result, but also it can effectively reduce the calculation and improves the calculation efficiency.

CT images generally saved into the DICOM format, which is exclusively used in medical science. The scale color depth various among different CT machines. The images we process normally of 512*512 pixels, the color depth is 16 bits, that is the light and shade degree of each pixel is represented as an integer value range from 0 to 4000.

3.2 ROI Setting

In order to determine two pathological changes area Ω_0, Ω_1, which needs symmetry calculation, during the process of algorithm research, we select rectangle as the shape of area. The range of the area is M*N*K, manually select the central point P0, P1 of the area.

The range of the cuboid area is determined by $(M_0,N_0,K_0),(M_1,N_1,K_1)$ and the central point $(x_0,y_0,z_0),(x_1,y_1,z_1)$.

3.3 Algorithm

First, we optimize the algorithm equation (2) for calculation of symmetry point. Using the calculation algorithm, for each different P, only by get the value of T each time, can we get the $S(P, P_0, P_1)$.

Second, for each different P, calculate S(P) which is the symmetry point of P. And also calculate the $S(\Omega_0)$ of the Ω_0 of ROI to formulate a three dimensional matrix in one time.

Then calculate the Ω_0 of P_0, P_1 and the data of their corresponding symmetry point matrix $S(\Omega_0)$.

Finally, if the symmetry point extend the boundary, let $f(S(P,P_0,P_1))=-3000$ to increase the numerator and the denominator in the calculation equation (also it can be set as –1, because the image pixel of DICOM is between 0 and 4000). That is to increase the value of asymmetry degree. Since the symmetry point extend the boundary, it is obvious that the difference of the image asymmetry increased. And the more difference, the bigger asymmetry degree.

4 Experiment and Results

The experiment is based on the medical images provided by the Eye Ear Nose and Throat Hospital of Shanghai Fudan University. Figure 4.1 shows the CT images of normal eardrum area and pathological eardrum area. The eardrum shown in upper area of the figure is normal; the eardrum shown in lower area is with cholesteatoma illness.

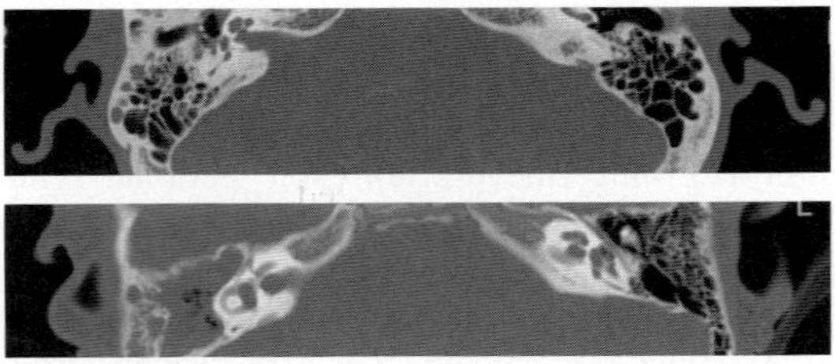

Fig. 4.1. The CT images of eardrum

4.1 Selection of Calculation Parameters

In the following, we will list the relationship between the time and value of r, which is corresponding to a particular point and the d. During the process of calculation, when the value d gets larger, the fewer value of r, and it take more time to calculate. The calculation result is more ideal when the value r get larger, and for the consideration of speed, so select d<=8 is more suitable, generally the value d=4.

4.2 Selection of ROI Area

When the selected ROI area is too small, generally the calculated result of r is also small, so the ROI area can't be too small. By experiment, the value of M, N, K is at the range of 30-50 is most suitable. When the value of M, N, K is too big (for example, above 55), not only the calculation will be too time consuming but also the asymmetry point will break bounds, which can result out the boundary error in calculation, but when the value of M, N, K is below 30, we can find the value of r is too small.

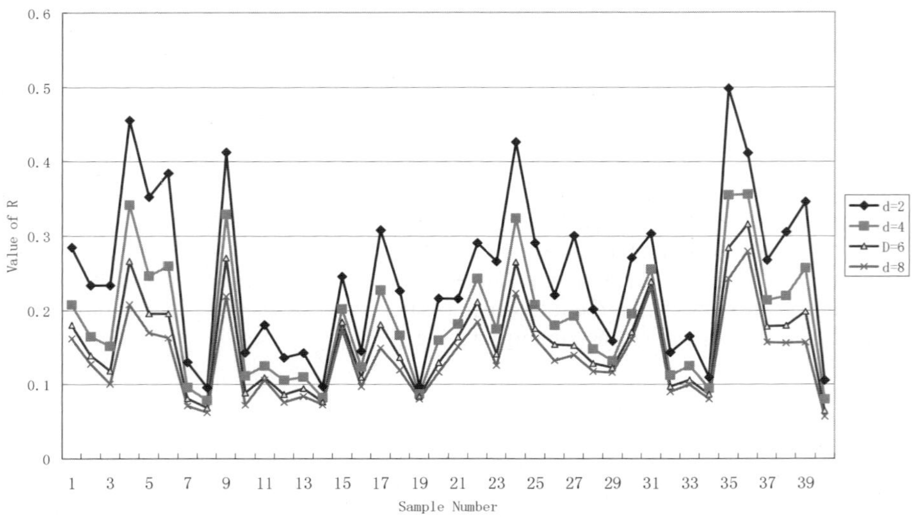

Fig. 4.2. The effect of precision to the algorithm

4.3 Results

During the process of calculation, when the value d gets larger, the fewer value of r, and it take more time to calculate. The calculation result is more ideal when the value r get larger, and for the consideration of speed, so select d<=8 is more suitable, generally the value d=4. Figure 4.2 shows the effect of precision to the algorithm.

Through a lot of calculation, we have obtained following results as shown in Table 4.1. the test sample consists of 40 people, among them 11 people are patients, and other 29 persons are healthy persons. When the threshold r is 0.2, the diagnose accuracy of healthy of normal people is 100%, when the threshold is 0.158, the diagnose accuracy of healthy people is 72.41%. The misdiagnose rate is 27.59%. The diagnose accuracy of patient is 72.73%, and the misdiagnose rate is 27.27%.

Table 4.1. Analysis of the obtained results

	Diagnose accuracy rate	Misdiagnose rate
Normal	72.41%	27.59%
Patient	72.73%	27.27%

From the data obtained from experiment, the calculation of asymmetry degree has some medical significances is real medical diagnose practice.

5 Conclusions

We propose an asymmetry-computing algorithm for cholesteatoma based on 3-D eardrum CT images. Based on this algorithm, the authors provide a complete numerical calculation method and computing framework. This algorithm has been successfully applied to real 3-D eardrum CT images. By a lot of data experiments, we have obtained the statistical result of asymmetry degree. By utilizing the asymmetry degree value, it can be used to aid disease diagnose and can greatly improves the diagnose accuracy. So the algorithm proposed in this paper has practical value in clinic diagnoses of cholesteatoma. Further works will focus on studying how to automatically segment the CT images and improve the efficiency of the algorithm.

Acknowledgments. This work is supported by National Natural Science Foundation of China(No. 60373071).

References

1. Vetter, T., Poggie, T., Bulthoff, H.H.: The importance of symmetry and virtual views in three-dimensional object recognition [J]. Current Biology 4, 18–23 (1994)
2. O'Mara, Owens, R.: Measuring Bilateral Symmetry in Digital Images. IEEE Tencon-Digital Signal Processing Applications, 151–156 (1996)
3. Glerup, N.: Asymmetry Measure in Medical Image Analysis [D]. IT University of Copenhagen, Copenhagen (2005)

4. Shen, L., Ford, J., Makedon, F., Saykin, A.: Hippocampal shape analysis: Surface-basedrepresentation and classi_cation. SPIE Medical Imaging - Image Processing (2003)
5. Gerig, G., Muller, K.E., Kistner, E.D., Chi, Y-Y., Chakos, M., Styner, M., Lieberman, J.A.: Age and treatment related local hippocampal changes in schizophrenia explained by a novel shape analysis method. In: Ellis, R.E., Peters, T.M. (eds.) MICCAI 2003. LNCS, vol. 2879, pp. 653–660. Springer, Heidelberg (2003)
6. Tuzikov, A., Colliot, O., Bloch, I.: Brain Symmetry Plane Computation in MR Images using Inertia Axes and Optimization [C]. In: Proceedings of 16th ICPR International Conference on Pattern Recognition, Quebec, pp. 516–519 (2002)
7. Bookstein, F.: Morphometric tools for landmark data: Geometry and biology [M]. Cambridge University Press, Cambridge (1991)
8. Desloge, R.B., Carew, J.F., Finstad, C.L., Steiner, M.G., Sassoon, J., Levenson, M.J., Staiano-Coico, L., Parisier, S.C., Albino, A.P.: DNA analysis of human cholesteatomas. Am. J. Otol. 18, 155–159 (1997)

Moving Target Detection and Labeling in Video Sequence Based on Spatial-Temporal Information Fusion

Shiwei Ma, Zhongjie Liu, Banghua Yang, and Jian Wang

School of Mechatronic Engineering and Automation, Shanghai University,
Shanghai Key Laboratory of Power Station Automation Technology,
Shanghai 200072, P.R. China
swma@mail.shu.edu.cn

Abstract. A novel method of moving target detection based on spatial-temporal information fusion was proposed. It combines the temporal properties within multiple frames and the spatial information in single frame of a moving target in video image sequence by using time-domain recursive filtering for background recovering and spatial image segmentation for separation of target from background. Together with follow-up morphological filtering and blob analysis, the area of an interested moving target can be detected and labeled by bounding box. Its application to swimmer tracking in underwater swimming video image sequence manifested the effectiveness of the method, which is expected to be applied to the implementation of underwater automatic tracking swimming video system.

1 Introduction

The detecting and labeling of moving target in video sequence is of importance in many applications, such as sports training, safety monitoring, etc. For example, in order to improve the quality of training of sportsmen by using computer and digital video image processing techniques, it is quite necessary to detect and analyze a player's various movement properties from real-time sport video. At present, the main existed approaches of moving target detection are background difference [1], time difference (frame difference) [2] and optical flow [3].

For the method of optical flow, since it involves sophisticated calculations, it is hard to meet real-time requirement without the support of specific hardware devices. The method of background difference is easy to be implemented since it can pick-up moving target from background by comparison of input image with background image. However, in this method, the used camera must be in condition of rest, hence it is not adequate to fully understand the target image and is vulnerable to noise and abruptness in environment.

The method of time difference is robust to the change of environment, since it assumes that there should be a certain degree of difference of moving speed between target and background, and the information of grey scale or gradient of temporal difference image can be employed to obtain the moving information from the deviation

K. Li et al. (Eds.): LSMS 2007, LNCS 4688, pp. 795–802, 2007.

of two or three consecutive frames. However, the changes between different frames can't be reflected on all of the boundaries of target. Therefore, by using time difference alone, the obtained profile may not be continuous and hence fail to accurately detect the moving target. So, the spatial information of moving target in single frame image should be considered.

In this paper, a novel method of moving target detection and labeling in video image sequence based on spatial-temporal information fusion was proposed. Related theory analysis and its application to swimmer detection and labeling in underwater swimming video are presented.

2 Spatial-Temporal Fusion Based Method

The proposed method is schematically depicted in Fig.1, which combines the temporal properties within multiple frames and the spatial information in single frame of a moving target. In temporal, time-domain recursive filtering can be employed to update background model of video image sequences. In space, image area segmentation methods can be employed to separate moving target from background in one frame by using the properties of color, gradation, edge or texture. Based on these processing, logical AND operation can be carried out to spatial target binary image with temporal target binary image. Further, together with follow-up morphological filtering and blob analysis [4], the area of moving target can be effectively detected and tracked by bounding box in video image sequences.

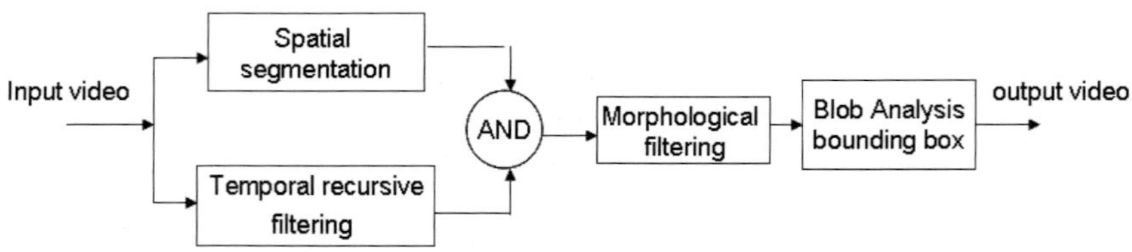

Fig. 1. Procedure of spatial-temporal fusion based method

2.1 Temporal Recursive Filtering

Considering the continuous change of scenery and noise caused by the moving of object within video, the background model must be recovered or updated continuously. Generally, the change speed of moving target is much faster than that of background in motion video. It means that in frequency domain, the background's signal is in lower frequency band, whereas the target's signal in high frequency band. Based on this principle, time-domain recursive low pass filter can be employed to estimate the properties of background [5].

The mathematic form of this filtering is given as:

$$f_a(i,j,t)=f_b(i,j,t-1)+k(i,j,t)*(g(i,j,t)-f_b(i,j,t)) \tag{1}$$

$$f_b(i, j, t) = f_a(i, j, t-1) \tag{2}$$

$$g(i, j, t) = f(i, j, t) + n(i, j, t) \tag{3}$$

Where, i and j means the horizontal coordinate and vertical coordinate respectively, and t is time. And g means the original frame consisted of real image f and noise n. f_b gives the estimation of background at the $(t-1)$ instance, f_a the prediction of background at t instance.

The function $k(i, j, t)$ is used to control the prediction error. In the proposed method, it is defined as a binary profile function of interested target, which can be adjusted automatically according to practical application and the moving speed of target. It becomes the method of adjacent frame difference with $k(i, j, t) = 1$, and of static background difference with $k(i, j, t) = 0$. When $k(i, j, t) = 0.5$, one can use the average value of sequences to update the background model. The structure of this filter is given in Fig.2.

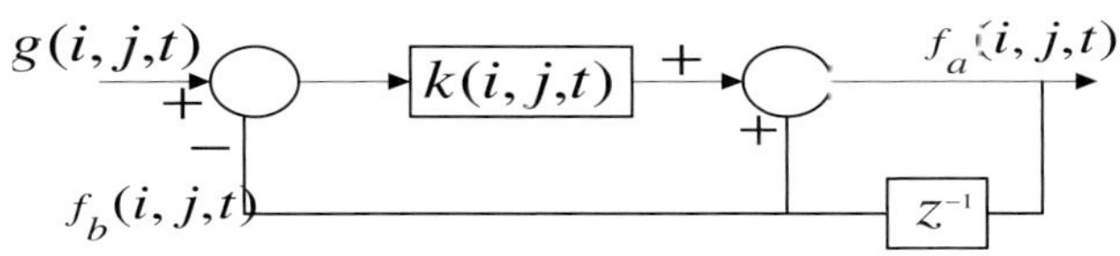

Fig. 2. Structure of time-domain recursive low pass filter

In order to adapt the change of input image sequence, $k(i, j, t)$ can be expressed as

$$k(i, j, t) = \rho \bullet M(i, j, t) + (1 - M(i, j, t)) \bullet \beta \tag{4}$$

Where, $M(i, j, t)$ gives the binary profile of moving target and can be obtained by adaptive threshold segmentation of current frame and the predicted background differential image. And the constants ρ and β determine the adaptation of background detecting, taking value within 0~1. The value of ρ must be small enough to guarantee the separation of moving target from background sequence, and the value of β should larger than ten times of the value of ρ so as to guarantee the operation of algorithm. Generally, they can be taken as $0.001 \le \rho \le 0.01$, $0.01 \le \beta \le 0.1$.

Hence one can see that, by using time-domain recursive filtering, one can not only reduce noise but also predict background. Then, by differentiating the current input frame with the background image predicted, the area of moving target can be detected.

2.2 Spatial Segmentation

Spatially, identical areas in one image frame can be segmented by using the properties of color, gradation, edge or texture. It can be implemented with the method based on profile information and the method based on area information. The result of segmentation operation is a set of componential areas in one image, commonly it is a part of moving target in video image sequence.

In the proposed method, we employ the color property of target for segmentation by sampling and classification of the target in image of YUV color space. With this processing, pixels can be classified and target can be separated from background.

2.3 Morphological Filtering and Blob Analysis

Since video target detection algorithms are pixel based, the foreground pixels within binary image obtained by the above processing are not necessarily came from true moving target, some of them may be caused by noise. Hence, the obtained binary image should be further processed. On the one hand, scattered pixels belonging to moving target should be connected; on the other hand, noisy pixels should be removed.

In the proposed method, some of morphological filtering operations such as etching, dilating, opening and closing are employed to remove small noise areas, at the same time to fill up the voids within the area of moving target caused by failure of detection. Then, blob analysis based connectable area identification is used to remove the foreground whose area is less than prescribed threshold. For blob information in image, one need to aggregate blobs belonging to identical object so as to convert blob information into target information and hence to obtain the information of an identical target.

3 Applications to Swimming Video

In swimming, underwater tracking camera system must be used so as to obtain accurate swimmer's moving video in water. However, up to now, practical systems are only hand priming or half-automatic device. Therefore, the automatic underwater tracking camera system must be development. In which, the study of effective methods for detecting and labeling of swimmer target in real-time swimming video image sequence is crucial.

For underwater swimming video, considering the requirement of real-time, continuous change of scenery and noise caused by water bloom and wave, it is difficult to detect swimmer target by using the method of optical flow and static background difference. Analyzing the properties of underwater swimming video image sequences, it is known that the interface sections of foreground and background often reveals obvious changes between different frames caused by the motion of swimmer. So, the temporal properties within multiple frames and the spatial information in single frame of moving swimmer in underwater video image sequence should be considered.

In this study, the proposed spatial-temporal fusion method was applied to swimmer detection in real-time underwater swimming video. Furthermore, based on blob analysis, by using some of the crucial points in boundary of connectable area, we can label the area of a single swimmer with bounding box, which has the vector form as:

$$z = (r_0, c_0, h, w)^T \tag{5}$$

The bounding box is shown in Fig.3 which gives its coordination in image. It can provide necessary information for further swimmer tracking.

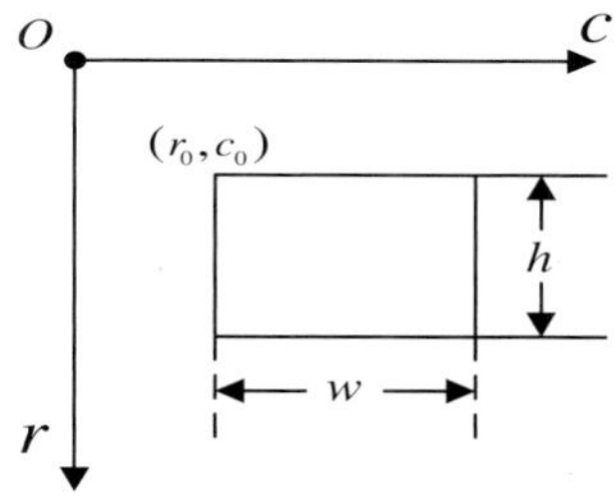

Fig. 3. Coordinates of bounding box in image

4 Experimental Results and Discussions

Experiment was carried out to real-time swimming video image sequences. The results manifested the effectiveness of the proposed method used in this field.

The original video was taken from published website, www.swim.ee. Where, the resolution of a single frame image is 352×288, frame frequency is 25fps. Due to page limits, the following only gives experimental results and discussions to its 19~22 consecutive four frames grey images. These original four frames images are given in Fig.4.

As a comparison, first use time-domain recursive low pass filter to update background, by setting $\rho = 0.008$ and $\beta = 0.08$. The obtained binary images are given in Fig.5 with arrangement corresponding to Fig.4. It can be found that, though entire swimmer's profiles were obtained, heavy noisy components in the vicinity of profiles still existed, by using time-domain filter along.

However, by using the proposed spatial-temporal fusion based processing method, the swimmer's profiles can be effectively detected and most of the noisy components in the vicinity of profiles, as shown in Fig.6, compared with Fig.5. In the experiment, color property based classification processing was used for spatial image segmentation and time-domain recursive low pass filtering was used to update background. After these processing, morphological filtering was used to remove small noise areas, at the same time to fill up the voids within the area of swimmer.

Further, blob analysis was carried out to the above obtained binary images as shown in Fig.6. Table 1 gives the values of bounding box coordinates used for labeling of swimmer area in underwater video sequence, as shown in Fig.7. For other single swimmer in this video sequence, we also obtained satisfactory experimental results.

Table 1. Values of bounding box coordinates of 19~22 frames

number of frame	19	20	21	22
r_0	122	120	118	118
c_0	66	58	52	48
h	112	111	109	114
w	190	209	220	226

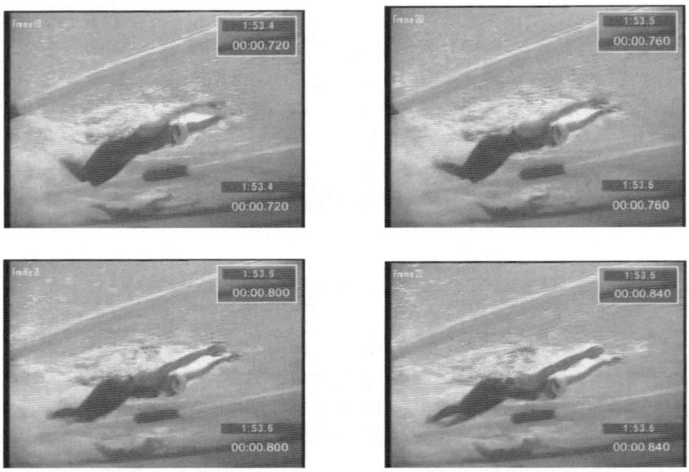

Fig. 4. Original swimming video images (form left to right and top to bottom are 19～22 frames respectively

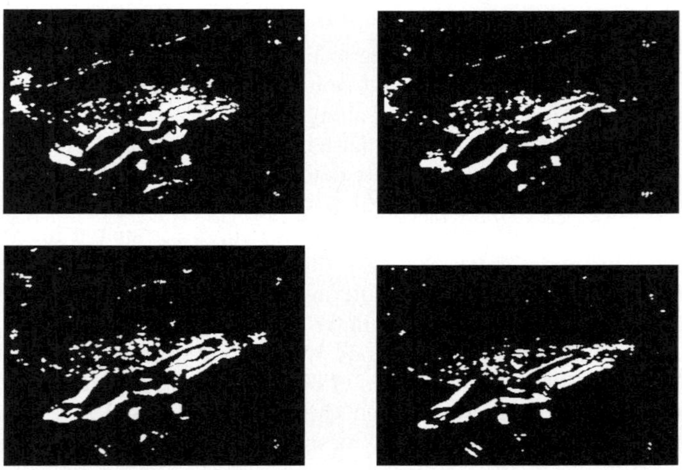

Fig. 5. Processed results by using recursive filtering along

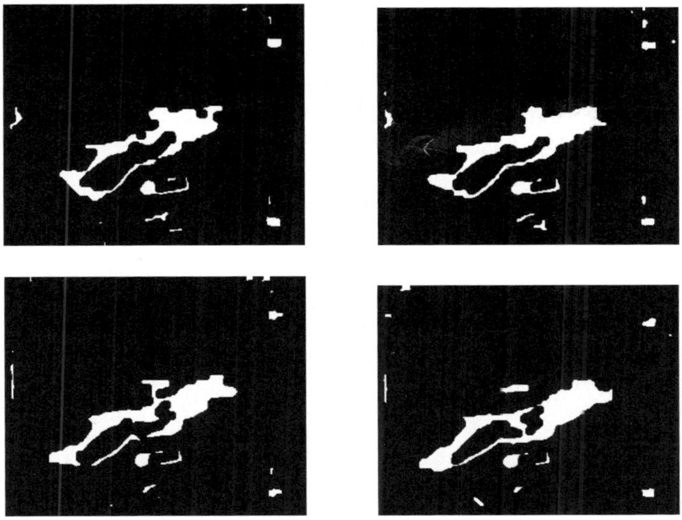

Fig. 6. Processed results by using temporal and spatial fusion method for swimmer detection

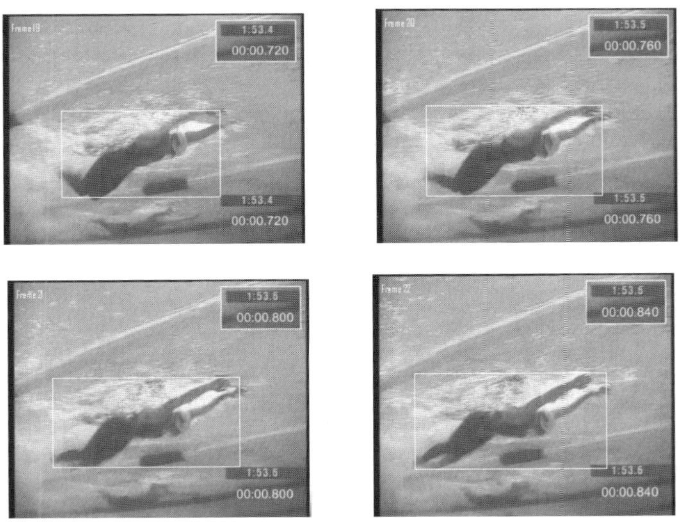

Fig. 7. Results of labeling by bounding box based on the detected results in Fig.6

These experiments manifested the effectiveness of the proposed method. It can provide important information for further swimmer tracking techniques, such as predictive tracking based on Kalman filtering. So, the method is useful for the implementation of underwater automatic tracking swimming video camera system used for swimming training.

5 Conclusions

The proposed temporal-spatial fusion based method combined the temporal properties within multiple frames and the spatial information in single frame of a moving target, and can be used for moving target detection and labeling in real-time video image sequence. The spatial image segmentation can be implemented by using color based classification of target area, and temporal background recovering can be implemented by using time-domain recursive low pass filter. Together with follow-up morphological filtering and blob analysis, the area of an interested moving target can be detected and labeled by bounding box. Its applications in swimming video and related experiments results manifested the effectiveness of this method.

Acknowledgements. The research is supported by Shanghai Leading Academic Discipline Project (T0103).

References

1. Anderson, C., Burt, P., van der Wal, G.: Change detection and tracking using pyramid transformation techniques. In: Proceedings of SPIE-Intelligent Robots and Computer Vision, vol. 579, pp. 72–78 (1985)
2. Wren, C., Azarbayejani, A., Darrell, T., Pentland, A.: Pfinder: Real-time tracking of the human body. IEEE Transactions on Pattern Analysis and Machine Intelligence 19(7), 780–785 (1997)
3. Barron, J., Fleet, D., Beauchmin, S.: Performance of optical flow techniques. International Journal of Computer Vision 12(1), 42–77 (1994)
4. Lipton, A., Fujiyoshi, H., Patil, R.: Moving Target Classification and Tracking from Real-time video. In: Proceedings of IEEE Workshop on Application of Computer Vision, Princeton, NJ, October 19-21, 1998, pp. 8–14. IEEE Computer Society Press, Los Alamitos (1998)
5. Pascual, J., Neucimar, J., Barros, Ricardo, M.L.: Background Recovering in Outdoor Image Sequences: An Example of Soccer Players Segmentation. Image and Vision Computing 24(4), 363–374 (2006)
6. Yujian, S., Xin, H., Zhihang, H.: Study on the method of determination for K coefficient in time-domain recursive filter. Journal of Systems Engineering and Electronics 23(3), 61–65 (2001)

Author Index

Printing: Mercedes-Druck, Berlin
Binding: Stein+Lehmann, Berlin

Lecture Notes in Computer Science

For information about Vols. 1–4415

please contact your bookseller or Springer

Vol. 4743: P. Thulasiraman, X. He, T.L. Xu, M.K. Denko, R.K. Thulasiram, L.T. Yang (Eds.), Frontiers of High Performance Computing and Networking ISPA 2007 Workshops. XXIX, 536 pages. 2007.

Vol. 4742: I. Stojmenovic, R.K. Thulasiram, L.T. Yang, W. Jia, M. Guo, R.F. de Mello (Eds.), Parallel and Distributed Processing and Applications. XX, 995 pages. 2007.

Vol. 4708: L. Kučera, A. Kučera (Eds.), Mathematical Foundations of Computer Science 2007. XVIII, 764 pages. 2007.

Vol. 4707: O. Gervasi, M.L. Gavrilova (Eds.), Computational Science and Its Applications – ICCSA 2007, Part III. XXIV, 1205 pages. 2007.

Vol. 4706: O. Gervasi, M.L. Gavrilova (Eds.), Computational Science and Its Applications – ICCSA 2007, Part II. XXIII, 1129 pages. 2007.

Vol. 4705: O. Gervasi, M.L. Gavrilova (Eds.), Computational Science and Its Applications – ICCSA 2007, Part I. XLIV, 1169 pages. 2007.

Vol. 4703: L. Caires, V.T. Vasconcelos (Eds.), CONCUR 2007 – Concurrency Theory. XIII, 507 pages. 2007.

Vol. 4697: L. Choi, Y. Paek, S. Cho (Eds.), Advances in Computer Systems Architecture. XIII, 400 pages. 2007.

Vol. 4689: K. Li, X. Li, G.W. Irwin, G. He (Eds.), Life System Modeling and Simulation. XIX, 561 pages. 2007.

Vol. 4688: K. Li, M. Fei, G.W. Irwin, S. Ma (Eds.), Bio-Inspired Computational Intelligence and Applications. XIX, 805 pages. 2007.

Vol. 4684: L. Kang, Y. Liu, S. Zeng (Eds.), Evolvable Systems: From Biology to Hardware. XIV, 446 pages. 2007.

Vol. 4683: L. Kang, Y. Liu, S. Zeng (Eds.), Intelligence Computation and Applications. XVII, 663 pages. 2007.

Vol. 4681: D.-S. Huang, L. Heutte, M. Loog (Eds.), Advanced Intelligent Computing Theories and Applications. XXVI, 1379 pages. 2007.

Vol. 4671: V. Malyshkin (Ed.), Parallel Computing Technologies. XIV, 635 pages. 2007.

Vol. 4666: M.E. Davies, C.J. James, S.A. Abdallah, M.D Plumbley (Eds.), Independent Component Analysis and Blind Signal Separation. XIX, 847 pages. 2007.

Vol. 4664: J. Durand-Lose, M. Margenstern (Eds.), Machines, Computations, and Universality. X, 325 pages. 2007.

Vol. 4649: V. Diekert, M.V. Volkov, A. Voronkov (Eds.), Computer Science – Theory and Applications. XIII, 420 pages. 2007.

Vol. 4647: R. Martin, M. Sabin, J. Winkler (Eds.), Mathematics of Surfaces XII. IX, 509 pages. 2007.

Vol. 4644: N. Azemard, L. Svensson (Eds.), Integrated Circuit and System Design. XIV, 583 pages. 2007.

Vol. 4641: A.-M. Kermarrec, L. Bougé, T. Priol (Eds.), Euro-Par 2007 Parallel Processing. XXVII, 974 pages. 2007.

Vol. 4639: E. Csuhaj-Varjú, Z. Ésik (Eds.), Fundamentals of Computation Theory. XIV, 508 pages. 2007.

Vol. 4638: T. Stützle, M. Birattari, H.H. Hoos (Eds.), Engineering Stochastic Local Search Algorithms. X, 223 pages. 2007.

Vol. 4628: L.N. de Castro, F.J. Von Zuben, H. Knidel (Eds.), Artificial Immune Systems. XII, 438 pages. 2007.

Vol. 4627: M. Charikar, K. Jansen, O. Reingold, J.D.P. Rolim (Eds.), Approximation, Randomization, and Combinatorial Optimization. XII, 626 pages. 2007.

Vol. 4624: T. Mossakowski, U. Montanari, M. Haveraaen (Eds.), Algebra and Coalgebra in Computer Science. XI, 463 pages. 2007.

Vol. 4619: F. Dehne, J.-R. Sack, N. Zeh (Eds.), Algorithms and Data Structures. XVI, 662 pages. 2007.

Vol. 4618: S.G. Akl, C.S. Calude, M.J. Dinneen, G. Rozenberg, H.T. Wareham (Eds.), Unconventional Computation. X, 243 pages. 2007.

Vol. 4616: A. Dress, Y. Xu, B. Zhu (Eds.), Combinatorial Optimization and Applications. XI, 390 pages. 2007.

Vol. 4613: F.P. Preparata, Q. Fang (Eds.), Frontiers in Algorithmics. XI, 348 pages. 2007.

Vol. 4600: H. Comon-Lundh, C. Kirchner, H. Kirchner (Eds.), Rewriting, Computation and Proof. XVI, 273 pages. 2007.

Vol. 4599: S. Vassiliadis, M. Berekovic, T.D. Hämäläinen (Eds.), Embedded Computer Systems: Architectures, Modeling, and Simulation. XVIII, 466 pages. 2007.

Vol. 4598: G. Lin (Ed.), Computing and Combinatorics. XII, 570 pages. 2007.

Vol. 4596: L. Arge, C. Cachin, T. Jurdziński, A. Tarlecki (Eds.), Automata, Languages and Programming. XVII, 953 pages. 2007.

Vol. 4595: D. Bošnački, S. Edelkamp (Eds.), Model Checking Software. X, 285 pages. 2007.

Vol. 4590: W. Damm, H. Hermanns (Eds.), Computer Aided Verification. XV, 562 pages. 2007.

Vol. 4588: T. Harju, J. Karhumäki, A. Lepistö (Eds.), Developments in Language Theory. XI, 423 pages. 2007.

Vol. 4583: S.R. Della Rocca (Ed.), Typed Lambda Calculi and Applications. X, 397 pages. 2007.

Vol. 4580: B. Ma, K. Zhang (Eds.), Combinatorial Pattern Matching. XII, 366 pages. 2007.

Vol. 4576: D. Leivant, R. de Queiroz (Eds.), Logic, Language, Information and Computation. X, 363 pages. 2007.

Vol. 4547: C. Carlet, B. Sunar (Eds.), Arithmetic of Finite Fields. XI, 355 pages. 2007.

Vol. 4546: J. Kleijn, A. Yakovlev (Eds.), Petri Nets and Other Models of Concurrency – ICATPN 2007. XI, 515 pages. 2007.

Vol. 4545: H. Anai, K. Horimoto, T. Kutsia (Eds.), Algebraic Biology. XIII, 379 pages. 2007.

Vol. 4533: F. Baader (Ed.), Term Rewriting and Applications. XII, 419 pages. 2007.

Vol. 4528: J. Mira, J.R. Álvarez (Eds.), Nature Inspired Problem-Solving Methods in Knowledge Engineering, Part II. XXII, 650 pages. 2007.

Vol. 4527: J. Mira, J.R. Álvarez (Eds.), Bio-inspired Modeling of Cognitive Tasks, Part I. XXII, 630 pages. 2007.

Vol. 4525: C. Demetrescu (Ed.), Experimental Algorithms. XIII, 448 pages. 2007.

Vol. 4514: S.N. Artemov, A. Nerode (Eds.), Logical Foundations of Computer Science. XI, 513 pages. 2007.

Vol. 4513: M. Fischetti, D.P. Williamson (Eds.), Integer Programming and Combinatorial Optimization. IX, 500 pages. 2007.

Vol. 4510: P. Van Hentenryck, L.A. Wolsey (Eds.), Integration of AI and OR Techniques in Constraint Programming for Combinatorial Optimization Problems. X, 391 pages. 2007.

Vol. 4507: F. Sandoval, A. Prieto, J. Cabestany, M. Graña (Eds.), Computational and Ambient Intelligence. XXVI, 1167 pages. 2007.

Vol. 4501: J. Marques-Silva, K.A. Sakallah (Eds.), Theory and Applications of Satisfiability Testing – SAT 2007. XI, 384 pages. 2007.

Vol. 4497: S.B. Cooper, B. Löwe, A. Sorbi (Eds.), Computation and Logic in the Real World. XVIII, 826 pages. 2007.

Vol. 4494: H. Jin, O.F. Rana, Y. Pan, V.K. Prasanna (Eds.), Algorithms and Architectures for Parallel Processing. XIV, 508 pages. 2007.

Vol. 4493: D. Liu, S. Fei, Z. Hou, H. Zhang, C. Sun (Eds.), Advances in Neural Networks – ISNN 2007, Part III. XXVI, 1215 pages. 2007.

Vol. 4492: D. Liu, S. Fei, Z. Hou, H. Zhang, C. Sun (Eds.), Advances in Neural Networks – ISNN 2007, Part II. XXVII, 1321 pages. 2007.

Vol. 4491: D. Liu, S. Fei, Z.-G. Hou, H. Zhang, C. Sun (Eds.), Advances in Neural Networks – ISNN 2007, Part I. LIV, 1365 pages. 2007.

Vol. 4490: Y. Shi, G.D. van Albada, J. Dongarra, P.M.A. Sloot (Eds.), Computational Science – ICCS 2007, Part IV. XXXVII, 1211 pages. 2007.

Vol. 4489: Y. Shi, G.D. van Albada, J. Dongarra, P.M.A. Sloot (Eds.), Computational Science – ICCS 2007, Part III. XXXVII, 1257 pages. 2007.

Vol. 4488: Y. Shi, G.D. van Albada, J. Dongarra, P.M.A. Sloot (Eds.), Computational Science – ICCS 2007, Part II. XXXV, 1251 pages. 2007.

Vol. 4487: Y. Shi, G.D. van Albada, J. Dongarra, P.M.A. Sloot (Eds.), Computational Science – ICCS 2007, Part I. LXXXI, 1275 pages. 2007.

Vol. 4484: J.-Y. Cai, S.B. Cooper, H. Zhu (Eds.), Theory and Applications of Models of Computation. XIII, 772 pages. 2007.

Vol. 4475: P. Crescenzi, G. Prencipe, G. Pucci (Eds.), Fun with Algorithms. X, 273 pages. 2007.

Vol. 4474: G. Prencipe, S. Zaks (Eds.), Structural Information and Communication Complexity. XI, 342 pages. 2007.

Vol. 4459: C. Cérin, K.-C. Li (Eds.), Advances in Grid and Pervasive Computing. XVI, 759 pages. 2007.

Vol. 4449: Z. Horváth, V. Zsók, A. Butterfield (Eds.), Implementation and Application of Functional Languages. X, 271 pages. 2007.

Vol. 4448: M. Giacobini (Ed.), Applications of Evolutionary Computing. XXIII, 755 pages. 2007.

Vol. 4447: E. Marchiori, J.H. Moore, J.C. Rajapakse (Eds.), Evolutionary Computation, Machine Learning and Data Mining in Bioinformatics. XI, 302 pages. 2007.

Vol. 4446: C. Cotta, J. van Hemert (Eds.), Evolutionary Computation in Combinatorial Optimization. XII, 241 pages. 2007.

Vol. 4445: M. Ebner, M. O'Neill, A. Ekárt, L. Vanneschi, A.I. Esparcia-Alcázar (Eds.), Genetic Programming. XI, 382 pages. 2007.

Vol. 4436: C.R. Stephens, M. Toussaint, D. Whitley, P.F. Stadler (Eds.), Foundations of Genetic Algorithms. IX, 213 pages. 2007.

Vol. 4433: E. Şahin, W.M. Spears, A.F.T. Winfield (Eds.), Swarm Robotics. XII, 221 pages. 2007.

Vol. 4432: B. Beliczynski, A. Dzielinski, M. Iwanowski, B. Ribeiro (Eds.), Adaptive and Natural Computing Algorithms, Part II. XXVI, 761 pages. 2007.

Vol. 4431: B. Beliczynski, A. Dzielinski, M. Iwanowski, B. Ribeiro (Eds.), Adaptive and Natural Computing Algorithms, Part I. XXV, 851 pages. 2007.

Vol. 4424: O. Grumberg, M. Huth (Eds.), Tools and Algorithms for the Construction and Analysis of Systems. XX, 738 pages. 2007.

Vol. 4423: H. Seidl (Ed.), Foundations of Software Science and Computational Structures. XVI, 379 pages. 2007.

Vol. 4422: M.B. Dwyer, A. Lopes (Eds.), Fundamental Approaches to Software Engineering. XV, 440 pages. 2007.

Vol. 4421: R. De Nicola (Ed.), Programming Languages and Systems. XVII, 538 pages. 2007.

Vol. 4420: S. Krishnamurthi, M. Odersky (Eds.), Compiler Construction. XIV, 233 pages. 2007.

Vol. 4419: P.C. Diniz, E. Marques, K. Bertels, M.M. Fernandes, J.M.P. Cardoso (Eds.), Reconfigurable Computing: Architectures, Tools and Applications. XIV, 391 pages. 2007.

Vol. 4416: A. Bemporad, A. Bicchi, G. Buttazzo (Eds.), Hybrid Systems: Computation and Control. XVII, 797 pages. 2007.